Machine Dreams

Economics Becomes a Cyborg Science

This is the first crossover book into the history of science written by a historian of economics, pioneering the combination of a number of disciplinary and stylistic orientations. It shows how what is conventionally thought to be "history of technology" can be integrated with the history of economic ideas, focusing on the history of the computer. The analysis combines Cold War history with the history of the postwar economics profession in America, revealing that the Pax Americana had much to do with the content of such abstruse and formal doctrines as linear programming and game theory. It links the literature on "cyborg science" found in science studies to economics, an element missing in the literature to date. The treatment further calls into question the idea that economics has been immune to postmodern currents found in the larger culture, arguing that neoclassical economics has surreptitiously participated in the deconstruction of the integral "Self." Finally, it argues for a different style of economics, namely, an alliance of computational and institutional themes, and challenges the widespread impression that there is nothing besides American neoclassical economic theory left standing after the demise of Marxism.

Philip Mirowski is Carl Koch Professor of Economics and the History and Philosophy of Science at the University of Notre Dame, Indiana. He teaches in both the economics and science studies programs at the university. The author of *More Heat than Light* (Cambridge University Press, 1989) and editor of *Natural Images in Economics* (Cambridge University Press, 1994) and *Science Bought and Sold*, Professor Mirowski has also written more than 100 articles in academic journals. He has held visiting positions at the University of Massachusetts, Yale, the Sorbonne, Erasmus University Rotterdam, and the University of Modena.

Machine Dreams

Economics Becomes a Cyborg Science

PHILIP MIROWSKI
University of Notre Dame

CAMBRIDGE
UNIVERSITY PRESS

CAMBRIDGE UNIVERSITY PRESS
Cambridge, New York, Melbourne, Madrid, Cape Town, Singapore, São Paulo

Cambridge University Press
The Edinburgh Building, Cambridge CB2 2RU, UK

Published in the United States of America by Cambridge University Press, New York

www.cambridge.org
Information on this title: www.cambridge.org/9780521772839

First published 2002

A catalogue record for this publication is available from the British Library

Library of Congress Cataloguing in Publication data

Mirowski, Philip, 1951–
Machine dreams : economics becomes a cyborg science / Philip Mirowski.
p. cm.
Includes bibliographical references and index.
ISBN 0-521-77283-4 – ISBN 0-521-77526-4 (pb.)
1. Economics. 2. Cybernetics – Economic aspects – History. 3. Economic history – 1945. 4. Technological innovations – Economic aspects – History – 20th century. 5. Computer science – History. I. Title.
HB71 .M636 2001
330.1–dc21 00-054703

ISBN-13 978-0-521-77283-9 hardback
ISBN-10 0-521-77283-4 hardback

ISBN-13 978-0-521-77526-7 paperback
ISBN-10 0-521-77526-4 paperback

Transferred to digital printing 2006

To those of the younger generation
who resist the confident blandishments of their teachers:
in them resides the hope for the future of economics.

Contents

Figures and Tables

FIGURES

TABLES

Acknowledgments

The following people have done me the profound favor of reading the manuscript all the way through and providing detailed comments: Mie Augier, Judy Klein, Kyu Sang Lee, Esther-Mirjam Sent, and Roy Weintraub. Although undoubtedly none are completely satisfied with the final product, it would have been much worse without their efforts. Without the help of Tori Davies, it would never have seen the light of day.

The Economics Departments at the Universities of Modena and Trento in Italy gave me the opportunity to try out parts of this manuscript as lectures during visiting stints, as did the Australian History of Thought Society, and for those memorable occasions I should like to thank David Lane (Modena), Steve Keen (HETSA-Sydney), and Axel Leijonhufvud and Vela Velupillai (Trento).

The following colleagues and students have made helpful comments over the five-year genesis of this book: Bruce Caldwell, Avi Cohen, John Davis, Paul Edwards, Duncan Foley, Matt Frank, Wade Hands, Michael Hutter, Albert Jolink, Gerard Jorland, Lily Kay, Alan Kirman, Matthias Klaes, Robert Leonard, Donald MacKenzie, Perry Mehrling, Andy Pickering, Andres Rius, Abu Rizvi, Koye Somefun, and Matt Weagle. They have all influenced the project in ways they perhaps do not fully recognize; but I do. Seminar participants at the University of Chicago, University of California–San Diego, Duke University, the London School of Economics, University of Illinois, York University, the University of Venice, and the Sorbonne also left their imprint.

Portions of this book have appeared in previous incarnations in *Social Studies of Science* (1999) and in John Davis, ed., *The New Economics and Its History* (1998).

I would like to thank the following: Kenneth Arrow for permission to quote the Arrow Papers; Marina von Neumann Whitman for permission to quote the von Neumann files; permission was also granted by Gregory Taylor and Nicolae Savoiu to reprint Figure 2.2 from *Models of Com-*

putation and Formal Languages; Institute Libraries and Special Collections, MIT Libraries, for permission to use material from the Norbert Wiener Papers; Rare Book, Manuscript and Special Collections, Duke University Library, for permission to quote from the Oskar Morgenstern Papers; Harvard University Archives for permission to quote the papers of Edwin Bidwell Wilson; Manuscripts and Archives, Yale University Library, for permission to cite the Tjalling Charles Koopmans Papers; and Tate Gallery for permitting reproduction of Jacob Epstein's *The Rock Drill.*

Abbreviations

ARCHIVAL SOURCES

CFYU	Cowles Foundation Archives, Yale University
CORE	Archives, Center for Operations Research, Economics and Econometrics, Université Catholique de Louvain, Louvain-la-neuve, Belgium
EWHL	Edwin Bidwell Wilson Collection, Harvard University Archives
HHPC	Harold Hotelling Papers, Columbia University Library Manuscripts Collection
HSCM	Herbert Simon Papers, Carnegie Mellon University Archives
JMLA	Jacob Marschak Papers, Young Library, University of California, Los Angeles
JTER	Jan Tinbergen Papers, Erasmus University, Rotterdam
KAPD	Kenneth Arrow Papers, Perkins Library, Duke University
MPRL	Michael Polanyi Papers, Regenstein Library, University of Chicago
NGRD	Nicholas Georgescu-Roegen Papers, Perkins Library, Duke University
NSMR	National Air and Space Museum, RAND Oral History Project, Washington, D.C.
OMDU	Oskar Morgenstern Diaries, Perkins Library, Duke University
OMPD	Oskar Morgenstern Papers, Perkins Library, Duke University
RANY	Rockefeller Foundation Archives, Sleepy Hollow, New York
SCOP	Computer Oral History Project, Smithsonian Museum of American History, Washington, D.C.

TKPY	Tjalling Koopmans Papers, Sterling Library, Yale University
VNLC	John von Neumann Papers, Library of Congress, Washington, D.C.
WMIT	Norbert Wiener Papers, Massachusetts Institute of Technology Archives
WWOC	Warren Weaver interviews, Oral History Collection, Columbia University

OTHER ACRONYMS

2P0Σ	Two-person zero-sum (game)
AAA	Autonomous Artificial Agents
AMP	Applied Mathematics Panel
ASWORG	Anti-Submarine Warfare Operations Research Group
C^3I	Command, control, communications, and information
DARPA	Defense Applied Research Projects Agency
ICBM	Intercontinental Ballistic Missile
MAD	Maxwell's Augustinian Demon
MOD	Manichaean Other's Demon
NBER	National Bureau of Economic Research
NDRC	National Defense Research Committee
NUTM	Nondeterministic Universal Turing Machine
OMDU	Oskar Morgenstern Diaries, Duke University
ONR	Office of Naval Research
OR	Operations research
ORO	Operations Research Office, Johns Hopkins University
OSRD	Office of Scientific Research and Development
SAGA	Studies, Analysis, and Gaming Agency, Pentagon
SAGE	Semi-Automatic Ground Environment
SRG	Statistical Research Group
SRL	Systems Research Laboratory, RAND
TGEB	*Theory of Games and Economic Behavior*, by John von Neumann and Oskar Morgenstern
TIMS	The Institute of Management Studies
TM	Turing Machine
UTM	Universal Turing Machine
USAF	United States Air Force

1 Cyborg Agonistes

> A true war story is never moral. It does not instruct, nor encourage virtue, nor suggest models of proper human behavior, nor restrain men from doing the things they have always done. If a story seems moral, do not believe it. If at the end of a war story you feel uplifted, or if you feel that some small bit of rectitude has been salvaged from the larger waste, then you have been made the victim of a very old and terrible lie.
>
> Tim O'Brien, *The Things They Carried*

The first thing you will notice is the light. The fluorescent banks in the high ceiling are dimmed, so the light at eye level is dominated by the glowing screens set at intervals throughout the cavernous room. There are no windows, so the bandwidths have that cold otherworldly truncation. Surfaces are in muted tones and mat colors, dampening unwanted reflections. Some of the screens flicker with strings of digits the color and periodicity of traffic lights, but most beam the standard Day-Glo palette of pastels usually associated with CRT graphics. While a few of the screens project their photons into the void, most of the displays are manned and womanned by attentive acolytes, their visages lit and their backs darkened like satellites parked in stationary orbits. Not everyone is held in the thrall of the object of their attentions in the same manner. A few jump up and down in little tethered dances, speaking into phones or mumbling at other electronic devices. Some sit stock still, mesmerized, engaging their screen with slight movements of wrist and hand. Others lean into their consoles, then away, as though their swaying might actually have some virtual influence upon the quantum electrodynamics coursing through their station and beyond, to other machines in other places in other similar rooms. No one is apparently making anything, but everyone seems nonetheless furiously occupied.

ROOMS WITH A VIEW

Where is this place? If it happened to be 1952, it would be Santa Monica, California, at a RAND study of the "man-machine interface" (Chapman et al., 1958). If it were 1957, then it could be only one place: the SAGE (Semi-Automatic Ground Environment) Air Defense System run by the U.S. Air Force. By 1962, there were a few other such rooms, such as the SAGA room for wargaming in the basement of the Pentagon (Allen, 1987). If it were 1967 instead, there were many more such rooms scattered across the globe, one of the largest being the Infiltration Surveillance Center at Nakhom Phanom in Thailand, the command center of U.S. Air Force Operation Igloo White (Edwards, 1996, pp. 3, 106). By 1977 there are many more such rooms, no longer only staffed by the military, but also by thousands of employees of large firms throughout the world: the SABRE airline reservation system of American Airlines (patterned upon SAGE); bank check and credit card processing centers (patterned upon that innovated by Bank of America); nuclear power station control rooms; the inventory control operation of the American Hospital Supply Corporation (McKenney, 1995). In 1987 a room like this could be found in any suburban shopping mall, with teenagers proleptically feeding quarters into arcade computer games. It might also be located at the University of Arizona, where "experimental markets" are being conducted with undergraduates recruited with the help of money from the National Science Foundation. Alternatively, these closed rooms also could just as surely be found in the very pinnacles of high finance, in the tonier precincts of New York and London and Tokyo, with high-stakes traders of stocks, bonds, and "derivatives" glued to their screens. In those rooms, "masters of the universe" in pinstripe shirts and power suspenders make "killings" in semiconscious parody of their khaki-clad precursors. By 1997, with the melding of home entertainment centers with home offices and personal computers via the Internet (a lineal descendant of the Defense-funded ARPANET), any residential den or rec room could be refitted as a scaled-down simulacrum of any of the previous rooms. It might be the temporary premises of one of the burgeoning "dot-com" start-ups which captured the imaginations of Generation X. It could even be promoted as the prototype classroom of the future. Increasingly, work in America at the turn of the millennium means serried ranks of Dilberts arrayed in cubicles staring at these screens. I should perhaps confess I am staring at the glow now myself. Depending on how this text eventually gets disseminated, perhaps you also, dear reader, are doing likewise.

These rooms are the "closed worlds" of our brave new world (Edwards, 1996), the electronic surveillance and control centers that were the nexus of the spread of computer technologies and computer culture. They are closed in any number of senses. In the first instance, there is the obviously artificial light: chaotic "white" sunlight is kept to a minimum to control the frequencies and the reactions of the observers. This ergonomically controlled environment, the result of some concerted engineering of the man-machine interface, renders the machines "user-friendly" and their acolytes more predictable. The partitioning off of the noise of the outer world brings to mind another sort of closure, that of thermodynamic isolation, as when Maxwell's Demon closes the door on slower gas molecules in order to make heat flow from a cooler to a warmer room, thus violating the Second Law of Thermodynamics. Then again, there is the type of closure that is more directly rooted in the algorithms that play across the screens, a closure that we shall encounter repeatedly in this book. The first commandment of the trillions of lines of code that appear on the screens is that they *halt*; algorithms are closed and bounded and (almost) never spin on forever, out of control.

And then the rooms are closed in another fashion, one resembling Bentham's Panopticon: a hierarchical and pervasive surveillance, experienced as an automatic and anonymous expression of power (Foucault, 1977). The range of things that the occupants of the room can access – from your medical records to the purchases you made three years ago with your credit card, from your telephone calls to all the web pages you have visited, from your genealogy to your genome – consistently outstrips the paltry imagination of movies haunted by suggestions of paranoid conspiracies and fin-de-siècle science run amok (Bernstein, 1997). Just as surely as death is the culmination of life, surveillance raises the specter of countersurveillance, of dissimulation, of *penetration*; and closure comes increasingly to resemble prophylaxis. The language of viruses, worms, and a myriad of other creepy-crawlies evokes the closure of a siege mentality, of quarantine, or perhaps the tomb.

The closure of those rooms is also radically stark in that implacable conflicts of global proportions are frequently shrunk down to something far less than human scale, to the claustrophobic controlled play of pixilated symbols on screens. The scale of phenomena seems to have become distended and promiscuously distributed. As the computer scientist Joseph Weitzenbaum once said, the avatars of artificial intelligence (AI) tend to describe "a very small part of what it means to be a human being and say that this is the whole." He quotes the philosopher (and cheerleader for AI) Daniel Dennett as having insisted, "If we are to make further progress in Artificial Intelligence, we are going to have to give up our awe of living things" (in Baumgartner & Payr, 1995, p. 259). The quickest way to divest oneself of an awe for the living in the West is to imagine oneself surrounded instead by machines. Whatever may have once been imagined the rich ambiguity of multiform experience, it seems enigmatic encounters and inconsistent interpretations can now only be expressed in this brave new world as *information*. Ideas are conflated with things, and things like computers assume the status of ideas.

And despite the widespread notion that as the global reach of these rooms has been stretched beyond the wildest dreams of the medieval magus or Enlightenment philosophe, the denizens of the modern closed rooms seem to have grown more insular, less experienced, perhaps even a trifle solipsistic. Closed rooms had held out the promise of infinite horizons; but the payoff has been . . . more closure. Who needs to venture any more into the inner city, the outer banks, the corridors of the Louvre, the sidewalks of mean streets? Travel, real physical displacement, has become like everything else: you need special reservations and a pile of money to go experience the simulacra that the screen has already conditioned you to expect. More annual visitors to Boston seek out the mock-up of the fictional bar "Cheers" than view Bunker Hill or Harvard Yard. Restaurants and museums and universities and corporations and Walden Pond are never quite as good as their web sites. Cyberspace, once a new metaphor for spatial orientation, comes to usurp motion itself. No, don't get around much any more.

WHERE THE CYBORGS ARE

Is this beginning to sound like just another pop sociology treatise on "being digital" or the "information superhighway" or "the second self" or denunciation of some nefarious cult of information (Roszak, 1994)? Calm your fears, dear reader. What the world needs now is surely *not* another panegyric on the cultural evils of cyberspace. Our whirlwind tour of a few clean, well-lighted places is intended to introduce, in a subliminal way, some of the themes that will structure a work situated more or less squarely within a distinctly despised genre, that of the history of economic thought. The novelty for most readers will be to cross it with an older and rather more popular form of narrative, that of the war story. The chronological succession of closed rooms is intended to serve as a synecdoche for a succession of the ways in which many economists have come to understand markets over roughly the same period, stretching from World War II to the end of the twentieth century. For while these closed rooms begat little models of closed worlds, after the fashion of Plato's Cave, the world as we had found it has rapidly been transubstantiated into the architecture of the rooms.[1] Modes of thought and machines that think

[1] Michel Foucault, 1977, pp. 211, 216: "the mechanisms [of disciplinary establishments] have a certain tendency to become 'de-institutionalized,' to emerge from the closed fortresses in which they once functioned and circulate in a 'free' state; the massive, compact disciplines are broken down into flexible methods of control, which may be transferred and adapted.

forged in British and American military settings by their attendant mobilized army of scientists in the 1940s rapidly made their way into both the natural and social sciences in the immediate postwar period, with profound consequences for both the content and organization of science.

The thesis that a whole range of sciences has been transformed in this manner in the postwar period has come to have a name in the literature of the history and sociology of science, primarily due to the pioneering efforts of Donna Haraway: that name is "cyborg science." Haraway (1991; 1997) uses the term to indicate something profound that has happened to biology and to social theory and cultural conceptions of gender.[2] It has been applied to computer development and industrial organization by Andy Pickering (1995a; 1997; 1999). Ian Hacking (1998) has drawn attention to the connections of cyborgs to Canguilhem and Foucault. Explication of the cyborg character of thermodynamics and information theory was pioneered by Katherine Hayles (1990b), who has now devoted prodigious work to explicating their importance for the early cyberneticians (1994; 1995a; 1999). Paul Edwards (1996) provided the first serious across-the-board survey of the military's conceptual influence on the development of the computer, although Kenneth Flamm (1988) had pioneered the topic in the economics literature of industrial organization.

. . . One can [thus] speak of the formation of a disciplinary society in this movement that stretches from the enclosed disciplines, a sort of social quarantine, to an indefinitely generalizable mechanism of 'panopticism.' "

2 "Nineteenth century scientists materially constituted the organism as a laboring system, structured by a hierarchical division of labor and an energetic system fueled by sugars and obeying the laws of thermodynamics. For us, the living world has become a command, control, communication, intelligence system in an environment that demands strategies of flexible accumulation. Artificial life programs, as well as carbon-based life programs, work that way" (Haraway, 1997, p. 97).

"I am adamant that the cyborg, as I use the term, does not refer to *all* kinds of artifactual, machinic relationships with human beings. . . . I am very concerned that the term cyborg be used specifically to refer to those kinds of entities that became historically possible around World War II and just after. The cyborg is intimately involved in histories of militarization, of specific research projects with ties to psychiatry and communications theory, behavioral research and psychopharmacological research, theories of information and information processing. It is essential that the cyborg is seen to emerge out of such a specific matrix. In other words, the cyborg is not 'born,' but it does have a matrix!" (Haraway, 2000, pp. 128–29).

"A cyborg is a cybernetic organism, a hybrid of machine and organism, a creature of social reality as well as fiction. . . . The cyborg is a creature in a post-gender world; it has no truck with bi-sexuality, pre-oedipal symbiosis, unalienated labour, or other seductions to organic wholeness through a final appropriation of all the powers of the parts into a higher unity. In a sense, the cyborg has no origin story in the Western sense – a final irony, since the cyborg is also the awful apocalyptic *telos* of the West's escalating dominations of abstract individuation, the ultimate self untied at last from all dependency" (Haraway, 1991, pp. 149–51).

Steve Heims (1991) documented the initial attempts of the first cyberneticians to reach out to social scientists in search of a Grand Unified Teleological theory. Evelyn Fox Keller (1995) has surveyed how the gene has assumed the trappings of military command; and Lily Kay (1995; 1997a) has performed the invaluable service of showing in detail how all the above played themselves out in the development of molecular biology. Although all of these authors have at one time or another indicated an interest in economic ideas, what has been wanting in all of this work so far is a commensurate consideration of the role of economists in this burgeoning transdisciplinary formation. Economists were present at the creation of the cyborg sciences, and, as one would expect, the cyborg sciences have returned the favor by serving in turn to remake the economic orthodoxy in their own image. My intention is to provide that complementary argument, to document just in what manner and to what extent economics at the end of the second millennium has become a cyborg science, and to speculate how this will shape the immediate future.

Just how serious has the cyborg incursion been for economics? Given that in all likelihood most economists have no inkling what "cyborgs" are, or will have little familiarity with the historical narrative that follows, the question must be confronted squarely. There are two preliminary responses to this challenge: one short, yet readily accessible to anyone familiar with the modern economics literature; and the other, necessarily more involved, requiring a fair complement of historical sophistication. The short answer starts out with the litany that every tyro economist memorizes in his or her first introductory course. Question: What is economics about? Answer: The optimal allocation of scarce resources to given ends. This catechism was promulgated in the 1930s, about the time that neoclassicism was poised to displace rival schools of economic thought in the American context, and represented the canonical image of trade as the shifting about of given endowments so as to maximize an independently given utility function. While this phrase still may spring effortlessly to the lips – this, after all, is the function of a catechism – nevertheless, pause and reflect how poorly this captures the primary concerns of neoclassical economists nowadays: Nash equilibrium, strategic uncertainty, decision theory, path dependence, network externalities, evolutionary games, principal-agent dilemmas, no-trade theorems, asymmetric information, paradoxes of noncomputability. . . . Static allocation has taken a back seat to all manner of issues concerning agents' capacities to deal with various market situations in a cognitive sense. It has even once again become fashionable to speak with confidence of the indispensable role of "institutions," although this now means something profoundly different than it did in the earlier heyday of the American Institutionalist school of economics. This is a drastic change from the 1930s through the 1950s, when it was taboo to speculate about mind, and all marched proudly under the banner of behaviorism; and society was thought to spring fully formed from the brow of an isolated economic man. So what is economics *really* about these days? The New Modern Answer: The economic agent as a processor of information.

This is the first, and only the most obvious, hallmark of the epoch of economics as a cyborg science. The other attributes require more prodigious documentation and explication.

THE NATURAL SCIENCES AND THE HISTORY OF ECONOMICS

The other, more elaborate, answer to the query concerning the relevance of cyborgs for economics requires some working familiarity with the history of neoclassical economics. In a previous book entitled *More Heat than Light* (1989a), I argued that the genesis of the supposed "simultaneous discovery" of neoclassicism in the 1870s could be traced to the enthusiasm for "energetics" growing out of the physics of the mid-nineteenth century. As was admitted by William Stanley Jevons, Leon Walras, Vilfredo Pareto, Francis Edgeworth, and Irving Fisher, "utility" was patterned on potential energy in classical mechanics, as were their favored mathematics of extremum principles. Their shared vision of the operation of the market (and the mind of the agent, if they were willing to make this commitment) was avowedly *mechanical* in an eminently physical sense of that term. Their shared prescription for rendering economics a science was to imitate the best science they knew, right down to its characteristic mathematical formalisms. It was a science of causality, rigid determinism and preordained order; in other words, it was physics prior to the Second Law of Thermodynamics, a science most assuredly innocent of the intellectual upheavals beginning at the turn of the century and culminating in the theories of quantum mechanics and statistical thermodynamics.

Some readers of that volume demurred that, although it was undeniably the case that important figures such as Jevons and Walras and Fisher cited physics as an immediate source of their inspiration, this still did not square with the neoclassical economics with which economists were familiar in the twentieth century. Indeed, a book by Bruna Ingrao and Giorgio Israel (1990) asserted that the impact of physics upon neoclassical economics was attenuated by the 1930s, precisely at the moment when it underwent substantial mathematical development and began its serious ascendancy. Others have insisted that a whole range of orthodox models, from the modern Walrasian tradition to game theory, betray no inspiration whatsoever from physics. The historiographical problem that these responses highlight is the lack of willingness to simultaneously examine the history of economics and the history of the natural sciences as jointly evolving historical entities, and not as fixed monolithic bodies of knowledge driven primarily by their internally defined questions, whose interactions with other sciences can only be considered as irrelevant rhetoric in whatever era in which they may have occurred. If you avert your gaze from anything other than the narrowly conceived entity called the "economy," then you will never understand the peripatetic path of American economics in the twentieth century.[3]

This book could thus be regarded as the third installment in my ongoing project to track the role and impact of the natural sciences on the structure and content of the orthodox tradition in economics, which is perhaps inaccurately but conventionally dubbed "neoclassical."[4] The first installment of this history was published in 1989 as *More Heat than Light*, and was concerned with the period from classical political economy up to the 1930s, stressing the role of physics in the "marginalist revolution." The second installment would comprise a series of papers coauthored over the 1990s with Wade Hands and Roy Weintraub,[5] which traced the story of the rise to dominance of neoclassical price theory in America from early

[3] "It will occasionally turn out that some piece of economics is mathematically identical to some piece of utterly unrelated physics. (This has actually happened to me although I know absolutely nothing about physics.) I think this has no methodological significance but arises because everyone playing this sort of game tends to follow the line of least mathematical resistance. I know that Philip Mirowski believes that deeper aspects of mainstream economic theory are the product of a profound imitation of nineteenth-century physical theory. That thesis strikes me as false, but I would not claim expert knowledge" (Robert Solow in Bender & Schorske, 1997, pp. 73–74). Other advocates of economic mumpsimus can be found in Niehans, 1990; Myerson, 1999; Baumol, 2000.

[4] As I tried to insist in *More Heat* (1989a, chap. 3) and *Natural Images in Economic Thought* (1994a), this should not be taken to imply that the commerce in metaphors and concepts moves only in a single direction, from the natural to the social sciences. Inspiration is a two-way street, with both physics and biology making sustained use of economic concepts. Yet, inevitably, this reverse appropriation is felt to lead rapidly to relativist or postmodern tendencies, and has been persistently ignored by historians of science, with the possible exception of the Darwin industry. Twice burned is thrice shy; and one can usually only effectively address one constituency at a time. For better or for worse, I presume in this text that the reader is interested in how economics was bequeathed its modern fascination for information and therefore tend to concentrate on traffic in only one direction.

[5] The narrative would begin with Henry Ludwell Moore's attempt to revive an empirically relevant mathematical theory of demand based on modern physics around the turn of the century, discussed in Mirowski, 1990. It continues with the innovations of Moore's student Henry Schultz and the statistician Harold Hotelling, recounted in Hands & Mirowski, 1998. The reaction to the duo and the development of three different American schools of neoclassicism is described in Hands & Mirowski, 1999. The attempt by one of these schools, associated with the Cowles Commission, to maintain it had relinquished the dependence on physics is recounted in Weintraub & Mirowski, 1994. Parenthetically, it is the philosophical position of this latter school, only one sect within the broad church of "neoclassicism," that informs the historical account given in Ingrao & Israel, 1990.

in the century up through the 1960s. The present volume takes up the story from the rise of the cyborg sciences, primarily though not exclusively during World War II in America, and then traces their footprint upon some important postwar developments in economics, such as highbrow neoclassical price theory, game theory, rational expectations theory, theories of institutions and mechanism design, the nascent program of "bounded rationality," computational economics, "artificial economies," "autonomous agents," and experimental economics. Because many of these developments are frequently regarded as antithetical to one another, or possibly movements bent upon rejection of the prior Walrasian orthodoxy, it will be important to discern the ways in which there is a profound continuity between their sources of inspiration and those of the earlier generations of neoclassical economics.

One source of continuity is that economists, especially those seeking a scientific economics, have always been inordinately fascinated by *machines*. François Quesnay's theory of circulation was first realized as a pump and some tubes of tin; only later did it reappear in abstract form as the *Tableau économique*. Simon Schaffer has argued that "Automata were apt images of the newly disciplined bodies of military systems in early modern Europe. . . . Real connections were forged between these endeavors to produce a disciplined workforce, an idealized workspace, and an automatic man" (1999, pp. 135, 144). It has been argued that the conception of natural order in British classical political economy was patterned on the mechanical feedback mechanisms observed in clocks, steam engine governors, and the like (Mayr, 1976). William Stanley Jevons, as we shall discuss in Chapter 2, proudly compared the rational agent to a machine. Irving Fisher (1965) actually built a working model of cisterns and mechanical floats to illustrate his conception of economic equilibrium. Many of those enthralled with the prospect that the laws of energy would ultimately unite the natural and social sciences looked to various engines and motors for their inspiration (Rabinbach, 1990). However, as Norbert Wiener so presciently observed at the dawn of the Cyborg Era: "If the seventeenth and the early eighteenth centuries are the age of clocks, the later eighteenth and nineteenth centuries constitute the age of steam engines, the present time is the age of communication and control" (1961, p. 39). Natural order for economists coming of age after World War II is still exemplified by a machine, although the manifestation of the machine has changed: it is now the computer. "It may be hard for younger economists to imagine, but nearly until midcentury it was not unusual for a theorist using mathematical techniques to begin with a substantial apology, explaining that this approach need not assume that humans are automatons deprived of free will" (Baumol, 2000, p. 23). Cyborg love means never having to say you're sorry. Machine rationality and machine regularities are the constants in the history of neoclassical economics; it is only the innards of the machine that have changed from time to time.

There is another, somewhat more contingent common denominator. The history of economics has been persistently swept by periodic waves of immigrants from the natural sciences. The first phase, that of the 1870s through the turn of the century, was the era of a few trained engineers and physicists seeking to impose some analytical structure on the energetic metaphors so prevalent in their culture. The next wave of entry came in the 1930s, prompted both by the Great Depression's contraction of career possibilities for scientists, and the great forced emigration of scientists from Europe to America due to persecution and the disruptions of war. Wartime exigencies induced physicists to engage in all sorts of new activities under rubrics such as "operations research." We shall encounter some of these more illustrious souls in subsequent chapters. The third phase of scientific diaspora is happening right now. The end of the Cold War and its attendant shifts in the funding of scientific research has had devastating impact on physics, and on the career patterns of academic science in general (Slaughter & Rhoades, 1996; National Science Foundation, 1995; Ziman, 1994). Increasingly, physicists left to their own devices have found that economics (or perhaps, more correctly, finance) has proved a relatively accommodating safe haven in their time of troubles (Pimbley, 1997; Baker, 1999; Bass, 1999; MacKenzie, 1999). The ubiquitous contraction of physics and the continuing expansion of molecular biology has not only caused sharp redirections in careers, but also redirection of cultural images of what it means to be a successful science of epochal import. In many ways, the rise of the cyborg sciences is yet another manifestation of these mundane considerations of funding and support; interdisciplinary research has become more akin to a necessary condition of survival in our brave new world than merely the province of a few dilettantes or Renaissance men; and the transformation of economic concepts described in subsequent chapters is as much an artifact of a newer generation of physicists, engineers, and other natural scientists coming to terms with the traditions established by a previous generation of scientific interlopers dating from the Depression and World War II, as it is an entirely new direction in intellectual discourse.

Finally, there is one more source of the appearance of continuity. As I argue in Chapters 4 and 5, the first hesitant steps toward economics becoming a cyborg science were in fact made from a position situated squarely within the Walrasian tradition; these initially assumed the format of augmentation of the neoclassical agent with some capacities to deal with the fundamental "uncertainty" of economic life. The primary historical site of this transitional stage was the RAND Corporation and its ongoing contacts with the Cowles Commission. Part of the narrative momentum of the story recounted herein will derive from the progressive realization that cyborgs and neoclassicals could not be so readily yoked one to another, or even cajoled to work in tandem, and that this has led to numerous tensions in fin-de-siècle orthodox economics.

ANATOMY OF A CYBORG

So who or what are these cyborgs, that they have managed to spawn a whole brood of feisty new sciences? A plausible reaction is to wonder whether the term more correctly belongs to science fiction rather than to seriously practiced sciences as commonly understood. For you, dear reader, it may invoke childhood memories of *Star Wars* or *Star Trek*; if you happen to be familiar with popular culture, it may conjure William Gibson's breakthrough novel *Neuromancer* (1984). Yet, as usual, science fiction does not anticipate as much as reflect prior developments in scientific thinking. Upon consulting the *Cyborg Handbook* (Gray, 1995, p. 29), one discovers that the term was invented in 1960 by Manfred Clynes and Nathan Kline in the scientific journal *Astronautics* (Clynes and Kline, 1995). Manfred Clynes, an Austrian émigré (and merely the first of a whole raft of illustrious Austro-Hungarian émigrés we shall encounter in this book), and one of the developers of the computed axial tomography (CAT) scanner technology, had been introduced to cybernetics at Princeton in the 1950s and was concerned about the relationship of the organism to its environment as a problem of the communication of information. As he reports, "I thought it would be good to have a new concept, a concept of persons who can free themselves from the constraints of the environment to the extent that they wished. And I coined this word cyborg" (Gray, 1995, p. 47), short for *cybernetic organism.* In a paper presented to an Air Force–sponsored conference in 1960, Clynes and Kline assayed the possibilities of laboratory animals that were augmented in various ways in the interest of directly engaging in feedback stabilization and control of their metabolic environment. The inquiry attracted the attention of the National Aeronautics and Space Administration (NASA), which was worried about the effects of long-term exposure to weightlessness and artificial environments in space. NASA then commissioned a cyborg study, which produced a report in May 1963, surveying all manner of technologies to render astronauts more resilient to the rigors of space exploration, such as cardiovascular modules, hypothermia drugs, artificial organs, and the like.

This incident establishes the precedence of use of the term in the scientific community; but it does little to define a stable referent. In the usage we favor herein, it denotes not so much the study of a specific creature or organism as a set of regularities observed in a number of sciences that had their genesis in the immediate postwar period, sciences such as information theory, molecular biology, cognitive science, neuropsychology, computer science, artificial intelligence, operations research, systems ecology, immunology, automata theory, chaotic dynamics and fractal geometry, computational mechanics, sociobiology, artificial life, and, last but not least, game theory. Most of these sciences shared an incubation period in close proximity to the transient phenomenon called "cybernetics." Although none of the historians just cited manages to provide a quotable dictionary definition, Andy Pickering proffers a good point of departure:

> Cybernetics, then, took computer-controlled gun control and layered it in an ontologically indiscriminate fashion across the academic disciplinary board – the world, understood cybernetically, was a world of goal-oriented feedback mechanisms with learning. It is interesting that cybernetics even trumped the servomechanisms line of feedback thought by turning itself into a universal metaphysics, a Theory of Everything, as today's physicists and cosmologists use the term – a cyborg metaphysics, with no respect for traditional human and nonhuman boundaries, as an umbrella for the proliferation of individual cyborg sciences it claimed to embrace. (1995a, p. 31)

So this definition suggests that military science and the computer became melded into a Theory of Everything based on notions of automata and feedback. Nevertheless, there persists a nagging doubt: isn't this still more than a little elusive? The cyborg sciences do seem congenitally incapable of avoiding excessive hype. For instance, some promoters of artificial intelligence have engaged in wicked rhetoric about "meat machines," but, indeed, where's the beef? After all, many economists were vaguely aware of cybernetics and systems theory by the 1960s, and yet, even then, the prevailing attitude was that these were "sciences" that never quite made the grade, failures in producing results precisely because of their hubris. There is a kernel of truth in this, but only insofar as it turned out that cybernetics itself never attained the status of a fully fledged cyborg science but instead constituted the philosophical overture to a whole phalanx of cyborg sciences. The more correct definition would acknowledge that a cyborg science is a complex set of beliefs, of philosophical predispositions, mathematical preferences, pungent metaphors, research practices, and (let us not forget) paradigmatic *things*, all of which are then applied promiscuously to some more or less discrete preexistent subject matter or area.

To define cyborg sciences, it may be prudent to move from the concrete to the universal. First and foremost, the cyborg sciences depend on the existence of the computer as a paradigm object for everything from metaphors to assistance in research activities to embodiment of research products. Bluntly: if it doesn't make fundamental reference to "the computer" (itself a historical chameleon), then it isn't a cyborg science. The reason that cybernetics was able to foresee so much so clearly while producing so little was that it hewed doggedly to this tenet. And yet, there has been no requirement that the science necessarily be about the computer per se; rather, whatever the subject matter, a cyborg science makes convenient use of the fact that the computer itself straddles the divide between the animate and the inanimate, the live and the lifelike, the biological and the inert, the Natural and the Social, and makes use of this fact in order to blur those same boundaries in its target area of expertise. One can always recognize a cyborg science by the glee with which it insinuates such world-shattering questions as, Can a machine think? How is a genome like a string of binary digits in a message? Can life-forms be patented? How is information like entropy? Can computer programs be subject to biological evolution? How can physicists clarify the apparently political decision of the targeting of nuclear weapons? Can there be such a thing as a self-sufficient "information economy"? And, most disturbingly, what is it about you that makes "you" really you? Or is your vaunted individuality really an illusion?

This breaching of the ramparts between the Natural and the Social, the Human and the Inhuman, is a second and perhaps the most characteristic attribute of the cyborg sciences. Prior to World War II, a surfeit of research programs attempted to "reduce" the Social to the Natural. Neoclassical economics was just one among many, which also included Social Darwinism, Kohler's psychological field theory, Technocracy, eugenics, and a whole host of others.[6] However, the most important fact about all of these early profiles in scientism was that they implicitly left the barriers between Nature and Society relatively intact: the ontology of Nature was not altered by the reductionism, and controversies over each individual theory would always come back sooner or later to the question of "how much" of Society remained as the surd of Naturalism after the supposed unification. With the advent of the cyborg sciences after World War II, something distinctly different begins to happen. Here and there, a cyborg intervention agglomerates a heterogeneous assemblage of humans and machines, the living and the dead, the active and the inert, meaning and symbol, intention and teleology, and before we know it, Nature has taken on board many of the attributes conventionally attributed to Society, just as Humanity has simultaneously been rendered more machinelike.

[6] Sorokin, 1956, provided an early catalog. More modern compendia include Cohen, 1994; Mirowski, 1994a; Hodgson, 1993; Degler, 1991; Hawkins, 1997.

Whereas before World War II, the drive for unification always assumed the format of a take-no-prisoners reductionism, usually with physicists unceremoniously inserting their traditions and formalisms wholesale onto some particular sphere of social or biological theory, now it was the ontology of Nature itself that had grown ambiguous. Not just the bogeyman of postmodernism challenged the previous belief in an independent Nature: the question of what counts as Natural is now regularly disputed in such areas as artificial life (Levy, 1992; Helmreich, 1995), cognitive science (Dennett, 1995), and conservation ecology (Cronon, 1995; Soule & Lease, 1995; Takacs, 1996). Interdisciplinarity, while hardly yet enjoying the realm of Pareto-improving exchange, now apparently takes place on a more multilateral basis. For instance, "genes" now unabashedly engage in strategies of investment, divestment, and evasion within their lumbering somatic shells (Dawkins, 1976); information and thermodynamic entropy are added together in one grand law of physical regularity (Zurek, 1990); or inert particles in dynamical systems "at the edge of chaos" are deemed to be in fact performing a species of computation.

This leads directly to a third signal characteristic of cyborg sciences, namely, that as the distinction between the Natural and the Social grows more vague, the sharp distinction between "reality" and simulacra also becomes less taken for granted and even harder to discern (Baudrillard, 1994). One could observe this at the very inception of the cyborg sciences in the work of John von Neumann. At Los Alamos, simulations of hydrodynamics, turbulence, and chain reactions were one of the very first uses of the computer, because of the difficulties of observing most of the complex physical processes that went into the making of the atomic bomb. This experience led directly to the idea of Monte Carlo simulations, which came to be discussed as having a status on a par with more conventional "experiments" (Galison, 1996). Extending well beyond an older conception of mathematical model building, von Neumann believed that he was extracting out the logic of systems, be they dynamical systems, automata, or "games"; thus manipulation of the simulation eventually came to be regarded as essentially equivalent to manipulation of the phenomenon (von Neumann, 1966, p. 21). But you didn't have to possess von Neumann's genius to know that the computer was changing the very essence of science along with its ambitions. The computer scientist R. W. Hamming once admitted:

> The Los Alamos experience had a great effect on me. First, I saw clearly that I was at best second rate. . . . Second, I saw that the computing approach to the bomb design was essential. . . . But thinking long and hard on this matter over the years showed me that the very nature of science would change as we look more at computer simulations and less at the real world experiments that, traditionally, are regarded as

> essential. . . . Fourth, there was a computation of whether or not the test bomb would ignite the atmosphere. Thus the test risked, on the basis of a computation, all of life in the known universe. (in Duren, 1988, pp. 430–31)

In the era after the fall of the Berlin Wall, when the Los Alamos atomic weapons test section is comprised of primarily computer simulations (Gusterson, 1996), his intuition has become the basis of routinized scientific inquiry. As Paul Edwards (1996) has observed, the entire Cold War military technological trajectory was based on simulations, from the psychology of the enlisted men turning the keys to the patterns of targeting of weapons to their physical explosion profile to the radiation cross sections to anticipated technological improvements in weapons to the behavior of the opponents in the Kremlin to econometric models of a postnuclear world.

Once the cyborg sciences emerged sleek and wide-eyed from their military incubator, they became, in Herbert Simon's telling phrase, "the sciences of the artificial" (1981). It is difficult to overstate the ontological import of this watershed. "At first no more than a faster version of an electro-mechanical calculator, the computer became much more: a piece of the instrument, an instrument in its own right, and finally (through simulations) a stand-in for nature itself. . . . In a nontrivial sense, the computer began to blur the boundaries between the 'self-evident' categories of experiment, instrument and theory" (Galison, 1997, pp. 44–45). While the mere fact that it can be done at all is fascinating, it is the rare researcher who can specify in much detail just "how faithful" is that particular fractal simulation of a cloud, or that global climate model, or that particular Rogetian simulation of a psychiatrist (Weizenbaum, 1976), or that particular simulation of an idealized Chinese speaker in John Searle's (1980) "Chinese Room." It seems almost inevitable that as a pristine Nature is mediated by multiple superimposed layers of human intervention for any number of reasons – from the increasingly multiply processed character of scientific observations to the urban culture of academic life – and as such seemingly grows less immediate, the focus of research will eventually turn to simulations of phenomena. The advent of the computer has only hastened and facilitated this development. Indeed, the famous "Turing Test" (discussed in Chapter 2) can be understood as asserting that when it comes to questions of mind, a simulation that gains general assent is good enough. In an era of the revival of pragmatism, this is the pragmatic maxim with a vengeance.

The fourth hallmark of the cyborg sciences is their heritage of distinctive notions of order and disorder rooted in the tradition of physical thermodynamics (a topic of extended consideration in the next chapter).

Questions of the nature of disorder, the meaning of randomness, and the directionality of the arrow of time are veritable obsessions in the cyborg sciences. Whether it be the description of information using the template of entropy, or the description of life as the countermanding of the tendency to entropic degradation, or the understanding of the imposition of order as either threatened or promoted by noise, or the depiction of chaotic dynamics due to the "butterfly effect," or the path dependence of technological development, the cyborg sciences make ample use of the formalisms of phenomenological thermodynamics as a reservoir of inspiration. The computer again hastened this development, partly because the question of the "reliability" of calculation in a digital system focused practical attention on the dissipation of both heat and signals; and partly because the computer made it possible to look in a new way for macrolevel patterns in ensembles of individual realizations of dynamic phenomena (usually through simulations).

The fifth hallmark of a cyborg science is that terms such as "information," "memory," and "computation" become for the first time *physical* concepts, to be used in explanation in the natural sciences. One can regard this as an artifact of the computer metaphor, but in historical fact their modern referents are very recent and bound up with other developments as well (Aspray, 1985; Hacking, 1995). As Hayles (1990a, p. 51) explains, in order to forge an alliance between entropy and information, Claude Shannon had to divorce information from any connotations of meaning or semantics and instead associate it with "choice" from a preexistent menu of symbols. "Memory" then became a holding pen for accumulated message symbols awaiting utilization by the computational processor, which every so often had to be flushed clean due to space constraints. The association of this loss of memory with the destruction of "information" and the increase of entropy then became salient, as we shall discover in Chapter 2. Once this set of metaphors caught on, the older energetics tradition could rapidly be displaced by the newer cybernetic vocabulary. As the artificial life researcher Tom Ray put it: "Organic life is viewed as utilizing energy . . . to organize matter. By analogy, digital life can be viewed as using CPU to organize memory" (in Boden, 1996, p. 113). Lest this be prematurely dismissed as nothing more than an insubstantial tissue of analogies and just-so stories, stop and pause and reflect on perhaps the most pervasive influence of the cyborg sciences in modern culture, which is to treat "information" as an entity that has ontologically stable properties, preserving its integrity under various transformations.

The sixth defining characteristic of the cyborg sciences is that they were not invented in a manner conforming to the usual haphazard image of the lone scientist being struck with a brilliantly novel idea in a serendipitous academic context. It is a historical fact that each of the cyborg sciences traces its inception to the conscious intervention of a new breed of science manager, empowered by the crisis of World War II and fortified by lavish foundation and military sponsorship. *The new cyborg sciences did not simply spontaneously arise; they were consciously made.* The usual pattern (described in Chapter 4) was that the science manager recruited some scientists (frequently physicists or mathematicians) and paired them off with collaborators from the life sciences and/or social sciences, supplied them with lavish funding along a hierarchical model, and told them to provide the outlines of a solution to a problem that was bothering their patron. Cyborg science is Big Science par excellence, the product of planned coordination of teams with structured objectives, expensive discipline-flouting instrumentation, and explicitly retailed rationales for the clientele. This military inspiration extended far beyond mere quotidian logistics of research, into the very conceptual structures of these sciences. The military rationale often imposed an imperative of "command, control, communications, and information" – shorthand, C^3I – upon the questions asked and the solutions proposed. Ultimately, the blurred ontology of the cyborg sciences derives from the need to subject heterogeneous agglomerations of actors, machines, messages, and (let it not be forgotten) opponents to a hierarchical real-time regime of surveillance and control (Galison, 1994; Pickering, 1995a; Edwards, 1996).

The culmination of all these cyborg themes in the military framework can easily be observed in the life and work of Norbert Wiener. Although he generally regarded himself as an antimilitarist, he was drawn into war work in 1941 on the problem of antiaircraft gunnery control. As he explained it in 1948, "problems of control engineering and of communication engineering were inseparable, and . . . they centered not around the techniques of electrical engineering but around the more fundamental notion of the message. . . . The message is a discrete or continuous sequence of measurable events distributed in time – precisely what is called a time series by statisticians" (1961, p. 8). Under the direction of Warren Weaver, Wiener convened a small research group to build an antiaircraft motion predictor, treating the plane and the pilot as a single entity. Because the idiosyncrasies of each pilot could never be anticipated, prediction was based on the ensemble of all possible pilots, in clear analogy with thermodynamics. In doing so, one had to take into account possible evasive measures, leading to the sorts of considerations that would now be associated with strategic predictions, but which Wiener saw as essentially similar to servomechanisms, or feedback devices used to control engines. Although his gunnery predictor never proved superior to simpler methods already in use, and therefore was never actually implemented in combat, Wiener was convinced that the principles he had developed had much wider significance and application. His report on the resulting statistical work, *Extrapolation, Interpolation and Smoothing of Stationary Time Series* (1949), is considered the seminal theoretical work in communications theory and time series analysis (Shannon & Weaver, 1949, p. 85n). Yet his manifesto for the new science of *cybernetics* (1961) had even more far-reaching consequences. Wiener believed his melange of statistical, biological, and computational theories could be consolidated under the rubric of "cybernetics," which he coined from the Greek word meaning "steersman." As he later wrote in his biography, "life is a perpetual wrestling match with death. In view of this, I was compelled to regard the nervous system in much the same light as a computing machine" (1956, p. 269). Hence military conflict and the imperative for control were understood as a license to conflate mind and machine, Nature and Society.

Although many of the historians (Haraway, Pickering, Edwards, et al.) I have cited at the beginning of this chapter have made most of these same points about the cyborg sciences at one time or another in various places in their writings, the one special aspect they have missed is that the early cyberneticians did not restrict their attentions simply to bombs and brains and computers; from the very start, they had their sights trained upon *economics* as well, and frequently said so. Just as they sought to reorient the physical sciences toward a more organicist modality encompassing mind, information, and organization, they also were generally dissatisfied with the state of the neoclassical economic theory that they had observed in action, especially in wartime. Although the disdain was rife amongst the cyborg scientists, with John von Neumann serving as our star witness in Chapter 3, we can presently select one more quote from Wiener to suggest the depths of the dissatisfaction:

> From the very beginning of my interest in cybernetics, I have been well aware that the considerations of control and communications which I have found applicable in engineering and in physiology were also applicable in sociology and in economics. . . . [However,] the mathematics that the social scientists employ and the mathematical physics they use as their model are the mathematics and mathematical physics of 1850. (1964, pp. 87, 90)

ATTACK OF THE CYBORGS

It is always a dicey proposition to assert that one is living in a historical epoch when one conceptual system is drawing to a close and another rising to take its place; after all, even dish soaps are frequently retailed as new and revolutionary. It may seem even less prudent to attempt the sketch of such a scenario when one is located in a discipline such as economics, where ignorance of history prompts the median denizen to maintain that the wisdom *du jour* is the distilled quintessence of everything that has ever gone before, even as he conveniently represses some of his own intellectual gaffes committed in his salad days. Although the purpose of this volume is to provide detailed evidence for this scenario of rupture and transformation between early neoclassicism and the orthodoxy after the incursion of the cyborgs, it would probably be wise to provide a brief outline up front of the ways in which the cyborg sciences marked an epochal departure from rather more standard neoclassical interpretations of the economy. The bald generalizations proffered in this section are documented throughout the rest of this volume.

As we have noted, economists did not exactly lock up their doors and set the guard dogs loose when the cyborgs first came to town. That would have gone against the grain of nearly seventy years of qualified adherence to a model of man based upon the motion of mass points in space; and anyway it would have been rude and ungracious to those physical scientists who had done so much to help them out in the past. Economists in America by and large welcomed the physicists exiled by war and persecution and unemployment with open arms into the discipline in the 1930s and 1940s; these seemed the sorts of folks that neoclassicals had wanted to welcome to their neighborhood. The first signs of trouble were that, when the cyborgs came to town, the ideas they brought with them did not seem to conform to those which had represented "science" to previous generations of economists, as we shall recount in Chapters 5 and 6. Sure, they plainly understood mechanics and differential equations and formal logic and the hypothetico-deductive imperative; but there were some worrisome danger signs, like a nagging difference of opinion about the meaning of "dynamics" and "equilibrium" (Weintraub, 1991), or suspicions concerning the vaunting ambitions of "operations research" and "systems analysis" (Fortun & Schweber, 1993), or wariness about von Neumann's own ambitions for game theory (Chapter 6). For reasons the economists found difficult to divine, some of the scientists resisted viewing the pinnacle of social order as the repetitive silent orbits of celestial mechanics or the balanced kinetics of the lever or the hydraulics of laminar fluid flow.

If there was one tenet of that era's particular faith in science, it was that logical rigor and the mathematical idiom of expression would produce transparent agreement over the meaning and significance of various models and their implications. This faith, however, was sorely tested when it came to that central concept of nineteenth-century physics and of early neoclassical economics, energy. When the neoclassicals thought about energy, it was in the context of a perfectly reversible and deterministic world exhibiting a stable and well-defined "equilibrium" where there was no free lunch. The cyborg scientists, while also having recourse to the terminology of "equilibria," seemed much more interested in worlds where there was real irreversibility and dissipation of effort. They seemed less worried whether lunch was paid for, because their thermodynamics informed them that lunch was always a loss leader; hence they were more concerned over why lunch existed at all or, perhaps more to the point, what functions did lunch underwrite that could not have been performed in some other manner? For the cyborgs, energy came with a special proviso called "entropy," which could render it effectively inaccessible, even when nominally present; many arguments raged in this period how such a macroscopic phenomenon could be derived from underlying laws of mechanics that were apparently deterministic and reversible.

The premier language that had been appropriated and developed to analyze macroscopic phenomena in thermodynamics was the theory of probability. The cyborg scientists were convinced that probability theory would come to absorb most of physics in the near future; quantum mechanics only served to harden these convictions even further. By contrast, neoclassicals in the 1920s and 1930s had been fairly skeptical about any substantive role for probability within economic theory.[7] Because they had grown agnostic about what, if anything, went on in the mind when economic choices were made, initially the imposition of some sort of probabilistic overlay upon utility was avoided as a violation of the unspoken rules of behaviorism. Probability was more frequently linked to statistics and, therefore, questions of empiricism and measurement; an orthodox consensus on the appropriate status and deployment of those tools had to await the stabilization of the category "econometrics," something that did not happen until after roughly 1950. Thus once the cyborg sciences wanted to recast the question of the very nature of order as a state of being that was inherently stochastic, neoclassical economists were initially revulsed at the idea of the market as an arena of chance, a play of sound and fury that threatened to signify nothing (Samuelson, 1986).

These two predispositions set the cyborg sciences on a collision course with that pursued by neoclassical economics in the period of its American ascendancy, roughly the 1940s through the 1960s. Insofar as neoclassicals believed in Walrasian general equilibrium (and many did not), they thought its most admirable aspect was its stories of Panglossian optimality

[7] This argument is made in Mirowski, 1989c. A very good example of an influential text that provided a justification for hostility to probability in the 1920s is Knight, 1940; although its modern interpretation seeks to interpret it as warrant for a reconciliation of calculable "risk" and incalculable "uncertainty" within the orthodoxy. For further consideration, see Emmett, 1998.

and Pareto improvements wrought by market equilibria. Cyborg scientists were not averse to making use of the mathematical formalisms of functional extrema, but they were much less enamored of endowing these extrema with any overarching significance. For instance, cyborg science tended to parse its dynamics in terms of basins of attraction; due to its ontological commitment to dissipation, it imagined situations where a plurality of attractors existed, with the codicil that stochastic considerations could tip a system from one to another instantaneously. In such a world, the benefits of dogged optimization were less insistent and of lower import, and thus the cyborg sciences were much more interested in coming up with portrayals of agents that just "made do" with heuristics and simple feedback rules. As we have seen, this prompted the cyborg sciences to trumpet that the next frontier was the mind itself, which was conceived as working on the same principles of feedback, heuristics, and provisional learning mechanisms that had been pioneered in gun-aiming algorithms and operations research. This could not coexist comfortably with the prior neoclassical framework, which had become committed in the interim to a portrayal of activity where the market worked "as if" knowledge were perfect, and took as gospel that agents consciously attained preexistent optima. The cyborg scientists wanted to ask what could in principle be subject to computation; the neoclassicals responded that market computation was a fait accompli. To those who complained that this portrait of mind was utterly implausible (and they were legion), the neoclassicals tended to respond that they needed no commitment to mind whatsoever. To those seeking a new theory of social organization, the neoclassicals retorted that all effective organizations were merely disguised versions of their notion of an ur-market. This set them unwittingly on a collision course with the cyborg sciences, all busily conflating mind and society with the new machine, the computer.

Whereas the neoclassicals desultorily dealt in the rather intangible, ever present condition called "knowledge," the cyborg scientists were busy defining something else called *information*. This new entity was grounded in the practical questions of the transmission of signals over wires and the decryption of ciphers in wartime; but the temptation to extend its purview beyond such technical contexts proved irresistible. Transmission required some redundancy, which was given a precise measure with the information concept; it was needed because sometimes noise could be confused with signal and, perhaps stranger, sometimes noise could boost signal. For the neoclassicals, on the other hand, noise was just waste; and the existence of redundancy was simply a symptom of inefficiency, a sign that someone, somewhere, was not optimizing. The contrast could be summed up in the observation that neoclassical economists wanted their order austere and simple and their a priori laws temporally invariant, whereas the cyborg scientists tended to revel in diversity and complexity and change, believing that order could only be defined relative to a background of noise and chaos, out of which the order should temporally emerge as a process. In a phrase, the neoclassicals rested smugly satisfied with classical mechanics, while the cyborgs were venturing forth to recast biology as a template for the machines of tomorrow.

These sharply divergent understandings of what constituted "good science" resulted in practice in widely divergent predispositions as to where one should seek interdisciplinary collaboration. What is noteworthy is that while both groups essentially agreed that a prior training in physics was an indispensable prerequisite for productive research, the directions in which they tended to search for their inspiration were very nearly orthogonal. The most significant litmus test would come with revealed attitudes toward biology. Contrary to the impression given by Alfred Marshall in his *Principles*, the neoclassical economists were innocent of any familiarity with biology and revealed minuscule inclination to learn any more. This deficiency did not prevent them from indulging in a little evolutionary rhetoric from time to time, but their efforts never adequately took into account any contemporary understandings of evolutionary theory (Hodgson, 1993), nor was it ever intended to. In contrast, from their very inception, the cyborg scientists just knew in their prosthetic bones that the major action in the twentieth century would happen in biology. Partly this prophecy was self-fulfilling, because the science managers both conceived and created "molecular biology," the arena of its major triumph.[8] Nevertheless, they saw that their concerns about thermodynamics, probability, feedback, and mind all dictated that biology would be the field where their novel definitions of order would find some purchase.

Another agonistic field of interdisciplinary intervention from the 1930s onward was that of logic and metamathematics. Neoclassical economists were initially attracted to formal logic, at least in part because they believed that it could explain how to render their discipline more rigorous and scientific, but also because it would provide convincing justification for their program to ratchet up the levels of mathematical discourse in the field. For instance, this was a major consideration in the adaptation of the Bourbakist approach to axiomatization at the Cowles Commission after 1950 (Weintraub & Mirowski, 1994). What is noteworthy about this choice was the concerted effort to circumvent and avoid the most disturbing aspects of metamathematics of the 1930s, many of which

[8] See Chapter 4 on Warren Weaver and the interplay of cyborg science and postwar science policy in America, and more generally the work of Lily Kay (1993, 1995, 1997a, 1997b) and Robert Kohler (1991).

revolved around Gödel's incompleteness results. In this regard, it was the cyborg scientists, and not the neoclassicals, who sought to confront the disturbing implications of these mathematical paradoxes, and turn them into something positive and useful. Starting with Alan Turing, the theory of computation transformed the relatively isolated and sterile tradition of mathematical logic into a general theory of what a machine could and could not do in principle. As described in the next chapter, cyborgs reveled in turning logical paradoxes into effective algorithms and computational architectures; and, subsequently, computation itself became a metaphor to be extended to fields outside of mathematics proper. While the neoclassical economists seemed to enjoy a warm glow from their existence proofs, cyborg scientists needed to get out and *calculate*. Subsequent generations of economists seemed unable to appreciate the theory of computation as a liberating doctrine, as we shall discover in Chapter 7.

THE NEW AUTOMATON THEATRE

Steven Millhauser has written a lovely story contained in his collection *The Knife Thrower* (1998) called "The New Automaton Theatre," a story that in many ways illustrates the story related in this volume. He imagines a town where the artful creation of lifelike miniature automata has been carried far beyond the original ambitions of Vaucanson's Duck or even Deep Blue – the machine that defeated Gary Kasparov. These automata are not "just" toys but have become the repositories of meaning for the inhabitants of the town:

> So pronounced is our devotion, which some call an obsession, that common wisdom distinguishes four separate phases. In childhood we are said to be attracted by the color and movement of these little creatures, in adolescence by the intricate clockwork mechanisms that give them the illusion of life, in adulthood by the truth and beauty of the dramas they enact, and in old age by the timeless perfection of an art that lifts us above the cares of mortality and gives meaning to our lives. . . . No one ever outgrows the automaton theatre.

Every so often in the history of the town there would appear a genius who excels at the art, capturing shades of human emotion never before inscribed in mechanism. Millhauser relates the story of one Heinrich Graum, who rapidly surpasses all others in the construction and staging of automata. Graum erects a Zaubertheatre where works of the most exquisite intricacies and uncanny intensity are displayed, which rival the masterpieces of the ages. In his early career Graum glided from one triumph to the next; but it was "as if his creatures strained at the very limits of the human, without leaving the human altogether; and the intensity of his figures seemed to promise some final vision, which we awaited with longing, and a little dread."

And then, at age thirty-six and without warning, Graum disbanded his Zaubertheatre and closed his workshop, embarking on a decade of total silence. Disappointment over this abrupt mute reproach eventually gave way to fascinations with other distractions and other artists in the town, although the memory of the old Zaubertheatre sometimes haunted apprentices and aesthetes alike. Life went on, and other stars of the Automata Theatre garnished attention and praise. After a long hiatus, and again without warning, Graum announced he would open a Neues Zaubertheatre in the town. The townsfolk had no clue what to expect from such an equally abrupt reappearance of a genius who had for all intents and purposes been relegated to history. The first performance of the Neues Zaubertheatre was a scandal or, as Millhauser puts it, "a knife flashed in the face of our art." Passionate disputes broke out over the seemliness or the legitimacy of such a new automaton theater.

> Those who do not share our love of the automaton theatre may find our passions difficult to understand; but for us it was as if everything had suddenly been thrown into question. Even we who have been won over are disturbed by these performances, which trouble us like forbidden pleasures, secret crimes. . . . In one stroke his Neues Zaubertheatre stood history on its head. The new automatons can only be described as clumsy. By this I mean that the smoothness of motion so characteristic of our classic figures has been replaced by the jerky abrupt motions of amateur automatons. . . . They do not strike us as human. Indeed it must be said that the new automatons strike us first of all as *automatons*. . . . In the classic automaton theatre we are asked to share the emotions of human beings, whom in reality we know to be miniature automatons. In the new automaton theatre we are asked to share the emotions of the automatons themselves. . . . They live lives that are parallel to ours, but are not to be confused with ours. Their struggles are clockwork struggles, their suffering is the suffering of automatons.

Although the townsfolk publicly rushed to denounce the new theater, over time they found themselves growing impatient and distracted with the older mimetic art. Many experience tortured ambivalence as they sneak off to view the latest production of the Neues Zaubertheatre. What was once an affront imperceptibly became a point of universal reference. The new theater slowly and inexorably insinuates itself into the very consciousness of the town.

It has become a standard practice in modern academic books to provide the impatient modern reader with a quick outline of the argument of the entire book in the first chapter, providing the analogue of fast food for the marketplace of ideas. Here, Millhauser's story can be dragooned for that purpose. In sum, the story of this book is the story of the New Automaton Theatre: the town is the American profession of academic economics, the classic automaton theater is neoclassical economic theory, and the Neues Zaubertheatre is the introduction of the cyborg sciences into economics. And Heinrich Graum – well, Graum is John von Neumann. The only thing missing from Millhauser's parable would a proviso where the military would have acted to fund and manage the apprenticeships and workshops of the masters of automata, and Graum's revival stage-managed at their behest.

2 Some Cyborg Genealogies: or, How the Demon Got Its Bots

> The human body is a machine which winds its own springs. It is the living image of perpetual motion.
>
> Joseph la Mettrie

> In Gibbs' universe order is least probable, chaos most probable. But while the universe as a whole . . . tends to run down, there are local enclaves whose direction seems opposed. . . . Life finds its home in some of these enclaves. It is with this point of view at its core that the new science of Cybernetics began its development.
>
> Norbert Wiener

A few years ago, the historian of science Adele Clark sought to see the world through a child's eyes and pose the question, "Mommy, where do cyborgs come from?" (in Gray, 1995, p. 139). As one often would when interrupted by a child's nagging questions, the overpowering temptation is to brush it off with a glib answer, perhaps something dismissive like: "Other cyborgs." But if one instead begins to appreciate the responsibilities of parental obligation, the temptation shifts to a different plane, that of providing a nominally correct but semantically empty answer, perhaps something concerning the "enclaves" about which Wiener wrote. Those enclaves sound an awful lot like incubators: they are closed rooms where pressure and temperature are maintained at homeostatic levels conducive to embryonic life. Yet someone must be there to maintain that temperature and pressure; they had to convey the germs of life there from outside the closed room; and someone else had to install the boxes there in the first place. Genealogies are a tricky business, almost as tricky as life itself, frequently threatening to succumb to entropic degradation.

For instance, I have no doubt that one of the main prerequisites for the debut of cyborgs in our century was the drive to control and manipulate the "quality" as well as the quantity of our own offspring, as Clarke and other feminists maintain. Cyborg aspirations have been a notoriously masculine province, with everything that implies about the struggle to wrest reproduction away from the female body. We have hinted above in Chapter 1 that fascinations with automata easily date back to the Enlightenment. Yet the overwhelming importance of World War II and its impact upon the organization and funding of the sciences was also crucial, as we shall learn in the subsequent chapters. Shifting locations of physics and biology within implied cultural hierarchies of knowledge, as well as the rise and fall of theories within the sciences, cannot be discounted as relevant. One should not forget that, in many ways, cyborgs were themselves spawned by dramatic transformations over the longer term in the bureaucratic organization of firms and corporations, as they struggled to exert some semblance of control over their sprawling structures and increasing complexity (Beniger, 1986; Martin Campbell-Kelley in Bud-Friedman, 1994). And then, I am sure, there are those who will seek to explain the appearance of the cyborgs as pure artifact of technological imperatives: the steam engine, the telegraph wire, the electricity grid, the radio, the magnetron, and, of course, the computer.

The problem with all these family trees is not so much that they are mistaken but that they are too limited, hemmed in by their individually parochial perspectives, each a partial account of something that looms well beyond their range of vision. Cyborgs by their very nature (if they can be even spoken about in such an idiom) tend to trample over such simple causal accounts and their disciplinary bailiwicks. Male versus female, pure versus applied, military versus civilian, conceptual versus technological, abstract versus concrete, mechanism versus organism, evolution versus teleology, success versus failure – nothing stays sacred for very long wherever cyborgs come loping into the neighborhood. The question provoked by the advent of cyborgs should perhaps rather be rephrased: what is a machine, that a living being might know it; and what is a (wo)man, that it can be comprehended by a machine?

The chapter epigraph – a juxtaposition of quotations from la Mettrie and Wiener – captures in a phrase the tensions between some older and newer approaches to the problem of the nature of life and the meaning of machines. In the eighteenth century, when Vaucanson constructed his mechanical duck (Riskin, 1998) and mechanical flautist, and John Merlin constructed his dancing automatons (Schaffer, 1996), they were intended to exemplify and embody the virtues of regularity, order, and harmony through the mechanical integration of smoothly interacting parts, subject to a central power source, a tangible artifact that "illustrated and thus reinforced the general world view of determinism" (Mayr, 1986, p. 119). Clocks first disciplined the laboring body and then became indistinguishable from it (Thompson, 1967). Life was regarded as essentially similar to other orderly natural phenomena, such as the solar system or the well-functioning hierarchical state. This pervasive determinism might create problems here and there for notions of freedom and will, as Kant and others noticed; but the dispute (such as it was) was carried out entirely within the unquestioned context of the presumption of a timeless dynamics. Consequently, the metaphor of balance ruled most eighteenth-century attempts at mathematical discourse (Wise, 1993). This mechanical mind-set was initially given a big boost by the institution of the energy concept in physics in the mid-nineteenth century, and, as I have argued (Mirowski, 1989a), the version that inscribed these laws as extremum principles subject to external constraints provided the immediate inspiration for the genesis of neoclassical economics.

The fact that the clockwork mind-set was distended beyond all recognition in the later nineteenth century has been noted by many historians: "the eighteenth century machine was a product of the Newtonian universe with its multiplicity of forces, disparate sources of motion, and reversible mechanism. By contrast the nineteenth century machine, modeled on the thermodynamic engine, was a 'motor,' the servant of a powerful nature conceived as a reservoir of motivating power. The machine was capable of work only when powered by some external source, whereas the motor was regulated by internal, dynamical principles, converting fuel into heat, and heat into mechanical work. The body, the steam engine, and the cosmos were connected by a single unbroken chain of energy" (Rabinbach, 1990, p. 52). The law of the conservation of energy, which had initially dangled the delectable promise of a grand unification of all physical and mental phenomena, instead ushered in an era in which conversions of all manner of energetic phenomena brought in their wake unforeseen loss, degradation, and dissolution. Steam engines never gave back as good as they got; this was the fundamental insight of Sadi Carnot's 1824 essay on the motive power of fire, now credited as one conceptual source of the Second Law of Thermodynamics. Controversies over the rate of cooling of the earth insinuated that even the clockwork heavens were really running down, heading for a Heat Death undreamt of in the heavenly world of the eighteenth-century philosophers.[1] Yet it seemed it was not just the solar system, or the British Empire, that was rushing headlong into a state of decline. If energy really were conserved, then what happened to the body, which displayed irreversible corruption

[1] As late as 1905, Ernest Rutherford maintained, "science offers no escape from the conclusion of Kelvin and Helmholtz that the sun must ultimately grow cold and this earth will become a dead planet moving through the intense cold of empty space" (in Smith & Wise, 1989, p. 551). Chapters 15 and 16 of Smith and Wise, 1989, discuss this literature in detail. It should be noted that these calculations were also used to attack Darwinism in this period. See Desmond & Moore, 1991, chap. 38.

and decrepitude? If the decline really was inevitable, then how was it that life did nonetheless persist, much less triumph? Was this merely the sacrifice of the individual organism for the benefit of the group? Or was it the triumph of *intelligence* over brute nature? In what sense was the theory of evolution opposed to the physical dictates of thermodynamics? Was Nature our nurturing mother, or was it demonic, malevolent, fundamentally hostile to human persistence? The skirmish between the believers in progress and the cassandras of a scientific prognosis of decay was one of the major cultural dislocations bequeathed to the twentieth century. It would extend even unto the realms of thought, which also had its fin-de-siècle prophets of decline (as it does in our own). Could a Golden Age really be followed by an Age of Dross, where the achievements of the former were lost to the mists of memory of the latter? These, I believe, were some of the primal anxieties that first gave rise to cyborgs.

In this volume we take it as gospel that one cannot abstract away the questions of the nature of life and its similarity to machines from questions concerned with the nature of thought, the telos of human striving, and the potential law-governed character of human provisioning and social organization. In other words, quotidian distinctions automatically made by almost all Western writers between physics, biology, psychology, and economics are rather unavailing when it comes to understanding the rise of the cyborg sciences in the twentieth century. This is not to propound the strong claim that all valid science is or should be unified (it is not); nor is it to make the rather ineffectual (but true) observation that transfer of metaphors has been fast and furious between these areas of inquiry in the same period. Rather, it is intended to document the thesis that comprehension of the trajectory of any one of these disciplines requires one to transgress the boundaries of the others with the same abandon and impudence as the cyborgs themselves.[2] Nevertheless, there is no disguising the fact that the overriding motivation of this volume is to produce an understanding of the making of the shape of the twentieth-century economics profession, along with its kit bag of orthodox doctrines generated in America around midcentury – a tall order, but still not quite the cyborg saga pitched at the most rarified levels of Olympian detachment favored by philosophers of science and some science studies scholars.[3] Here instead we attempt to provide in this volume a history of economics *from the cyborg point of view*. In order to

[2] "I would like to suggest that the history of computer science – if and when it comes to be written – will establish a new and different paradigm for history writing. It may indeed rid us of certain stereotypes common to the history of science, with its overemphasis on the history of theoretical physics" (Metropolis, 1992, p. 122).

[3] If names must be attached, exemplars may be provided by Dennett, 1995; Haraway, 1997; and Latour, 1999.

simulate that perspective, we find that in this chapter we have to zig and zag between early computer history, nineteenth-century British economics, Maxwell's Demon, mathematical logic, conceptions of biological metabolism and evolution, cybernetics, and then back to the abstract theory of computation. Our staggered syncopated account is motivated by the fact that these component themata often developed out of phase with one another, sometimes getting decoupled for a generation or more, and then abruptly coming back into contact, with profound effects for all concerned. While this will occur time and again in our subsequent narrative, it may help offset the threatened vertigo to make a few brief preliminary statements about the origins of neoclassical economics, the tradition that would serve as the basis for the eventual construction of the economic orthodoxy in America.

The pall of turn-of-the-century gloom in the physical sciences having to do with dissipation and death had a curiously delayed impact on the nascent discipline of economics. Indeed, if one narrowly confines one's attention to the school of neoclassical economics, just showing its first stirrings in various works in the 1870s, it is almost completely absent. Whereas the early neoclassical progenitors had paid close attention to mechanics for their inspiration, they betrayed no signs of appreciation for thermodynamics, and the attempts of Marshall and others to take a few lessons on board from biology were garbled, ineffectual, and quickly forgotten (Hodgson, 1993). By neglecting many of the trends in the sciences outlined in this chapter, the neoclassical tradition of scientific discourse managed to maintain a pristine determinism and an untrammeled optimism nearly unparalleled in the contemporary histories of many of the other sciences. By occupying what it saw as the side of the Angels, it missed out on an appreciation for the activities of some very lively Demons. It is a prerequisite for understanding the vicissitudes of the neoclassical orthodoxy in the twentieth century to keep in mind that the paradigm of the constrained optimization of utility split off from its physics inspiration just prior to the elaboration of the Second Law of Thermodynamics (Mirowski, 1989a, pp. 389–90); this goes some distance in explaining its persistent blindness to problems of dissipation, the fragile nature of knowledge, and problems of temporal orientation until at least the middle of the twentieth century, if not later.[4] Another way of phrasing

[4] "We may safely accept as a satisfactory scientific hypothesis that the doctrine so grandly put forward by Laplace, who asserted that a perfect knowledge of the universe, as it existed at any moment, would give perfect knowledge of what was to happen thenceforth and forever after. Scientific inference is impossible, unless we regard the present as the outcome of what is past, and the cause of what is to come. To the perfect intelligence nothing is uncertain" (Jevons, 1905, pp. 738–39).

this observation is to say that neoclassical economics was (by choice or chance) generally excluded from the cyborg genealogy outlined in this chapter after an initial promising encounter, thus rendering their rendezvous in the period after World War II (the subject of Chapters 4–6) all that much more dramatic. In order to document this missed opportunity, in this chapter we first demonstrate that the histories of economics and the computer were initially intimately intertwined, only then to become separated by a *cordon sanataire* by the end of the nineteenth century. These separate strands were then brought back together in the twentieth century by John von Neumann, the subject of our next chapter. The remainder of this chapter is taken up with some prerequisite background in the history of thermodynamics and computation, if only so that the reader may better recognize the return of the cyborg in subsequent encounters.

THE LITTLE ENGINES THAT COULD'VE

When one begins to entertain the notion that "machines are good to think with," one could potentially mean any number of conflicting statements. It could signify merely that, bolstered by machines of all sorts, it is just easier for you and I to soar on wings of thought. A warm building, a reading light, printed books, a mechanical pencil: all grease the wheels of intellectual endeavor. A deeper signification might be that the very exemplars of thought in the modern world have come to take their bearings from experience with machines. For instance, by the seventeenth century, "The taking apart of clockwork became an illustration of that process known as *analysis*" (Mayr, 1986, p. 84). How much truly useful mathematics, in the first instance, has been prompted by machines? And a third signification, the one so dissonant that sends shivers up the spine of every humanist, is that the machines themselves can do the thinking for us. This, of course, is a veiled reference to something called "the computer."

It is striking that conventional histories of the computer nowadays start with people we would consider *economists*, before settling down to recount the accomplishments of farsighted engineers and perspicuous logicians. Take, for instance, the popular history by Martin Campbell-Kelley and William Aspray. Their narrative starts out with the problem of the production of various mathematical tables and almanacs, such as tables of logarithms and nautical almanacs. There we are ushered along to make the acquaintance of Gaspard de Prony (1755–1839), a professor of mathematical analysis at the Ecole Polytechnique and director of a Bureau de Cadastre in Paris dedicated to the production of detailed maps and tables under the just-instituted novel metric system. De Prony found the logistics of the operation daunting, but later claimed to have discovered the inspiration for its solution in reading – of all things – Adam Smith's *Wealth of Nations*. As he put it in his own retrospective account: "I came across the chapter where the author treats of the division of labor; citing, as an example of the great advantages of this method, the manufacture of pins. I conceived all of a sudden the idea of applying the same method to the immense work with which I had been burdened, and to manufacture logarithms as one manufactures pins."[5] Whereas academic mathematicians were recruited to organize and prescribe the work to be performed, unemployed hairdressers (displaced by the obsolescence of aristocratic hairstyles in the French Revolution) were tapped to perform the tedious and repetitive calculations. Complex and involved calculations were resolved down into their simplest arithmetic components and then farmed out to low-skilled and cheaper labor. These hairdressers were the first computers, according to Campbell-Kelley and Aspray, because the term "computers" originally referred not to machines but to people assigned these tasks of calculation.

Actually, the identification of de Prony as a direct progenitor of the computer is not an intemperate flight of fancy of historians; he was distinctly identified as such by someone who is readily conceded by all and sundry as deserving that designation, namely, Charles Babbage.[6] Babbage was a mathematician and a polymath who could not readily be confined to any particular disciplinary box, although his writings on economics are indispensable for understanding his greatest claim to fame, the design and abortive construction of the Difference Engine, and the sketch of the later Analytical Engine. The story of these engines, retailed in many places (Swade, 1991; Bowden, 1953), is the quest by Babbage to mechanize the production of mathematical tables beginning in 1821. He managed to arrange a subvention from the British government to underwrite the production of Difference Engine No. 1 in 1823, and was engaged in supervising its construction until the process ground to a halt in 1833. Numerous obstacles arose to the completion of the engine, ranging from

[5] This account appears in Campbell-Kelley & Aspray, 1996, pp. 11–12. The best source on de Prony and his work can be found in Grattan-Guinness, 1990; the subsequent quotation from de Prony is taken from p. 179 of that article.

[6] Charles Babbage (1791–1871). Babbage's own "autobiography," really a rather disjointed set of memoirs, was published as *Passages from the Life of a Philosopher* (1864, reprint, 1994). The passage on de Prony can be found in the *Economy of Machinery*, chap. 20. Modern biographical sources are Swade, 1991; Hyman, 1982; Schaffer, 1994, 1996. A brief attempt to summarize Babbage's importance as an economist may be found in Rosenberg, 1994. As Campbell-Kelley and Aspray insist, "Babbage's unique role in 19th century information processing was due to the fact that he was in equal measure a mathematician and an economist" (1996, p. 13).

technological glitches arising from the extrafine tolerances demanded by his designs for the cogs and toothed gears, to disputes with his mechanic Joseph Clement, to Babbage's own irascible character, including a predisposition to continually redesign and improve upon the original plans to the detriment of the exigencies of bringing the process to a swift conclusion. It was a characteristic flaw of Babbage's temperament that, when the project broke down, not only did he have the hubris to dun the government for more funds, but had the indisputably bad judgment to concurrently hint that the crisis had provoked him to rethink the entire design of the engine in 1834 and conceive of a genuine all-purpose calculator (unlike the purpose-built Difference Engine, which was geared only to perform one form of arithmetic operation) called the Analytical Engine. Peel's government understandably began to worry about throwing good money after bad and pulled the plug on the entire project. Babbage's Analytical Engine, which existed only as a disheveled set of drawings, is now considered to have displayed all the major components of a modern von Neumann architecture for the stored-program computer roughly a century before the fact. It was never built in his lifetime, although a working version of Difference Engine No. 2 was constructed for exhibition at the London Science Museum in 1990.[7]

Abundant evidence motivates the proposition that Babbage was a protocyborg well before the fact. For instance, the tenets of his economics informed his design of the Analytical Engine, and vice versa, a fact Babbage never hesitated himself to point out. His constant kibitzing on fine points of the construction of the Difference Engine led him to tour all manner of factories in England and abroad searching for improved machine technologies; his expertise in knowledge of factories thus obtained then inspired the composition of *On the Economy of Machinery and Manufactures* (1832; reprint, 1989b), an unsurpassed tour de force on the social and technological organization of factory production in the Industrial Revolution. The major theme of the *Economy of Machinery* is a revision of Adam Smith's account of the division of labor, where for Babbage the great virtue of dividing labor is separating out the dreary low-skill components of any job and lowering the pay commensurately for those unfortunate enough to qualify to do it. Marx leaned heavily on Babbage's text, only to take the insight one step further away from praise: the sociotechnical organization of the factory was for him the means by which the exploitation of labor guaranteed the maintenance of the rate

[7] Difference Engine No. 2 has 4,000 parts excluding the printing mechanism, is 7 feet high, 11 feet long, and weighs 3 tons. It is purportedly capable of carrying out its calculations to 31 places of accuracy. Modern photographs can be perused in Swade, 1991. The machine on display was not in operation when I visited there in 1999.

of profit. Babbage, of course, took the insight in a different direction: in chapter 20 on the "Division of Mental Labours" he maintained that human thought was as susceptible to principles of rational organization as the weaving of brocade cloth. As he put it later, "there is no reason why mental as well as bodily labour should not be economized by the aid of machinery" (1994, p. 81). His was a call to prosthesis by means of the reduction of thought to mechanism, but equally a proscription to fragment the laboring body into modules, and then to recombine the components according to the convenience and dictates of the factory master. After his disappointing experience with his London craftsmen, he sought to banish the laborer more drastically out of the loop in his Analytical Engine, and ended up contemplating the calculation of recursive functions, which he referred to as the machine "eating its own tail." The very architecture of the Analytical Engine, now acknowledged as the first stored-program computer, thus constituted a projection of a more perfect factory. At one juncture, Babbage conceded that one of the greatest difficulties in the design and construction of his Difference Engine was the combination of many diverse modalities of motion within a single machine; he then continued, "It instantly occurred to me that a very similar difficulty must present itself to a general commanding a vast army" (1994, p. 133). Countess Ada Lovelace, in her 1843 account of the Analytical Engine, only rendered this implication more plain: "The Analytical Engine is an embodying of the science of operations . . . this engine [is] the executive right hand of abstract algebra" (in Babbage, 1989a, 3:118, 121). For these reasons, Bowden (1953, p. 11) suggests that Babbage should equally be considered as having anticipated the field of Operations Research (OR, whose history is recounted in Chapter 4).

But the cyborg anticipations do not stop there. Babbage was equally fascinated by automata, and had purchased and refurbished a naked mechanical dancer he had seen at Merlin's Museum as a child (Schaffer, 1996), which he called his "Silver Lady." He ruefully recounts in his autobiography that the Silver Lady was displayed in one room of his residence and an incomplete part of the Difference Engine in another adjacent room; and that there was no contest among the bulk of his visitors as to which machine more thoroughly captivated his guests (1994, p. 320). This experience convinced him that hoi polloi would only be impressed with machines if they suitably mimicked some activity that an average human would recognize as resembling their own. What could that be for a device essentially devoted to calculation? After some rumination, Babbage decided that the most humanlike thing that a thinking machine like his Analytical Engine could conceivably do was – play a game! "I selected for my test the contrivance of a machine that should be able to play a game of purely intellectual skill successfully; such as tic-tac-to, drafts, chess, etc. . . . I soon arrived at a demonstration that every game of skill is susceptible of being played by an automaton" (1994, pp. 349–50). True to form, Babbage concocted grandiose plans to charge the public a fee to play his machine in some common game of skill and thereby fund the further construction of his engines; but nothing concrete ever came of it. Unfortunately, he never published his demonstration of the susceptibility of games to being played by automata; for had he done so, the rest of this history as recounted in subsequent chapters would surely never have turned out the way it did. One can only shiver at his prescient recognition that games and thinking machines go together like a horse and carriage; and that a general theory of games would surely follow. As he wrote, "As soon as the Analytical Engine exists, it will necessarily guide the future course of science" (1994, p. 103).

Cyborg enthusiasms practically coruscated off Babbage's fascination with intelligence in all its forms. Babbage spent a fair bit of effort on cryptography and decipherment, convinced as he was that "every cypher can be decyphered" (1994, p. 174). He was an inveterate inventor, coming up with a "black box" recorder for monitoring the condition of railway tracks, occulting lights for ship communications, and a device for delivering messages using aerial cables. He realized that computational complexity could be measured along time and space dimensions, and that one might be traded off for some advantage in the other (p. 94). He anticipated that efficiency in computation would require an ability of the machine to "foresee" the sequence of calculations some steps ahead, now a standard practice in software engineering (p. 86). He invented a "Mechanical Notation" that would facilitate the design of complex components of the engine – in clear anticipation of software programs – and then made the astounding leap: "It [also] applies to the description of combat by sea or by land. It can assist in representing the functions of animal life" (p. 109). Indeed, one of the more underresearched aspects of Babbage's influence upon nineteenth-century thought has to do with Darwinian evolution. Babbage in his *Ninth Bridgewater Treatise* had a long passage in which he described how miracles (or what seemed to most people like miracles) could be compared to long machine programs that produced protracted observed regularities, only to shift abruptly into some novel behavior due to a preprogrammed switch in algorithms. Babbage's point was that determinism rules and that, in any event, humans were generally incapable of judging otherwise. Charles Darwin was apparently impressed with these opinions, and they had some bearing on his construction of the origin of species as conformable to natural law without divine intervention (Desmond & Moore, 1991, chap. 15).

It is nearly impossible to read Babbage today without suffering the chill shock of cyborg recognition, incongruously encrusted in Victorian verbiage. Clearly I am myself incapable of resisting falling into this Whig practice. But the question remains as to the extent to which his contemporaries understood his ideas; or even whether the quantum of influence he had on any of the protagonists in the rest of our narrative surpassed minuscule ceremonial obeisance. I think it is fair to say that Babbage was not merely underappreciated by his contemporaries but that he actually tended to alienate them by his behavior. Earlier on in his career, he had been a fellow traveler of a movement to render political economy a more inductive endeavor, but he rapidly suffered a falling out with his compatriots (the movement failed for numerous other reasons).[8] Later in his career, he became more famous for his legal crusade against street musicians (which "destroys the time and energies of all the intellectual classes of society by its continual interruptions of their pursuits" [Babbage, 1994, p. 253]) and his attacks on the Royal Society than for anything substantial that he had accomplished. Subsequent generations do not seem to have derived much more in the way of inspiration or illumination from his writings. None of the major protagonists of our subsequent chapters seem to have had more than a passing acquaintance with Babbage's ideas before their own contributions: not von Neumann, not Wiener, not Weaver, not Turing, and certainly none of the wartime generation of neoclassicals.[9] The causes of his obscurity for the natural scientists would seem relatively straightforward; and we shall leave the fine points of that phenomenon for future historians of the computer to puzzle out. His utter irrelevance for subsequent economists after Marx bears somewhat more concerted scrutiny, however.

One possible explanation of the deliquescence of Babbage's themes from the hearts and minds of economists would be traced to the fact that his leitmotiv of the centrality of the division of labor to economics had itself subsequently suffered irreversible decline, first in late British classical political economy, but then with a vengeance in neoclassical theory. When the latter shifted analytical focus from the social structure of the workplace and the class structure of society to the internal mental configurations of the generic economic agent, the very idea of a distributed architecture of organization and control, predicated upon irreducible cognitive differences amongst people, simply evaporated. Indeed, the telos

8 On the Whewell group of British mathematical economics, and the machinations behind the formation of Section F of the British Association for the Advancement of Science, see Henderson, 1996; James Henderson in Mirowski, 1994a.

9 Campbell-Kelley & Aspray (1996, pp. 70–71) relate the story of Howard Aiken stumbling across a dusty piece of Babbage's difference engine at Harvard in 1936; but this seems a relatively inconsequential connection to the more momentous sequence of events leading up to the construction of the digital computer in the 1940s.

of the neoclassical program has been to portray every economic agent as exactly alike (a tendency documented at length in Chapter 7) – a situation hardly conducive to the promulgation of the doctrine of the division of labor. But because this did not happen all at once, there must have been something else about the neoclassical program that served as a prophylactic against Babbage's cyborg enthusiasms. One way to explore this conundrum further is to examine briefly the one early neoclassical economist who did seek to appropriate Babbage as a significant influence: William Stanley Jevons.[10]

Jevons, whose primary posthumous claim to fame is as one of the inventors of neoclassical price theory, was better known in his own time as a writer of popular textbooks on logic and the philosophy of science. Under the influence of the mathematician Augustus De Morgan, Jevons became an avid supporter of the idea that logic should be treated as a formal, even mechanical, set of operations. He also was the first significant critic of George Boole's *Laws of Thought* in the early 1860s. The subterranean connections between nineteenth-century logic and economics were much more evident for Jevons than they are for us now: for instance, Jevons expends some effort in his textbook *Pure Logic* in reprimanding Boole for his own discussion of an attempt to render Nassau Senior's earlier definition of economic wealth a logical syllogism within his system (Jevons, 1890, p. 62). Of course, this would imply that he would also identify Babbage as an important precursor and, indeed, he did:

> It was reserved for the profound genius of Mr. Babbage to make the greatest advance in mechanical calculation, by embodying in a machine the principles of the calculus of differences. . . . [In the] Analytical Engine, Mr. Babbage has shown that material machinery is capable, in theory at least, of rivalling the labours of the most practised mathematicians in all the branches of their science. Mind thus seems to impress some of its highest attributes upon matter, and thus to create its own rival in the wheels and levers of an insensible machine. (1890, pp. 140–41)

All of this would therefore seem to argue that Jevons would have taken up the baton for a cyborg science of economics, but in fact, quite the opposite occurred.

[10] William Stanley Jevons (1835–82): British Mint, Sydney, Australia, 1854–59; Owens College, Manchester, 1859–76; University of London, 1876–80. Jevons's relationship to logic and computers was a mainstay of the history of computation literature but, until very recently, absent from the large secondary literature on Jevons in the history of economic thought. Some honorable exceptions are Grattan-Guinness, 1991; Mosselmans, 1998; Mays & Henry, 1973. The paper by Maas (1999) appeared after this chapter was written and diverges appreciably from the interpretation proffered herein.

Jevons nurtured grandiose ambitions to propound a unified conception of science predicated on what he also called "laws of thought." These laws were truly promiscuous, encompassing both things and their signifiers: "logic treats ultimately of thoughts and things, and immediately of the signs which stand for them. Signs, thoughts and exterior objects may be regarded as parallel and analogous series of phenomena, and to treat any one of these three series is equivalent to treating either of the other series" (1905, p. 9). Jevons sought to parlay this ontologically flattened world into the proposition that mathematics should be reduced to logic, but as any number of commentators have noted, he did not even approach success in this endeavor and was ignored in the subsequent logicist line of Frege and Russell and Whitehead.[11] Indeed, it clarifies an otherwise tangled situation to see that Jevons effectively tried to minimize the role of algebra in logic, and concomitantly to revise both Boole's notation and his project, which in turn led him to assume some rather curious positions with regard to machine cognition and, consequently, to his own version of machinelike economics in the form of his theory of the final degree of utility.

Rather than become embroiled in his obsolete notation and baroque principles, it will be quicker to examine briefly the physical instantiation of his system of laws of thought in baywood and brass rods, or his "logical piano," a direct descendant of his Logical Abacus.[12] Jevons begins the paper by nodding toward Babbage's engines, but then bemoaning the fact that no such comparable machines for mechanizing logical inference had yet been constructed. Here we observe the curious paradox that, although Jevons purported to believe that mathematics is logic, and acknowledged that Babbage's Analytical Engine could in principle perform any mathematical calculation, he does not draw the conclusion that the Analytical Engine is precisely just such a logic machine. (He might perhaps have been distracted by the fact that Babbage's Difference Engine was hard-gear engineered for numbers.) Instead, Jevons effectively assumed the position that logic and calculation are separate procedures, each requiring

[11] See, for instance, Schabas, 1990, p. 64; Mosselmans, 1998, pp. 86–87. Hailperin (1986, p. 119) calls Jevons's work a "marked regression in standards of formality." In this regard, it is necessary to warn the reader of pervasive flaws in the article by Buck & Hunka, 1999, which so thoroughly misrepresents Jevons's actual logical position that it then finds it plausible to claim he produced an early "mechanical digital computer."

[12] These are described in Jevons's paper "On the Mechanical Performance of Logical Inference," read before the Royal Society on January 20, 1870, and published in the *Transactions of the Royal Society*, 160 (1870): 497–518. We quote here from the reprint in his *Pure Logic* (1890). One should note the close proximity in time to the composition of *Theory of Political Economy* (1871; reprint, 1970), as well as the irony of the venue, which Babbage had grown infamous for disparaging. A diagram of the piano may be found in Buck & Hunka, 1999, where it is also claimed a working model of the piano was donated to the Oxford Museum for the History of Science.

altogether separate devices. Next, he revealed his own understanding of logic was essentially combinatorial and intensional, based on attributes and qualities, rather than extensional, as later versions based on classes and sets, and thus incompatible with what later became known as the predicate calculus. Consequently, he asserts that propositions and their negations can be substituted and combined according to his three "basic" principles of identity, noncontradiction, and the law of the excluded middle. The stage is thus set for a machine to accept a small number of propositions as input on piano keys (in Jevons's piano just four: A,B,C,D and their negations) and some specified logical connectives (+, OR, "full stop") and a "clear" operation; and then to display through a "window" whether any given statement constructed out of those propositions conforms to the built-in principles of his version of logic. To use one of Jevons's own examples (1890, p. 103), undoubtedly intended to captivate the reader with the author's broad-minded cosmopolitanism, let A = negro, B = fellow-creature, and C = suffering, then inputting A = AB into the machine, we discover that AC = ABC, or that "A suffering negro is a suffering fellow-creature." Jevons was quite proud of his little contraption, writing, "I have since made a successful working model of this contrivance, which may be considered *a machine capable of reasoning*, or of replacing almost entirely the action of the mind in drawing inferences" (1890, p. 120; emphasis in original).

Inadvertently, in many ways his machine serves to illustrate the profound divergence of Jevons's ambitions from those of Babbage. Babbage would be awarded the Gold Medal of the Astronomical Society in 1825 because "Mr. Babbage's invention puts an engine in the place of the computer" (in Schaffer, 1994, p. 203): that is, he replaced the human with a machine in certain identifiable tasks, thus extending the prosthesis of industrial technology into the realm of human thought. He took this mandate so seriously that he became bogged down in the technological minutiae of gear ratios and power transmission. Furthermore, neither Babbage nor his faithful amanuenses Lady Ada Lovelace or Luigi Menabrea ever suggested that the engine was an adequate representation of what happened in the human mind: it was incapable of dealing with semantics and flummoxed by originality.[13] Jevons, by contrast, misconstrued Babbage's intentions as aimed at creating a *rival* to the human mind "in the wheels and levers of an insensible machine." Jevons consequently cribbed together a spindly contraption of rods and levers from some odds and ends of wood and metal, capable of performing substantially less in the way of inference than anything Babbage had ever

[13] Both memoirs by Lovelace (1842) and Menabrea (1843) are reprinted in Babbage's *Works* (1989), vol. 3.

built, and immediately proceeded to conflate it with the totality of operations of the human mind, imagining it "replacing almost entirely the action of the mind in drawing inferences" (1890, p. 120). Of course, its principles of operation were so transparently trivial that these assertions were immediately recognized for the puffery and intemperate hyperbole that they actually were. Many logicians, among them Charles Sanders Peirce and John Venn, were relatively disparaging about this simple combinatorial device (Gardner, 1968, pp. 104–9). Boole and de Morgan were also less than enthusiastic. Undoubtedly in reaction to such skepticism, Jevons in his 1870 exposition was forced to admit that he did *not* "attribute much practical utility to this mechanical device"; indeed, "The chief importance of the machine is of a purely theoretical kind" (1890, pp. 170–71).

Because Jevons was grappling with issues of logic, simulation, and calculation, it is all too easy to paint him in retrospect as a budding cyborg; but now we can begin to entertain the notion that he was no such creature. The way to undertake the evaluation is to observe what he himself made of his machine. As a serious device for calculation, it was useless, as Jevons had been forced to admit. Not only would it run afoul of combinatorial explosions in even the most common concatenation of logical propositions, but, in the end, it was an instantiation of his own flawed system of logic (and *not* Boolean algebra, as has been suggested in some careless histories). It was equally superfluous as a pedagogical device, if only because it was not a machine for *testing* logical propositions; nor did it actually teach handy procedures for making logical inferences. So what was it good for? Here we should have recourse to Jevons's faith in unified science and the importance of analogy for transporting the concepts of one developed science to its lesser developed cousins: as Mosselmans so deftly puts it, "the substitution of similars leads to a project of unified science" (1998, p. 89).

It seems that Jevons really did believe in the practical similarity of logic and things or, to put it another way, that mind was directly reducible to matter, and therefore it followed that his logical piano provided a superior model for the mind. Thus what could not be retailed successfully in baywood and brass to logicians found a more receptive audience as political economy. The machine was ultimately used to fashion an abstract model of the economy: "The importance of the machine is of a purely theoretical kind." Jevons reworked the levers of his logical piano into a theory of market equilibrium of equalized final degrees of utility explicitly patterned on the equilibrium law of the lever (1970, pp. 144–47). "The theory of the economy, thus treated, presents a close analogy to the science of statistical mechanics, and the laws of exchange are found to resemble the laws of equilibrium of a lever" (1970, p. 144). The projection of a machine onto a portrait of mind could not have been more baldly wrought; but what is of particular salience for our narrative is that it was accomplished in such a ham-fisted fashion. The laws of thought so avidly sought in political economy turned out not to be his putative laws of logic, his propositions of the form AB = ABC, but rather the palpable *physical* laws of motion of the machine, which inscribed the laws of logic in bevels and levers.[14] To grasp the problem, consider a more modern parallel category mistake that misguidedly sought the "laws of thought" inscribed in modern computers in Kirchoff's Law or Ohm's Law, simply due to the observation that the personal computer runs on electricity.

The confusions and errors over the relationship of logic to mathematics, of the relationship of syntax to semantics, and of the conceptual relationship of physical principles to the design of the machine that was predicated upon them, all argue that Jevons's economic project was crippled by numerous self-inflicted wounds from its inception. But more deleterious were the divergences from the cyborg project nascent in the writings of Babbage. First and foremost, Babbage was concerned to formulate effective procedures of computation, and not to pursue some goal of abstraction entirely removed from everyday experience and workable inference. Jevons may not be guilty of being the provocateur of a persistent infatuation of the neoclassical school with noneffective mathematical procedures (described in Chapter 6), but he certainly contributed his modest moiety to this practice. Indeed, one can easily detect in Jevons's departure from Babbage the first in a whole sequence of neoclassical offspring of purely phantasmagoric machines held to no particular standards of construction or implementation, and having nothing whatsoever to do with hardworking machines encountered in real social life, much less with real physics. An early instance of this surfaced in an unpublished paper of Alfred Marshall entitled "Ye Machine" first published in 1994 but probably written in 1867. In this manuscript, Marshall simply equates mind with mechanism without apology or motivation, and proceeds to imagine a pleasure machine comprised of all manner of prodigious wheels and "bands" and boilers. Because the neoclassical fancy grew to Promethean proportions, the impossible was merely confused with the abstract in the machine dreams of Marshall: "If the

[14] Of course, the saga of Jevons's peregrinations from the 1862 "Brief Account" of his theory to the full statement in *Theory of Political Economy* of 1871 was determined by much more than the simple analogy to the lever. The other major influences involved the early reception of psychophysics in the British context, as well as the larger energetics movement first described in Mirowski, 1989a. The one historian who has best sorted out the complicated relationship of Jevons to various trends in the natural sciences is Michael White (in Mirowski, 1994a; forthcoming). This account would unfortunately draw us too far afield from our cyborg narrative and is regrettably bypassed here.

wheels etc. of the machine be sufficiently numerous, it must of course have infinite power" (1994, p. 122). Clearly Marshall had never ventured very close to any machine that would be recognized by contemporary engineers. As the neoclassical economists proceeded to merrily postulate their imaginary pleasure machines, then "of course" they would possess infinite computational capacity. Historians have tended to forget that one pivotal context of the rise of neoclassicism in the British scene was a silly season of the 1870s where all and sundry felt comfortable engaging in the loosest and gleefully undisciplined speculation in arguing out the question "Are We Automata?" in the quarterlies and the reviews.[15]

But Jevons rejected the cyborg manifesto in other ways as well. Although Jevons is sometimes credited with introducing probabilistic considerations into economics, what leaps out at the modern reader is his studious avoidance of treatments of randomness in any way resembling those being proposed in contemporary thermodynamics. Instead of exploring the stochastic regularities of aggregates of actors, he boldly presumes that the "trading body" can be analytically described as identical with his version of the individual. Probability densities are not explicitly written down, nor are moments of distributions accessed. A privileged direction of time is completely absent in his theory. Furthermore, in his rush to inscribe levers in the mind, he entirely neglected the greater importance of thinking machines as *prostheses*, as implements to aid and assist in the activities of humans. Babbage understood, as Jevons did not, that machines would first have to prove themselves useful in some human endeavor before they might be upgraded from their status as tools to something that might eventually serve to redefine the very quiddity of rationality. In a sense, Jevons jumped the gun, producing an utterly useless machine, and then proceeding to shout from the rooftops that it could think and was therefore indistinguishable from a brain. And finally, Babbage firmly rooted his machine design in the experience of the division of labor of factory organization, whereas Jevons repudiated the tradition of political economy that aimed to explain the social organization of

[15] The controversy was given a boost by the address to the BAAS and subsequent essay by Thomas Henry Huxley (1874), which reads as a rather more cogent version of what is regarded as wickedly avant-garde today when written by a Daniel Dennett. We can observe from the essays of Jevons and Marshall that comparisons of economic actors to automata had been prevalent for at least a decade before this. Other contributions to the nineteenth-century automata debate are James, 1879; Carpenter, 1875a, 1875b. Parenthetically, the distressing lack of appreciation for these issues sometimes leads historians of economics to make anachronistic and embarrassing claims for their subjects: "Cybernetic feedback is an interesting feature of Marshall's Machine and of his economic agent" (Raffaelli, 1994, p. 79). If there ever could have been an anti-Wiener *avant la lettre*, he would have resembled Alfred Marshall.

production, in favor of a vision of the economic process that was solipsistically confined within a single consciousness, or at best, a "trading body" that was indistinguishable from a solitary economic man. Everything about machines that could have lent a patina of legitimacy to the project had been neglected or repressed, which helps explain why Jevons found himself making outlandish claims in his *Theory of Political Economy* about the power of mathematical expression per se to render economics a scientific research project.

The point of summarily drumming Jevons out of the cyborg Hall of Fame is not to besmirch his title as the progenitor of neoclassical economics but rather to raise the possibility that the birth of neoclassical microeconomic theory in the 1870s marks the (temporary) suppression of cyborg themes in economics and that this divarication can be illustrated by his own work. Indeed, there is no further interaction between the theory of the computer and the profession of economics until we encounter John von Neumann in the 1940s. This silence is deafening; if the "Marginalist Revolution" did not itself bring about the divorce of economics from computation by means of the interpolation of fantastical machines, then something else equally pervasive will have to be accessed by a more perceptive researcher in order to explain how economics and computers grew to have such a strained and separate existence from roughly the 1870s till the1950s.

In any case, in order to follow the cyborgs, especially due to their banishment from economics, it now becomes necessary to shift our attention for the remainder of the chapter to the natural sciences, and their fin-de-siècle fascination with demonology.

ADVENTURES OF A RED-HOT DEMON

Someday, when the history of cyborgs finds its Emile Meyerson (or, better yet, its Ken Burns), it will become commonplace to appreciate that the taproots of cybernetics and the theory of computation were buried deep in the loam of thermodynamics. It was thermodynamics, as we have already hinted, that was firmly situated in the vanguard in the destruction of the previous clockwork conceptions of order that had been part and parcel of the eighteenth-century scientific world view. It was also thermodynamics that eventually pointed toward a novel and user-friendly replacement for the concept of Natural Order, one that would exfoliate throughout the natural and human sciences in the twentieth century. The reason that this has not been the subject of extended commentary in science studies was that the path along which thermodynamics wrought its magic was unprecedentedly indirect and circumforaneous: in bald summary, thermodynamics begat a novel world view by means of an extended detour through biology and the construction of the computer, with some brief side trips into quantum mechanics. This story is prodigious, too vast to explore comprehensively in a book devoted to economics. Here we merely provide some indicators to landmarks along the way that will prove pivotal in later chapters. For readers whose curiosity is piqued by the incongruity of something as prosaic as a steam engine giving rise to an alternative account of the Meaning of It All, we direct them to some of the more suggestive texts in the history of science that bolster this account.[16]

For those who long for a good old-fashioned linear plot line, it is possible, although ultimately misleading, to suggest the overall function of our summary in this section of part of the history of thermodynamics within the context of the larger narrative of this book. Thermodynamics threw down the gauntlet to any comforting account of mankind's place in the universe: what did it mean that we were stuck in natural circumstances that were inexorably running down? Although many answers were bruited about in the intervening years, two ripostes in particular warrant our attention. The first answer was that something like "intelligence" stands as a bulwark against the inexorable grind of entropic dissolution; and, luckily enough, this was something mankind possessed in abundance. This first line of defense is recounted herein as the saga of Maxwell's Demon. Eventually the quest to render intelligence commensurate with entropy rapidly dovetailed with the development of the computer, with consequences we explore in detail in the remaining sections of this chapter, as well as in the next. But this particular "solution" to the problem of the Meaning of Life harbored within itself a very nasty possibility, first noticed by Norbert Wiener, the coiner of the term "cybernetics": If mankind is endowed with this new kind of "intelligence," this ability to process information about the world, then why wouldn't it also be the case that Nature also possessed that capacity? And if Nature also had the capacity to exert intelligence, then how would we ever know whether Nature was merely indifferent to the fate of mankind or, more distressing, was both malevolent and misleading with regard to the supposed triumph of mankind over dissolution? Wiener inadvertently raised the issue that perhaps "intelligence" was not the vaunted solution to man's privileged place in the universe that it had first seemed; later thinkers merely reinterpreted malevolence as the natural state of intelligence. This subjection of Nature to a hermeneutics of suspicion was later easily extended to Society.

[16] The tour would begin with the works of Stephen Brush (1978, 1983) and Charles Ruhla (1992), continue with the anthology of original papers in Leff & Rex, 1990, consult Georgescu-Roegen (1971) and then venture into the twentieth-century history with Evelyn Fox Keller (1995) and Katherine Hayles (1990a).

Conversely, the second answer to the challenge of thermodynamics was generated within the precincts of biology, although it also was eventually disseminated throughout the rest of the sciences. Here an appeal was made to the phenomenon of "evolution" as providing the offset to the problem of entropic dissolution, primarily by making use of entropy to produce order. This jiujitsu trick of using the opponent's strengths to defeat them has proved perennially appealing down to the present; yet the problem attendant to this approach over the same period was to come up with some plausible story about how these laws of biology could apparently trump the laws of physics without openly violating them. For some, this would take the format of contrasting "open systems" as dynamos for conjuring order out of chaos, in invidious contrast to their closed counterparts. For many others, recourse to the novel discourse of "intelligence" – that is, the first answer – held the promise of circumventing the incompatibility. Both answers prove indispensable for understanding the subsequent development of twentieth-century economics.

It goes without saying that this caricature of a plot summary in no way stands as a substitute for a thick and textured history of thermodynamics. Indeed, even this paltry summary leaves out the most important aspect of science organization in the twentieth century, the vast watershed of World War II. The wartime mobilization of scientists provided the occasion for the final drawing together of all these diverse skeins of scientific inquiry – thermodynamics, the computer, cybernetics, intelligence, cryptanalysis, the logic of deception – into a single massive research program. It is often overlooked that the wartime push to produce the atomic bomb and its immediate postwar offspring, the thermonuclear bomb, highlighted the centrality of thermodynamics in all these disciplines (Rhodes, 1986, p. 249). The problem of whether the deuterium and tritium would "burn" or fizzle out was one of the central questions of this crash program, and provoked the calculations that were the very first task of the electronic stored-program computer. The gun-aiming problems that inspired cybernetics were heavily implicated in the Air Force's concern over the relative vulnerability of bombers. Intelligence and cryptanalysis also took their cue from models taken from thermodynamics and assumed heightened importance in a world where the superpowers were both poised nervously upon their atomic hair triggers. After World War II cyborgs and their entropic concerns were to be found everywhere. It was no longer the heavenly world of the eighteenth-century philosophers, but instead twentieth-century America: a closed world sleek with dread and heavy with doom. It was the world of a John von Neumann; and, in an unexpected twist, it also turned out to be a world in which neoclassical economics managed to thrive.

Thermodynamics for Dummies and Demons

Scientists often bemoan the difficulties of explaining the laws of thermodynamics and the concept of entropy to laypeople; even philosophers of science tend to eschew thermodynamics in favor of glitzier fields such as quantum mechanics or molecular biology.[17] It does seem odd that the Second Law of Thermodynamics, so apparently elementary in its guise as the statement that "*heat cannot by itself pass from a colder to a warmer body*," would throw up such insuperable barriers to comprehension. Nonetheless, the issues are fraught with controversy, not the least because of their implications for the direction of time, the nature of life, and the future of the universe. Economists have been blessed with a lucid expositor of these issues in the person of Nicholas Georgescu-Roegen, who in a classic book (1971) sought, with indifferent success, to instruct the discipline in their implications and significance for economics. The reader in need of a four-page primer could do no better than to consult Georgescu's entry for "entropy" in the *New Palgrave* (1987, pp. 153–56) for a quick refresher course. However, Georgescu's gloss on entropy was misleading in at least one idiosyncratic sense: he was implacably opposed to the extensions of the entropy concept in the direction of its interpretation as a measure of "information," and therefore he missed out on one of the most consequential aspects of the cyborg genealogy in the late twentieth century. Indeed, that the unusually philosophically sophisticated economist Georgescu-Roegen could not become reconciled to the major thrust of the cyborg sciences in the late twentieth century will shortly constitute yet another bit of evidence that neoclassical economists and cyborgs don't mix.[18]

Rather than embark upon a dreary and protracted tutorial in thermodynamics, it may prove more entertaining to structure our narrative of cyborg themes around an actual protagonist, a picaresque tale of noble quests and temporary setbacks, albeit with an unusually tiny *picaro*. The hero of our yarn will be that Gremlin with No Name, the minuscule master of molecular motion known as Maxwell's Demon. The story of the Demon has been told before in many contexts, but none that accords him his rightful place as the progenitor of cyborg science.

The tale begins with Rudolf Clausius coining the term "entropy" in 1854 for the changes in a quantity of heat energy divided by absolute

[17] An honorable exception to this trend, and a very insightful introduction to many of the thorny problems involved, is Sklar, 1993. Other good historical introductions to thermodynamics are Brush, 1983; Earman & Norton, 1998.

[18] Compare, for instance, Georgescu-Roegen's treatment of Maxwell's Demon as "empty" (1971, pp. 188ff.) with the following section.

temperature in the thermal equilibrium of an ideal heat engine. Due to the restrictions of the second law about converting heat into work, Clausius noted that entropy was always positive in an irreversible system and drew the portentous conclusion that, "The energy of the universe is a constant. The entropy of the universe tends to a maximum." The meaning of this statement was understandably obscure; and the British physicist James Clerk Maxwell took it upon himself to attempt to render its significance more clear. In order to insist that the Second Law of Thermodynamics only displayed a statistical regularity, he first broached the idea of the Demon in a letter to Peter Tait in 1867 and to the world in his 1871 *Theory of Heat*. William Thomson was responsible for popularization of the Demon in the 1870s, as well as endowing him with the patronymic:

> The word "demon," which originally in Greek meant a supernatural being, has never been properly used to signify a real or ideal personification of malignity. Clerk Maxwell's "demon" is a creature of imagination having certain perfectly well defined powers of action, purely mechanical in their character, invented to help us understand the "Dissipation of Energy" in nature. . . . He cannot create or annul energy; but just as a living animal does, he can store up limited quantities of energy, and reproduce them at will. By operating selectively on individual atoms he can reverse the natural dissipation of energy, can cause one-half of a closed jar of air, or of a bar of iron, to become glowingly hot and the other ice cold. (Thomson in Leff & Rex, 1990, p. 5)

Because heat was being reconceptualized as molecular motion, Maxwell and others believed that the second law could be reversed, at least in principle at the molecular level, permitting heat to flow from a cooler to a warmer body. In his text, Maxwell imagined a nimble-fingered homunculus stationed at a door partitioning off a cooler from a warmer gas. Because the cooler gas was composed of faster and slower molecules, the Demon could wait till one of the faster molecules was headed toward the door, quickly to whip it open letting the molecule pass to the warmer gas, and then close it with even more precise alacrity to prevent the back migration of faster molecules. In this manner the Demon could make heat flow from a cooler to a hotter body and violate the second law. This possibility rapidly assumed significance in the Victorian context all out of proportion to its tiny protagonist. We can observe from Thomson's quotation that the issue of the similarity of the Demon to "living animals" insinuated itself from the very beginning. Nature was read in tooth and claw, but the predator had now undergone transmutation: "The general struggle for existence of animate beings is therefore not a struggle for raw materials . . . nor for energy which exists in plenty in any body in the form of heat (albeit unfortunately not transformable), but a struggle for entropy, which becomes available through the transmission of energy from the hot sun to the cold earth" (Boltzmann, 1974, p. 24). But more to the point, the second law was represented in the popular mind by the running down of the world engine and the eventual Heat Death of the entire Universe. Note well: the dissipation of heat had become grimly conflated with death, for both the individual organism and the universe as a whole. This called forth further metaphorical calamities, because circumvention of the second law meant that (at least in principle) a perpetual motion machine might not be such a fatuous notion. Cheating death was one thing; but cheating capitalism by getting something for nothing was quite another.

From this point onward, it becomes apparent that the Demon took on a life of his own. Laymen often think that science comprises forbidding formulas and colorless facts; but little folklore heroes like Maxwell's Demon are often just as important for the elaboration of scientific thought. The fact that an illustrious parade of physicists, from Maxwell to Kelvin to Erwin Schrödinger to Leo Szilard to Brillouin to Rolf Landauer to Charles Bennett have been enchanted by such a fanciful conceit is proof of that observation. As Leff & Rex (1990, p. vii) have noted, the Demon himself has repeatedly been pronounced dead in physics, only to be resurrected in some new disguise. The twin images of the Demon and universal dissolution have stalked the dreams of physicists in the intervening century, spurring their search for a place for life in the seemingly hostile universe. It was the Demon that pointed the way out of an intolerable cul-de-sac; as Keller (1995, p. 52) explained, "between the promise of progressive development in Darwin's theory and the threat of inexorable decay and dissolution in the laws of thermodynamics, Maxwell foresaw a third possibility in the nature of molecular structures – neither of progress nor of decay but of stability."

Over time, for many the Demon came to stand for the triumph of Life over Death (even if only temporarily); the beauty of this synecdoche was the way it left the essential character of life indeterminate, a tabula rasa upon which future scientists could inscribe their own obsessions. What was it that allowed life to prevail, to maintain homeostasis and growth in the face of entropic degradation? The answers ranged from Henri Bergson's *élan vital* to Walter Cannon's *Wisdom of the Body* (1939); but we focus on two that constituted the direct inspiration of the cyborg sciences: the image of life as feeding voraciously upon pockets of low or negative entropy, and the conception of life as the intelligent maintenance of "memory" in the face of dissipation. Molecular genetics, the theory of information, cybernetics, and even John von Neumann's theory of automata all found their point of departure at the portal manned by the versatile Demon.

By all accounts, Leo Szilard was the smithy who next stoked the furnaces in the workshop of Demonology.[19] Szilard, perhaps best known for facilitating America's drive to build the atomic bomb early in World War II and then later for his attempts to control the bomb once it emerged from Pandora's box, was one of a small but significant group of scientific refugees who moved restlessly between physics, biology, and economics all their lives. While at the University of Berlin in the 1920s, he was part of an intense discussion group on economics, which encompassed John von Neumann, Eugene Wigner, and Jacob Marschak, figures who loom large in our subsequent narrative (Lanouette, 1992, p. 76). His paper that resuscitated the Demon grew out of a seminar on statistical mechanics that Szilard, von Neumann, Wigner, and Denis Gabor took from Einstein in 1921; Szilard wrote "On the Decrease of Entropy in a Thermodynamic System by the Intervention of Intelligent Beings" in 1922 in response to the lectures, though it was not published until 1929. Although the ideas therein were treated with respect, it was more due to von Neumann's endorsement than anything Szilard did with them subsequently that caused them to be taken up and developed in the postwar period.[20] Szilard, characteristically, grew bored with the topic and never did anything further on it.

The objective of Szilard's 1929 essay was to "find the conditions which apparently allow the construction of a perpetual motion machine of the second kind, if one permits an intelligent being to intervene in a thermodynamic system" (translated in Leff & Rex, 1990, pp. 124–33). The imaginary apparatus described in the essay, in which a single molecule is made to do work on a piston, should not detain us here: it was, rather, the interpretation superimposed on the device that set off reverberations throughout the decades. For Szilard suggested that one might save the second law from violation by associating the measurements made by the Demon of the velocities of molecules with their own production of entropy. As one friend only half-facetiously glossed the paper, "Thinking produces entropy" (in Lanouette, 1992, p. 64). With the usual panache of the physicist, Szilard insisted that "ignorance of the biological

[19] Leo Szilard (1898–1964): Ph.D. in physics, University of Berlin, 1922. A book-length biography is Lanouette, 1992; but see also Rhodes, 1986. Lanouette reports that Szilard apparently considered doing a second doctorate in economics early in his career. Szilard's role in the history of economics, unfortunately overlooked by Lanouette, is discussed in Chapters 3 and 5.

[20] "Von Neumann gave one of the earliest statements of a form of Szilard's Principle" (Earman & Norton, 1999, p. 6). The pivotal role of the first generation of cyborg scientists in propagating the Demon is a topic that has so far eluded historians of science. To give just one example, it was Warren Weaver who first alerted Leon Brillouin to the importance of Szilard's paper (Lanouette, 1992, p. 64).

phenomenon need not prevent us from understanding that which seems to us the essential thing." And that essential thing was *memory*: the first use of that term in a physical sense, but not the last. "We show that it is a sort of memory facility, manifested by a system where measurements occur, that might cause a permanent decrease of entropy and thus a violation of the Second Law of Thermodynamics, were it not for the fact that the measurements themselves are necessarily accompanied by the production of entropy" (Leff & Rex, 1990, p. 124). The fact that this analytical move became possible close on the heels of the inception of the "sciences of memory" in the late nineteenth century (Hacking, 1995) would appear no coincidence.

What Szilard had done was not so much amend the actual physical laws or discover anything new – after all, the essay merely described a *Gedankenexperiment* – and he brushed aside all the serious physical considerations that might be used to compromise the capacities of the smaller-than-life character, such as ignoring the nature of the Demon's thermal equilibrium with the gas, or the physical character of the signals it used to locate the molecules, or the energy dissipation of its own nervous system. What Szilard did accomplish was to initiate the investment of "information" with a thinglike character, so that one could talk about its genesis and destruction in much the same way one would talk about entropy. This notion of the interaction of the observer with the phenomenon through the act of measurement was prescient, given that the Copenhagen interpretation of quantum mechanics still lay in the future when Szilard wrote the essay.[21] The more immediate result was a novel construction of the meaning of life: memory, which facilitated the apparent bootstrapping of free energy, plus observation, which reimposed the Natural discipline of scarcity by imposing a cost. While the nods in the direction of the conflation of life with cognition were flattering to a certain self-image of privileged humanity, the yoked team of memory plus observation nonetheless still left life pretty much where it had been stranded before in physics: it was inexplicable.

At this particular juncture, following the Demon in all his disguises becomes devilishly difficult; but it is noteworthy that everywhere the

[21] The Copenhagen interpretation of quantum mechanics, and Szilard's role in von Neumann's understanding of the doctrine, is discussed in Chapter 3. Some modern commentators also see Szilard's paper anticipating cybernetics: "Szilard's observation that an inanimate device could effect the required tasks – obviating the need to analyze the complex thermodynamics of biological systems – was a precursor to cybernetics. . . . [However,] it is unclear whether the thermodynamic cost is from measurement, remembering or forgetting" (Leff & Rex, 1990, p. 16). It should be stressed, however, as Lily Kay has reminded me, that Szilard never himself actually makes use of the term "information." As we shall learn, it was von Neumann who promoted that linkage.

Demon went, a major cyborg science soon was sure to go. There was no end of sites where one might encounter the Demon from the 1950s onward. One stream of thought recast the problem of the reinstatement of order as largely one of the possible reversibility or irreversibility of computation itself, aligning itself with mathematical logic and the theory of computation. This line of inquiry begins with John von Neumann and Alan Turing and is more recently associated with the names of Rolf Landauer and Charles Bennett (Leff & Rex, 1990). A second line of inquiry attempted to treat biological evolution as though it were itself an entropic process (Brooks & Wiley, 1988). A third direction of inquiry led more explicitly into molecular genetics and the treatment of DNA as a "code" (Yockey, 1992; Kay, 1997a, 2000). A fourth skein of research led directly to Norbert Wiener's cybernetics and Claude Shannon's theory of information, the subject of the next section. Although this scattering of Demon tracks into geographically separate and intellectually diverse territories is potentially baffling, the proliferation of tiny footprints does give some indication of the extent to which thermodynamics progressively became entrenched in the mind-set of the postwar sciences. As Wiener once wrote, "It is easier to repel the question posed by Maxwell's Demon than to answer it" (1961, p. 57); precisely for that reason, many of the most daring and innovative sciences undertook to confront it in the postwar period.

Evelyn Fox Keller in her *Refiguring Life* (1995) has done a wonderful job in tracing all the metaphors that went into the making of the "gene" and linking them to the questions posed by Maxwell's Demon (esp. pp. 59–75). By her account, the text that crystallized out the connection between entropy and the gene for many was Erwin Schrödinger's pamphlet *What Is Life?* (1967 [1944]). While it has now become commonplace to deride the lecture as having little in the way of solid content (Gould, 1995), it does appear that the generation that ran the race to discover the double helix all read it avidly and found it inspirational, from Watson and Crick on down (Judson, 1979, pp. 244–45). Schrödinger situated the question of his title in a tradition which had become conventional wisdom by his time: "What then is that precious something contained in our food which keeps us from death? . . . What an organism feeds upon is negative entropy" (1967, p. 76). It was convenient that this "sucking orderliness from the environment" was perfectly compatible with a market system, in that no one seemed to be getting away with something for nothing; however, it had not provided an adequate guide to the definition of life up to that point. It did not distinguish Life from a waterspout, say, or a black hole. This required a particular leap of imagination: "An organism's astonishing gift of concentrating a 'stream of order' on itself and thus escaping decay into atomic chaos . . . seems to be connected with the presence of 'aperiodic solids,' the chromosome molecules" (p. 82).[22]

The major inspiration that Schrödinger bequeathed to molecular biology, as Keller points out, was to swap the Demon as homunculus for the Demon as "code," a "one-to-one correspondence with a highly complicated and specified plan of development." No more worries would intrude about the troublesome size or corporeal thermodynamics of our smaller-than-life homunculus; a molecule is surely small enough to avoid most of that. Yet this was simultaneously a molecule of prodigious organizational capacity: "Since we know the power this tiny central office has in the isolated cell, do they not resemble stations of local government dispersed throughout the body, communicating with each other with great ease, thanks to the code that is in common to all of them?" Szilard's "intelligence" had taken on a new connotation, just in time for the Second World War. The "communication" was of a curious sort (since all the cells apparently possessed the same instructions), but this did not prove a metaphorical hindrance, because armies and governments were more interested in controlling the flow of information rather than any two-way dialogue. This subsequently encouraged what Keller has called the doctrine of the "master molecule" in molecular biology, the privileged proteins wherein "the ordinary laws of physics do not apply."

The influence of Maxwell's Demon on biology did not propagate solely or even primarily through Schrödinger; another major conduit was through cybernetics and Shannon's theory of communication (Kay, 1995). A straightforward summary of the theses of cybernetics and information theory can tend to be misleading in recounting this history; for whereas both doctrines experienced a surge of enthusiasm in the 1950s, it is now commonplace to dismiss them as all show and no substance. For instance, Judson writes, "Information theory made no difference to the course of biological discovery: when the attempt was made to apply it . . . the mathematical apparatus of the method produced comically little result" (1979, p. 244). Similar assessments of information theory have been written in many other fields, most notably, economics.[23] Lily Kay's evaluation of its

[22] For the sake of faithfulness to the science, it should be noted here that in classical thermodynamics there would be no such thing as the "entropy of the organism," because living beings are open, nonequilibrium entities, whereas classical thermodynamics dealt with closed systems at thermal equilibrium. In the modern context, it is doubtful whether such a phenomenon could be subject to measurement. See, for instance, Leon Brillouin in Buckley, 1968, p. 153. Nevertheless, these qualms do not seem to have hindered generations of those seeking to reconcile life to the laws of thermodynamics.

[23] See Chapter 6 for the response of selected neoclassical economists to information theory. For dismissals of cybernetics in other fields, see (for psychology) Machlup & Mansfield, 1983, p. 417; for qualms about information theory in the social sciences, see Anatol Rapoport in Buckley, 1968.

intellectual consequence was even more devastating: "if information theory is invoked to bestow scientific legitimacy on the new representations of heredity . . . its semantic value must be relinquished; and so must the textuality of the genome, since the genetic communication system conveys no meaning" (1997a, p. 28). Yet it would be intellectually premature and historically unwise to write off peremptorily either Shannon's information theory or cybernetics as unimportant for science, simply because their specific mathematical formalisms in their 1950s manifestations did not serve to consolidate all theoretical questions in either the natural or social sciences under a small set of abstract doctrines (Heims, 1991, chap. 12), or perhaps alternatively due to the conviction that whatever of value they may have expressed has now been absorbed into more legitimate disciplines (Machlup & Mansfield, 1983, pp. 39–41). Their effect was simultaneously more diffuse but also more far-reaching than that.

It is possible to make the case that Maxwell's Demon played the role of the Pied Piper in recruitment and promotion for the cyborg sciences, at least in the twentieth century. The allure was the promise of rendering physics, and especially thermodynamics, the means by which all sorts of previously inaccessible phenomena might be brought under the purview of science. The Devil managed to get all the good tunes, including:

- The fascination with the myriad of ways that violations of the Second Law of Thermodynamics might be bound up with a physical definition of "information" began with Szilard but devolved to the master impresario John von Neumann. This way lay the genealogy of the computer.
- The Demon not only conjured the wizardry of information, but concomitantly focused attention on issues of command and control. "A Demon unable to follow its planned operations consecutively cannot produce work to begin with. It can no more produce work than Brownian motion can" (Shenker, 1999, p. 358). So it was not enough to know something to violate the second law: organization and control came to be more intimately bound up with thermodynamics. This way smoothed the path to the novel discipline of operations research, the topic of Chapter 4.
- Due to the recurring problematic of the ability of life to transcend dissipation and decay, theories of evolution increasingly became inflected with thermodynamical formalisms and concepts. For instance, R. A. Fisher's reformulation of evolution was more or less inspired by the statistical mechanics of his time (Depew & Weber, 1995, chap. 10).
- These three themes were knitted together in an enticing way in Norbert Wiener's manifesto *Cybernetics* (1961 [1948]), which ended up becoming more than just the sum of its components, as

we shall explore in the next section. Life conceived as a struggle against chaos and dissipation was transmogrified into a struggle against a wily and deceptive opponent, giving rise to a thermodynamics of suspicion, and a framework for cryptanalysis. Here we enter the realm of economics and game theory.

DEMONS WHO CAME IN FROM THE CODE: CYBERNETICS

"Cybernetics" was an important etymological way station for cyborgs, if not their actual point of departure; and "information theory" has stood as something much more powerful than a simple adjunct to telecommunications engineering or systems theory.[24] "Cybernetics" has proved difficult to define, even for its enthusiasts in the immediate postwar period (Bowker, 1993; Pickering, 1998). Sometimes jokingly defined as "The sciences of life in the service of the sciences of death," this phrase did not begin to capture the layers of ambition and ambivalence that supported the quest. Cybernetics for the generation after World War II became the public philosophy of a whole set of scientific practices growing out of the wartime mobilization of research; and it turned out to be all the more effective because its primary spokesman, Norbert Wiener, was publicly identified with an antimilitarist position. It started off with Wiener positing a set of technologies to restrain entropy and chaos through feedback and later was transmuted into theories of self-organization, where entropy would itself under certain circumstances give rise to "higher" levels of order. Information theory, per contra, became the obscure mathematical rationale for the reification of one of the central concepts of the cyborg sciences, especially in the new technologies of command, control, and communication. Although it is true that neither managed to attain the status of a fully fledged cyborg science in and of itself, together they did foster an environment in which the subsequent cyborg sciences could flourish.

Cybernetics began as a science of a certain class of machines but rapidly and inadvertently became the vehicle for a unified science of people and things. There had been previous attempts to pattern theories of people and society upon the natural sciences, but what cybernetics encouraged was the blurring of all such distinctions, treating mind as essentially no different from generic machines. Paul Edwards (1996, pp. 178–79) has provided a suggestive roster of the immediate heritage of cybernetics for psychology, something that will gain gravitas as we venture further into

[24] For instance, Marvin Minsky once wrote: "The era of cybernetics was a premature anticipation of the richness of computer science. The cybernetic period seems to me to have been a search for simple, powerful, general principles upon which to base a theory of intelligence" (1979, p. 401).

the postwar history of economics in subsequent chapters: (1) a fundamental belief in the existence of "internal" cognitive processes and modules, encompassing perception, memory, and, of course, calculation; (2) a revival of emphasis on experimentation with human subjects; (3) a predisposition to portray organisms as active and creative, especially with respect to goals and expectations; (4) a bias toward regarding mental structures as innate; (5) a strong commitment to computation as a metaphor for all thought; and (6) cognition treated as symbolic information processing.

Once we have stabilized (at least temporarily) the referent for that slippery term "cybernetics," the time has arrived to return to our tiny protagonist.

From MAD to MOD

What rendered Maxwell's Demon such a compelling character? Why should it be a matter of moment whether a patently artificial homunculus could or could not reverse the Second Law of Thermodynamics? That is one question that percolates as a subtext throughout many of Norbert Wiener's writings.[25] Once thermodynamic degradation had been linked to questions of information and control, as had been initiated by Szilard, then the Demon was strategically positioned to become a personification of the Human Condition, at least in Wiener's view. Thermodynamics was therefore poised to be parlayed into a general theory of the Natural and the Social. This is rendered most transparent in his popular book, *The Human Use of Human Beings*:

> [Humans are] playing against the arch enemy, disorganization. Is this devil Manichaean or Augustinian? Is it a contrary force opposed to order or is it the very absence of order itself? The difference between these two sorts of demons will make themselves apparent in the tactics to be used against them. The Manichaean devil is an opponent, like any other opponent, who is determined on victory and will use any trick of craftiness or dissimulation to obtain this victory. In particular, he will keep his policy of confusion secret, and if we show any signs of beginning to discover his policy, he will change it in order to keep us in the dark. On the other hand, the Augustinian devil, which is not a power in itself, but the measure of our own weakness, may require our full resources to

[25] A somewhat unsatisfactory biography of Norbert Wiener (1896–1964) is Masani, 1990; a fascinating attempt to compare and contrast Wiener to the major protagonist of our narrative, John von Neumann, is Heims, 1980; but it still must be said that the definitive biography of either figure remains to be written. Much more entertaining are Wiener's own two installments toward an autobiography, *Ex-Prodigy* (1953) and *I Am a Mathematician* (1956). Besides these sources, the following section is based on materials in the Wiener archive, WMIT.

> uncover, but when we have uncovered it, we have in a certain sense exorcised it. . . . The Manichaean devil is playing a game of poker against us and will readily resort to bluffing . . . as von Neumann explains in his *Theory of Games*. . . . Nature offers resistance to decoding, but it does not show ingenuity in finding new and indecipherable methods for jamming our communication with the outer world. (1954, pp. 34–36)

This quotation announces the moment from which Maxwell's Demon should be recognized as having spawned a *Doppelganger*, although I believe the actual birth date should probably be celebrated a decade or so earlier. Before this date, Maxwell's Augustinian Demon (or MAD), one and the same picaresque demon described in the preceding section, struggled mightily against randomness in the implacable laws of nature, paralleling mankind's own struggle to wrest improved efficiency from the steam engine. The early cyberneticians readily appreciated this version of the Demon: machines could output more information than was input into their design by making use of other random information, in the same way a single man could set in motion an entire factory without violating the laws of energy conservation by bootstrapping the free energy of the coal burning in the furnace (Asaro, 1998). Various arguments could be broached as to whether a suitably equipped MAD could in fact ultimately prevail over Nature; but there was no doubt that the opponent played fair: the laws of nature were fixed, eternal, and implacably indifferent to mankind and demons alike.

However, the introduction of "information" and "memory" into the struggle, as Szilard had done, had subtly destabilized the very fairness, or at least the ontology, of the "game" against Nature. What guarantees could be tendered or ground rules confirmed when calculation, recollection, and intentionality were inserted into the equation? After all, Wiener's other trademark doctrine was that we are all feedback machines, be they electronic, mechanical, or organic. If homunculi and other humans could exhibit intelligence, then in principle so could inanimate Nature, at least in its manifestation as machines. But then came the troublesome thought that perhaps Nature was not so very indifferent to us after all; if we could exhibit cunning and guile in our interrogation of Nature, just as our nimble demon deployed these virtues in interrogating individual molecules concerning their speeds and directions, then what was Nature busily doing to us? At just this juncture, World War II and its unprecedented mobilization of the natural sciences intervened to drive the question home.

It has not yet been fully appreciated that the profound shift from the nineteenth-century sciences of energy to the twentieth-century sciences of information and control pivoted precisely on this revision of the conception of Nature from passive obstruction to our plans and projects to Nature as potentially dangerous and deceptive opponent. This first surfaced in the wartime discourse concerning machines. All the combatants in World War II had come to share the experience during the war of dealing with an Enemy Other, a powerful, yet wily and devious Other, whose rationality had to be somehow deduced from past observed actions interspersed with randomized feints and intercepted coded messages studded with random noise, rather than through direct unmediated interrogation of experience. Statistics became the device through which signal could be precipitated out of noise. However, if disorder could be offset in this way through the exercise of intelligence, as MAD seemed to do, couldn't it also have been initially *created* by a malevolent intelligence? For some, the very existence of atomic weapons began to resemble the sort of Faustian bargain concocted by a demonic intelligence bent on the wholesale destruction of the human race. For others, familiarity with systems approaches elevated the concept of unintended consequences to a natural state of affairs. And if disorganization could be considered a calculated phenomenon, then perhaps might a similar sort of demon, augmented with a *strategic* capability, possess sufficient capacity to sort it out? In other words, perhaps Maxwell's Augustinian Demon wasn't really up to snuff when it came to the brave new postwar world of communication and intelligence? Wouldn't it take a newer, more strategically savvy demon to measure up to the Natural onslaught of disorder?

Wiener in the foregoing quotation strove to reassure us that no such dire predicament had come to pass; but the bulk of his writings belie such confidence. Instead, his work is studded with forebodings and premonitions of the consequences of the shift to a cybernetic sensibility. Was the prediction of the evasive maneuvers of an aircraft really so very different from predicting the motions of a molecule in an enclosed gas? Was the mind really an agglomeration of statistical feedback algorithms similar to those found in a torpedo? Was the escalation of atomic terror really so different from the operation of a servomechanism? Was bluffing in chess really so different from the collapse of the wave packet in quantum mechanics? In a phrase, we have met the Enemy, and it is us. This led to the chilling thought that entropic degradation was the Natural outcome of the only sort of fair game to be played in this or any other world. As Wiener asked at the end of his life: "Can God play a significant game with his own creature? Can *any* creator, even a limited one, play a significant game with his own creature? In constructing machines with which he plays games, the inventor has arrogated to himself the function of a limited creator" (1964, p. 17). After this realization, the Manichaean Other's Demon (MOD) made his appearance, inhabiting a different sort of closed world, struggling to confine the Other between his cross hairs. It came down to Hobson's choice: either we are trapped in a mug's game with the deck always stacked to win a meaningless victory, or else our machines will inevitably outgrow our plans and intentions, slough off their subordinate status, and become more like Nature itself: duplicitous, random, evolving, and pitted against our interests. Entropy becomes identical with agonistic strife for Wiener. This, I believe, exposed the connection in Wiener's oeuvre between his fascination with entropy and his dread of the trends in social theory, which he continually aired in his popular writings. Katherine Hayles (1999, p. 111) has correctly insisted that Wiener feared his beloved cybernetics would annihilate the Enlightenment individual as the captain of the soul, and thus he preached that the social sciences should be sequestered off from the natural sciences, if only to protect ourselves from the bleak anomie of the natural world.

But MOD could not be so effortlessly exorcised or quarantined. Enthusiasts for cybernetics brought the natural and social sciences in ever more intimate intercourse, and the very thing that Wiener perennially feared has most certainly come to pass (as discussed in detail in Chapter 7). For every technology of surveillance and control, there arises a parallel technology of simulation and deception. In consequence, relations between the natural and the social sciences have never been the same since. Operations research, the "strategic revolution" in economic thought, organizational theories patterned on military hierarchies, the economic agent as information processor, exchange as the crafty victimization of the inept, and much else in this volume flows directly from this innovation.

Norbert Wiener

All manner of scientists were led by the war to reinterpret thermodynamics in a strategic direction in one arena or another from the 1940s onward; and we shall repeatedly encounter them in our subsequent narrative: John von Neumann, Warren Weaver, Alan Turing, Claude Shannon, John Holland, Ross Ashby, Patrick Blackett, Kenneth Arrow, Leonid Hurwicz, Michael Rabin, Philip Morse, Herbert Simon, Fischer Black, Kenneth Boulding, and Jacob Marschak, among others. However, there was no figure who synthesized as many of the different competing strands of disciplinary approaches into a tidy philosophical raison d'être quite as well as Norbert Wiener. He was the first to sense that the demands of MOD in combination with the development of the computer would lead not only to profound intellectual upheaval, but also what he liked to think of as a "Second Industrial Revolution" transforming the social landscape. Other scientists foresaw the shift of the real action in the immediate future into the precincts of biology and fantasized about the future of the computer; but it was Wiener who gave those dreams a name, a core metaphor, and a rationale. It is worthwhile to recall that the subtitle to his 1948 bestseller *Cybernetics* was: "Control and Communication in the Animal and the Machine."

Wiener, like so many of the mathematically inclined cyborg scientists of the immediate postwar generation, began his studies in philosophy and logic. He was not completely enamored of the logicist program, as he tells us: "when I studied with Bertrand Russell, I could not bring myself to believe in the existence of a closed series of postulates for all logic, leaving no room for arbitrariness in the system" (1956, p. 324). Nevertheless, Wiener did find some corners of mathematics that did resonate with his preanalytic vision of a chaotic universe, and thus he became a mathematician. As one of his biographers, Steve Heims put it: "in conversation [Wiener] spoke about chance, irrational impulses, and stochastic processes, as if to him these phenomena were close to each other" (1980, p. 155). This vision comes through quite clearly in his writings: "We are swimming upstream against a great torrent of disorganization which tends to reduce everything to the heat-death of equilibrium and sameness described in the second law of thermodynamics. . . . we live in a chaotic moral universe. In this, our main obligation is to establish arbitrary enclaves of order and system" (Wiener, 1956, p. 324). Wiener initially thought he accomplished this feat by applying the theory of Lebesgue integration to the description of Brownian motion, clarifying the ergodic theorem (1961, chap. 2), and developing generalized harmonic analysis and the Wiener-Hopf equations. After "an apprenticeship in ballistic computation in World War I" (1956, p. 227), he joined the faculty at MIT in 1919, becoming one of their intellectual stars by the beginning of World War II. In the late 1920s, he worked closely with Vannevar Bush on his differential analyzer, which was an early analog computation device. The connection with Bush was fortuitous, not only because of the prescient familiarization with engineering problems of computation, but also because Bush would shortly become one of the three main organizers of the American mobilization of the scientists' war effort, along with Warren Weaver and James Conant.[26]

Wiener was very anxious to participate in the war effort in some academic fashion. He began by proposing a "mechanical method for encoding and decoding messages" (1956, p. 239); but, as it happened, MIT's track record in assisting cryptography was dismal, at best (Burke,

[26] On Bush, see Reingold, 1995; Edwards, 1996, pp. 46–48; Owens, 1994. An excellent biography of Bush is Zachary, 1997; but compare also his autobiography *Pieces of the Action* (1970). For a biography of Conant, consult Hershberg, 1993. The shape of the World War II mobilization is discussed in Chapter 4, focusing on the central role of Bush and Warren Weaver.

1994). By 1940, however, Wiener had an idea that would inform all his work until the end of his career, even if it never actually resulted in an effective military device: he would use electrical networks and his work on probability theory to predict several seconds in advance the position of an aircraft trying to evade antiaircraft fire (Galison, 1994). The predictor began as a simulation on paper, which conveniently made use of Bush's analyzer for the calculations. As early as 1941, Wiener and his associates realized they would have to formalize the behavior of the pilot as part of the problem: "We realized that the 'randomness' or irregularity of an airplane's path is introduced by the pilot; that in attempting to force his dynamic craft to execute a useful maneuver, such as straight line flight or 180 degree turn, the pilot *behaves like a servo-mechanism*, attempting to overcome the intrinsic lag due to the dynamics of his plane as a physical system, in response to a stimulus which increases in intensity with the degree to which he has failed to accomplish his task."[27] Pilots used to talk about their sense of resistance of their machine to certain maneuvers as being the handiwork of "gremlins"; but Wiener instead saw MOD at work. All of a sudden, the outlines of a new science fell into place.

> At the beginning of the war the only known method of tracking an airplane with an anti-aircraft gun was for the gunner to hold it in his sights in a humanly regulated process. Later on in the war, as radar became perfected, the process was mechanized . . . and thus eliminated the human element in gun-pointing. However, it does not seem even remotely possible to eliminate the human element as far as it shows itself in enemy behavior. Therefore, in order to obtain as complete a mathematical treatment as possible of the overall control problem, it is necessary to assimilate the different parts of the system to a single basis, either human or mechanical. Since our understanding of the mechanical elements of gun pointing appeared to us far ahead of our psychological understanding, we chose to try and find a mechanical analogue of the gun pointer and the airplane pilot. . . . We call this negative feedback. (1956, pp. 251–52)

The logical elision between the servomechanical character of correcting the flaps of an aircraft and the correction of the orientation of a gun attempting to bring down the aircraft was more than a mathematical trick or a crude attempt to eliminate the "human element"; for Wiener, it constituted a wholesale reconceptualization of the human-plus-machine. The whole ensemble of airplane-gunner could be treated as a conceptual system from the viewpoint of MOD, if the process of evasion-targeting

[27] Norbert Wiener, "Summary Report for Demonstration," National Defense Research Committee Contractor's Technical Report, June 10, 1942, NDCrc-83, quoted in Galison, 1994, p. 236.

could be regarded as a set of "communications" subject to feedback correction. In effect, this deconstructed what could only appear to the untutored as an intrinsically physical process of airfoils and explosions; the physical and the human both had to undergo ontological metamorphosis into "messages with noise" in order to be combined into a new synthesis. This is the meaning of Wiener's later mantra that, "The physical identity of an individual does not consist in the matter of which it is made. . . . Organism is opposed to chaos, to disintegration, to death, as message is to noise" (1954, pp. 101, 95).

The beauty of the literal conflation of men and machines is that it pointed the way forward for the construction of new machines. Wiener and his assistant Julian Bigelow began a laboratory where they might "put our ideas into metal almost as fast as we could conceive them" (1956, p. 248). They first used Bush's differential analyzer to predict the position of an evading aircraft a few seconds into the future but realized that the superior speed of electronics might require a digital computer. Also, to model pilot reactions better, they set up a brace of simulation experiments where subjects attempted to train a light spot under the sway of a deliberately sluggish control stick onto a moving target. This was the beginning of the cyborg predilection for linking simulations with human experimentation to render the "man-machine interface" more harmonious, rendering the human more machinelike and vice versa. They made substantial progress with their predictor, but when it was tested against other, much less elaborate and less expensive predictors, it was inferior to one developed by Hendrik Bode at Bell Labs.[28] So, in one sense, Wiener's cyborg brainchild was a failure by 1943. But, in a larger sense, it was only just the beginning.

The turning point came with an essay by Wiener and his collaborators Arturo Rosenbleuth and Bigelow on "Behavior, Purpose and Teleology" (Rosenbleuth, Wiener, & Bigelow, 1943), in which they proposed a program of unified behaviorism to encompass all phenomena of intentionality under the formalism of prediction and feedback. "The term purposeful is meant to denote that the act or behavior may be interpreted as directed to the attainment of a goal – i.e., to a final condition in which the behaving object reaches a definite correlation in time or space with respect to another object or event" (p. 18). Whereas his antiaircraft predictor had not panned out as a gun guidance system, perhaps it could stand instead as a first step in the understanding of the brain itself. This conviction received a boost from the appearance of a paper by McCulloch

[28] Hendrik Bode (1905–82): B.A., Ohio State, 1924; Ph.D. in physics, Columbia, 1935; Bell Labs, 1926–67; professor of systems engineering, Harvard, 1967–?. For the Bode predictor beating out Wiener's, see Galison, 1994, p. 244; Millman, 1984.

and Pitts (1943), which related a simple model of the neuron to the abstract computational power of an ideal computer.[29] There seemed to be a critical mass of co-workers in the war effort who shared just these sorts of convictions, enough for Wiener to join forces with John von Neumann and Howard Aiken (an early Harvard computer pioneer) to convene a meeting of the newly minted Teleological Society in January 1945 (Heims, 1980). The prior military connections of this nascent movement did not leave it starved for encouragement: "We are also getting good backing from Warren Weaver, and he has said to me this is just the sort of thing that Rockefeller should consider pushing. In addition, McCulloch and von Neumann are very slick organizers, and I have heard from von Neumann mysterious words concerning some thirty megabucks which is likely to be available for scientific research. Von Neumann is quite confident he can siphon some of it off."[30]

Wiener's vision not only energized the nascent computational community; it also informed an emergent community concerned with prediction within the field of statistics. In Wiener's opinion, "the nervous system and the automatic machine are fundamentally alike in that they are devices which make decisions on the base of decisions they have made in the past" (1954, p. 33). This meant that the types of rationales that had underpinned statistical inference in the past – say, R. A. Fisher's doctrines (Wiener, 1961, p. 62) – had to be rethought and remolded into the new approach. Unlike earlier inchoate notions of the "information" contained in a sample, Wiener proposed to substitute a notion of information predicated upon an entropy measure inspired by thermodynamics. Wiener then projected well-known linear least-square projections (which he had used in the gunnery predictor) into the frequency domain, making good use of his earlier work on Brownian motion and the Wiener-Hopf theorem (Whittle, 1983, chap. 6; Wiener, 1961, chap. 3). The result was a recasting of the prediction problem as one of extracting "signal" from "noise" in a communication channel – just the sort of thing a wily MOD might do. This work was written up as a classified report in wartime – called, with characteristic American insensitivity, the "yellow peril" due to its cover (Wiener, 1956, p. 262) – and published in declassified form after the war as the *Extrapolation, Interpolation and Smoothing of Stationary Time Series* (1949). As with everything else Wiener did, the mathematics was valued more as a means to a larger end. "The work I did on the statistical

[29] This sentence refers to the Turing machine, discussed in the next section. The relationship of the McCulloch-Pitts paper to von Neumann is discussed in Chapter 3. Wiener (1961, pp. 12–13) briefly discusses McCulloch-Pitts. This history of the McCulloch-Pitts paper is nicely illuminated in Anderson & Rosenfeld, 1998.

[30] Wiener to Arturo Rosenbleuth, January 24, 1945, quoted in Aspray, 1990, p. 316.

treatment of anti aircraft control has led eventually to a general statistical point of view in communications engineering . . . this is passing over to less orthodox fields such as meteorology, sociology and economics" (1956, p. 255).[31]

Wiener's own trajectory diverged substantially from those of most of his colleagues in the wartime mobilization toward the end of the war. He had forsaken participation in the Manhattan Project, and when he learned of the devastation of the bomb dropped on Hiroshima, he was shaken to his very core. Around that time he resolved not to have anything further to do with the military organization and funding of science, which, if anything, only grew more pervasive at the end of the war. For example, he refused to be included on a distribution list for RAND documents and memoranda.[32] This isolated him from most of the major developments in computer engineering in the immediate postwar period, as well as the innovations in social science recounted in Chapter 6; but he seemed not to mind, withdrawing even more into biological and medical applications of his statistical and feedback ideas. Much of his collaborative work was concentrated on areas like hearing aids, artificial limbs and ataxia, and heart arrhythmia. He once wrote that these were all exercises in pursuit of a "general philosophy of prosthesis" (1956, p. 287). Yet his increasing feelings of scientific and political isolation in the Cold War era shunted him in an ever more philosophical direction, first instantiated by his book *Cybernetics* (1948) and his later more explicitly popular books *Human Use of Human Beings* (1950) and *God and Golem* (1964). In Wiener's mature philosophical phase, his version of MOD was inflated into a Theory of Everything, from code breaking to physiological homeostasis to capitalism to brains as self-organizing systems. All really seductive philosophical systems are monistic, and thus Wiener's fit this mold, collapsing differences between the animate and inanimate, organism and machine, signal and message, the Natural and the Artificial.

It seems that one motivation for his philosophical turn was that his insistence that "humans do not differ from machines" had come back round to haunt him, in that it took but little imagination to see that once this prognostication was taken seriously, it would open up breathtaking new vistas of exploitation and inhumanity on the part of those who were

[31] Wiener here was more prescient than even he, prognosticator extraordinaire, could realize. This quotation is intended to point to the subsequent rise of "rational expectations" theory in neoclassical economics, a topic briefly touched upon tangentially here but covered in detail in Sent, 1998.

[32] See Alex Mood to Wiener, September 18, 1952, box 10, folder 155, WMIT. See also the correspondence with Oskar Morgenstern in box 5, folder 75, where a suspicious Wiener asks whether Morgenstern wants copies of his wartime statistics papers to be employed in further war work.

bent on becoming the new Demons of society. If cybernetics were to be a technology of *control and communication*, then it was a foregone conclusion this technology would be patterned on the practices of control already exercised in the military and in the factory. "From the very beginning of my interest in cybernetics, I have been well aware that the considerations of control and communication which I have found applicable in engineering and in physiology were also applicable in sociology and in economics" (1964, p. 87). But Wiener was not happy with any of the enthusiasms expressed for the insights of cybernetics by contemporary sociologists or economists. Indeed, he was repelled by what he knew of neoclassical economics. As early as 1936, he had resigned his membership in the Econometrics Society, offering as his explanation "misgivings regarding the possibilities of employing more than elementary statistical methods to economic data."[33] The more he learned about neoclassical economics, the less he liked it. The following passage from his *Cybernetics* was characteristic of his attitude:

> [O]ne of the most surprising facts about the body politic is its extreme lack of efficient homeostatic processes. There is a belief, current in many countries, which has been elevated to the rank of an official article of faith in the United States, that free competition is itself a homeostatic process: that in a free market the individual selfishness of the bargainers . . . will result in the end in a stable dynamic of prices, and with [*sic*] redound to the greatest common good . . . unfortunately, the evidence, such as it is, is against this simple-minded theory. The market is a game. . . . The individual players are compelled by their own cupidity to form coalitions; but these coalitions do not generally establish themselves in any single determinate way, and usually terminate in a welter of betrayal, turncoatism, and deception, which is only too true a picture of the higher business life. (1961, pp. 158–59)

He spent inordinate numbers of pages in his popular works criticizing game theory, as I document here, and corresponded with social scientists more to his liking about the dire consequences of von Neumann's social mathematics. Some of the strongest language was aired in a letter to Wiener by Gregory Bateson:

> Two years ago – or was it three – at one of the Cybernetic meetings, I said that I wished you or some other mathematician would build a branch of mathematics to describe the propagation of paranoid premises. You instantly replied that no such branch of mathematics could ever be constructed because it would run into the Theory of Types. . . . The problem is fundamental to theory and urgent in a world which stinks. . . . No doubt in short time perspective the Rand people are right and have

[33] Norbert Wiener to Charles Roos, April 12, 1936, and response, box 3, folder 45, WMIT.

'useful' advice to offer on such subjects as intercepting hostile aircraft or bamboozling hostile diplomats . . . but it still stinks. In the long run, the application of the theory of games can only propagate the theory by reinforcing the hostility of the diplomats, and in general forcing people to regard themselves and each other as Von Neumannian robots.[34]

Perhaps then it comes as less of a surprise to realize what Wiener wrote near the end of his life: "Thus the social sciences are a bad proving ground for the ideas of cybernetics" (1964, p. 92).

Wiener's relationship with that other cyborg visionary, John von Neumann, is a sensitive barometer of his postwar conundrum. As Steve Heims (1980) makes abundantly clear, Wiener and von Neumann started out sharing many interests and approaches, even to the point of working on many of the same formal problems; nevertheless, by the ends of their respective careers, one could not find two thinkers more diametrically opposed concerning the significance and legitimacy of their respective enterprises. While von Neumann became more deeply embroiled in the military organization of science (as described in Chapter 3), Wiener withdrew. As von Neumann slid Right, Wiener tacked Left. Von Neumann became the consummate political insider in the 1950s, while Wiener sulked as the outsider and public gadfly. The leitmotiv of von Neumann's work became that of the maintenance of the complexity of organization and the use of randomness to produce further organization, whereas that of Wiener remained the control of randomness and dissolution through feedback. While Wiener was willing to associate his trademark fright-Demon, the MOD, with von Neumann's game theory, he increasingly signaled his disaffection with game theory throughout his postwar career. In the first edition of *Cybernetics* (1948), he already implicitly commented on the underdetermination of the "stable set" solution concept.[35] In the second edition he added another chapter critiquing game theory from the

[34] Gregory Bateson to NW, September 22, 1952, box 10, folder 15, WMIT. Biographical information on Bateson can be found in Harries-Smith, 1995; on the relationship to Wiener, see Heims, 1980, pp. 307–9. The timing of this letter and its relationship to "the Rand people" and the Nash game theory generated just before it are quite stunning, as the reader of Chapters 6 and 7 will come to realize.

[35] "Where there are three players, and in the overwhelming majority of cases, when the number of players is large, the result is one of extreme indeterminacy and instability. The individual players are compelled by their own cupidity to form coalitions; but these coalitions do not establish themselves in any single, determinate way, and usually terminate in a welter of betrayal, turncoatism and deception. . . . Naturally, von Neumann's picture of a player as completely intelligent, completely ruthless person is an abstraction and a perversion of the facts" (Wiener, 1961, p. 159). Note well that the "welter of betrayal" is precisely the sort of situation an MOD is supposed to confidently rectify; so this is an implicit criticism that that game theory is not sufficiently cybernetic. Von Neumann's stable set is described in Chapter 3.

point of view of machines that could learn to play games, with a long discursus on chess (1961, chap. 9). In *Human Use* he warns of the fallacies of using game theory to organize actual military strategies in a Cold War environment (1954, pp. 181–82). In *God and Golem* he asked the pointed question of von Neumann, "Can *any* creator, even a limited one, play a significant game with his own creature?" (1964, p. 17). Thus whereas Wiener sought the deliverance of his science in the mind-numbing complexities of the neuron, something made by God and not by man, von Neumann sought to move away from the particularities of the biological organism toward an abstract mathematical theory of an artificially constructed automaton.

In the other direction, one notes from the tone of their correspondence that von Neumann was patiently suffering what he regarded as Wiener's eccentricities.[36] It seems von Neumann was miffed that Wiener was getting credit for proselytizing for the connection between entropy and information, something that von Neumann believed he had appreciated much earlier (Heims, 1980, p. 208). This showed up in von Neumann's noticeably restrained review of *Cybernetics* for the *Scientific American*:

> The author is one of the protagonists of the proposition that science, as well as technology, will in the near and in the farther future increasingly turn from problems of intensity, substance and energy to problems of structure, organization, information, and control. . . . The book's leading theme is the role of feedback mechanisms in purposive and control functions. . . . Several students of the subject will feel that the importance of this particular phase of automat-organization has been overemphasized by the author. . . . The reviewer is inclined to take exception to the mathematical discussion of certain forms of randomness in the third chapter of the book. . . . The technically well-equipped reader is advised to consult at this point some additional literature, primarily L. Szilard's work. . . . There is reason to believe that the general degeneration laws, which hold when entropy is used as a measure of the hierarchic position of energy, have valid analogs when entropy is used as a measure of information. On this basis one may suspect the existence of connections between thermodynamics and the new extensions of logics. (von Neumann, 1949, pp. 33–34)

[36] "What you wrote me about my journalistic contacts and their appurtenances has all the virtues of a cross-word puzzle. What is the 'labor article'? . . . I have been quite virtuous, and had no journalistic contacts whatever (except with a Mr. McDonald from *Fortune* who asked me some quasi-technical questions on the theory of games – but that pew is in a different church). . . . I hope I need not tell you what I think of 'Cybernetics,' and, more specifically, of your work on the theory of communications: We have discussed this many times." John von Neumann to NW, September 4, 1949, box 7, folder 104, WMIT.

Von Neumann also grew progressively disenchanted with what he regarded as premature conflations of the brain and the computer, especially as he became more embroiled in their actual design; most of these barbs were aimed obliquely at Wiener, for whom the ontological equivalence of the brain and computers was a central tenet of cybernetics. The further that Wiener sought to direct cybernetics toward neuroscience and psychology, the more von Neumann distanced himself from the project. The sources of their disagreement came out much more starkly in their correspondence:

> Our thinking – or at any rate mine – on the entire subject of automata would be much more muddled than it is, if these extremely bold efforts [of Pitts and McCulloch, Pitts, Wiener and Rosenbleuth] – with which I would like to put on one par the very un-neurological thesis of R. [*sic*] Turing – had not been made. Yet, I think that these successes should not blind us to the difficulties of the subject. . . . The difficulties are almost too obvious to mention. . . . To understand the brain with neurological methods seems to me as hopeful as to want to understand the ENIAC with no instrument at one's disposal that is smaller than 2 feet across its critical organs, with no methods of intervention more delicate than playing with a fire hose (although one might fill it with kerosene or nitroglycerine instead of water) or dropping cobblestones into the circuit. Besides the system is not even purely digital. . . . And it contains, even in its digital part, a million times more units than the ENIAC.[37]

It seems clear that von Neumann held an entirely different appreciation for the computer as a machine, in contrast to Wiener; but that he also bore a different conception of the implications of thermodynamics than Wiener. For Wiener, entropy was ultimately something oppressive, something to be channeled and neutralized; for von Neumann, it was a fruitful source of inspiration for his conception of abstract automata:

> Such a theory remains to be developed, but some of its major characteristics can already be predicted today. I think that it will have as one of its bases the modern theory of communications, and that some of its most essential techniques will have a character very near to Boltmannian thermodynamics. It should give a mathematical basis for concepts like "degree of complication" and of "logical efficiency" of an automat (or of a procedure).[38]

The mathematical theory of automata, which we shall argue was the culminating fact of von Neumann's career, never held much of an allure for his inverted interlocutor. Wiener, for his part, tried making some jokes

[37] Von Neumann to NW, November 29, 1946, box 5, folder 72, WMIT.

[38] Von Neumann to Jeffress, "Abstract of a Paper by von Neumann," [1948?], box 19, folder 19, VNLC.

about the public reaction to machines enjoying sex, with reference to von Neumann's theory of self-reproducing automata; von Neumann did not think the joke at all funny (Heims, 1980, p. 212). In his autobiography, Wiener even went so far as to make some jaundiced remarks about von Neumann's approach to weather prediction as insufficiently appreciative of the stochastic nature of prediction (1956, p. 259), which, for Wiener, was tantamount to complaining that von Neumann lacked an appreciation for his life's work.

Claude Shannon and Information Theory

The testy relationship between the two cyborg visionaries was partially mediated by a third vertex of the cyborg triangle, Claude Shannon.[39] Compared with the profound sophistication of the Hungarian wizard and the Harvard ex-prodigy, Shannon would play the role of the aw-shucks midwestern tinkerer. Unusually, Shannon combined an interest in symbolic logic with a background in electrical engineering. On landing a job at MIT helping Vannevar Bush with his differential analyzer, he wrote a 1938 master's thesis showing that electronic switching circuits could be described by Boolean algebra, a fact now taken as transparently obvious in modern digitalized society. On Bush's advice, he stayed on to do a Ph.D. in mathematics at MIT, taking some time off to learn some genetics at Cold Spring Harbor. His 1940 thesis tried to do for genetics what he had just done for electronic circuits; but the attempt was premature. In 1940 a research fellowship permitted a year of residence at the Institute for Advanced Study, where he came in close contact with von Neumann. This led to Shannon serving as a consultant to the National Defense Research Committee to do fire-control work, like so many others in this narrative. In a widely circulated story, it is reported that von Neumann told Shannon to link his nascent theory of information to thermodynamics: "You should call it 'entropy' for two reasons: First, the function is already in use in thermodynamics under that name; second, and more importantly, most people don't know what entropy really is, and if you use the word 'entropy' in an argument you will win every time!" (Tribus in Machlup & Mansfield, 1983, p. 476). In 1941 he took a job at Bell Laboratories, where he was assigned to work on cryptography research on Project X, the development of a speech encoder that quantized the waveform and added a digital signal before transmission. There he collaborated with Alan Turing in 1943 on cryptography problems but also on their ideas about human brains and computers (Hodges, 1983, p. 249).

[39] Claude Shannon (1916–2001): Ph.D. in mathematics, MIT, 1940; Princeton & NDRC, 1940–41; Bell Labs, 1941–56; professor at MIT, 1958–78. Biographical information can be found in Kay, 1997a; Lucky, 1989; and Shannon, 1993. For an obituary, see *Science*, April 20, 2001, p. 455.

It is not often noted that Shannon's first ideas on information theory were published in a 1945 classified Bell Labs memorandum "A Mathematical Theory of Cryptography." Indeed, it is hard to differentiate what later became known as information theory from cryptology: "The work on both the mathematical theory of communications and cryptology went forward concurrently from about 1941. I worked on both of them together and I had some of the ideas while working on the other. I wouldn't say one came before the other – they were so close together you couldn't separate them."[40] At the urging of his supervisor Hendrik Bode (the very same inventor of the winning antiaircraft gunnery predictor), he published a declassified version of his information theory in the *Bell System Technical Journal* in 1948, linking it to earlier work done at Bell Labs by H. Nyquist and R. Hartley (Aspray, 1985). In the July 1949 *Scientific American*, Warren Weaver published an article popularizing his own synthetic interpretation of the work done by Wiener, Shannon, and others on information theory, which he had directly been organizing and funding during the war through his activities on the Applied Mathematics Panel, the Rockefeller Foundation, and the National Defense Research Council. Wilbur Schramm, a communications theorist at the University of Illinois, then arranged to package the articles by Shannon and Weaver together as *The Mathematical Theory of Communication* (1949), the format in which Shannon's information theory is still available to most modern readers. It may be prudent to keep in mind the odd circumstances of publication, because Weaver's ideas should not be confused with those of Shannon.

The relationship of Shannon's information theory to Wiener is a bit difficult to interpret. Wiener claimed that he was not acquainted with Shannon's work during his education at MIT (1961, p. 140), though this hardly rings true, given both of their overlapping responsibilities for Bush's differential analyzer. Wiener asserted something like separate but equal priority for the entropy definition of information, according Shannon priority for the "discrete" versus his "continuous" definition (1956, p. 263). The situation is further muddied by the presence of Weaver, who was continually shuttling back and forth between the various cryptographic, computer, and gun-control problems in his capacity as research coordinator during the war. After all, Wiener's work was directly "under the supervision of Dr. Warren Weaver" (1956, p. 249). Weaver took it upon himself to referee priority in a footnote to the Illinois volume (Shannon & Weaver, 1963, p. 3), but the sentiments ascribed to Wiener are not very convincing. Shannon, for his part, had access to Wiener's "yellow peril" in the early 1940s at Bell Labs (Schement & Ruben, 1993, p. 45).

[40] Shannon interview, quoted in Kay, 1997a, p. 51. See also J. Rogers & R. Valente in Schement & Rubin, 1993, p. 39; Edwards, 1996, p. 200.

We now know that Wiener, Shannon, and Weaver were jockeying for priority hard on the heels of the publication of *Cybernetics*. Shannon wrote Wiener in October 1948 to remind him that he had already provided a mathematical expression for information that was the negative of Wiener's expression, and that perhaps Wiener's account of Maxwell's Demon in that book was not all that sound.[41] Weaver rapidly followed up with a letter to Wiener in December asking him how much priority he would cede to Shannon.[42] In letters to third parties, Wiener always stressed that it was Wiener himself who had been seeking to provide a general theory of the living organism, whereas he portrayed Shannon as a mere technician working on the limited problems of Bell Telephone (Kay, 1997a, p. 52). Yet, given Shannon's early ambitions for biology, this slur was unwarranted.

But maybe all the hairsplitting about priority ultimately doesn't matter, because it should be regarded as another instance of the MOD demon making his nefarious presence felt. After all, no participant was altogether candid about his motivations. As Shannon is reported to have said: "I started with Hartley's paper and worked at least two or three years on the problems of information and communications. That would be around 1943 or 1944; and then I started thinking about cryptography and secrecy systems. There is a close connection; they are very similar things, in one case trying to conceal information, in the other case trying to transmit it."[43] The blurring of the distinction between natural randomness and conscious intentional evasion, entropic control and stealthy dissimulation, and communication and misinformation is the hallmark of MOD.

Von Neumann was certainly correct in anticipating that conflating information and entropy would result in mass confusion; Shannon's information concept has to be one of the most misunderstood notions on the planet, perhaps second only to the theory of relativity or the "laws" of supply and demand. Yet the irony is that this has been its strength, for it has been Shannon's information theory, and not Wiener's, that became the touchstone of postwar cyborg science. One should realize that

[41] "I consider how much information is *produced* when a choice is made from a set – the larger the set the *more* information. You consider the larger uncertainty in the case of a larger set to mean less knowledge and hence *less* information. The difference in viewpoint is partially a mathematical pun. . . . In connection with the Maxwell demon, I have the intuitive feeling that if the gas is in complete statistical equilibrium with the demon's eye (for both matter and radiation) then the demon will not be able to measure anything useful regarding the speeds and positions of the particles." Claude Shannon to Norbert Wiener, October 13, 1948, box 6, folder 85, WMIT.

[42] Weaver to NW, December 21, 1948, box 6, folder 87, WMIT.

[43] Oral history with Robert Price, reported by Rogers & Valente in Schement & Ruben, 1993, p. 42.

Shannon does not use the notion of information in any of its colloquial senses, and diverges from the interpretation pioneered by Szilard; but, nevertheless, the diffusion of Shannon's notions have come to color the vernacular understanding over time. The best description I have ever encountered is that it is all a matter of "a statistical relation between signs"; but, then, that makes it sound as though it were some sort of game one foists upon children to keep them quiet on a long car trip. Technically, it is the theory of a "transducer," a device capable of decoding and recoding strings of symbols as inputs and outputs, one that "may have an internal memory so that its output depends not only on the present input symbol but also on its past history" (Shannon & Weaver, 1949, p. 57). Possibly it is more felicitous to imagine the situation as one of those classic experiments in telepathy: there is a source of "information" arranged by some external experimenter (a pack of cards), a transmitter (here, the putative telepath), a signal (a black circle on a card), a channel (the ethereal medium?), possibly some noise (static in the astral plane; distractions for the telepath), and a receiver (a talented "sensitive" in another room). Shannon would imagine that it is a job of the telepath to send the image of the circle to the other room, devoid of any concern for their meaning or significance. Indeed, it is the role of the experimenter to control all the important semantic characteristics of the symbols so they do not influence the experiment. The experimenter "chooses" which symbols to transmit at random, guided by full knowledge of the prior distribution of the inscribed cards. Shannon then asserts that it is possible to formalize the "amount of information" conveyed by the signal solely by the probability of its "choice"; "semantic aspects of communication are irrelevant to the engineering problem" (p. 31), or so he claims.

Shannon followed Hartley (1928) in suggesting that the "amount of information" in a transmitted message should be inversely proportional to the probability of its "choice" by the experimenter. One can bend in all sorts of epistemic pretzels by telling stories about high-probability symbols already being "expected" and therefore lacking the element of surprise; but I personally think this insinuates an element of intentionality to the receiver which is not justified by the mathematical setup. It is better to just phrase the question as to whether the "sensitive" can be statistically regarded as "just wildly guessing," or if there is an argument that some communication has successfully taken place. The next problem is to pick some specific function of the probabilities p as defining "information," and Shannon chose the logarithmic function in base 2:

$$H(p) = -\log(p) = \log(1/p)$$

In his essay, Shannon provides some rather breezy justifications for this choice (Shannon & Weaver, 1949, p. 32), such as the information content of two independent events is just the sum of their individual information measures, "two punched cards should have twice the storage capacity of one for information storage," and the log function is continuous and monotonic so that it is easy to take limits,[44] as well as recourse to some "intuitive feelings." Maximum expected information would occur in a situation where the a priori probability of all the symbols would be the same, or $p = 1/N$, $H(p) = \log N$. In all other instances, the expression for expected information content was the same as the conventional expression for entropy in thermodynamics – the one major motivation that Shannon mentioned only later in the text (p. 51):

$$H = \sum_{i=1}^{n} p_i \log \frac{1}{p_i}$$

In the thermodynamic interpretation, increased entropy implies a greater degree of disorder or randomness. Shannon seems to twist this into a paradox by equating greater randomness with less redundancy in transmission, so higher entropy means higher expected information content. It is an implication of this view that a string of gibberish like "qxvy/zz2j" could be said to embody more "information" than, say, "Mississippi." Before this makes one dizzy, try and translate that statement into the proposition that a randomized deck of cards in the context of a controlled experiment will help us learn more about the efficacy of our telepath than one with some discernible order. Another way of seeing how the metaphor of entropy does not carry over entirely unaltered into Shannon's version is to realize that thermodynamic entropy is a measure of the number of ways the unobserved (and therefore probabilistic) microdynamics of molecules can make up a measurable macrostate, like temperature. In Shannon's version, there is no macro-micro distinction, only a given probability of a particular symbol showing up, and a measure of the likelihood of strings of symbols. This is often rephrased by suggesting that Shannon's entropy is about "choice" of symbols (Hayles, 1990a, p. 54), an interpretation first popularized by Weaver: "information, in communication theory, is about the amount of freedom of choice we have in constructing messages" (Shannon & Weaver, 1949, p. 13). This invocation of "choice" is extremely dubious – isn't it perhaps sneaking intention and semantics back in through the back door? – but will prove to be significant when we come to consider the cyborg incursion into orthodox economics.

Actually, the definition so far tells us very little about anything we might like to know. In this world where it appears that meaning has been leached out of information, there is not much sensible that can be said about its

[44] Not to mention avoiding Wiener's complicated resort to measure theory.

role in the process of communication. But wait: not much, but something nonetheless. Shannon was concerned, as was his employer, AT&T, about the theory of the capacity of various sorts of communication channels, such as telephone cables or television signals. One set of theorems that Shannon produced described what an efficient code would look like (*vide* its provenance in cryptography) and demonstrated that if a channel without "noise" had a transmission capacity of C bits per second, it would not be possible to transmit information at a rate greater than C/H (Shannon & Weaver, 1949, p. 59). A second set of theorems imagines the existence of a "noise" that, because it has a random character just like the posited symbol set, cannot be filtered out according to some fixed deterministic principles. He proposed that his definition of information could be used to define an equation which would read: "amount sent + noise = amount received + missing information." He then proved a theorem that said that a noisy channel with capacity C is capable, with suitable coding, of transmitting at any rate less than C with a vanishingly small probability of error. This thesis was initially greeted with stunned disbelief by communications engineers at the time. Here we must distinguish between entropy of the source and entropy of the channel. In effect, the theorem assured them that there existed a way to use some of the received bits of information to locate and correct the errors due to noise, and that if the fraction of bits devoted to this task is greater than the channel entropy H, then it should be possible to find and correct virtually all the errors. Here again we can observe MOD at work: a demon taught to neutralize the devious codes of the Enemy now trains his strategic knowledge on Nature to defeat the forces of dissolution and disorder.

Shannon's theory of information has proved immensely fruitful in communications and computer engineering, in everything from compression codes to error-correction algorithms to encryption packages (Lucky, 1989). It has also set in motion one of the most farcical trains of misconceptions and misunderstandings in the modern history of the sciences, namely, that "information" is a palpable thing with sufficient integrity to be measured and parceled out like so many dollops of butterscotch. It is often asserted that Shannon himself tried to prevent everyone and their cousin from projecting their own idiosyncratic connotations of "information" (and, for that matter, "entropy") on the mathematics; and, to be fair, there is evidence that he did raise the yellow flag here and there.[45]

[45] For instance, there is his intervention in the Eighth Macy Conference in 1951: "It seems to me that we can all define 'information' as we choose; and, depending upon what field we are working in, we will choose different definitions. My own model of information theory, based mainly on entropy, was framed precisely to work with the problem of communication. Several people have suggested using the concept in other fields where in many

Nevertheless, a vast scrum of scholars was quick off the mark to co-opt Shannon information theory back into their own fields, ranging from psychology and neurology to molecular biology, traffic engineering, and (lest we forget) economics (Tribus in Machlup & Mansfield, 1983). In some fields like psychology, it experienced a pronounced boom and bust, having been evacuated from the journals by the 1970s (p. 493). In tandem with cybernetics, it did not so much disappear, as mutate into a second, and now third, generation of research, sometimes called "autopoesis," sometimes "cognitive science," or perhaps something else (Hayles, 1994, 1999).

The question I should like to raise is whether Shannon was really so very intensely concerned that his various mathematico-metaphorical flights of fancy not be improperly highjacked by a gaggle of irresponsible scholars, or whether he was just engaging in a kind of standoffishness, which most physicists and engineers feel is their birthright when confronted with the despised breed of social scientists. After all, he did participate in three of the "Macy Conferences," which were the lineal descendants of the meeting of Wiener and von Neumann's original "Teleological Society," and were intended to spread the gospel of cybernetics to those working in the fields of psychology and anthropology (Heims, 1991). More importantly, one need only cast a glance over the rest of his career to realize Shannon was no narrow humble technologist just trying to solve a few practical problems for AT&T. In 1950 he published a paper on "Programming a Computer for Playing Chess." In the very first paragraph, Shannon manages to make an elision between "machines which will handle the routing of telephone calls" to "machines for making strategic decisions in simplified military operations" to "machines capable of logical deduction" (1950, p. 256). His presentations at the Macy Conferences concerned an electronic "rat" that could supposedly navigate a maze in good behaviorist fashion. When a coworker at Bell Labs built a rudimentary set of circuits to "guess" the strategy of an opponent at a simple coin-toss game, Shannon was inspired to build a simpler version, and to pit his machine against the other one.[46]

cases a completely different formulation would be appropriate. . . . If you are asking what does information mean to the user of it and how it is going to affect him, then perhaps [MacKay's] two number system might be appropriate" (in von Foerster, 1952, pp. 207–8). Or for a harsher warning, see p. 22: "I don't see too close a connection between the notion of information as we use it in communication engineering and what you are doing here. . . . I don't see how you measure any of these things in terms of channel capacity, bits, and so on."

[46] "It was necessary to construct a third machine to act as a referee, and to dole out the random numbers. . . . Hagelbarger observed that if you told people that you were conducting an experiment in probability, no one was interested. But if you gave people

He participated in the earliest attempts to elaborate on von Neumann's approach to automata and the theory of computation (Shannon & McCarthy, 1956); and was one of the co-organizers of the 1956 Dartmouth conference credited in retrospect as the birthplace of "Artificial Intelligence" as a discipline (Edwards, 1996, p. 253). It is reported that, as a hobby, he even built little juggling automata out of clockwork and electronics, and his own chess-playing machines. On the debit side, he published little more on his information theory after the 1950s. So here was someone who thought and acted just like the other cyborgs in good standing: his goal has persistently been to either erase or transcend the distinctions between humans and computers; and that is why his semantically impoverished theory of information was still essentially thought to apply equally to machines and to people; and it should be recognized that his "information theory" was part and parcel of a project to build a Theory of Everything out of some probability theory and computational metaphors. This is documented directly in a recent interview:

> *Omni:* In the Fifties you criticized people for applying your ideas to fields other than communications. . . . Are you as skeptical now as you were then about such attempts?
>
> *Shannon:* . . . It's possible to broadly apply the term *information theory* to all kinds of things, whether genetics or how the brain works or this and that. My original ideas were related to coding information for transmission, a much narrower thing. But some of these applications may be valid. For example, animals and humans transmit information along nerve networks. . . . It is a noisy, redundant system.
>
> *Omni:* Do you agree with Norbert Wiener's denial of any basic distinction between life and nonlife, man and machine?
>
> *Shannon:* . . . I believe in evolutionary theory and that we are basically machines but of a very complex type. . . . We are the extreme case: a natural mechanical device.[47]

Perhaps Shannon's position relative to the efflorescence of work done in the name of "information theory" might be elucidated by his own

the idea that two thinking machines were dueling to the death, then everyone was excited" (Lucky, 1989, p. 54). The relevance of "dueling computers" to conceptions of rationality in the cyborg sciences is further explored in Chapter 7.

[47] Shannon interview, *Omni* magazine, August 1987, pp. 64, 66. In this article we also learn that Shannon has tried his hand at the mathematical modeling of stock prices, but has never published his ideas. Nonetheless, he did give talks on the topic at MIT, and claims to have amassed a substantial fortune in the stock market. A similar interest may be implied by a popular piece done on Shannon by John Horgan in the January 1990 *Scientific American*. There it is written: "What is information? Sidestepping questions about meaning, Shannon showed that it is a measurable commodity."

notion of "equivocation": this was the term he coined for the amount of information contributed to the message decoded by a receiver by the noise induced in the transmission channel. Now, under the inspiration of MOD, it was true that Shannon generally regarded this "surplus" information as something unwanted, something to be extirpated; but that was a semantic judgment that he made, which could not be defended from within his own system, with its proleptic self-denying ordinance about meaning. This peremptory judgment that all noise was bad, especially in a formalism that treated message and noise as ontologically identical, was Shannon's very own "ghost in the machine."[48] Weaver, for one, did not necessarily see it the same way; nor, as we have suggested, did von Neumann; and a subsequent generation of cyberneticians reveled in the novelty and surplus meaning induced by noise (Hayles, 1990a, p. 56; 1994). From this vantage point, Shannon himself was "equivocal": his importation of thermodynamics as a measure of "information" was a message badly infected by a species of noise – the chattering of all those camp followers lusting after a mathematical science of information, yet also the ineradicable surplus meaning of metaphor – but a noise that itself assumed some surplus moiety of meaning in the very process of transmission. In the end, it appears Shannon wanted information to look more and more like something that had a curious kind of integrity independent of the cognitive makeup of the receiver, or the intentions of the sender.

Hence, all the profuse endless denials that Shannon information has anything whatsoever to do with semantics or meaning, or that it *should* be so restricted, or deductions that it must therefore necessarily be disqualified from general relevance to anything outside of communications engineering, themselves seem to have utterly missed the point.[49] After all, what Shannon accomplished was not all that different from what Gödel had done before him: in order to develop metaprinciples about what was and was not possible within a formal system, one had to abstract away from the meaning of statements through the technique of Gödel-numbering, and then discuss the formal manipulation of those resulting numbers. System self-reference usually requires a modicum of semantic repression. Shannon's symbol transducers performed the same function, although perhaps his unwillingness to become too embroiled in issues of software prevented Shannon from searching out his own recursive paradoxes. Because he "believed in evolution," it was quite possible for

[48] This presumption that noise was bad was itself an artifact of the military origins of the cyborg sciences. On the tremendous din of mechanized battle and the difficulties of maintaining military chain of command and communication, see Edwards 1996, chap. 7.

[49] The reader will be asked to further entertain this thesis in the novel context of some controversies between Cowles and RAND in Chapter 6.

Shannon to see brains and genes as all processing his kind of information, precisely because (by some accounts) evolution has no final meaning or purpose. Likewise, for cyborgs like Shannon, what is it that constitutes the ultimate telos of capitalism? After all, to reduce intentionality and semantics to mechanics is the first tenet of the Cyborg Creed.

THE DEVIL THAT MADE US DO IT

Thus far this volume has subjected the reader to repeated references to computational metaphors and calculating machines but almost no reference to formal computational *theory*. Partly this has been due to the fact that no one we have surveyed at any length – Szilard, Wiener, Shannon – had actually done anything on the topic. (Von Neumann, as usual, deserves a category of his own in this regard.) Yet there abides a much more substantial reason, namely, that a vast proportion of the history still remains unwritten. Modern fascination with the computer has tended to fasten upon the grosser aspects of hardware or, worse, the buccaneers who built the companies whose logos grace our laptops (Campbell-Kelley & Aspray, 1996; Ceruzzi, 1998). But "the computer" is a most curious intersection of a motley set of traditions, from abstract mathematics to electrical engineering to weapons development to operations research to software engineering to artificial intelligence to television broadcasting to molecular biology; and its hybrid character flaunts all standard distinctions made between "pure science" and "technology" (Mahoney, 1988). Some seem to believe that the separation of software from hardware could serve to enforce that pristine purity: the former the province of an abstract "computer science," the latter the preserve of the engineers. Although we concentrate on the former in this section, it should become increasingly apparent as we go along that this aspiration is forlorn.[50] Indeed, cyborgs constantly are able to transgress the Social and the Natural because the computer repeatedly transgresses the science-technology boundary.

Michael Mahoney has recently cautioned us that it will prove exceedingly tricky to write the history of something called "computer science," if only because the academic discipline has experienced great difficulties in its attempt to coalesce around some unified formal doctrine in the twentieth century (1997). Not only did the physical hardware rarely sit still for very long; but the traditions that were accessed to "explain"

[50] "Although we commonly speak of hardware and software in tandem, it is worth noting that in a strict sense the notion of software is an artifact of computing in the business and government sectors in the 1950s. Only when the computer left the research laboratory and the hands of the scientists and engineers did the writing of programs become a question of production" (Mahoney, 1988, p. 120).

formally the action of the computer have turned out to bear little relevance to the actual design and operation of those very real machines. In practice, the logical tradition of Turing and von Neumann was developed in parallel but rarely informed computer design; and this was also the case for a second field of the algebraic theory of electronic circuits growing out of the early work of Claude Shannon. Mahoney suggests that this unsatisfactory situation was clarified when Michael Rabin and Dana Scott (1959) adopted some of von Neumann's hints about automata theory and the central importance of randomness and reworked them into what has become the de facto orthodoxy in the pedagogy of computer science, as one can readily observe from perusal of any modern textbook in the theory of computation (Lewis & Papadimitriou, 1981; Davis, Sigal, & Weyuker, 1994; Taylor, 1998). Yet, even here, it is not uncommon to hear the complaint that "theoretical" computer science has had precious little bearing upon the lively practice of imagining and building machines. This also seems to be the case for those whose efforts are absorbed in the production of useful software: "If computers and programs were 'inherently' mathematical objects, the mathematics of the computers and the programs of real practical concern had so far proved to be elusive. Although programming languages borrowed the trappings of mathematical symbolism, they did not translate readily into mathematical structures that captured the behavior of greatest concern to computing" (Mahoney, 1997, p. 632). Even if Mahoney is right, and someday we shall have a superior context-sensitive history of computer science, one that rivals some of the finest work done on the history of the natural sciences, it will still be necessary for the present reader to endure a brief introduction to the Standard Version of the history of computational analysis, if only because some of the concepts emanating therefrom, such as effective computability and the halting problem and hierarchies of computational complexity, will play important roles in our subsequent narrative of the recent history of economics.

The canonical account of computer science that Mahoney calls into question is one where the theory of computation was said to arise out of some currents in formal mathematical logic in the 1930s (Herken, 1988; Mahoney, 1990). The search for unassailable logical foundations of mathematics had begun with Gottlob Frege's attempt to reduce arithmetic to logic in the 1880s, which had in turn led to paradoxes highlighted in the work of Bertrand Russell and Georg Cantor (Grattan-Guinness, 1994a, pp. 600–34; 2000). Proposals to reestablish the logical foundations on firmer principles were propounded by Russell and Whitehead (1910) and David Hilbert's "formalist" program (Hilbert & Ackermann, 1928). Among Hilbert's procedures was to pose what was widely called the *Entscheidungsproblem*: did there exist a general procedure for deciding whether a statement belonging to a given axiom system has a proof within that system?[51] This had the effect of focusing the attention of mathematical logicians, especially those of the Göttingen school, to what were often called "mechanical" procedures for proof and inference. The formalist program was mortally wounded by the 1931 proof of Kurt Gödel that for any formal system adequate for the expression of number theory or arithmetic, assertions could be found that were not decidable within that formal system. Among these undecidable propositions was one that stated that the given formal system itself is consistent, one of Hilbert's desiderata. Gödel's proof was built on a technique of assigning "Gödel numbers" to statements within a formal system and constructing the proof as the erection of effective procedures for computing Gödel numbers for specific propositions from the Gödel numbers of the axioms. Here was an instance where logical decision procedures were brought increasingly closer to algorithms for calculation.

What does all this have to do with computers? Whenever we encounter Demon tracks, we know we are getting warmer. How could MOD possibly have anything to do with an arid abstract subject like mathematical logic? The clue to this connection has to do with the existence of the "undecidable" at the very heart of mathematics, uncovered by a quartet of authors in 1936: Alonzo Church, Stephen Kleene, Emil Post, and Alan Turing (Gandy, 1988). The problem, stated in crude intuitive terms, was to ask to what extent would the undecidable wreak havoc with the project of mathematics to provide rigorous foundations to rational inference. Was the problem simply one of natural noise, which could be offset by the exertions of something like MAD, or were there more devious and dangerous fifth columnists undermining mathematics from within? Was the problem of the undecidable somehow linked to the intentions of the constructor of the axiomatic system, therefore requiring extra *strategic* considerations in the restoration of order to impending chaos? I argue that strategic considerations were indeed present, and they were neutralized by the expedient of reducing the Opponent to a machine of known

[51] Hilbert's formalist program, as well as the impact of Gödel upon the ideas of von Neumann, is discussed in further detail in Chapter 3. It is interesting to note that von Neumann initially in 1927 regarded the mechanization of decision rules as threatening the project of mathematics: "As of today we cannot in general decide whether an arbitrary well-formed formula can or cannot be proved from the axiom schemata given below. And the contemporary practice of mathematics, using as it does heuristic methods, only makes sense because of this undecidability. When the undecidability fails then mathematics, as we now understand it, will cease to exist; in its place there will be a mechanical prescription for deciding whether a given sentence is provable or not" (quoted in Gandy, 1988, p. 67). Clearly by 1948 he had come to adopt an entirely different attitude toward a "machine theory of mathematics."

algorithmic capacities. Although all four authors were regarded as coming up with essentially the same answers in retrospect, I believe it was Alan Turing who most clearly perceived the Demonic implications of the undecidable, and for that reason his work (and not Church's lambda-calculus or Post's grammars) became the primary inspiration for the theory of computation, as well as a stimulus for the actual construction of computers in the 1940s.

Alan Turing

The Demon's tracks are clearly perceptible in Turing's later writings:

> There is a remarkably close parallel between the problems of the physicist and those of the cryptographer. . . . We might say then that insofar as man is a machine he is subject to very much interference. In fact, interference is the rule rather than the exception. He is in frequent communication with other men, and is continually receiving visual and other stimuli which themselves constitute a form of interference. It will only be when the man is "concentrating" with a view towards eliminating these stimuli or "distractions" that he approximates a machine without interference. (Turing, 1992, p. 118)

The cyborg manifesto is surely encrypted in this paragraph. Turing began with the premise that Gödel had demonstrated that men and women can invent mathematics that cannot be rendered decidable within some a priori set of axioms. Nevertheless, if man is really a machine, then we can provisionally equate undecidability with "noise" and provide a formal account of what is decidable in mathematics by "concentrating" our attention on algorithmic mechanical thought. Humans, of course, seem to be able to do some amazing things other than to "calculate"; but that turns out to be just distraction, the irreducible surd that constitutes interference to machine inference. Nevertheless, one cannot disparage the noise as entirely superfluous in a world that believes in Darwin rather than Bishop Paley: "I believe this danger of the mathematician making mistakes is an unavoidable corollary of his power of sometimes hitting upon an entirely new method. This seems to be confirmed by the well known fact that the most reliable people will not usually hit upon really new methods" (Turing, 1996, p. 256). The undecidability of who or what is really reliable is the central defining preoccupation of Turing's work.

Turing, it seems, was engaging in surreptitious autobiography here. He was an inveterate misfit and an outsider, an odd combination of schoolboy and head-in-the-clouds academic; perhaps this is one reason he has been the subject of one of the truly great scientific biographies of the twentieth century (Hodges, 1983) and a play, *Breaking the Code*. His suicide by eating a cyanide-dipped apple after having been prosecuted under the

Gross Indecency Act for homosexuality undoubtedly adds dramatic catharsis to what is already an action-packed narrative of assisting in decrypting the German Enigma code machine at Bletchley Park during World War II. Yet it is some other facets of the prodigious range of his cyborg interests that concern us in the present context.

The 1936 essay on computable numbers has earned him a permanent place in the pantheon of computation, as well as his work in conjunction with Max Newman on the wartime Colossus (the machine that broke the Enigma codes), which is credited in some quarters as the first electronic programmable computer. Less well known is his work on probability theory, mathematical biology (morphogenesis), information theory, and game theory. In 1943 Turing visited at Bell Labs and consulted frequently with Shannon about their shared interests. Hodges (1983, p. 251) relates an incident that reveals the extent to which they held opinions in common:

> Alan was holding forth on the possibilities of a "thinking machine." His high-pitched voice already stood out above the general murmur of well-behaved junior executives grooming themselves for promotion within the Bell corporation. Then he was suddenly heard to say: "No, I'm not interested in developing a *powerful* brain. All I'm after is just a *mediocre* brain, something like the President of American Telephone and Telegraph Company." The room was paralyzed while Alan nonchalantly continued to explain how he imagined feeding in facts on prices of commodities and stocks, and asking the machine the question "Do I buy or sell?"

The cyborg resemblance is even closer to John von Neumann, a similarity noted by numerous authors (Hodges, 1983, pp. 519, 556; Mirowski, 1992). Perhaps unsurprisingly, their careers intersected at numerous points, ranging from an encounter just prior to the paper on computable numbers in 1935 (Hodges, 1983, p. 95) to Turing's stint at Princeton from 1936 to 1938 and his declining of an offer to stay on as von Neumann's research assistant (pp. 144–45), to von Neumann's expanding appreciation of the formalism of the Turing machine from the late 1930s onward (p. 304). It is entertaining to speculate what might have happened if von Neumann had not met Oskar Morgenstern, and instead Turing's subsequent collaborations with the Cambridge economist David Champernowne had ripened beyond inventing and playing "war games" and chess. Not only did Turing write about the problems of the mechanization of the playing of chess (1953) around the time that this became seminal for the birth of artificial intelligence (see Chapter 7), but he also composed some unpublished papers on game theory in the later 1940s (Hodges, 1983, p. 373). Games were not restricted to the diversions of the poker table but also played a pivotal role in the definition of machine intelligence for Turing.

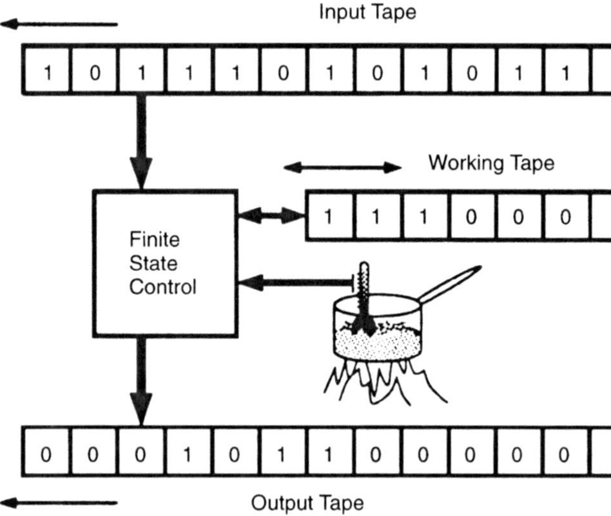

Figure 2.1. Turing machine. *Source*: Crutchfield, 1994c, p. 24.

Turing's work on computation was inspired by lectures on the foundations of mathematics by Max Newman at Cambridge in the spring of 1935.[52] Newman presented logic from a Hilbertian perspective, discussing proof techniques as subjected to "chess-like" rules of logical deduction, and finished the course with Gödel's recent theorems. Turing became captivated with the *Entscheidungsproblem* and came up with a way of showing there was no algorithm possible for solving the problem, thus sinking the last floundering ambitions of the formalist program. This was based on a novel, almost childlike conceptualization of what a human being did when he performed a calculation. Turing took the daring metaphorical leap of rendering a "mechanical calculation" as actually taking place on something that very much resembled a typewriter circa 1930 (see Figure 2.1). (The conflation of man and machine was already apparent at this early stage. One of the reasons why Turing's version of computability came to dominate those of Church and others is precisely

[52] The following account (as so much else in this section) is based primarily upon Hodges 1983, pp. 96–104. Other accounts of the Turing Machine can be found in Cutland, 1980; Feynman, 1996; Sommerhalder & van Westrhenen, 1988; Epstein & Carnielli, 1989; Gandy, 1988; Martin Davis in Herken, 1988; Taylor, 1998. The curious can play with a Turing Machine on their own PC using Barwise & Etchemendy, 1993. The only biographical source for Maxwell Newman (1897–1984) is Hilton, 1986. Max Newman again surfaces in our narrative in Chapter 3; he should be acknowledged as one of the great unsung early mediators of the intersection of computation and game theory.

because the subsequent technological trajectory of the personal computer initially came to combine the external features of the typewriter and the tickertape.)

Turing imagined a machine that could read, erase, and write on an infinite tape, one symbol per unit square. On appeal to the limited capacities of the human computer, Turing restricted the number of squares the machine could read per step, the distance through which the machine's scanner could move per step, and (most critically) the number of "states of mind" that the machine could assume upon reading a symbol. A purely deterministic machine inscribed one symbol for each state and each symbol read.[53] Calculation was reduced to following a fixed table of rules, one for each eventuality; "computable numbers" were those strings of symbols produced by a fixed definite algorithm. The numbers could include irrational numbers such as $\pi = 3.1416\ldots$, with infinite decimal expansions; this was the origin of the requirement of an infinite tape. While the work of the machine on this infinite decimal would never end, it would arrive at a fixed digit decimal place at a fixed finite time at a fixed finite place on the tape, so the number was still deemed "calculable." This construct became known as a Turing machine (TM).

Once his machine dream had progressed this far, Turing realized he could make use of a trick that has become standard in logic and computer science, which is known as Cantor's diagonal argument. Cantor used it to show that there were "more" real numbers than integers, even though both were treated as sets of infinite dimension. Turing used it to show that one could always define another "uncomputable number" from the set of computable numbers. Not stopping there, he then demonstrated that, following the technique of "Gödel numbering," one could index each table of instructions by a number and dream a more imperious dream of a universal Turing machine (UTM), which, fed one of these TM index numbers, would precede to *mimic* or simulate the operation of the identified TM. In effect, the UTM was MOD all over again: the demon had the power to *decrypt* the hidden power of an individual machine and turn it to his own advantage, extracting order out of an apparently

[53] Slightly more technically, a deterministic Turing Machine consists of the following components: an unbounded memory consisting of an infinite number of tape cells of fixed capacity; a machine head whose functions are limited to sequential read, write, and motion along the tape; and a finite control unit consisting of a finite set of instructions. In each given state of the machine, the instructions allow for one and only one action. If the memory tape is not infinite, then we instead have a machine of lesser capacity, called a "finite state machine." If the Turing Machine is augmented with the ability to "choose" between more than one instruction to execute at each state, then we have a Non-deterministic Turing Machine (NTM). NTMs are generally regarded as "guess and test" algorithms and prove central to the definition of computational complexity.

meaningless string of symbols. In mechanizing the process of Cantor's diagonal argument, however, there was a catch. He showed rigorously that there was no way that one could tell if a UTM would produce an appropriate number upon the input of a given number string. It could very well simply continue grinding away forever, never outputting any number at all. In the subsequent jargon of computational theory, it was said that in this eventuality the Turing machine failed to "halt": there existed numbers that were rigorously definable but uncomputable. As Feynman (1996, p. 68) put it, "if you try to get a UTM to impersonate itself, you end up discovering there are some problems that no Turing machine – and hence no mathematician – can solve." This possibility was patterned upon Gödel's own proof of the existence of undecidable propositions.

This 1936 result of Turing's sounds unremittingly negative and gloomy, but the miracle of ensuing events is that it instead proved increasingly fruitful, not the least for pivotal figures such as von Neumann, Warren McCulloch and Walter Pitts, and Michael Rabin. Robin Gandy (1988, p. 90) suggests four conceptual innovations that paved the way for the construction of a real-world universal device: the restriction of elementary steps to simple fixed-length instructions; the universal machine as a stored program device; the placing of conditional and unconditional instructions on equal footing; and the easy adaptation to binary storage and operation. Of course, one can quibble with any of these prognostications and their attribution to Turing; and as we have already admitted, the relationship between the mathematical theory of computation and the physical engineering of the computer has never been tightly synchronized or reciprocally coordinated. Nevertheless, it is clear that Turing managed to give expression rigorously to the stunningly simple and potentially Protean nature of the abstract computer, separating out what could be accomplished in principle from what could only be dreamed of. There would be no more cavalier attributions of "infinite powers" to ideal machines. This thesis, sometimes called the "Church-Turing thesis," is that any rigorous definition of "effective calculation" can be shown to be equivalent to the Turing machine formalism.[54] The Church-Turing thesis will be put to repeated use in the following chapters.

Turing machines as mathematical entities have become central to the development of many diverse intellectual disciplines in the ensuing years.

[54] A different expression is given by Gandy, 1988, p. 83: "Any function which is effectively calculable by an abstract human being following a fixed routine is effectively calculable by a Turing machine – or equivalently, effectively calculable in the sense defined by Church – and conversely." The reason it is a "thesis" rather than a "theorem" is that there is no closed definition of the intuitively vague notion of human calculation. For more on this, see P. Galton in Millikan & Clark, 1996, and Taylor, 1998, pp. 395–405.

Some, like their use in the foundations of theories of formal grammars and linguistics and theories of computational complexity, are briefly surveyed in the next section. Others, such as their use as a basic building block in von Neumann's theory of automata, make their appearance in Chapter 3. In still others, it has provided the point of embarkation for whole fields of inquiry. One field that bears profound relevance for the future of modern economics is the cyborg science of artificial intelligence, which owes its genesis to the Turing machine.

Alan Turing was not only interested in developing theorems in mathematical logic; he also wanted to shake up common presuppositions about what it meant to "think." Although he indicated this intention in a number of places, he managed to raise the most hackles with his famous article in *Mind* in 1950.[55] In this article, he proposed an index of machine intelligence with what he called "the imitation game," although, perhaps misleadingly, it is now called by all and sundry the "Turing Test." In this game, a contestant confronts two curtains, both hooked up to the ubiquitous typewriter. The initial object is to deduce from a set of questions posed on the keyboard to the other contestants which person behind the curtain is a man and which a woman (again, a thinly disguised reference to his own predicament). Then, Turing suggests that one of the hidden players be replaced by a machine. If the machine manages to win the game, Turing asks, What bearing would this have on the question whether or not machines can think? His answer: "I believe that in about fifty years" time it will be possible to programme computers . . . to make them play the imitation game so well that an average interrogator will not have more than 70% chance of making the right identification after five minutes of questioning" (1992, p. 142). The prodigious amount of gender dissimulation on MUDs (multiple user domains) on the modern Internet testifies that Turing's prediction was right on the mark (Turkle, 1995).

One of the more festive aspects of Turing's legacy is the Loebner Prize competition held at Dartmouth College, which has turned the Turing Test into a public tournament. At the Loebner 2000, ten judges faced ten computer terminals and were told that there was at least one human and one computer program "behind" them. The judges were allotted fifteen minutes of typed conversation with each one before deciding which was which. At this event, not one person mistook a computer for a human

[55] Reprinted in Turing, 1992. This article grew out of a controversy at Manchester with Michael Polanyi, the famous philosopher of "tacit knowledge." Whereas Polanyi wanted to rescue science from the planners and the military by insisting that some aspects of research could not be codified and planned, Turing was bent upon dispelling any mystery about human thought. Turing, of course, was undisputedly a representative participant in militarily planned science.

being; however, from the judges' scores it turned out there was a one in five chance of mistaking a human being for a computer, itself a commentary on the impact of a half century of experience with Turing's offspring (Harrison, 2000).

It would appear obvious that, once again, we encounter the demon in one of his lesser nefarious disguises. Here, the thermodynamics of MOD has been transmuted once more, only now into the question of whether intelligence should not be treated as isomorphic to the ability to deceive, to confound our attempts to *decode* the messages being sent from behind the curtains. This, of course, is why Turing employed the language of a "game," fully aware of its connotations linking it to von Neumann's game theory. He understood that intelligence was being accorded a "strategic" aspect: "It might be urged that when playing the 'imitation game' the best strategy for the machine may possibly be something other than imitation of the behavior of a man" (Turing, 1992, p. 135). The standard game-theoretic quandary of "He (it) thinks that I think that he thinks that . . ." makes its appearance. "The reply to this is simple. The machine . . . would deliberately introduce mistakes in a manner calculated to confuse the interrogator" (p. 148). The link forged to the Turing machine in the article (pp. 141–42) is that Turing wanted to assert that no specialized equipment would be needed for such a game, because the universality of the TM underwrote its relevance. If man is really nothing more than a machine, and a UTM can simulate any other machine, then it follows with all the inexorability of logic that a UTM can play the imitation game with the best of them.

The suggestion that simulation satisfied some sort of criteria for the definition of that notoriously slippery concept "intelligence" was a defining moment for the rise of the cyborg discipline of "artificial intelligence." This is not to say that much of anyone now thinks that the Turing Test is a legitimate index of machine intelligence; indeed, many decry it as an unfortunate and disastrous distraction away from more substantial issues (Ross Whitby in Millikan & Clark, 1996). But this again fails to comprehend the warchest for the keys. Turing's game set in motion a completely novel project to conflate man with machine, which was to raise the seemingly technical question: to what extent was 'mind' captured by the notion of a Turing machine? It has been argued that Turing accomplished this primarily through deft sleight-of-hand (Shanker, 1995), especially by appealing to behaviorist notions of "learning" as the mechanical following of given rules. Yet this simple ploy fostered a milieu where "thinking" could be portrayed as having raised itself by its own bootstraps out of randomness, in clear analogy with the supposed ascent of life out of primeval ooze, not to mention the ur-metaphor of thermodynamics: "Many unorganized machines have configurations such that if once that configuration is reached, and if the interference is thereafter appropriately restricted, the machine behaves as one organized for some definite purpose . . . [an] unorganized machine with sufficient units can find initial conditions which will make it into a universal machine with a given storage capacity" (Turing, 1992, pp. 118–19). The efforts of MOD to ward off the distractions of interference, cited at the beginning of this section, combined and conflated with evasion of the probing intelligence of the "interrogator" intent on sniffing out deviance, reenacted the struggle of life with entropic dissolution. Learning, development, evolution, and wartime evasive tactics were all conflated under the sign of the machine. And the only answer to the question of the meaning of it all is, "The Demon made us do it."

Turing's Machine and Turing's Test stand as the two premier icons of the cyborg sciences in the late twentieth century. They represent the two orthogonal directions of development of research strategies in the cyborg-inflected disciplines, although they were demonstrably not separate and distinct in Turing's own writings; as such they play an important role in the subsequent chapters of our narrative. In the former case, we have a tradition that attempts to extract the formal properties of a computational entity as a prelude to making machine-independent analyses of the conditions under which computation can proceed in an effective and efficacious manner. This tradition informed John von Neumann's approach to the theory of automata, discussed in Chapter 3, as well as various attempts to create a formal account of computational complexity, described immediately in this chapter. It should nevertheless be acknowledged that a self-sufficient formal doctrine of computer science fully unified under the exemplar of the TM does not reign triumphant, except insofar as the uneasy coexistence of a set of fragmented fields is often countenanced within university structures.[56] Cohabiting cheek by jowl with the formal theory of computation mentioned earlier, there subsists the whole tradition of computer simulacra, or the widely disparaged culture of simulation, so characteristic of the cyborg era. It was the tradition best exemplified in its early manifestations by the work of Herbert Simon. This fundamental distinction between automata and simulacra has been almost uniformly obscured by practitioners and historians of the sciences alike and is covered in detail in Chapter 7. It will

[56] On the search for a unified field of computer science, see Mahoney, 1992, 1997. The field has actually been punctuated with periodic attempts to pronounce last rites over the TM. For instance, see Juris Hartmanis's retrospective on the 1960s (1979, p. 227): "The unrestricted Turing machine was dismissed as an unrealistic model and one joked about the 'Turing tarpit.'" More recently, Iain Stewart proclaims, "The Demise of the Turing Machine in Complexity Theory" (in Millikan & Clark, 1996). Nevertheless, most textbooks begin with the TM.

serve as a major chunk of our Rosetta Stone in trying to decode fin-de-siècle economics.

THE ADVENT OF COMPLEXITY

One remaining chunk of the legacy of the Turing machine will prove indispensable for evaluation of postwar economics; it is that vexed bit of jargon, namely "complexity." While Turing himself did not propose any formal index of the difficulty of computation conducted upon a TM, this rapidly became an issue once computers graduated from the status of rare playthings of scientists and defense analysts. One characteristic trademark of the cyborg sciences was to assert that they existed in order to pioneer a whole range of novel analyses which were neither too "simple," say like rational mechanics, nor too unwieldy, and thus only amenable to aggregative probability analyses.[57] Cyborgs imagined a whole intermediate class of phenomena, where patterns straddled the random and the deterministic, phenomena that had previously eluded efforts of scientific modeling.

This is where the UTM came in. Its claim to be truly universal, in that it could in principle imitate the operation of any other machine, rendered it attractive as a base line for the attempt to characterize different levels of the tractability of classes of problems. Although this would appear the extension of a continuum defined by Turing's poles of "computable" and "uncomputable" numbers, in practice the guiding metaphor more frequently appeared to be the "cost" of a unit calculation (Sommerhalder & Westrehen, 1988). In this sense, the inquiry which is retailed under the rubric of "computational complexity" in modern literature sports a very "economic" flavor, although perhaps not in the sense which would immediately occur to economists.

The literature on computational complexity is conventionally traced back to some pathbreaking work done in the 1950s by Michael Rabin and Dana Scott (Shasha & Lazere, 1995, pp. 68–88), and some simultaneous work done on formal linguistics by Noam Chomsky. In this first category of complexity measure, the "in-principle" computability of various problems is related to a hierarchy of automata of increasing computational power, with the Turing machine at the top of the hierarchy. For the remainder of this volume, we shall refer to this class of distinctions as "machine complexity." A second category of complexity measure is instead related to the amount of computational "resources" enjoyed by an abstract machine, usually but not exclusively posited to be a TM. These measures were brought to the wider attention of computer scientists in

[57] This theme was first rendered explicit in Warren Weaver's (1947) article "Science and Complexity." It is discussed in Chapter 4.

general by Juris Hartmanis in the middle 1960s. Interest in the field received a boost from some results proved by Scott Cook and Richard Karp in the 1970s, which made it possible to define equivalence classes of problems from a computational viewpoint. For the duration, we shall henceforth refer to this as "resource complexity." Rather than present this development as a historical narrative, as we have tended to do up until now, we shall simply organize our account of "machine complexity" around the Chomsky hierarchy (Taylor, 1998, chaps. 9–12), which will come back into play in Chapter 8.

Noam Chomsky defined an abstract language as a set of rules for producing strings of symbols from a finite alphabet. Some strings will qualify as words in good standing in the language, and others will not; the task he set for himself was to formalize how a speaker would "recognize" legitimate words and sentences, patterned upon the analogous situation of a machine "accepting" a symbol string. Some strings would prove easier to test than others, and the "complexity" of a language would be predicated upon various formal restrictions on the algorithmic procedures for the formation of new strings, called "production rules." Various grammars might therefore be taxonomized by how much machine power would be required in general by successful interlocutors of the language in question. The Chomsky hierarchy is schematized in Figure 2.2, with more complex grammars subsuming simpler grammars as proper subsets.

The most complicated languages in the Chomsky hierarchy are dubbed *recursively enumerable languages*. These are the set of all languages that can in principle be decoded, although something called the "word problem" suggests that the question whether any given word w belongs to any given language L is formally undecidable. The next lowest category of languages are the *context-sensitive languages*, the set of grammars for which all string productions $P \to Q$ where the length of Q is at least as long as the length of P. Each application of the production rule results in a string monotonically longer than the one started with, and this characteristic accounts for the lower computational complexity of the language. Each word in a context-sensitive language could only have as its antecedent a word of the same length or less, a finite set. Hence the parsing of the grammar is guaranteed to produce an answer in finite time. The next lower level of languages in the Chomsky hierarchy is the *context-free languages*. Here the string production $P \to Q$ is restricted to those instances where P is only a single nonterminal symbol. The designation context-free refers to the fact that in this grammar the substitution of symbol Q for symbol P is licensed for any occurrence of P whatsoever, that is, independent of context. The lowest class in the hierarchy comprises the *regular languages*, which are built up from very simple substitution of symbols in a sequential order.

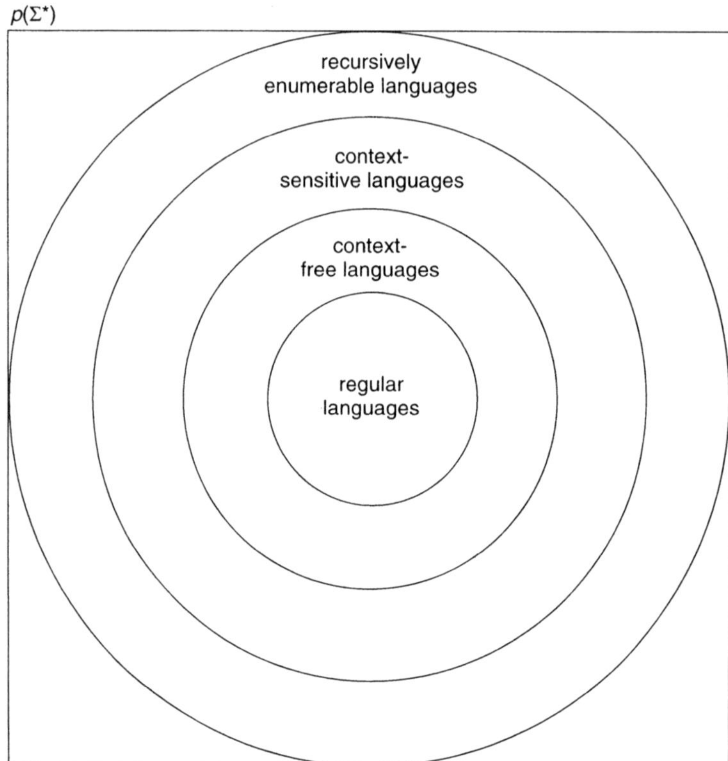

Figure 2.2. Computational hierarchy of languages. *Source*: Taylor, 1998, p. 644.

The link to the complexity of computation came with the demonstration by Kleene, Chomsky, and Rabin and Scott that the hierarchy of languages mapped directly into a hierarchy of machines of increasing computation capacity required to recognize these languages (Mahoney, 1997). The most rudimentary regular languages could be recognized by the most rudimentary automata – namely, a deterministic finite state machine with no memory. The somewhat more complex context-free languages would require a nondeterministic automata with pushdown-stack memory; this memory repository acts like a first-in, first-out inventory container. By contrast, context-sensitive languages would require a linear bounded automaton, one able to cope with more complex memory requirements but whose memory capacity is still bounded. Because parsing words in context-sensitive languages can never lengthen the word, the bounded memory will always suffice. Finally, the recursively enumerable language would require a full-fledged Turing machine, that is, an automaton with access to infinite memory.

In the theory of computation, "complexity" considerations may be divided into those which seek to develop ordered hierarchies of different "machines" or automata processing a fixed "language" to characterize increasing difficulty of computation, and those which operate with a single fixed machine, generally a specially equipped Turing machine, to allocate various types of algorithms to a hierarchy of complexity classes. The former approach, just described in the Chomsky hierarchy, is what we have called "machine complexity," an approach that is better regarded as envisioning a fixed machine burning up a certain quantum of resources in the course of completing its calculations. This duality of approaches – should we gauge complexity by progressively more souped-up machines or just by the amount of fuel consumed? – tends to mirror a certain ambiguity about the nature of the "costs" to which complexity theory is putatively addressed. Partisans of the first approach tend to be interested in the extent to which the machine metaphor might illuminate a phenomenon that is undergoing dynamic alteration through time, be it the capacity of a growing human to comprehend language, or the extent to which inert physical dynamical systems might be asserted to have attained the ability to perform calculations. This group harbors a vague notion of a trade-off between programmability, computational efficiency, and evolutionary adaptability (Robert Conrad in Herken, 1988). "Evolution" is then equated with scaling the complexity hierarchy with emergent novel solutions to the trade-off (although this might not accord with some understandings of the Darwinian theory).[58] Partisans of the second approach – and here it should be stressed this encompasses the vast majority of contemporary computer scientists – instead regard the primary objective as the taxonomy of difficulty of computation of broad classes of algorithms relative to an agreed-upon general machine model that does not change (with some minor exceptions). For this second group, "costs" do not refer to actual monetary expenses, but rather some abstract "scarce resources" commensurate with the abstract character of the UTM. Because this approach is much more widespread, we need only conduct a brief flyover of the issues.

How to compare the "difficulty" of all algorithms, even when restricted to the "same" UTM? Computer scientists have settled on two, not entirely commensurable indices of "costs," namely, time complexity and space

[58] Those anxious to cut directly to the conclusion will realize this explains why we have adopted this approach in Chapter 8. An evolutionary economics should stress the diversity of the entities embarking upon computation in order to describe analytically their change through time.

complexity. Because these are dictated by the abstract model of the UTM, they do not refer to real clock time or physical real estate; rather, computations are more "complex" if they require a greater number of "operations" (sequential processors operate on periodic cycles) or a greater length of tape (which translates into greater memory usage). For purposes of comparison, a large number of assumptions must be imposed to render algorithms quantifiable along these axes. Most will be neglected here for purposes of exposition, but it is important to note that measures of complexity refer to "worst-case scenarios" and not specific or average case instantiations of the algorithms at issue.[59] The complexity classes then start with an index of the size of the input string, say "n," to the machine, and ask what happens to the time (or space) requirements as n gets large, in the eventuality that the computation does halt. At this juncture, the injunction to stick to a fixed machine model is violated slightly, in that the same algorithm is imagined inserted in two different Turing machines, one the conventional "deterministic" machine (DUTM), and another, augmented with an "oracle" or source of randomness, dubbed a "non-deterministic" Turing machine (NUTM). The "oracle," as its name implies, is a source of "guesses" for the machine unrelated to the operation of the algorithm. Classes of algorithmic or resource complexity are then defined relative to machine type and order of increase in time or space required to halt given an increase in input size n.

The most famous complexity classes are denominated in terms of time requirements (Garey & Johnson, 1979; Taylor, 1998, chap. 8). An algorithm is said to belong to class **P** if it can be solved by a DUTM in time that is a polynomial function of n. An algorithm is said to belong to class **NP** if it can be solved by a NUTM in time that is a polynomial function in n. One of the most pressing open problems in computational theory is the question of whether class **P** does or does not equal class **NP** or, in other words, whether the oracle inherently can handle more difficult problems than the purely deterministic machine (Sommerhalder & Westrehen, 1988, p. 283; Thagard, 1993). It is widely believed that the classes are different, and that **NP** problems are inherently more complex because a NUTM can "guess" at solutions and then simply "check" them

[59] For this reason alone, not to mention that we are always dealing with asymptotic situations, it is imperative that "complexity" measures never be confused with "efficiency" indices. Small problems may very well be more efficiently solved by algorithms that would qualify as computationally intractable in the limit. An illustration of this fact of interest to economists is the simplex algorithm for the solution of linear programming problems, which has been proved to be of exponential complexity in the arguments but seems to work quite satisfactorily in actual applications (Sommerhalder & Westrehenen, 1988, p. 312).

in polynomial time; but there are as yet no formal proofs. This, as the reader may begin to tire of hearing, is the appearance of the Demon in yet a further disguise: the "equivocation" of randomness may actually assist in speeding up the computation. The significance of these complexity classes is that many in the computer science community tend to treat the desideratum "solvable in polynomial time" as essentially tantamount to "feasible in practice." Other complexity classes defined in an analogous manner for space constraints are **PSPACE** and **NPSPACE**.

The reader should not derive the impression from this unpardonably telegraphed summary that the theory of computational complexity is either complete or uncontentious. It is neither, for "complexity," like "computability," is an intuitive notion, which has been cashed out in a number of formal proposals over the past half century, all of which are still actively under negotiation. Yet, "complexity" is the cyborg trademark par excellence, and we shall meet it over and over in our subsequent account of economics and its close encounters with cyborgs of a keening kind.

3 John von Neumann and the Cyborg Incursion into Economics

> At Princeton, where in 1933 von Neumann at 29 became the youngest member of the newly established Institute for Advanced Study, the saying gained currency that the Hungarian mathematician was indeed a demigod but that he had made a thorough, detailed study of human beings and could imitate them perfectly.
>
> Richard Rhodes, *The Making of the Atomic Bomb*[1]

Our explicit narrative of the constitution of modern economics begins with John von Neumann because I believe, with the benefit of a little additional hindsight and the provision of some previously neglected evidence, he will come to be regarded as the single most important figure in the development of economics in the twentieth century. It would initially

[1] Rhodes, 1986, p. 109. Numerous works seek to describe von Neumann from various vantage points, but no synthetic work begins to capture the full range of the remarkable man's talents, motivations, ideas, and activities. Perhaps no historian can hope to master the full panoply of his genius; but because he was uncommonly aloof, von Neumann left little in the way of human material for the biographer. The next sentence of Rhodes's text reads: "The story hints at a certain manipulative coldness behind the mask of bonhomie von Neumann learned to wear, and even Wigner thought his friendships lacked intimacy." Geoff Bowker, in a review on some historical essays of game theory, noted, "We never get inside von Neumann's head as we do Morgenstern's – but then von Neumann's is a difficult nut to crack. Despite his other successes, [William] Aspray (1990) has also failed here, and when [Steve] Heims (1980) compared von Neumann and Wiener, it was Wiener who came to life: von Neumann remained a distant dabbler in technologies of death" (1994, pp. 239–40). The reader should especially be warned concerning the popular biography of von Neumann by Macrae (1992). This book manages to combine unabashed ignorance about every technical subject that occupied von Neumann, from mathematics to economics to physics, with an ill-repressed tendency to project the author's persona upon the biographical subject. Von Neumann, whatever his faults, certainly deserves better. Nevertheless, these texts, in combination with material taken from the John von Neumann papers at the Library of Congress (VNLC), constitute the main sources of information accessed for this chapter.

appear I am not alone in this conviction. Roy Weintraub (1985, p. 74), suggests, "Von Neumann's [1937] paper is, in my view, the single most important article in mathematical economics." Mohammed Dore (in Dore, Chakravarty, & Goodwin, 1989, p. 239) asserts that "John von Neumann *changed* the way economic analysis is being done." Nicholas Kaldor ventured, "He was unquestionably the nearest thing to a genius I have ever encountered" (ibid., p. xi). Jurg Niehans's textbook (1990, p. 393) states flatly, "In the second quarter of the twentieth century, it happened for the first time that a mathematical genius made fundamental contributions to economic theory." Given the spread of game theory throughout the core microeconomics curriculum since 1980, it would appear a foregone conclusion that von Neumann should be revered as the progenitor of that tradition and, thus, of microeconomic orthodoxy at the end of this century. So it would seem that his central location within the economic canon is secure, and the reasons for this status are widely understood.

Appearances are deceiving. The primary counterevidence is that there exists a fair amount of hostility to according von Neumann any such exalted status in the Pantheon. The prosecution's star witness here is Paul Samuelson. Never one to forget a slight, he recounts an incident from the early 1940s in which von Neumann insisted at a seminar at Harvard that progress in economics would require a mathematics different from that which derived from the time of Newton. Samuelson challenged him then and remains defiantly unrepentant now: "except for the philosophical complications introduced by games involving more than one person, I do not honestly perceive any basic newness in this so-called non-physics mathematics" (in Dore et al., 1989, p. 112). Even those demonstrably more sympathetic to game theory, such as Kenneth Binmore – a figure whom we shall encounter at greater length in Chapter 7 – have openly disparaged the importance of von Neumann and Morgenstern's *Theory of Games and Economic Behavior* (1944): "I have read the great classic of game theory from cover to cover, but I do not recommend the experience to others! Its current interest is largely historical. . . . they abandoned the attempt to specify optimal strategies for individual players" (1992a, p. xxix, 12). In his history textbook, Niehans finds he must repeatedly contradict the programmatic statements of the demigod: "The substantial insights, however, remained sparse. The hoped-for revolution did not materialize" (1990, p. 399). And at the end of this century, some game theorists have more aggressively sought to portray von Neumann's initiative as a dead end: "the Nash program opened the door to the questions of information economics, while the von Neumann program led away from it" (Myerson, 1999, p. 1075).

Conversely, and more to the point, previous writers who would betray little hesitation in elevating von Neumann to canonical status are vexed by some trouble in agreeing upon what precisely he did to deserve the laurels. Some praise the 1937 paper for its pioneering use of fixed point theorems and convexity arguments to provide existence proofs, a practice that indeed became central to the Walrasian tradition in America in the 1950s. Others laud the 1937 paper as a revival of classical value theory, which subsequently found further expression in Sraffian economics. In some quarters, he is most fondly remembered for his axiomatization of the theory of expected utility (Fishburn, 1989). For others, he was the prophet of duality approaches in price-quantity models. Some regard the 1944 book as the first expression of a level of mathematical rigor woefully absent from previous studies of economics but finally attained in the postwar period with the recruitment of talent from the physical sciences and mathematics (Debreu, 1983a); this would encompass the Hilbertian specialty of use of convex polygons and separating hyperplanes (Sonja Brentjes in Grattan-Guinness, 1994, 2:830). Still others would point to his encouragement of linear programming (George Dantzig in Lenstra, Kann, & Schrijver, 1991), or his 1928 proof of the minimax theorem. Of course, each of these innovations is important in its own right, and bears great significance for the particular tradition within economics that seeks to appropriate the hallowed name of von Neumann for its extended lineage. And yet, in this book, none of these conventional genealogies even begin to adequately explain the ultimate significance of John von Neumann for economics. That is because a thoroughgoing understanding of his work, one that extends its purview beyond the 1928 and 1937 papers and the 1944 book with Morgenstern, will reveal that von Neumann was not at all interested in shoring up anyone's existing tradition of economics: not the classical-Sraffian model, not the Walrasian tradition, not a psychology of measurable utility functions, and certainly not the contemporary academic orthodoxy of Nash equilibria in noncooperative game theory. And it is a travesty to try and recruit him to some sort of Bourbakist crusade to invest economics with a purity of rigorous abstraction that holds formal expression as the ne plus ultra of intellectual accomplishment.[2] Things are indeed often rarely what they seem.

Perhaps the most distressing thing for a fin-de-siècle economist to accept is that this figure, this demigod who cut through Gordian knots with his terrible swift sword, irreversibly transforming the discipline of economics, was not really all that concerned with economics as an intellectual tradition or field of discursive practice. It was, for von

[2] The only historian of economics to my knowledge who has lodged a protest against this travesty is Rashid (1994). The only economist who has perceived that von Neumann's mature program for mathematical economics might have diverged dramatically from the trajectory of modern neoclassical theory is Albin (1998, p. xv).

Neumann, always a sideline, an avocation, a detour away from matters more pressing and innovations more epochal. On any scale of consequence, the formalization of the theory of quantum mechanics, participation in the design of the atomic bomb, and the development of the electronic computer far outstripped in importance anything any economist managed to say or do in the middle of the century; moreover, thermodynamics, formal logic, quantum mechanics, and biology also occupied higher rungs in the hierarchy of von Neumann's own personal estimation. Nevertheless, von Neumann developed the conviction over time that economics stood badly in need of revision and reconceptualization; what changed over the course of his writings was the intended shape and contours of this revision. That, in turn, was a product of von Neumann's own shifting concerns throughout his life. It was as if a castle could capture a pawn in passing, rather than vice versa.

Without attempting to impose any hint of inevitability, the trend of von Neumann's attitudes about economics might be summarized as follows. Early in his career, in the 1920s and early 1930s, his tangential interest in economics was confined to correcting some of its more egregious errors of reasoning, and essaying the possibility of subjecting it to the discipline of the Hilbert program of axiomatization and formalization. Much of this interest derived from encounters with some protagonists who did care more deeply than he about the conceptual problems of economics and who managed to turn his attention briefly in that direction. As Leonard (1995) correctly insists, the early 1928 paper on game theory had essentially no connection with any of these economic concerns, and therefore should not be enshrined as a member in good standing of this sequence. By the mid-1930s, von Neumann had conceived of a fairly strong aversion to neoclassical price theory; but by then he had also lost faith in the Hilbert program of formalist foundations for the applied disciplines. By his own account, 1937 was a sort of watershed, "and it was through military science that I was introduced to the applied sciences. Before this I was, apart from some lesser infidelities, essentially a pure mathematician, or at least a very pure theoretician. Whatever else may have happened in the meantime, I have certainly succeeded in losing my purity."[3] Attitudes

[3] As quoted in Aspray, 1990, p. 26. As this language attests, 1937 marked multiple ruptures in von Neumann's life. As we document here, it was the beginning of his consultation relationship with the military, which came to dominate the rest of his life. It was also the year in which his first wife Mariette Kovesi took their baby daughter Marina away and left him to marry Horner Kuper (Heims, 1980, p. 178; Macrae, 1992, p. 161). Von Neumann married Klara Dan in 1938, after he helped to arrange her own divorce in Budapest and her exit visa. He also became a naturalized United States citizen in 1937. His relationship to his religious heritage is also interesting. His brother Nicholas reports that the entire family converted from Judaism to Catholicism after their father's death in 1929, "for the sake of

toward purity and impurity of motives are the cryptographic keystream that permit us to unriddle the enigma of von Neumann's work and decipher the half-submerged history of postwar economics.

After his expanding economy model of 1932 (presented to the Vienna Menger Colloquium in March 1934, and published in 1937 [von Neumann 1945]), von Neumann did essentially nothing more on economics until the discussions with Oskar Morgenstern commenced in 1940 (Morgenstern, 1976a, p. 808), leading ultimately to their collaboration on *Theory of Games and Economic Behavior* (1944). This collaboration falls rather neatly on the far side of the watershed in von Neumann's work between the pure and the impure, and correlates with his proliferating connections with the military. The nature of that collaboration has been the subject of much examination and some controversy (Mirowski, 1991; 1992; Rellstab, 1992; Leonard, 1992; 1995; Shubik, 1992; Dimand & Dimand, 1996; Schmidt, 1995b). Our primary source for the present interpretation of von Neumann's evolving attitudes toward economics in the context of that collaboration has been the extensive diaries of Oskar Morgenstern.[4] The years of the composition of that book might be regarded as taking place within the second phase of von Neumann's encounter with economics, with its publication roughly marking the boundary of that period.

Far from terminating his brush with economics, as it is often intimated in retrospective histories seeking their denouement in 1944, he subsequently initiated a previously undisclosed number of interventions in the work of economists until his death in 1957. Many of these interventions were decisive in shaping the contours of the postwar orthodoxy of mathematical economics. One purpose of this volume is to document his activities and the reverberations of those interventions, be they direct or indirect, as they echo down the past fifty years. Although there exists no further explicit programmatic statement from his pen, there does exist substantial evidence that von Neumann had extensively clarified his vision of a progressive economics by the early 1950s; in a word, economics in his view was slated to become yet another instantiation of the sciences of

convenience, not conviction" (box 34, folder 3, VNLC). This may go some distance in explaining von Neumann's prickly response to a query about his fleeing Nazism: "I first came to the US in January 1930 and, since conditions in Europe at that time were, in the main, normal, I would not consider myself a refugee scientist" (letter to Maurice Davie, May 3, 1946, box 3, folder 6, VNLC). Locating the watershed in 1937, rather than during World War II, as does Heims, may have some relevance for a better understanding of game theory.

4 These extensive and previously underutilized diaries will be subsequently referenced in the text by the indication OMDU. They are housed with his papers (OMPD) in the Duke University Archives.

information processing, which he was busily propagating across the landscape: another automata self-replicating in his wake.

John von Neumann, not his various epigones, initiated this transformation in economics. Hence, the period from roughly 1943 to his death constitutes a third phase in von Neumann's approach to economics; it is no coincidence that it coincides exactly with von Neumann's prodigious labors serving as midwife to the electronic stored-program computer. This third and final phase, which grows in importance with each passing decade, has received absolutely no attention from historians of economics and inadequate attention from historians of science. Once one breaks free from the idea that the minimax solution, or the expanding economy model, or fixed point theorems, or the *Theory of Games*, or von Neumann–Morgenstern expected utility was the ultimate terminus ad quem and full fruition of von Neumann's economic quest, one is finally free to ask, *Where was he going? What sort of economics did he seek to promote?* Thus, when we say that von Neumann was the most important figure in American economics in the twentieth century, we mean that, more than any other single actor, he was responsible for the conditioning and promotion of economics as one of the cyborg sciences. And, stranger still, precisely because of his imposing stature amongst the cohort of American mathematical economists at midcentury, even those revulsed by his vision of economics as a cyborg science found their own doctrines inexorably structured in opposition to his own and therefore falling equally under the sway of his agenda.

ECONOMICS AT ONE REMOVE

Is the child always the father to the man? We should at minimum be wary of that maxim in the case of cyborgs, who more often than not flaunt the ambiguities in their genetic inheritance and taxonomic categorization in order to deny any certifiable patrimony. Von Neumann himself coined a phrase for this kind of ambiguity, "a topological version of the truth." Margittai Neumann János Lajos, born in Budapest on 28 December 1903, was the son of the director of one of Hungary's leading banks, the Magyar Jelzalog Hitelbank. His younger brother, Nicholas Vonneumann, suggests that finance and business theory were common topics of discussion around the family dinner table (in Glimm, Impagliazzo, & Singer, 1990, p. 20); also, he feels that the defeat of the Central Powers in World War I, followed by the overthrow of the Karolyi government and the brief institution of a Soviet Republic, had a profound impact upon young John.[5] It does seem

[5] See, in particular, "John von Neumann as Seen by His Brother," private mimeo, 1987, box 34, folder 3, VNLC. There Nicholas Vonneumann recounts the brothers playing a game of John's invention upon graph paper, simulating battles. "The aim was to demonstrate and

that von Neumann harbored a lifelong aversion to Russia. As he admitted after World War II, "I am violently anti-Communist, and I was probably a good deal more militaristic than most. . . . My opinions have been violently opposed to Marxism ever since I can remember, and quite in particular since I had about a three-months taste of it in Hungary in 1919" (in Aspray, 1990, p. 247). The depths of his contempt was demonstrated by his public advocacy of a preemptive first strike against Russia with nuclear weapons as early as 1950 – that is, less than a year after their demonstration of the possession of an atomic bomb. Something resembling a game theoretic logic is evident in statements like, "If you say why not bomb them tomorrow, I say why not today? If you say today at 5 o'clock, I say why not one o'clock?" (Heims, 1980, p. 247). Elsewhere, J. Robert Oppenheimer reported he once said in conversation, "I don't think any weapon can be too large" (in Dyson, 1997, p. 78). He was openly contemptuous of those who sought political control of atomic weapons and, at least initially, believed that world domination by the United States was the logical consequence of the atomic bomb.[6]

The purpose of beginning with politics, and some rather brutal politics at that, is to establish some base line concerning John von Neumann's attitudes toward economics. There is some evidence that he had very early on briefly considered a business career (Heims, 1980, p. 43), and, unlike many other academic intellectuals, he never disparaged the business world or political success or pecuniary pursuits. His friend Stanislaw Ulam reports that, "He was given to finding analogies between political problems of the present and of the past. . . . [He] seemed to take a perverse pleasure in the brutality of a civilized people like the ancient Greeks. For him, I think it threw a certain not-too-complementary light on human nature in general" (1976, pp. 80, 102). It is reported (Heims, 1980, p. 327) that he once said, "It is just as foolish to complain that people are selfish as it is to complain that the magnetic field does not increase unless the electric field has a curl. Both are laws of nature." In this sense, one can observe that he had an instinctive affinity for the Stoic distrust of the motives and moral competence of individuals, combined with a belief in the beneficial

practice ancient strategies." The family left its home in Budapest during the Soviet period, returning in 1919 with the advent of the Horthy government (Heims, 1980, pp. 45–47). For a political history of this period, see Rabinbach, 1985. Macrae reports the interesting tidbit that von Neumann's father Max was also an amateur chess expert.

6 See Tjalling Koopmans to Jacob Marschak, September 1, 1945, box 92, file: Koopmans, JMLA: "Regarding the atomic energy developments, he [von Neumann] did not regard intervention and control as politically possible. The logic of that position is that of the two remaining alternatives, mutual destruction, or world domination by one power, the latter is the lesser evil, and does become relatively desirable. Privately he did not object to that inference being drawn from his judgment regarding the political impossibility of control by agreement."

outcomes of free competition. The one passage wherein he praises the invisible hand comes in the context of a discussion of the problem of subjecting science to social planning, but it captures the general "Austrian" flavor of his outlook: "a large part of mathematics which became useful developed with absolutely no desire to be useful, and in a situation where nobody could possibly know in what area it would become useful; and there were no general indications that it would ever be so. . . . This is true for all of science. Successes were largely due to forgetting completely about what one ultimately wanted, or whether one wanted anything ultimately. . . . And I think it extremely instructive to watch the role of science in everyday life, and to note how the principle of *laissez faire* has led to strange and wonderful results."[7] Moreover, it is fairly evident from his work in many different fields that he was never opposed to what would now be called a "methodological individualist" ontology throughout his career, whether the issue was the treatment of measurement in quantum mechanics, the characterization of rationality in game theory, or the von Neumann architecture for the computer.

The ingrained hostility to Marxism, the belief in laissez-faire, the implicit acceptance of methodological individualism, and the privileging of mathematics would all seem to suggest an inclination to favor the neoclassical version of economics; but here is where the syllogism fails. The (admittedly scant) evidence from what we are calling the "first period" of his career, roughly 1926 through 1937, is that whenever he encountered neoclassicism, he found it wanting. Economics was of course rarely a topic of explicit concern, because his lot had been definitively cast with mathematics and the physical sciences once he left Budapest in 1921. He spent the years 1921–23 in Berlin, attending lectures by Albert Einstein and Fritz Haber, and associating with a stellar circle of Hungarian scientists such as Leo Szilard, Eugene Wigner, Denis Gabor, and Michael Polanyi. Encouraged by his father to get a practical education, he enrolled in the Eidgenossische Technische Hochschule in Zurich to study chemical engineering from 1923 to 1925, resulting in a diploma in 1926; in his "spare time" he managed to pursue pure mathematics with George Polya and Herman Weyl. He was also awarded a doctorate in mathematics from the University of Budapest in 1926. The years from 1925 to 1930 were spent between the University of Göttingen, the epicenter of Hilbert's school of

[7] Von Neumann, 1961–63, 6:489–90. In case this might be seen as contradictory to what has been said earlier about the planned character of the cyborg sciences, we should reiterate that this story about "disinterested science" was promulgated by all the major figures promoting the military subsidy and management of American science in the immediate postwar period, starting with Vannevar Bush's *Science: The Endless Frontier*. The complex relationship of the public *ethos* of free science to the military mobilization of scientists and, more explicitly, to the cyborg sciences, is explored in Chapter 4.

mathematics, and the University of Berlin, where he was appointed Privatdozent in 1927 (Begehr et al., 1998). He was granted a half-time appointment in mathematics at Princeton in 1930, returning to Berlin to teach half of each year until he was granted a full-time appointment at the Institute for Advanced Study (or, as he liked to say, the Institute for Advanced Salaries) in Princeton in 1933. He remained in the IAS position until his death in 1957.

The first bit of evidence we have of any expressed opinions about economic theory comes from a retrospective account by the economist Jacob Marschak. In a letter to Michael Polanyi, he wrote:

> Yes, I remember those encounters in Berlin, 1926, most vividly. Of the participants, I cannot identify the Indian physicist. The others were: yourself, Szilard, Wigner, Neumann. We were sitting at an oval table and I recall how v Neumann was thinking aloud while running around that table. And I remember the issue. I was talking, in the "classical" Marshallian manner, of the demand and supply equations. Neumann got up and ran around the table, saying: "You can't mean this; you must mean inequalities, not equations. For a given amount of a good, the buyer will ofeer [*sic*] *at most* such-and-such price, the seller will ask *at least* such-and-such price!" This was (later?) pointed out by another mathematician, Abraham Wald, perhaps in the "Menger Semiar [*sic*]" in Vienna, and certainly in 1940 in USA.[8]

There are many fascinating aspects about this poorly documented event. First, it shows that physicists and physical chemists in Berlin were independently arguing about many of the issues that surfaced in the formalization of economics a decade or more later in Vienna and elsewhere (Lanouette, 1992, pp. 76–77). Marschak was probably the only legitimate "economist" at the table, although he himself had started out as a mechanical engineer. The roster of participants is also retrospectively credible, because both Leo Szilard and Michael Polanyi went on to write essays on economics later in their careers.[9] Second, this is clearly the earliest indication we have of a motive which von Neumann may have had in producing his expanding economy model; and it is all the more striking that the neoclassical Marschak himself apparently gave no indication of noticing this. While some scholars have sought to link the expanding economy model to the prior work of Robert Remak, another Privatdozent in mathematics at the University of Berlin (Kurz &

[8] Jacob Marschak to Michael Polanyi, July 8, 1973, box 12, folder 8, MPRL. The subsequent role of Jacob Marschak in the promulgation of neoclassical price theory in America is discussed in Hands & Mirowski, 1999, and in Chapter 5.

[9] Szilard's contributions are mentioned in Lanouette, 1992, p. 320. Polanyi's work in economics is discussed in Mirowski, 1997, 1998a.

Salvadori, 1993), it was never very plausible that von Neumann would have himself gone to all the trouble of writing a paper just to "correct" a colleague without mentioning him by name.[10] It was much more likely that the Hungarian clique of physicists and close friends in Berlin were discussing these issues on a periodic basis and that the dissatisfaction nurtured there with conventional neoclassical approaches eventually expressed itself in desultory papers at various stages in their respective careers in the natural sciences.

This latter interpretation is supported by a second anecdote recounted by Nicholas Kaldor as happening in Budapest in the early 1930s (he is not more specific as to the date).

> One day he expressed an interest in economics and he asked me whether I could suggest a short book which gives a formal mathematical exposition of prevailing economic theory. I suggested Wicksell's *Uber Wert, Kapital und Rente*. . . . He read it in a very short time and expressed some skepticism of the "marginalist" approach on the grounds it gives too much emphasis to substitutability and too little to the forces which make for mutually conditioned expansion. I also remember that following upon his reading of *Uber Wert, Kapital und Rente* he wanted to look at the original Walrasian equations. He told me afterwards they provide no genuine solution, since the equations can result in negative prices (or quantities) just as well as positive ones – whereas in the real world, whilst commodities can be "free goods" (with zero prices), there is nothing analogous to *negative* prices. (in Dore et al., 1989)

This resonates with the first anecdote, in the sense that von Neumann was once more offended by the paucity of mathematical rigor and the treatment of price-quantity functions as unique mappings giving rise to a single equilibrium solution. This theme would be revisited in the other two "phases" of his career, albeit with diminished urgency. Unlike many other intellectuals during the Great Depression, he was never attracted to macroeconomic themes or schemes of wholesale reform as much as he was to questions of the *logical* character of prices and the restrictions which must be imposed upon them to formalize their operations.

The third anecdote, by contrast, dates from after the construction of the expanding economy model but still well within the 1937 boundary of what we would consider the first period of von Neumann's economic

[10] By 1932, when the paper was apparently written, John von Neumann was already deemed a mathematical genius to be reckoned with, whereas Remak languished in relative obscurity and died in Auschwitz in 1942. A brief description of Remak appears in Begehr et al., 1998, pp. 127–28. It is entirely possible that Remak's papers of 1929 and 1933, corresponding to the period of his teaching at Berlin, were responses to the Berlin round-table discussions of economics such as the one reported here.

apprenticeship. In February 1934 the director of the Institute for Advanced Study, Abraham Flexner, sent von Neumann a book, which, as he confessed, "I have no idea whether it will be of interest to you or your associates." It was a book claiming to apply the formalisms of thermodynamics to the theory of value.[11] It would seem that Flexner was cognizant of another lifelong conviction of von Neumann, namely, that *thermodynamics* would provide the inspiration for the formalization of a number of previously intractable sciences. What is intriguing about this incident is that the book was critical of the "Lausanne mathematical school" as not having based its abstractions upon sufficiently scientific concepts. His extended comments permit us to see how he would differentiate himself from critics who might superficially share some of his motives:

> I think that the basic intention of the authors, to analyze the economic world, by constructing an analogical fictitious "model," which is sufficiently simplified, so as to allow an absolutely mathematical treatment, is – although not new – sound, and in the spirit of exact sciences. I do not think, however, that the authors have a sufficient amount of mathematical routine and technique to carry this program out.
>
> I have the impression that the subject is not yet ripe (I mean that it is not yet fully understood, which of its features are the essential ones) to be reduced to a small number of fundamental postulates – like geometry, or mechanics. The analogies with thermodynamics are probably misleading. The authors think, that the "amortization" is the analogon to "entropy." It seems to me, that if this analogy can be worked out at all, the analogon of "entropy" must be sought in the direction of "liquidity." . . . The technique of the authors to set up and deal with equations is rather primitive, the way f.i. in which they discuss the fundamental equations (1), (2) on pp. 81–85 is incomplete, as they omit to prove that 1: the resulting prices are all positive (or zero), 2: that there is only one such solution. A correct treatment of this particular question, however, exists in the literature. . . . I do not think that their discussion of the "stability of the solutions," which is the only satisfactory way to build up a mathematical theory of economic cycles and of crises, is mathematically satisfactory.
>
> The emphasis that the authors put on the possibility of states of equilibrium in economics seems to me entirely justified. I think that the importance of this point has not always been duly acknowledged. (in Leonard, 1995, pp. 737–38)

Once again we observe von Neumann's trademark themes: the call to heed the strictures of modern mathematics, the admonitions to worry

[11] The book was Georges Guillaume's *Sur le fundements de l'economique rationelle* (1932). The incident is recounted in greater detail in Leonard, 1995.

about existence and uniqueness of solutions, as well as the insistence upon restrictions to positive prices. What is novel about this last incident is that it now hints at a firmer opinion about the state of contemporary economics: somehow, the fundamental postulates are missing or not well understood – a not altogether uncommon opinion during the Great Depression. This, it goes without saying, implies that economics was not (yet?) a candidate for the style of axiomatization, which von Neumann himself had by then wrought for set theory, *Gesellschaftsspiele* ("games of society" or "parlor games," and not, as the English translation has it, "games of strategy"), and quantum mechanics. Because he would not have regarded himself as someone ideally placed to rectify this situation, it was a tacit admission that he was not then seeking to reform or recast economics. What is striking, however, is his brusque dismissal of the legitimacy of the thermodynamic analogy. This combination of attitudes can only be understood when set in the larger context of von Neumann's full portfolio of research in the period of the 1920s and 1930s.

PHASE ONE: PURITY

When confronted with von Neumann's research accomplishments, even limited to his early career, it is impossible not to be daunted by the list of seemingly disparate topics represented in his published papers. In the period up until 1932, there are papers on logic, quantum mechanics, and game theory; and the mathematical papers can be further categorized under the rubrics of the spectral theory of Hilbert space, ergodic theory, rings of operators, measures on topological groups, and almost periodic functions.[12] And yet a few observations could be made to render them a little less diverse and unrelated. The first concerns the overwhelming influence of David Hilbert in the choice of topics and techniques. Every one of these areas was a site of active research in Göttingen during von Neumann's sojourn there; and, further, his angle of attack can often be traced to an active research community congregated around Hilbert or his program. Furthermore, von Neumann often produced his breakthroughs by transporting the findings of one research congregation into the problem situation of another. Some of the more obvious instances were his use of Hilbert space to axiomatize and generalize the matrix mechanics, which had been just developed at Göttingen in 1925 (Mehra & Rechenberg, 1982, chap. 3); the extension of the theory of operators in Hilbert space used in the quantum mechanics to the general theory of rings of operators (now called "von Neumann algebras"); the appeal to statistical mechanics in

[12] The complete bibliography through 1932 of forty-seven papers and two books is very conveniently provided in Aspray, 1990, pp. 357–59. A brief characterization of the work appears on ibid., pp. 13–15. See also Begehr, 1998, pp. 100–3.

macroscopic quantum mechanics leading to the ergodic theorem; the work on groups as a response to Hilbert's Fifth Problem and some aspects of quantum phenomenology (J. Segal in Glimm et al., 1990); and, not the least, the extension of the axiomatization of sets and the matrix mechanics of bilinear groups to the axiomatization of games. The choice of mathematical topics can often be found in some prior physical theory or philosophical dispute being argued out in Göttingen in the 1920s. How, then, was it that von Neumann believed that he maintained his "purity" of mathematical approach?

Von Neumann's friend Stanislaw Ulam had one explanation of this curious maintenance of purity in the midst of physical application. "It was only, to my knowledge, just before World War II that he became interested not only in mathematical physics but also in more concrete physical problems. His book on quantum theory is very abstract and is, so to say, about the grammar of the subject. . . . it did not, it seems to me, contribute directly to any truly new insights or new experiments" (1980, p. 95). But that opinion misconstrues the subtleties of the ambitions of Göttingen in the 1920s, and the attractions of the Hilbert program for von Neumann.

It seems that the spirit of Hilbert's formalist program in metamathematics both defined the purity and certainty of mathematical endeavor, but simultaneously sanctioned its characteristic forays into the treacherous domains of "applied" knowledge, allowing it to emerge unscathed from phenomenological (and psychological) confusion in the encounter. Much of this has been explained in admirable detail in the recent work of Leo Corry (1997b). Hilbert's program was a response to the enigmas and antinomies unearthed by Georg Cantor's work on set theory in the 1880s (Dauben, 1979). The keystone to the formalist response was the method of axiomatization, instantiated by Hilbert's own axiomatization of Euclidean geometry in 1899. In contrast to the logicist program of Russell and Whitehead, Hilbert did not believe mathematics could be "reduced" to logic. Rather, the axiomatic method was the epitome of the process of abstraction, the one characteristic practice of mathematicians which was self-justifying, and could be projected outward onto empirical domains without becoming complicit in their own uncertainties, paradoxes, and confusions. As Hilbert wrote in 1922:

> The axiomatic method is indeed and remains the one suitable and indispensable aid to the spirit of every exact investigation no matter in what domain; it is logically unassailable and at the same time fruitful; it guarantees thereby complete freedom of investigation. To proceed axiomatically means in this sense nothing else than to think with knowledge of what one is about. While earlier without the axiomatic method one proceeded naively in that one believed in certain rela-

> tionships as dogma, the axiomatic approach removes this naivete and yet permits the advantages of belief. (in Kline, 1980, p. 193)

Thus mathematics for Hilbert should not be regarded as some species of a posteriori distilled factual knowledge (perhaps in the manner of Mill), but rather as the process of the formal manipulation of abstract symbols leached of all intuitive reference. Because mathematics was the one field of human endeavor that could demonstrate the legitimacy of its own foundations, namely, that its inferences were well defined, complete, and consistent, it could then be used to formalize and axiomatize other external bodies of knowledge, exploring the question of whether they also exhibited those virtues. As Davis and Hersh (1981, pp. 337, 335) put it:

> Hilbert's program rested on two unexamined premises; first, the Kantian premise that *something* in mathematics – at least the finitary part – is a solid foundation, is indubitable; and second, the formalist premise, that a solidly founded theory about formal sentences could validate the mathematical activity of real life. . . . This program involved three steps. (1) Introduce a formal language and formal rules of inference, so that every "correct proof" of a classical theorem could be represented by a formal derivation, starting from axioms, with each step mechanically checkable. . . . (2) Develop a theory of the combinatorial properties of this formal language, regarded as a finite set of symbols subject to permutation and rearrangement. . . . (3) Prove by purely finite arguments that a contradiction, for example 1 = 0, cannot be derived within the system.

The resemblance of the formalist program to the structure of "games" has often been noted (Corry, 1997b, p. 84). Davis and Hersh (1981, p. 336) write, "As the logicist interpretation tried to make mathematics safe by turning it into a tautology, the formalist interpretation tried to make it safe by turning it into a meaningless game." The flaw in this interpretation resides in the adjective "meaningless." As Corry points out, Hilbert's axiomatics were intended to be applied to already well-established mathematical entities in "concrete" as well as abstract settings, for the purpose of cleaning up the logic of their explanatory structures, and demonstrating completeness, consistency, independence, and simplicity of the fundamental axioms. The task was to attain logical clarity, test for correctness of inferences, compare sets of axioms for appropriateness to task, and purge possible contradictions. This explains Hilbert's lifelong fascination with physics, in his view an ideal field for the deployment of this axiomatic practice on a par with geometry. If axiomatization resembled a game, it was by no means "meaningless," but rather a retrospective evaluation of the state of a research program, with the intention of further interventions in future research. It would be more faithful to the spirit of Hilbert to think of the game as a microcosm of a complex and unwieldy reality, exaggerating and highlighting its internal structure – say, *Monopoly* as a caricature of real estate investment in 1930s New York. In his early career up until 1905, Hilbert in his lectures proposed the extension of axiomatization not only to classical mechanics and thermodynamics, but also to the probability calculus, the kinetic theory of gases, insurance mathematics, and even (!) psychophysics (Corry, 1997b, p. 179).

Von Neumann's early work on logic followed essentially the same approach as Hilbert (Moore, 1988, p. 119). Three of von Neumann's early papers provide an axiomatization of set theory (1961–63, 1:35–57), and in the words of Stanislaw Ulam (1958, pp. 11–12), they "seem to realize Hilbert's goal of treating mathematics as a finite game. Here one can divine the germ of von Neumann's future interest in computing machines and the mechanization of proofs." While there certainly is a connection, it is neither as direct nor as unmediated as Ulam suggests. The more immediate connection is to the first paper on the theory of games, presented to Hilbert's Göttingen seminar on December 7, 1926.

It has now become widely accepted that von Neumann's early work on game theory was prompted by questions thrown up by Hilbert's program of the formalization of mathematics and *not* by any special concern with economics or even with the social practices of gaming themselves.[13] Earlier, Hilbert had brought Zermelo to Göttingen in order to work on the foundations of logic (Corry, 1997b, p. 124). Von Neumann in the 1920s was working within the tradition of Zermelo-Fraenkel set theory; and his Göttingen colleague Ernst Zermelo had previously delivered an address to the International College of Mathematicians in 1912 "On the Application of Set Theory to Chess." In the address, he insists his object is "not dealing with the practical method for [playing] games, but rather is simply giving an answer to the following question: can the value of a particular feasible position in a game for one of the players be mathematically and objectively decided, or can it at least be defined without resorting to more subjective psychological concepts?" (1913, p. 501). Zermelo provided a proof that chess is strictly determined, in the sense that either white or black can force a win, or both sides can at least force a draw.[14] Various mathematicians, all in contact with one another and including Denes König, Lazlo Kalmar, and von Neumann, were engaged in 1926–27 in improving Zermelo's proof

[13] To my knowledge, this case was first argued in Mirowski, 1992, pp. 116–19. It is presented with some elaboration in Leonard, 1995. Dimand and Dimand (1996) are the stubborn holdouts, simply ignoring the weight of evidence.

[14] Robert Aumann's *Lectures on Game Theory* (1989) begins with this proof.

from the viewpoint of confining the axioms (principally finiteness) to those sanctioned by the Hilbert program.

The other distinctly Hilbertian project in Göttingen in the later 1920s was the elaboration of quantum mechanics, principally in the matrix mechanics format of Max Born, Werner Heisenberg, and Pascual Jordan (Mehra & Rechenberg, 1982; Cushing, 1994). Von Neumann wrote his first paper on the topic in collaboration with Hilbert and Lothar Nordheim, and produced two more papers in 1927 (1961–63, 1:151–235) that sought to deploy axiomatization in order to settle a raging controversy about the correct approach to quantum mechanics: which approach was "better," Göttingen's matrix mechanics or Erwin Schrödinger's wave mechanics? Von Neumann's answer was that both were equivalent because they could both be derived from an improved (but more abstract) formalism of operators on vectors in Hilbert space. These papers were revised and extended in an axiomatic treatment of the topic in his first book published in 1932, *Mathematical Foundations of Quantum Mechanics* (1955); although remaining controversial as proof of "no hidden variables" in quantum mechanics, this formalism has set the pattern for the preponderance of treatments of quantum mechanics down to the present (Baggott, 1992).

One of the primary reasons why von Neumann persists as such an inscrutable figure in the historiography of science is that his work in logic, game theory, mathematics, and physics is frequently treated in individually splendid isolation. Nowhere is this more evident than his work on quantum mechanics, which has been greeted with disdain by many physicists (Cushing, 1994; Moore, 1989; Beller, 1999), who regarded it as "a luxury, an example of striving for mathematical exactness for its own sake" (Redei, 1996, p. 509; Feyerabend, 1999, p. 157), and dazed incomprehension by other, lesser mortals. This is unfortunate, because it is in the work on quantum mechanics that most of the themes that von Neumann returned to throughout his life are most evident; and, moreover, it is one of the few areas in which explicitly *philosophical* commitments were indispensable to the project being pursued, and therefore received some semi-explicit treatment.

Themes Out of Quantum Mechanics

Although it would be awkward to attempt a summary of all the issues bedeviling quantum mechanics in the 1920s, we can provide a list of the theses most salient to the subsequent development of game theory and of the theory of computation and automata, in the form of an abbreviated bill of particulars.

1. The first message of *Mathematical Foundations* is that Hilbert's axiomatic method, far from being simply a method of

redescription of physical theory, can actually resolve important theoretical controversies in physics once and for all. The controversy at issue here was the existence of a "causal" account of quantum mechanics versus the "acausal" interpretation associated with Niels Bohr, now called the Copenhagen interpretation (Cushing, 1994; Torretti, 1999). Von Neumann purported to provide an abstract mathematical proof that no "hidden variable" theory could reproduce the results attributed to existing quantum mechanics. Because Göttingen was a stronghold of the Copenhagen interpretation, this was hailed as resounding defeat of its opponents. (We shall pass on any attempt to pronounce whether von Neumann's proof was actually *used* by others in this manner. On this issue, see Peat, 1997, p. 125, and Cushing, 1994, p. 131.)

2. The Copenhagen interpretation argued, crudely, that electron states existed as simultaneous possibilities until the act of observation caused the "collapse of the wave-packet" to a classical particle. Bohr's principle of "complementarity" suggests that because quantum particles have no intrinsic properties independent of the measuring instrument, and seemingly contradictory descriptions, such as the famous wave-particle duality, they must coexist in a complementary fashion to encompass the phenomenon. "The Copenhagen interpretation essentially states that in quantum mechanics we have reached the limit of what we can know" (Baggott, 1992, p. 87). Probability considerations were not artifacts of ignorance of the observer; they were inherent in the quantum phenomena. This interpretation was disturbing to many physicists, because it seemed to suggest that the observer has the capacity to exercise appreciable powers over the constitution of the reality which is to be probed. Von Neumann, it should be noted, was an adherent of this view.
3. Although it may have been implicit in previous work, von Neumann's method stressed the explicit use of state vectors to provide a unique representation of the events to be covered by the theory. The method of positing states was often prosecuted in a manner analogous to the way Hilbert treated choice of axioms. The method of the partitioning of states was then used to define the statistics of the problem.[15]

[15] See the chapter by J. Segal in Glimm, Impagliazzo, & Singer, 1990, esp. p. 155. The spread of treatment of "information" as the partitioning of a universe of states is examined further in Chapter 6.

4. Due to the Göttingen background of matrix mechanics and the initial development of the theory of Hilbert spaces, matrix bilinear forms, eigenvalues, and matrix algebra were generally entertained as more familiar technique than would have been the case in the rest of the physics community of the 1920s. "Hilbert was excited about the advantages and the insight afforded by the vectorial formulation of Eulerian equations. Vectorial analysis as a systematic way of dealing with phenomena was a fairly recent development that crystallized towards the turn of the century" (Corry, 1997b, p. 146). While von Neumann himself opposed the use of matricies in quantum mechanics in favor of linear operators, he was certainly intimately familiar with the bilinear form in which Heisenberg's version of the theory was cast, and was a master of its idiosyncrasies. One observes this facility in both the first game theory paper and in the subsequent expanding economy model.[16]
5. Von Neumann's book ventured beyond the immediate concerns of the developers of quantum mechanics, in that he believed the theory had implications for macroscopic collections of quantum particles. The act of measurement in quantum mechanics transforms an ensemble of N indistinguishable particles into a mixture of (as many as) N particles in different eigenstates. This indicates a place for a quantum statistical mechanics, in which the collapse of the wave packets produces an increase in entropy. Thus, familiar standard Boltzmannian concerns about time asymmetry of ensembles of particles should also apply to the time irreversibility of quantum measurement (Baggott, 1992, pp. 175–76). This concern would result in von Neumann's work on the ergodic theorem.
6. If physical measurement is irreversible in an entropic sense, then it makes sense to ask what it is that could offset the rise in entropy in the act of measurement. The answer, taken directly from the paper by Leo Szilard (1929), is that *information* is key to understanding the analogue to Maxwell's Demon in this novel quantum

[16] Matthew Frank reminds me of the differences as well as the similarities between game theory and quantum mechanics as formulated by von Neumann: "The bilinear forms of the game theory paper are all bounded, over real vector spaces, and skew-symmetric – whereas his key innovations in the quantum-mechanical domain were dealing with unbounded operators, with complex vector spaces, and with symmetric forms. . . . the central issue of the game theory paper (finding a saddle point) is pretty foreign to the spectral theory and other key problems of quantum mechanics" (e-mail, September 25, 1999). Others seem somewhat more enthusiastic about the parallels, however; see, for instance, Meyer, 1999.

environment (von Neumann, 1955, p. 369). Indeed, "The discussion which is carried out in the following . . . contains essential elements which the author owes to conversations with L. Szilard" (1955, p. 421). This is the initial significant link to the traditions of thermodynamics and information encountered above in Chapter 2. Von Neumann's fascination with the idea that "thinking generates entropy" dates back to a seminar on statistical mechanics taught by Einstein to Szilard, himself, Wigner, and Gabor in Berlin in 1921 (Lanouette, 1992, p. 58).

7. It is also important to see where von Neumann goes "beyond Copenhagen" in his 1932 book. The orthodox Copenhagen interpretation in that era refused to commit to what it was about the act of observation that caused the collapse of the wave packet. The equations of quantum mechanics were not themselves irreversible; and it would be difficult to argue that quantum mechanics did not also govern the behavior of the macroscopic measuring instrument, or indeed the human observer. In contrast to his Copenhagen-inclined comrades, von Neumann tentatively grasped the nettle and suggested that the wave function of a quantum particle collapses when it interacts with the consciousness of the observer (Torretti, 1999, sec. 6.3.2). Because pointer-readings were registered with photons, and the impact of photons on the retina was a perfectly good quantum phenomena, as was the signal from the retina to the brain, it was only there that the wave function encountered a system that was not itself governed by just one more wave function. Thus, the problem of the nature of consciousness was raised with disarming immediacy in von Neumann's physics.
8. In this work, von Neumann already expresses impatience with conventional theories of subjectivity that claim to reduce consciousness to purely mechanical processes. He wrote: "the measurement or the related process of the subjective perception is a new entity relative to the physical environment, and is not reducible to the latter" (1955, p. 418). But this position itself gave rise to further problems and paradoxes having to do with intersubjective agreement (Heims, 1980, p. 135). What if two rival consciousnesses each experience a collapse of the wave packet? Will they necessarily agree? Von Neumann essentially finessed this problem by having the second observer treat the first as no different from any other inanimate object or measuring device; but this appears to contradict the previously quoted statement, and does not banish the problem in any event, as his friend Eugene Wigner realized and later made famous as the "Paradox

of Wigner's Friend" (Baggott, 1992, pp. 187–88). Hence, the problem of intersubjective agreement when "each observer influences the results of all other observers" was already as pressingly insistent in the quantum mechanics as it was to become in the game theory; indeed, it was raised there first.

It is possible to argue that the first four elements (1–4) of this philosophical program had already been manifest in the 1928 paper on the theory of games, as well as the scant work on economics in this early phase of von Neumann's career. The foundational role of the Hilbert program of axiomatics (theme 1) for the 1928 paper has already been canvassed; and it is just as clearly signaled in the later *Theory of Games* (1964 [1944], p. 74). Yet, just as in the case of quantum mechanics, this option is not paraded about as formalization for its own sake: it is thought to have real consequences for theoretical controversies. The social theory at issue here is, not coincidentally, reminiscent of the Copenhagen interpretation (theme 2): it has to do with the place of probability in social phenomena.[17] The cumulative message of the 1928 paper is that probability is ontic rather than epistemic, only in this instance not due to the behavior of electrons, but rather because the canons of rational play demand recourse to randomization. "Although in Section 1 chance was eliminated from games of strategy under consideration (by introducing expected values and eliminating 'draws'), it has now made a spontaneous reappearance. Even if the rules of the game do not contain any elements of 'hazard' . . . in specifying the rules of behavior for players it becomes imperative to reconsider the element of 'hazard.' The dependence on chance . . . is such an intrinsic part of the game itself (if not of the world) that there is no need to introduce it artificially by way of the rules of the game" (1959, p. 26). This refers to the guarantee of the existence of a minimax strategy in a two-person zero-sum game if "mixed strategies" or randomization of individual play is allowed. "Strategies" are themselves defined as state vectors (theme 3), while the game itself is fully characterized as a set of states and their consequences. The proof of the central theorem, that the minimax solution coincides for both players, is premised upon the mathematics of matrix bilinear forms (theme 4). Von Neumann himself admitted later that he used the same fixed point technique for the expanding economy model as he did for the minimax theorem (1964, p. 154n).

If one sets out to acknowledge the parallels, then the striking aspect of this first period of von Neumann's endeavors that fairly jumps out from

[17] The question of in what manner, or even whether at all, probability ought to play a significant role in economics was a divisive and unsettled issue in the 1920s. On the possible objections within economics, see Knight, 1940; Mirowski, 1989c.

our enumeration is that, although the first brace of concerns (1–4) are brought into play, the second brace of considerations (5–8) are *not* brought over into the game theoretic context, and therefore the treatment of the active role of consciousness is notably, if not painfully, absent. This does not imply, however, that they were in some sense irrelevant, as the following examples show. Theme 5: Although the state characterization of games would seem to be naturally suited to elaboration of a Boltzmannian "ensembles" approach, there is no evidence of any such interest in the 1920s.[18] Theme 6: Since the question of irreversible play does not arise, neither does the question of the role of information in the reduction of entropy. That does not mean that "information" is not mentioned at all in 1928; rather, relative to the conceptualization of the quantum mechanics, the effect of information is directly *suppressed*. This is achieved by collapsing it into a prior, unspecified exhaustive enumeration of states: "it is inherent in the concept of 'strategy' that all the information about the actions of participants and the outcomes of 'draws' a player is able to obtain {NB-PM} or to infer is already incorporated in the 'strategy.' Consequently, each player must choose his strategy in complete ignorance of the choices of the rest of the players" (von Neumann, 1959, p. 19). Themes 7 and 8: One can already see where this species of treatment of consciousness and intersubjectivity might potentially lead. Because the opponent has in a very precise sense been objectified and neutralized, the gamester might as well be playing a finite state machine as a putatively sentient opponent. That machines might be the ideal instantiation of game-theoretic rationality would only loom large in von Neumann's research in the third and final phase of his career.[19]

In retrospect, and returning to the 1920s, there was something distinctly odd about a project that purported to be an investigation of "the effects which the players have on each other, the consequences of the fact (so typical of social happenings!) that each player influences the results of all other players, even though he is only interested in his own" (von Neumann, 1959, p. 17), but ended up subsuming and neutralizing all those influences under the rubric of some independently fixed external conditions. Yet the phenomenon of cognitive interplay of active conscious entities has already encountered a similar obstacle in his quantum mechanics, in the paradox of Wigner's Friend. The problem was not the advocacy of a fairly strict methodological individualism with a myopically

[18] This situation is noted by von Neumann himself in the second phase of his work. See von Neumann & Morgenstern, 1964, pp. 400–1.

[19] The thesis that game theory finds its logical apotheosis in thinking machines is the main topic of Chapters 7 and 8. Here we observe the initial glimmerings of the tradition of cyborg economics.

selfish actor – a doctrine that von Neumann generally endorsed, as we previously observed – but rather the portrayal of mutual interdependence of interest as a one-way mechanical process, and not "subjective perception [as] a new entity relative to the physical environment" (theme 8). Much of this derived directly from the mathematical convenience of positing global invariance of the payoff, the "conservation of value," or the "zero-sum" condition. Consequently, information and its vicissitudes had no effective purchase in the early theory of games, and even in the area in physics in which he had ventured "beyond Copenhagen" (theme 7), it was attenuated. It was not altogether absent, however; in a coda to the section on the three-person case, he wrote: "the three-person game is essentially different from a game between two persons. . . . A new element enters, which is entirely foreign to the stereotyped and well-balanced two person game: struggle" (p. 38).

This construction of the notion of "struggle" is important for understanding precisely how the early theory of games does and does not diverge from the work in quantum mechanics, especially because it looms ever larger in significance in the next phase of von Neumann's forays into economics. Von Neumann painted the two-person game as possessing an intrinsically random component (due to mixed strategies), but nevertheless exhibiting fundamentally determinate solutions in strategic play and outcomes. He achieved this perplexing portrayal primarily by negating any efficacy of consciousness in minimax play. Moving to three players constitutes a substantial qualitative leap, however, because in von Neumann's opinion, two players may gang up on the third, thus reducing this case to the two-person setup. The two-person play is still determinate, but it is indeterminacy in the shape and behavior of the coalition – that is, the conscious interaction with another sentient being – that prevents overall sharp and determinate results. Consciousness and communication, banished in the original definition of strategy, threaten to reemerge with a vengeance in the eventuality that the very definition of a "player" and a "payoff" could become subsumed under the province of the strategic manipulation of the agents. It is noteworthy that the word "struggle" is introduced at precisely this juncture – a word rarely associated with "purity," and one that fairly squirms with discomfort at the inability to produce general solutions to games of arbitrary payoffs and numbers of players.

In 1928, therefore, this nascent "theory of games" was poised at the cusp between parlor games and something played with a bit more gusto; between describing a trivial pursuit and formalizing something done a bit more for keeps. It was perched precariously between quantum mechanics and thermodynamics, between purity and danger, between formal certainty and empirical distress, between a human and a machine, between detached contemplation and active engagement, along with its progenitor. It remained there, poised, stranded, for a decade or more. Its progenitor did not.

PHASE TWO: IMPURITY

As we mentioned at the beginning of this chapter, John von Neumann himself identified his life and work as having changed irreversibly some time around 1937. Some of these changes were personal. For instance, two of the friendships which had the greatest impact on his intellectual career, with Stanislaw Ulam and Oskar Morgenstern, date from this watershed.[20] The year 1937 marked the beginning of wider ceremonial recognition: the American Mathematical Society named him Colloquium Lecturer in 1937 and bestowed the Bocher Prize in 1938. Other changes had more to do with the opening out of his concerns to larger and more diverse audiences – and here the military dominated all the other supplicants clamoring for his attention. The experience of war leaves no one unscathed; but it affected von Neumann in a distinctly peculiar way. Other changes were intellectual, which he characterized as a slide from "purity" to something less pristine, but in fact involved aloofness (if not rejection) toward many of his earlier enthusiasms. Again cognizant of the dangers of imposing a telos a posteriori to something that was in reality much more disheveled, it is difficult to regard this period, roughly 1937–43, as anything other than a transitional phase. Only when von Neumann found the computer, everything else tended to fall into place as part of a grand unified conception of the place of mathematics, and hence human thought, in the universe. Unfortunately for the comprehension of von Neumann's impact upon the discipline of economics, the text self-advertised as most directly pertinent to economics as it unfolded in the subsequent scheme of things, *Theory of Games and Economic Behavior*, falls squarely within this transitional phase. His role in the writing of the book was finished by December 1942, and by all accounts von Neumann got interested in computing in 1943.

Immediately upon becoming a U.S. citizen in the watershed year 1937, von Neumann began to consult with the military at the Ballistics Research

[20] See note 3. The first of these friends was Stanislaw Ulam, who was appointed his assistant at the Institute for Advanced Study in 1936. "Stan was the only close friend von Neumann ever had. Von Neumann latched on to Stan and managed to be as close to him as often as possible" (Rota, 1993, p. 48). For his autobiography, see Ulam, 1976. The second was Oskar Morgenstern, whom he first met in fall 1938 (Rellstab, 1992, p. 90). There exists no useful biography of Morgenstern, although his own personal account of his collaboration with von Neumann was published (1976a). One can't help but notice that these were two *mitteleuropaisch* expatriates, of upper-class backgrounds, both deeply disaffected from their new American surroundings.

Laboratory operated by the Army Ordnance Department in Aberdeen, Maryland, apparently through the intercession of Oswald Veblen (Aspray, 1990, p. 26; Goldstine, 1972, p. 177). The early date tends to belie the claim that brute wartime emergency first turned von Neumann in the direction of the military. So does the speed with which he became attached to a multifarious collection of military boards, governmental committees, and research units.[21] It appears that von Neumann had become very involved with questions of turbulence and fluid flow in the mid-1930s, and was particularly fascinated by the meaning and significance of the Reynolds number, which demarcated the phase transition between laminar and chaotic turbulence.[22] This expertise justified a consulting relation concerning the treatment of shock and detonation waves; and this, in turn, explained his role in arguing for an implosion design on the first nuclear bomb (Rhodes, 1986). Technical and mathematical expertise rapidly became parlayed into a general advisory capacity, with von Neumann visiting various military research sites and counseling the allocation of effort and resources across endeavors. As we shall observe in the next chapter, he was instrumental in blurring the line between technical weapons consultant and military strategist and, in the process, spreading the new gospel of "operations research," conducting research in the area from at least 1942. It will prove important later in this volume to realize that his conception of Operations Research diverged from the previous British conception of operations analysis, eventually combining thermodynamics, optimization, game theory, and the computer into a prodigious

[21] The following is a partial list from a sheet entitled "J. von Neumann Government Connections as of February 1955," box 11, folder 3, VNLC:

1940–date	Member, Ballistic Research Laboratories Scientific Advisory Committee, Aberdeen Proving Ground
1941–46	Navy Bureau of Ordnance
1943–46	Manhattan District, Los Alamos
1947–date	Naval Ordnance Laboratory
1948–date	RAND Corporation, Santa Monica
1948–53	Research and Development Board, Washington
1949–54	Oak Ridge National Laboratory
1950–date	Armed Forces Special Weapons Project
1950–date	Weapon Systems Evaluation Group, Washington
1951–date	Scientific Advisory Board, U.S. Air Force
1952–54	U.S. Atomic Energy Commission
1952–date	Central Intelligence Agency

[22] See Ulam, 1980, p. 93. Indeed, Ulam & von Neumann, devoted extended consideration to the tent map $F(x) = 4x(1 - x)$ on the interval [0,1], which is now the mainstay of elementary introductions to what is sometimes called "chaos theory." The inspiration of fluid dynamics is discussed in Galison, 1997, while the link to simulation is surveyed in Weissert, 1997.

new scientific discipline absorbing literally thousands of technically trained workers; it eventually had incalculable implications for the future of the economics profession as well. The transition correlates rather precisely with his transfer out of pure mathematics; Goldstine (1972, p. 175) estimates by 1941 applied mathematics had become his dominant interest.

The smooth traverse from pure to applied pursuits was not negotiated by many other mathematicians during World War II (Owens, 1989), so it is pertinent to ask what paved the way in his particular instance. Certainly von Neumann's vaunted talent for polymathy would loom as the dominant explanation; but there is also the issue of the character of his break with the Hilbert program and its conception of the relative standings of formalism and application. The key event here is the announcement of Gödel's incompleteness results in axiomatic set theory at Königsburg in September 1930 and the cumulative effect they had upon von Neumann's later work. As is now well known, Gödel regarded himself as participating in the Hilbert program and began his career by proving the completeness of first-order logic, only to undermine the program by demonstrating the existence of formally undecidable propositions in his classic 1931 paper. It is perhaps less well known that von Neumann was one of the very first to appreciate the profundity of these results (John Dawson in Shanker, 1989, pp. 77–79; Dawson, 1997, p. 69). "The incompleteness theorems so fascinated von Neumann that on at least two occasions he lectured on Gödel's work rather than his own" (Dawson, 1997, p. 70). It appears that von Neumann was so impressed that he had Gödel invited repeatedly to give lectures at the Institute for Advanced Study – for the academic year 1933–34, again in October 1935, and in the fall of 1938 (Dawson in Shanker, 1989, p. 7; Dawson, 1997, p. 123). These visits were trenchant for von Neumann. One effect they had is that he went around telling his Princeton students that he did not bother to read another paper in symbolic logic thereafter (Kuhn & Tucker, 1958, p. 108). That was certainly hyperbole; on a number of occasions he returned to the incompleteness results in the last phase of his career.

Further, the fascination with Gödel's work did not end there. When Arthur Burks wrote to Gödel to ask him which theorem von Neumann had intended in his informal citation in his 1949 Illinois lectures, Gödel responded (in von Neumann, 1966, p. 55):

> I think the theorem of mine which von Neumann refers to is not that on the existence of undecidable propositions or that on the lengths of proofs but rather the fact that a complete epistemological description of a language A cannot be given in the same language A, because the concept of the truth of sentences of A cannot be defined in A. . . . I did not, however, formulate it explicitly in my paper of 1931 but only my Princeton lectures of 1934.

One might expect that epistemological obstacles such as these might have had a bearing upon the ambitions, if not the structures, of a formalism like the theory of games. For such an arid topic, it is amazing that this work did seem to resonate with certain themes that subsequently rose to prominence in applied work, especially for the military. Gödel's theorems certainly did have some bearing on the military quest for a science of command, control, communication, and information, if only because they were couched in terms richly suggestive of cryptanalysis. In seeking an arithmetization of syntax, Gödel subjected not only the symbols of a mathematical language, but finite sequences of such symbols, and sequences of such sequences, to being assigned numerical codes in an effective, one-to-one fashion. His theorems then queried whether some predictably terminating algorithm could determine whether a given number was code for a sequence of formulas, and hence could serve as a "proof" for a given proposition. His breakthrough was to demonstrate that this sharpened notion of "provability" was not testable in that manner. Of course, the theorems also played a large role in bringing mathematical logic to bear on problems of mechanizing computation, as described in the previous chapter.

So, for von Neumann the mathematician, the leitmotiv of the period of the mid-1930s was one of absorbing the implications of Gödel's results, and incorporating them into a recalibrated sense of the efficacy and aims and purposes of mathematics, and reconciling them with his newfound alliance with the military. What could rigor and formalization do for us after Gödel? Von Neumann's interactions with Alan Turing at Princeton in the years 1936–38 were part and parcel of this process. He uncharacteristically acknowledged this philosophical turnabout in his famous 1947 lecture "The Mathematician," which reads as an apologia for the years spent wandering in the desert, for what we have dubbed phase two (1961–63, 1:6):

> My personal opinion, which is shared by many others, is, that Gödel has shown that Hilbert's program is essentially hopeless. . . . I have told the story of this controversy in such detail, because I think that it constitutes the best caution against taking the immovable rigor of mathematics too much for granted. This happened in our own lifetime, and I know myself how humiliatingly easy my own views regarding the absolute mathematical truth changed during this episode, and how they changed three times in succession!

His appreciation of Gödel's theorems only grew more avid and nuanced through time. In a 1951 presentation of the Einstein Award to Gödel, he marveled: "Kurt Gödel's achievement in modern logic is singular and monumental – indeed it is more than a monument, it is a land mark which will remain visible far in space and time. Whether anything comparable to it has occurred in the subject of modern logic in modern times may be debated. . . . Gödel's results showed, to the great surprise of our entire generation, that this program, conceived by Hilbert, could not be implemented."[23]

Because the Hilbert program had not stood as some purely abstract dogma that had little bearing upon mathematical practice or mathematical ambitions, the rejection of Hilbert's project implied some repudiation of its attendant prior approaches to the physical sciences, as well as some bedrock philosophical beliefs. Perhaps the main symptom of the dissolution of the Hilbert program for von Neumann was that he no longer looked to scientific formalisms that were "well developed" or subjects that were conventionally "ripe" for axiomatization, in contrast to the opinion expressed in the letter to Flexner quoted in the previous section. Von Neumann instead began to show a proclivity to strike out into neglected or ill-understood areas, using mathematics as a heuristic rather than as an instrument of consolidation. For instance, von Neumann came in this period to be dissatisfied with the Hilbert space representation of quantum mechanics (Redei, 1996). He collaborated with F. J. Murray on a brace of papers on rings of operators, and with Garrett Birkhoff on a different approach to quantum mechanics called "quantum logic." The other area he revisited for reevaluation in the late 1930s was his earlier ideas on the formalization of rationality. This assumed two differing guises: the first, a tentative inquiry into the physiology of brains; and the second, renewed efforts to formalize the rational play of games.

The evidence that we have on these two related themes comes primarily from a remarkable brace of correspondence with the Hungarian physicist and family friend Rudolf Ortvay (Nagy, Horvath, & Ferene, 1989). These letters are often cited to prove that von Neumann was contemplating the nature of the brain and the idea of the computer well before his documented encounter with computational limitations in war research in 1943; but they also indicate the ways in which he was triangulating on a revised notion of rationality from his previous work in quantum

[23] "Statement in connection with the first presentation of the Albert Einstein Award to Dr. K. Gödel, March 14, 1951," box 6, folder 12, VNLC. Von Neumann here gave one of the best short descriptions of Gödel's achievement ever written. "The accepted modes of reasoning in mathematical logic are such that anything can be rigorously inferred from an absurdity. It follows, therefore, that if the formal system of mathematics contained an actual contradiction, every mathematical theorem would be demonstrable – including the one about the absence of inner contradictions in mathematics. . . . Gödel's result accordingly that the absence of such contradictions is undecidable, provided there are no contradictions in mathematics, or to be more exact: In the formalized system identified with mathematics. In other words, he established not an absolute, but a relative undecidability of a most interesting and peculiar kind."

mechanics, logic, and game theory. In March 1939 von Neumann wrote Ortvay, "I am deeply interested in what you write about biology and quantum theory and the subject is very sympathetic to me, too. . . . I have been thinking extensively, mainly from last year, about the nature of the 'observer' figuring in quantum mechanics. This is a kind of quasi-physiological concept. I think I can describe it in the abstract sense by getting rid of quasi-physiological complications" (in Nagy et al., 1989, p. 185). On Christmas 1939 Ortvay responded to von Neumann, describing neurons as, "a highly sophisticated switchboard system, the scheme of which we do not know. The task would be to propose theorems about this" (p. 186). Von Neumann answered Ortvay on 13 May 1940, "I cannot accept that a theory of prime importance which everybody believes to be elementary, can be right if it is too complicated, i.e., if it describes these elementary processes as horribly complex and sophisticated. Thus, I am horrified by the present-day quantum chemistry and especially biochemistry" (p. 187). Ortvay responded on May 30, "I think that it will be possible to find a proper and simple theoretical approach to the operation of the brain. The teleological concept of variation principles was widely spread in the eighteenth century and it was used in a very naive way at times so that it became discredited later. . . . Another remark: don't you think that classical mathematics developed along with the rationalization of the physical world and made this development possible. By now, mathematics far exceeds this framework (set theory, topology, mathematical logic), and can be expected to contribute to the rationalization of psychology and social sciences" (p. 187).

On January 29, 1941, Ortvay wrote, "Once again I read through your paper on games and would be interested in having your new results published, preferably in a popular manner. I liked your paper very much at the time, and it gave me hope that you might succeed in formulating the problem of switching of brain cells. . . . This model may resemble an automatic telephone switchboard; there is, however, a change in the connections after every communication" (p. 187). Finally, encapsulating the metaphorical shifts between machines, brains, and social entities, Ortvay wrote on February 16, 1941, "these days everybody is talking about organization and totality. Today's computing machines, automatic telephone switchboards, high-voltage equipment such as the cascade transformer, radio transmitter and receiver equipment, and also industrial plants or offices are technical examples of such organizations. I think there is a common element in all these organizations which can be the basis for an axiom. . . . I believe that once it is possible to identify clearly the essential elements of an organization such as these, this would give us a survey of possible forms of organization" (p. 188). From the modern vantage point, we could regard Ortvay as urging a revival of the nineteenth-century tradition of the interplay of economic organization, machine metaphors, and logic, which flourished briefly with Babbage and (to a lesser extent) Jevons.

Corresponding with Ortvay, a physicist, von Neumann felt comfortable to range freely between the problem of the observer in quantum mechanics to the interactions of neurons in the brain to the possibility of an electronic switching machine capturing some aspects of thought to the interactions of humans in games to the promise of a generic social science of organizations, all predicated upon a "new" mathematics. The paper on games that Ortvay references in January 1941 is most likely the draft dated October 1940 of a manuscript entitled "Theory of Games I: General Foundations."[24] It is now known that von Neumann had once more turned his attention to the theory of games, probably in 1937, prior to his encounter with Oskar Morgenstern (Rellstab, 1992). The surviving draft manuscript gives some clues as to the reasons for once again taking up the topic. First, significantly, there is still absolutely no reference to economics whatsoever. If this is "applied mathematics," then it resides at a stage not very different from the type of "application" made in the 1928 paper. Players and their objectives are treated in the most abstract manner, with no concessions to didactic context. Second, the manuscript begins by repeating the 1928 results, and then asserting that extension to the n-person case is possible through the intermediary of the "value function" v(S). There he wrote:

> the set function v(S) suffices to determine the entire strategy of the players i = 1, . . . , n in our game. . . . v(S) describes what a given coalition of players (specifically: the set S) can obtain from their opponents (the set S˜) – but fails to describe how the proceeds of the enterprise are to be divided among the partners. . . . the direct "apportionment" by means of the f [individual payoffs] is necessarily offset by some system of "compensations" which the players must make to each other *before* coalitions can be formed. These "compensations" should essentially be determined by the possibilities which exist for each partner in the coalition S to forsake it and join some other coalition T.[25]

[24] See box 15, OMDU. Part II of the manuscript was dated January 1941, which was probably too late for Ortvay to have also received it.

[25] "Theory of Games I," pp. 5–6, box 15, OMDU. The table of contents for this paper reads:

1. General scheme of a game. The 2-person game and its strategies. Previous results concerning the 2-person game. Value of a 2-person game.
2. The n-person game. Coalitions. The function v(S). Intuitive discussion of v(S). The characteristic properties of v(S). Construction of a game with a given general v(S).
3. Equivalence of games, i.e., of v(S)'s. The v(S) as a partially ordered vector space. Normal forms for a game. Essential and inessential games.
4. Intuitive discussion of the 3-person game. Exact definitions for the n-person game, based on the above intuitive discussion: Valuations, domination, solutions.

Third, one can observe the nascent resurgence of the Boltzmannian approach to the solution of n-person games. The set function v(S) indexes macroscopic equivalence of games, even though they are comprised of complexes of payoffs ("energies") that differ at the microscopic level. This, by analogy, promises the prospect of general macroscopic laws of games: the entire game is putatively determined by the exogenously set levels of payoffs, even though coalition payoffs are determined in an endogenous manner. An ordering on valuations is defined, and a solution is asserted to be a set of valuations in which no two included valuations dominate each other, and every external valuation is dominated by at least one valuation within the set. This definition became the "stable set" solution concept for games in the 1944 book. Fourth, for this to be tractable, it is still necessary to insist that the formation of "coalitions" is merely an artifact of the payoffs; permissible configurations of "compensations" are somehow set in stone *before* the coalitions can be formed. The problem situation of 1928 has not been solved, as much as it has been finessed in the very same way that it had been done in the coalition-free solution of the two-person minimax. Coalitions, which had threatened to insinuate the consciousness of the observer into the constitution of the observed in 1928, were now without further ado demoted to the status of combinatorial readings of preset inanimate measurement devices. "Which 'winning' coalition is actually formed, will be due to causes entirely outside the limits of our present discussion" ("Theory of Games I," p. 13, OMPD). Coalitions now become like wave packets, superpositions of possibilities which only "collapse" down to observable social organizations with actual play. The static treatment of the measurement process and the Paradox of Wigner's Friend in the quantum mechanics – the Copenhagen "problem of the observer" – still bedevils the theory of games at this juncture.

It is difficult to see how von Neumann believed that this manuscript was a substantial improvement over the earlier work, in the sense that all games were now purportedly subsumed under a general formal approach, much less that anyone would regard it as solving any real-life problems. In a series of lectures delivered at the University of Washington on game theory in the summer of 1940, the only applications mentioned are to poker and chess; yet he was well aware that his game theory did not practically serve to illuminate the actual play of these games, either. There is little indication that anyone else regarded it as a substantial contribution to pure mathematics, except on the strength of his growing reputation. For

5. Existence of one-element solutions: Inessential game, unique solution.
6. Determination of all solutions of the essential 3-person game. Graphical method.
7. Intuitive discussion of the results of 6. The intuitive notions of proper and improper solutions.

himself, it may have represented a fleeting glimpse of how the self-organization of information and communication through the process of struggle could ideally result in a kind of order;[26] but it was precisely the soul of those interactive and informational aspects of the human phenomena that his formalism tended to suppress. He seemed to be casting about for a concrete social phenomenon that could provide guidance and inspiration for the *useful* development of the theory of games, upon having relinquished the Hilbertian quest to capture the abstract quintessence of the category "game." It stood naked and unadorned, a formalism in search of a raison d'être, a situation that was rectified by the intervention of Oskar Morgenstern.

Regrettably, we cannot describe in this volume the saga of Morgenstern's own intellectual quest in anywhere near the detail which justice would require, if only because of our overriding commitment to the pursuit of the cyborg genealogy.[27] It is agreed by most commentators that Morgenstern himself did not make any substantial contribution to the mathematical side of the theory of games, but that it would nevertheless be a mistake to disparage his contribution on that account. Morgenstern's great virtue was to share many of von Neumann's enthusiasms and, better, to provide game theory with the cachet of a "real-life application," which it had so clearly lacked up till that point. His skepticism about the legitimacy of existing neoclassical theory, his fascination with the new physics,[28] his dabbling in logic, and his lifelong concern about the psychology of prediction and valuation in interdependent circumstances rendered him an almost ideal sounding board for von Neumann's own project at this juncture. Throughout his life, Morgenstern was someone who sought to collaborate with mathematicians on his vision for an improved economics; and he eventually trained his hopes and ambitions on von Neumann over the course of 1940.

[26] This is suggested by a conversation with von Neumann reported by Bronowski (1973, p. 432): "Chess is not a game. Chess is a well-defined form of computation. You may not be able to work out the answers, but in them there must be a solution, a right procedure in any position. Now, real games, he said, are not like that at all. Real life is not like that. Real life consists of bluffing, of little tactics of deception, of asking yourself what is the other man going to think I mean to do. And that is what games are about in my theory." This passage is interesting in that it illustrates the very transitional meaning of "games" in themselves, relative to the earlier Hilbert formulation.

[27] See, however, Mirowski, 1992; Leonard, 1995; Martin Shubik in Weintraub, 1992; Shubik, 1997.

[28] An indication of his interest is a report in his diary of a discussion with Neils Bohr at the Nassau Club in February 1939, where it is reported that Bohr said that "physics is too often taken as a model for the philosophy of science." February 15, 1939, OMDU.

In the period April–May 1940, according to the Morgenstern diaries, he and von Neumann had extensive discussions concerning game theory. Von Neumann left for the West Coast in summer 1940, during which he reported making progress on games of four and more persons. On his return, discussions resumed in earnest about the possible relationship of game theory to specific economic topics. A September diary entry recounts conversations about the notion of an "imputation" and the attribution of marginal productivity. "N. says that he can also get by without homogeneous functions. We want to continue this. He finds that one of his functions bears a striking resemblance to an important function in thermodynamics. It also provides great analogies to higher and lower energies, etc." (September 20, 1940, OMDU). Rejection of neoclassical doctrines are also repeated subjects of comment. "Lunch with vN. I told him about some criticisms of the theory of indifference curves and he agreed with me. He is also convinced that marginal utility does not hold. I shall work all that into the Hicks review" (October 19, 1940, OMDU).

On a number of occasions during this period Morgenstern brought informational and cognitive considerations to von Neumann's attention. "Yesterday I showed him the contract curve; it has a bearing upon his games because it makes a difference who goes first. Moreover, I showed him the causal relationships in BB's [Bohm-Bawerk's] price theory where, as one can easily see, one obtains different results, depending upon the assumed knowledge of the other's position. It differs from a game because profit is possible" (November 9, 1940, OMDU). This passage is significant because it points to the few economic authors (Edgeworth, Bohm-Bawerk) who had attempted some quasi-informational amendments to prewar neoclassical price theory by recognizing possible indeterminacies; it ventures beyond von Neumann's prior position that all strategies are uniformly known to all players a priori; and it clearly locates the zero-sum assumption as one of the mainstays of the neutralization of the cognitive efficacy of the opponent. It is equally noteworthy that von Neumann counters Morgenstern's "empirical" concerns with some of his own renewed struggles with quantum mechanics. The diary reports on January 21 that von Neumann explained his paper on quantum logics to Morgenstern. "And that disturbs me deeply, because it means that far-reaching epistemological conclusions must be drawn. The nice, comfortable division of science into logical and empirical components fails. That is immense. 'Everything comes from quantum mechanics.' And it leads again to the foundational debates." He continues, "I suspect that in economics it would be necessary to introduce some sort of new mode of reasoning. For example, a logic of desire. This again leads to [Karl] Menger.[29] The problem with this is I perceive it as necessary, but I fear it eludes me. In all probability, this is because I never had the necessary mathematical training. Sad" (January 21, 1941, OMDU).

Even though von Neumann was being increasingly drawn by war work into areas such as "gases and ballistics" (May 17, 1941, OMDU), he nevertheless finally decided to actively collaborate with Morgenstern on a text concerned with "Games, Minimax, bilateral monopoly, duopoly" (July 12, 1941, OMDU) and related issues in July 1941, in effect abandoning his previous draft manuscript "Theory of Games" of 1940–41. With the assistance of von Neumann, Morgenstern drafted a first chapter of a work they tentatively entitled, "Theory of Rational Behavior and of Exchange."[30] Apparently one of the earliest thorny issues prompted by the collaboration (as opposed to von Neumann's lone elaboration of mathematical concepts) was the attitude and approach to the neoclassical doctrine of value as individual "utility." A "logic of desire," a formalism "mainly concerned with quasi-psychological or even logistical concepts like 'decisions,' 'information,' 'plans,'" and so forth, would seem to be

[29] This is a reference to Karl Menger's attempt to formalize ethics in his *Moral, Wille und Weltgestaltung* (1934). The importance of Menger's model for early game theory is discussed by Leonard, 1998.

[30] This is the title reported in a prospectus for the book in a letter to Frank Aydelotte dated October 6, 1941, box 29, folder 5, VNLC. The prospectus divides the work into three sections: unmathematical exposition of the subject, mathematical discussion, and application of the theory of games to typical economic problems. Only the first part was included with the prospectus, while von Neumann cited his "1940–41 notes" as constituting the core of the second part. "§1 is a discussion of the possibility and the (desirable) spirit of the application of mathematics to economics. . . . §3 contains our views on the notion of 'utility.' The 'numerical utility,' as developed in §3.3, is based on considerations which – we think – have been overlooked in the existing literature." Their motives are stated with exemplary simplicity in the attached prospectus: "It is attempted to analyse the similarities and dissimilarities between economic and social theory on one hand, and the natural sciences – where mathematical treatment has been successful – on the other hand, and to show that the differences – the existence of which is not denied – are not likely to impede mathematical treatment. It is proposed to show, however, that the present mathematico-economical theories follow the pattern of mathematical physics too closely to be successful in this qualitatively different field. The theory of economic and social phenomena ought to be primarily one of rational behavior, i.e., mainly concerned with quasi-physiological or even logistical concepts like "decisions," "information," "plans," "strategy," "agreements," "coalitions," etc. A discussion which comes up to the standards of exact discussions – as maintained in the more advanced sciences (the natural sciences) – must primarily find "models" for this type of problem, – i.e., simple and mathematically completely described set-ups which exhibit the essential features of economic and social organizations as outlined above. The authors believe that the concept of such models for economic and for social organizations is exactly identical with the general concept of a game."

precisely what the neoclassicals thought they had already advocated and brought to a fevered pitch of perfection; but this presumption was what both members of the duo rejected. I believe there subsisted a tension between von Neumann and Morgenstern over the reasons that each respectively thought "utility" was untenable (Mirowski, 1992); but since Morgenstern deferred to the great mathematician in all things, this is difficult to sort out.[31]

In any event, it seems that Morgenstern finally convinced von Neumann that they must proceed tactically by means of the conciliatory move of phrasing the payoffs in terms of an entity called "utility," but one that von Neumann would demonstrate was cardinal – in other words, for all practical purposes indistinguishable from money – so long as one remained committed to application of the expectation operator of the probability calculus to combinations of strategies and payoffs. It went without saying that the one major contribution of the 1928 theory of games that hinged on precisely this construct – the insistence upon the intrinsically probabilistic character of play – had to be retained in the refurbished research program. This, therefore, is the meaning of the curiously inverted comment in the final text that "We have practically defined numerical utility as being that thing for which the calculus of mathematical expectations is appropriate" (von Neumann & Morgenstern, 1964, p. 28). In the mind of Morgenstern, this constituted an attack on the fractious discussion in the 1930s and 1940s of the nature and significance of interdependence in neoclassical demand theory (Hands & Mirowski, 1998): "In the essay more remarks are going to be made about the measurability of utility. It turns out that one can easily make genuine suppositions of complimentarity (completely reasonable and possible suppositions) by which the whole indifference curve analysis becomes untenable. Instead of this the preferences of the individual are partially ordered sets" (August 11, 1941, OMDU). This became known subsequently as the axiomatization of von Neumann–Morgenstern expected utility, the formal derivation of which was added as an appendix to the second edition of *Theory of Games and Economic Behavior* (*TGEB*) in 1946 (1964, pp. v, 617–32), although the outline of the proof was constructed in 1942 (April 14, 1942, OMDU).

Because "von Neumann–Morgenstern expected utility" was the one concept in the book that many in the economics profession latched onto

[31] However, much of Morgenstern's diary entries, as well as the one substantial letter from von Neumann to Morgenstern during the process of composition (letter, October 16, 1942, box 23, OMDU), are concerned with the adequate or appropriate treatment of utility in various game contexts. This seemed to be one of Morgenstern's few areas where he could make analytical comments; the mathematical formalization of games was left entirely to von Neumann.

with alacrity in the period immediately following the publication of *TGEB*, it is important to try and gain some perspective on its place in von Neumann's oeuvre. It must be said that he had no objections to the idea there was some sort of individual ordering of personal preference; as we suggested earlier in the chapter, it resonated in a low-key manner with his methodologically individualist proclivities. Moreover, he also was moderately favorably inclined toward the use of extrema in identifying rest points over preferences; selfishness, as he said, was a law of nature. Yet, arrayed against those motivations were other, rather more compelling reasons *not* to accept utility as any sort of reasonable characterization of psychology.

First, as we have noted, he regarded the mathematics of utility as a pale and inadequate imitation of classical mechanics, verging on obsolescence. He had repeatedly hinted throughout his writings that entropy, and not energy, held the key to a modern representation of the processing of information. Second, the text of *TGEB* insists that it will have no truck with the subjectivist theory of probability, but will remain firmly committed to the frequentist interpretation (von Neumann & Morgenstern, 1964, p. 19). This merely reiterated a position that von Neumann advocated throughout his career that probabilities were ontic, and not epistemic. Third, the text explicitly wears its heart on its sleeve, in the sense that the word "utility" is treated as just a placeholder for "money," something transferable, numerical, and fungible in that everyone is presumed to want in unlimited amounts (pp. 15–16). Because payoffs are effectively denominated in its currency as part of the rules, and in magnitudes independent of play, people are presumed to know what they are getting as outcomes for each move and countermove before the pieces hit the board; *pace* Morgenstern, that aspect is never allowed to become a matter of interpretation. Whereas the theory of games was indeed intended to make some stabs in the direction of a logic of strategy, the theory of utility was in no way regarded as a necessary or credible component of that project.[32] Fourth, there was the contemporary "pragmatist" riposte that people act "as if" they had utility functions, even though this was not a serious theory of psychology; what is interesting about von Neumann is that he was less and less inclined to make these sorts of arguments as time went on. As we shall observe repeatedly in subsequent chapters, the wizard of computation betrayed no inclination to privilege the Turing Test or to simulate rational behavior in humans;

[32] "Von Neumann was here recently. He stated that he did not agree with Wald in applying minimax ideas to the treatment of uncertainty." Kenneth Arrow to Jacob Marschak, August 24, 1948, box 154, JMLA.

rather, his quest was for a formal theory of rationality independent of human psychology. This distinction only grew more dominant in phase three of his research trajectory. Fifth, and most crucially, the final phase of von Neumann's career, from 1943 until his death in 1957, was heavily explicitly occupied with the study of the brain and the formalization of rational calculation via the instrumentality of the computer. At no time in this period did von Neumann return to "utility" in order to complement, illustrate, or inform those objectives. Utility theory simply dropped from sight.

Themes Out of Game Theory

Given the vast outpouring of literature on games, especially since the 1980s, it is sobering to realize how very few authors can be bothered actually to describe what von Neumann and Morgenstern wrote in 1944 or to take the trouble to distinguish their project from what subsequently became the orthodoxy in game theory.[33] Of course, if one believes that, "while the publication of von Neumann and Morgenstern's book attracted much more attention, it was Nash's major paper on noncooperative equilibrium that truly marked the beginning of a new era in economics" (Myerson, 1996, p. 288), then such exercises are rendered nugatory, and von Neumann's own ideas are erased. Perhaps the most egregious elision encouraged by this confusion is the historical evidence that von Neumann explicitly rejected that subsequent Nash program. The net effect of this selective amnesia, intentional or not, is that everything in the original 1944 text that foreshadowed cybernetics and computational architectures has been peremptorily banished, and the context of its immediate (non)reception in economics dissipated, like the shards and remnants of a bad dream. But troubling dreams of reason have a way of recurring, especially in a book devoted to cyborg reveries. Here we restrict ourselves to providing a summary of game theory of vintage 1944.

The point of departure that any summary worth its salt must confront is the iconoclastic tone of the text. Much of this language can be attributed to Morgenstern, who bore primary responsibility for sections 1–4 and 61–64, but there is every indication that von Neumann concurred with

[33] Examples of the unthinking conflation of von Neumann with subsequent game theory would seem endemic in the science studies literature (Edwards, 1996; Mendelsohn & Dalmedico in Krige & Pestre, 1997; Fortun & Schweber, 1993). The one honorable exception here is Christian Schmidt (1995a, 1995b). I explicitly recount his theses in Chapter 6. The attempt to jury-rig a continuous line of development from Cournot to Nash (thus downgrading von Neumann's work), exemplified by the writings of Roger Myerson, is critically examined in Leonard, 1994a. The rise of Nash equilibria in game theory is covered in detail in Chapter 6.

most of the methodological pronouncements concerning economics. There is, for instance, the call for a "new mathematics" to displace the Newtonian calculus, because it "is unlikely that a mere repetition of the tricks which served us so well in physics will do so for the social phenomena too" (von Neumann & Morgenstern, 1964, p. 6). Here the duo distanced themselves from the earlier vintage of physics envy, which had so dominated the history of neoclassical economics. Nevertheless, it would be rash to interpret this as a self-denying ordinance when it came to physical metaphor, especially in a book that mentions quantum mechanics on four separate occasions and ventures numerous comparisons of individual theoretical points to classical mechanics not enumerated in the index. Instead, what is rejected is the imitation of classical mechanics exemplified by the "Lausanne School" (p. 15) since "there exists, at present, no satisfactory treatment of the question of rational behavior" (p. 9). The remedy proffered consisted of the generic formalism that is intended to replace it, the construct of the game: "For economic and social problems games fulfill – or should fulfill – the same function which various geometrico-mathematical models have successfully performed in the physical sciences" (p. 32). Some reasons are suggested for why various mathematical exercises had not been successful in economics prior to their work, the major reason being the dearth of formal rigorous proofs (p. 5). All of this, we should realize by now, is fairly standard fare coming from von Neumann.

Caution must be exercised, however, in interpreting what "rigor" now means in the world after Gödel. There are only two full-scale axiomatization exercises in all of *TGEB*: the first, already mentioned, is the axiomatization of expected utility; the second is the formal axiomatization of the concept of a "game" (pp. 73–79). The first exercise is not fundamental to the project but rather appended to the second edition as an afterthought; and the second purports to "distill out" the intuitive fundamentals of a game without necessarily being committed to any specific interpretation, still after the fashion of Hilbert (p. 74). The distinction between syntax and semantics becomes abruptly twinned, with the introduction of the concept of a "strategy" as a complete formal plan of play that is independent of the information or interpretation imposed by either the player or his opponents (p. 79), leaning on the draconian restriction of the model to one-time-only static play to lend this some legitimacy. The assurances that this exercise would ultimately provide impeccable warrants of perfect generality and consistency are muted, however; and this is the first taste of the legacy of Gödel. The Hilbertian goals of freedom from contradiction, completeness, and independence of the axioms are indeed mentioned (p. 76), but completeness is confessed to be beyond reach, if only because there appears to be no unique object or attribute that formally characterizes all games.[34] Unlike the case of geometry or classical mechanics or quantum mechanics, there did not then exist a well-developed science of "games" which could provide sufficient guidance in this matter: and metaphors were never thought to be appropriate candidates for axiomatization. Furthermore, the other two goals are merely treated as throwaways, exercises for the reader, and not classed among the essential results of the project. Hilbert's grand program persists in *TGEB* in a stunted and withered form.

The "transitional" character of the text is illustrated by the fact that the axiomatization exercise does not really serve to underwrite any of the asserted "generality" of the theory of games. That comes instead from the systematic attempt to reduce all possible games to some elaboration or recasting of the central model, which von Neumann identified as the two-person zero-sum (2P0Σ) game of the 1928 paper. If all games, suitably reconfigured, were really of the 2P0Σ form, then the minimax concept would still constitute the generic solution algorithm for all games. The book is structured as an attempt to cement this case by moving first from two-person to three-person to four-person zero-sum games, only then to loosen up the zero-sum payoff assumption to incorporate general n-person non-zero-sum games. The way this was accomplished was to retain the innovation of the unpublished "Theory of Games" manuscript of 1940, namely, using the uninterpreted concept of "coalitions" in order to collapse the 3-person game to a "coalition vs. odd-man-out" game, a four-person to a three-person, and a promissory note to continue the process for five persons and more. If the generalization to larger numbers of players was successful, then the non-zero-sum game could be transformed into a zero-sum game through the mathematical artifact of the introduction of a "fictitious player" (pp. 506ff.), turning an N-person non-zero-sum formalism into an (N + 1)-person zero-sum game. It is evident that von Neumann believed that this process of the "imbedding" larger more complex games into simpler, more tractable games by adding

[34] In von Neumann & Morgenstern, 1964, p. 76n3, von Neumann brushes off this problem as one that is shared by axiomatizations of mechanics as well. It is curious that he does not indicate the general problem of paradoxes of self-reference, which is the more relevant issue in this instance. The text then continues to suggest that the real benefit of axiomatization is to demonstrate "that it is possible to describe and discuss mathematically human actions in which the main emphasis lies on the psychological side. In the present case the psychological element was brought in by the necessity of analyzing decisions, the information on the basis of which they are taken, and the interrelatedness of such sets of information. . . . This interrelatedness originates in the connection of the various sets of information in time, causation, and by the speculative hypotheses of the players concerning each other" (p. 77). The problem with this auxiliary justification is that it is an *interpretation*, and a poorly grounded one at that. The treatment of the cognitive status of the opponent is part of the problem, and not part of the solution, in *TGEB*.

or consolidating "dummy players" was the major contribution to a general theory at this stage. This is signaled by one of the primary citations of quantum mechanics (and, implicitly, thermodynamics) in the text:

> The game under consideration was originally viewed as an isolated occurrence, but then removed from this isolation and imbedded, without modification, in all possible ways into a greater game. This order of ideas is not alien to the natural sciences, particularly to mechanics. The first standpoint corresponds to the analysis of so-called closed systems, the second to their imbedding, without interaction, into all possible greater closed systems. The methodological importance of this procedure has been variously emphasized in the modern literature on theoretical physics, particularly in the analysis of the structure of Quantum Mechanics. It is remarkable that it could be made use of so essentially in our present investigation. (pp. 400–1)

Nonetheless, the process of "imbedding" or stepwise reduction to 2P0Σ games is not as direct or unmediated as the quotation suggests. The introduction of "coalitions" necessitates mathematical innovations that were not present in the simpler theory, namely the concept of "domination." "Thus our task is to replace the notion of the optimum . . . by something which can take over its functions in static equilibrium" (p. 39). Because a payoff now incorporates the stipulations of an uninterpreted "agreement," it is now dubbed an "imputation," with all the connotations of intentionality and contestability that it implies. An imputation x is said to "dominate" another imputation y "when there exists a group of participants each one of which prefers his individual situation in x to that in y, and who are convinced that they are able, as a group – i.e., as an alliance – to enforce their preferences" (p. 38). A "solution" to an $N > 2$ person game is a set S of imputations such that: (1) no y contained in S is dominated by an x contained in S; and (2) every y not contained in S is dominated by some x contained in S. When this is extended to the $(N + 1)$-person non-zero-sum case, it became known later as the "stable set" (Gilles, 1953).

How did this diverge from the more familiar constrained optimum of mechanics and neoclassical economics? The first consideration is that the ordering defined by domination will not generally be transitive, an attribute "which may appear to be an annoying complication" (von Neumann & Morgenstern, p. 39). Second, the solution will generally be a set containing numerous members; the theory will not generally be capable of identifying which specific realization is selected. Rather than regard this multiplicity and indeterminacy – especially relative to the original minimax – as somehow symptomatic of the incoherence of the abstraction or imperfection of the formalism, von Neumann and Morgenstern opted to try and turn it into a virtue. In a revealing change of metaphor, they exhort the reader to "forget temporarily the analogy with games and think entirely in terms of social organizations" (p. 41n). A solution is then compared with an "established order of society" or "accepted standard of behavior," which, it is hinted, comes to be instituted through causes entirely outside the purview of those captured by the game formalism. Because coalitions already have taken on this character, solutions also cannot be prevented from appealing to this external entity as an uncaused first cause. All of these considerations culminate in a troubling admission, at least to a mathematician: existence and uniqueness of the stable set have not been proved. Indeed, "We have not given a definition of S, but a definition of a property of S – we have not defined the solution but characterized all possible solutions" (p. 40). So even in chapter 4, the ambivalence (and some would say, "struggle") with respect to the Hilbert program of formalist foundations is evident.[35]

It is consequently important to understand how a "Theory of Rational Behavior and Exchange" (September 22, 1941, title) became a "Mathematical Theory of Economic Behavior" (December 24, 1941), and then, the final and rather more chastened "Theory of Games *and* Economic Behavior." As the formalism of the game was extended to greater numbers of players and non-zero-sum situations, the very notion of rationality experienced subtle shifts. In part, this was due to von Neumann more directly having to confront the uses and disuses of symmetries in order to render the problem tractable. In the original two-person version, symmetry between players was elided by the zero-sum assumption, itself the specification of a rather strong symmetry in payoff space: one person's gain is exactly the other's loss. This made it possible to plan strategic play regardless of whether the opponent was "different" in any relevant sense, or even, as von Neumann suggested, irrespective of his rationality. Once coalitions were introduced when $N > 2$, similarities and differences of players could no longer be thrust aside as irrelevant, even when the zero-sum condition was upheld. Imposition of symmetric imputations by fiat would simplify solutions and reduce their multiplicity, but because the reasons for formation of coalitions were obscure in the first place, and would in most cases hinge upon subtle differences between players,

[35] "All these considerations illustrate once more what a complexity of theoretical forms must be expected in social theory. Our static analysis alone necessitated the creation of a conceptual and formal mechanism which is very different from anything used, for instance, in mathematical physics. Thus the conventional view of a solution as a uniquely defined number or aggregate of numbers was seen to be too narrow for our purposes, in spite of its successes in other fields. The emphasis seems to be shifted more towards combinatorics and set theory – and away from the algorithm of differential equations which dominate mathematical physics" (von Neumann & Morgenstern, 1964, p. 45).

asymmetric solutions could no longer be ruled out (p. 315). But this raised the vexing and threatening issue of how asymmetry and a generic "rationality" were to be reconciled, one with another. Thus had von Neumann stumbled upon a paradox of symmetry and rationality which has continued to bedevil economics for more than a century.[36]

This paradox came home to roost when the theory was finally presented in its full "generality" of the (N + 1)-person non-zero-sum case. Because the fictitious player was appended as a mathematical artifact, the "dummy" clearly broke player symmetry, it became necessary to describe the meaning of a stable set for the "real players," and this proved difficult. "The best one can say of it is that it seems to assume the effective operation of an influence which is definitely set to injure society as a whole (i.e., the totality of all real players). Specifically in this case domination is asserted when all players of a certain group (of real players) prefer their individual situation . . . if the remaining (real) players cannot block this arrangement, and if it is definitely injurious to society as a whole. . . . The reader will have noticed by now that [this solution concept] is of a rather irrational character, but nevertheless not altogether unfamiliar" (p. 523). While this may indeed have resonated with von Neumann's own rather jaundiced view of humanity, it was becoming more and more difficult to maintain the pretense that this was a formalization of unalloyed commonsense "rationality," at least in any vernacular sense.

At this juncture, we can come to see that *TGEB* is better understood as a tentative exploration of various paradoxes of certain definitions of "rationality" and, as such, stands shoulder to shoulder with the other two such contemporaneous explorations which von Neumann held in high esteem: those of Kurt Gödel and of Alan Turing. This project had begun with the very notion of a mixed strategy, that is, the portrayal of pure randomness of activity as evidence of pure strategic rationality under certain circumstances (p. 52). Far from avoiding it as an intractable paradox, von Neumann found the prospect of possible reconciliation of such antipodes exhilarating, just as he had done with his response to the Copenhagen interpretation. Now, with the extension of game theory to the problem of "coalitions," von Neumann had ventured further into antitheses to explore the possibly further paradoxical nature of a "strategy" conceived as a "program." A strategy was defined as "a plan which specifies what choices he will make in every possible situation, for

[36] This problem, of how a truly comprehensive rationality could coexist with others of its ilk, without reducing the world to a mere solipsistic excrescence, haunts the annals of modern economics, from Nash game theory to rational expectations macro to the modern "no-trade theorems" such as Milgrom & Stokey, 1982. The generic philosophical problem was first cogently discussed by Esther-Mirjam Sent in her 1994 thesis, revised in 1998. The problem for Nash equilibria is covered in Chapter 6.

every possible information which he may possess at that moment in conformity with the pattern of information which the rules of the game provide for him in that case" (p. 79). Even though the formalism in *TGEB* dealt only with static one-shot play, which by construction ruled out induction, learning, and most other cognitive functions, and had further saddled the players with an implausibly mechanical utilitarian psychology, von Neumann was fascinated by the idea that, even in this impoverished world, strategic thinking would nevertheless encourage the conceptual redefinition of players (through coalitions) and payoffs (through "compensations"), leading to solutions that would not conform to simple physically dictated global extrema.

Moreover, one observes from the definition of strategies that the key intermediate term that created these possibilities was *information*. In "the process of trying to define 'good' 'rational' playing . . . the problem is to adjust this 'extra' signaling so that its advantages – by forwarding or withholding information – overbalances the losses which it causes directly" (p. 54). But because the very calculation of net benefit was itself a strategic variable, except in certain restrictive cases (like the 2P0Σ game), the search for a solution was slowly mutating into search for a mutually supportive set of information processors constructing the parameters of their game on the fly, a configuration that need not produce an outcome foreordained in the absence of all ratiocination. "Therefore the conduct of affairs of this coalition – the distribution of the spoils within it – is no longer determined by the realities of the game – i.e., by the threats between partners – but by the standard of behavior. . . . It must be re-emphasized that this arbitrariness is just an expression of the multiplicity of stable standards of behavior" (p. 417). Hence this (admittedly cartoonish version of) thought in action could give rise to something novel, something more complex, something not simply a reflection of underlying physical givens. It was seemingly paradoxical, in that the outcome could be "rational" from the viewpoint of players involved, but perhaps not from some external, detached vantage point. This portrayal of society as something other than a bald epiphenomena of natural "constraints" was diametrically opposed to the Lausanne school as well as the general physicalist orientation of economics in that era (Mirowski, 1989a). This theme, that "rationality" gives rise to certain organizational structures, which then refer back to themselves in a recursive fashion, redefining the prior notion of rationality in a more complex way, was leitmotiv of the last decade of his life.[37] It

[37] The way that von Neumann's themes of information processing, complexification, and strategic considerations came together in his attitudes toward game theory is nicely revealed in a report of comments made at a conference on game theory held at Princeton in February 1955: "Von Neumann then outlined the program of a new approach to the cooperative game by means of some (not yet constructed) theory of the *rules* of games.

marked the transition from logical inference as an invariant state to calculation as a program or process – or, better yet, from thermodynamics to cyborg science.

Having identified those aspects of *TGEB* which were genuinely innovative, we must acknowledge that the "economic" content of the book was exceedingly thin. The "applications" section occupies only thirty-two pages (pp. 555–86), with its prime candidate being Bohm-Bawerk's horse market. Not only was the choice of an obscure Austrian price-theoretic homily (bearing the fingerprints of Morgenstern) essentially orthogonal to any contemporary concerns of orthodox neoclassical theory, but the most charitable thing that could be said about the analysis was that it actually expanded the range of price indeterminacy relative to Bohm-Bawerk's theory (p. 564). Issues of monopoly and monopsony were also mentioned but with little in the way of substantial clarification. If game theory really were poised to revolutionize the world of economics in 1944, then one would have been sorely hard-pressed to say how this might happen. Precisely for this reason, in combination with its basically dry prose style (when you could find the prose), all the hoopla and overheated rhetoric that greeted its publication, including a front-page review in the *New York Times Book Review* of March 10, 1946, practically cries out for historical explanation.[38] Much of this will be more immediately explained by von Neumann's other, higher-profile activities during the war and immediately afterward and *not* about his attitudes concerning economics.

PHASE THREE: WORLDLINESS

The year 1943 heralded the emergence of John von Neumann as a player on a global stage in two Promethean dramas: the Manhattan Project, and the development of the electronic computer. The two events were, of course, intimately related. The manuscript of *TGEB* had to be rather hurriedly completed in late December 1942 because he had become attached to the Navy Bureau of Ordinance in September 1942, which had required a move to Washington, D.C.; in early 1943 the work took him to England, where he had his first encounter with problems of large-scale

A class of *admissible extensions* of the rules of the non-cooperative game would be defined to cover communication, negotiation, side payments, etc. It should be possible to determine when one admissible extension was 'stronger' than another, and the game correspondingly more cooperative. The goal would be to find a 'maximal' extension of the rules. A set of rules such that the non-cooperative solutions for that game do not change under any stronger set. If this were possible, it would seem the ideal way in which to solve the completely cooperative game" (in Kuhn & Tucker, 1958, p. 103).

[38] Box 17, folder 3, VNLC. See also the glowing reviews by Hurwicz (1945), Stone (1948), Marschak (1946).

calculations (Aspray, 1990, p. 28). Upon his return, he was enlisted as a consultant for the atomic bomb project at Los Alamos, where his expertise in shock waves and hydrodynamics was brought to bear on the implosion design for the device (Rhodes, 1986, p. 480; Galison, 1997, pp. 694–98). The equations for the hydrodynamics of this kind of explosive were not amenable to analytic solution and had to be approximated by numerical calculation or experimental simulation. This requirement, in conjunction with needs for intensive computation of ordinance firing tables at the Aberdeen Ballistics Research Laboratory, had alerted many in the research community to be on the lookout for more efficient computational devices and schemes. Notably, von Neumann early on approached Warren Weaver for help in learning about existing machines for automating the weapons calculations.[39] Curiously, Weaver put him in touch with everyone working in the area except for the researchers at the Moore School at the University of Pennsylvania, who since 1943 had been constructing an all-electronic calculator for Army Ordinance, dubbed ENIAC (Aspray, 1990, pp. 34–35). The story of how von Neumann encountered Herman Goldstine on a train platform in Aberdeen in August 1944 (Goldstine, 1972, p. 182) and first learned of ENIAC is an appointment with destiny that has swelled to mythic proportions in the interim. It certainly ushered von Neumann into avid collaboration on the design of the next-generation computer EDVAC, in the process composing the world's first comprehensive description of the design of an electronic stored-program computer. The "First Draft of the Report on the EDVAC" (in von Neumann, 1987) was written in spring of 1945, and rapidly became the design bible of the nascent computer community.

The year 1943 not only marked the birth of the computer and the bomb, but it also was the year in which von Neumann's lifelong interest in rationality took a decidedly biological turn. Soon after its publication, von Neumann read a paper by Warren McCulloch and Walter Pitts on the logical calculus as expressed by a simplified model of the neuron (1943). Von Neumann was reportedly immensely impressed with this paper, because it linked his burgeoning interest in Turing's ideas about computation and the long-standing obsession with the nature of rationality, combined with a demonstration of how the brain could be structured, at least in principle. From 1943 onward von Neumann developed numerous contacts with the neurophysiological and medical communities, in an attempt to canvass what was known about the brain. By late 1944, von Neumann thought there was sufficient interest and substantive results in

[39] Warren Weaver was a Rockefeller Foundation officer who was in charge of the wartime Applied Mathematics Panel at the time of this incident. Weaver's career, so pivotal for the development of the cyborg sciences, is described in Chapter 4.

the area to collaborate with Norbert Wiener and Howard Aiken in convening a meeting at Princeton on a new field "as yet not even named." The meeting was closed because so much of the work was subject to secret classification, restricted to a handpicked set of scientists and engineers concerned with "communications engineering, the engineering of computing machines, the engineering of control devices, the mathematics of time series in statistics, and the communication and control aspects of the nervous system" (in Aspray, 1990, p. 183). Wiener proposed the group be dubbed the "Teleological Society," although later he was responsible for coining the more specialized term "cybernetics." "Teleology is the study of the purpose of conduct, and it seems that a large part of our interests are devoted on the one hand to the study of how purpose is realized in human and animal conduct and on the other hand how purpose can be imitated by mechanical and electrical means."[40]

It is no coincidence that von Neumann's angle of attack on the problem of purposive conduct shifted so dramatically after 1943. It is hard to believe that he was altogether satisfied with his recent game theory results, which, after all, were themselves initially intended to revolutionize the study of human purposive (read: strategic) conduct. Although nominally concerned with the processing of information, the formal games model did not adequately take the act of information processing into account; whereas from a mathematician's vantage point, it was devoid of much in the way of novel mathematics; nor did it really come to grips with the developments in logic which had rocked the field in the 1930s. On top of all that, there was the nagging problem of the lack of an existence proof for the stable set; and then there were no really successful applications which could be pointed to. The formalism was proving to be a hard sell to the economics community in the immediate aftermath of publication.[41]

[40] Aiken, von Neumann, and Wiener to Goldstein [*sic*], December 28, 1944, quoted in Aspray, 1990, p. 183. On the series of interdisciplinary conferences largely inspired by Wiener and von Neumann's cybernetic enthusiasms, see Heims, 1991. It is noteworthy that the only person invited to these conferences who even remotely could qualify as an economist was Leonard Savage, and that would be a stretch.

[41] Some of the reactions can be gleaned from the Morgenstern diaries. For instance, after presenting the theory to a hometown seminar at Princeton, Morgenstern reported in an entry dated May 2, 1944: "Lutz was hostile and ironic. It appears that he perceives the disagreement the most because he has vested interests in theory. This will be the reaction of most theoreticians." After a visit to Harvard, he wrote on June 8, 1945: "None of them has read *The Theory of Games* and no one has said anything besides Haberler, who has not read any further. But they will not be able to ignore it because the mathematicians and physicists there, as elsewhere, ought to be very enthusiastic about it." After a subsequent visit to Harvard, where he gave a seminar on game theory, he recorded on March 4, 1946: "The economists are partially hostile and partially fascinated because the physicists so strongly and masterfully rule over it." On December 20, 1947: "Ropke even said later that game theory was Viennese coffeehouse gossip." And on June 13, 1947, after

Even the economists at the Cowles Commission, whom von Neumann had initially targeted as ideal candidates for proselytization through his acquaintance with Tjalling Koopmans, proved recalcitrant in absorbing the content, as opposed to the mathematical form, of the new doctrine or strategic rationality, becoming instead mesmerized by the allure of the appendix on cardinal expected utility. By contrast, *TGEB*'s residual dependence upon an outdated utilitarian psychology could only grate upon scientists willing to delve into serious neurobiology or cognitive psychology. This is not to claim that he callously abandoned his progeny after 1944; indeed, in this period he was rather primarily responsible for fostering a home for the development of game theory in the postwar defense analysis community. But nevertheless, the unmistakable withdrawal from any further work on game theory after 1944, with only a few minor exceptions, coincides exactly with his turn toward the theory of computation, brain science, and the theory of automata. Some of this neglect can be attributed to his increasingly onerous military obligations; but not all of it. After all, von Neumann's towering legacies to the twentieth century, computers and atomic weapons and the theory of automata, all date from this period. As argued in Chapter 8, the theory of automata ultimately stands as his most profound contribution to economics, and his crowning bequest to the cyborg sciences. What he chose to work on in those last hectic years was an expression of a personal judgment about his potential legacy and place in history.[42]

Unlike in the cases of the previous two phases of his career, no single book sums up von Neumann's intellectual concerns in a manner that he felt sufficiently confident himself to prepare for publication. Indeed, in his third period the inhumanly busy von Neumann took up the practice of having someone else act as amanuensis during some lecture and allowing the lucky scribe to be listed as coauthor. The closest candidate to canonical text for his third-period concerns is the *Theory of Self-Reproducing Automata* (1966); but that is actually a series of lecture notes posthumously edited and completed by Arthur Burks, and it leaves the

a lecture on game theory at the Perroux Institute, he wrote: "Allais opposed: we had *not* disproved there was a social maximum for free competition(!) . . . Nobody has even seen the book. The copy I sent to Perroux has not yet arrived." After a talk at Chicago on game theory, the entry for January 1, 1948, reads: "It is clear from what he says that Schumpeter has never read the book." The entry for October 8, 1947, reads: [Von Neumann] says [Samuelson] has murky ideas about stability. He is no mathematician and one should not credit him with analysis. And even in 30 years he won't absorb game theory." After a talk at Rotterdam on October 30, 1950: "They had heard of game theory, but Tinbergen, Frisch, etc. wanted to know nothing about it because it disturbs them." All quotations from OMDU.

[42] "Von Neumann perhaps viewed his work on automata as his most important one, at least the most important one in his later life" (Goldstine, 1972, p. 285).

distinct impression of quitting just as it is getting started. Yet the work on computer architectures, probabilistic logics, pseudorandomness, and the theory of automata are all of a piece; as Aspray put it, von Neumann "thought that science and technology would shift from a past emphasis on the subjects of motion, force, energy, and power to a future emphasis on the subjects of communication, organization, programming and control" (in von Neumann, 1987, p. 365). Indeed we have already encountered this cyborg refrain in Wiener, Turing, Shannon, and elsewhere. It takes as its pole star the central question: "To what extent can human reasoning in the sciences be more efficiently replaced by mechanisms?" (von Neumann, 1987, p. 318). Never one to regard this as solely a question of pragmatic engineering, von Neumann trained his predilection for abstraction upon the computer as a vehicle for computation and the computer as a metaphor for organization and self-control.

The choice of the term "automata" was itself revealing. The *Oxford English Dictionary* defines "automaton" as "1. Something which has the power of spontaneous motion or self-movement. 2. A living being viewed materially. . . . 4. A living being whose actions are purely involuntary or mechanical." Von Neumann undoubtedly intended each and every connotation as components of his project. While distancing the theory from robots, drones, wind-up toys, mechanical ducks, and Turkish chess-players, he started out by conflating self-motion and self-regulation in the earlier *kinematic* model of an automaton floating in a pond of prospective parts (1966, p. 82). Hence the earliest conceptualization is redolent of a certain nostalgia for boys and their toys.[43] The demands of abstraction rapidly transmuted "self-motion" into "self-regulation" and thence to information processing, with the model recast into the format of cellular automata. The "cells" were an abstract grid or lattice, with each cell assuming a finite set of "states" subject to rules written in terms of both own- and neighbor-states. The latter's resemblance to board games like Nim or checkers was salutary, in that the toys revealed their provenance in the earlier theory of games; but the language also simultaneously indicated that living creatures were to be seen as composed of purely mechanical subsystems.

The theory of automata was to claim as its subject any information-processing mechanism that exhibited self-regulation in interaction with the

[43] "At first he conceived of his device as three-dimensional and bought the largest box of Tinker Toys to be had. I recall with glee his putting together these pieces to build up his cells. He discussed his work with Bigelow and me, and we were able to indicate to him how the model could be achieved two-dimensionally. He thereupon gave his toys to Oskar Morgenstern's little boy Karl" (Goldstine, 1972, p. 278). A popular account of the structure and emergence of cellular automata out of something closer to toys can be found in Levy, 1992, chap. 2.

environment, and therefore resembled the structure and operations of a computer. But beyond that pedestrian exploration of parallels, von Neumann had a vision of a formal theory of ever widening ambit that would eventually establish thermodynamics as the preeminent basis of biological and social phenomena. The inspiration dated back to Szilard's 1929 paper, itself the result of discussions with von Neumann in 1921–22: thermodynamics would finally encompass thought. MAD would finally be tamed by von Neumann's various mathematical enthusiasms. This would begin with Shannon's use of the entropy-analogy in his theory of information; it would continue onward to incorporate Turing's universal machines to discuss the conditions for attaining the status of logical universality in the manipulation of information. Experience had shown that information processors could be constituted from widely varying substrata, from vacuum tubes to McCulloch-Pitts neurons to clanking analogue devices; it was the task of a theory of automata to ask, What were the necessary requisites, in the abstract sense, for self-regulation of an information processor? Conditional upon the answers to that question, the theory would then extend Turing's insights to inquire after the existence of a "universal" constructor. Biology and thermodynamics made a reappearance at this juncture, when the theory would ask, Under what conditions could the universal constructor reconstruct a copy of itself? The problems of self-reference identified by Gödel here "naturally" came to the fore. The logical prerequisites for self-replication that would resist entropic degradation in a process inherently temporal and fraught with randomness were thus brought within the ambit of the theory of automata, recapitulating a theme dating back to Szilard's 1929 paper. The introduction of time's arrow could actually serve to *dissolve* previous paradoxes of logic.[44] Von Neumann's vast ambition then hinted at a final apotheosis of the theory of automata, namely, research into the logical prerequisites for information processors to be capable of creation of successors logically more complicated than themselves. In searching for the conditions under which simple automata gave rise to increasingly complex automata, mathematicians would then finally have blazed the trail to a formalized logical theory of evolution.

Von Neumann's collaborators on automata theory promptly realized that there was some sort of connection between his later preoccupations

[44] "There is one important difference between ordinary logic and the automata which represent it. Time never occurs in logic, but every network or nervous system has a definite time lag between the input signal and the output response. . . . it prevents the occurrence of various kinds of more or less vicious circles (related to 'non-constructivity,' 'impredicativity' and the like) which represent a major class of dangers in modern logical systems" (von Neumann, 1987, p. 554). The structure of von Neumann's automata theory is reprised in Chapter 8.

and his earlier theory of games. One of the most perceptive, Arthur Burks, wrote:

> Automata theory seeks general principles of organization, structure, language, information and control. Many of these principles are applicable to both natural and artificial systems. . . . von Neumann's logical design of a self-reproducing cellular automaton provides a connecting link between natural organisms and digital computers. There is a striking analogy with the theory of games at this point. Economic systems are natural; games are artificial. The theory of games contains the mathematics common to both economic systems and games, just as automata theory contains the mathematics common to both natural and artificial automata. (in von Neumann, 1966, p. 21)

Although this quotation certainly displays the characteristic cyborg willingness to scramble promiscuously the Natural and the Social, it does not altogether succeed in getting at the crux of the analogy. As we have noted, *TGEB* did not manage to explicate or encompass much of any recognizable "economy"; and it is not at all clear that von Neumann personally got very far in actually formalizing any living biological system before his death. Yet, the attraction of games for the subsequent pioneers of automata theory was undeniable. Claude Shannon, Alan Turing, Warren McCulloch, Ross Ashby, Max Newman, and a whole host of others wrote about machines playing games at some time during their careers. It was almost as though "games" (although perhaps not precisely von Neumann's own theory of games) were one of the cleanest instantiations of the type of activity an abstract automata might engage in. Indeed, given the unanalyzed prejudice that games were what automata were good at, the reactions of some of the first-generation cyborg theorists to von Neumann's game theory is one of the better barometers of its perceived flaws and drawbacks in the 1940s.

A very good example of this phenomenon is one of the key British facilitators of computer development, Max Newman.[45] In 1935 he was the Cambridge mathematician whose lectures on mathematical logic and Gödel's theorems inspired Alan Turing's classic 1936 paper on computable numbers (Hodges, 1983, pp. 90–95) and provided the connections to Princeton, which nurtured the earliest British-American efforts on computation. In 1942 he was also at Bletchley Park, organizing the effort to

[45] Maxwell Newman, FRS (1897–1984): originally born "Neumann" (though apparently no relation to John von Neumann); degrees in mathematics, University of Cambridge, 1921, and University of Vienna, 1923; fellow of St. Johns, Cambridge, 1923–45; university lecturer, 1927–45; visiting fellow, Princeton, 1928–29, 1937–38; Bletchley Park Code and Cipher School, 1942–45; professor of mathematics, University of Manchester, 1945–64. For a biography, see Hilton, 1986. For his work on early British computers, see Bowden, 1953.

build the electronic "Colossus" to crack the German Enigma code. After the war he accepted a chair in pure mathematics at Manchester University and, like von Neumann, promptly proceeded to renounce any pretensions to purity by obtaining a grant from the Royal Society to build a computer at Manchester (Hodges, 1983, pp. 340–43; Bowden, 1953). In some fascinating correspondence with von Neumann in early 1946, he moved effortlessly between discussions of how computers could be used as heuristic devices – presciently predicting that they could be used for "testing out, (say), the four-colour problem or various theorems on lattices, groups, etc. for the first few values of *n*" – to problems of computer architecture to criticisms of the theory of games.[46] In an elaborate set of notes, Newman expressed his discomfort with the interpretation of the stable set as a standard of behavior. "Your definitions perform the surprising feat, – I should have thought impossible –, of arriving at definite solutions of the bargaining problem by pure analysis of the profit motive, without getting involved in 'degrees of intelligence,' or other arbitrary 'psychological' assumptions. Isn't it throwing away a great deal of the point and subtlety of this analysis to allow taboos and such things in after all?" He also proposed an amendment to the *TGEB* definition of domination, measuring the extent of the domination by the attractiveness of the change to the least attracted member of the coalition. Von Neumann responded on March 19, 1946:

> I realize, that the emergence of "discriminatory" solutions may cause one to worry and to hesitate, particularly if there are very many of them and in a very amorphous complexity. Yet, I cannot quite see, that one should make up one's mind already now to reject them. After all, they may correspond in reality to stable forms of social organization. Besides, even the "main" solutions include elements of arbitrary discrimination, and I suspect that the hard-and-fast distinction which exists in the 3-person case . . . will get less and less precise as one goes through increasing numbers of participants. . . . I admit that this is vague, especially since it is not proven that even with my "wide" definition all games have solutions.

Newman answered in a handwritten missive attached to a letter dated April 14:

> I believe it would be better to discuss definitions if their *motivations* were made more explicit, in the form of assumptions about the behavior of players in bargaining. . . . There are two points in particular on which I feel in need of further prudence. (1) To what extent is enlightened self-interest supposed to be the sole governing motive of the players? . . . It seems to me that pure self-interest of single players, however

[46] This correspondence is in box 5, folder 15, VNLC.

> sophisticated, cannot lead to the idea of a solution as a "convention" of behavior, but that class-interest, or "solidarity" of set S, might do so. . . . (2) I cannot bridge the gap between the theory of the valuation function v developed in Chs. II & III . . . and the idea of an *imputation* as the final end of bargaining. . . . In fact in the general formulation of 30.1.1 the *play* of the game seems rather to have faded out, and I have the impression that some other more subtle mechanism for determining an actual division of profits is envisaged. If that is so, what is the meaning of v(S)?

Newman was trying to nudge von Neumann in the direction of clarification of motives, interpretations, and the role of play as a performative activity – perhaps some version of what was more commonly known at that time as "psychology"; but these were exactly the directions that he had coolly resisted ever since the early work on quantum mechanics, and topics he would continue to avoid in his later writings. In their place, von Neumann nurtured a vision of explanation that was much more willfully mechanical and hierarchical. In this "teleological" world, simple microlevel, rule-governed structures would interact in mechanical, possibly random, manners; and out of their interaction would arise some higher-level regularities that would generate behaviors more complex than anything which could be explained at the level of micro entities. The complexity of these emergent macro properties would be formally characterized by their information-processing capacities relative to their environments. The higher-level structures would be thought of as "organisms," which in turn would interact with each other to produce even higher-level structures called "organizations." This, I believe, helps explain why von Neumann continued to favor the solution concept of the "stable set" long after it had lost its allure for the rest of the game theory community. The mathematics of cellular automata more aptly served as paradigmatic for the types of explanation von Neumann sought in his travels amongst the engineers, neurophysiologists, biologists, cyberneticians, operations theorists (and, not to be forgotten, military patrons). Rival attempts to ground human purposive behavior in more conventional psychology, be they behaviorist or Freudian, he treated with ill-concealed disdain.[47]

The immediate postwar period was one where the first generation of cyborg scientists struggled with what would be regarded as legitimate approaches to psychology; and these often were brought out into the

[47] One can observe this in his response to a request from Robert Clark, May 12, 1955: "Do you believe – or have you ever considered – that the theory of games and economic behavior may be applied to the long-continued interrelations of small groups of human beings in family life and psycho-therapy?" In a brusque letter dated May 16, von Neumann replied, "I am too ignorant of the relevant circumstances to have an opinion." Box 3, folder 2, VNLC.

open in their opinions about game theory. Norbert Wiener, who was avid in his enthusiasm for direct modeling of the brain, nevertheless was desirous of quarantining society off from mechanical modeling (Heims, 1980). This manifested itself, amongst other things, as a hostility toward von Neumann's brand of game theory.[48] Alan Turing, on the other hand, became interested in flattening all hierarchical distinctions between the individual and society, and therefore went on a behaviorist crusade in the late 1940s to shock all and sundry with the thesis that machines were "intelligent" (Hodges, 1983, pp. 424–25). In order to argue that machines could think, he found himself reprising his own argument for the universality of Turing machines by insisting that all thinking was indistinguishable from computation (Shanker, 1995). Because his "psychology" remained relatively unsophisticated, he felt himself drawn to game theory around 1948 (Hodges, 1983, p. 373; Alan Turing in Bowden, 1953; Turing, 1992), that is, after it has been more or less abandoned by von Neumann.

It is important to differentiate sharply von Neumann's later position from all of these other permutations, especially because they become perilously more entangled with the subsequent rise of "cognitive science," "artificial intelligence," and other cyborg sciences. For von Neumann, the theory of automata constituted an abstract general theory of information processing and the evolution of complexity of organization; however, it was definitely *not* a surrogate or replacement for the field called psychology. The more he learned about the brain, the more he became convinced that the computer did not adequately capture either its structure or its mode of function. As Aspray (1990, pp. 208–9) points out, his final undelivered Silliman lectures, published as *The Computer and the Brain* (1958), surveyed the various ways in which the von Neumann architecture was dissimilar to the brain. The objective was not to go in search of a better psychology but rather that "a deeper mathematical study of the nervous system . . . will affect our understanding of the mathematics itself that are involved" (p. 2). This also explains his valedictory statement on the prospects for a science of economics:[49]

[48] "Even in the case of two players, the theory is complicated, although it often leads to the choice of a definite line of play. In many cases, however, where there are three players, and in the overwhelming majority of cases, when the number of players is large, the result is one of extreme indeterminacy and instability. The individual players are compelled by their own cupidity to form coalitions; but these coalitions do not generally establish themselves in any single, determinate way, and usually terminate in a welter of betrayal, turncoatism and deception, which is only too true a picture of the higher business life" (Wiener, 1961, p. 159). Wiener's attitudes toward von Neumann's theory are outlined in Chapter 2.

[49] This quotation is from an address to the National Planning Association, December 12, 1955; it can be found in von Neumann, 1961–63, 6:100. It is interesting to note that von

> There have been developed, especially in the last decade, theories of decision-making – the first step in its mechanization. However, the indications are that in this area, the best that mechanization will do for a long time is to supply mechanical aids for decision-making while the process itself must remain human. The human intellect has many qualities for which no automatic approximation exists. The kind of logic involved, usually described by the word "intuitive," is such that we do not even have a decent description of it. The best we can do is divide all processes into those things which can be better done by machines and those which can be better done by humans and then invent methods by which to pursue the two.

Themes out of Automata Theory

It would be fair to say that the connections between von Neumann's second and third phases of thought were never spelled out in his lifetime, particularly when it came to economics. We have already mentioned the unfinished state of his theory of automata; and, further, there was his uncharacteristic practice of having developed automata theory in isolation from any collaborative enterprise (Goldstine, 1972, p. 285; Aspray, 1990, p. 317n67). While the exercise of outlining a Neumannesque "economics of automata" will of necessity sport a speculative air, it is nonetheless crucial for our subsequent narrative, as well as an understanding of one of the most suppressed aspects of von Neumann's history, namely, his hostility to the game-theoretic equilibrium concept proposed by John Nash in 1950.[50]

I believe the best way of making sense of the evidence from von Neumann's last decade is to regard game theory as being progressively displaced as a general theory of rationality by the theory of automata. Again, this did not dictate a complete repudiation of a subfield of mathematics called "game theory," nor did it preclude a judicious word here and there to the military authorities about the benefits of encouraging its development. The clear preeminence of Princeton as a breeding ground for game theorists would itself belie the possibility of any open renunciation. Nevertheless, the widespread dissatisfaction with the 1944 book cited earlier most likely led to the conviction that the original angle of approach to rationality was faulty, encouraging too many distracting detours into endless thickets of psychology, a deviation from which no

Neumann reiterates his complaint from the 1930s as to the greatest obstacle to a science of economics: "I think it is the lack of quite sharply defined concepts that the main difficulty lies, and not in any intrinsic difference between the fields of economics and other sciences" (p. 101).

[50] This hostility to the Nash program in game theory is documented in Chapter 6.

mathematician ever emerged rigorously unscathed. How much more in character for von Neumann to cut any nettlesome ties to the muddled preoccupations of the economists, circumvent the repeated objections of his cyborg colleagues, and ratchet the level of abstraction up a notch to a general theory of automata: a mathematics that (*pace* Burks) truly applied to the Natural and the Social, the live and the lifelike, uniformly and impartially. This theory would apply indifferently and without prejudice to molecules, brains, computers, and organizations. Amidst this generality, the architecture of computers would stand as paradigmatic and dominate the inquiry, if only because, "of all automata of high complexity, computing machines are the ones we have the best chance of understanding. In the case of computing machines the complications can be very high, and yet they pertain to an object which is primarily mathematical and which we understand better than we understand most natural objects" (von Neumann, 1966, p. 32). Of course, in accessing the argument that we can usually better understand something that we ourselves have constructed, the distinction between artifact and naturally occurring entities became blurred in passing. Machines for the first time become truly generic entities.

In the transition between game-theoretic rationality and automata, some aspects of the formalism are preserved, whereas others are amended or summarily abandoned. Most obviously, where game theory tended to suppress formal treatment of communication and the role of information, the theory of automata elevates them to pride of place. The very idea of a roster of possible moves in the game is transmuted into an enumeration of machine states. Strategies, a foundational concept of game theory, now become internally stored programs. However, under the tutelage of Gödel and Turing, exhaustive prior enumeration of strategies, as well as the idea of a generic defensive minimax, is abandoned as implausible: "in no practical way can we imagine an automaton which is really reliable. If you axiomatize an automaton by telling exactly what it will do in every completely defined situation you are missing an important part of the problem. . . . Rather than precautions against failure, they are arrangements by which it is attempted to achieve a state where at least a majority of all failures will not be lethal" (1966, pp. 57–58). Randomness has been displaced from a condition of rational play to an inherent condition of the incompleteness of strategies and the imperfect reproduction of automata. The rather drastic implication of this insight is that the gyrocompass of von Neumann's general theory of games, the reduction of all games to the 2P0Σ game, must be abandoned. This constitutes the final sloughing off of the original ambitions of the Hilbert program.

The very notion of a "solution" now lurches in the direction of process rather than outcome, in keeping with the new stress on evolutionary language. The formalization of successful process is equated with Turing computability, which, as we have seen, underlies the entire notion of hierarchies of automata. Curiously enough, the previous "normal form" representation of a game, if indeed still at all legitimate, would tend toward a summary interpretation of the parallel processing of programs; but von Neumann had decided early on that parallel processing architectures were unwieldy and had opted definitively for serial processing. Consequently, the graph-theoretic extensive form of the game, originally treated as an unloved ugly duckling in the 1944 book (von Neumann & Morgenstern, 1964, pp. 77–78), now returns in triumph as the programmer's flow diagram and the finite automaton state graph.[51] Indeed, such a full specification of the game-automaton becomes de rigueur because of the paramount importance of informational aspects: specificity trumps elegance; impurity comprehensively triumphs over purity.

The notions of "strategy" and "struggle" also undergo some very interesting transubstantiations in this third phase. In *TGEB*, the problem of one player "finding out" another's strategy is a major consideration in the motivation of the solution concept: opponents must be assumed to know the "theory of games" as well as the original player, so that ambage and deceit become a reason why one should ignore the peccadilloes and idiosyncratic psychology of the opponent (von Neumann, 1964, p. 148). Now, however, Turing's demonstration of the existence of a Universal Turing machine that can simulate any other Turing machine changes the very meaning of "finding out" the other's program. The UTM serves as a base-line characterization of the rationality of the opponent; but the "halting problem" (Arbib, 1969, p. 149; Cutland, 1980, p. 102) suggests that there is no effective procedure for computing a general a priori answer to the question: Will the opponent succeed? Whereas randomization was introduced into the early theory of games in order to render the player opaque to the opponent and guarantee the existence of a minimax, randomization is now rather treated as an *oracle* (Cutland, 1980, p. 167) to augment the computational capacity of the Turing machine. Finally, in the 1944 book the "dummy player" was introduced in order to encompass more complex games into the zero-sum rubric; that analytic move is rendered no longer necessary. Undaunted,

[51] See Arbib, 1969, pp. 101–5. The close relationship between the extensive-form game tree, the flow chart of a computer program, the neutron-scattering simulation of the detonation of the atomic bomb, and the organization chart of a military hierarchy is discussed under the rubric of "operations research" in the next chapter.

the "dummy" still lives on in the theory of automata, only now as one fundamental building block from which all rationality is built up. Machines augment human rationality when they are accorded commensurate status with human ratiocination; they are the "dummies" which herald the next quantum leap in hierarchical complexity through their prosthetic interaction with humans.

Game-theoretic notions of struggle and interdependence soldier on in the theory of automata, but now they are rendered subordinate to the larger project of a computational approach to interaction. For instance, von Neumann does not treat individual automata as self-consistent monads; he is willing to entertain the idea that there could be "antagonism between parts."

> There are some indications (I do not wish to go into them in detail now) that some parts of the organism can act antagonistically to each other, and in evolution it sometimes has more the character of a hostile invasion of one region by another than of evolution proper. . . . It has already happened (and it is, of course, just by the introduction of automata into mathematics that it begins to happen) that you are no longer thinking about the subject, but thinking about an automaton which would handle the subject. It has already happened in the introduction of mass production into industry that you are no longer producing the product, but you are producing something which will produce the product. The cut is, at present, never quite sharp, and we still maintain some kind of relation with the ultimate thing we want. Probably the relationship is getting looser. It is not unlikely that if you had to build an automaton now you would plan the automaton, not directly, but on some general principles which concern it, plus a machine which could put these into effect, and will construct the ultimate automaton and do it in a way that you yourself don't know any more what the automaton will be. . . . I think that if the primary automaton functions in parallel, if it has various parts which may have to act simultaneously and independently on separate features, you may even get symptoms of conflict. (in Jeffress, 1951, pp. 109–10)

Here, in compact form, we observe the apotheosis of strategic reasoning as the decentering of the supposed integrity of the rational actor, all expressed in the mixed idiom of military and industrial command and control. Here the erstwhile lockstep regimentation of means to ends is "probably getting looser," while the automaton issues a declaration of independence from the creator, and automata themselves engage in conflict as well as cooperative coalitions. MAD becomes MOD as we face the brave new world of apocalyptic conflict. It is difficult not to detect the Cold War in all this agonistic closed-world discourse. Yet, far from evoking the Western nightmare of the Frankenstein scenario, von Neumann proposed in its place a reverie of coevolution of mechanisms and objectives, a conception of competition far removed from the facile optimizations of the Lausanne school or, indeed, later orthodox game theorists. Such dreams would, in the future, become the calling card of the cyborg researcher.

In the final analysis, thermodynamics and the Demons encountered in Chapter 2, and not the theory of games, reasserted its centrality in the theory of automata. The idea of entropic degradation was paralleled in the idea of the tendency toward degeneration in the self-reproduction of automata. Von Neumann thought he would find something like a "Reynolds number" in this theory; a parameter for "complexity" possessing a critical value below which the degeneration proceeds unchecked, but above which the system would bootstrap itself into a plane where automata could produce more complex offspring. The key to offsetting entropic degradation was the processing of *information*, the theme pioneered by Szilard in 1929. Sometimes von Neumann regarded the computer itself as a heat engine, with speed and capacity of components the analogue to temperature differentials between sources and sinks (von Neumann, 1987, p. 387). If there was something specifically "economic" about all this, it was because of the "economic" themes previously built into the laws of thermodynamics and not due to any brand of economic theory of which he had been made aware over the course of his career.[52] Only toward the end of his life did von Neumann's idea of economics as the science of emergent complexity through the hierarchical development of social organizations begin to find its voice. Yet the cancer spreading through von Neumann's bones stilled the voice prematurely.

There is evidence that von Neumann sought to interest a few selected economists in these computational themes in the third phase of his career,

[52] This explains one of the more bizarre incidents in von Neumann's later career, his attempt to interest the Cowles Commission in the idiosyncratic work of Andrew Pikler, someone who attempted to import thermodynamic formalisms wholesale into economics. See von Neumann to Jacob Marschak, April 25, 1947, box 94, JMLA:

"I want to tell you that I have talked to him about his ideas in considerable detail and that I have come to the conclusion that the model constructions with which he likes to illustrate them are not very essential but that there is probably a very healthy nucleus in his way of using statistical ensembles and transition probabilities; i.e., in applying to the fluctuations of money those concepts which have turned out to be appropriate to gas and chemical kinetics. I think that his work deserves interest and encouragement and that he could get it in the milieu of the CC much better than anywhere else, even if his interests form somewhat of an angle with the CC's main direction of approach." It goes without saying the Cowles economists did not welcome Pikler with open arms. For more on Pikler's work, see Mirowski, 1989a, pp. 389–90; Pikler, 1951, 1954.

evidence distributed throughout our subsequent chapters. The roll call of those touched by his enthusiasms is really quite astounding, including Tjalling Koopmans, Friedrich von Hayek, Jacob Marschak, Gerard Debreu, Merrill Flood, David Novick, Alain Lewis, Michael Rabin, Herbert Simon, Robert Aumann, and John Nash; and those are merely the more significant protagonists in our saga. The unintended consequences of von Neumann's ambitions for economics are to be found in his interventions in the history of the Cowles Commission. To give the reader a foretaste of what will be understood in retrospect as one of the most consequential encounters in the history of twentieth-century American economics, we reproduce here an excerpt from a letter from Tjalling Koopmans of the Cowles Commission to von Neumann dated January 18, 1952.[53] Enclosed with the letter is a draft of Gerard Debreu's "Saddle Point Existence Theorems," which would become one of the landmarks of the postwar Walrasian economic program. There is no evidence that von Neumann evinced much interest in it. Instead, Koopmans refers to "the question you asked me in Boston, as to the possibilities of large scale computation involving economic data and designed to answer questions of economic policy." Koopmans, just coming down off a chill at Cowles toward the whole idea of a successful program of econometric empiricism, was none too encouraging about this prospect. But he did respond more warmly to another hint of von Neumann:

> I should like to mention another way in which the study of computation technique is relevant to economics. The design and method of operation of computation equipment can be regarded as problems in the theory of organization which in some respects are greatly simplified. Since the parts of the machine (unlike, as you pointed out, the cells of the human organism) have no separate objectives or volitions, all complications arising from this circumstance in actual organizations are absent. It is therefore possible to concentrate on the more mechanical elements of organization, such as transmission times, capacities for memory or transmission of arithmetical operations, etc. If with the help of the computation analogy the more mechanical aspects of organization can be studied first, then we will be ready to take on the more difficult aspects associated with the diversity of objectives.

This passage, only slightly garbled, indeed contains the germ of that vision of economics von Neumann held in the third phase of his career. Koopmans, as was his wont, was more concerned to reserve a place for

[53] Box 5, folder 4, VNLC. The influential relationship of von Neumann to Koopmans and Cowles is discussed in detail in Chapter 5, as is the larger context of this letter.

"volitions" as he understood them; but he nevertheless appreciated the significance of the computer as an information processor for a reconceptualization of the very nature of the economy. The extent to which the neoclassical economists at Cowles were willing to entertain this vision, much less take it on board as the Magna Charta of cyborg science in economics, is the subject of the remainder of this book.

4 The Military, the Scientists, and the Revised Rules of the Game

It takes a war to make an industrialist out of a physicist.

Merle Tuve[1]

WHAT DID YOU DO IN THE WAR, DADDY?

It is quite the spectacle to observe how postwar economists, those hard-boiled beady-eyed realists, so eager to unmask the hidden self-interest lurking behind every noble sentiment, undergo a miraculous transubstantiation when the topic turns to their own motivations. When summoned to reflect on their personal successes, they regularly cite such lofty goals as the alleviation of pain, the augmentation of the general welfare, the abolition of injustice, and the advancement of human understanding (Szenberg, 1992). It is on its face a singularly amazing accomplishment, as if some new Augustine had unearthed the philosopher's stone capable of conjuring agape out of avarice, leaving him alone zaddick in a non-zero-sum world. It would be too much of a distraction from our present itinerary to inquire exactly how the prestidigitation is accomplished in every case, or indeed to even ask whether the individuals in question truly believe it deep down in the recesses of their psyches; but, nevertheless, it will serve to explain one very striking lacuna in the modern treatment of the history of economics. No one seems to want to ask the quintessential economic question about the modern economics profession – Who pays? Qui bono?

In this respect the historians of the physical sciences have been simultaneously more bold and more incisive. There now exists a fairly large literature tracing the evolution of the funding of the natural sciences in the United States over the course of the twentieth

[1] Quoted in Reingold, 1991, p. 300.

century.[2] In a nutshell, it suggests that the pursuit of American science in the nineteenth century was very uneven and lacking in any consistent or sustained means of support. Whereas Europeans had patterned their treatment of indigenous scientists as akin to a hereditary aristocracy or meritocratic bureaucrats deserving of subsidy, Americans had nothing but suspicion and skepticism about those harboring intellectual pretensions. This was manifest in the political vicissitudes of the few governmental agencies that sought to support research, such as the U.S. Geological Survey, the Bureau of Entomology, and the Weather Service. The situation began to improve incrementally around the turn of the century, with the institution of some industrial research laboratories in various large corporations formed in the first great wave of mergers and industrial concentration just prior to that period. The primary function of these laboratories, however, was not so much the uncovering of new products or processes for their patrons as it was a strategic device to defeat competitors through patent coverage and ward off antitrust actions by the government (Reich, 1985; Mowery & Rosenberg, 1998; Mirowski & Sent, 2001). "Pure science" was honored more in the breach than in the laboratory; science patronage was dominated by strategic considerations in America from square one. Furthermore, any industrial connection to science in the early twentieth century was intimately bound up with a commitment to "scientific management" and Taylorism. Indeed, one could make a case that one major function of American science in the early twentieth-century corporate context was to provide the tools, the instrumentalities, and a rationale for the burgeoning mass of middle managers which had been one of the more troublesome unintended consequences of the great concentration of corporate wealth and its induced innovation of alternative corporate governance structures in the American context. Science (or, more correctly, engineering) provided the basis for many of the new industries in electrical, chemical, metal-working, and automotive fields; and now science was importuned for solutions to the crisis of control of far-flung factories and offices (Beniger, 1986).

While the captains of industry had ventured to provide employment for a few scientists in their research laboratories, they did not initially lavish their attentions to the production of improved science in American society.

[2] A survey of the highlights of this literature would include Reingold, 1991, 1995; Kevles, 1995; Forman, 1987; Kay, 1993, 1998; Galison & Hevly, 1992; Reich, 1985; Dennis, 1994; Michael Dennis in Krige & Pestre, 1997; Kragh, 1999. A standard summary from the viewpoint of science policy is Morin, 1993. The work of Donna Haraway and Andy Pickering, cited in Chapter 1, first linked this change to the rise of the cyborg sciences.

Universities were growing and proliferating in this period, but by and large they did not treat research as their primary responsibility, nor could they take it upon themselves to fund such expensive activities as scientific laboratories. In many fields, the only dependable route to advanced study and graduate degrees passed through the new German universities, which were structured to encourage research through academic competition and government subvention. By contrast, grants to researchers in the United States, such as they were, had been regarded as something akin to poor relief, parsimonious temporary expedients to tide over underprivileged academics. The next phase in the transformation of American science was initially a function of the turn-of-the-century concentration of corporate wealth, but grew out of an altogether different unintended consequence of the vanity of the robber barons.

For reasons of both tax avoidance and public perception, industrialists such as Andrew Carnegie and John D. Rockefeller were persuaded to donate some moiety of their corporate windfalls to endow large philanthropic organizations such as the Carnegie Institution (1911) and the Rockefeller Foundation (1913). These nominally eleemosynary organizations were founded and run by businessmen who had pioneered the new business methods and managerial practices of the megacorporations that had been the engines of their wealth; for reasons of expedience as well as experience, they sought to import these hierarchical managerial practices into their funding of academic science. Instead of administering small grants and fellowships, which were deemed too insignificant and too uneven in productivity to warrant the attention of program officers, these foundations decided to innovate new forms of patronage, and to subject science to some of the rationalization devices that had revamped the American corporation (Kohler, 1991). One program was aimed at "institution building," and the biggest beneficiaries were Princeton, Stanford, MIT, Cal Tech, Chicago, and Harvard. A few favored private (but not state) universities were encouraged to nurture the role of the academic entrepreneur, mimicking the captains of industry who had provided their seed capital. Another innovation was the targeting of specific disciplines or research areas that were deemed by the foundations as ripe for development. Grants became patterned on the business instrument of *contracts* for specified projects instead of the previous handouts patterned upon poor relief. Peer review was reconstituted as special scientific review panels, based on the precedent of corporate boards of directors (and sometimes just as convenient a figleaf for the operating officers, as some suggested). Other innovations in the planning and funding of science involved the inevitable bureaucratic complement to the various projects and initiatives, such as application forms, progress reports, and other paper trappings of impersonal accountability for funds provided.

Historians now acknowledge that a few philanthropic organizations like Rockefeller and Carnegie pioneered the modern American system of science planning, although this is not the way it is conventionally portrayed in popularizations, which imagine science as a kind of natural Brownian motion of individual random genius. All of the major figures of the next phase of our saga, namely, the military mobilization of science in World War II, served their apprenticeships in this quasi-private quasi-public netherworld of science policy in the runup to the Great Instauration of 1941. The important thing is to view the massive military reorganization of American science in midcentury as essentially *continuous* with the prior corporatist innovations pioneered at the beginning of the century. The leaders of this movement themselves managed to combine corporate, foundation, and military ties. Vannevar Bush, after helping found the Raytheon Corporation, teaching electrical engineering at MIT, developing the early analog computer, and doing cryptographic research for the Navy (Burke, 1994), served as president of the Carnegie Institution, all the while supervising the nation's research efforts in wartime as czar of the Office of Scientific Research and Development (OSRD) (Zachary, 1997; Bush, 1970). James Conant "lived through a low-tech preview of Los Alamos" by engaging in poison gas research for the government in World War I, only to attempt to run a private chemical firm while employed as a Harvard junior faculty member in the interwar period; catapulted to fame as president of Harvard in 1933, he served under Bush as head of the National Defense Research Committee (NDRC) during World War II, where among other responsibilities he urged "operations research" upon the armed services, and was the primary advocate for the Manhattan Project (Hershberg, 1993). In 1947 Conant created the course "Nat Sci 4" at Harvard to teach the uninitiated the "tactics and strategy of science," which had been honed in wartime; one of his first assistants in the course was Thomas Kuhn (Hershberg, 1993, p. 409; Fuller, 2000). Hendrik Bode, the electrical engineer at Bell Labs who served as Claude Shannon's mentor, and the designer of the antiaircraft device that beat out the version by Norbert Wiener, also pioneered concepts of systems engineering and "technical integration" to better coordinate research at Bell with the economic needs of Western Electric (Fortun & Schweber, 1993, p. 599). John von Neumann, as we have documented, had become the consummate insider, moving effortlessly between foundations, corporations, and military research units from the 1930s onward. Warren Weaver, Grandmaster Cyborg extraordinnaire, someone we have already encountered repeatedly in the narrative, abandoned an academic career as a mathematics professor at Cal Tech and Wisconsin to become natural sciences program officer in 1932 and later vice-president at Rockefeller, member of many corporate boards, and subsequently director of a major portion of the mathematical military research effort during World War II.

The superficial lesson to be drawn from this dense web of interlocking directorates is that, at least in America, postwar science policy was itself developed in close conjunction with both military and corporate imperatives; and, further, nowhere was this more apparent than in the case of the cyborg sciences. A more profound appreciation might suggest that impressions of hierarchical planned organization of research and the cognitive structure of inquiry itself began to bleed and intermingle one with another, with untold consequences for the subsequent development of the sciences.

The primary thesis of this chapter is that the military usurpation of science funding in America in World War II; the rise of *theories* of science planning, organization, and policy; the rise of the cyborg sciences; and the rise to dominance of neoclassical economic theory within the American context are all different facets of the same complex phenomenon, and they need to be understood from within this larger perspective (Mirowski & Sent, 2001). It is now commonly recognized that, although neoclassical economic theory was inspired by developments in physics in the later nineteenth century (Mirowski, 1989a), only from the 1940s onward has American economics assumed its characteristic modern format and scientific pretensions, which have persisted more or less intact down to the present. The American orthodoxy became more formal, more abstract, more mathematical, more fascinated with issues of algorithmic rationality and statistical inference, and less concerned with the fine points of theories of collective action or institutional specificity.[3] We maintain that this turn was not foreordained but instead was part and parcel of the other trends listed here. In brief, economists themselves did not actively assist in the creation of the cyborg sciences, but rather were drafted into the ranks of the cyborg sciences (with all that implies about voluntary acquiescence) as a consequence of the overall wartime reorganization of science in America. Furthermore, it was through this extremely oblique route that physicists came to play an indispensable role in the conceptualization and stabilization of neoclassical economics for the *second* time in its curious history (the first being the aforementioned "marginalist revolution"). Therefore, both science studies scholars and historians of economics, should they be concerned with how we got to the "end of history" at the end of the millennium, will neglect the history of military mandates, changing science policy, as well as the novel field of "Operations Research"

[3] Various versions of this characterization can be found in Niehans, 1990; Ingrao & Israel, 1990; Morgan & Rutherford, 1998; Robert Solow & David Kreps in Bender & Schorske, 1997; Yonay, 1998.

at their peril. In World War II, physicists and their allies, recruited to help deploy new weapons of mass destruction, participated in the *reorganization* of science patronage and management by conceiving a novel *theory of organization* inspired by physics and (latterly) the theory of the computer, which was in turn subsequently imperfectly absorbed and revised by a key subset of economists into a variant of the man-machine cyborg celebrating *market organization* within the neoclassical tradition. Problems posed by military management of science were reprocessed into a mechanistic science of management, and then once again fused into an unstable amalgam with the previous neoclassical tradition.

Because many will disparage or discount the importance of the military connection for everything that follows, I should make an attempt to clarify the present approach. There does exist a rather debased genre of historical narrative which uncovers some aspect of military patronage in science, and then flaunts it as prima facie evidence of the dubious or untrustworthy character of the research performed. This was a rather common trope amongst those who bewailed the existence of the "military-industrial complex," although it is now somewhat less commonplace in the world after the fall of the Berlin Wall.[4] The juxtaposition of "tainted science" with the supposedly virtuous variety – itself, significantly, a refrain pioneered by Bush, Weaver, and their compatriots – is emphatically *not* the intention in which the military patrimony of the cyborg sciences is to be invoked here. To make my position painfully precise: there is no *necessary* connection between military funding (or, indeed, any funding) and any notion of "tainted knowledge." Rather, my objective is to highlight the specific manner in which shifts in science funding and organization fostered an intellectual sea change in the way in which issues of communication, command, control, and information – the military mantra of C^3I – came to dominate the continuum of scientific thought in the postwar period. In this, we reiterate the insight of Peter Buck that World War II "gave many academics their first taste of working within rather than merely consulting for large operating agencies. The realities of bureaucratic life and politics taught them that power and influence went together with the right to define problems for others to study. That insight shaped one of the most familiar features of the postwar social scientific landscape . . . to ground their empirical and applied research on basic theoretical principles" (in Smith, 1985, p. 205). And these principles were all

[4] Examples might be found in Gray, 1997; Melman, 1971, 1974; Noble, 1979; Everett Mendelsohn in Krige & Pestre, 1997. Pickering 1997 is a warning against this tendency. For those concerned about a suspicion of paranoia in the ascription of efficacy to the military in its pervasive role behind the scenes, we consider this issue once more in Chapter 6.

the more effective if they conformed to modern high-tech mathematical models of randomness and organization, models largely pioneered and promulgated by John von Neumann.

One of the more curious aspects of this Great Transformation is the fact that, however much one might regard such problems of communication and control in hierarchical organizations as somehow intrinsically economic and social, given their origins in the business sector, it was not the economists but instead a curious motley of physical scientists and engineers who made the breakthrough investigations in the novel approaches to organization and initially staked out this area as their own. Chapter 2 has described how researchers in the areas of thermodynamics and mathematical logic, people such as Szilard, von Neumann, Wiener, Turing, and Shannon, had begun to converge on ideas of information, memory, feedback, and computation by the 1930s. These ideas found their physical culmination in the digital electronic computer, as much a product of wartime hothouse development as the atomic bomb or radar. Many of these scientists were subsequently called upon in wartime to assist in taming problems of military intelligence gathering and strategic control, to bring about the rationalization of what had become an unwieldy sprawling hydra of military initiatives and technological innovations. The successes of what soon came to be called "operations research," "systems analysis," and a host of lesser neologisms was just as much a function of these newer developments in the physical sciences as was the computer; and the two were often invoked in the same breath as mutually definitive.

In order to mobilize the sciences adequately for the war effort, the military with the assistance of Bush, Conant, Weaver, von Neumann, and some lesser luminaries like P. M. S. Blackett, Philip Morse, Karl Compton, Lee DuBridge, Franklin Collbohm, John Williams, and Ivan Getting, acted with consummate coordination to revamp science planning, funding, and organization. Almost contemporaneously, the military brought in a number of civilian scientists to deploy the scientific method in order to streamline and rationalize communications, control, intelligence, and operations at a level unprecedented in military history. After the war, instead of dissolving and dismembering the jerry-built wartime structures, a half decade of wrangling over the political shape of a civilian National Science Foundation allowed the military to consolidate its control over postwar science funding and organization, as the cyborg sciences contemporaneously consolidated their positions within the academy. But did the military dictate the outlines of the postwar Big Science model, or did the scientists discipline the postwar military in their own image? Undoubtedly, it is better to think of scientists and the military as mutually constituting each other through bilateral initiatives (Pickering, 1995a);

but before we run the risk of becoming mired in blowsy platitudes of everything depending on everything else, it would be prudent to identify some actors who did *not* participate in the initial constitution of the cyborg sciences. The most notable odd man out in these wartime proceedings was the American economist, not in the sense of some blanket exclusion from the war effort, but rather in his exclusion from active participation in the reorganization of science planning as well as his initial isolation from the developments of the sciences of information and control.

In this way the question, Who pays? becomes relevant to the present narrative. The Rockefeller Foundation had earlier sought to reorient the economics profession toward what it deemed was a more "scientific" direction in the 1920s, but that initiative had been abandoned as ineffective (Craver, 1986b; Kay, 1997b). As the cyborg scientists were busily rationalizing the prosecution of wars hot and cold, various and sundry economists were recruited to the war effort, and found themselves in the uncomfortable position of being tutored by natural scientists in the mysteries of strategy, information processing, stochastic search, and the economics of control. Some turned out to be quick understudies, whereas others tended to recoil from some of the newer doctrines. For their own part, many cyborg scientists conceived of a disdain for the achievements of the social sciences in this interval, as documented in the preceding chapters. After the war, the prolongation of the military funding regime thus tended to skew sharply the distribution of those economists chosen to participate in the brave new world of Big Science, especially in the direction of those willing to make a separate peace with operations research (OR). Enthusiasts for the brave new world aborning during World War II, such as Paul Samuelson – "It has been said that the last war was the chemist's war and that this one is the physicist's. It might equally be said that this is an economist's war" (Samuelson, 1944a, p. 298) – quickly learned to bite their tongue in the immediate postwar period. Disputes over the legitimacy of military organization of science also threatened to have pervasive consequences for the social sciences. The exigencies of Cold War military classification, combined with a certain reticence in admitting the extent to which the topics explored were dictated by military needs, have served to stifle the discussion of this class of causes of the postwar stabilization of the neoclassical orthodoxy in America until now.[5]

[5] A lesson that has been brought home repeatedly to the present author in the decade in which he has researched this volume in various archives listed in the appendix to this volume is to be aware of the extent to which archival collections have been purged of military evidence – for example, vitas conveniently omit military reports and publications, correspondence with military funders is destroyed or sequestered.

The story remains sprawling and shapeless, at least in part, because it has been so thoroughly repressed in the interim. In this chapter, we make a start at describing how the exigencies of war brought together some physical scientists and some economists under the auspices and patronage of the military, resulting in profound unintended consequences for the shape of the postwar neoclassical orthodoxy in America. The story commences with some cyborg themes underpinning the reorganization of science in World War II; continues with the birth of operations research during the war and its bifurcation into British and American variants; shifts to the crisis of mathematical neoclassical theory in America in the 1930s; and concludes with the deliverance of neoclassical economics from this impasse by the good graces of its encounters with OR, to such an extent that the three main schools of neoclassical economics in postwar America (Hands & Mirowski, 1998) can be correlated with their distinct attitudes and relationships to the variant versions of OR. In short, this is the saga of the reorganization of the neoclassical wing of the American economics profession by the war.

RUDDLED AND BUSHWHACKED: THE CYBORG CHARACTER OF SCIENCE MOBILIZATION IN WORLD WAR II

The names of Warren Weaver (1894–1978) and Vannevar Bush (1890–1974) are not often mentioned in the history of the social sciences. Perhaps development economists may have heard of Weaver's role in the Rockefeller promotion of the "Green Revolution" in postwar Third World agriculture, whereas those with a generalist's interest in history may recall that Bush had organizational responsibility for the development of the atomic bomb. In the larger culture, Bush is more often than not misremembered as progenitor of things for which he was not actually responsible: he was by no stretch of the imagination the "father of modern computing" (Zachary, 1997, p. 262), nor was he even remotely responsible for the idea of the internet or hypertext, any more than H. G. Wells could be held responsible for the idea of the atomic bomb. Bush was, however, the author of what became the canonical justification for state patronage of scientific research in America in the postwar period, *Science: The Endless Frontier* (1945). Weaver, if anything, remains forever yoked in the public mind with Claude Shannon as an expositor of information theory, although, as we note in Chapter 2, he had very little to do with its conceptual development or elaboration. Other than a 1929 textbook on electromagnetic field theory coauthored with Max Mason, Weaver's actual scientific output was rather thin (Rees, 1987). But that did not prevent him from having profound impact on the shape of postwar American thought.

These confusions and misattributions are unfortunate, but they also become comprehensible when one comes to understand that Bush and Weaver were the first of the modern science managers, and, as managers often do, they frequently garner the credit for work that is done by those in their employ. Their overriding historical importance lay in their capacity as patrons and organizers of the cyborg sciences, and as facilitators and popularizers of the achievements of the military-science interface. The historical irony is that these deeply conservative men, preternaturally hostile to government control of the economy and instinctively wary of the corrupting power of the military, ended up presiding over the greatest expansion of government-funded and military-directed scientific research in the history of the country and, furthermore, found themselves taking credit for many of its perceived successes.

Beating around the Bush

Vannevar Bush was a tinkerer and inventor who managed to parlay a modest career in electrical engineering into a path to the pinnacle of wartime science administration. With a degree from MIT, he began a dual career as an academic at Tufts College and the head of research for the American Research and Development Corporation, a lineal precursor of Raytheon. During World War I, Bush attempted to develop a submarine detector, but was rebuffed by the Navy, an experience that marked his attitudes for life (Zachary, 1997, pp. 36–38; Burke, 1994, p. 31). After the war he moved to MIT's department of electrical engineering, rapidly becoming dean of engineering and vice-president of the institute. In this period he developed the "differential analyzer," a mechanical analogue computer for solving differential equations. Many of the more stellar figures of cyborg science spent long hours of apprenticeship learning how to nurse along Bush's analyzer, from Claude Shannon to Norbert Wiener to Philip Morse. Bush's first contacts with Weaver came in soliciting Rockefeller support for improvements of the device, and Weaver decided to support Bush instead of researchers at Harvard and Iowa State (Burke, 1994, p. 78). Never one to become long engrossed in a single machine, Bush in the 1930s was also involved in a quest for optical retrieval of information from high-speed microfilm funded by the FBI and firms like National Cash Register and Kodak; as well as a secret project for the Navy on a device to mechanize certain aspects of cryptanalysis. A disturbing trend emerged in the 1930s, with Bush promising completed machines with impossibly short deadlines, and then reneging on delivery, only to hop imperiously to a different funding source and a reconfigured machine project (Burke, 1994). Part of the problem seemed to be his penchant for analogue devices over digital, and another was his partiality toward microfilm to solve problems of memory storage. None of this permanently harmed his credibility, however, because his skills as an administrator were increasingly becoming apparent to those seeking to develop the role of science manager. In 1938 Bush attempted to lure research managers of corporate laboratories into closer cooperation with each other and with academic scientists; in 1939 Bush was appointed to the lucrative presidency of the Carnegie Institution, with its large portfolio of research support and its key location next to centers of power in Washington.

In an extremely unlikely turn of events, the inveterate elitist and anti–New Deal Bush managed to get access to Franklin Delano Roosevelt and to convince him to authorize the creation in June 1940 of a "National Defense Research Committee" of unprecedented autonomy: "a direct line to the White House, virtual immunity from Congressional oversight and his own line of funds" (Zachary, 1997, p. 112). Rejecting the idea of government labs to research new weapons, Bush opted in favor of the foundations' innovation of research contracts to be written with academics and industrial labs. This gave him and his committee, which he had personally handpicked, enormous latitude over what projects and which researchers would receive government funds. If World War I had been a chemists' war, Bush believed he would preside over a physicists' and engineers' war. Bush himself admitted that the formation of the committee was "an end run, a grab by which a small company of scientists and engineers, acting outside established channels, got hold of money and authority for the program of developing new weapons" (Zachary, 1997, p. 116; Bush, 1970, pp. 31–32).

With war formally declared, Bush expanded his burgeoning domain by getting Roosevelt to authorize creation of the Office for Scientific Research and Development (OSRD) as an umbrella organization in control of the NDRC in mid-1941, which endowed him with a line-item Congressional appropriation and the ability to contract to build hardware, as well as pursue research in weapons and devices (Rau, 1999, chap. 2). In partial violation of Bush's own proscription against the creation of "government" labs, the NDRC decided to capitalize upon British developments of radar technology by siting the institution of the "Radiation Laboratory" (a name designed to mislead the curious, especially when universally shortened to "Rad Lab") at his home institution of MIT, thus beginning a prodigious flow of wartime funds to that entity. The appearance of favoritism apparently bothered even Bush, who "repeatedly asked for some sort of blanket immunity against future conflict-of-interest claims" (Zachary, 1997, p. 136). Ultimately, 90 percent of all OSRD academic funds went to only eight universities, while the MIT Rad Lab garnered 35 percent of this total all to itself. It was an unprecedented channeling of massive subsidy through a very few institutions, continuing the prewar practice of the foundations' bias toward "institution building," and it set a precedent for much of postwar science management.

Although Bush was wary of the personal political jeopardy which such a concentration of funding and effort entailed, he did not regard it as a generic or structural political problem for scientists as a group, mainly because he harbored a rather jaundiced view of democracy and its relationship to science. Doling out contracts of staggering immensity to old acquaintances with no guarantees or surety was not a dubious procedure fraught with pitfalls, because they were unquestionably the "best men" in his estimation; and science was definitely not a democracy. As his biographer reports, Bush possessed "a view of American society that pitted sober pragmatic elites against the untutored volatile masses" (Zachary, 1997, p. 324), and this view encompassed the military as well as the man in the street. His earlier prewar experience of wrangling over Navy contracts had especially soured him on that branch of the service, and he wreaked his revenge early in the war by "defeating" one of his strongest critics in the person of Admiral Bowen (Sapolsky, 1990, pp. 16ff.) and bestowing control over the atomic bomb project to the Army.[6] Nevertheless, the vast expansion of scientific research he oversaw could only happen within a military framework, and that meant that military brass had to call the shots. This created a paradox: how could the pragmatic elites exercise their superior talents and free-ranging inquiries while eluding both military and political encumbrances?

The instrumentality of research "contracts" with scientists at their universities was supposed to slice through the knotty contradiction (Bush, 1970, pp. 78–79), but in the largest projects, such as the Rad Lab and the Manhattan Project, that was little more than a public relations ploy. For instance, it was Army general Leslie Groves who vetted and appointed Robert Oppenheimer scientific head of Los Alamos, and not vice versa (Rhodes, 1986, pp. 448–49). The extent and nature of the final control of the military over the scientists and their research was continually a matter of irritation and contention, only muted by the shared determination during wartime to repress frictions and cooperate in the complete and rapid destruction of the enemy.

Over time, Bush came round to the conviction that the British had stumbled across another means to insulate the scientific elite from military

[6] See Zachary, 1997, p. 203. Bush's irritation with the Navy shows up repeatedly in his memoirs (1970). The intricacies of interservice rivalries go far to explain much of the shape of postwar military subsidy of science, although it is one that would distract us too far afield here. For instance, Bush's defeat of Bowen came back to haunt him in the postwar period, with Bowen "developing a clone of the NDRC within the Navy" (Zachary, 1997, p. 127), namely, the Office of Naval Research. Another example is the effort of the Air Force to influence weapons procurement and wrest atomic control from the Army through the institution of Project RAND. Interservice rivalry should always be promoted ahead of any putative altruistic motives when discussing the funding of postwar "pure science."

meddling and yet simultaneously impose scientific good sense upon the unruly military brass through their innovation of "operations research." It is standard to trace the origins of OR to the British scientific rationalization of antiaircraft and antisubmarine warfare in the 1930s – a genealogy to be examined in detail in the next section – and to credit Vannevar Bush with encouraging the American services to consider instituting similar cadres in 1942 (Trefethen, 1954, p. 12; Zachary, 1997, pp. 172–74; Bush, 1970, pp. 91–93). Actually, the story was much more complicated than that (Rau, 1999).

When Bush set up the NDRC, he was a partisan of the doctrine that the scientists and engineers should stick to the things that they knew best; namely, to the weapons and other physical devices that were the fruits of their research, and leave to the military the actual prosecution of the war. When word began to filter back from Britain about a newfangled set of doctrines called "operational research" then assisting the military in using radar to fight submarines and bombers, he gave vent to some of his prejudices about nonscientists:[7]

> [It] would be undesirable for NDRC to become closely identified with such matters. This is for the general reason that the NDRC is concerned with the development of equipment for military use, whereas these [OR] groups are concerned with the analysis of its performance, and the two points of view do not, I believe, often mix to advantage. . . . The type of man to be used in such work is very different from the type needed for experimental development. Analysts, statisticians and the like are not usually good developmental research men.

Yet, others were more enthusiastic about the promise and prospects of OR, and in 1942 units began to appear in the Army Air Force, the Signal Corps, and the Navy Bureau of Ordnance, initially outside the purview of OSRD. Whereas Bush was inclined to tar these activities as nonscientific, it became obvious that many of the research units were headed by physicists of repute; and, further, Bush's deputies James Conant and Warren Weaver were inclined to favor these forays into questions of strategy and tactics. The latter not only believed that scientific discoveries could help the war effort but also that a "science of war" was not an oxymoron. By 1943 Bush came round to the position that room should be made for OR to be included under the OSRD umbrella, either in such faits accompli as the Applied Mathematics Panel, or more directly, as with the Office of Field Services. The latter office alone employed 464 scientists by the war's end, one-third of whom were physicists.

[7] V. Bush to Moses & Lee, May 29, 1942, National Archives, quoted in Rau, 1999, pp. 135–36.

Under the pretext of helping the soldier in the field successfully deploy the new contraptions that were pouring out of the labs funded by OSRD, these American "scientific consultants" rapidly came to extend their purview to questions of strategy and tactics that were nominally the preserve of military officers. Enjoying the same privileges of military officers, they were free of the obligations of subordination to the chain of command, not to mention risks to life and limb, because Bush insisted that no OSRD scientist in the field would actually enlist in the military. As he candidly admitted, OSRD scientists wore officers' uniforms without insignia or service designation (1970, p. 113), and "My rank in the hierarchy was never defined, but it certainly was not minor" (p. 111). Hence for Bush, OR eventually constituted a device for physical scientists to wrest a modicum of control over strategy and tactics away from their less savvy military patrons, *the better to maintain control over their own research*, taking the money while remaining aloof from the chain of command, without seeming to undermine military authority (Zachary, 1997, p. 160). For science czars like Bush and Conant, the "scientific method" became of necessity an abstract virtue detached from any specific scientific practices because some such protean capacity had to underwrite the pretentions of the scientists to rationalize such diverse and "unscientific" military activities as submarine stalking, saturation bombing, or the "best" way to mine a harbor. The midcentury attitude that "science" was the universal solvent of all social conundrums was a pretension that itself derived from the need to offset the potentially deleterious consequences of the military funding on the direction of science.

Perhaps incongruously, it was as the postwar defender of an independent "pure science" capability that deserved no-strings-attached public funding that Bush attained popular fame, after his 1945 government report *Science: The Endless Frontier*. In that government report, he justified a general "best science" approach controlled by a few insiders by claiming that technological downstream developments would more than compensate the cost of their autonomy. The ensuing chequered history of the National Science Foundation – first blocked by a political pincers movement between those seeking greater accountability and those who wanted to preserve the wartime military structures of science organization; then finally instituted five years later as a relatively small and ineffectual agency in practice still subordinated to military control and funding of Big Science – is often read as the defeat of Bush's plans for postwar science (Kevles, 1995; Reingold, 1991). Subsequent reconsideration of this episode, while acknowledging the adversarial orientation of many in the military toward Bush, adopts a more nuanced approach, researching the ways in which Bush's "ability to juggle his support for both civilian and military management of research illustrated the thinness of the distinction, especially after the war" (Zachary, 1997, p. 470).

The funding and direction of science was somehow to be exacted from the military and corporate sector without bearing any responsibility for the allocation of those resources or for accountability for the consequences. The path from pure science to economic prosperity was never charted with any seriousness because Bush could not be bothered with such utilitarian concerns; indeed, he was not even concerned with the impact of his favored "best science" approach on the education of a new generation of scientists, or on the structure of his bulwark universities.[8] It would appear that Bush believed societal benefits would flow equally and indifferently from civilian, industrial, or military science; he never once entertained the idea that "pure" science could easily disappear up its own naval into obscure irrelevance, or that military requirements for research could so diverge from the industrial concerns that they would lose all contact with one another. The ultimate irony is that the figure who midwifed Big Science in America never gave a serious thought to the effects of social organization upon the content and biases of research, probably because he projected his own self-image as lone tinkerer upon his ideal-type Scientist.

This obliviousness was nowhere more apparent than in his attitudes toward the computer, the one technology that did owe its genesis to the OSRD and its military progeny (Flamm, 1988). Bush's earliest crypt-analysis machines for the Navy were crippled not only by their engineer's bias toward analogue devices, but also by Bush's insensitivity to issues of coding, information, and logical organization.[9] In September 1940, when Norbert Wiener proposed to his old friend that NDRC fund research into a binary computer as part of the gun control project, Bush nixed the project as possessing insufficient immediate application (Zachary, 1997, pp. 265–66). The Army in 1943 underwrote support of one of the first vacuum-tube calculators for ballistics at the Moore School of the University of Pennsylvania, a machine that metamorphosed into the later pathbreaking ENIAC; but experts within Bush's OSRD opposed the

[8] "After 1945, the increase in federal funding and the increasing impact of federally supported research centers threatened the hegemony of the universities and their departments. By the 1950s there was talk of a researcher's allegiance to their discipline, rather than their university, and about the neglect of teaching by the research-minded" (Reingold, 1995, p. 313).

[9] "Bush did not realize all the negative consequences of tape (serial) processing, or of a coding scheme without hierarchy and without code position significance. . . . Bush's ideas about coding were very primitive. He was so oriented towards a system for experts in particular fields that he thought he could ignore all the complexities that tormented professional cataloguers and indexers" (Burke, 1994, p. 190).

project (Flamm, 1988, p. 48). In 1949 Bush wrote von Neumann that, "After looking at the computing machine situation in the country as a whole I am rather fearful that there are some programs that have been instituted at great expense that will never go through to completion or, on the contrary, more machines will be finished than can be serviced or used" (Zachary, 1997, p. 274). Instead of actively coordinating the burgeoning military support of the development of electronic computers (largely at von Neumann's instigation), Bush was occupied in 1945 writing science fiction articles for *Atlantic Monthly* and *Life* magazines on an imagined "memex," a microfilm-based personal television device where scientific articles could be called up by keyboard and cross-referenced according to personal associations keyed in by the user. The contrast with active computer researchers was characteristic of Bush's approach to science. Whereas other cyborg scientists like von Neumann and Wiener and Shannon were achieving real computational breakthroughs by treating information like a thing, the body like a feedback device, and computer architectures like the organization charts of bureaucracies, Bush was still mired in the nineteenth-century image of a lone genius at his desk making idiosyncratic connections between flocculent bits of knowledge, with little concern spared for how the feat would be accomplished, who would bear responsibility for maintenance of the mega-database, or what purposes it would serve. For Bush, the memex was a wistful solution to the problem of the anomie and dissociation of the division of labor of Big Science, and the consequent sheer inability to be an intellectual jack-of-all-trades, a technological fix to a social transformation that he himself had largely helped to bring about.[10]

It is important at this stage to recognize that the therapeutic virtues of the scientific method did not seem to Bush to work their magic on the "social sciences," even though much of what both he and his operations research cadres were doing with the military would easily qualify as social analysis or social theory. His own ambivalence about statisticians has been illustrated already. This, implicitly, drove a wedge between previous social science and the advent of OR, a gulf that turned out to be significant. As

[10] Nevertheless, Bush has been adopted as patron saint of what might be "right cyborgism," the belief that computers can serve to liberate entrepreneurial instincts and support individual freedom. In direct contradiction to this vision, he might also be regarded as a precursor to William Gibson's dystopic cyberspace fantasies (1984; Goodwin & Rogers, 1997). In an unpublished essay of 1959, he dreamed of a souped-up version of the memex, in which a mind amplifier would sense brain activity and participate in forging associations without the tiresome intervention of motor or sensory skills (Zachary, 1997, p. 400). One is reminded of Gibson's description of "a deranged experiment in Social Darwinism, designed by a bored researcher who kept one thumb permanently on the fast-forward button."

the first American textbook on OR put it, with the charming air of the naïf, "Large bodies of men and equipment carrying out complex operations behave in an astonishingly regular manner, so that one can predict the outcome of such operations to a degree not foreseen by most natural scientists" (Morse & Kimball, 1951, p. 7). For Bush, things mostly worked by themselves; and when they didn't, there providentially existed a sort of pragmatic managerial expertise, but one vested in those who had naturally risen to positions of power in the economy; so there was no need for codified lawlike propositions about social phenomena. His favorite put-down of the military brass was: "This kind of organization would not be tolerated a week in a manufacturing concern producing bobby pins" (Zachary, 1997, p. 310). His contempt for the social sciences was made explicit both during the war, when he blocked their participation in OSRD cadres, and afterward, when he fought against any movement to include the social sciences in the funding mandate of any postwar National Science Foundation. Bush insisted that scientific research was not motivated by social needs, though he sometimes found he had to shade this denial in political contexts (p. 440); if anything, he believed "society" should cheerfully subordinate its unregenerate interests to the needs of scientists. Belying all the consequences of his own prodigious efforts in the arena of science management, he persisted in portraying the scientist in his writings as a lonely maverick, following his muse wherever it might lead; and, as for the rest of humanity, the only reason to subject them to empirical study was to take unscrupulous advantage of their painfully evident deficiencies of reason. As Bush once railed against postwar advertisers: "What are they now going to sell? Nothing but hogwash, and the public seems to fall for it. . . . As long as the public is gullible, satisfied with hokum, why move? My wonder is that the public seems to have unlimited capacity to swallow fantasy" (p. 387).

Warren Weaver, Grandmaster Cyborg

Warren Weaver, by contrast, enjoyed a much more sanguine opinion of humanity and its foibles and, perhaps for that reason, was better suited to operate down in the trenches with scientists, rather than serve as remote bureaucratic figurehead like Bush, or as a high-profile political operative like James Conant. He was much more the consummate science *manager*, the hands-on assessor of specific research projects, the forger of transdisciplinary research networks, the conjurer of new fields out of inchoate fragmented developments (Kohler, 1991, p. 301). It was he whom the movers and shakers sought out to coordinate their wartime research programs, perhaps because he had a superior working understanding of the intellectual issues involved; Weaver, like Kilroy in World War II, had been everywhere that was going anywhere; he was the anonymous entity behind the lines who left his mark on most of the nascent cyborg sciences. Precisely for that reason, the C^3I paradigm of military research is something most transparently inscribed in his work: "it may be reasonable to use a still broader definition of communication, namely, one which includes the procedures by means of which one mechanism (say, automatic equipment to track an airplane and to compute its probable future position) affects another mechanism (say, a guided missile chasing this airplane)" (Shannon & Weaver, 1949, p. 3). More than anyone else, it is Warren Weaver whom we have to thank for opening doors to the closed rooms for the cyborgs to find their way into the fin-de-siècle world; and yet, of all the wartime science czars such as Bush, Conant, or von Neumann, he is the only one to not have attracted a biographer.[11]

Warren Weaver took degrees in civil engineering and mathematics from the University of Wisconsin, there becoming the protégé of the physicist Max Mason. After some World War I work on dogfights at the Army Air Corps and a stint at Cal Tech, he spent the decade of the 1920s on the mathematics faculty at Wisconsin. Mason subsequently brought him to New York in 1932 to head up the newly created division of natural sciences at the Rockefeller Foundation, just when the Depression was forcing the foundation to contract and consolidate its grants. Although bereft of background in the area, Weaver managed to reorient the program toward his own vision of a "molecular biology," which encompassed mathematical and physiological approaches to psychology (Kohler, 1991). As he confessed in his memoirs, "I started with genetics, not because I realized in 1932 the key role the subject was destined to play, but at least in part because it is a field congenial to one trained in mathematics" (1970, p. 69). One historian has contextualized this reorientation of the Rockefeller philanthropies away from physics and towards "molecular biology" in the 1930s as part and parcel of its initiative to produce a new model "Science of Man": genetics was deemed a more respectable underpinning for the social sciences than the field of eugenics (Kay, 1993).[12]

[11] The only partial exception is Rees, 1987, which is singularly uninformative on many issues. More attention is paid in Kohler, 1991, although only in the narrow context of a description of Rockefeller Foundation activities. Weaver published a relatively colorless and bland set of memoirs later in life (1970). Enhanced insight into some of his career is provided by the oral history in WWOC.

[12] Scholars have yet to correlate this shift in Rockefeller priorities with its initiatives in economics in the 1930s, described in Craver, 1986b. Fisher, 1993, seems to neglect most issues relevant to economics. It is unfortunate that the social sciences have yet to find their Kohler or Kay.

It would do Weaver a disservice to stress his success in a single field like molecular biology, however earthshaking, because his real forte lay in dissolving the inherited barriers between disciplines. Indeed, his doctrine of transdisciplinarity could be regarded as a direct artifact of the innovations in science funding and management going on at Rockefeller in the 1930s, for to preclude the overarching intellectual precedence of the "discipline" was to deny that Rockefeller had any special funding commitment to any discipline (Kohler, 1991, p. 277). Instead, foundation officers could freely pick and choose individual scientists for encouragement, tinker with career trajectories, reconfigure research networks more to their own liking, and appeal for legitimation to "peer evaluations" on advisory boards after the fact. Weaver followed his own nose and became enthused when he caught a whiff of the combination that would prove so heady in the postwar years: usually, some physicist or engineer who had mathematized something that would conventionally be regarded as the province of biology or psychology, perhaps by accessing issues of thermodynamics, probability, computation, and electrical engineering.

For precisely these reasons, the Weaver connection with Bush dated back well before World War II. One of his first tours of duty after coming on board at Rockefeller in 1932 was a survey of all computer projects in the nation; and, consequently, he supported Bush's project to improve the differential analyzer with substantial funding in the depths of the Depression (Burke, 1994, p. 78). Reciprocally, when Bush was appointed head of the NDRC, Weaver was one of his first choices to head a crucial section area, that of antiaircraft fire control (Weaver, 1970, p. 77), and later he was tapped to be supervisor of the mathematical and scientific instrument section. It was in his role as chief of the Applied Mathematics Panel (AMP) of the NDRC that he makes his first pivotal appearance in this narrative. (The second comes with his presence at the creation of the RAND Corporation.) Indeed, Weaver was ideally placed to mediate between the fire control and mathematics: the technologies envisioned were automated gun-laying devices, high-speed photoelectric counters for ballistics tests, and instruments to monitor atomic emissions; unlike Bush, Weaver opted in each case for electronic digital solutions. In this capacity, Weaver initiated coordination and control of electronics research at private firms, such as National Cash Register, RCA, Eastman Kodak, and Bell Labs, as well as at wartime universities (Burke, 1994, p. 207). This preemption of interfirm rivalries and government-business distinctions paved the way for many fire control developments to proceed with alacrity into cryptanalytic devices, commercial servomechanisms, and early computers like the ENIAC. Weaver was also responsible for bringing Norbert Wiener into the gun control project, thus redirecting his interests toward cybernetics (Wiener, 1956, p. 249). Weaver hired von Neumann to work on shock waves;[13] and he was also strategically waiting there when von Neumann began to turn his attention to computation. The earliest evidence of von Neumann's sustained interest in the computer comes in a letter to Weaver in January 1944, soliciting his expertise about the various computing technologies that could be commandeered in the effort to build the atomic bomb (Aspray, 1990, p. 30). In response, Weaver delegated a subordinate to scare up some IBM equipment to assist von Neumann. It was also Weaver who first put von Neumann in touch with the various researchers across the country building prototypes of the electronic computer.[14]

Weaver admits in his memoirs that the fire control section was rather obviously bifurcating into "hardware" and "mathematical" directions as early as 1942; "problems we worked on sometimes related to, and were preliminary to, the design of devices; often they related to the optimum employment of devices; and sometimes they were still of a broader character, concerned with tactical or even strategic plans" (1970, p. 86). When the OSRD was reorganized in 1942, with fire control devices shifted to a new "Division 7" and the creation of the new agency, the Applied Mathematics Panel, which Weaver was asked to lead, Weaver made most of the personnel choices for the AMP, in consultation with Bush and Conant.[15] Whereas the panel was originally conceived as a way of incorporating the efforts of the American mathematics community into the war effort, the irony ensued that it mostly ended up excluding them. However much such distinguished mathematicians as Marshall Stone, Marston Morse, and Griffith Evans claimed they could effortlessly straddle the divide between the pure and the applied, Weaver saw that it took something else to subordinate oneself to planned and coordinated science. "It is unfortunately true that these conditions exclude a good many mathematicians, the dreamy moonchildren, the prima donnas, the a-social geniuses. Many of them are ornaments of a peaceful civilization; some of them are very good or even great mathematicians, but they are certainly a

[13] See von Neumann to Weaver, July 23, 1943, box 7, folder 12, VNLC. His choices for assistant on the project were Leonard Savage, S. Heims, or R. M. Thrall.

[14] The rapid transition from game theory to the computer for von Neumann is nicely illustrated in the letter accompanying the presentation copy of *Theory of Games and Economic Behavior* to Weaver: "I am exceedingly obliged to you for having put me in contact with several workers in the field, especially with Aiken and Stibitz. In the meantime, I have had a very extensive exchange of views with Aiken, and still more with the group at the Moore School in Philadelphia, who are now in the process of planning a second electronic machine. I have been asked to act as the advisor, mainly on matters connected with logical control, memory, etc." Von Neumann to Weaver, November 1, 1944, box 7, folder 12, VNLC.

[15] Rees, 1988; Owens, 1989. This was confirmed by Mina Rees, herself brought into the AMP by Weaver. See Rees interview, March 16, 1969, p. 10, SCOP.

severe pain in the neck in this kind of situation" (Weaver quoted in Owens, 1989, p. 296). Weaver turned out to be much more in tune with those of mathematical bent but coming from diverse applied backgrounds, those willing to trample disciplinary imperatives – that is, those more like himself – especially since he interpreted the mandate of the panel as growing organically out of the problems of devices like gun control and bomb targeting.

Hence Weaver increasingly directed the panel toward problems that had been pioneered by the British operational researchers, problems disdained as trivial by the bulk of the mathematicians.[16] His introduction to the practice of operational analysis came as a member of the second American scientific mission assigned to consult with the British OR contingent in March 1941 (Weaver, 1970, p. 89). He had become an enthusiastic proponent of the British procedures by 1942 and was instrumental in convincing Bush that they were a legitimate field deserving of OSRD support. By 1943 the AMP was already splitting its time equally between weapons development and operational analysis (Rau, 1999, p. 295). At this juncture Weaver decided that economists were to be recruited into the AMP, with the consequence of these favored few getting their first introduction to cyborg themes, an event which deserves closer examination.

Weaver's close association with the computer persisted even as he was switching hats and chairmanships in the mid and late 1940s. In the spring of 1945, issues of computer support at the NDRC had come to a head, and Weaver asked von Neumann to write a report for the AMP to survey the past achievements and future prospects of electronic digital computing machines (Apsray, 1990, p. 240). When von Neumann decided to have his own computer built at the Institute for Advanced Study in 1946, Weaver was one of the first to be approached for funding (p. 56). Few have appreciated that von Neumann's "biological turn" (discussed in Chapter 3) coincides rather neatly with Weaver's accelerated proselytizing for the importance of "molecular biology" for the future of science. It was in late 1944 that von Neumann briefly banded together with such figures as Norbert Wiener, W. Edwards Deming, Warren McCulloch, and Herman Goldstine to form the Teleological Society, the forerunner to the cybernetics meetings (Heims, 1991). Although the nascent congregation

[16] Mac Lane, 1989, p. 508: "Scientific war research, like other scientific activities, is not immune from nonsense; especially because of the pressure of the work it is possible to set up problems which look superficially sensible, but which turn out to be either hopeless of solution or meaningless in application. This tendency is especially strong when the problem comes to the scientist through a long chain of channels." Mac Lane was director of the Applied Mathematics Group at Columbia in 1944–45, and admits, "As director, I often found myself in disagreement with Warren Weaver" (p. 501).

did not coalesce immediately, Wiener wrote in a letter to another of the participants: "We are also getting good backing from Warren Weaver, and he has said to me this is just the sort of thing that the Rockefeller should consider pushing."[17]

Weaver's postwar career was no less significant for the wider nurture of the cyborg sciences. On his return to full-time residence at the Rockefeller Foundation, he began to ponder the impact of the computer upon the structure of postwar culture. His meditations led to at least three conclusions. First, and nearest to home, "I believed that the computer might have a truly profound effect upon the whole structure of mathematics, demanding a revised foundation which from the outset recognized discontinuity and discreteness as natural and necessary aspects of quantification" (1970, p. 106). Although he did little to develop the mathematics of this vision himself, in the world of chaos, fractals, and cellular automata, we might regard this conviction as prescient. Second, Weaver took Shannon's information concept and popularized it across the vast gamut of physical and social sciences. As we noted in Chapter 2, Weaver's interpretation of the mathematical concept was not identical to that of Shannon, but this actually served to make it more palatable in many far-flung contexts, not the least of which was economics. One can sample Weaver's OR-flavored version in the following:

> To be sure, this word *information* in communication theory relates not so much to what you *do* say, as to what you *could* say. That is, information is a measure of the freedom of choice when one selects a message. . . . The concept of information applies not to the individual messages (as the concept of meaning would), but rather to the situation as a whole, the unit information indicating that in this situation one has an amount of freedom of choice, in selecting a message, which is convenient to regard as a standard or unit amount. (Shannon & Weaver, 1963, pp. 8–9; emphasis in original)

Whereas Shannon at least maintained the facade of restricting the purview of his theory to communications engineering, Weaver threw caution to the winds, extolling the theory's generality of scope, classical simplicity, and power (p. 25). The evidence offered for these statements was a trifle contrived, because the "contributions" that Shannon's information had made to cryptanalysis and computer architectures up to that juncture were better understood as artifacts of the conditions of its genesis. Nevertheless, Weaver did manage to convey a feeling of boundless promise: "this analysis has so penetratingly cleared the air that one is now, perhaps for

[17] Norbert Wiener to Arturo Rosenbleuth, January 24, 1945; quoted in Aspray, 1990, p. 316.

the first time, ready for a real theory of meaning. . . . Language must be designed (or developed) with a view to the totality of things that man may wish to say; but not being able to accomplish everything, it too should do as well as possible as often as possible. That is to say, it too should deal with its task statistically" (p. 27).

Third, and perhaps most important of all, Weaver wrote an article in 1947 that pioneered the notion that science had crossed a watershed sometime after 1900, away from problems of "simplicity" and into the arena of what he called "disorganized complexity." The harbinger of the new approach was statistical mechanics. Probability theory had come to sweep physics; and now "the whole question of evidence and the way in which knowledge can be inferred from evidence are now recognized to depend on these same statistical ideas, so that probability notions are essential to any theory of knowledge itself" (1947, p. 538). But science had gone from one extreme (say, two or three variables of stripped-down rational mechanics) to another (say, Gibbsean ensembles), leaving a vast middle region of phenomena untouched. He suggested, "these problems, as contrasted with the disorganized situations with which statistics can cope, show the essential feature of *organization*. . . . They are all problems which involve dealing simultaneously with a *sizable number of factors which are interrelated into an organic whole*" (p. 539, emphasis in original). He called this range of phenomena problems of "organized complexity," thus bestowing a name to a concept that the cyborg sciences would claim for their own in the postwar period. Indeed, Weaver identified "complexity" as the premier frontier of the sciences of the future slightly before John von Neumann gave it mathematical expression in his theory of automata. As we observed in Chapter 2, "complexity" has become one of the watchwords of the computer sciences as well.

If this realm of organized complexity sounded suspiciously like dressed-up OR in Weaver's account, trailing an economic resonance in train, Weaver was not loath to acknowledge it. "On what does the price of wheat depend? . . . To what extent is it safe to depend on the free interplay of such forces as supply and demand? To what extent must systems of economic control be employed to prevent the wide swings from prosperity to depression?" (p. 539). Existing economics, with which Weaver was familiar from his wartime experience on the AMP, was essentially brushed aside as having little to offer to clarify these problems of organized complexity. Instead, "Out of the wickedness of war have come two new developments that may well be of major importance in helping science to solve these complex twentieth-century problems" (p. 541). And what hath the war wrought? One new tool was the computer, wrote Weaver, and the other was "the mixed-team approach of operations analysis." In other words, wartime science management plus the computer as exemplar would dissolve the intractability of organized complexity, which had left the field unexplored for so long.

This may have seemed excessively intangible and imprecise to most readers in 1949 – Weaver had no concrete examples of problems solved to proffer at that point – but he was nonetheless busily putting his prescriptions into practice. Unlike Bush, in public Weaver was not half so contemptuous of the military funding of science. He worried that abruptly shutting down the OSRD hard upon the war's conclusion would necessarily imply a drastic contraction in all funds for science and would thus inadvertently scuttle the civilian control of science planning and management, which had been so carefully built up during the war. Immediate passage of a bill for civilian science funding would "give the whole national science foundation idea a black eye and two lame legs. It would take years to build up again the situation of increasing confidence which now exists between scientists and universities, on the one hand, and the government on the other."[18] The solution as he saw it was to insinuate lots of big and little OSRDs throughout the postwar government and the military, thus installing Rockefeller practices throughout the corridors of public patronage of science. Consequently, Weaver became a ubiquitous "civilian" representative on military research advisory boards.

For instance, Weaver participated in the initial institution of the Office of Naval Research (ONR), which became the largest funder of science in the immediate postwar period. He found the new ONR so to his liking that he once said that it was like being in the Rockefeller Foundation but without the need to deal with troublesome trustees (Reingold, 1995, p. 302). He sat on the board of the original RAND Corporation and sought to recreate within the Evaluation of Military Worth Section of RAND another version of the wartime Applied Mathematics Panel (Jardini, 1996, p. 83). To that end, he installed one of his protégés, John Williams, to head RAND's nascent mathematics division. At RAND's first conference on social science, he explained RAND's mandate was to inquire "to what extent it is possible to have useful quantitative indices for a gadget, a tactic, or a strategy, so that one can compare it with available alternatives and guide decision by analysis" (Kaplan, 1983, p. 72). Later on in his career he was named vice-president of the Sloan Foundation, which then played a major role in the institution of the nascent field of "cognitive science" (Gardner, 1987). Always eschewing the limelight, he consistently kept the money flowing and the intellectual firepower aimed at his conception of the problems of "organized complexity," and for this he deserves to be remembered as the Maxwell's Demon of cyborg science. He kept the door open and let a "few quick

[18] Weaver to Naval Research Advisory Committee, December 20, 1946; quoted in Zachary 1997, p. 470.

ones" get through, bringing hot new ideas from thermodynamics, electrical engineering, and computation into disciplines previously impervious to their charms, generally shaking things up.

OPERATIONS RESEARCH: BLIPKRIEG

Operations research languishes as the unloved orphan of the history of science, even though some writers close to events had argued it "was one of the chief scientific features of the war" (Crowther & Whiddington, 1948, p. 91). To many, at first glance it looks suspiciously like a social science, but the history of economics has been written as though it were absent, or at best, irrelevant.[19] Historians of physics (e.g., Kevles, 1995, chap. 20; Pickering, 1995a; Fortun & Schweber, 1993) have paid more attention but only in passing, declining to describe any of the intellectual content or trace much in the way of its tangled disciplinary affiliations. Historians of nuclear strategy (Kaplan, 1983; Herken, 1987; Sherry, 1995) tend toward more enhanced appreciation of its importance but also shy away from content. For a short while after World War II, it was frequently dealt with in the larger genre of war anecdotes (Clark, 1962; Thiesmeyer, 1947). More recently, OR has been lionized from the viewpoint of the history of technology (Hughes, 1998), demonized from the vantage of the sociology and politics of science (Mendelsohn, 1997), and interrogated as to its immediate wartime origins (Rau, 1999; Zimmerman, 1996; Hartcup, 2000). There are, of course, internalist practitioner histories (Trefethen, 1954; Holley, 1970; Waddington, 1973; Miser, 1991; Dorfman, 1960; Johnson, 1997), written primarily to supply a sense of doctrinal continuity and progress for their chosen field, which commit the opposite offense of ignoring anything that might suggest the conditions of production and ongoing transformation of the discipline. But, more tellingly, it is obvious that the practitioners themselves have suffered identity crises verging on neuroses, struggling again and again to define themselves to themselves. In order to come to appreciate better the quandaries of attempting to write a history of something that resists acknowledgment of any central tendencies, we begin with some relatively conventional narrative accounts, but then retell the story in ever widening outward spirals, ending up by design where the cyborg rubber met the neoclassical road.

OR in Some Standard Narratives

The standard synopsis can begin with a textbook definition by Ellis Johnson, an eminent physicist and practitioner of the early discipline: OR "is the prediction and comparison of values, effectiveness and costs of a

[19] A partial exception to this generalization is Leonard, 1991; although it misunderstands the dynamics of OR by phrasing the question as one of what the economists had to "offer" the military. Another is Rider, 1992.

set of proposed alternative causes of action involving man-machine systems" (Ellis Johnson in McCloskey & Trefethen, 1954, p. xxiii). Other than the man-machine business at the end, this sounds suspiciously like some version of generic neoclassical economics, where bean counters assiduously compare the allocation of scarce means to given ends. If that was all there was to it, then most nineteenth-century industrial planning exercises, such as use of statistics of freight traffic to reduce recourse to empty freight cars on return trips, would count as OR.[20] Yet, if these and other bureaucrats like them were really just deploying constrained optimization without realizing it, whyever did they feel impelled to constitute themselves as a separate entity, the "Operational Research Society of America" in 1952? Far from being a small coterie of enthusiasts trapped in some backwater of history, this organization grew to encompass 5,000 dues-paying members within the first decade of its existence. In 1995, when the Operations Research Society of America merged with the Institute of Management Sciences, the umbrella organization boasted more than 11,700 members.[21]

A rather simple and straightforward account is sometimes retailed concerning this train of events. OR, as this story goes, was simply the growth of an applied arm of economics or social science in general. As another illustrious progenitor, Patrick Blackett, put it, it was nothing more than "social science done in collaboration with and on behalf of executives" (Blackett, 1962, p. 201). First industrial firms, and then perhaps the military, got so big and so unwieldy that they were forced to cast about for help with their problems of command and coordination; fortuitously, some social scientists appeared on the scene and proceeded to solve some of their problems (Hay, 1994). Because most of the problems had to do with the discipline of human beings, in this narrative OR is little more than an extension of the Taylorist movement of "scientific management" at the turn of the century (Fortun & Schweber, 1993, pp. 620–25; Rider, 1994, p. 841; Waring, 1991). With the founding

[20] Beniger (1986, p. 231) makes this case for the work of Daniel McCallum on the Erie Railroad in the 1850s. On the next page, he reaches further back to Charles Babbage as an earlier and more illustrious anticipator of OR, as have other writers quoted in Chapter 2.

[21] The modern answer to the question "What is OR/MS?" may be found on the INFORMS web site at <www.informs.org/Join/Orms.hmtl>. There we find that members of this organization "aim to provide rational bases for decision making by seeking to understand and structure complex situations and to use this understanding to predict system behavior and improve system performance. Much of this work is done using analytical and numerical techniques to develop and manipulate mathematical and computer models of organizational systems composed of people, machines and procedures." The promiscuous enumeration of organizational building blocks testifies to the ongoing cyborg inspiration.

of postwar business schools and management curricula, OR ended up just a convenient umbrella under which to conduct the kinds of analyses that had already become commonplace in more advanced sectors of the social sciences.

This Topsy story stumbles on the historical specifics of the genesis of the discipline. First and foremost, it was natural scientists, and primarily physicists, who were responsible for the creation, conduct, and codification of OR in World War II. Fortun and Schweber do take note of this, but suggest it was merely an artifact of the historical accident that physicists were more familiar with probability and statistics in the 1940s than were their social scientific counterparts.[22] Second, there is the problem broached in the previous section that it would be imprudent to claim that the military "wanted" or actively sought out civilian advice on how to conduct the war; the same could probably be said for the captains of industry with their cadre of research scientists. It was not incidental that the application of OR was all bound up with struggles over the modality of the funding of scientific research and the ability of the paymasters to dictate the types of research pursued, much less encourage it at all. It might make more sense to follow Weaver, Conant, and the other science czars in approaching OR as a strategic innovation in the relationship of the scientist to the military patron. Who would be gracious enough to help whom, and with what activities, was part of the outcome of the process, and not part of the given problem situation. Third, the very idea that there was a generic "scientific method" to be deployed in advising clients, while certainly a popular notion in the era after the bomb, was repeatedly challenged by the fact that there was no obvious unity or quiddity to what an operations researcher seemed to do.[23] Was he merely a humble

[22] This unfounded assertion that physicists were more familiar with statistics than any other disciplinary matrix in this time period is extremely implausible. Indeed, it was as a reservoir of statistical experts that economists were first brought into contact with OR, as described in the next section. However, the undeniable fact that it was *physicists*, and not economists, who innovated OR is one of the most naggingly disturbing facts confronting those who wish to account for its genesis. For instance: "A more understandable and defensible definition of operations research is a sociological one. OR is a movement that, emerging out of the military needs of World War II, has brought the decision-making problems of management within the range of large numbers of natural scientists. . . . The operations researchers soon joined forces with mathematical economists who had come into the same area – to the mutual benefit of both groups" (Simon, 1960, p. 15).

[23] The ubiquitous gnashing of teeth and rending of garments over the exact content of OR are among the more fascinating phenomena of this history, to which future historians need to attend more carefully: for one commentator, OR "is not a subject-matter field but an approach or method. And, even after a study of hundreds of examples of work classified as OR, it is by no means clear just what the method is other than it is scientific" (Dorfman,

Baconian collector of quantitative facts, grist for the statistician's mill? No one went so far as to suggest that it was indispensable that highly qualified physicists were needed to do that sort of mundane work. Was he instead a theorist of social organizations, bringing a well-developed formal theory to clarify interpersonal relationships and smooth over rough spots in psychological conflicts? This came closer to some of the cherished self-images of the operations researcher, although they would strenuously deny knowing anything specific about psychology or social theory. Perhaps, then, it was all just a matter of possession of rare mathematical expertise in which everyone else was deficient – namely, the ability to make arguments *rigorous* through the instrumentality of formal expression.

This comes closest to the way operations research is presented in both internal documents and in many retrospective accounts. In this view, OR is what mathematicians (or mathematically inclined physicists) found themselves doing when pressed into military service. In that case, the content of OR is best displayed by enumerating the mathematical formalisms that found their way into the operations analyst's tool kit. Robin Rider (1992, 1994) opts for this route, and characterizes OR as consisting of game theory, symbolic logic, (Shannon) communications theory, linear and dynamic programming, queuing theory, Monte Carlo estimation methods, and theories of production and inventory control. Robert Lilienfeld (1978, p. 104), less concerned to portray OR as predicated solely on narrow mathematical techniques, gives the following expanded list: linear programming, queuing theory, simulation techniques, cost-benefit analysis, time series and cross-section statistical estimation, operational gaming patterned upon war games, time and motion study, and network analysis. One modern OR textbook (Winston, 1991) covers linear programming, network models, nonlinear programming, theories of decision making under uncertainty, game theory, inventory models, markov chain models, queuing theory, and simulation techniques. Once again, the fin-de-siècle reader might be forgiven for thinking the overlap with the tool kit of the modern mathematical economist is broad enough to warrant at least some curiosity.

Nevertheless, this definition of OR as a branch of mathematics founders on the inconvenient fact that mathematicians would rarely, if ever, deign to accord it that honorific status. There was nothing particularly novel or even erudite about the sorts of mathematics being used by the original OR cadres: bits of rudimentary probability theory, some statistical inference, some computational approximations to integrals,

1960, p. 575). Warren Weaver himself acknowledged this to Bush: "The term 'Operational Research' is actually being applied to activities of very different characters; and certain observations and conclusions concerning OR are confused and vague." Weaver to Bush, February 25, 1943; quoted in Rau, 1999, p. 279.

some partial differential equations (Morse & Kimball, 1951; Fortun & Schweber, 1993, p. 665). Few if any conceptual unifying threads ran through the individual calculations and exercises. Professional mathematicians participating in the OR cadres could be fairly scathing about the types of mathematics going on under their aegis: "one famous document purported to study the relative merits of different operations by methods of mathematical economics used to compute the relative loss in manpower to the enemy and to us. It is the author's opinion that this particular document represents the height of nonsense in war work."[24]

Rather than defining OR as a "part of mathematics," a better characterization of events would be that some mathematicians were dragooned during the war into mixed disciplinary units dominated by physicists and directed to participate in a motley of activities lumped together under the OR rubric; but once the war was over, they abandoned those activities for the bracing precincts of Bourbaki with relief, renouncing much of their own prior participation. The physicist administrators of those cadres often kept up their participation after the war, but as the OR community increasingly organized its own academic disciplinary identity, physicists too eventually migrated out of the area. Under the dual imperatives of finer specialization and refined expertise, the academic face of the OR discipline did become more enamored of mathematical formalisms for their own sake, but by then the process was entirely cut off from the community of mathematicians.[25] To outsiders (and clients) it may have looked like a forbidding branch of mathematics; but to most mathematicians, it was a hodgepodge not worth sustained attention.

OR as Boundary Work

I would like to suggest that the reason there is as yet no useful synoptic history of OR is that no historian has yet seen fit to frame OR as the harbinger of the practice of "boundary work" (Gieryn, 1995) and the attendant "interdisciplinarity" that has come to be the hallmark of the military-induced organization of so many fin-de-siècle sciences. While it is

[24] Saunders Mac Lane in Duren, 1989, pp. 505–6, quoting his own report on the Applied Mathematics Group at Columbia written immediately after the war. His more contemporary attitudes toward OR had not mellowed over time: "These cases may also indicate that the problems with the Military-Industrial complex may have started back then. The later overeager manufacture of 'World models' . . . may have had an origin in the hurried work of WWII in operations analysis" (p. 506). Another participant is a bit less withering: "Is OR mathematics? Nowadays, the practitioners insist that it is a separate discipline, and I guess by now it is. It is certainly not now taught in departments of mathematics. But, it grew out of mathematics." Barkley Rosser in Duren, 1988, p. 304.

[25] Philip Morse complained that the journal renamed *Operations Research* in the 1970s looked "more like advanced mathematics texts, rather than of journals of physical science." Morse Papers, box 21, MIT; quoted in Rider, 1992, p. 232.

certainly germane to inquire "Why was OR innovated by physicists?" it is woefully insufficient to suggest that their "centrality is accidental" (Fortun & Schweber, 1993, p. 627), or due simply to their wartime competence and self-confidence.[26] OR should instead be located by a process of triangulation between the natural sciences, economics, and the nascent field of science policy; furthermore, the computer was the tapis upon which the moves were carried out, which is the single most important fact rendering these developments so central to the inception of the cyborg sciences. Again, physicists wanted to be paid by the military but not be *in* the military; physicists wanted to do social research for the military, but not *be* social scientists; physicists wanted to tell others what to do, but not be *responsible* for the commands given. To be granted these dispensations, they somehow had to innovate new roles balancing this delicate combination of engagement and aloofness from the chain of command. OR turned the humble role of consultant into a fully fledged "discipline," with everything that implies. The reason that OR became so important after the war was not due to any particular technical innovation or bit of mathematical wizardry; rather, it was the workshop where the postwar relationship between the natural scientists and the state was forged, and, inadvertently, the site where neoclassical economics became integrated into the newfound scientific approach to government, corporate management, and the very conceptualization of society as a cybernetic entity. The fact that OR could boast no uniform roster of practices did not mean that it was incapable of having profound effect upon the intellectual content of academic disciplines such as economics, psychology, and even computer science. To complicate matters further, the situation was subsequently occluded by the intervention of cultural factors: OR looked different when constituted in Britain, or America, or on the European continent, or in the former Soviet Union; and much of this can be traced to the vicissitudes of World War II, divergent state policies towards science, and ensuing local interactions with culturally variant conceptions of economics.

The origins of OR can be traced to the British incorporation of scientists into the military command structure with the Tizard committee and Blackett's "Circus"; but no historian to my knowledge has made the attendant observation that its further development was effectively

[26] Certain passages in Fortun & Schweber, 1993, p. 631, seem to give the impression that physicists were recruited because they were naturally suited to be jacks-of-all-trades. If that were the case, then the physicists were not willing to serve in that capacity for very long. Rather, such claims should be understood as their assertions of expertise in the "scientific method" sanctioned them to intervene in any management functions that they deemed warranted their intervention.

frustrated in Britain by the fact it was regarded as the Trojan Horse of the "social planning of science" movement: all of the major figures of OR in Britain doubled as advocates of science planning and government funding, including Patrick Blackett, J. D. Bernal, A. V. Hill, and Solly Zuckerman. This prompted a swift and negative reaction on the part of a group of scholars located outside of the war effort, including Michael Polanyi, Friedrich von Hayek, John Baker, and Arthur Tansley, to found the "Society for the Freedom of Science" for the purpose of protecting science from corruption through government planning.[27] It is quite possible that this overt political hostility toward the seeming alliance of military organization and funding with socialist politics stunted the subsequent development of OR in Britain, or at least its all-important links with the military, and had the unintended consequence of decoupling the computer from all sorts of useful early alliances, even though British technical computer expertise was roughly commensurate with American knowledge during the war.

No such obstacles stood in the way of OR in America. Somewhat incongruously, in the land of laissez-faire the military entanglement in science planning and funding provoked no similar hue and cry to preserve and protect the virtue of scientists that occurred in Britain.[28] (Did the fact that the threat of socialism seemed more remote actually promote the planning of science? Or was it simply the indigenous prewar experience

[27] The best source on the British controversy over science funding is McGuckin, 1984. The connection between OR, the politics of economic planning, and the rise of the social studies of science in Britain is a topic we unfortunately cannot develop in further detail here. Discussions about "tacit knowledge," social epistemology, and freedom in science organization, however, would greatly benefit from taking cognizance of the interplay of trends in economics, hostility to computer metaphors, and visions of the history of science in a figure like Michael Polanyi. On this point, see Mirowski, 1997. The intellectual cross-currents often made for strange bedfellows, as when the Society for Freedom in Science reprinted a pamphlet by Warren Weaver on "Free Science" (McGuckin, 1984, p. 286). The possible interplay between the rapid decline of the British computer industry and the relatively stunted status of British OR is another topic that cries out for analysis, but which we must regretfully pass by here. It is tempting to note that James Fleck (1987) has argued that even though Britain was awash with early cyberneticians, Artificial Intelligence was strangely retarded in Britain relative to its status in the United States, then only to be championed by "outsiders" to the computer community, and that this may have had something to do with an unwillingness to accept or the unavailability of military funding in Britain.

[28] The relationship of OR to science policy and computer development in both continental Europe and the former Soviet Union would be an indispensable input to evaluation of these hypotheses. Unfortunately, work on continental OR is very scarce; and language barriers barred my inadequate attempts to find out something about Soviet cybernetics and operations research. Some suggestive work on the history of cognitive science in France is Chamah, 1999; Dupuy, 1994.

with planning innovated by the foundations?) Shepherded primarily by John von Neumann and John Williams, the purview of American OR was expanded to include new mathematical techniques like game theory, Monte Carlo estimation, and linear programming; competence in computer programming and simulation modeling became the American trademark of the crossover physical scientists like Philip Morse, George Kimball, George Gamow, William Shockley, John Bardeen, Hendrik Bode, Herman Kahn, Ivan Getting, and a whole host of others. Just as gun control was transubstantiated into cybernetics, OR became subsequently transmuted into systems analysis; and the drive to subsume organizational reform under the umbrella of quantification and computer expertise led inexorably to another search for a Theory of Everything every bit as ambitious as the holy grail of a unified field theory. As Philip Morse mused in retrospect, "None of us was the same as we had been in 1939, and we suspected that our lives would not be a return to prewar days. Whether we liked it or not, physics and physicists had transformed warfare and would inevitably have a profound influence on the peacetime" (1977, p. 213).

OR as a Leaner and Meaner Economics

If we are to gauge the impact of this development upon economics, it is necessary once again to sketch the trajectory of OR, only now from the alternative vantage point of economics. As is acknowledged in most retrospectives, OR began with the institution of the Tizard Committee on air defense in Britain in 1935 (Clark, 1962). The use of the newly invented radio wave detectors, or "radar," appeared to demand the recruitment of scientists to adequately understand and use the devices in air defense (Buderi, 1996). The earliest deployment of what the British (in their own inimitable fashion) would call "operations analysis" was in addressing the problem of using radar to aim antiaircraft guns (Blackett, 1962, p. 207). British research produced no effective hardware for gun laying, the first applicable device only constructed subsequently by the Americans in the Wiener–Bell Labs competition described in Chapter 2 (i.e., in the very incident that inspired Wiener's cybernetics). Nevertheless, Patrick Blackett's "Circus" began in 1940 as the Anti-Aircraft Command Research Group; it consisted of three physiologists, four physicists, two mathematicians, one army officer, and one surveyor. As Blackett is once reported to have said, "From the first, I refused to be drawn into technical midwifery" (Clark, 1962, p. 146); he conceived his remit as strategy and tactics, and not trouble-shooting temperamental machines. They were initially attached to the British Royal Air Force, which had existed as a separate military service since 1918; this should be contrasted with the situation in the United States, where the Army Air Forces were only established in 1941. Blackett later shifted to the Coastal Command, developing models for U-boat warfare.

A standard internalist account notes that, "psychologists and social scientists appear to have been missing from the early groups" (Trefethen, 1954, p. 11). This is especially curious, because it was only when the circus ventured beyond its nominal expertise in electronics into areas of military tactics that it achieved its first noteworthy successes. These encompassed directives such as changing the depth of explosion of air-dropped depth charges, altering bombing runs to increase damage, painting planes white to further elude visual identification from subs, and so on. What is noteworthy is that the "methodology" employed to arrive at these prescriptions resembled nothing so much as a pragmatic version of Marshallian economics, an identification noted by Blackett himself.[29] Blackett even posited his own "equilibrium theorem," "analogous to the use of the 'virtual work' theorem in mechanics," that "an intelligently controlled operation of war, if repeated often enough . . . will tend to a state where the yield of an operation is a maximum" (1962, pp. 187, 186). Nonetheless, as yet economists were not included as part of the "mixed teams" prosecuting this method. Further, access to computational devices seemed to be restricted to the separate cryptanalysis section at Bletchley Park in Britain. British science, lacking experience with planned integration into larger teams of coordinated researchers fostered by the foundations in the 1930s, seemed to project their string-and-sealing-wax ethos of research onto operations analysis itself.[30]

There is some dispute as to precisely who imported the British ideas to America, but by 1942 there were at least four very active OR groups in the

[29] "[T]he variational methods discussed below . . . are rather nearer, in general, to many problems, say, of biology or of economics, than to most problems of physics" (Blackett, 1962, p. 177). In general, a generic "yield" (of kills, targets hit, etc.) was posited, and then a rough-and-ready production function was written Y = F(a,b,c, ...) for the various causes involved in the activity. Crude data were collected on the derivatives of the function dY/da, and strategies were adjusted to optimize yield. That this application of simple rational mechanics to social or war phenomena bore strong resemblances to Marshallian neoclassical economics is no accident, as readers of *More Heat than Light* will realize. Indeed, early British OR can be regarded as a recapitulation of events in the later nineteenth century when physicists-engineers imported their versions of rational mechanics into economics.

[30] There is an irony about the British versus American conceptions of OR, which can be noted but not explored here. In 1934 Blackett scanned individual cloud chamber pictures by hand to search for subatomic events. By the 1960s in America, Blackett's other brainchild came back to reorganize and rationalize particle physics, with OR techniques employed to industrialize the search for cloud chamber events at Berkeley. On the latter, see Galison, 1997, pp. 402–5.

United States: the Operational Analysis Section of the Army Air Force; a unit at the Naval Bureau of Ordnance called the Mine Warfare Operations Research Group (MWORG), one at the Army Signal Corps, and another constituted under NDRC Division 6 at the Chief of Naval Operations under the acronym ASWORG (AntiSubmarine Warfare Operations Research Group) at MIT's Radiation Laboratory.[31] Significantly, a prominent member in his role as consultant of three out of the four groups from the beginning was John von Neumann. Von Neumann's participation was decisive for the shape that OR assumed in America, not to mention smoothing the way for OR to become a dominant concern in research units as the Applied Mathematics Panel, of which he was also a member.[32] He introduced game theory as an adjunct of OR, lending mathematical firepower and intellectual clout to the wartime concern over "strategy"; he forged the lasting links with the computer as tool and as exemplar of organizational rationalization; he endorsed the promise of linear programming to various military clients; he supplied the bridge to information theory; and he arranged for the OSRD to found training units in OR at Princeton and the Rad Lab, seeking out those physicists and engineers inclined to do applied social science.

Thus it is no accident (to put it mildly) that this list of mathematical accoutrements of OR parallels so closely von Neumann's own interests in the 1940s. He was the first to insist that the formalisms of early game theory and early linear programming resemble one another.[33] It was nowhere near odd that Stanislaw Ulam and von Neumann had recourse to the word "game" when describing indifferently yet simultaneously formal logic, social organizations, and the Monte Carlo simulation technique (Galison, 1997, p. 762). Wouldn't it be the most natural thing in this world to apply Monte Carlo techniques to approximate the solution to an especially thorny tactical game (Ulam, 1956, p. 63) and then realize this is just another form of computer simulation? Serendipitously, the extensive form description of the play of a game as a "tree" reproduced in Figure 4.1 could be construed to resemble equally the diagram of a neutron-scattering problem in atomic bomb design, the flow chart of a

[31] See, for instance, Fortun & Schweber, 1993, p. 603; Trefethen, 1954, p. 12; Johnson, 1997b, p. 897; Bush, 1970, p. 92; Hay, 1994. The nature of these units in the earliest period is covered in Rau, 1999. ASWORG later became the Center for Naval Analyses.

[32] See Trefethen, 1954, p. 15; Fortun & Schweber, 1993, p. 603; Rau, 1999, p. 164. Von Neumann's "role, not only during the war but after its conclusion, was unique; for he was a consultant or other participant in so many government or learned activities that his influence was broadly felt" (Mina Rees in Duren, 1988, p. 277).

[33] See the letter of von Neumann to Lee Ohlinger, April 10, 1950, where he admits that solving linear equations, eigenvalue problems, ordinary least squares, linear programming, and minimax solutions for a zero-sum, two-person game with n strategies are "really" all the same mathematical problem. Box 6, folder 1, VNLC. See also Schwartz, 1989.

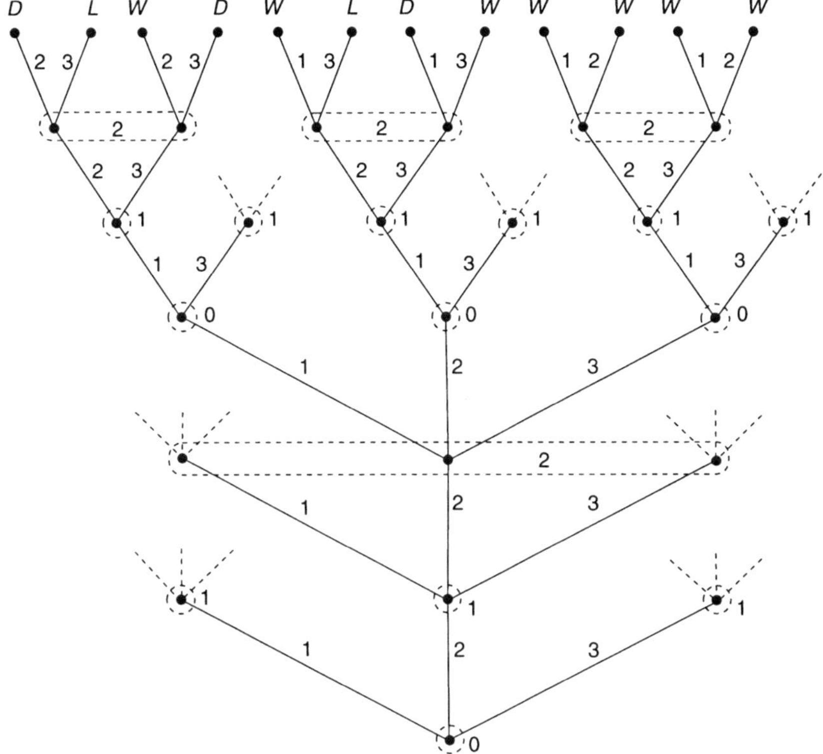

Figure 4.1. Game tree. *Source*: Luce & Raiffa, 1957, p. 46.

computer program, and (inverted) the organization chart of a military unit or a multidivisional corporation.[34] These were the kind of synergies that von Neumann (at least initially) found could be expressed in the idiom of OR. It was formalism transcendent of impurity, metaphorical promiscuity gone worldly, machines regimented for action.

By the early 1950s von Neumann was too much encumbered with governmental responsibilities and corporate consulting to participate in the founding of professional societies such as the Operations Research Society of America and TIMS (Institute of Management Studies); but that does not mean he was not responsible for much of the shape of postwar American OR – and, by implication, much of American fin-de-siècle economics – in any event. It is now well known that game theory was not,

[34] "I think I have now come near to a procedure which is reasonably economic at least as a first try: I will treat the Ph[otons] by the old 'blackjack' method" (von Neumann quoted in Galison, 1996, p. 465). "In a way, operations research was like applying quantum mechanics to military operations" (Riordan & Hoddeson, 1997, p. 104). Actually, it was usually closer to thermodynamics.

with very few exceptions, developed within the ambit of the immediate postwar economics profession, American or otherwise. Funding and recruitment for budding game theorists therefore had to be sited somewhere else if the tradition were not to be stifled prematurely; and OR was the vehicle through which von Neumann kept game theory on life support in this period. Philip Morse, a leader of early OR, acknowledged this: "OR has its own branch of mathematics, called the Theory of Games, first developed by von Neumann and since worked on intensively by Project RAND and other military operations groups" (quoted in Rider, 1992, p. 236). The military knew it needed OR; John von Neumann told the military it consequently needed game theory; on the strength of his recommendation, the military believed it – for a while.

Thanks largely to von Neumann's encouragement and Weaver's behind-the-scenes efforts, the Air Force think tank RAND became the showcase of what OR American-style could accomplish in the 1950s; and not coincidentally, it also stood as the world's largest installation for scientific computing in the same period (Edwards, 1996, p. 122). It was at RAND that OR openly embraced the postwar cyborg. There, problems of rationality and organizational efficiency became conflated with problems of computer design and programming. It was also at RAND that OR assumed its more ambitious postwar trappings: "Bombing accuracy, gunnery practices, maintenance concepts, supply and inventory problems, these were [World War II–era] topics. . . . they were overtaken and subordinated to much larger issues of weapons choice, strategic doctrine, procedures of R&D, methods of ensuring interservice cooperation in combat conditions" (Perry in Wright & Paszek, 1970, p. 114).

It was equally at RAND that American science policy was first yoked to an "economics of innovation" (Hounshell, 1997a), thus demonstrating the inseparability of science policy and OR. The individual armed services, for their part, felt compelled to have their own extensive internal OR units, if only just to "answer" the studies and prognostications coming out of RAND (Holley, 1970). In an ever widening spiral of escalation, OR moved into the Department of Defense itself with Robert McNamara's "whiz kids" in the early 1960s (Herken, 1987; Kaplan, 1983; Shapley, 1993). It was only in the Vietnam War that some of the ardor for OR and a science of war began to noticeably cool (Twomey, 1999).

It will be crucial for subsequent arguments to come to understand that, even during World War II, but certainly by the 1950s, American OR had diverged so profoundly from the British variant that fault lines began to appear within the discipline. For instance, Ivan Getting informs us, "it is clear that the British Mission of Tizard in suggesting gun-laying did not intend that the American effort should involve the concept of automatic tracking. In fact it was quite clear that the British were opposed to the approach" (1989, p. 108). We might then extrapolate this hostility toward automated tracking forward to an analogous postwar suspicion about Wiener's cybernetics. Or indeed, as American OR shifted its focus from humble consultancy concerning already existing weaponry to concocting military strategy for weapons that did not yet exist, many of the progenitors of British OR such as Blackett began to speak out openly against the necromancy of the American discipline. These splits were not merely political: they were also conceptual, revealing an uneasiness about the direction of the nascent cyborg sciences, which was relatively absent in the United States.

As one might expect from the ubiquitous participation of John von Neumann, OR American-style would not be all that favorably inclined toward the methods characteristic of neoclassical economics, in stark contrast to its British incarnation.[35] What had begun as a British hands-on procedure stressing field work and tacit knowledge achieved through observation in wartime (Morse & Kimball, 1951, p. 141) quickly became transformed into a more abstract, aloof, and theoretical discipline. However, abstraction came to mean something quite particular in American OR. Constrained optimization was viewed as relatively trivial, except in more complex situations with inequalities as constraints. Probability models were patterned increasingly on thermodynamics: "I could see interconnections between statistical mechanics and OR" (Morse, 1977, p. 311). Minimax concepts were imported from game theory, and elaborate stochastic models became the calling card of the operations researcher. The enemy was portrayed as receiving "messages" from military offensives, processing the information, and responding in kind, entertaining counter-counter-countermeasures. The intrusion of nuclear weapons only intensified this trend. As Morse complained as early as 1950, "The problems the [OR] groups were asked to work on became more tinged with interservice politics, and the data underlying their evaluations became ever more vague. . . . Calculations of nuclear war capabilities became less and less scientific and more and more astrological" (1977, p. 259). As the dangers grew progressively more intangible, where the U.S. Navy could pose as immediate a threat as the KGB to the U.S. Air Force, the mantra of the "reduction of uncertainty" grew more insistent: "many organizations saw in such new management tools as game theory, decision analysis, OR and econometric model building a way to substitute predictable formal and analytical skills for managerial (and command) ones that could never be perfectly known" (Rochlin, 1997, p. 190). But the effect of turning the focus

[35] "The SSRC gave me a rough time when I told them I believed that OR differed both in technique and in subject matter from the social sciences on one hand and economics on the other" (Morse, 1977, p. 290).

inward, of looking less and less at concrete weapons and repeatable events and more and more at the problems of information, blue-sky organizational alternatives, and virtual feints, resonated quite well with the diffusion of the computer throughout all levels of the military command structure. RAND pioneered the practice of running training simulations on the same computer equipment that would be used to direct actual nuclear war. Simulations and gaming were consequently elevated to roughly equivalent status with analytical solutions to models, not only out of necessity, but also out of a revised conception of the nature of abstraction (Edwards, 1996, pp. 119–20). Abstraction became separate from simple quantification on one hand, and Bourbakist monastic communion with Platonist reality on the other; it was instead constituted as a novel professional ethos. Further, OR created whole new identities for middle management and swelled to fill new niches. Like postwar corporations, the military grew to suffer from bureaucratic bloat. Whereas 65 percent of Army personnel were combat soldiers in World War II, by 1990 the proportion was down to 25 percent. Wars had become huge logistical nightmares, conceived as vast cybernetic problems of command, control, communication – information cascading over hordes of quasi-stochastic actants. "Blipkrieg" was more than a pastime; it had become a legitimate vocation.

Where were the American economists when all this was going on? Here the plot not only thickens but positively curdles.

THE BALLAD OF HOTELLING AND SCHULTZ

To understand what neoclassical economists were doing during World War II, one needs to know what they were doing *before* the war and, in particular, during the Great Depression. Because no consensus history of American economics in the Depression and postwar period exists, my coauthor Wade Hands and I have sought elsewhere to provide a narrative of the rise to dominance of the neoclassical orthodoxy in the period from roughly 1930 to 1980 (Hands & Mirowski, 1998, 1999). In impossibly terse summary, the narrative there suggests that there was no dominant orthodoxy in American economics prior to World War II, although the indigenous strain of American Institutionalism held some key strategic outposts at Columbia and Wisconsin. This situation stood in sharp contrast to Britain, where a Marshallian orthodoxy held confident sway from the beginning of the century. The Depression destabilized the American situation profoundly, provoking some key figures (like our two present protagonists, Hotelling and Schultz) to search for an improved scientific basis for neoclassical economics. They failed; but out of their failure, combined with some external influences traceable to World War II, a robust economic orthodoxy did arise like a phoenix in the immediate postwar period. One of the novelties of our interpretation of this period is that this orthodoxy did not succeed out of monolithic consistency or logical clarity, but rather by adopting multiple, variant positions with regard to a key doctrine, the nature of neoclassical price theory. These three versions of American neoclassicism, described in elaborate detail in our other writings, can be dubbed the Chicago doctrine, the Cowles approach, and the MIT style.

It would take us too far afield to describe these events and doctrines in detail; for that the reader is directed to our cited works. But (as explained in Chapter 1) the present book is intended as the third in a trilogy explaining the impact of the natural sciences upon economics, and it is right at this point – First Contact of neoclassicals with the fledgling cyborg scientists – that we need to begin our description of the major players on the economists' team with greater precision and care. This becomes all the more imperative when we come to appreciate that the distribution between the three versions of American orthodoxy had quite a bit to do with the reactions of key individuals in each school to wartime experiences with OR and their differential response to the failures of Hotelling and Schultz. To reprise the observation of Peter Buck, the politics of interdisciplinary research confronted these economists with the overwhelming importance of theory. Moreover, local conditions dictated the *shape* of the theory pursued. Crudely, Chicago economics derived from experience at the Statistical Research Group of the Applied Mathematics Panel, whereas the Cowles Commission was predicated upon lessons learned from RAND; MIT economics, perhaps the least affected of all by the war, was nonetheless contingent upon experience at the Radiation Lab. It will become apparent in the course of our narrative that these three locations coincide exactly with the three preeminent centers of operations research in America in the 1940s. But beyond that crucial observation, this account also serves as a prerequisite for an understanding of how and why the next generation of neoclassical economists ultimately were repelled by the cyborg sciences, instead consolidating their newfound competencies into a dominant academic discipline remaining relatively hostile to cyborg themes, at least until the 1980s. For purposes of tracking those whom we identify as the major protagonists in this saga, we shall therefore take up the narrative thread with two 1930s economists who became the obligatory passage points for the subsequent development of mathematical economics in America, Harold Hotelling and Henry Schultz.

Desperately Seeking Demand

In the black-and-white world of Depression America, there were numerous economists seeking to deploy the "Law of Supply and Demand" to confront the appalling array of problems facing the nation: crop failures, widespread unemployment, falling price levels, the concentration of economic power in large trusts, calls for a national economic planning, and so forth. Yet in (say) 1932 there were a paltry few economists who thought that the ubiquitous disarray and patent failures of economics could be promptly resolved by an abstract reconsideration of the scientific foundations of the demand function, conditional upon acquiescence that one should remain within the general framework of individual constrained maximization of utility, and pursue the consequences of the general interdependence of the prices, quantities, and incomes that were the observable artifacts of these calculations. On the contrary, there subsisted a large contingent of the profession who believed experience had effectively repealed those putative "laws"; and there was an equally large proportion who believed that existing Marshallian theory could adequately account for contemporary events (Barber, 1985, 1996). Both segments of the profession would have regarded the idea that what was most urgently required was a more rigorous restatement of the mathematical relationship of constrained optimization to demand functions and the injection of a more formalized brace of statistical procedures to fortify economic empiricism with something akin to dumbfounded disbelief. Nevertheless, this was precisely the project initiated by Harold Hotelling in his 1932 paper on the "Edgeworth Taxation Paradox" and prosecuted with singular intensity in tandem with Henry Schultz until Schultz's death in a car accident in November 1938.

Hotelling and Schultz discovered in each other the perfect complement to their own respective, independently chosen research program in economics. Schultz had been guided down the road of empirical estimation of something called "demand functions" by his mentor Henry Ludwell Moore at Columbia, who had conceived of a deep distrust of Marshallian economics; Schultz had subsequently come round to the position while at the University of Chicago that formal demand theory had to be based on Walrasian-Paretian mathematical models, in opposition to his mentor (Mirowski, 1990). His fervent desire to render his project more consonant with contemporary understandings of the scientific method led him to look to physics for inspiration and to champion the physicist Percy Bridgman's (1927) philosophy of "operationalism." Schultz was searching for a tractable mathematical economics that would acknowledge pervasive interdependence at the market level, yet one you could actually use to diagnose real problems of agricultural production and sales in the 1930s – a statistical empirical economics to stand in contrast with the casual empirical observations of the British Marshallians. Schultz also became the founder of a statistical laboratory operation at the University of Chicago, which anticipated the organizational structure of Big Science – large teams of divided labor engaged in elaborate calculations and data gathering – in some respects. In order to ground his mathematical exercises firmly, he required some sort of guarantee or warrant that the individually estimated demand functions really meant something in a world of pervasive interpenetration of causes; he was convinced that there must be underlying dependable "laws" behind the possibly spurious observable functions of prices and quantities.

Harold Hotelling, trained in physics and mathematics at Princeton, also became engaged in estimating agricultural demand functions at Stanford's Food Research Institute in the later 1920s. His initial interest was in Fisherian hypothesis testing (an enthusiasm shared with Schultz), but he also was rapidly brought round to an appreciation for the mathematical neoclassical theory that buttressed this effort, although his favorite authors tended more toward Cournot and Edgeworth. Called to the Columbia economics department to replace Henry Ludwell Moore in 1931, he rapidly became embroiled in efforts to produce a mathematical theory that would explain and underwrite the legitimacy of demand functions; not incidentally, he believed that such a foundation would be relevant to political issues of evaluation of welfare and the diagnosis of causes of the Depression. In common with Schultz, he harbored nothing but disdain for the British Marshallians with their easy appeals to ceteris paribus and their smatterings of mathematics. Due to his relatively limited prior background in economics (Darnell, 1990, pp. 3–4) and his obviously superior mathematical talent, he enjoyed rather greater degrees of freedom than Schultz in his understanding of the physics analogy and in his willingness to amend the Walrasian-Paretian organon in order to derive a rock-solid demand function. This was nowhere more evident than in his 1932 article on the "Edgeworth Taxation Paradox," published in the *Journal of Political Economy*. If Edgeworth could demonstrate that the imposition of a tax on one good might actually lower its equilibrium price in tandem with that of another related good, then what could be said with any conviction in the neoclassical tradition? Hotelling believed he had found a model that would produce more dependable laws of demand than those previously cited.

Schultz was the editor at the *Journal of Political Economy* assigned to the Hotelling paper, and this contact initiated an intensive correspondence about the statistical consequences of his new model foundations for demand. As described in detail by Hands and Mirowski (1998), Hotelling had proposed two novel ways to "derive" a demand curve: one (subsequently ignored) from a cumulative normal density function, and the other from an unconstrained optimization of the quantity $U - \Sigma(p \times q)$ (utility minus the sum of quantity purchased at equilibrium prices). The latter, which Hotelling called his "price potential" model in direct analogy with the treatment of the motion of a particle in classical mechanics, would guarantee downward-sloping demand functions and straightforward welfare indices even in a world of pervasive interdependence of prices and quantities. While it possessed other virtues (which we have elaborated elsewhere), it also bore the drawback that the "income" or "budget" term was not treated as an entity fixed independently of the equilibrium outcome. This bothered Schultz, but, to his credit, he chose to collaborate with Hotelling to explore various justifications for the novel treatment of the budget in the constrained maximization problem.[36] Schultz, after all, was searching for some fundamental theory of interdependent demand curves, and this seemed a viable formal candidate, even though it was not to be adequately found in Walras or Pareto.

The novel treatment of utility optimization also gave rise to some other symmetry (or "integrability") restrictions on demand, which Schultz found unpalatable. Nonetheless, Schultz immediately set out to test these symmetry restrictions in his agricultural price data, only to discover that they were widely violated. Hotelling and Schultz then proceeded to discuss all the various auxiliary hypotheses that one might bring to bear to understand the failure, with Hotelling directing his efforts primarily toward pure theory, while Schultz tended to cast about evaluating statistical and data problems. During this hiatus, the Slutsky symmetry conditions were brought to their attention – and this is crucial for our later story – by Milton Friedman, who had taken an M.A. at Chicago in 1933, had been studying under Hotelling at Columbia in 1934, and was to return to Chicago in 1935 to help Schultz write up his next brace of demand

[36] Here experience dictates I must signal that this was not "simply" an appeal to the neoclassical special case of the Marshallian constant marginal utility of money, nor was it straightforwardly a special case to be rigidly restricted to some separate sphere of "production." These issues are discussed in detail in Hands & Mirowski, 1998. For the present, it suffices to insist that Hotelling and Schultz were engaged in a process of negotiation over the meaning and significance of the mathematical budget term, which would imply revision of a whole array of other theoretical terms in tandem, including but not restricted to the measurability of utility, the treatment of money, the significance of complementarity, the nature of pervasive interdependence, the meaning of scarcity, and so on. This becomes clear in their detailed correspondence over the problem; see box 1, folder: Henry Schultz correspondence, HHPC. One example is found in the letter of Schultz to Hotelling, November 9, 1933: "Of the two alternatives which you propose for 'overcoming' the difficulty, I prefer – at least for the time being – the one which assumes the utility is measurable, and that there exists a total utility function. I do not wish to define the connection in terms of a third commodity such as money, for the reason that the third commodity may itself have utility and then the difficulty is with us again. The difficulty might perhaps be overcome if we could develop criteria for telling us whether the *entire group* of commodities are competing or completing."

estimates.[37] Hotelling acknowledged the Slutsky equations as another way to underwrite observed demand curves, by means of the more "Walrasian" route of maximizing an individual utility function given an independent income constraint; however, its drawbacks centered on the severely diminished ability to provide guarantees for welfare theorems, uniformly guaranteed negative slopes for demand curves, and so forth. Schultz then decided to redouble his efforts to estimate well-founded demand curves, this time testing the Slutsky and Hotelling symmetry conditions as rival hypotheses. To his dismay, both conditions appeared to be equally contradicted by the data.

Here the saga of Hotelling and Schultz might superficially seem to draw to a close, but with a denouement that was satisfying to neither of the protagonists. Schultz wrote up the results of his decade-long search for the Walrasian principles underlying demand curves in his 1938 book *Theory and Measurement of Demand.* It is not often realized (because it is not often read) that the book was closer to a swan song than a manifesto for an empirical mathematical Walrasian economics, a dirge for the theory of demand: Schultz bravely reported the empirical debacle in detail, and then produced a litany of excuses why things had not worked out as hoped. Paul Samuelson, who had attended Schultz's class in Chicago from 1934 to 1935, had also been closely involved in trying to tidy up the unwieldy mess. "Already in 1936, on a trip back to Chicago, I had unsuccessfully tried to interest Henry Schultz in Jacobian conditions for rigorous aggregation (functional dependencies of separability). Also, I spent countless hours pursuing testable regularities that aggregate demand data must theoretically obey" (1986, p. 223). Milton Friedman had helped scour the estimates for empirical infelicities. But it was all for naught, it seems. The book ended with a promissory note, which bore little in the way of substantive promise. Capping frustration with tragedy, Schultz died in a car accident in California in November 1938, just as the book appeared.

Hotelling's Retreat

The reaction of Hotelling to the empirical disappointments and the loss of Schultz was no less unexpected and presents more of an explanatory challenge to the historian. Initially, Hotelling showed every sign of wishing

[37] Milton Friedman to Harold Hotelling, March 6, 1935, box 1, folder: Milton Friedman correspondence, HHPC. See, for instance, the interview with Hammond (1993, p. 222): "Henry Schultz was a good mechanic, but he wasn't really very smart. . . . [he] had a great deal of influence on me, primarily by recommending me to Hotelling at Columbia [laughter]. . . . His book on the *Theory and Measurement of Demand* is a good book – even the parts I didn't write [laughter]."

to pursue his initial "price potential" program further.[38] He continued to teach a graduate course in mathematical economics at Columbia; but the war and its own demands intervened. On April 15, 1943, he was formally appointed Consultant to the Applied Mathematics Panel of the National Defense Research Council, in recognition of his work in putting together the Statistical Research Group at Columbia. This event, described in detail in the next section, is the first time cyborgs crossed the path of Hotelling and his students: they had been recruited to do OR for the military. On the whole, for Hotelling it was not a happy encounter. Even though Hotelling would seem to have been ideally sympathetic to the statistical bent of the cyborgs, and concurred with their respect for the physics, he permanently disengaged himself immediately from OR upon the termination of the war. There also exists some evidence of personal friction with some cyborg representatives, and in particular von Neumann, from around that time.[39]

[38] Indeed, it seems that for Hotelling, the price potential model had become equated with the very idea of "rational action" *tout court*. Evidence for this comes from the unpublished note entitled "On the Nature of Demand and Supply Functions," undated, folder: Mathematical Economics, box 26, HHPC: "'Rational Actions' may be taken to mean a system of demand functions such that a 'potential' U exists with $p_i = \partial U/\partial q_i$. Such demand & supply functions may well be taken as central, all others being treated as more or less causal deviations, often of only temporary importance. But U may be a function not only of the q's but of their time- or space-derivatives. Thus non-static conditions may arise, e.g., irreplaceable resources. Also, each person's U may depend upon the consumption of others (emulation; competitive display; but also less wasteful types of activity, as when in intellectual cooperation a particular subject occupying the focus of attention of a group may, advantageously to society, be pushed). The statistical determination of $\partial p_i/\partial q_j$, which equals $\partial p_j/\partial q_i$ for 'rational action,' involves a least-squares solution & ideas of correlation which generalize ordinary calculus of correlation by replacing individual variables by matrices. These matrices will, moreover, by symmetric [*sic*], giving rise to interesting theory."

As late as 1939, he still included in his self-description of his research program in mathematical economics: "the nature of demand and supply functions when the behavior of buyers and sellers is rational; the incidence of taxation in the case of related commodities, with a disproof of certain classically accepted notions in this field." Vita included in letter to P. C. Mahalanobis, January 26, 1939, box 4, HHPC.

[39] George Dantzig describes his first presentation of the idea behind linear programming as follows: "After my talk, the chairman called for discussion. For a moment there was the usual dead silence; then a hand was raised. It was Hotelling's. . . . This huge whale of a man stood up in the back of the room, his expressive fat face took on one of those all-knowing smiles we all know so well. He said: 'But we all know that the world is non-linear.' Having uttered this devastating criticism of my model, he majestically sat down. And there I was, a virtual unknown, frantically trying to compose a proper reply. Suddenly another hand in the audience was raised. It was von Neumann. . . . Von Neumann said: 'The speaker titled his talk linear programming and carefully stated his axioms. If you have an application that satisfies the axioms, well, use it. If it does not, then don't,' and then he sat down" (Dantzig in Lenstra et al., 1991, p. 45). Von Neumann's role in the genesis of

Hotelling's ties to economics were further attenuated when, immediately after the war, he was recruited by the University of North Carolina to set up an Institute of Mathematical Statistics. His intellectual preoccupations, already drawn during the war back toward the physical sciences from which he had began, now progressively gravitated toward fields that were striving to incorporate newer developments in mathematical statistics into their own research practices, such as psychology and medical diagnosis. So it would seem that we have another instance of a peripatetic mathematical savant, bringing his portable expertise to one field, only to abandon it abruptly to enrich another. Hotelling certainly never wrote down any further substantial economic model after Schultz's death, even though his academic career extended for another two decades. But doubts about his attitude toward the subsequent developments in neoclassical price theory linger. He certainly had plenty of opportunities to vent them, if he so wished. He continued to be treated with the greatest respect by the community of mathematical economists, as demonstrated by his later election as a Fellow of the Econometric Society and the first Distinguished Fellow of the American Economics Association (although he had never been a member: see Darnell, 1990, p. 9).

A survey of all the evidence suggests that Hotelling simply stopped reading any economics by 1940. Requests by others to comment on subsequent developments were met with reiterations of points he had made before in the 1930s. When asked to write a letter of recommendation for Tjalling Koopmans's move to Yale in 1955, he managed to avoid citing or discussing any of his work in economics.[40] His class notes also suggest that he almost exclusively restricted his lectures to problems tackled in his 1930s papers.[41] It would seem that with the loss of his prime collaborator and interlocutor, Schultz, he "had no one to talk to" – a characterization that often crops up in the retrospective memoirs of others. Although this situation is often taken to mean that he was bereft of students and

linear programming is discussed in the next chapter. Hotelling never expressed an opinion in print concerning the merits of von Neumann's game theory, but it is interesting to note that in 1947 he ventured the opinion that humans were unlikely to achieve true randomization on their own, a conviction that would certainly impugn the very idea of mixed strategies in game theoretic solutions.

40 Hotelling to Lloyd Reynolds, September 24, 1954, box 5, Misc. correspondence "R," HHPC.

41 See the comments of Kenneth Arrow in Feiwel, 1987b, p. 640: "As to the influence of his work, well, at the time [1941–42] he was not pushing economics, so there was no direct influence. The course was on his papers of the 1930s in economics, but it was not current work." Or see the comment of Tibor Scitovsky in Colander & Landreth, 1996, p. 209: "At Columbia [in 1939] no one said or asked anything about Keynes. Hotelling and Wald were in a very different field." Hotelling was nevertheless the main inspiration in the early career of Kenneth Arrow, as discussed in the next chapter.

colleagues possessing adequate training in mathematics at Columbia, such a construction flies in the face of reality. It was precisely the students of Hotelling (and Schultz) who were responsible for the stabilization of the postwar neoclassical orthodoxy and the creation of at least three versions of price theory to break the impasse encountered by the duo and described in *Theory and Measurement of Demand.* One need only gather the names together in one place to begin to doubt that interpretation: Milton Friedman, Kenneth Arrow, and Robert Dorfman and, at one remove (through Schultz), Paul Samuelson and George Stigler. However frustrated Hotelling may have been with his Institutionalist colleagues at Columbia, lack of surrounding talent cannot explain the clear decision to disengage from economic theory.

There is one other possible explanation: Hotelling never entirely repudiated his 1932 "price potential" model, and his wartime experience at the AMP soured him on all the alternatives. He continued to teach his personal model well into his career and late in life conceived an interest in experimental work to measure individual utility functions directly – an eventuality more in tune with the 1932 model than with the behaviorist denial of any psychological basis of preferences so rife in that period.[42] All signs point to a belief in his own conception of straightforward quantitative welfare indices and less bothered by "proper" treatments of the budget constraint, which could then be regarded as a mere auxiliary consideration. But after the war, nobody was left who shared Hotelling's particular vision of a scientifically legitimate neoclassical price theory.

Curiously enough, there remains one way in which Hotelling's experience with OR may have subtly changed his attitude toward the goals of neoclassical theory. Whereas in the 1930s discussions with Schultz utility functions were still treated as "recoverable" from actual economic behavior, there is some evidence that in the late 1940s they took on the trappings of an "ideal type" rationality not to be found in any actual experience. In a rare commentary on some disputes over integrability theory in the 1940s, he wrote:

> Preference and demand functions and consumer's surpluses are commonly understood to refer to people's actual preferences and choices. Sometimes people do not make their choices rationally and consistently.... in wartime a national food controller has at his disposal

[42] Hotelling to R. G. D. Allen, February 7, 1966, box 2, Misc. correspondence "A," HHPC: "I have plans gradually becoming more definite, for a book of a special sort on mathematical economics, consisting chiefly of numerical illustrations of utility or indifference functions with solutions worked out showing curves as free from singularities as possible and providing illustrations of effects of various kinds of excise and other taxes."

> much information about the science of nutrition . . . and can apportion them among people in a way that will give them better health and welfare than as if each could freely choose for himself. If we are to consider such systems of planning and allocation, or the consumer's choices that would result from improved information on their part, then it is appropriate to take as preference or utility functions something based not on what consumers have been observed to do, but rather on what they ought to do if they were entirely rational and well-informed. (1949, p. 188)

Although Hotelling never himself followed up on this theme of the "virtual" rational consumer, predicated on an ideal state of "information," using computationally intensive techniques like linear programming, imitating the operations researcher, and circumventing all the vexing problems of empirical verification, it did show up in the work of his most illustrious student, Kenneth Arrow.

Hotelling was, of course, fully aware of the Slutsky decomposition of demand into income and substitution effects from 1935 onward; he subsequently taught that version of the model in his courses, as well as his own 1932 model. Repeated postwar references to "measuring welfare" might seem ambiguous in this ambidextrous context, unless they are juxtaposed to his 1932 convictions about the necessary character of the integrability relations.[43] Yet the postwar neoclassical consensus, largely created by the illustrious roster of his students, coalesced around the doctrine that one must start with Walras and Slutsky, and nowhere else, to become an orthodox "mathematical economist" in good standing, not to mention a strident rejection of the measurability of utility and a hermeneutics of suspicion concerning welfare. We should now come to see this as a localized cultural prejudice, a specifically American phenomenon, one that seems to have had some relationship to prior exposure to the Hotelling-Schultz dialogue of the 1930s.

SRG, RAND, RAD LAB

Recently it has become fashionable to ask to what extent neoclassical economics has contributed to the study of war;[44] the answer, more often

[43] Hotelling, 1932b, p. 452: "To the doubts whether utility is objective, the reply may be made that the demand and supply functions are objective things, and that if utility is defined as an integral or other functional of those functions, it too is objective. In this sense, utility has the same legitimacy as a physical concept such as work or potential, which is the line integral of force, provided certain integrability conditions are satisfied. The weakness of discussions of utility which start on a psychological basis are those of treatments of force which start from muscular exertion."

[44] See, for instance, Doti, 1978; Leonard, 1991; Goodwin, 1991; Sandler & Hartley, 1995; de Landa, 1991.

than not, has been: not much. Our contention is that the question is ill-posed, and should be inverted: how much did attempts to study war determine the postwar shape of neoclassical economics? Most of the major protagonists in our postwar narrative would strenuously protest that such a phenomenon was essentially unthinkable. My favorite dismissal, bordering upon mendacity, comes from Nobelist George Stigler: "Wartime throws up many economic problems, but the problems do not succeed in holding the interest of economists" (1988, p. 63). If this statement holds the smallest mustard seed of truth, it is because *active* consulting for the military was *mostly* conducted under the rubric of "operations research" for the first three decades after World War II; and postwar OR had been hived off as a disciplinary structure separate from academic economics, so that "Defense Management" never appeared under "Economics" in course catalogs, although many of the personages described in the rest of this volume maintained dual affiliations. Nevertheless, military and quasi-military units such as the Office of Naval Research, the Atomic Energy Commission, National Advisory Committee for Aeronautics (later NASA), Defense Applied Research Projects Agency (DARPA), and RAND (to only name the largest) constituted the primary source of all support for academic science in the immediate postwar period, including mathematical economics. It has been estimated that in 1949 the ONR alone was funding 40 percent of the nation's "pure" or academic contract research (Nanus, 1959, p. 105). At MIT, the ONR alone accounted for half of all sponsored research in the period 1947–50.[45] And this represented a period of "drawdown" from the empyrean heights of expenditure characteristic of wartime mobilization.

It might be retorted that it didn't really matter where the money came from; as long as the researchers were free to engage in whatever research captured their ivory tower fancies, choosing topics as "useless" and conclusions as farfetched from the perspective of their patrons as they wished, then no bias or influence could be inferred. Precisely at this point the historian must resist the vague blandishments of the memoir, the blanket reassurances of the reminiscence, the bland complacency of the science manager. For as we have seen, Weaver and Bush were busily advocating the complete and utter freedom of scientists to choose their own research programs with no strings attached, all the while constructing the most targeted and micromanaged scientific research effort in the history of the United States. Historians of physics have begun to inquire how military priorities shaped the face of postwar physics (Forman, 1987; Mendelsohn,

[45] F. Leroy Foster, "Sponsored Research at MIT," unpublished manuscript, MIT Archives.

1997; Kragh, 1999), and more recently we find research into its impact on other natural sciences (Oreskes, 2000; Kay, 2000); it would seem incumbent upon historians of the social sciences to follow their lead. The path is especially treacherous and daunting because the participants were willing to submit to the regime of secrecy and classification in order to benefit from the largesse; and thus they were demonstrably willing to continue to shield themselves from inquiry behind a labyrinth of classification and selective silence. Even a half century after the fact, the standard skills of the historian are often not adequate or sufficient to disinter the full remains of the postwar cyborg.

Although undoubtedly the better way to demonstrate the mesmerizing power of military patronage would be to track the careers of individual economists, we shall opt in the remainder of this chapter to concentrate instead on a few selected research units organized for military purposes: Hotelling's Statistical Research Group (SRG) of the Applied Mathematics Panel at Columbia, the RAND Corporation of Santa Monica, and the Rad Lab at MIT in Cambridge, Mass.[46] Not unexpectedly, all three bear the hallmarks of Warren Weaver's plans and aspirations for postwar science. The three are of course interrelated in their research mandates; but from the vantage point of the history of economics, they were also different enough to impress a distinctive stamp upon three major postwar approaches to neoclassical price theory. The SRG was responsible in many ways for the peccadilloes of the postwar "Chicago school" of political economy, whereas RAND had a profound influence on the evolution of the Cowles Commission in the 1950s; and the Rad Lab impressed some of its research practices upon members of the postwar MIT economics department. Wade Hands and I have argued that the symbiosis of these three distinct approaches to neoclassical price theory was constitutive of the postwar orthodoxy of American economics. If one can accept that proposition, then this chapter implies that military patronage was crucial in defining the shape of postwar economics.

[46] This selection omits other important units and agencies, both during wartime, such as the Research and Analysis Branch of the OSS, the War Production Board, the Enemy Objectives Unit of the Economic Warfare Division of the State Department (Leonard, 1991), and postwar, such as the CIA, the Weapons Systems Evaluation Group of the Joint Chiefs of Staff, and the ONR itself. A very interesting history of the OSS by Barry Katz (1989) suggests that when different economists were thrown together with humanists, as they were in the OSS (rather than natural scientists in the OR units covered in this chapter), then their subsequent careers in economics tended to display a rather greater skepticism toward the neoclassical penchant for scientism. The principles of selection and omission of economists in this chapter are primarily the dual criteria of relevance to the postwar stabilization of the "three school" configuration of neoclassical economics, and the relationship to the nascent cyborg sciences. Much remains to be done.

The Statistical Research Group/vAMP

In the spring of 1942 Warren Weaver decided that the AMP should convene a group to work on statistical problems of ordnance and warfare at Columbia under the principal leadership of Harold Hotelling.[47] Upon consultation with Hotelling, W. Allen Wallis [48] was chosen to run the day-to-day activities of the group. Wallis had been a student of Hotelling in 1935–36, after studying under Schultz at Chicago, and was at that time teaching statistics at Stanford. It appears that Weaver envisioned convening one of his mixed-team OR groups built around a joint nucleus of mathematical statisticians and engineers, and to this end suggested members Julian Bigelow (later to join Norbert Wiener in gun control work) and Leonard Savage. But the primary disciplinary orientation of Hotelling and Wallis was still at that juncture inclined more toward economics, especially considering Hotelling's 1930s pipeline to the Chicago department, which accounted for the subsequent recruitment of Milton Friedman and George Stigler for the group. Other principal staff members of significance were Edward Paxson, Abraham Wald, Meyer Girshick,[49] Jack Wolfowitz, Churchill Eisenhart, and Herbert Solomon (Wallis, 1980, p. 324), most of whom were recruited expressly for their statistical expertise. There was another "Statistical Research Group" nominally based at Princeton under Samuel Wilks, another Hotelling student, which counted Frederick Mosteller and John Davis Williams among its numbers; apparently there was free movement between the two groups.[50]

The tasks initially assigned to the SRG bore a close resemblance to those generally carried out by British OR units, leavened by the gun-

[47] Box 12, file: Statistical Research Group, HHCP.

[48] Wilson Allen Wallis (1912–98): B.A. in psychology, University of Minnesota, 1932; post-graduate education at University of Chicago, 1932–35, and Columbia, 1935–36; economist on the National Resources Commission, 1935–37; instructor of economics at Yale, 1937–38; assistant professor of economics at Stanford, 1938–46; research associate at NBER, 1939–41; research director at SRG, 1942–46; professor of business statistics and economics at University of Chicago, 1946–62; special assistant to President Eisenhower, 1959–61; president of the University of Rochester, 1962–82; and under secretary of state for economic affairs, 1982–89.

[49] Meyer Abraham Girshick (1908–55): emigrated from Russia to the United States, 1922; B.S., Columbia, 1932; student of Hotelling at Columbia, 1934–37; statistician, U.S. Department of Agriculture, 1937–44; SRG, Columbia, 1944–45; Bureau of Census, 1946–47; RAND, 1948–55; professor of statistics, Stanford, 1948–55. For biographical information, see Arrow, 1978. Also, author (with David Blackwell) of *Theory of Games and Statistical Decisions* (1954).

[50] The combined AMG and SRG at Columbia constituted a little more than half of the entire budget of the AMP during the war, totaling $2.6 million, which gives some indication of its size. Smaller contracts were given to Samuel Wilks and Merrill Flood at Princeton, the latter working on the B-29 bomber (Owens, 1989, p. 288).

aiming concerns of Weaver's fire control panel. They included evaluating gun effectiveness in air dogfights, comparing bomb sights, calculating pursuit curves for homing torpedoes, estimating the probability of aircraft damage with random flak and with Merle Tuve's newly developed proximity fuze, and gauging B-29 vulnerability. Some of these problems were evaluated by electronic calculation devices designed by Bigelow. Although relations with the Navy seemed to be most cordial, and this tended to bias the work in certain directions over the course of the war, there was no cumulative project or area of expertise which emerged out of the SRG. Of the 572 reports and memoranda produced by the unit, the one identifiable idea or technique that attracted further interest and development in the postwar period was the theory of sequential analysis in determining optimal sample size, a doctrine later generally attributed to Abraham Wald (Anderson et al., 1957; Wallis, 1980; Klein, 1999b). Thus, economists were not so much sought out as ideal team members for an OR unit, or for their social theoretic expertise as such; in this case, rather, they were incorporated into the SRG through a series of accidents, as individuals for whom Hotelling would vouch for as having a good grounding in statistical theory and practice. Because "orthodox" doctrines of statistical inference had yet to gel (Gigerenzer & Murray, 1987), the line between the identity of statistician, mathematical probabilist, and economist had not yet become as sharp as it would in the postwar era.

Although no profound engineering or theoretical breakthroughs on a par with the atomic bomb or radar occurred at the SRG, it was nevertheless the occasion for the consolidation of what later became known as the "Chicago school" of economics in the postwar period. When Milton Friedman returned along with Wallis to Chicago in 1946, and Leonard Savage was brought to Chicago in 1947, the groundwork was laid for the SRG in Exile on the shores along Lake Michigan. When Allen Wallis was appointed dean of the Chicago Business School in 1956, he lured George Stigler to the Walgreen Professorship in 1958, completing the reunion and cementing the Chicago school (Shils, 1991). Along with Friedman's student Gary Becker, this group came to share a relatively coherent response to the problems of neoclassical price theory that had been brought into prominence by the work of Hotelling and Schultz in the 1930s (Hands & Mirowski, 1998). This position has been described (Hammond, 1996; Hirsch & de Marchi, 1990; Reder, 1982), as has its specific relationship to neoclassical price theory (Hands & Mirowski, 1999). Briefly, to risk caricature of this position, its first commandment is that the market always "works," in the sense that its unimpeded operation maximizes welfare. Its second commandment is that the government is always part of the problem, rather than part of the solution. The third commandment is that the demand curve is the rock-bottom fundamental entity in price theory, and that attempts to "go behind" the demand curve in order to locate its foundations in the laws of utility or "indifference," as had been the objective of Hotelling and Schultz, were primarily a waste of time and effort. Even the attempts to subject individuals to psychological experiments in order to map their indifference surfaces – a movement given impetus during the war – would not serve to further the economists' quest "to obtain exact knowledge of the quantitative relation of consumer expenditures to prices and incomes for the purpose of predicting the effect of changes in economic conditions on the consumption of various commodities" (Friedman & Wallis, 1942, p. 188). This did not counsel wholesale repudiation of something like utility, however, because it "may be a useful expository device"; nevertheless, one should not "mistake the scaffolding set up to facilitate the logical analysis for the skeletal structure of the problem." This later became Friedman's notorious methodological doctrine of neoclassical theory as convenient fiction: "we shall suppose that the individual in making these decisions acts *as if* he were pursuing and attempting to maximize a single end" (1966, p. 37).

Our own objective here is not to maximize the faithful description of the Chicago approach but to point out its clear relationship to the activities of the SRG. As already suggested, the SRG, not noted for any epoch-making breakthroughs in methodology or theory, was basically the extension of wartime British OR (plus some more advanced statistical techniques) to a motley of problems in the American armed services. The Chicago school of economics, with its rough-and-ready pragmatism about the nature of the underlying objective functions, *was little more than Blackett's OR imported back into economics.* Recall the standard practices of Blackett's circus: look for "quantifiable" variables, even when the problem resists quantification. Then posit a simple "as if" objective function and collect data on the supposed partial derivatives of the function. Assume repeated operations tend to a maximum of the function. Marvel at the extent to which "large bodies of men and equipment behave in an astonishingly regular manner." Treat the set of designated causes as comprising a closed system, invoking ceteris paribus if necessary to avoid the fact the list cannot be exhaustive. Use statistics to paper over the uncertainties and unknowns of the problem as portrayed, both for the economist and the agent, and to churn out implementable predictions for the client. Keep psychology out of it. Remember OR is just "social science done in collaboration with and on behalf of executives"; don't let the analysts think they can run things better than those actually in charge. The resemblance is capped by Milton Friedman's championing of a "Marshallian" approach to price theory (1949): that is, a *British* version of economics, in contrast to the Continental Walrasian approach, or the indigenous Institutionalist statistics of the National Bureau for Economic Research (NBER), where Friedman had worked on collecting data on incomes of professionals before joining the SRG (Friedman & Kuznets, 1945).

Is it just too farfetched to think that these neoclassical economists learned some of their trademark practices and theoretical ambitions from the natural scientists who staffed wartime OR units? Let us examine one relatively specific instance. A prerequisite for understanding the Chicago school of economics is to see that it shared a specific stance toward the treatment of income and income effects in neoclassical price theory (Hands & Mirowski, 1999). Briefly, the viewpoint of Schultz in the 1930s was that income effects created difficult problems for neoclassical price theory, especially with regard to the constitution of "normal" demand functions. The Chicago response of the 1940s and 1950s was to assert that, for all practical purposes, income effects don't matter or else wash out, thus preserving the analytical priority of demand functions in partial equilibrium exercises. The rationale behind this response was that "incomes" didn't really exhibit the invariance in real life that they were accorded in the Slutsky theory of income and substitution effects; Friedman had decided in his study of professional incomes (Friedman & Kuznets, 1945, chap. 7) that there were "transitory" and "permanent" components, and that these would be very difficult to separate in practice. The "transitory" component could loom quite large; so the most legitimate sort of demand curves would be of the "income compensated" variety, with income itself treated as an average. The intricate causal interconnections between demands and incomes in neoclassical theory were rendered putatively harmless by portraying incomes as virtual statistical entities.

So what has all this to do with the SRG? Quite a bit, according to Friedman's own testimony

> One article of mine that in a very important sense grew almost entirely out of the work of the SRG was the article which I wrote in 1953 on "Choice, Chance and the Personal Distribution of Income." It traced directly to our work on the proximity fuze. One element in the work on the proximity fuze was the attempt to approximate the time distribution of bursts. The proximity fuze had two impulses, one forward and one backward. As a result there were generally two modes in the distribution. We treated this as the sum of two separate distributions. . . . The resemblance of those distributions of bursts to income distributions got me started to thinking about whether the same method could not be used to describe income distributions, and that result is directly and immediately reflected in the article I referred to. (in Wallis, 1980, p. 329)

So forward and retrograde radar triggers became jolts of permanent income superposed with transitory random shocks; and probabilistic considerations were recruited to "explain" the naturalness of the income distribution and to dissolve a vexing problem of interdependence of income feedbacks in price theory. Any feedback of price changes on income changes would thenceforth be considered as a second order of smalls by construction. Here we explicitly encounter Pickering's cyborg Theory of Everything, layering gun-aiming technologies across the academic board. Likewise, other specific models of protocybernetic feedback learned on the job would osmose into formal economic models at Chicago; to trace the filiations would require a painstaking census to trawl the hundreds of reports (many still classified after all these years) to enumerate how often this happened. Yet much more than individual specific models were absorbed at the SRG; there was also a broader style of interdisciplinary research that could be picked up from the can-do hubris of the OR teams:

> A lesson I learned from this experience is that one can become an expert in a narrow field with astonishing rapidity. One subject I worked on was the vulnerability of aircraft to various kinds of firepower (20 mm. cannon, .50-caliber machine guns, etc.). Within six months after our group began work on this subject, we were consulted by other war-research agencies on the details of aircraft vulnerability. One day I would be measuring a secretary to estimate how many square feet of target a seated pilot made, and a short time later I would be gravely discussing that number with another research group. (Stigler, 1988, p. 62)

Just remember that the next time you encounter a Chicago paper on the economics of suicide, or cigarette addiction, or the "economics of the family," or the trajectory of optimum saving over the life cycle. You don't need combat experience to sell yourself as an expert on war.

The legacy of British OR in Chicago economics resides not only in some of its trademark doctrines but also in the sorts of innovations it opted to set itself against. For example, there was never any substantial impact of the computer upon Chicago doctrines; it rarely made an appearance in any metaphorical incarnation, and skepticism reigned regarding its use as a tool, especially in elaborate econometric exercises. The Chicago school was never very receptive to any of John von Neumann's innovations, be it the expanding economy model or game theory or automata theory, tending to subject these formalisms to a frigid silence. It displayed very little interest in simulation exercises, or in fledgling attempts to conduct experiments with human subjects. If it ever did make reference to something akin to signal extraction, such as George Stigler's "economics of information" (1961), which characteristically ignored every possibly relevant cyborg innovation, from Shannon's information concept to Wiener's feedback to Weaver's notions about complexity. Instead, the economic agent was frequently conflated with a Chicago economist: believing in a partial equilibrium model of the world, the consumer carried out simple inductive statistical exercises to augment the unerringly accurate information provided by the market. Science didn't need any coordination, and neither did the participants in the marketplace. The greatest threat to order and progress was posed by renegade scientists who stubbornly refused to learn this lesson. And they found their bête noire just down the Midway, in the form of the Cowles Commission.

People who think of neoclassical economics as a monolithic system of beliefs relentlessly drummed into generations of docile students have great difficulty comprehending the depths of animosity between the Chicago economics department and the Cowles Commission, especially in the period just following World War II. They seemingly shared so much – economic theory as constrained optimization, the primacy of the individual and her wants, the insistence that the neoclassical tradition was the *only* scientific game in town, the publishing in the same journals, even occupying the same quarters for a short time – whatever could so aggravate tensions until Cowles decamped for Yale in 1955? The battles were often fought out on the terrain of "method" and philosophy; there were of course clashing conceptions of the legitimacy of macroeconomics; many would rightly point to a political subtext; and there was the issue of who would be willing to pay for abstract mathematical lucubrations, as well as the dispute over the appropriate version of neoclassical price theory (Hands & Mirowski, 1999). Yet all of this, however accurate, misses one of the fundamental differences between these two wings of neoclassical theory: in essence, the battle can be understood as the clash between older "British" OR and the new-fangled American OR. It was one of the first indications of close encounters of a cyborg kind. It ultimately boiled down to the difference between the SRG and the RAND corporation.

RAND/OR

The existence of RAND floats about the fringes of consciousness of much of the economics profession. There is a *RAND Journal of Economics*, fulsome acknowledgments of RAND support in the first footnote of some classic articles, RAND stints on the curricula vitae of colleagues, even a RAND graduate curriculum. Yet few stop to accord its ubiquity a second thought, and there has been almost no curiosity about RAND in evidence in the community of historians of economics.[51] Its significance has been a watchword in the history of Cold War policy and increasingly recognized in the history of the sciences. These disciplinary impediments to understanding are unfortunate, because RAND was the primary intellectual influence upon the Cowles Commission in the 1950s, which is tantamount to saying RAND was the inspiration for much of the advanced mathematical formalization of the neoclassical orthodoxy in the immediate postwar period. But beyond that, RAND was also the incubator for cyborgs inclined to venture out into the worlds of management, the military, and the social sciences. Systems analysis, artificial intelligence, and the discipline of software engineering all enjoyed their first stirrings there; game theory found its life support system there in those all-too-critical early years. RAND itself was constructed to break down disciplinary barriers between the natural and the social sciences and to spread the gospel of complexity, after the example promulgated by Warren Weaver and wartime OR; it was not always successful in its ambitions for conjuring this metascience, as evidenced by outbreaks of disputes between the economists and other defense analysts, and the fragmented blinkered local conceptions of the significance of RAND prevalent so long after the fact. Nevertheless, many of the figures we shall subsequently discuss have testified that their RAND experience was the turning point in their intellectual careers; that alone justifies a careful look.

Project RAND was originally the resultant of two sets of events in 1945: the desire of certain Air Force figures to maintain something like a postwar research organization and scientific capacity resembling that which they had enjoyed during wartime (and were in danger of losing with Bush's drastic demobilization of the OSRD) by locking in some major expenditure commitments in the medium-term horizon; and the desire of the aircraft industry to maintain a research and development pipeline to retain their lucrative military contracts (Raymond, 1974). The fact that it began as a subsidy to an aircraft producer betokens the historical accident of a close relationship between Air Force general Hap Arnold and Donald Douglas as much as it did a foray into innovation of new forms of science

[51] The sole exception has been Leonard, 1991. The standard institutional history of RAND is Smith, 1966, although it is very dated, and much more concerned with the perspective of the science manager deciding what the Air Force should do with its unruly mutant offspring, a "nonprofit" corporation. A good history from the vantage point of the evolution of nuclear strategic doctrines is Kaplan, 1983, which does not delve very deeply into the other activities of RAND. The early years of the corporation are treated in some detail in Collins, 1998. Edwards (1996) insisted upon the importance of RAND for the history of the computer in America, and Jardini (1996) sets new standards of excellence for the history of RAND in the social sciences. Hounshell (1997b) provides an overview of the range of fields researched in the early years of RAND.

management; indeed, Vannevar Bush initially opposed the move to site Project RAND at Douglas, because it would threaten to undermine the role of the university in science. These "efforts represented a determination on the part of Arnold and others in the professional military not to be dependent upon scientists and scientific institutions beyond their control" (Collins, 1998, p. 67). The vague original mandate of Project RAND concerned study of "the broad subject of intercontinental warfare, other than surface, with the object of recommending to the Army Air Forces preferred techniques and instrumentalities for this purpose," although specifics tended to center on further study of the V-1 and V-2 rockets (Smith, 1966, p. 41). The project was put under the direction of Douglas Aircraft engineers Arthur Raymond and Franklin Collbohm, and physically situated at the Douglas Aircraft plant. Contrary to the impression conveyed in some retrospectives, the mandate for research had been conceptualized more broadly than mere weapons system engineering from the very beginning, encompassing the social sciences and abstract mathematical inquiry as well as computational innovation.[52] The paradigm was clearly the operations research team, which had become so ubiquitous during the war. This was essentially guaranteed by the fact that Collbohm had worked under Weaver doing OR for the Air Force during the war, and that "Weaver's ideas . . . would provide the basis for an alternative strategy for integrating the military and the civilian" (Collins, 1998, p. 222). "OR was a tool for military managers to enhance rational decision making in integrating civilian scientific and technological resources. Managerial and political ends were primary" (Collins, 1998, p. 175).

Something about OR, however, did not thrive so well in an industrial setting, at least in the immediate postwar period; the persona of the

[52] This point is made forcefully by Jardini, especially by unearthing some very early documents. One is a December 1945 draft by Collbohm entitled "R&D Contract: Long Range Air Power," which states: "one of the most important points in an evaluation of the proposed system and its components involves a study of its economics. We must get the most effectiveness considering the number of units, accuracy, destructive area, vulnerability and indirect effects as well as direct effects upon our peacetime economy" (in Collins, 1998, p. 114). Another is a presentation by Arthur Raymond dated August 1947. It states that RAND would be concerned "with systems and ways of doing things, rather than particular instrumentalities, particular weapons, and we are concerned not merely with the physical aspects of these systems but with the human behavior side as well. Questions of psychology, of economics, of the various social sciences, so-called, are not omitted because we all feel that they are extremely important in the conduct of warfare" (in Jardini, 1996, p. 32). Nevertheless, as of 1949, the RAND research staff consisted of 50% engineers and mathematicians, 14% physicists, 14% computer scientists, and only 3% economists (p. 37). On the other hand, the largest component of the *consultant* staff consisted of economists (35%), followed by physicists (16%) and mathematicians (14%).

OR analyst was generally someone who stood outside and apart from hierarchies and chains of command, and not just another branch in the organization chart, another suit spouting corporate culture. Project RAND researchers were having trouble getting proprietary information from Douglas's competitors, and they didn't like the research atmosphere; Douglas, for its part, thought RAND was hurting its chances in landing Air Force contracts; industry did not believe the early RAND portfolio was resulting in substantial production contracts. The practice dictated by Douglas of subcontracting out research also seemed to dilute any coherent intellectual identity that would be RAND's lifeblood. Hence, in 1948 Project RAND was divorced from Douglas, moved to its own quarters in downtown Santa Monica, and was incorporated in California as a curious hybrid entity, a private nonprofit (but nonphilanthropic) corporation, with Collbohm as its president from 1948 to 1967.

Given the amorphous character of the founding mandate, and the inevitable contest over control of the fledgling organization, it was perhaps not unexpected that a number of figures attempted to assert their own priorities on the unit. There were those, such as Collbohm, who would have preferred a team of hardware engineers, with some systems analysis resembling that done in the aircraft industry thrown in for good measure. His vision was embodied in early projects such as the space satellite feasibility study and the incorporation of titanium alloys into aircraft bodies, and the pursuit of technologies bequeathed by the war, like jet engines, microwave communications, and atomic weapons design. However, Collbohm was a person who took a distinctly hands-off attitude toward the initiation of projects.[53] He spent a lot of his time placating the Air Force brass, who wanted very specific answers to its own short-term preoccupations; for the first decade or so of RAND, such requests were easily fended off. Then there was Warren Weaver, the *éminence grise* of Rockefeller and the OSRD, who was looking to preserve the wartime project of an across-the-board mathematical science of war. Weaver first coined the term "military worth" and pushed to have an interdisciplinary section structured around the concept. Weaver also articulated the notion that a "science of war" was really just a cyborg Theory of Everything in disguise: "The distinction between the military and the civilian in modern war is . . . a negligible distinction. . . . It may even be, for example, that the distinction between war and peace has gone by the board" (Weaver quoted in Collins, 1998, p. 253). Weaver imposed his objective of

[53] See, for instance, Robert Belzer interview, July 23, 1987, p. 21, NSMR: "the management committee would kick it around and usually they would try to steer the discussion . . . so that Frank didn't make the decision. Somebody would volunteer and say, 'Let John [Williams] and I work this out.' . . . There was a defined attempt to keep the thing from developing to the point where Frank would make a command decision."

preserving the continuity of an independent site for development of OR by having RAND hire his protégé John Davis Williams[54] from the Statistical Research Group, who then in turn became the contact for a veritable pipeline for recruits from the Columbia AMP and Princeton to Santa Monica, including Edwin Paxson,[55] Ed Hewitt, Olaf Helmer,[56] and Meyer Girshick. Helmer in particular signaled the beginning of a RAND practice of hiring formal logicians to do OR. When it came time to decide the composition of the "Evaluation of Military Worth Section" at RAND, Weaver was there closeted with Williams and Helmer in a December 1946 meeting, along with consultants Samuel Wilks and Frederick Mosteller (Jardini, 1996, p. 84). Not surprisingly, it was decided to model the unit upon the AMP. Weaver, in giving the plenary at the September 1947 RAND Conference of the Social Sciences, explained to the gathered handpicked group who were being looked over for potential recruitment what it was that he thought RAND would be doing. "I assume that every person in this room is fundamentally interested in and devoted to what

[54] John Davis Williams (1909–64): B.S., Arizona, 1937; graduate studies in mathematics, Princeton, 1937–40; statistical consultant, 20th U.S. Army Air Force, 1941–42; SRG, Columbia, AMP, 1943–45; RAND staff, 1946–64. See Kaplan, 1983, pp. 62–68, Robert Specht interview, June 26, 1989, and Robert Belzer interview, July 23, 1987, NSMR. Specht interview, p. 7: "John was a disciple of Warren Weaver's. Warren Weaver was a guide to him, and I'm sure it was John who was responsible for RAND getting into artificial intelligence." See also Mood, 1990, p. 40: "John Williams spent the war years as a dollar-a-year scientist with the military and the NDRC. Afterwards, he was influential in persuading the Air Force and particularly General Hap Arnold to create the RAND Corporation . . . John became head of the RAND Mathematics department and at once urged many of his old associates to join him." Collins (1998, p. 166) calls Williams "Collbohm's closest advisor."

[55] Edwin W. Paxson (1913–79): Ph.D. in mathematics, California Institute of Technology; assistant professor, Wayne State University, 1937–42; scientific advisor, U.S. Army Air Force Air Proving Ground Command, 1942; technical aide, AMP, 1943–45; consultant, U.S. 8th Air Force in England, 1944–45; consultant, U.S. Strategic Bombing Survey, 1945–46; codirector of mathematics, U.S. Naval Ordnance Test Station, 1946–47; RAND staff, 1947–79. Paxson was briefly Williams's boss at Naval Ordnance, before their roles were reversed at RAND. Paxson's relative anonymity in treatments of game theory is explained by James Digby as follows: Paxson "published largely in classified form. Because he liked to use specific examples involving nuclear weapons designs, most of his important early work was 'Restricted Data' and has not been declassified" (1989, p. 4). See also Kaplan, 1983, pp. 86–89.

[56] Olaf Helmer (1910–76): Ph.D. in mathematics, University of Berlin, 1934; Ph.D. in logic, University of London, 1936; New School for Social Research, 1936; research assistant, University of Chicago, 1937–38; professor of mathematics, University of Illinois, 1938–41; CCNY, 1941–44; Princeton, SRG, AMP, 1944–46; RAND staff, 1946–68; Institute for the Future, 1968–73; USC, 1973–76. Some biographical information can be found in Rescher, 1997, who asserts, "He was impatient of detail – and the active writing up of research was for the most part left to his collaborators." Helmer's relationship to Kenneth Arrow is briefly described in Chapter 5.

can broadly be called the rational life," he told them (Kaplan, 1983, p. 72). That suited some economists in the room just fine, Charles Hitch among their number; Hitch was tapped to head up the new economics department.

With Williams and Paxson on board, the agenda of a third party became dominant. Both stood very much in awe of John von Neumann. Williams had become acquainted with von Neumann when he was at Princeton. While Paxson was still at the Naval Ordnance Test Station in October 1946, he had corresponded with von Neumann about the application of game theory to naval tactics (Jardini, 1996, p. 50). In what appears to be the earliest detailed attempt to cast specific military scenarios into a formal game framework, Paxson proposed a destroyer-submarine duel where payoffs are denominated in amounts of shipping saved or lost, a combinatorial problem of coalitions of interceptors resisting an attack group, and the choice of strategies as choice between differently configured weapons systems. When Paxson came to RAND in early 1947, he set about to mathematize aerial bombing systems analysis along strategic lines, which included the construction of a special Aerial Combat Research Room to simulate geometric and game-theoretic aspects of maneuvers and evasions. Paxson also conducted a 1948 summer study on "The Theory of Planning in Relationship to National Security," casting weapons choice in a game theoretic idiom (Collins, 1998, p. 286). Through their efforts, the Neumannesque version of OR described in this chapter came to dominate the culture at RAND. By June 1947 Williams and Paxson decided that the computational demands of these and other simulations were too daunting and, after consultation with von Neumann, argued that RAND should obtain its own EDVAC-style computer, even though only one original existed at that time. A machine resembling the one von Neumann was having constructed at the Institute for Advanced Study was erected at RAND in 1953 and named – what else? – the "Johnniac" (Gruenberger, 1979).

Von Neumann was certainly accustomed to military contractors clamoring for his advice but here was ambition on a scale that could not escape even his sorely taxed attention. It seems that around 1947 he decided that here was a golden opportunity to push his agenda not in a piecemeal fashion, as tended to happen in his consultations with the AMP or Los Alamos or the Moore School, but in a concerted and systematic manner. Ensconced in Santa Monica was a ready and willing core group poised to work on game theory, linear programming, Monte Carlo simulation, and machine computation, and it showed every sign of willingness to hire further mathematicians and other academics to do the same. That it was pledged to subordinate its research to the needs of the Air Force was no drawback; von Neumann had already decided that this kind of "applied mathematics" left more than adequate room for maneuver. Moreover, here was a group who had previously specialized in mathematical logic, but now sought connections to real-world questions of computation and strategy. The concentration of specialist logicians was virtually unprecedented in any academic setting anywhere in the world.[57] He foresaw that he would be able to set the research agenda without any of the nuisance of academic students and colleagues (even by the extraordinarily light standards of the IAS), and even avoid dealing directly with the military paymasters. Offered a no-strings consultantship with RAND in December 1947, he accepted. As John Williams wrote in the cover letter, "Paxson, Helmer and all the rest are going to be especially nice to me if I succeed in getting you on the team!"[58] It didn't take long for von Neumann to decide that if the economists were not going to cotton up to game theory, then it could find a home at RAND, along with his other mathematical preoccupations. Warren Weaver noticed, too. "I was awfully pleased on a recent visit in southern California, to find that two or three good men in Project RAND are working seriously on your theory of games. This seems to me so very obviously desirable that I was glad to find out they were busy at it."[59]

Given that game theory is treated in popular accounts as almost synonymous with RAND (Poundstone, 1992; Kaplan, 1983; Hounshell, 1997b), it comes as a bit of a shock to realize its trajectory at RAND traced a meteoric rise and an Icarian fall, all in the space of little more than a decade (Leonard, 1991; Jardini, 1996). Indeed, it may have

[57] Albert Wohlstetter interview, July 29, 1987, NSMR: "I was surprised, however, to find that RAND, this organization that worked for the Air Force, was publishing a research memorandum by [Alfred] Tarski . . . I was living in Santa Monica . . . we just ran into Abe Girshick, Olaf Helmer and Chen McKinsey on the street, and they were overjoyed to see us. Mathematical logic was a very, very small world. There were only a little over a dozen mathematical logicians before the war in the United States, and two jobs in mathematical logic. For the rest, you had to teach either calculus, as Chen McKinsey did, or philosophy or something of that sort." The role of RAND in the funding and encouragement of a particularly American style of analytical philosophy is a subject that still awaits its researcher. RAND, for instance, funded Tarski's work on decision algebras in this period; see George Brown interview, March 15, 1973, p. 29, SCOP.

[58] Williams to von Neumann, December 16, 1947, box 15, VNLC. The letter informs von Neumann he would be retained on the identical terms as Warren Weaver and Sam Wilks: a retainer for services unspecified: "the only part of your thinking time we'd like to bid for systematically is that which you spend shaving."

[59] Warren Weaver to von Neumann, February 25, 1948, box 7, folder 12, VNLC. The same letter states, "I have been very much out of touch, for some time, with the progress and developments of your own computer project. If you have any little statement . . . that tells where you are and where you are going, I would also be very much interested to see that." And, chatty Cyborg that he was, he closes with a query: "have you ever seen any estimate of the total number of possible games of chess?"

shadowed von Neumann's own level of enthusiasm, with a slight lag.[60] Nevertheless, it packed enough pyrotechnics to shape the next half century of game theoretic ideas, as we shall recount in the next chapter. From Nash equilibria to "psychological" experimental protocols, from "scratchpad wars" to computer simulations, from dynamic programming to evolutionary dynamics, from "rational decision theory" to automata playing games, all constitute the fin-de-siècle economic orthodoxy: in each case, RAND was there first. More than ninety researchers produced memoranda on games at RAND from 1946 to 1962, and the roster reads like a *Who's Who* of postwar game theory. Yet the observation that von Neumann was responsible for the marriage of game theory and operations research at RAND does not even begin to exhaust his influence on the outlines of postwar economic orthodoxy.

Perusal of the list of consultants who passed through the portals of RAND in its first years reveals a distinct pattern in the composition of economists. Even though many of the staff just mentioned had connections to the wartime AMP at Columbia, very few of the economists who passed through the mathematics section at RAND looked much like the economists who had staffed the SRG. (The Economics Division at RAND, now a separate entity, was a different story.) Early on, it became clear that analysts like Williams, Paxson, and Helmer were not all that enamored of the sorts of empirical "statisticians" that eventually populated the Chicago school. They wanted theorists more like

[60] One of the very few papers on game theory von Neumann published post-*TGEB* was a RAND memorandum, which reveals the shifting balance of concern toward computational issues and away from innovation in the theory of games per se. The set of circumstances leading up to this work is described in the George Brown interview, SCOP. See Brown & von Neumann, "Solutions of Games by Differential Equations," reprinted in von Neumann, 1961–63, vol. 6), which provides a constructive proof of existence of good strategies for two-person, zero-sum games. See also the correspondence with J. C. McKinsey, February 16 and 18, 1949, box 17, folder 6, VNLC, and the letter to Warren Weaver, March 1, 1948: "I have spent a good deal of time lately on trying to find numerical methods for determining 'optimal strategies' for two person games. I would like to get such methods which are usable on electronic machines of the variety we are planning, and I think that the procedures that I contemplate will work for games up to a few hundred strategies." Box 7, folder 12, VNLC.

The von Neumann papers have numerous pleas by RAND figures trying to get him more involved in their projects and problems with game theory. In a letter shifting his payments from a retainer to "when actually employed" basis, Williams mentions "We intend to make major efforts on the applications of game theory" and pleads, "If you were really to pour your torrent of energy into these subjects for a while, there would probably be a handsome pay off." Williams to von Neumann, December 27, 1948, box 15, VNLC. See also Alexander Mood to von Neumann, October 1, 1951, ibid.: "Since we have not had much luck luring you here during the past few years, we have talked the powers that be into doubling your daily rate."

themselves, with backgrounds in mathematical logic and the natural sciences, and an appreciation for rigor in proofs. If war was just another problem in logic, then so too was politics and economics; and machines were proving to be unsurpassed in manipulations of logic. Hence they sought out people who really believed deep down that rationality was algorithmic and did not treat cognitive considerations in what they considered a cavalier manner. In effect, they wanted to recruit a different breed of economist, and with the help of von Neumann, they found their target populace in the Cowles Commission.[61]

Two observations on the Cowles Commission are relevant to the themes dominating this chapter. The first revisits the question, Who pays? The answer now tendered is that RAND picked up the tab for Cowles at a vulnerable period when its other major sources of support were waning. The second observation recalls that, as always, the question of who pays is never separable from the sort of work that is paid for. In the case of Cowles, the subvention from RAND, and the larger military initiative in the commissioning of economic research generally, coincided almost exactly with a profound sea change in the type of research being done at Cowles. As we argue in Chapter 5, it is no coincidence that the careers of Tjalling Koopmans, Jacob Marschak, Leonid Hurwicz, and Kenneth Arrow took a noticeable turn away from the earlier quest for an econometric validation of neoclassical theory and toward a reconceptualization of the "rational" economic agent as an information processor. Not only were the standards of rigor recalibrated at Cowles to aspire to the mathematical practices of a John von Neumann, but the very *content* of the notion of rationality and the very *referent* to cutting-edge natural science took its cue from his visionary project. Of course, it is prudent to remind ourselves periodically that taking a cue and recapitulating a performance are two entirely different things.

Here we need briefly review the problem situation of Cowles in the immediate postwar era. The Cowles Commission had not come through the war in very good shape (Hands & Mirowski, 1999). Its intellectual fortunes had begun to revive with the accession of Jacob Marschak to the research directorship of Cowles in 1943 – the very same Marschak we encountered in Chapter 3 who had participated in the Berlin seminar with von Neumann, Szilard, and Polanyi in 1926.[62] Marschak had himself run

[61] "In personal conversations with the author Robert Kalaba and Lloyd Shapley have independently remarked that there were intensive discussions at RAND, in the early 1950s, about the appropriate mathematization of economics" (Velupillai, 1996, p. 267).

[62] The primary biographical source on Jacob Marschak (1898–1977) is Arrow, 1986. For other views, see Radner, 1984; 1987. Marschak's work in price theory is discussed in Hands & Mirowski, 1999. Marschak's career is briefly described in the next chapter.

a small seminar on mathematical economics and econometrics during the early phases of the war at the New School in New York, and brought the program of "structural econometrics" to Cowles at Chicago as the centerpiece of its revived research efforts. The major thrust of this effort was to carry on Schultz's program of the empirical verification of neoclassical general equilibrium theory, while avoiding the "pitfalls" of demand analysis which Marschak thought he had detected. It is the postwar invention of such doctrines as the "probability approach" to econometric theory, the identification problem, the correction of structural equations estimations bias through simultaneous equations techniques, and the genesis of the large-scale Keynesian macromodel for which Cowles is fondly remembered by its historians;[63] but, with only one exception (Epstein, 1987), none recount the fact that Marschak's program had run into both internal and external obstacles by the later 1940s. Internally, the program of the empirical verification of the Walrasian system of price theory had hit numerous conceptual snags (Hands & Mirowski, 1999). Externally, it was getting harder and harder to convince any funding source that the increasingly elaborate statistical precautions against error were producing anything that wasn't already known from far less elaborate and arcane statistical techniques. Cowles's primary rival, the then-bastion of Institutionalist economics, the National Bureau of Economic Research (NBER), was growing increasingly skeptical of commission's strident claims for sole possession of scientific legitimacy.

The Cowles Commission had been heavily dependent on two funding sources: the Alfred Cowles family and the Rockefeller Foundation. Their only other substantial source, the NBER, was increasingly withdrawing its support for the single wartime external grant to study price controls, as can be observed in Table 4.1. Indeed, what comes across most forcefully is the utter dependence of the wartime Cowles Commission upon the continuing beneficence of Alfred Cowles and his kin.

The problem that Marschak faced during the war was that two out of three of his sources of funding were being jeopardized by the ongoing friction with the NBER, and the Cowles family was not savoring the prospect of having to shoulder the entire burden. The Rockefeller program officers were hearing disparaging comments about the Cowles Commission from some prominent researchers at the NBER, particularly

[63] Given the number of retrospectives penned about Cowles, it is truly astounding how none of them begin to capture the primary historical determinants of this most important organization for the history of American economics. The attention of most standard accounts, such as Christ, 1952; 1994; Hildreth, 1986; Morgan, 1990; and Klein, 1991, tends to stress innovations in econometrics, or perhaps Keynesian macro, which were arguably the least important events at Cowles. Cowles without neoclassical microtheory and operations research is truly Elsinore without the Prince, and the absence of Alfred Cowles himself leaves out the Ghost.

Table 4.1. Cowles Commission Budget, 1939–46, Calendar Year Basis, dollars

Summary of Surplus Account, July 20, 1945

Category	Income	Expense	Surplus	Cumulative surplus
1939*	28,056.24	27,607.90	448.34	1,428.96 [deficit]
1940†	28,552.44	21,702.24	6,850.20	5,421.24
1941	22,788.74	19,832.65	2,956.09	8,377.33
1942	20,778.00	17,657.16	3,120.84	11,498.17
1943	22,601.84	16,482.92	6,118.92	17,617.09
1944	21,234.23	20,824.51	409.72	18,026.81

Report January 1, 1945‡

Funds received	*1942*	*1943*	*1944*
Rockefeller Foundation	11,175	6,000	0
National Bureau	2,400	2,400	1,200
Cowles Commission	4,850	8,750	3,650
Expenditures			
Directors	1,768.34	5,750	5,500
Research staff	6,910.58	10,739.68	3,421.91
Clerical staff	1,397.42	2,249.58	1,323.49
TOTAL	10,579.28	19,473.95	10,543.25

Report 1944

Donations				
Alfred Cowles	5,000			
Other Cowles	12,000			
Total		20,000		
Sale of books	34.23			
NBER Price Control Study		1,200		
TOTAL INCOME			21,234.23	
Expenses				
Salary: Marschak (half-time)		3,750.00		
Total salaries			16,157.78	
Publishing costs			1,734.99	
Grant to Econometrics Society			1,000	
TOTAL EXPENSES				20,824.51
Surplus for 1944		409.72		

Report 1945

Donations		
A. Cowles	5,000	
Total Donations		20,000

(continued)

Table 4.1 *(continued)*

Grants					
SSRC Univ. Chicago		8,263.33			
Sale of books		1,403.61			
Interest		239.92			
TOTAL INCOME				30,075.02	
Salaries: JM (half-time)			3,875.00		
Research assoc.					
T. Anderson			291.67		
HT Davis			1,000.		
L Hurwicz			1,678.08		
L. Klein			750.00		
T. Koopmans			166.66		
D. Leavens			3,465		
Total salaries				22,118.31	
Book pub (Katona)				1,630.40	
Grant to Econometrics Society				1,000.00	
Conference expenses				965.00	
TOTAL EXPENSES				27,813.79	
Surplus for 1945		2,261.23			
Report 1946					
Donations					
A. Cowles	9,343.75				
Total donations		27,705.75			
Sale books		964.60			
Interest				212.44	
TOTAL INCOME					29,090.37
Salaries: JM (4125), T. Anderson (204), HT Davis (500), D. Leavens (3637)[§]					
Total salaries		17,931.89			
Book pub.		1,023.03			
Travel & conferences		2,155.84			
Loss on sale of securities		649.70			
TOTAL EXPENSES					24,660.73
Net surplus	4,429.64				

* Former contributions of Cowles family amounted to $27,000 a year, but were reduced to $20,000 a year beginning 1941.

† Income for this year was high because it included part of Mr. Cowles's personal contribution due in previous years but not paid since we were not in need of cash. Dickson Leavens, Treasurer.

‡ Separate document on "Committee on Price Control and Rationing" letterhead, Marschak and Katona codirectors, calendar year basis.

§ Note explains Koopmans & Hurwicz shifted to SSRC grant, now treated as off-budget.

Source: Box 148, folder: Cowles Commission Treasurer Reports, JMLA.

Wesley Clair Mitchell and Arthur Burns, whom they held in the highest respect.[64] The Price Control Study was turning out to be more trouble than it was worth.[65]

Nonetheless, Marschak did manage to coax a further $7,500 out of Rockefeller, which he used to hire Tjalling Koopmans; thus his ambitions outran fiscal prudence, and the commission persisted in dire financial straits. As Willits reported in a Rockefeller internal memo in January 1947:

> [Marschak] is in a rather tight situation one year from now. He is spending at the rate of $50,000 a year, $20,000 of which comes from the Cowles family (not from endowment, but by funds raised by Mr. Cowles) and $10,000 from Rockefeller Foundation. Mr. Cowles has agreed to raise $20,000 a year for, I think, three years. Their surplus saved by Yntema will be exhausted by the end of 1947.... He has gone to President Hutchins, and they told him to get out a sales document, he being referred to the food institute document, – the perusal of which made him as a scientific man sick.... M says that the NBER is only telling what occurred, is not trying to pose theories and test them. Cowles Commission does not have facilities to prepare its own data, therefore depends upon the Government and the NBER.[66]

The heavy reliance on the NBER and, by extension, Rockefeller, for funding and data sources was just too galling to be suffered much longer, especially given the reservations that were being broached about the commission at Chicago. This was the oft-neglected background to the frequently misunderstood "Measurement without Theory" controversy, which broke out into open warfare between Cowles and the NBER in 1947–49.

The devolution of the research directorship from Marschak to Koopmans in July 1948 marked the escape route out of the intolerable bind; it was brought about by means of the cultivation of a new patron and a new research program. That patron was RAND. Koopmans's ties

[64] Marschak to Louis Wirth, February 8, 1944: "there seems to be a difference of opinions and sympathies between the approach here and that used by Mitchell and Kuznetz [*sic*] at the National Bureau.... The combination of economic theorists and statisticians at Chicago is interested in a closer connection between statistics and theory. It seems that [Joseph] Willits has been advised by representatives of the 'their school' hence this difficulty in understanding some of our terms." CFYU. Wirth was a sociologist at Chicago and a member of both the SSRC and the Cowles Advisory Committee. This incident is discussed further in Mirowski, 1989b.

[65] See the letter of Jacob Marschak to George Garvy, August 31, 1943; box 100, folder: Demand Analysis letters, JMLA: "Nearly one-half the Commission's own funds and about two-thirds of the outside grants are absorbed at present by the Price Control Study."

[66] Interview notes, Joseph Willits, January 21, 1947, RANY.

to von Neumann (described in Chapter 5) now began to pay off for Cowles.[67] Like other sciences blessed by the postwar largesse, the Cowles budget ballooned to $153,000 a year by 1951, with RAND covering 32 percent of the total, and the Office of Naval Research another 24 percent.[68] The new research program was dictated primarily by von Neumann's version of OR, or at least as much of it as Cowles thought it could reconcile with its ongoing unshakable theoretical commitments to Walrasian general equilibrium theory. The individual contracts ranged from military applications of linear programming to the shape of the American economy after an atomic war; but the overall transformation wrought by RAND upon Cowles was abandonment of Marschak's (and thus Schultz's) program of econometric estimation and validation of neoclassical price theory in favor of Koopmans's and Arrow's abstract mathematical reconceptualization of the neoclassical economic agent (see Chapter 5). At the bidding of von Neumann, the neoclassicals at Cowles were exhorted to entertain the virtues of game theory and linear programming and the computer; and, in return, the Cowles economists proposed that what neoclassicals had to offer to RAND and the world of OR was something called "decision theory." The nitty-gritty questions that shaped the negotiations involved the types of mathematics to be used in specific models (Weintraub, 1985, pp. 74–77), but the issues at stake were much larger than that. The negotiations over what this new-fangled decision theory could possibly be and "how much" each side would accept of the other's aims and preconceptions were delicate and protracted, – the subject of the next two chapters – but the consequences for economics proved profound, because it set the terms of how cyborgs made their debut in economics.

Unlike the previous case of the SRG-Chicago connection, there is no earthly way to summarize in a page or two the conceptual and technical impact that RAND and American OR had upon the distinct brand of neoclassical economics that was developed at the Cowles Commission in the

[67] Contact was initiated in January 1948, initially through Koopmans. See box 93, File: Koopmans, JMLA. "Before this, I had heard of the Rand Project only by rumor, and even now I am still very unclear as to the purposes of the project." Tjalling Koopmans to Ed Paxson, March 1, 1948, box 93, JMLA.

[68] The budget data come from "Application to Ford Foundation," September 17, 1951, CFYU. See also interview report, LCD with Koopmans, Simpson, and Marschak, March 21, 1951, RANY. "Under a contract for Rand Corporation, for example, they are working on a theory of resource allocation for the Department of Defense; this is chiefly a mathematical formulation designed to develop a regional system for decentralizing decision-making and responsibility in a complex organizational structure . . . another is a study for Rand on how to determine the ramifications in the defense program of technological changes. They are negotiating with the Office of Naval Research for a grant to carry forward their work on the general theory of organization."

1940s and 1950s. The consequences ranged from something as superficial as the recasting of the idiom of mathematical expression into a characteristically modern vernacular to something so deep-seated as the reconstitution of the very subject of economic analysis. The task of beginning to do it justice will occupy most of the rest of this volume. It turned out to be all the more potent because the Version according to Cowles came to represent the rarefied heights of "high theory" to the economics profession, well into the 1980s. The way we begin to summarize the metamorphosis in this chapter is that, if Chicago came to represent Good Old-Fashioned OR (GOFOR?) and operated as if von Neumann had never existed, then Cowles was clearly the site where the entire gamut of von Neumann's legacy was at least confronted and grappled with by economists.[69]

From scattered references, it appears that von Neumann at the end of his life believed that some combination of American-style OR plus computational models of economic organization constituted the wave of the future for economics, slated to displace the neoclassical models he had never much liked. He imagined a world of mechanical logical prostheses, in lieu of the ultimate Cyborg Dream, where the machines would themselves become the economic actors. In one of his last public talks entitled "The Impact of Recent Developments in Science on the Economy and Economics" (1961–63, 6:100–1) he stated:

> There have been developed, especially in the last decade, theories of decision-making – the first step in its mechanization. However, the indications are that in this area, the best that mechanization will do for a long time is to supply mechanical aids for decision-making while the process itself must remain human. The human intellect has many qualities for which no automatic approximation exists. . . . For trying out new methods in these areas, one may use simpler problems than economic decision-making. So far, the best examples have been achieved in military matters.

This idea that a new theory of social rationality had been pioneered in the military context of American OR was also expressed in correspondence late in his life. For instance, when John McDonald wrote in 1955 asking him in what sense there had been a "break-through in decision theory which is likely to affect practical decision making in the next 25 years," von Neumann responded that "In some forms of industrial and military operations analysis, it does so already."[70]

[69] As I argue in the next section, the MIT school represents an uneasy intermediary case, trying to both accept and deny von Neumann in a single analytical movement.

[70] John McDonald to von Neumann, May 9, 1955; von Neumann to McDonald, May 19, 1955; both in box 1, VNLC.

The economists at Cowles, while not actively disputing von Neumann's vision for OR and economics during his lifetime, wanted to claim a stake and a legacy for the neoclassical model within that tradition. For anyone who had been part of a large-scale OR operation in wartime, it was more than a little difficult to assert that economists had had much of anything to do with the institutionalization and elaboration of OR. The economists associated with RAND all realized this; and many of them were not afraid to admit it in writing. Thomas Schelling wrote about OR in World War II, noting that it "was distinctly *not* the application of economics by economists. . . . [OR] had a distinctive amateur quality about it" (in Cooper et al., 1958, p. 221). Koopmans, for one, realized this: "The question naturally arises how it came about that a spirit of progress on problems central to economics had to await an influx of talent from other fields." His own answer at first seemed to ignore the implied rebuke, and to try and embrace the interlopers as One Big Happy Family: "physical scientists who have entered OR have looked on it as an extension of the methods of the physical sciences to a new range of problems. On the other hand, impressed by the pervasiveness of valuation, economists have looked at the new field as an application of economic thinking in a wider range of circumstances. . . . Both views are valid" (1991, p. 184). But this validity would be legitimate if both sides deferred to the other's conceptions of OR; yet there was very little reconciliation involved in the eventual Cowles response. Instead of a computational theory of organization that distinguished human brains from computers, Cowles preserved its neoclassical price theory by recasting its a priori commitment to utilitarian psychology as though it were best described as the operation of a virtual computer. Donald Patinkin, one of the Cowles researchers, put it with admirable succinctness:

> Indeed we can consider the individual – with his given indifference map and initial endowment – to be a *utility computer* into which we "feed" a sequence of market prices and from whom we obtain a corresponding sequence of "solutions" in the form of specified optimum positions. In this way we can *conceptually* generate the individual's excess demand functions. (1965, p. 7)

By repositioning the computer metaphor and preserving individual constrained optimization within the Walrasian organon, while putatively "solving" the problem of excess demands, "decision theory" was propounded to dissolve the impasse bequeathed the neoclassical program by Hotelling and Schultz. As a not inconsiderable sidelight, it could be sold to the military as relevant to their problems of command, control, communications and information – that is, if RAND could be persuaded to buy it.

Rad Lab/subMIT

The Radiation Laboratory at MIT has enjoyed a number of detailed histories dealing with individual accomplishments in recent times (Guerlac, 1987; Budieri, 1996; Leslie, 1993), as well it might, since it was the conviction of many that radar won the war, whereas the atomic bomb merely ended it; and wartime radar development emanated from the Rad Lab. Nevertheless, there exists no comprehensive history covering the full range of its activities and encompassing all its participants. At its apogee in 1945, it was comparable in size and resources with the Manhattan Project, with roughly 3,800 employees and an annual budget of around $13 million. Just as with the Manhattan Project, attention has more or less been monopolized by the hardware aspects of the research, to the neglect of attendant mathematical and operational research. Although we cannot survey all the ways in which OR was represented at the Rad Lab (especially bypassing the Anti-Submarine Warfare Operations Research Group, or ASWORG), we will here concentrate upon Division 8, the gun-aiming and servomechanisms section, under the wartime leadership of Ivan Getting, because of its possible influence upon the postwar MIT school. "Ralph Phillips had in his stable of theoreticians probably the most distinguished group of mathematicians anywhere in the country: L. Eisenbud, Witold Hurewicz, W[alter] Pitts and for a while Paul Samuelson" (Getting, 1989, p. 193).

The military problem of the utilization of radar was not a simple problem of getting microwave hardware to work properly, although it started out that way. Indeed, one of the early motivations behind the constitution of Division 8 was that Getting had noticed that the Rad Lab radar devices and the Bell Labs gun director (described in Chapter 2) could not be made to work well together; this need for a species of integration of projects over and above mere technological considerations is sometimes cited by historians as the beginning of "systems analysis" (Johnson, 1997, p. 33). The benefits of radar were intrinsically bound up with traditional military issues of strategy and tactics, especially when used in aircraft and submarine detection, and then in gun-laying systems. It is instructive how early in the war issues of radar jamming, or "spoof" (pulsed replies calculated to be read as large concentrations of nonexistent aircraft), or "chaff" (aluminum foil strips dropped to confuse radar readings) became rapidly integrated into the toolbox of the radar expert (Budieri, 1996, pp. 192ff.). Each new technological advance would quickly be met with another evasive countermeasure confounding the new technology, leading some military brass to bemoan the dawn of the new regime of "wizard war." Late in the war, the three innovations of SCR-584 radar, the Bell Labs gunnery predictor, and the proximity fuze were finally beginning to be integrated for automatic use against the V-1 rocket, although the war ended before this combined cybernetic defense mechanism was fully put into action.

All these developments dictated that radar research would encroach on issues of tactics and strategy, and therefore the Rad Lab numbered among its research portfolio the workers in Division 8. It appears to have encompassed conventional topics as pursuit curves and optimal gunnery techniques; we have the testimony of insiders that "most of the ideas for applying radar were from the Rad Lab people themselves and not from the military."[71] What is noteworthy is that this unit is not mentioned in any of the standard histories of OR in World War II; it seemed to be situated off the beaten track in the conventional rounds of OR experts, including Philip Morse's ASWORG, considered in retrospect one of the founding units of American OR. More curiously, it was not on the itinerary of Warren Weaver, and John von Neumann did not list it in his roll call of consultancies. Indeed, there is some evidence that the Rad Lab also maintained an array of distinct OR units, including a set of advisers at the Eighth Air Force headquarters separate from and in competition with the OR directorate (McArthur, 1990, pp. 150–51).

The import of this separate existence in wartime was that Division 8 of the Rad Lab operated in relative isolation from the major wartime innovations of American OR. It did not sport any of the broad interdisciplinary ambitions of the latter, nor, curiously enough, did it seem overly exercised about the implications of issues of command, communications, and the man-machine interface, which were coming to dominate the expert prognostications of other civilian scientists. This seems all the more curious, in that Division 8 was directly concerned with servomechanisms and automated gun-aiming devices, and included among its staff the Walter Pitts of neural network fame. Nevertheless, perhaps due to the inclinations of Getting, it seems to have more or less restricted itself to producing and testing breadboard models of integrated radar-aimed gun control devices (Getting, 1989, chaps. 8–11). When using radar to aim antiaircraft fire,

> the tracking was characterized by jerkiness or jitter, resulting in part from the fluctuations of the reflected signal strength and in part from the preponderant reflections from the aircraft shift from one part of the plane to another. This jitter is important because fire control requires the prediction of the movement of the target during the time of flight of the bullet and such accurate prediction requires accurate angular rate

[71] Interview with Charles and Katherine Fowler by Andrew Goldstein, June 14, 1991, p. 8. For interview, see <www.ieee.org/history_ce . . . tories/transcripts/fowler.html>.

determination. Clearly the instantaneous rates had to be averaged or smoothed. The optimization of this smoothing became one of the basic contributions of the Theory Group. (1989, pp. 112–13)

In 1943 Getting arranged for the Rad Lab to produce an intermediate-range gun-aiming device for heavy guns on naval vessels, called the Gun Director Mark 56. This is significant because we know that Samuelson joined Division 8 rather late in the war, on the staff of the Rad Lab only in 1944–45, and was assigned to work on MK-56. We possess one declassified example of the sorts of work in which Samuelson was engaged in his Rad Lab report no. 628 (1944b), the calculation of a smoothing filter for angular position errors for actual data from a prototype MK-56 test firing. A few observations can be made about what was essentially a very rote exercise of tweaking the operation of a complicated bit of automated gun aiming. First, Samuelson was being employed as a glorified statistician, and not in any capacity as an economist, much the same as most of the other figures mentioned in this chapter. Second, he was not part of the "hands-on" approach of British OR, but rather sequestered behind the lines dealing almost exclusively with a machine in isolation from its intended surroundings. Third, in this report Samuelson nevertheless does not adopt a concertedly stochastic approach but rather a quasi-physicalist one, suggesting repeated integration of the electrical current output of the torque converter as a smoothing solution to the problem of error.

The task of this chapter is to inquire briefly into the relevance of this wartime experience and the subsequent postwar military connections of Paul Anthony Samuelson for his own writings and for the outlines of the third postwar school of orthodox neoclassical theory, the MIT school.[72] From many perspectives, this is a much more difficult task than relating the military experience of either the Chicago School or the Cowles Commission to their postwar configurations. The trouble derives from many quarters: that Samuelson has been extremely protective of his own legacy, and therefore has not opened his archives to historians; that Samuelson has been quick to point out the military connections of other economists but loath to discuss his own;[73] that MIT itself owes its postwar success to the fact that it was, as one scholar has put it, "a university polarized around the military" (Leslie, 1993), but so far research has been

[72] Paul Anthony Samuelson (1915–): B.A., Chicago, 1935; Ph.D., Harvard, 1941; National Resources Planning Board, 1941–43; Rad Lab, 1944–45; professor, MIT, 1940–86; RAND consultant, 1948–65. I (1993) have discussed some of the peculiarities of Samuelson's career; but the better attempt to understand Samuelson's role in postwar microtheory is Hands & Mirowski, 1999.

[73] "Kenneth Arrow's two Nobel-stature contributions to economics were financed by the Office of Naval Research and the RAND Corporation" (Samuelson, 1986, p. 867); see also the "economist's war" quote earlier in this chapter.

confined to just a few of the natural sciences; and finally, that economists have been much less inclined to acknowledge the existence of an MIT school of economics in the same way that they would readily concede the existence of the Chicago and Cowles schools. These are all severe obstacles to making the case that MIT economics was beholden to the military for its postwar dominance at the same level as that found at Chicago or Cowles; and, undoubtedly, many readers will greet this proposition with skepticism. In lieu of devoting elaborate resources to try and explore this thesis further, I simply proffer a few observations which might spark further research when conditions become more propitious.

Perhaps the best way to understand Samuelson's place in the postwar neoclassical orthodoxy is as someone who has been most strident in his insistence upon the "scientific" character of neoclassicism, but simultaneously someone who has been obsessed with the idea that there might be a "third way" out of the Hotelling-Schultz impasse, somehow located between the aggressive "Marshallianism" of the Chicago school and the full-blown Bourbakist general equilibrium formalisms of the Cowles Commission. This putative third way was always a very delicate entity, composed of a host of seemingly self-contradictory propositions; and it is the opinion of the present author that it was the first of the three-school configuration of postwar American neoclassicism to fall apart, as early as the 1960s. Yet, through a series of fortuitous events (such as his 1948 textbook), it was Samuelson's version of neoclassicism that first captured the attentions of the immediate postwar economics profession, and it was he who came to exemplify the brash self-confident face of the "new" mathematical economics. Samuelson's self-promotion as a technically skilled mathematician, in combination with his professed fascination with thermodynamic analogies, rendered him a suitable candidate for military patronage; and indeed, his Rad Lab connections served him well in his subsequent capacity as RAND consultant and Summer Studies participant. Nevertheless, it seems inescapable that Samuelson appears least influenced by any of the signal innovations of OR, be it of the outdated British variety or the cyborg heritage of von Neumann. Incongruously, Samuelson managed to coauthor a commissioned text for RAND on the promise of linear programming with Robert Dorfman,[74] an important figure in American OR, and his protégé Robert Solow,[75] a text that strains

[74] Robert Dorfman (1916–): B.A., Columbia, under Hotelling, 1936; Ph.D., Berkeley, 1950; BLS, 1939–41; OPA, 1941–43; operations researcher, U.S. Air Force, 1943–50; associate professor, Berkeley, 1950–55; professor, Harvard, 1955–87. Biographical information can be found in Dorfman, 1997, where he writes: "My specialty was bombing tactics, though I had never actually seen a bomb or been aboard a military aircraft of any kind" (p. xiv).

[75] Robert Solow (1924–): B.A., Harvard, 1947; Ph.D., Harvard, 1951; professor, MIT, 1949–95; RAND consultant, 1952–64.

to deny any linkages to game theory and the computer, and conjures an imaginary lineage where linear programming merely "resolv[es] the problems of general equilibrium left unsolved by Walras" (Dorfman, Samuelson, & Solow, 1958, p. 7; also Samuelson, 1966, p. 496). Unlike Chicago, which simply ignored von Neumann and the cyborgs, and Cowles, which struggled mightily to assimilate them to the neoclassical organon, it is hard to regard MIT as doing anything other than actively trying to purge economics of any residual trace of cyborgs, which it knew was already there.

Whereas the fully fledged doctrines regarding price theory of Chicago and Cowles have been tied to their wartime experiences, no such similar connection can be made for Paul Samuelson's contribution to neoclassical price theory, the doctrine dubbed "revealed preference," simply because it was published in its first incarnation in 1938, well before Samuelson went to MIT in 1940 and began work at the RAD Lab in 1944. As is well known, Samuelson proposed the theory originally to transcend the Hotelling-Schultz impasse by rejecting ordinal utility theory and putting the observable content of neoclassicism "up front" in the assumptions rather than "out back" in the econometric consequences.[76] Later, in 1948, the theory was reinterpreted as solving the problem of reconstructing an individual's utility map from observable market behavior, something that contradicted Samuelson's earlier denunciation of any and all psychological premises. Finally, in 1950, it was admitted that revealed preference theory was the "observational equivalent" of ordinal utility theory, which had been codified in the interim by the Cowles contingent. We do not wish here to question whether this latter admission left the theory without raison d'être, or, as Wong (1978, p. 118) puts it, "What is the problem to which the revealed preference theory is a proposed solution?" Rather, we simply aim to point out that the revealed preference theory was born of an earlier conception of thermodynamics, one that preceded the developments outlined in Chapter 2. The claims for its avant-garde status were rapidly outstripped by events, and with this phenomenon the context of the cyborg sciences becomes relevant. Thus, although Samuelson could have initially appeared to be participating in the larger transformation of the sciences occurring in the twentieth century, closer scrutiny would reveal his progressive hostility toward those very same trends.

Samuelson has acknowledged on a number of occasions his indebtedness to his teacher Edwin Bidwell Wilson and his lectures on

[76] The history of the doctrine of revealed preference is carefully related and dissected in Wong, 1978. The description of the shifting of the empirical content as a reaction to Hotelling and Schultz can be found in Hands & Mirowski, 1999.

Willard Gibbs's equilibrium thermodynamics.[77] The use of the Gibbs conditions as an analogy for stability conditions on joint changes of price and quantity is revealed in some correspondence from early 1938 and led to his championing of the "Le Chatelier Principle" as an aspect of the general logic of maximization.[78] What is striking about the uses made of thermodynamics in this theory is the way they are completely and utterly uncoupled from any considerations of randomness, or the distinction between micro determinism and macro regularity, or any clear analogue for time's arrow of the second law based on prior specification of a first law: in other words, Samuelson's understanding of thermodynamics managed to bypass all the standard epistemological conundrums that had become the old standbys of the discussion of Maxwell's Demon from Szilard's paper onward. Instead, Samuelson appeared to take the thermodynamic analogy as a blanket warrant to ignore most of the thorny problems of the microfoundations of observable macrolevel phenomena (Samuelson in Colander & Landreth, 1996, p. 162) without appeal to probabilistic limit theorems. This attitude also extended to his well-known penchant for defending comparative statics exercises without engaging in a full-blown formal stability analysis of a Walrasian general equilibrium system (Weintraub, 1991, p. 56). One could easily conclude that Samuelson was not really interested in participating in the Cowles project at all – no cognitive commitments, no problems of information processing, no commitment to rigorous interdependence of markets, no attempt at a computational interpretation of dynamics – and certainly his results were not ultimately deemed to extend to the levels of generality that the Cowles workers had subsequently attained, which is one reason that they tended to recede in importance in the 1960s. He acknowledged this in an oblique way with a gentle (for him) backhanded comment about Bourbakism (1986, p. 850). For all his putative concern over dynamics, he ultimately attempted to transcend the Hotelling-Schultz impasse by indulging in the Chicago trick of repressing income effects and hence the very essence of market processes. As Frank Hahn later noted, "he gave no account of

[77] Samuelson, 1986, pp. 850, 863; 1998; Paul Samuelson in Colander & Landreth, 1996, p. 163; Weintraub, 1991, pp. 59–61. Edwin Bidwell Wilson (1879–1964): Ph.D. in mathematics Yale, 1901; professor, Yale, 1900–6; MIT, 1907–22; Harvard, 1922–45; ONR consultant, 1948–64. Biographical information may be found in Hunsaker & Mac Lane, 1973; his relevance to Samuelson is discussed briefly in Mirowski, 1993. His relationship to Samuelson is illuminated by some correspondence in the Wilson archives, EWHL.

[78] See P. A. Samuelson to E. B. Wilson, January 25, 1938, HUG 4878.203, Personal Correspondence, EWHL. This "Principle," which played a minor role in the history of physics, was much more closely linked to French Taylorist developments earlier in the century. This deceptive appeal to physics deserves closer scrutiny than it has received so far in the history literature.

the model of the price mechanism. . . . The excess demands are written as functions of the prices only, that is, endowments are not included. And so it came about that he did not consider processes in which endowments are changing in the process of exchange, nor did he discuss this matter" (in Brown & Solow, 1983, p. 47).

Moreover, Samuelson never revealed any curiosity about the emergence of order from molecular chaos, which was a hallmark of American OR. Thus, the first attribute of his program to be contrasted with the cyborg sciences was its relatively antiquated understanding of the multiple meanings and significance of thermodynamics, relative to that prevailing elsewhere in the contemporary sciences.

The second signal attribute of Samuelsonian economic theory was his hostility to the work of von Neumann, as already mentioned in Chapter 3. He rarely missed an opportunity to subject von Neumann to some invidious comparison or another (1977, p. 883). But more to the point, Samuelson never accorded game theory much respect. This shows up in his retrospective comments but also early on in the survey of linear programming that was commissioned by the RAND Corporation (Dorfman, Samuelson, & Solow, 1958, esp. p. 445), where it is treated as having "no important applications . . . to concrete economic problems." Rather than regarding linear programming as a novel development, Samuelson has persistently used linear analysis to subsume it and the von Neumann expanding economy model under the umbrella of neoclassical production functions, which led in turn to the debacle of the Cambridge capital controversies. For a while, he even opposed the von Neumann–Morgenstern expected utility theory, although he did an about-face on that position when he later became an advocate of one special version of "stochastic economics."

The third characteristic attribute of Samuelson's work is the nuanced way in which he invokes Maxwell's Demon. However much he likes to quote Robert Solow's dictum that "probability is the most exciting subject in the world" (1977, p. 488), it is not so clear how much Samuelson feels the same way. As he has acknowledged, he regarded the fascination with Wiener processes as fundamentally antithetical to the very notion of an economic law governing prices (1977, pp. 471ff.). From Maurice Kendall's paper of 1953 to Benoit Mandelbrot's work in the 1960s, he thought modeling prices as stochastic processes was an attack on neoclassical price theory; and he was provoked to produce his 1965 "Proof" that neoclassically anticipated prices would tend to fluctuate randomly. The idea that an analyst could extract "information" out of price data offended his own understanding of classical equilibrium thermodynamics, which we have already noted was formative for his own model of price formation. He later wrote: "any subset of the market which has a better ex ante knowledge of the stochastic process that stocks will follow in the future is in effect possessed of a 'Maxwell's Demon' who tells him how to make capital gains from his effective peek into tomorrow's financial page reports" (1986, p. 485). Thus the centrality of information was banished from Samuelson's world, and this extended to his own notions of the role of statistics. We noted that the role of "statistician" was the entrée for both Chicago and Cowles into the brave new world of OR; but this was emphatically not the case with MIT. Samuelson himself exhibited a lifelong aversion to econometrics; and moreover, the MIT attitude toward econometric estimation – basically, any baroque procedure was justified as proof of mathematical virtuosity, but any fool could obtain an empirical fit: it was the model that counts – was one of the debilitating practices that set up the fall from grace of econometrics in the 1980s.

There is a fourth phenomenon, toward which we can only gesture. Of all postwar American economists, Samuelson seems to have felt most acutely the possible conundra of a neoclassical economics that seeks to imitate physics. He seems to have inherited this appreciation from his mentor Wilson, who had written on this problem on a number of occasions.[79] For von Neumann and his cadre of American Operations Researchers, the solution was not to deal in analogies but rather to forge a transdisciplinary science of complexity, a cyborg science applicable indifferently to social and natural phenomena. Samuelson and the MIT school seem never to have found this alternative worth comment, much less exploration. Rather, he has pioneered the refrain – echoed by Solow (in Bender & Schorske, 1997) and repeated as recently as 1998 – that it is pure coincidence that the same mathematical formalisms are used in physics that underpin neoclassical economics. Playing coy footsie with the natural sciences is one of the least intellectually attractive aspects of the MIT school of economics.

With these four characteristics in mind, we can perhaps now come to see how Samuelson's relative isolation from American OR, both in the Rad Lab and in the postwar configuration of military funding, may have encouraged a lifelong aversion to cyborg themes in economics. Samuelson has effectively been out of sympathy with most of the developments in the modern sciences outlined in Chapters 2 and 3. In contrast to the position of von Neumann, randomness was something to be borne as a trial, and not a creative principle in itself. One passively accepts the noise as part of the landscape; it does not participate in the constitution of the signal. Samuelson was forced in later life to acknowledge the existence of

[79] See, for instance, E. B. Wilson to Joseph Willits, January 5, 1940: "What are you going to do to protect the social sciences from the natural scientists? Have you read Julian Huxley's Washington speech?" Correspondence, 1940, folder I–Z, EWHL.

Maxwell's Augustinian Demon, but he could never bring himself even to admit to the existence of the Manichaean Other Demon. In this regard he never advanced as far as his MIT colleague Norbert Wiener. There was no possibility that randomness was an artifact of deception, or the strategic feints of the Other; randomness was at worst just another aspect of the State of Nature. The actual processing of information was never entertained, because, as a card-carrying "operationalist," Samuelson believed he had eschewed all cognitive commitments. Indeed, the theory of revealed preference conflated observed behavior with internal intentionality; any disjuncture between the two would reduce his theoretical edifice to rubble. Samuelson even took it upon himself to crusade against the version of "evolution" emanating from RAND (1986, pp. 697ff.). Computation never became an issue in Samuelson's work, because the entire question of who or what is carrying out the computation had been thoroughly repressed.

Thus we were bequeathed not one but three wildly divergent versions of neoclassical theory. We have only begun to hint at their distinctions and disjunctions in this section. Chicago, ever proud of its conservatism, hewed to Good Old Fashioned Operations Research, and got a workable rough-and-ready version of neoclassical theory, which served its purposes well. Cowles, wanting to be more *au fait*, grappled with the newer cyborg sciences at RAND. Last, MIT tried to have it both ways: there was nothing really new under the sun, but its epigones were nonetheless busy trying to co-opt the latest mathematical techniques to demonstrate that they were the most *au courant* of scientists. Not the least of the ironies besetting their project was that they were housed in one of the biggest nests of cyborgs to be found anywhere on the planet – the MIT of Whirlwind, Lincoln Labs, Bolt Berenek & Newman, and the Media Lab.

5 Do Cyborgs Dream of Efficient Markets?

When I think of it, it's not such a great distance from communist cadre to software engineer. I may have joined the party to further social justice, but a deeper attraction could have been to a process, a system, a program. I'm inclined to think I've always believed in the machine.

Ellen Ullman, *Close to the Machine*

FROM RED VIENNA TO COMPUTOPIA

As I sit here staring at a laptop, it seems one of the most histrionic statements in the world to insist that computers have a politics. It would appear that I could effortlessly bend this machine – small, portable, and eminently personal (though perhaps a little too warm and nowhere near cuddly) – to whatever idiosyncratic purposes I might harbor. However much it may now sport a tabula rasa on its user-friendly face, that was definitely not the way it was viewed in the first half of the century. An economist in the runup to the 1950s would have almost inevitably associated a fascination with the existence of something called a computer with a leftish inclination to believe that market economies could be controlled through some form of scientific planning. Although this political connotation was left out of the account of the rise of the three postwar schools of neoclassical economics in the previous chapter, its continued omission would be unpardonable. Because the confluence of computation, mathematical economics, and state planning is a key theme in the histories of the Cowles Commission and RAND, a basic prerequisite for understanding the cyborg incursion into economics is a passing acquaintance with the perfervid question of "market socialism" in the 1930s and 1940s.

The "socialist calculation controversy," like so much else in this cyborg narrative, began in Vienna.[1] Ludwig von Mises wrote an influential

[1] We now have a number of excellent accounts of the controversy, especially by Austrian economists who have rescued the major points at issue from subsequent misrepresentation by neoclassical economists. See Lavoie, 1985, 1990; Steele, 1992; Caldwell, 1997a; Thomsen, 1992; Chaloupek, 1990.

and provocative paper in 1920 asserting that comprehensive rational socialist planning of the economy was bound to fail, in response to a claim by the economist and philosopher Otto Neurath that wartime experience had demonstrated it was feasible to do away with market valuations and substitute "natural" units and centralized calculations. The problem, as Mises saw it, was that doing away with market valuations in money terms would render all rational calculation not just difficult, but essentially impossible. This stark equation of socialism with irrationality drew numerous responses and retorts, the thrust and parry of which we must necessarily bypass here, but whose ultimate consequence was a redoubled effort to clarify what sort of calculations could or should be carried out within markets; this in turn lent considerable impetus to the project of the elaboration of a rigorous mathematical economics in Vienna in this era. It also drew into the fray a number of figures who would subsequently play major roles in American economics, including Joseph Schumpeter, Wolfgang Stolper, and Karl Polanyi. But we shall be more interested in the way that the controversy had a way of acting as a magnet for cyborgs.

It has yet to be fully appreciated that many of the key figures in the history of the postwar Cowles Commission cut their eye teeth on this controversy. Jacob Marschak's first published article (1923) was a response to the Mises broadside. Leonid Hurwicz, who joined Cowles in 1942, subsequently constructed a separate interpretation of the controversy. The bellwether case of Kenneth Arrow is covered in detail shortly. This pattern was presaged in the 1930s at Cowles, with the best-known socialist response in the Anglophone world written by Oskar Lange (originally 1936–37, reprinted with Fred Taylor, 1938), produced just as both he and Cowles were moving to the University of Chicago. Lange's *Economic Theory of Socialism* embraced the mathematical model of Walrasian general equilibrium as a faithful representation of market operation and, in that light, found it relatively easy to substitute a central planning authority for the mythical Walrasian auctioneer. If a trade coordinator could calculate excess demands and broadcast revised notional prices, then why couldn't a central planning authority provide accounting prices to socialist managers? Mises's insistence upon the centrality of money was brushed aside as tangential. His arguments in favor of his brand of socialism won Lange much notoriety, but also left him deeply unhappy with the state of the Chicago department. Lange resigned his position in the Chicago department and Cowles in 1945 to return to his native Poland, but not before he had arranged for Marschak to assume the research directorship of Cowles from Theodore Yntema in 1943. It was for reasons such as these that mathematical Walrasian theory had assumed a dangerous "pinkish" cast in the eyes of other members of the Chicago economics department such as Frank Knight, and this conviction was the stage set for the ongoing hostility of the Chicago school to Walrasian general equilibrium in the postwar period. As Donald Patinkin (1995, p. 372) noted, "it was the socialist Oskar Lange who extolled the beauties of the Paretian optimum achieved by a perfectly competitive market – and Frank Knight who in effect taught us that the deeper welfare implications of the optimum were quite limited."

The vicissitudes of the socialist calculation controversy briefly raised all sorts of partially submerged issues but without really salvaging any of them, probably due to the highly charged political setting, the wartime disruption of careers, and the fact that just too many thickly overlapping philosophical controversies were being played out simultaneously. For instance, Lange accused Mises of advocating "the institutionalist view . . . that all economic laws have only historico-relative validity" (Lange & Taylor, 1938, p. 62); the Cowles habitus of baiting Institutionalists, while getting an early workout, was clearly unwarranted in this instance. Then there was the issue of the definition of "market socialism," the objective of which Lange seemed to equate with the artificial *simulation* of what markets putatively already did; beyond the issue of sheer redundancy, of course this distorted the rather more widely held opinion that markets and centralized controls might augment one another's operation in various ways (let us ignore for the moment the really important question of the exercise of political power), or at least subsist in some sort of symbiotic relationship. The Walrasian model was treated in an unmotivated manner as a faithful description of market operation, although this would be one of the least likely doctrines to be held dear by socialists in the Great Depression. What "competition" would signify in this Walrasian framework would persistently prove egregiously obscure.[2] And then there was the disciplinary question of precisely what the mathematical models were intended to accomplish: were they merely demonstrations of the possibility of existence of some equilibrium situation, or were they abstract exemplars of what it meant to conduct an economy in a "rational" manner, or were they more like computational tools to be used in the actual prosecution of coordination and control? None of these questions was adequately clarified in the subsequent debate, mainly because

[2] The following quotation reveals Hayek's awareness of this point, as well as the fact that it was intimately connected to the planning of science controversy in the British context: "Until quite recently, planning and competition used to be regarded as opposites. . . . I fear the schemes of Lange and Dickinson will bitterly disappoint all those scientific planners who, in the recent words of B. [*sic*] M. S. Blackett, believe that, 'the object of planning is largely to overcome the results of competition' " (1948, p. 186). This refers to Lange's belief that the benefits of competition had been obviated by the rise of monopolies and had to be restored by a central planner.

they were rarely tackled directly – and this is the valuable insight of the modern Austrian school of economics cited earlier – but nevertheless, two interesting developments can be traced back to these controversies. One consequence, possibly unintended, was to elevate to awareness the very problems of economic calculation *qua* computation. The other was to initiate a shift of the center of gravity of economic conceptions of the "function" of the market away from earlier neoclassical notions of exchange as brute allocation – that is, as physical motion in commodity space mirroring the movement of goods from one trader to another – and toward the image of the market as a conveyor and coordinator of "knowledge" or "information" between agents. The pivotal agent provocateur in disseminating the germs of these cyborg themes in both cases was Friedrich Hayek.

As early as 1935 Hayek was arguing against those who proposed a Walrasian model of central planning that "the mere assembly of these data" needed to prosecute the calculation "is a task beyond human capacity" (1948, p. 156); but moreover, "every one of these decisions would have to be based on a solution of an equal number of simultaneous differential equations, a task which, with any of the means known at the present [1935], could not be carried out in a lifetime." By 1940 he was trading dueling quotes from Vilfredo Pareto with Lange, reminding him that, even if one accepted the point that "values in a socialist society would depend on essentially the same factors as in a capitalist society, . . . Pareto, as we have seen, expressly denied that they could be determined by calculation" (1948, p. 183). Hence at least some of the premier disputants in this controversy had arrived at the curious impasse of suggesting that the salient difference between capitalism and socialism was that the former accomplished some (virtual?) computations that the latter could not. Lange attempted to meet this objection by suggesting a trial-and-error algorithm would work better in a socialist economy than in a capitalist context, since it could itself be optimized to produce an answer in a smaller number of iterations (Lange & Taylor, 1938, p. 89). Hayek retorted that, "This seems to be much the same thing as if it were suggested that a system of equations, which were too complex to be solved by calculation within a reasonable time and whose values were constantly changing, could be effectively tackled by arbitrarily inserting tentative values and then trying about until the proper solution were found."[3] It is difficult in retrospect not to regard this as a thinly disguised dispute over what computers could and could not do, predating any

[3] Hayek, 1948, p. 187. Here he inadvertently anticipated the technique of Monte Carlo simulation, unfortunately in a context where he wanted to suggest that such a thing was impossible.

awareness of the fact of their existence: a conjuration of Turing without his Machine. Lange later in life explicitly took this position to an extreme conclusion:

> Were I to rewrite my essay today, my task would be much simpler. My answer to Hayek and Robbins would be: so what's the trouble? Let us put the simultaneous equations on an electronic computer and we shall obtain the solution in less than a second. The market process with its cumbersome tatonnements appears old-fashioned. Indeed, it may be considered a computing device of the pre-electronic age. (1967, p. 158)

Again in 1935 Hayek was precociously promoting the term "information" as one of the central instrumentalities of market coordination. The cognitive problems of economic interactions had long been a trademark Austrian preoccupation; for instance, Hayek's colleague and successor as director of the Institut für Konjunkturforschung in Vienna, Oskar Morgenstern, had already written extensively about the problems of foresight and statistical prediction and had insisted on an essential indeterminacy in most economic endeavors (Mirowski, 1992). Nonetheless, it was within the specific context of the socialist calculation controversy that Hayek found himself appealing directly to this nebulous "thing" that the market processed but the central planner lacked: "The information which the central planning authority would need would also have to include a complete description of all the technical properties of every one of these goods. . . . But much of the knowledge that is actually utilized is by no means 'in existence' in this ready made form" (1948, pp. 154–55).

The need to refute "market socialists" such as Lange thus led directly to the initial landmark revision of the image of market functioning away from static allocation and toward information processing. The clearest statement of this momentous shift can be found in Hayek's "Use of Knowledge in Society," which has been quoted repeatedly as the manifesto (in retrospect) of the Cyborg Revolution.[4]

> What is the problem which we try to solve when we try to construct a rational economic order? On certain familiar assumptions the answer is simple enough. *If* we possess all the relevant information, *if* we can start out from a given system of preferences, and *if* we command complete knowledge of available means, the problem which remains is purely one of logic. . . . This, however, is emphatically *not* the economic problem which society faces. . . . The peculiar character of the problem of a

[4] See Simon, 1981, p. 41; Gottinger, 1983, p. 113; Makowski & Ostroy, 1993; Smith, 1991a, pp. 221ff.; Lavoie, 1990; Weimer & Palermo, 1982.

> rational economic order is determined precisely by the fact that the knowledge of the circumstances of which we must make use never exists in concentrated or integrated form but solely as dispersed bits of incomplete and frequently contradictory knowledge which all the separate individuals possess. The economic problem of society is thus not merely a problem of how to allocate "given" resources . . . it is a problem of the utilization of knowledge which is not given to anyone in its totality. (1948, pp. 77–78)

This bold proposal to change the very subject of academic economics did not come at a propitious moment. Hayek had just produced the popular political tract *The Road to Serfdom* (1944), which warned that various schemes to extend state control of the economy would lead to the types of totalitarian regimes resembling those which the allies had just defeated; few in the West were willing to make that particular leap, particularly given the patent recent success of governments in mobilizing the war economy. Keynesianism was being embraced by the avant-garde of economic advisors in the Anglophone world, and consequently Hayek was shunned, with a reputation as having been soundly routed by John Maynard Keynes and his Cambridge Circus (Caldwell, 1997b; Colander & Landreth, 1996). This impression was perhaps inadvertently reinforced by Hayek's leaving the London School of Economics for an appointment outside of the economics department (in the Committee for Social Thought) at the University of Chicago in 1949. But more significantly, Hayek's railing against "scientism" during the war had banished him outside the charmed circle of those ambitious to extend the "scientific method" to social theory and planning (Mirowski, 1989a, pp. 354–56). As commonly happens in such incidents, competing conceptions of the very nature of "science" were at stake in the controversy, and not untrammeled Luddism, although this passed unrecognized at that time.

In an attempt to clarify his theses about the role of the market in transmitting information, Hayek was propelled into consideration of some of the characteristic themes, and intermittently the actual neighborhood, of the cyborg theorists. Soon after his 1945 manifesto, he turned his attention to psychology (1982, p. 289), working on the manuscript that became *The Sensory Order* (1952). The spectacle of an economist with no evident credentials in contemporary psychology presuming to write on the fundamentals of the topic was rather less common in that era than it is in our own; even with his prior record of popular notoriety and substantial theoretical accomplishments, Hayek could not garner much in the way of respect or attention from experts in the field, and the isolation shows in his final product. However, three decades later, he did claim kinship with one very special brain theorist:

> I wasn't aware of [von Neumann's] work, which stemmed from his involvement with the first computers. But when I was writing *The Sensory Order*, I reasoned that I could explain to people what I was doing. Usually I found it very difficult to make them understand. And then I met John von Neumann at a party, and to my amazement and delight, he immediately understood what I was doing and said that he was working on the same problem from the same angle. At the time his research on automata came out, it was too abstract for me to relate it to psychology, so I couldn't really profit from it; but I did see that we had been thinking along very similar lines. (1982, p. 322)

By his own admission, Hayek was incapable of appreciating von Neumann's mathematical enthusiasms. He may also have been oblivious to the withering skepticism trained in his direction by his former colleague Morgenstern.[5] Thus he never really attained membership in good standing of the Cyborg Club, although his prognostications did become increasingly Teleological as time went on. Perhaps it is better to think of Hayek as someone who filtered various cyborg themes into economics at second- and third-hand, motivated to search them out by his prior commitment to the metaphor of the market as a powerful information processor. Once given a voice, however, his ambition for a theory of a computational marketplace could easily be turned round to provide further metaphorical inspiration for many cyborgs in good standing.[6] This certainly explains his status in some quarters as the premier political philosopher of the Information Age.

The ambition of *The Sensory Order* is to explain how the brain works, but not in the manner of a Turing or a McCulloch and Pitts but rather as (in his own words) "a ghost from the 19th century" (1982, p. 287). Much taken with Ernst Mach's "principle of the economy of thought," Hayek conceived of the nervous system as engaged in a process

[5] The Morgenstern diaries are replete with disparaging remarks about Hayek. There is a certain *Rashomon* quality to his description of what is very probably the very same party to which Hayek refers above: "Hayek is here for 10 days. I gave a party. . . . I heard him in a Col. after a dinner. I can't stand it any longer. Neither him, nor his opponents; that's no science. I could predict all his answers" (entry April 19, 1945, OMPD). The tantalizing entry for May 22, 1946, reads: "Need to write more another time, especially about 'Schmonzologie' [Idle-talkology] (Johnny about Hayek's presentation).' The entry for May 23, 1946: 'In the evening Hayek gave the Stafford Little Lecture on competition. Weak, 'literary' and scholastic. The students were very critical . . . what is done by others doesn't interest him. He has never started reading our Games, but he is "against it." What he said about epistemological questions was very primitive."

[6] One of the most significant computational theorists to acknowledge Hayek's influence was Frank Rosenblatt of *Perceptron* fame (1958). An interesting attempt to make the case that "For Hayek, society is a complex automaton in von Neumann's sense" can be found in Dupuy, 1996; see also Dupuy, 1994.

of classification of sense impressions, a process distributed across the gamut of neurons and undergoing continuous amplification and revision. In this schema, the map of prior associations plays the role of memory (absent the computational concern with its capacity), but because knowledge is treated as relational and distributed, it does not generally conform to a symbolic or propositional format. For Hayek, this accounts for the "tacit" character of most expertise, the very attribute that would thwart "information" from being centrally managed and controlled. Indeed, in a reversal of what would later become the norm in economics, his prior image of the structure of the market provided the metaphoric template for his theory of the structure of the brain, something he openly acknowledged: "In both cases we have complex phenomena in which there is a need for utilizing widely dispersed knowledge. The essential point is that each member (neuron, or buyer, or seller) is induced to do what in the total circumstances benefits the system. Each member can be used to serve needs of which it doesn't know anything at all. . . . knowledge is utilized that is not planned or centralized or even conscious" (1982, p. 325). This image of passive associations formed under a bombardment of stimuli fits rather uneasily into the larger Austrian tradition, as has been noted by some sympathetic commentators (Smith, 1997), because it leaves little room for the conscious subject to intentionally alter her own cognition: the entrepreneur is banished here just as surely as she had been in the neoclassical theory of the firm.

Nevertheless, Hayek stands out squarely as one of the earliest prophets of the importance of "complexity" in economics. His version of complexity, never quite spelled out in its entirety, did seem to be related obliquely to notions of Kolmogorov complexity theory: "The minimum number of elements of which an instance of a pattern must consist in order to exhibit all the characteristic attributes of the class of patterns in question appears to provide an unambiguous criterion" (1967, p. 25). However, there is no question about the original source of his ideas in this area: here he repeatedly credits (who else?) Warren Weaver (1967, 4n, 26n). For Hayek, complexity was just another contrapuntal line in his trademark chorus that knowledge could never be centralized and regimented. In later life, he repeatedly stated that only entities of a higher level of complexity could encompass and thus "understand" entities of a given lower level of complexity; thus, while the mind could adequately comprehend simpler systems, it could never provide a comprehensive account of its own operation. This, for instance, is how he came to (mis)interpret Gödel's incompleteness results: "It would thus appear that Gödel's theorem is but a special case of a more general principle applying to all conscious and particularly all rational processes, namely the principle that among their determinants there must always be some rules which cannot be stated or even be conscious" (1967, p. 61).

While misconstruing Gödel had become a pastime nearly as popular as lawn tennis at midcentury, at least amongst economists, Hayek's attempts to access cyborg themes to insist upon the limits of knowledge and its organic interdependence helps shed light on what some have called "Hayek's transformation" (Caldwell, 1988). His pronounced shift away from equilibrium theory and toward a process description of market operation was accompanied by a shift from what started out as a deep distrust of engineers to something approaching admiration. What might initially appear to be an inexplicable about-face, from the Cassandra decrying "scientism" in economics to the late advocate of something verging upon systems theory,[7] going so far late in life as an embrace of even the evolutionary metaphor, becomes much more comprehensible when read as an intuitive elaboration of the incompatibilities of the nascent cyborg sciences with the heritage of neoclassical economics in midcentury. This recourse to "computation without computers" was altogether inseparable from its political context: namely, the vertiginous turn toward Walrasian theory that the dispute between capitalism and socialism had taken within the economics profession in that era.

Be it attributed to lightness in recursive function theory or darkness at tune, Hayek's effective subsequent isolation from cyborg discourse crippled his ability to make any substantial contributions to their development. For instance, however much Hayek wanted to portray knowledge as distributed across some structure, he maintained an implacable hostility to statistical reasoning throughout his career, a stance that effectively blocked any further comprehension of perceptrons and connectionist themes. Indeed, the formal template for his understanding of neurons was his own previous *Pure Theory of Capital*, a deeply flawed attempt to explicate the Austrian theory of capital using simple deterministic geometric models.[8] His innocence of formal logic was pervasive: it constituted one of the motivations behind Morgenstern's scorn. In a deeper sense, he was more akin to a romantic poet than a software engineer, in that he wanted to maintain that there was something inviolate and ineffable about rationality, something that could never be reduced to an algorithm or mechanical device. This is why, on the one hand he can sound so very modern – "the price and market system is in that sense

[7] "In the first few years after I had finished the text of [*Sensory Order*], I made an effort to complete its formulations. . . . I then endeavored to elaborate the crucial concept of 'systems within systems', but found it so excruciatingly difficult that in the end, I abandoned [it]" (Hayek, 1982, p. 290).

[8] See his rather revealing admission of this point (Hayek, 1982, p. 291).

a system of communication, which passes on (in the form of prices, determined only on the competitive market) the available information that each individual needs to act, and to act rationally" (1982, p. 326) – and yet, on the other, continue to treat this "information" as something ethereal, impervious to all attempts at codification, analysis, and control, and in the end, fleetingly evanescent.

It may seem unfair to subject Hayek to the Rules of the Code, because his goals and intentions need not have uniformly overlapped with the cyborgs; but then, this misses the point of the comparison. Whatever Hayek the human being may have wanted, Hayek the historical figure has been retrospectively thrust into a ceremonial pigeonhole which he was ill-fitted to occupy. This contradiction is captured in an implicit contrast between Hayek and Shannon in one of his late documents. Shannon, as we have mentioned in Chapter 2, sought to endow the theory of automata with substance by constructing a mechanical mouse that was capable of negotiating a maze, which could itself be sequentially reconfigured by the observer. Hayek rejected the implication that this exercise accomplished anything in the way of understanding of rodent cognition. His advocacy of the ultimate untransmissibility of lived experience, echoing Rickert and Dilthey, led him to claim, "in order to understand what a rat will do and why it does it, we would in effect have to become another rat. Need I spell out the consequences that follow from this for our understanding of other human beings?" (1982, p. 293). In effect, Hayek had renounced both the formal theory of abstract logical automata and the pragmatic technics of simulation, and yet clung to the language and metaphor of the nascent cyborg sciences.

Such delicate discriminations were nowhere in evidence in the 1940s and 1950s in academic economics. No one as yet had a clear idea of what it would mean to take a computational approach to economics at that juncture. What mattered to the economists whom we shall encounter in the next section was that Hayek was the acknowledged major protagonist in the socialist calculation controversy, that he was there on the ground at Chicago from 1949 onward, and that Cowles had become the de facto standard-bearer for the Walrasian pro–"market socialist" position in the same period. Furthermore, John von Neumann and his enthusiasms were encountered everywhere the ambitious Cowlesmen turned. Everyone was speaking cyborg without really being fully aware of it – or at least until they were rudely awakened to that fact.

THE GOALS OF COWLES, AND RED AFTERGLOWS: GETTING IN LINE WITH THE PROGRAM

Once the question has been broached about computers, symmetry demands it should equally be applied to neoclassical economics: does neoclassical economic theory embody a specific politics? A century of uneasy bromides have done little to seriously clarify the issue, perhaps because of a tendency to confuse the various levels of specificity at which an answer might be tendered and defended. At its most abstract and general, the response must surely be: a few equations imperfectly cribbed from rational mechanics need have no special affinity with any political position whatsoever, as long as sufficient ingenuity is expended in their interpretation and elaboration. Yet, at a slightly less elevated level, the question as to whether individual neoclassical economists put forth arguments that it supported various political programs cannot be answered in any other manner but the affirmative. Walras thought it underwrote a limited conception of state ownership of land; Pareto thought it resonated with a sour cynicism about the circulation of elites; Hayek began by thinking it helped explain an antistatist position; Milton Friedman believed it the economics of Dr. Pangloss. But this raises the further question: is there anything intrinsic to the doctrine that biases it toward one political orientation or another? This question can only be addressed at the very specific level of individual cultural formations at well-demarcated historical junctures.

The story of the Cowles Commission is the tale of "high theory" in postwar American neoclassical economics, but despite numerous "official" accounts (cited in Chapter 4), the most significant aspects of its heritage to the economics profession still remain shrouded in oblique shadows. The conventional accounts find it convenient to focus mainly on technical issues in statistical estimation of structural econometrics, but this only serves to paper over a critical rupture in the history of Cowles, a breach toward which we have begun to gesture in the previous chapter. In effect, the postwar Cowles Commission innovated the standards and practices for orthodox economic *theory* for the next generation of American economists to a much greater extent than it stabilized empirical practice. What is missing from the existing accounts is a narrative of how the key Cowles figures managed to forge an accommodation with their perilous postwar environment, producing a forward-looking economic doctrine for a refurbished economics profession, effectively complementing the dawn of the American Half Century. The three neglected components of this novel package were: the political reorientation of economic discourse, the cultivation of a new scientific patron, and the attempt to assimilate and tame some new conceptual developments emanating out of the cyborg sciences.

Seeing Red

In the immediate postwar era, Cowles was ground zero of Walrasian market socialism in America. Although Oskar Lange had departed by

1945, the remaining participants were fully aware that they held substantially more in common than some diffuse fondness for mathematical rigor: "we members of the Cowles Commission were seeking an objective that would permit state intervention and guidance for economic policy, and this approach was eschewed by both the National Bureau and the Chicago School" (Klein, 1991, p. 112). Jacob Marschak, research director from 1943 to 1948, had written, "I hope we can become social engineers; I don't believe we are much good as prophets" (1941, p. 448); his own commitment to socialism had been subject to private commentary by colleagues in the 1930s.[9] By 1945 Marschak was disputing with Michael Polanyi over the interpretation of the basic message of neoclassical economic theory:

> Briefly, there is no disagreement that a social optimum (or approximate optimum) in some sense which requires proper definition, will be reached if both the consumers and the hired factors of production were left free to maximize their satisfactions (leaving aside the question of unequal incomes and of a necessary minimum of "unwanted" but useful things such as schools). The disagreement arises on the question of the managers of enterprises: will an approximate social optimum be secured by individual managers maximizing their profits? This seems to be the case if the following two conditions are satisfied: 1) perfect competition; 2) perfect foresight. To the extent that these conditions are not fulfilled, a search for other methods becomes necessary.[10]

Tjalling Koopmans had been a member of a research unit in Holland gathered around Jan Tinbergen that sought to develop the technical means to prosecute rational economic planning. When he came to Cowles in 1944, he rapidly became immersed in the question of market socialism:

> I have been reading lately about welfare economics. After Lange's book on socialistic economics, which appeared before the war, Lerner has

[9] See the letters between Joseph Schumpeter and Ragnar Frisch on Marschak's nomination as a Fellow of the Econometrics Society, and Schumpeter's letter to Alfred Cowles on his appointment to the Cowles Commission in the Schumpeter Papers, Harvard University Archives. Schumpeter to Frisch, February 2, 1932: "You do me an injustice: I am not so narrow as to object to anyone because he is a socialist or anything else in fact. If I did take political opinion into consideration I should be much in favor of including socialists in our lists of fellows. In fact, I should consider it good policy to do so. Nor am I or have I ever been an anti-Semite. The trouble with Marschak is that he is both a Jew and a socialist of a type which is probably unknown to you: his allegiance to people answering these two characteristics is so strong that he will work and vote for a whole tail of them and not feel satisfied until we have a majority of them, in which case he will disregard all other qualifications." Schumpeter Papers, HUG/FP 4.8, box 1, Harvard University Archives.

[10] Jacob Marschak to Michael Polanyi, August 2, 1945, box 94, file: Polanyi, JMLA.

> published a textbook on "The Economics of Control." Then there is still Meade's "Economic Analysis and Policy." . . . This discussion does not get the attention it deserves in this country. The new stream, to which Lerner has also been converted . . . is that it does not matter whether the economic system is named capitalist or socialist. It all amounts to whether the pricing policy is based on marginalist principles, which will lead to the optimum allocation of resources. To my tastes, this discussion is still too academic, although it is refreshing in abandoning the old hobby horses. It is not a simple case to calculate the marginal cost of the transportation of a railroad passenger between Boston and New York, or even the "expected value" of it. Hoping to be able to contribute to a more concrete discussion, I have begun in my free time a study in which these ideas are applied to shipping.[11]

Lawrence Klein was recruited to Cowles, against the advice of his thesis advisor Paul Samuelson, by Marschak suggesting to Klein, "What this country needs is a new Tinbergen model" (in Feiwel, 1987a, p. 340). Arrow's first attempt at a Ph.D. thesis also involved recasting the Tinbergen model into the American context (1991b, p. 8). The Cowles Commission bore close ties to the postwar National Planning Association, whose publications were primarily the work of Arthur Smithies and Jacob Mosak, the latter an erstwhile Cowles researcher. Cowles also served as a magnet for outsiders interested in questions of economic planning, including one of the key protagonists from our cyborg narrative in Chapter 2:

> Leo Szilard, who was also a good friend of Marschak and who had done unusually noteworthy work in getting Jewish scientists out of Germany, of which Marschak was a part . . . used to come to Cowles quite regularly. He used to spend a lot of time in the evenings with [Klein], Hurwicz and others. He had very clever ideas about structuring an economy that was cycle-free. He demonstrated this by means of a game. . . . Szilard used to say to me, "I am going to prove to Hayek, von Mises, and the other free-marketeers, just why their ideas are wrong and what you need to do to get an economy that is cycle-free." (Klein in Feiwel, 1987a, p. 344; see also Lanouette, 1992, p. 320)

Thus there is no question but that the Cowlesmen were acutely aware of Hayek, even before he arrived on the Chicago scene. Indeed, Marschak was tapped by the University of Chicago Press to act as referee on Hayek's

[11] Tjalling Koopmans to Jan Tinbergen, July 18, 1945, JTER. I wish to thank Albert Jolink for facilitating access to this correspondence, and for the translation from the Dutch. See also the transcript of Questions about Koopmans supplied by his wife Truus, September 1991, file: Koopmans, HSCM. There she states (p. 5): "As far as I know, he had quite some sympathy towards socialism and regarded the economic experiment in the USSR with great interest."

book *The Road to Serfdom*; in the spirit of raising the tone of the debate, he counseled publication.[12] A goodly portion of Cowlesmen's abiding faith in planning was derived in equal measure from their (mostly) shared European backgrounds, their previous political travails, their prior training in the natural sciences, and their Old World cultural presuppositions, although as (again mostly) immigrants and refugees, they did find reconciliation of their beliefs with the folkways of the indigenous inhabitants of their newfound land a bit of a trial.[13] Perhaps the midwesterners, mired in their mugwump xenophobia, could not be expected so willingly to concede to the Cowlesmen the right to "plan" the economy of their newly adopted Arcadia. Of course, contemporary events did not conspire to make their ambitions any more likely: the Soviet A-bomb test of September 1949; the convictions of Communists under the Smith Act in October; the antics of Senator Joseph McCarthy in early 1950; and the invasion of South Korea by the North in June 1950.

The trouble for Cowles began with its skirmishes with the Chicago economics faculty, and especially with Milton Friedman and Frank Knight. Later on, Allen Wallis also became a formidable adversary.[14] In better circumstances this might have only been confined to a tempest in an academic teapot; but national politics also bore down upon Cowles's

[12] See "Report on the Road to Serfdom," box 91, folder H, JMLA. It states: "Since in this country the terms 'plan' and 'socialism' have often been used to include monetary and fiscal policies, social security, and even progressive income tax the American reader will possibly expect from Hayek a more concrete demarcation between what the book calls 'plan in the good sense' and the (undesirable) planning proper. In fact, the non-economic chapters . . . are more impressive than the economic ones."

[13] See, for instance, Tjalling Koopmans's tribute to Jacob Marschak, dated July 2, 1958: "When we moved to Chicago, I had just about gotten through the immigrant's first years of bewilderment about the new country, and had begun to feel at home, with a sense of newly acquired patriotism. When I expressed this to Jascha he told me he had gone through a similar process in three countries in succession! . . . There was a strong sense of mission and of standing together in the early postwar years of the Commission, which in retrospect is quite amusing. But it did bring out our best efforts in pursuing the goal as we saw it. With Klein, Hurwicz and others we battled for simultaneous equations and for the idea of econometrically guided policy, in the annual meetings of the professional societies and in our skirmishes with the National Bureau, as if the future of the country depended on it!" Box 16, folder 303, TKPY.

[14] See Koopmans memo on conversation with Bernard Berelson of the Ford Foundation, October 20, 1952: [Koopmans] "Explained that sometimes we communicated more effectively with mathematicians in other Social Science disciplines (Lazardsfeld, Coombs) than with economists at large. Explained some coolness to us in our own profession including our own campus from an incorrect imputation of bias toward planning or government interference (based on Lazardsfeld's advice to give our best hypothesis concerning Wallis' antagonism, but without mentioning Wallis)." Folder: Ford Foundation, box 99, JMLA.

aspirations in the immediate postwar period. Much has been written in a general way on the Red Scare and the McCarthy witch-hunts and their repercussions upon academic freedom in America (Selcraig, 1982; Schrecker, 1986), but consideration of the local fallout on specific academic disciplines and schools is still in its infancy. What is known is that Illinois was not a happy-go-lucky place to be if you harbored socialist sympathies in the late 1940s. For instance, the Broyles Commission in the Illinois legislature ominously threatened to institute loyalty oaths and vet educational institutions in 1949. As Herbert Simon wrote, "By 1948, Communists and supposed Communists were being discovered under every rug. . . . Any graduate of the University of Chicago, with its reputation for tolerance for campus radicals, was guaranteed a full field investigation before he could obtain a security clearance" (1991a, p. 118). Robert Hutchins, the president of the University of Chicago, was reasonably effective in his defense of academic freedom in these years;[15] yet some of the Cowlesmen, with their foreign passports and/or their socialist pasts, could be construed as having at least half a skeleton secreted in their South Side closets. Some members of Cowles were exposed to loyalty questions in an unfortunate incident at the University of Illinois in 1950–51 (Solberg & Tomilson, 1997). In an attempt to build up the economics department, Dean Harold Bowen and chairman Everett Hagen had set out to recruit various Cowles researchers to the department, including Leonid Hurwicz, Franco Modigliani, and Donald Patinkin. A local political dispute was escalated into some rather nasty Red-baiting in 1950, and an outside committee of businessmen dictated that Bowen should be removed from his administrative duties. The local threat of a higher-powered mathematical economics was conflated with the perceived threat of socialist tendencies in the ill-named Urbana, and Hagan, Hurwicz, Patinkin, and Modigliani resigned over the next two years.

Some other Cowles members had backgrounds that were deemed by the authorities as warranting political scrutiny. Kenneth O. May, a Cowles research consultant in 1946–47, was an openly declared member of the Communist Party (Simon, 1991a, p. 124); he left for a tenured position at Carleton College before the witchhunts and was ejected from Cowles in 1947.[16] Others weren't quite so fortunate. When Lawrence Klein moved to

[15] On Hutchins, see Dzuback, 1991, p. 201; compare the behavior of James Conant with regard to the defense of Harvard University faculty in Hershberg, 1993, chaps. 21–23.

[16] Kenneth May (1915–77): B.A. and M.A., Berkeley, 1936; Ph.D. in mathematics, Berkeley, 1946. See folder: Kenneth May, box 148, JMLA, where it becomes obvious that Marschak banished May due to his skcpticism over the worth of neoclassical theory and not his politics per se. For biographical information, see Jones et al., 1984.

Cowles in 1944, he also began to teach at the Abraham Lincoln School. As part of his activities there, he did join the Communist Party, although he claimed to have dropped out of its activities by 1947. He wrote a paper during that time period comparing Marx and Keynes, but Marschak refused to include it in the Cowles reprint series.[17] Even though Cowles was expanding, Klein's econometric models became viewed there as a liability after 1948, and he was not given one of the internal academic positions. One of his colleagues at the time, Kenneth Arrow, mused with a certain laconic understatement, "Lawrie obviously had trouble getting a job. He got a National Bureau Fellowship. They used to give it to people who could not get a job. I guess it was his politics" (in Feiwel, 1987b, p. 648). The observer might be tempted to contrast Klein with Arrow; according to a friend's description, "While in college he was a Norman Thomas socialist. But wherever he stood in the past, he has largely abstained from contemporary political involvements" (Lipset in Feiwel, 1987b, p. 693). After some other temporary assignments, Klein did land an instructor position at the University of Michigan in 1950, but his past continued to dog his progress. He was served with a subpoena to appear before the Clardy Committee in Michigan in 1953. The economics department did counter by proposing he be offered a full professorship, but without tenure; disheartened, Klein decamped to a visiting professorship at Oxford in 1954, never to return to Michigan (Brazer, 1982, pp. 221–27).

The late 1940s were not only a time of external dangers and lurking conspiracies, but also profound internal intellectual tumult at Cowles. The Marschak program at Cowles in the mid-1940s had essentially been an extension of the project of Henry Schultz in the 1930s: namely, the improvement of empirical estimation of Walrasian systems of demand equations by means of the innovation of new and enhanced statistical techniques, now augmented by the motivation of Lange-Lerner questions of planning and control. This was the accomplishment for which Cowles is still most fondly remembered, that of the invention of structural econometric estimation techniques (Hildreth, 1986; Christ, 1994; Epstein, 1987; Koopmans, 1950). The problem, as one of the participants put it

[17] Koopmans was especially scathing in his response to Klein's paper. See his memo to Marschak, December 10, 1946, box 148, folder: Klein, JMLA: "This paper is an attempt to sell the idea of econometric model building to adherents of Marxian economic doctrine. I shall explain in these comments why I believe that such attempts, including the present one, are harmful to the objectives of econometric model building. The main reason is that econometric research of the type in which we are engaged in is essentially based on modern theories of economic behavior. The way to sell it to anybody, including Marxian economists, is first to sell modern economic theory, and then to sell econometric methods. There are no short cuts."

in retrospect, was that "the whole statistical apparatus at the Cowles Commission did not pay off" (Klein in Feiwel, 1987a, p. 350). The Marschak angle of attack, which had initially been seconded by Koopmans, was to treat the previous empirical failures of the Walrasian model as predominantly "technical" problems that would yield to technical solutions: causal identification, statistical definitions of exogeneity, vigilance toward observational equivalence, corrections for simultaneous equations bias, and the like.[18] Unhappily, the prescribed "corrections" only served to multiply a whole raft of further logical elisions and undermotivated theoretical compromises, all in the service of arriving at parameter estimates, which, if anything, looked worse than the comparable estimates produced by the simpler and less computationally intensive old-fashioned single-equation ordinary least-squares techniques. Signs and magnitudes of some coefficients were still "wrong," and standard errors tended to balloon out of control. Prediction of key variables outside of sample periods displayed disappointing accuracy. And if the techniques were applied to a limited number of Keynesian macroeconomic equations, which were by definition more tractable, such as Klein's models I–III of the U.S. economy, both Cowles insiders and their critics complained that the estimated functions were not sufficiently grounded in rigorous "theory." It seemed as if the mountain had labored mightily to bring forth a mouse: Milton Friedman especially delighted in tormenting Marschak and Koopmans about their failure to solve any real scientific problems in economics, but Arthur Burns at the NBER also chimed in with disdain.[19] The vision of a statistically based macro-

[18] The relationship of the statistical program to prior issues of observable implications of the Walrasian system is treated in detail in Hands & Mirowski, 1999. As late as 1946, Marschak could write: "All is not well with static economics. . . . The more significant implications of this analysis (e.g., the famous Slutsky equation) have not yet been subjected to valid verification by the facts; but neither have they been disproved" (1946, p. 97). The context of this admission, in a review article of von Neumann & Morgenstern's newly appeared book on game theory, will shortly assume heightened significance.

[19] "Marschak has expended effort on what seems to me to be a sterile exercise in translation. . . . [his] principal complaint here is that Mitchell has not worked out numerous demand and supply schedules. Marschak writes with enthusiasm of what could be accomplished with such schedules. . . . But I do not think it would be easy to supply concrete instances of outstanding successes in Marschak's direction by others. Marschak describes a goal that may be attained some day, not one that has already been reached" (Burns, 1951, pp. 26, 30). Friedman's attack is discussed in detail in Epstein, 1987, pp. 108–10. In later retrospectives, some Cowlesmen tried to appropriate the external critiques of Cowles macromodels as evidence for their triumphal acceptance: "Keynes' bold severing of the connection with rational behavior was undermined by the intellectual need to understand behavior, which we interpret as explaining it in rational terms" (Arrow, 1991b, p. 3).

economic planning device in service of market socialism grew increasingly forlorn. The 1949 business cycle conference seemed particularly traumatic, as witnessed by these internally circulated postconference wrap-up notes:[20]

> Quotations from expert critics:
> Samuelson: Identification difficulties with serial correlation.
> Leontief: Many data, don't need hyper-refined methods.
> Smithies: Effect of policy of given direction and unknown magnitude.
> Metzler: "Theory" in Klein's work not sufficiently systematic.
> . . . .
> Hurwicz: We want as much of disaggregation *as we can afford.*
> Marschak: Traditional theoretical model not always the most useful to interpret data with (e.g., production function).
> Hurwicz: Much criticism incurred because we fully state our assumptions.
> Simpson: State objections and answer them.
> Hurwicz: Comment on existing writings on policy questions to state their implications and limitations.
> Marschak: Education of profession in use of simple models.
> Koopmans: Not fight too much.
> Christ: Need to show results.
> Hildreth: How anxious should we be to convince everyone?

Hildreth's question was perceptive: should Cowles researchers remain beholden to older notions of the sources of legitimacy of economic discourse or should they develop new roles for themselves? The Cowles idiom of arcane mathematical statistics was not exactly captivating the imaginations of the political masters whom they had sought to advise and educate; and the graybeards of the legitimate economics profession had begun to sense the hollowness of their program to provide the Walrasian system with impeccable empirical content and to grow wary of the political content of their Keynesian and Lange-Lerner exercises. The response of most Cowles researchers circa 1948 was to drop most of their work on the theory of statistical estimation in relatively short order, and turn instead to the cultivation of an altogether different approach to neoclassical theory. One after another, they adopted a style of argument bequeathed them by John von Neumann and his protégés and found a sympathetic patron for their reoriented work in the military, and particularly at RAND.

Koopmans on the Road to Damascus

The metamorphosis began just before the accession of Tjalling Koopmans to the research directorship of Cowles in July 1948, a job he would hold

[20] "Discussion to Size-up Results of Business Cycle Conference. Early December, 1949," CFYU. On this incident, see the commentary of Epstein, 1987, p. 111.

until just before Cowles moved to Yale in 1955.[21] Koopmans's regency coincides with the years of the construction of the New Order at Cowles. The Commission was reconstituted anew in 1948 as a self-governing body with its own executive committee and right to recommend tenure at Chicago independent of academic status in other departments. This enhanced autonomy was itself a reflection of the newly demonstrated capacity of the commission to attract lavish sources of outside funds, primarily from the military. Not all such sources were open and above board, even in Cowles's annual *Research Reports*: for instance, Cowles held a conference on the economic consequences of an atomic war for postwar urban centers in 1945, but no proceedings were ever published.[22] Cowles's wartime connections to members of the Metallurgical Lab (the site of the first controlled nuclear chain reaction [Rhodes, 1986, pp. 432–42]), and especially Leo Szilard, had given them an inside track into the early effusion of postwar research funding on all things "atomic." Projects on atomic energy that were made public, such as the 1946 study on the development of civilian atomic power under the direction of Marschak and Sam Shurr, had their funding routed through less obvious channels, such as the Social Science Research Council or the Life Insurance Association of America. But Cowles did not become just another promiscuous contract research house hawking a motley portfolio of empirical projects to a jumble of grant agencies; there was a clear direction toward which its research was reoriented, to such a degree that the new regime was already evident in the celebratory *Twenty Year Research Report* (Christ, 1952). In the guise of a history of the Commission, it announced that Cowles was putting a new face to the world, one that the text dubbed "Economic Theory Revisited, 1948–52." Yet far from déja vu, there were numerous tokens of a new departure. As part of the twentieth anniversary, it was decided that Cowles's original motto – *Science Is Measurement* – had become a bit of an embarrassment, given that most of the principals

[21] Tjalling Charles Koopmans (1910–85): M.A. in physics and mathematics, University of Utrecht, 1933; Ph.D. in mathematical statistics, University of Leiden, 1936; lecturer, Netherlands School of Economics, 1936–38; economist, League of Nations, Geneva, 1938–40; research associate, Princeton, 1940–41; Penn Mutual Life Insurance Co., 1941–42; statistician, Combined Shipping Adjustment Board, 1942–44; Cowles Commission, 1944–45; professor, University of Chicago, 1946–55; professor, Yale, 1955–85. There is no really reliable biographical source; but see Scarf, 1997.

[22] Koopmans to Jan Tinbergen, September 30, 1945, JTER. The circumstances surrounding the conference on Atomic Energy, essentially arranged by Leo Szilard, can be found in Kathren et al., 1994, pp. 757–66). See also Koopmans to Marschak, September 1, 1945, box 92, folder: Koopmans, JMLA: "Szilard and some of his colleagues have bad conscience about this world-shaking invention. Klein and I helped Szilard with some estimates of the real cost of spreading out the cities."

had ceased to do statistical work. Clifford Hildreth suggested a motto more in tune with the times and in line with the plan, and it was quickly replaced on their emblem: *Theory and Measurement*.[23] It was bland, noncommittal, erased inconvenient historical attachments and served to deflect curiosity concerning the real objectives of the New Model Cowles.

Tjalling Koopmans proved to be the ideal catalyst to bring about this Reformation at Cowles. He had earned a degree in quantum physics in Holland under the tutelage of the physicist Hendrik Kramers (Dresden, 1987). He explained to a meeting of the American Physical Association in 1979 why he had opted to switch careers:

> Why did I leave physics at the end of 1934? In the depth of the worldwide economic depression, I felt the physical sciences were far ahead of the social and economic sciences. What had held me back were the completely different, mostly verbal, and to me almost indigestible style of writing in the social sciences. Then I learned from a friend that there was a field called mathematical economics, and that Jan Tinbergen, a former student of Paul Ehrenfest, had left physics to devote himself to economics. Tinbergen received me cordially and guided me into the field in his own inimitable way. I moved to Amsterdam, which had a faculty of economics. I found that I benefited more from sitting in and listening to discussions of problems of economic policy than from reading the tomes. Also, because of my reading block, I chose problems that, by their nature, or because of the mathematical tools required, have similarity with physics.[24]

But the process of conversion and retooling was not without its dangers. Koopmans had worked for two years on a Tinbergen-style econometric model for the League of Nations in Geneva, an experience he found

[23] See "Motto file," CFYU. It contains a letter from a researcher from the Cowles *ancien régime*, Harold Thayer Davis to Koopmans, May 30, 1952: "The suggestion made to change the motto to 'Theory and Measurement' might be a very good solution to the problem, particularly since nothing is asserted and hence there can be no objection."

[24] The speech continues: "By way of example, as a result of war work in the British Shipping Mission in Washington, I worked on a problem of the best routing of empty ships from ports of cargo discharge to next ports of cargo loading. The problem turned out to be quite similar to Kirchhof's [*sic*] classical problem – solved a century earlier – of the distribution of electric currents in a network of conductors." Box 18, folder 333, TKPY. The analogy with Kirchoff's law is discussed in some detail in Koopmans, 1951, pp. 258–59. A slightly different account is provided by his wife Truus in her transcript found in HSCM: "Tjalling became interested in economics in his physics years. He had started to read Karl Marx's *Das Kapital* volume 1, his first economics book, in order to understand more about Marxism. He got more and more frustrated with abstract theoretical physics and wanted to make more applications especially to the reality of unemployment and social change. He thought that the careful use of probability and mathematics in physics theory could also be applied to economic data."

deadening, and from which he fled abruptly when Germany marched into the Netherlands in mid-1940. Upon disembarking in America, he discovered to his chagrin that academic jobs were extremely hard to find. With the assistance of Samuel Wilks and von Neumann, he settled in Princeton, doing various odd jobs as a research associate for Merrill Flood, at the School for Public Affairs of Princeton, and teaching night classes in statistics at New York University.[25] Finding this existence unbearably tenuous, and chafing at the prospect of having "sold my soul to the devil of empiricism," Koopmans embarked on a sequence of unsatisfactory jobs. For more than a year he worked in an econometric unit at the Penn Mutual Life Insurance Company, and then discontent once again propelled him to work for the British Section of the Combined Shipping Adjustment Board in Washington. There is no telling how long this peripatetic life of quiet desperation might have continued had not Marschak offered him a position on the staff at Cowles in 1943.[26] His vicissitudes illustrate the fact that there were very few career opportunities in mathematical economics outside of Cowles in America as late as the Second World War, and that the process of assimilation of foreign academics was very sporadic and uneven: any trivial bias in selection could potentially have profound repercussions.

The installation of Koopmans at Cowles, far exceeding the von Neumann–Morgenstern collaboration, constituted the primary channel through which von Neumann came to shape the postwar economics profession in America. We have observed in Chapter 3 that Marschak had been acquainted with von Neumann as early as the 1920s: but there is little evidence that he ever really took to heart the full complement of Neumannesque themes, or ever managed to engage von Neumann's attention while research director at Cowles. Koopmans was another matter. Here was a bona fide quantum physicist who could potentially appreciate the necessity to bring economics into line with developments in twentieth-century science; here was a specialist in statistics who respected

[25] See box 5, folder 4, VNLC, where appreciable evidence exists for his early contacts with von Neumann. See also the Tinbergen correspondence, JTER, where the "sold my soul" quotation comes from a letter dated December 18, 1940. For further information on Merrill Flood, see note 39 and Chapter 6. Koopmans, in box 16, folder 303, TKPY, reports that Samuel Wilks arranged the job with Merrill Flood. Wilks (1906–64) was a colleague of von Neumann's in the mathematics department at Princeton, and was responsible for its program in mathematical statistics. Wilks was an important member of the NDRC and the Applied Mathematics Panel during World War II. We have already encountered Wilks in the previous chapter as one of the early organizers of RAND, along with Warren Weaver and John Williams. This also cemented Koopmans's early ties to RAND.

[26] The letter of Koopmans sounding out Marschak for a job (August 22, 1943) and the favorable response (September 9, 1943) can be found in box 92, folder: Koopmans, JMLA.

the aspirations to rigor of modern mathematics. Furthermore, Marschak tended to think of his economic research as devoted to the service of peace and prosperity; but Koopmans shared with von Neumann a fascination with war. As Koopmans wrote to Marschak as early as September 1945:

> [Jacob] Viner's strong belief in competition made him think that, if only every country had this [atomic] weapon, each would be so afraid of retaliation, that no military warfare would take place anymore, and issues would be decided by economic warfare. It seemed to me that, at least, with this hypothesis errors of the second kind would be extremely costly. Szilard thinks the whole idea is a dangerous illusion. . . . It seems to me that the control of the military uses of economic power is now the most important political issue facing the planet.[27]

Koopmans was nearly obsessed with the bomb, perhaps not unreasonably, given what he was hearing from von Neumann, on one hand, and Szilard, on the other. Hence the finale of World War II did not prompt him to relax his war footing; nor did he think that should happen at Cowles. As early as March 1948 he wrote to Marschak: "If war comes in the near future. I cannot yet quite believe this, but we must think about what would be the best use of our resources. I have been thinking of giving empirical content to the systems of linear production functions by studying the operations of an economy under successive or simultaneous bottlenecks, to determine 'accounting prices' where money prices fail as guides to allocation."[28] By late March the anxiety was worse: "Finally, I am worried about the possibility of war, which would frustrate our present reorganizational effort. Unless we continue the group by securing a quantitative war economics project, for which cooperation with the Rand group will be helpful. Without a war project, the draft would blow us up."[29]

Von Neumann served as Koopmans's muse, and perhaps, his guardian angel in the 1940s. Although the date of their first meeting is uncertain, there is substantial evidence of regular contact with von Neumann well before Koopmans went to Cowles. In von Neumann's archives there are a brace of documents that suggest he aided in getting Koopmans on his feet soon after his arrival in America in summer 1940. Among these papers is a photostat of Harold Hotelling's letter to the counsel-general of the

[27] Koopmans to Marschak, September 1, 1945, box 92, folder: Koopmans, JMLA. The same letter then has a postscript, which goes on to relate news from a meeting with von Neumann on September 4 from 10:30–3:00, which, among other events, has von Neumann offering to let Cowles have some time on the computer newly set up at Aberdeen Proving Grounds.

[28] Koopmans to Marschak, March 17, 1948, box 92, folder: Koopmans, JMLA.

[29] Koopmans to Marschak, March 23, 1948, box 92, folder: Koopmans, JMLA.

Netherlands dated January 2, 1940, requesting that Koopmans be absolved from Dutch military service; another is a letter from Kramers dated February 8, 1939, plaintively asking if Koopmans is "completely lost for physics or not?"; a third is a vita and list of publications dating from mid-1941. A letter from Koopmans to von Neumann dated November 3, 1941, refers to discussions the two men had the previous summer about a proof of the distribution of a quadratic form in normal variables on the unit sphere.[30] Von Neumann also delegated Samuel Wilks to look after the refugee.

The relationship was clearly fortified after Koopmans went to Cowles. There were two reasons, one fortuitous, the other more substantial. The fortuitous reason was that von Neumann was forced to change trains in Chicago on his numerous trips between Princeton and Los Alamos, and Koopmans was privy to the attempts by others at the University of Chicago to take advantage of these precious lulls in the otherwise overtaxed von Neumann schedule.[31] The more serious reason was that, during the mid-1940s, there materialized an occasion for mutually advantageous exchange. Von Neumann was initially interested in finding some mathematically sophisticated economists to go to work on game theory, while Koopmans and Cowles were especially concerned to get his help with the computational problems of the statistical estimation of systems of equations, which had been the research focus at Cowles. Von Neumann gave two lectures on his game theory at Cowles in May 25–26, 1945 (Feiwel, 1987b, p. 20; Debreu, 1983a), and it was thus arranged that two (eventually three) Cowles members would provide favorable reviews of *TGEB* in major outlets of the American economics profession (Marschak, 1946; Hurwicz, 1945; Simon, 1945). In return, von Neumann sketched a method for computation of maximum likelihood estimates of a set of linear equations in a letter soon after his visit.[32]

[30] These letters can all be found in box 5, folder 2, VNLC.

[31] "We hear from Szilard that it happens more often that you pass through Chicago. Please be assured that we are profoundly grateful for any time you feel able to spare for a discussion with us." Koopmans to von Neumann, June 22, 1945, box 24, VNLC. See also Klein, 1991, p. 110.

[32] See "Copy taken from a letter of J. von Neumann, June 2, 1945," box 24, VNLC, where he discussed the problem of finding the maximum of the function [D(**X**) exp (−1/2 **T(XMX′)**)], where the boldface variables are matrices. This is discussed in a letter from Koopmans to A. H. Taub dated July 24, 1959, in the same folder. There Koopmans writes: "I remember very well the discussions I had with von Neumann. . . . There were several discussions in Chicago on the theory of games, just about at the time that the von Neumann Morgenstern book appeared. Reviews of that book by Hurwicz, Marschak, and Simon in various journals reflect in part the effects of these discussions. Finally, much later at the RAND Corporation, probably in the summer of 1952, I had some discussions with von Neumann about the so-called 'assignment problem.'"

Just as in real-life market exchanges, the transaction did not simply leave the parties equipoised in a Pareto-improved but fully anticipated and invariant state; rather, it initiated a process whereby the goals of Cowles were rather comprehensively revised; and perhaps a case of lesser influence could similarly be made in the reverse direction for von Neumann (though we decline to do so here). The occasion for the mutual regeneration was a realization that solving a game and the problem of maximizing a matrix likelihood function were not so much different questions as similar mathematical problems. These were the sorts of epiphanies in which von Neumann reveled but the kind that had been rather scarce in economics up to that point. What had struck Koopmans forcibly at that juncture was that his own earlier work on efficient tanker scheduling was yet another mathematical reformulation of the same problem; it was only at this stage that he began to glimpse some of the virtues of von Neumann's expanding economy model, which he (and the rest of the economics profession) had previously ignored.[33] Out of this conversion experience came the conviction that what was important about game theory was not necessarily all the verbiage (which Koopmans would have found indigestible in any event) about indeterminacy and strategic problems of information processing, but rather that it provided the paradigm for a rather more sophisticated type of maximization procedure employing a different set of mathematical tools. In the eyes of the Cowles researchers, that became the beacon guiding the quest for the reanimation of the Walrasian tradition, not through the previous Cowles doctrine of empirical testing but now rather through a recasting of interdependent maximization as an actual planning tool that mimicked the operation of an idealized "perfect" market. Socialist politics of a Lange-Lerner variety, adherence to the neoclassical creed, the desire for a more rigorous science, and the commitment to mathematical sophistication all came together in a tidy bundle of neatly reconfigured packages called, variously, "linear programming," "activity analysis," "decision theory," and "Arrow-Debreu general equilibrium."

This, of course, is where the military came into the picture. The older machine dreams of a socialist like Lange were simply nowhere on the radar screen in the immediate postwar America, fearful as it was of a Red under every bed. However, there was one place where comprehensive planning was not only being put into practice on a daily basis in postwar America but, even better, where dark suspicions about patriotism and aspersions

[33] See the letter of Koopmans to Robert Dorfman, October 31, 1972, which states "I heard von Neumann lecture in Princeton on his model, probably when I lived in Princeton, 1940–41, or soon thereafter, at the invitation of Oskar Morgenstern. To my everlasting frequent regret, I did not see the great importance of it." Box 11, folder 186, TKPY.

about foreign-sounding surnames were kept at bay, if not altogether banished. Whatever the wider doubts about the intellectual coherence of "market socialism," the American military was one place where unquestioned adherence to the virtues of the market cohabited cheerfully with the most vaunting ambitions of centralized command and control, without ever provoking any hand-wringing about conceptual consistency or soul-searching over freedom. Indeed, for researchers of a certain political stripe, the security clearance required for military consultants could be worn proudly as a badge of legitimation, serving to ward off a modicum of superficial distrust or the ubiquitous pall of McCarthyite suspicion. Conversely, the avatars of planning had something that the military might find useful. The military had come out of World War II with an appreciation for the role of the scientist in assisting with problems of command, control, and communications; operations research was regarded as the wave of the future. That the military had also emerged from the war as far and away the largest patron of postwar science did not go unnoticed. The shape of the New Order at Cowles was largely a process of bending the concerns of Walrasian economics to the contours of the military's requirements, coming to terms with operations research, and adjusting to the already existing wartime tradition of physical scientists doing social science at the behest of the command hierarchy.

Linear Programming

Koopmans had precociously developed some existing ties to the military through his own war work and acquaintance with the atomic scientists at Chicago, but once the connection was rejuvenated via von Neumann, vast new horizons were opened up to the Cowlesmen. The first manifestation was Koopmans's introduction to George Dantzig, the developer of the "simplex method" for solving linear programming problems.[34] Dantzig was an operations researcher for the Army Air Force at the Pentagon during World War II who then wrote a statistics thesis proving the existence of optimal Lagrange multipliers for a linear model under Jerzy Neyman

[34] The account in these two paragraphs is based on the following sources, which disagree with one another in certain minor respects: Dantzig, 1987, 1991; Dorfman, 1984; Lenstra et al., 1991; Schwartz, 1989 and a letter of Koopmans to Dorfman, October 31, 1972, box 11, folder 186, TKPY. All Koopmans quotes in the next paragraph are from this letter.

George Dantzig (1914–): B.A. in mathematics and physics, University of Maryland, 1936; Ph.D. in statistics, Berkeley, 1946; U.S. Army Air Force statistical control headquarters, 1941–46; mathematics advisor, U.S. Air Force, 1946–52; RAND consultant, 1947–51; RAND staff, 1952–60; professor of OR, Berkeley, 1960–66; professor of OR and computer science, Stanford, 1966–. Some biographical information is available from the interview in Albers, Alexanderson, & Reid, 1990.

immediately after the war. He was enticed to return to the Pentagon as advisor to the U.S. Air Force Comptroller in 1946 with the charge to explore the extent to which the military planning process could be mechanized using the newly developed electronic computers. At first drawn to Wassily Leontief's input-output matrices as a promising organizational framework, Dantzig decided that what the military really needed instead was an algorithm to choose between alternative activities oriented toward the same goal, preferably subordinate to some sort of optimization procedure. It was at this point that Dantzig decided to turn to some different economists and, in particular, to Koopmans, to learn what they had to say on the maximization of a linear function subject to linear inequality constraints.

Precisely how Dantzig had learned about Koopmans is not explained in any of the retrospective accounts, but what is not in doubt is that their prior shared knowledge of Neyman-Pearson statistics and the tanker-routing paper (restricted and not yet published)[35] signaled a close parallel of mathematical interests. Once Dantzig visited Koopmans at Cowles in June 1947, they eventually discovered that they also shared an avid interest in planning models and a fascination with the mathematical analogy between solving a routing problem and solving a likelihood function for a system of linear equations. "Koopmans at first seemed indifferent to my presentation, but then he became very excited – it was as if, in a lightening flash, he suddenly saw its significance to economic theory" (Dantzig in Albers et al., 1990, p. 74). Indeed, Koopmans became so enthusiastic that he arranged for Cowles associate Leonid Hurwicz to spend time with Dantzig at the Pentagon that summer to learn more about the nascent developments. He also gave Dantzig a copy of von Neumann's expanding economy paper, inducing him to quickly consult with von Neumann in October and cement the three-way collaboration. (It was

[35] Koopmans's first *public* presentation of his tanker work was a presentation at the International Statistical Conference in Washington, D.C. in September 1947 – that is, *after* the initial meeting with Dantzig. A brief note on the work appeared in an *Econometrica Supplement* in July 1947, but the first substantial explanation of Koopmans's 1942 tanker memo had to await the activity analysis volume in 1951. It was finally published in Koopmans's collected papers (1970). In fact, the problem had been independently "discovered" many different times before, dating back to Fourier in 1824; on the longer-term history, see Grattan-Guinness, 1994b. An essentially similar solution of the problem by Frank Hitchcock published in 1941 languished unappreciated until Dantzig and Koopmans mobilized the military in 1947. On the Russian development of linear programming, see Dorfman, 1984. In an interview with Albert Tucker found at ⟨libweb.princeton.edu.2003/libraries/firestone/rbsc/finding_aids/mathoral/pm04.htm⟩ Merrill Flood claimed that Koopmans had first gotten interested in the "48 States Problem" of Hassler Whitney while working for Flood on the Princeton Surveys and local school bus routing, which may have sown the seeds for the tanker scheduling problem.

expressly for his work on linear programming that Koopmans was jointly awarded the 1975 Nobel Prize in economics with the Russian Leonid Kantorovich.) Dantzig has often related his astonishment at von Neumann giving him an off-the-cuff personal lecture of more than an hour on how to solve a linear programming problem, when he had been unable to find any relevant papers in the mathematics literature. He reports von Neumann as reassuring him:

> "I don't want you to think I am pulling this all out of my sleeve on the spur of the moment like a magician. I have recently completed a book with Oskar Morgenstern on the theory of games. What I am conjecturing is that the two problems are equivalent." . . . Von Neumann promised to give my computational problem some thought and to contact me in a few weeks which he did. (1991, p. 24)

When confronted with this particular format of optimization, Von Neumann was reminded of his game theory and the computer; when Koopmans was confronted with the same problem, he saw only Walras. He later admitted: "Before that encounter, the only [linear programming] model I was aware of was the transportation model. . . . With regard to other uses of the production function in economics, my thinking was at the time traditional and in terms of the smooth neoclassical model. . . . I think George [Dantzig] was flabbergasted by my lack of interest in doing algorithmic development by myself. But don't forget, George, I was and am a 'classical' economist!"[36] This ability to find some common ground in the mathematics – most of which had *not* been generated by the economists – while steadfastly remaining committed to the neoclassical program is the key to understanding how Cowles responded in general to both von Neumann and the military. For Koopmans, linear programming, or as he tried to rename it with little success, "activity analysis," was an improved version of Lange, imagining the market as a vast *program* in which the prices that emerged from competitive markets came to resemble the dual variables in Dantzig's algorithm. Koopmans explicitly acknowledged the political subtext that, "Particular use is made of

[36] See note 33 for Koopmans to Dorfman, October 31, 1972. We should also take this occasion to caution the reader of the attempts by Albert Tucker and Harold Kuhn to retrospectively revise the historical record, something that will show up again in the next chapter. "Al Tucker, who had been reading the manuscript of my book *Linear Programming* asked me, 'Why do you ascribe duality to von Neumann and not to my group?' I replied, 'Because he was the first to show it to me.' 'That is strange,' he said, 'for we have found nothing in writing about what von Neumann has done.' 'True,' I said, 'but let me send you the paper I wrote as a result of my first meeting with von Neumann'" (George Dantzig in Albers et al., 1990, p. 76). For an interesting paper that challenges the standard account of the genesis of the Kuhn-Tucker theorem as itself historically dubious, see Kjeldsen, 2000.

those discussions in welfare economics (opened by a challenge of L. von Mises) that dealt with the possibility of economic calculation in a socialist society. The notion of prices as constituting the information that should circulate between centers of decision to make consistent allocation possible emerged from the discussions by Lange, Lerner and others" (1951, p. 3).

This vision was not restricted to practical allocation problems, but became central to the other conceptual triumph of Cowles in this period, the Arrow-Debreu existence proof for Walrasian general equilibrium (discussed in Chapter 6). Koopmans helped innovate Cowles's explicit novel riposte to Hayek, that the Walrasian model actually illustrated the "economy of information" that existed in a perfect market or in decentralized market socialism (1991, p. 22). The mantra was fashioned along the lines that the neoclassical agent worked solely on a need-to-know basis: purportedly he required access only to his very own supply set, his own personal preferences and the given system vector of prices (by whom it was left deliberately vague) in order to calculate his own optimal response. Because Koopmans had avoided algorithmic questions as beyond the pale, waxed agnostic on questions epistemological, and showed no interest in any historical market structure, this assertion was based on little more than the sheerest bravado; moreover, there would be almost no attempt at effective specification of this "simple" cognitive procedure at Cowles Mark II. Economists of all stripes, above all else, should have been wary of the smell of snake oil and the whining of perpetual motion machines that surrounded this early approach to "information" at Cowles: "the economies of information handling are secured free of charge. . . . It is therefore an important question of economic analysis to discern and characterize the situations, as regards both the technology of production and the structure of preferences, in which this costless economy of information handling is possible" (p. 23). But, bedazzled and in awe of the nascent computer, no one saw fit to complain. Indeed, the planning portrayal of the economy as one gigantic computer was deployed by Koopmans as a rhetorical resource to argue that the military encouragement of mathematical economics was a relatively benign phenomenon, and that no hidden agenda was being freighted in alongside the abstract prognostications.[37]

[37] "There is, of course, no exclusive connection between defense or war and the systematic study of allocation and planning problems. . . . If the apparent prominence of military application at this stage is more than a historical accident, the reasons are sociological rather than logical. It does seem that government agencies, for whatever reason, have provided a better environment and more sympathetic support for the systematic study, abstract and applied, of principles and methods of allocation of resources than private industry. There has also been more mobility of scientific personnel between government and universities, to the advantage of both" (Koopmans, 1951, p. 4).

The language of "programming" was fairly suffused with the military preoccupation with C^3I in the 1940s, something all too readily forgotten, now that computers have osmosed into every corner of modern life. When the first digital computers were invented, designing a physical logic circuit was treated as part and parcel of the design of the algorithm it would execute, and instructions were frequently patched with phone plugs rather than "written" in symbols. As Dantzig explained with his coauthor Wood, the distinction between the planning of an elaborate calculation and the planning of a social structure was likewise elided in the development of "linear programming." "It's purpose is close control of an organization. . . . Programming . . . may be defined as the construction of a schedule of actions by means of which an economy, organization or other complex of activities may move from one defined state to another" (in Koopmans, 1951, p. 15). The evocative language of the finite state automata was no rhetorical flourish here, as elsewhere. Yet there was another attractive feature of the word "programming," for which many participants at RAND shared an insider's appreciation. It was not at all obvious just how long the Air Force would turn a blind eye to the amount of "pure" or playful mathematical research that was being funded at RAND; and there was the converse problem that one could not be too explicit in the 1950s as to the breadth of purview of ambitions to "plan" any economic process. The beauty of that protean word "programming" was not only that it evoked the computer, but it would prove a convenient euphemism for any potentially controversial doctrine concerning control and power that had to be encapsulated with a certain opacity.[38] It became the most effective of the serried ranks of euphemisms for political phenomena at RAND, such as "organizational theory," "systems analysis," "logistics support," and "decision analysis" (Fisher & Walker, 1994).

The language of programming, and its polymorphously promiscuous referents, constituted a bridge spanning the concerns of the economists, the logicians, the engineers, and the operations researchers. This was not the first time the computer had served such a function, and neither would it be the last. Not only did the mathematics of linear programming resemble that of von Neumann's game theory; the very act of programming a problem freighted with it many of the resonances of playing

[38] "I felt I had to do something to shield Wilson and the Air Force from the fact I was really doing mathematics inside RAND Corporation. What title, what name should I choose? In the first place, I was interested in planning, in decision-making, in thinking. But planning is not a good word for various reasons. I decided therefore to use the word 'programming.' . . . it's impossible to use the word, dynamic, in a pejorative sense. . . . Thus I thought dynamic programming was a good name. It was something not even a Congressman could object to. So I used it as an umbrella for my activities" (Bellman, 1984, p. 159).

a game. As Paul Edwards reminds us, "all computer programming, in any language, is gamelike" (1996, p. 170). It was the structured manipulation of abstract symbols according to well-defined rules in order to arrive at preconceived outputs. Validity in such a reference frame comes to be assessed primarily in terms of internal coherence and consistency, and doing the mathematics came to loom increasingly large in significance relative to demonstrated competence in the task at hand. Strategies could be treated as subroutines, and, given the emphasis on mixed strategies in game theory, statistical inference came to be conflated with the mixing and matching of algorithms. Hence the language of "programming" fostered an atmosphere within which Cowles could initially appear noncommittal about whether it was concertedly engaged in actually doing game theory, or just baroque optimization, or perhaps something else equally machinelike.

But what was especially timely about the genesis of "linear programming" and "activity analysis" was that it was conjured right on the historic cusp of the separation of "software" from "hardware." The distinction itself was evoked by military exigencies: Merrill Flood claimed to have coined the term "software" in a 1946 War Department memo, so as to be able to distinguish the ballooning research cost items that could not be directly attributed to military hardware budgets (Ceruzzi, 1989, p. 249). As inscriptions of algorithms came to occupy a different conceptual plane than the machines they were being run on, it became more likely that inscriptions of mathematical models could themselves be conceptualized as "running" on a social structure like a mainframe, or that commanders would set the values of a few variables in the software while subordinates executed their commands on the organization hardware. "Both the organization leader who determines efficiency prices and the subordinate who determines the in- and outputs for his agency proceed by alternate steps: each reacts to the preceding move of the other. . . . The problem reduces to the mathematical one of finding the fastest iterative method for a computation" (Cowles Commission, 1951, p. 17). This metaphor neatly dovetailed with Koopmans's own visions for the future of economics: mathematical economists could aspire to be the antiseptic "software engineers" of the brave new world of economic planning, writing the algorithms for the manipulation of "efficiency" targets or "shadow" prices, while leaving all the messy institutional details of markets and organizations and history and politics to the lower-status hardware engineers (the "literary" economists? or the lawyers and politicians?).

This presumed hierarchy in the division of economic labor lay at the crux of the other major incident in Koopmans's *anno mirabilis* of 1947, the "Measurement without Theory" controversy with the NBER (Koopmans, 1947; Mirowski, 1989b). In that celebrated broadside which first earned him widespread notice outside the charmed circle of the mathematical elect, he consigned Wesley Clair Mitchell and Arthur Burns to playing the subordinate roles of Tycho Brahe and Kepler to his Newton. In a slight travesty of the actual history of science, it was made to seem as if renunciation of responsibility for the generation and processing of data was the hallmark of the nobility and objectivity of scientific endeavor. Yet there festered a minor inconsistency buried in Koopmans's blueprint for the brave new world, for, as we noted, he did not really care a fig for the actual algorithmic specifics of how one would calculate the linear program or generally process market data. In effect, his ambition was all the more vaulting, aspiring to the vocation of master programmer *sans* computer. This aspiration to be a software engineer somehow divorced from the tiresome process of writing, debugging, and implementing the code would eventually prove a spanner in the machine dream.

Slouching toward Santa Monica

The Cowles pipeline to the Air Force was not limited to a few select Pentagon brass or Santa Monica strategists. RAND, as we noted in the previous chapter, was a godsend for Koopmans and for Cowles. Koopmans's first patron at Princeton, the statistician Samuel Wilks, was one of the consultants, along with Warren Weaver and Frederick Mosteller, to decide the composition of the "Evaluation of Military Worth" section at RAND in 1947 (Jardini, 1996, p. 84). After the war, Wilks was a major figure in the Social Science Research Council, just one of the ways in which "civilian" and "military" funding of social science had become inextricably intertwined soon after V-J day (Mosteller, 1978). Merrill Flood, Koopmans's first temporary employer in America, having had himself been an Army consultant, was now employed at RAND to do game theory.[39] Meyer Girshick, who had been principal statistician for the NDRC during the war, had gone to work for RAND when it was still

[39] Merrill Meeks Flood (1908–): B.A., Nebraska, 1929; Ph.D. in mathematics, Princeton, 1935; instructor in mathematics, Princeton, 1932–36; director of statistical section surveys, Princeton, 1936–40; research associate in mathematics, 1940–45; Ordnance department, Frankford Arsenal, 1943–44; OSRD Fire Control Research Group, 1944; general staff, U.S. Army, 1946–47; RAND staff, 1949–52; professor of industrial engineering, Columbia, 1953–56; professor of industrial engineering, University of Michigan, 1956–68; principal scientist, Systems Development Corp., 1968–69; professor of education, University of Michigan, 1974–; president, TIMS, 1955; president, ORSA, 1961. Flood was a student of von Neumann at Princeton, and had read *TGEB* in manuscript, and spent part of World War II working on aerial bombing strategies. On this, see Poundstone, 1992, p. 68, and the interview with Albert Tucker in the series "The Princeton Mathematical Community in the 1930s" at ⟨libweb.princeton.edu.2003/libraries/firestone/rbsc/finding_aids/mathoral/pm04.htm⟩. Flood claimed to have coined the term "software," as mentioned earlier; he also claims to

a subsidiary of Douglas, roughly overlapping his time as a research consultant at Cowles. George Dantzig was consulting with RAND by mid-1947 (Jardini, 1996, p. 90). And then, of course, there was John von Neumann, working behind the scenes to shape and define the research program at RAND.

When RAND separated from Douglas Aircraft in 1948, around the same time that Koopmans engineered Cowles's reorganization, golden opportunities opened up for extensive patronage and cooperation with the newly "private" think tank. As Koopmans wrote in his earliest correspondence with Ed Paxson, "My main interest is in the topics described by you as 'one-sided' planning. I am interested in these as practical problems of planning, as well as theoretical problems in welfare economics."[40] It quickly became commonplace for Cowles researchers to make extended visits to RAND, usually over the summer. The roster of visitors in the first decade included Kenneth Arrow (beginning in 1948), Leonid Hurwicz, Theodore Anderson, Evsey Domar, Harry Markowitz, Jacob Marschak, Andrew Marshall, and Herbert Simon. But, more important, the RAND connection opened up the military coffers to lavish long-term funding for the brand of mathematical economics that Cowles wanted to explore, thus guaranteeing a unified program of growth and expansion. Substantial Cowles ties began in 1948, with two of Koopmans's working papers included in the RAND "research memoranda" series.[41] The first RAND research contract with Cowles was initiated in January 1949 under the rubric "Theory of Resource Allocation"; the first large contract with the Office of Naval Research on "Decision-Making under Uncertainty" commenced later in July 1951 (Christ, 1952, pp. 46–47). RAND, by contrast, regarded Cowles as the ideal platform from which to entice economists into the development of operations research, systems

have invented the term "linear programming" with John Tukey. His primary claim to fame has to do with the genesis of the "Prisoner's Dilemma" with Melvin Dresher, discussed in detail in Chapter 6.

In the interview with Tucker, Flood says, "when Sam [Wilks] heard from Tjalling Koopmans that Tjalling wanted to come to the United States, he came over to me and said, 'Is there any way you can do something about it?' I worked around for a while and finally found some money, though not much. I had him come to work at the Local Government Survey, and brought him over from the League of Nations. We were going to do a book together on time series theory." Koopmans often had occasion to depend on the kindness of strangers over the course of his career, in part because he restlessly sought to move on from wherever he was situated.

40 Koopmans to Ed Paxson, March 1, 1948, box 92, file: Koopmans, JMLA. The letter continues in suggesting that Herbert Simon, Carl Kaysen, and Kenneth Arrow also be included in RAND activities.

41 "Systems of linear production functions," February 10, 1948, RAND RM-46; "Remarks on reduction and aggregation," July 21, 1948, RAND RM-47.

analysis, and contract procurement evaluation for the military. To that end, RAND sponsored a conference on the applications of game theory to military tactics at Chicago and Cowles on March 14–15, 1949 (Jardini, 1996, p. 92), and a conference on linear programming in June 1949. The Cowlesmen rapidly learned to conform to the military's notions of openness as the concomitant of accepting their funds: the proceedings of the former have never been published, much less mentioned in the *Annual Reports*, whereas a vetted version of the latter did appear as Cowles monograph 13, *Activity Analysis of Production and Allocation* (Koopmans, 1951). The Cowlesmen began to get the Econometrics Society involved as well, jointly sponsoring (with TIMS) a symposium on game theory in September 1949, which was essentially a showcase for RAND, presided over by von Neumann and Wilks, and including a talk by von Neumann and papers by Girshick, Paxson, Dantzig, and John Tukey (*Econometrica*, 17 [1949]: 71–72).

So it would appear that in relatively short order, by 1951 at the latest, Cowles had gotten into line with von Neumann's cyborg program by looking seriously at game theory and linear programming, reorienting its efforts toward a general theory of organization patterned on the computer. Indeed, many Cowles-affiliated economists underwent something akin to conversion experiences at RAND,[42] setting out on a path that would earn them fame and (for many) Nobel Prizes. One after the other, Koopmans, Marschak, Hurwicz, Arrow, and Simon halted their prior pursuit (to varying degrees) of econometric empiricism and instead became theorists of rational choice, information processing, and optimal planning. As one of the exiles (and a non-RANDite) described it, somewhat ruefully: "many of my colleagues at Cowles got turned off from model-building and application at an early stage. Either they found it too much work, too tedious, or they found it not to work mechanically well enough. They went into other highly varied areas: Kenneth [Arrow] went into social choice, Koopmans went into activity analysis, Marschak went into team theory, Haavelmo went into more theoretical and speculative ideas about economic growth and economic philosophy, Anderson went back to his work on statistical inference, and Patinkin went into pure macro theory. Our team then fell apart" (Klein in Feiwel, 1987a, p. 342). The pattern Klein does not explore is that the dissolution of Cowles Mark I consisted

[42] See, for instance, Kenneth Arrow in Feiwel, 1987b, p. 648: "This was a very exciting period. I really learned a great, great deal. In many ways, certainly with respect to mathematical tools for equilibrium theory, the time at RAND was more stimulating than at Cowles. . . . It was not that I was unaware of game theory before I came to RAND, but there you really learned the technical aspects rather than just vague impressions." For a similar testimonial, see Simon, 1991a, p. 116.

of a distinct split between those working on a newly bequeathed program – namely, members of our roster for Cowles Mark II who developed ties with RAND – and those like Haavelmo or Patinkin or Klein himself, who were summarily spun out of the military orbit into unrelated research programs.

The exhilaration of the Mark II RAND cohort in having gone "beyond these frontiers [of traditional economics] into the theory of organization, the study of communication and the flow of information in society" is evident in their grant application to the Ford Foundation in September 1951.[43] As exemplars of bold transgressions and new departures, the document cited Arrow's Impossibility Theorem, von Neumann and Morgenstern's theory of games, John F. Nash's bargaining solution (1950), "work by Norbert Wiener and others on cybernetics," and "methods of programming the interdependent activities of a large organization, done by the Air Force." It crows that, "it is no longer necessary (and sometimes ineffective) to urge upon social scientists emulation of the example of the physical sciences in regard to rigor, clarity and method. The studies referred to in this section can stand the test of comparison in these respects, and in addition go to the heart of social science problems through logical and mathematical methods that are appropriate to these problems (rather than being borrowed from the physical sciences)." Economists are exhorted to experience the newfound freedom of the operations researchers to boldly go where no scientist has gone before: "the boundaries between the various social sciences are vague and unimportant. Thus, without a systematic intention to that effect, the Cowles Commission studies just described have led us to a situation where we no longer know whether we are working in economics, in organization theory, [or] in sociology."

Well, yes, but . . . We have all become inured to a certain modicum of hyperbole in grant proposals, but there is something about this story that is not quite straight; something is getting deflected so that the narrative can appear more sleek and linear. For instance, in the enumeration of exemplary achievements in its grant application, the Cowles contingent could not legitimately claim credit for any of them. The practical development of linear programming was really due to Dantzig; game theory and Nash bargaining theory were products of Princeton and RAND, as explained in the next chapter; there was little or no development of cybernetics to speak of anywhere at Cowles; and the nearest thing to a legitimate triumph, Arrow's Impossibility Theorem, was also more

43 "Application to the Ford Foundation by the Cowles Commission, September 17, 1951," mimeograph copy, CFYU. There is no author listed, but both content and protocol suggest it was Koopmans. All quotations in this paragraph are from this document.

appropriately regarded as a RAND product, as we outline in the next section. Furthermore, it was really only the last doctrine that was well and truly absorbed into the Cowles theoretical tradition. As for the Great Emancipation from Physics, this line is often bruited about in the modern history of economics (Koopmans, 1991; Ingrao & Israel, 1990; Debreu, 1991) and finds its echoes in an earlier clarion call issued by von Neumann and Morgenstern (1964, pp. 6, 32); but measured meditation upon the story as presented in this volume so far reveals that themes fresh out of physics still tended to play an overwhelming role in both the formalization and conceptualization of cybernetics, game theory, and information theory. Piling insult on injury, the assertion contradicts Koopmans's own retrospective testimony, quoted earlier, about the ways that physical inspiration had governed his own personal career in economics.

And as for the vaunted interdisciplinary freedom of the Cowles researchers, it was a freedom born of ignorance rather than knowledge, as freely admitted in the very same document: "With one possible exception (Simon) the present staff of the Commission can claim no special competence in the tomes of the social science literature, nor do we think that this should be a primary criterion in the selection of individual staff." In a handwritten note to Marschak accompanying a draft of this grant application, Koopmans wrote: "I feel we have gone about as far as we can justify in suggesting the social science relevance of what we do. We must be careful not to pass ourselves off as social scientists."[44] In the future, economists would blithely trample all over the topics asserted as the province of other social sciences; but no one would ever seriously mistake those economists for bona fide sociologists, psychologists, political scientists, or (after some time elapsed) operations researchers. The practical effect of the Cowles program was to "toughen up" the mathematical training of economists and thus repel anyone trying to trespass from another social science – to eject history from the curriculum and to renounce loudly any psychological commitments. (Herbert Simon was the one exception to this rule; but then, as argued in Chapter 7, he never had fully accepted neoclassical economics.)[45] What Cowles ultimately sought to do was to shore up the boundaries between neoclassical economics and the other social sciences; pending that, transcendental urge was

[44] Koopmans to Marschak, August 31, 1951, box 99, folder: Ford Foundation, JMLA.

[45] Koopmans's own attitudes toward Simon are best revealed in a letter congratulating him on reception of the Nobel Prize, October 16, 1978: "We are hoping, Herb, that you will see this as a challenge to draw economists more into the circle of people you are writing for. . . . [Your audiences] tend to be somewhat insular. You have called them the physicists of the social sciences. Harvey Brooks calls them the thermodynamicists of the s[ocial] s[ciences]." Folder: Koopmans, HSCM. We, of course, call them cyborgs.

reconceptualized as the periodic forays of the economic imperialist, bringing back home raw materials wrest forcibly from the natives as fuel for their stationary engine of analysis.

Cooling out the Mark: Koopmans and von Neumann

This brings us back to the letter from Koopmans to von Neumann quoted at the end of Chapter 3. Both its date (January 18, 1952) and its subject matter now take on added significance. Koopmans, as we now know, had been the prime conduit for von Neumann's influence on Cowles; the letter was written after Cowles had successfully been reengineered to promulgate the New Order, version Mark II. Von Neumann, as argued in Chapter 3, had by this time become thoroughly immersed in the major research program of the end of his career, his theory of automata. The original 1945 pact of intellectual reciprocity, the "exchange" of help with the computation of maximum likelihood estimation for a close reading of *Theory of Games and Economic Behavior*, had not panned out as either party had originally envisioned it. Both parties to the exchange had, in effect, cooled in their original enthusiasms regarding those particular topics, but still found themselves both being looked to for inspiration and guidance by large military-funded quasi-academic think tanks still nominally devoted to those particular subjects. The letter reveals the state of negotiations between two rather divergent visions of the future of economics, and for that reason it is worth (re)quoting at length:

> Enclosed I am sending to you a paper "Saddle Point Existence theorems" by Dr. Gerard Debreu of the Cowles Commission. This is an attempt to find weaker conditions, sufficient for the existence of a saddle point, than those contained in the literature. . . . I have also given some further thought to the question you asked me in Boston, as to the possibilities of large scale computation involving economic data and designed to answer questions of economic policy. In the first place, I feel that any such computations now proposed by Dantzig or by Leontief are definitely worth while experimenting on, for the development of computation methods as well as for the introduction of computability considerations in economic model construction. I do not know if we can expect other fruits besides these at this time. I continue to feel that further thought should be given to the definition of the data to be collected, to the sources and methods of data collection (stratified sample versus complete coverage), and to the formulation of the policy questions to which answers are sought, before we can feel that economics is really using the opportunities afforded by modern computation equipment to good effect.
>
> I should like to mention another way in which the study of computation technique is relevant to economics. The design and the method of operation of computation equipment can be regarded as problems in

> the theory of organization which in some respects are greatly simplified. Since the parts of the machine (unlike, as you pointed out, the cells of the human organism) have no separate objectives or volitions, all complications arising from this circumstance in actual organizations are absent. It is therefore possible to concentrate on the more mechanical elements of organization, such as transmission times, capacities for memory or transmission or arithmetical operations, etc. If with the help of the computation analogy the more mechanical aspects of organization can be studied first, then we will be more ready to take on the more difficult aspects associated with the diversity of objectives.[46]

Although no comparable programmatic letter from von Neumann's pen survives, it should be fairly clear that he was here promoting his "third period" concerns, suggesting that computers as prostheses and the theory of automata might have profound conceptual implications for economics. Here also is evidence that he was even evaluating the biological analogy – another hallmark of the nascent cyborg sciences. Koopmans had countered with his own agenda, which in many respects contradicted that of von Neumann, but could still be portrayed as sharing some common goals. At the top, the proffered Debreu paper was an early installment of what would become the other crowning achievement of 1950s Cowles, the Arrow-Debreu proof of existence of general Walrasian equilibrium. The significance in this context is that Cowles would shortly make use of one of von Neumann's prior proof techniques, in conjunction with some of the mathematics of his game theory, but only now to shore up the pre-existent Walrasian neoclassical program (especially after it had failed in its empirical incarnation in Cowles' own estimation), something that we have already argued would have left von Neumann cold.

Next, Koopmans signals a rather unenthusiastic evaluation of existing attempts to implement economic models on the computers available in that era. There is no evidence this evaluation was done on formal computational grounds. We have already observed that Koopmans remained indifferent to the computational implementation of linear programming; and Cowles had become notoriously contemptuous of Leontief's input-output models, which had also enjoyed lavish Air Force support.[47] Although it was not mentioned, Cowles had also abandoned any support or encouragement of large-scale macroeconometric models, such as

[46] Box 5, folder 4, VNLC. Part of this same letter was quoted at the conclusion of Chapter 3; but repetition is intended, so the reader might read the letter in two different lights.

[47] See the comments by Klein: "I found the Cowles group a little flippant about Leontief's work. . . . One day I met Kenneth [Arrow] in the hall and I said something favourable about the Leontief system. . . . Kenneth replied that he thought this was just an accounting system. To people at Cowles, accounting identities were something to take into account but they did not get one anywhere" (in Feiwel, 1987a, pp. 343, 349).

Klein's U.S. model. Koopmans cites the quality of the data as his primary motivation; but this seems somewhat disingenuous coming from a research organization which had already banished all its specialists in data collection under Marschak and had subsequently slashed support for econometric empiricism to the bone. The Cowles aspiration under the New Order Mark II was to cut itself free from all historico-temporal dependence upon the "data" – to transcend the mundane realities of specific nations, specific cultures, specific institutional structures. The "data" had given them nothing but trouble in the past, as had their status as strangers in a strange land; it was now time to leave all that behind.

The last paragraph of the letter is most revealing. Here apparently is an explicit cyborg conflation of the economic organization with the machine, dating back to the socialist calculation controversy. This was a hallmark of promise of linear programming; it also was a recurrent theme in von Neumann's theory of automata. Yet, just as in the case of the brain, it was von Neumann who insisted upon the *disanalogies* between the machine and the organization; and it was to be expected that the crux of the disanalogy lay in the "antagonism between parts" (Jeffress, 1951, pp. 109–10). This was the remaining vestige of his game theory, now transmuted into a tendency to stress issues of strategic interaction and information transmission within a computational framework. Koopmans's reaction is indicative of the path that was taken at Cowles for the next two decades. Although Koopmans demonstrated he was aware of various "mechanical elements of organization," such as Shannon's information theory (transmission times), hierarchies of automata (capacity of memory), and strategic considerations, Cowles would nonetheless proceed to give all of these considerations a wide berth in its future research. The Cowles position toward the "diversity of interests" of the components of an organization was subsequently rotated in a direction orthogonal to that of von Neumann's orientation. Sometimes, diversity was simply denied so as to treat the organization as if it were no different from a single individual. "Those studies [of Koopmans, Debreu, et al.] excluded from consideration the conflict of individual interests. . . . With the specific difficulty of the diversity of individual interests removed, it is in principle a simple matter to test the efficiency of an economic decision" (Cowles Commission, 1951, p. 16). More frequently, the "diversity" of interests was ensconced as a nonnegotiable desideratum of the Walrasian program but with the computational analogies entirely banished from consideration. This diversity would come to be interpreted in the immediate future not as strategic cognitive interaction, as in game theory, because those phenomena got mislaid along with the computational aspects, but rather as the nonnegotiable dictum that rigorous general equilibrium theory should allow for the maximum of freedom in characterization of abstract "rational" individual preference orderings.[48] In a manner that was rarely made explicit, the Walrasian market was deemed to be praised as accommodating almost any idiosyncratic "rational" preference (hence the formal fascination with "ordinal" utility, "lexicographic" preferences, and other fanciful psychological entities) purportedly without requiring much in the way of specification of informational or institutional competencies on the part of *Homo economicus*. The beauty of the Walrasian market for Cowles, at least initially, was that it supposedly transcended any need for prodigious individual capacities for computation. This curious doctrine was the residue of the original Cowles stake in the socialist calculation controversy by the later 1950s; it became known subsequently as the theory of "social choice."

Hence the Cowles turn to the Arrow-Debreu formalization of Walrasian general equilibrium in the 1950s as the core doctrine of neoclassical orthodoxy actually constituted a *rejection* of the cyborg vision of von Neumann, although even in retrospect it sports enough in the way of superficial resemblances and mathematical prejudices that few have seen fit to explicate the full extent of the divergence. From von Neumann's perspective, Cowles was just a very small cog in the larger military-industrial-academic complex; economists had not been major contributors to the wartime development of American OR, which he believed was transforming social science; and, in any event, he did not live to see what economics in America was destined to become. Von Neumann was used to brushing off those whom he felt were mired in outdated or unimaginative research agendas.[49] Conversely, the task that Cowles Mark II set itself was to accommodate Walrasian theory to the new mathematical orientation best represented by von Neumann, and to the novel funding regime promoted by the military to which it was intimately linked. What this was deemed to imply for Koopmans, Arrow, et al. was the preservation of substantial elements of doctrinal continuity with neoclassical economic theory of Hotelling and Schultz in the 1930s (e.g., the rejection of British Marshallianism, hostility to American

[48] The highly misleading way in which the values of freedom and diversity were ultimately accorded salience by neoclassical economists is the subject of Chapter 7.

[49] See von Neumann's brisk apology to Marschak for letting his membership in the Econometrics Society lapse: "I resigned from it because I assumed that membership didn't matter very much to anyone else and felt I had joined too many societies already. Besides, while I have always been, and still am, strongly interested in economics or mathematical economics or econometrics, I have not actually worked in this field since 1943 and expect to be tied up with other matters for the next few years. (This does not mean I do not sincerely hope to come back to the subject at some later time)." Von Neumann to Marschak, June 24, 1947, box 93, folder N, JMLA.

Institutionalism, openness to physics inspiration, resort to probability theory), while gingerly engineering some ruptures with deeply ingrained practices that had characterized economics in the prewar era. As just one example, the behaviorist disdain for mental entities would itself have to be repudiated if neoclassicism were to receive a booster shot of cyborg credibility. Close familiarity with the operation of market institutions on the ground, not to mention the data of realized prices and quantities, would also go by the boards. The paramount importance of RAND as the New Automaton Theatre wherein one could become schooled in the modern mathematical techniques and some of the colophon attitudes of OR was what attracted Cowles; nevertheless, RAND was also the site of many cyborg encounters of the second and third kind – namely, with the computer and with the culture of simulation.

EVERY MAN HIS OWN STAT PACKAGE: COWLES UNREPENTANT, UNRECURSIVE, AND UNRECUSANT

The aim of this chapter is to recreate the problem situation in neoclassical economics from the vantage point of someone situated at Cowles in midcentury. The dangers besetting this conclave might be viewed from the vertiginously global perspective of the Cold War, or pitched at a level as parochial as the ongoing friction with the Chicago school, or phrased as the need to constitute a self-confident neoclassical school of mathematical economics in a world skeptical of such arcane indulgences. Nonetheless, our guiding presumption of the first half of the present volume has been that Cowles Mark II eventually set the tone and timbre for the subsequent American half century of neoclassical economics. This thesis in itself, although hardly controversial, would require substantial qualification and clarification in ways that have already been discussed (Hands & Mirowski, 1999). The Cowles Walrasian project was forged and hardened through conflict and cooperation with neoclassical schools at Chicago and MIT, and a case could be made that in the 1950s and 1960s the latter schools were more prevalent in providing the day-to-day pedagogical and public face so critical for the disciplinary identity of economics. By contrast, the Cowlesmen tended to remain a bit more aloof and standoffish when it came to public rough-and-tumble; the closest thing to a "popularization" of their position came with Koopmans's 1957 *Three Essays in the State of Economic Science* (1991). Yet even this cool technocratic reserve was not sufficient to ward off the looming clash with the Chicago department; the threshold of tolerance was surpassed in 1954, and Cowles decamped and was reincarnated once more at Yale in 1955 as the "Cowles Foundation for Research in Economics." If this were a full-fledged history of postwar neoclassical microeconomics (which it most assuredly is not), then the narrative would have to be structured to recount the complex interplay of

Chicago, Cowles, and MIT in the ecology of academic economics and its intellectual jousts. In lieu of that, a few words concerning the fate of Cowles as an institutional entity may be in order, before we turn our attention once more back to cyborg themes.

Relaxin' at New Haven

If we provisionally dub this reincarnation at Yale as Cowles Mark III, it would seem that the third time was not the charm, for thereafter Cowles lost some modicum of the coherence and clarity of purpose it had struggled to maintain under Koopmans's guidance at Chicago. Koopmans alluded to this possibility, even as he was arranging the escape to New Haven: "Colleagues are friends to each other [at Yale] as well as colleagues, and there are no symptoms of doctrinaire attitudes visible yet. . . . The one possible drawback is that with the old world conviviality goes a certain exclusiveness, and not much of a feeling of equality before the god of science that rightly prevails in Chicago. So far I have found more of this among economists."[50] With James Tobin assuming the research directorship, Cowles came to resemble the other schools of economics to a much greater degree, augmenting its credibility but diluting its distinctiveness.[51] One after another, the other major protagonists of Cowles Mark II drifted off to other positions (though retaining consultant status) – Arrow to Stanford in 1949, Hurwicz to Minnesota in 1951, Debreu from Yale to Berkeley in 1960, Marschak to UCLA in 1960 – leaving Koopmans to preside over a hollowed shell, a remote figurehead for the next generation of Cowles researchers. Koopmans himself then periodically drifted off to other arenas, such as the newly formed International Institute for Applied Systems Analysis in Vienna, a Cold War think tank where he could mix with other natural scientists who wanted to do social theory without becoming social scientists (Hughes & Hughes, 2000, p. 418). With postwar travel budgets including airfare, far-flung pilgrimages to the god of science could always compensate for a lack of doctrinal fervor at home. Due perhaps to this inevitable diffusion, as much as to the homogenizing dynamic of the new situation at Yale, Cowles relaxed into much less of a crusading stance and experienced a commensurate diminution of its proselytizing activity. The Cowles Monograph series languished, and the reports of research activities were issued with less

[50] Koopmans to Marschak, October 21, 1954, box 152, JMLA.

[51] On this point, see the Gerard Debreu interview by E. R. Weintraub, May 4–5, 1992, on deposit at Duke University archives. On p. 58, Debreu agrees with Weintraub that Tobin altered their status as outsiders to the profession: "And I seem to find that Tobin was very much an insider and became director of Cowles Foundation, that helped to change that."

frequency. Furthermore, it seemed that the ascension of Cowles Mark III coincided closely with the passing of the internally perceived "Golden Age" at RAND, a theme broached briefly in Chapter 6.

Our own adopted stance herein is to view Cowles Mark II as the locus of a lively process of accommodation to the cyborg imperatives encountered at RAND – different in intellectual specifics for each one of the Cowles researchers, but sharing some broad general commitments – tempered so that it would appear for all the world as though there had been substantial continuity with the Walrasian program which had been inherited from Mark I.[52] Neoclassicals had to learn to stop worrying and *love* the machine. Cowles Mark III was then the product of that (temporarily) successful rapprochement. Because neither the Walrasian Credo nor the Cyborg Codex was a household word for the rest of the American neoclassical profession in the years immediately following World War II, and would not have been automatically held in high esteem even if it had been more familiar, its research agenda fostered the common impression of the Cowlesmen as situated well outside the mainstream of economic discourse in the 1950s – a situation frequently but mistakenly attributed solely to Cowles's evident fondness for arcane mathematical formalisms. With the move to Yale, the thaw set in: the cyborg themes were further attenuated, but simultaneously, the Walrasian creed had grown more respectable within the American profession; and the quality of graduate mathematical training and/or recruitment in economics had risen in the interim; and, consequently, Cowles Mark III had moved much closer to the unperturbed center of orthodox opinion by the time it became settled in New Haven. However, the net result was that much of what had been notably distinctive about von Neumann's vision for economics had been progressively sloughed off or rendered ineffectual in the interim, leaving the fortified but complacent neoclassical orthodoxy unprepared for a renewed cyborg onslaught starting in the 1980s. This Return of the Cyborgs is covered in Chapters 6 and 7.

Because it is central for our narrative to understand how the cyborgs were ultimately repelled at Cowles, our immediate task in the remainder of this chapter is to sketch out how rupture and continuity were negotiated at Cowles Mark II, to the neglect of subsequent developments in Cowles Mark III. This is depicted, first, as a brace of broad-brush characterizations of the neoclassical construction of information processing

[52] "Although the analyses have changed greatly, the questions remain relatively constant. In fact, at few times in the history of [economic] thought have there been radical innovations" (Arrow, 1991b, p. 3). This unwillingness to entertain the existence of ruptures in intellectual discourse is precisely one of the main motivations for the banishment of the history of economic thought within the postwar economics profession.

as inductive inference, linking it to Cowles's erstwhile enthusiasm for econometrics; then, as some brief indications of how these tensions were played out in the individual careers of Cowlesmen such as Koopmans, Marschak, Hurwicz, and Arrow; and, finally, as an inventory of some of the ways in which Cowles invariably adopted a conceptual stance opposed to RAND's enthusiasms, once the cyborg themes had threatened to undermine its fledgling strategy of accommodation to Walrasian doctrines. We provide three different views of the clash between Cowles and cyborgs precisely to drive home the thesis that this problem was *structural*, and not simply a matter of idiosyncratic disputes at the level of individual peccadilloes or personalities. Nevertheless, these tectonic shifts were covered with an overlay of Cold War politics. We therefore conclude this chapter just as it began, by returning to the question of politics, and observe that an intuitive statistician, just like the computer, also is freighted with political content; and that it was Kenneth Arrow who effectively brought the socialist calculation debate to a close by spelling this out. All that was left to do afterward was to relax in New Haven.

Homo economicus as Betting Algorithm

The political point of departure for Cowles Mark II was the question of the nature of a rational economic order, and the unexamined presumption that the issue could be posed primarily as one of logic, such as the "logic of rational choice." But politics now began to set the parameters for how far the Cowles juggernaut might travel. This sleek Walrasian engine of inquiry had run up hard against Hayek's taunts in the 1930s that this would only be the case "*if* we possess all the relevant information, *if* we can start out from a given system of preferences, *if* we command complete knowledge of available means." RAND had initially seemed to offer the promise of a crushing riposte to Hayek, chockablock as it was with mathematical logicians itching to rationalize the planning process for the military. The retaliatory reproach might go: Hayek was innocent of mathematical logic; Hayek had a nineteenth-century understanding of the brain; Hayek had alienated the physicists with his strictures on "scientism"; Hayek had misread the lessons of World War II. As Edward Bowles had testified before Congress in November 1945: "for the sake of safety, we will have a pre-planned integration . . . a planned economy, you might say" (in Collins, 1998, p. 88n). But there was a catch: Cowles had essentially sided with Lange, insisting that the Walrasian model captured whatever it meant to possess a rational economic order, be it manifest as a "perfect" market or an omniscient central planner. Now, under those circumstances, in what sense could the experts at RAND assist in fending off the complaint that Walrasian general equilibrium had no resources to illuminate, much less even enunciate, the more pressing economic problem of "the utilization of knowledge which is not given to anyone in its totality"?

Here is where Cowles Mark II sought out a golden mean between denying the problem altogether (as was done at the postwar MIT and Chicago economics departments) and essentially acknowledging the implied indictment of Walras (as we have suggested was von Neumann's position). The price of gaining access to RAND's expertise was that something had to be given up: return to the status quo ante of Marschak's Cowles Mark I was simply not an option. The individual Cowlesmen would consequently have to learn to relinquish their hard-won agnosticism about the nature of the brain of the economic agent: to abandon the infamous inscrutability of the Other. What was purchased (if one can speak in such crass terms) was a custom-made souped-up version of the economic agent as an information processor, which could be plug-compatible with the Walrasian model, resulting in a newer and shinier engine of inquiry capable indifferently of either singing the praises of markets or dancing to the tune of the economic planners. Less than a full-fledged Turing machine, it was nevertheless something more than the mass point coming to rest so redolent of nineteenth-century neoclassicism.

Although the apprenticeship at RAND and the brush with von Neumann goes quite some distance in explaining why the thematic of information processing was so "obvious" a direction in which economics was fated to turn in that era, it is not anywhere near so straightforward a task to provide a comprehensive explanation of why the bundling of Walrasian general equilibrium theory with the individual cognitive agent as statistical information processor was to prove so compelling that it grew to be treated at Cowles as an unquestioned first commandment of rigorous economic model building. Certainly none of our protagonists ever proffered more than an unapologetic bald fiat when it came to reciting the creed. At Cowles Mark II, *Homo economicus* was to be reconstructed not as an all-purpose computing machine, but rather as a canned statistical package that might be found on any of a range of mainframes in that era: perhaps SPSS, or SAS, or TSP. "Information" would come to be conflated initially with "sufficient statistics" in the manner of Neyman-Pearson; and once that eventually grew stale, others would insist instead on a mechanical Bayesian updating device. Yet the height of insularity at Cowles was the tendency to bruit about the notion that cognition as intuitive statistics had all been its own idea: "Statistical theory can be regarded as an economics of information. . . . It was Marschak's papers of 1954 and 1955 that made explicit the role of information in economic behavior and organization" (Arrow, 1991b, p. 15). The historical problem is not so much to marvel at this as an isolated event – on the contrary, treating "cognition as intuitive statistics" had became quite commonplace in the immediate American postwar social sciences (Gigerenzer & Murray, 1987) – but rather to begin to comprehend how the Cowlesmen could come to believe so fervently that this move was underwritten by the vision of von Neumann and cyborg developments at RAND and not see it as the retrograde move it so clearly turned out to be.

Much of this attitude must have been bound up with contemporary images of the putatively universal nature of the "scientific method," which itself arose out of the physicists' participation in the war effort. "Only after the inference revolution, when inferential statistics finally were considered an indispensable instrument and had to some extent mechanized inductive reasoning and decisions, did it become possible to reconsider the instruments as a theory of how the mind works" (Gigerenzer & Murray, 1987, p. 58). In many ways, the version of information processing so favored at Cowles initially bore a closer resemblance to "old-fashioned" British OR than it did to any innovations linked to von Neumann's computer and American OR. The trick in managing the straddle between Walras and the cyborgs was to hint at the brave new world of information processing – namely, the tradition sketched in Chapter 2 from thermodynamics to information to command and control and connectionist theories of the brain – but, when writing down the model, hew as closely to previous econometric formalisms as possible.[53]

There was, however, at least one salient wartime precedent for "guessing machines" being conflated with information processors, derived primarily from large-scale mechanized cryptanalysis, such as Turing's Colossus or Bush's Navy machine. Shannon's theory of "information" clearly bore the birthmarks of this genesis, as described in Chapter 2. Nonetheless, few cyborgs long rested content with statistics as a freestanding paradigm of machine cognition. If a lesson was to be learned from wartime experience with cryptanalysis, it was that inferential statistics alone were *never* sufficient to constitute an effective engine of effective induction. At Bletchley Park the British came to appreciate that the problem was "to capture not just the messages but the whole enemy communication system. . . . The Hut 3 filing system, therefore, had to mirror the German system as a whole. Only when this had been done could the Enigma decrypts yield real value – not so much in juicy secret messages, but in giving general knowledge of the enemy mind" (Hodges, 1983, pp. 196–97). This lesson was brought home time and again in areas such as pattern recognition,

[53] "I have had a glimpse of some neurological literature but got frightened by the word 'synapsis' and other specialties." Marschak to Calvin Tompkins, October 14, 1952, box 84, JMLA.

machine translation, and the whole range of military intelligence activities. The premature tendency to mistake a very small part of the problem of information processing for the whole, if only to reify some convenient mechanical procedure as a tidy "proof" of some optimality result or another, was stubbornly prevalent in America and became a noticeable weakness of Cowles's intellectual trajectory from this point forward. Here the fascination with the mechanics of statistical inference tended to defocus attention on the need for decryption to deal with issues of strategic deception, cognitive limitations, interpersonal semantic systems, and an accounting of the differential sources of randomness.

But cryptanalysis and parochial notions of *the* scientific method are still not sufficient to account for the fascination with intuitive statistics across the disciplinary board in the 1950s and 1960s. Here, following some hints (Goldstein & Hogarth, 1997), we trace the spread of the human betting machine back to – you guessed it! – some offhand ideas of John von Neumann. Briefly, it seems that he had influenced the course of postwar American psychology at least as much as American economics, and its consequences, especially at RAND, had an unanticipated feedback upon the course of events at Cowles.

A number of recent works now argue that World War II was a critical turning point in the history of American psychology.[54] Although there are many facets to this transformation, the one that concerns us here is the projection of the economistic metaphor of life as a "gamble" onto all manner of previously mentalistic phenomena such as decision making, judgment, and ratiocination. In part, this can be traced to a dynamic where psychologists first "toughened up" their discipline by the importation of various statistical procedures and algorithms, and then turned their "tools" into "theories" of cognitive capacities. Combined with the "cognitive revolution" against previous behaviorist doctrines, this transmutation rapidly led to portrayals of minds as congeries of statistical algorithms. Gerd Gigerenzer graphically captures this situation in his only partly satirical diagram, reproduced here as Figure 5.1.

The mind as intuitive statistician was not solely the artifact of the adoption of various statistical tools by academic psychologists; it was equally a function of the spread of OR throughout the postwar social sciences. In America, it occurred to any number of mathematically sophisticated researchers interested in psychology that John von Neumann had structured his entire game-theoretic formalism around the theme that most cognition could be reduced to a gamble of one sort or another. His early stress on mixed strategies in games subsumed all decision making

[54] See William Goldstein & Chrisman Weber in Goldstein & Hogarth, 1997; Edwards, 1996, chaps. 5–7; Gigerenzer, 1992; Capshew, 1999.

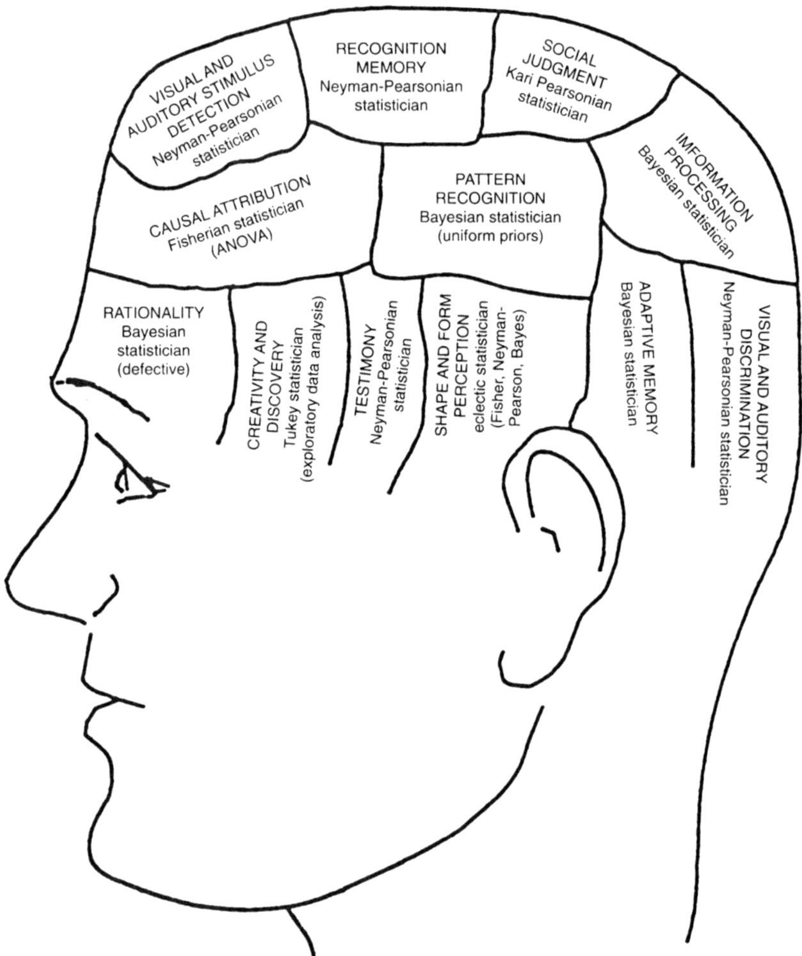

Figure 5.1. Cognition as intuitive statistics. *Source*: Gigerenzer, 1992, p. 336.

under the gambling motif; and, further, his axiomatization of the theory of expected utility seemed to provide the promise of a measurable and observable set of consequences to fortify this approach. Hence, just when the bulk of the American psychology profession began to project their statistical algorithms upon the mental landscape, a self-consciously scientific cadre of psychological researchers inspired by this understanding of von Neumann instituted the new field of "decision theory." A different approach to experimental psychology became attached to this field; and RAND in particular became one of the centers of this style of quantitative research. We cannot even begin to trace the subsequent vicissitudes of this field, from an early fascination with "utility" to a later disillusionment with the concept, and from the initial hostility with which it was greeted by mainstream economists to a later source of inspiration of the field of "experimental economics" (see, however, Chapter 8). All we need to stress for present purposes is that at RAND in the 1950s there was no clean separation between OR, "decision theory," psychology, and economics; and that one place where they all intersected was in the presumption that the modern scientific approach to cognition lay through the portrayal of mind as betting machine. The analytical choices made by the Cowlesmen should be situated squarely within this formation. Their agent as little econometrician assumed a gravitas and plausibility in such an environment.

What was it that Cowles Mark II stood for in the 1950s? This commitment to cognition as intuitive statistics was the pivotal move; at least five consequences flowed from this creed. The first and immediate task the Cowlesmen set themselves was thus to come up with some version of the economic agent as information processor that would neither mortally derange the Walrasian model nor make it appear that the rupture between the newer doctrines and older economic theories was catastrophic and irreversible. The task was urgent, and the prospect daunting, because no one had taken utility theory seriously as a viable *cognitive* theory since the third quarter of the nineteenth century. The turn-of-the-century Freudian ascendancy had ridiculed the pretension of the transparent self-knowledge of motivations and drives; and then the more recent computer models of cognition being broached by Turing, von Neumann, McCulloch, and their students were diametrically opposed to the tradition of preformationist hardwired transitive preferences. The method by which the Cowlesmen thus attacked the problem was to deflect their prior economic commitments from being applied to themselves (in the format of putative inductive rules of scientific method) and instead to project them onto their model of the neoclassically rational economic agent.

When you've got hold of a mass-produced hammer, all the world looks like a nail, so maybe it is not all that incongruous that even as they were themselves relinquishing the project of econometric empiricism, the Cowlesmen more or less uniformly decided that "information processing" dictated that rational economic man behaved like a little econometrician (and *not* a computer sporting a von Neumann architecture). It certainly conveniently prevented most of their prior statistical training from becoming obsolete. The displacement of the onus of bearing the brunt of statistical inference from themselves and onto the Walrasian trader, more than any other set of intellectual influences, explains why the most discussed aspect of the *Theory of Games and Economic Behavior* in the postwar economics literature was the addendum to the 1947 revised edition, where von Neumann provided an axiomatization for a cardinal utility that was linear in probabilities, and which is now called "von Neumann–Morgenstern expected utility." As we argued in Chapter 3, for von Neumann this was merely an expedient to mollify the economists and reconcile them to what he considered as the imperative for a cardinal value index for his theory of games; by no stretch of the imagination was the axiomatization itself intended as a serious theory of cognitive information processing. (To be fair, the hordes of aspiring decision theorists equally misunderstood von Neumann's real intentions.) For the Cowlesmen, per contra, it appeared as the slender lifeline thrown them by von Neumann; it would reel them in the direction of what they hoped would be the reconciliation of the theory of games with the Walrasian organon; it dangled an escape route through which the utilitarian agent could display something vaguely resembling cognitive interaction with his uncertain environment; its only immediate drawback being that its deduction of a "measurable" utility (up to a factor of proportion) might vitiate their own prior inclinations to believe that a "welfare economics" should only deal in terms of interpersonally incommensurate (and possibly inaccessible) preferences. But these were just the sort of puzzles that could keep mathematical economists busy with their yellow pads, exploring special cases until the military cash cows came home; and that is precisely what happened.[55] Hence the Cowlesmen rejoiced, and inflated all out of proportion this one way that *TGEB*, an explicitly antineoclassical document, could be tendentiously reinterpreted as one of the foundational texts of postwar neoclassical theory.[56]

A second Cowles commitment, really a trademark of the school, had been the conviction that (from its vantage point) the cavalier way in which the Chicago economists and the MIT contingent had dealt with income effects and market interdependencies in their microeconomic theory had been unconscionable, which is why Cowles had occupied the first line of defense of the Slutsky equation since the 1930s. This advocacy

[55] Arrow to Marschak, August 24, 1948: "As you say, there has been a good deal of discussion of utility and its measurability here at RAND. By now, I'm pretty much convinced that von Neumann has a good point in regard to measurability; as far as probability considerations, maximizing the expectation of utility is probably allright. Von Neumann was here recently. He stated he did not agree with Wald in applying minimax ideas to the treatment of uncertainty." Box 154, JMLA.

[56] "The work of von Neumann and Morgenstern and later of Savage restored the confidence of economists in expected utility theory by showing that the theory could be interpreted in ordinalist terms" (Arrow, 1991b, p. 15). Economists, including that author, were not so very confident about the virtues of expected utility in the 1950s: see Arrow, 1950. This concertedly antihistorical reading can also be found in Fishburn, 1989, 1991.

of extreme generality of approach to statistical inference did not pan out by the late 1940s, as we (and they) have indicated. Once more, with the advent of the Walrasian agent as proto–information processor, the strictures constraining the behavior of the economic scientist were instead displaced onto to the economic actor. Instead of treating the extent of interrelatedness of markets and individual preferences as an empirical issue, the Cowlesmen opted for maximum interrelatedness of prices and the bare minimum of structural content of preferences as an analytical desideratum for the "legitimate" cognitive characterization of the agent. This was made manifest as the dictum that all problems of information processing encountered by the agent had to be specified as problems of "imperfections of information" and decidedly *not* as problems of cognitive computational limitations.[57] The rationale behind this stubborn anticyborg principle was that if cognitive limitations really were intrinsically computational, then the economic agent would partition the problem space into nonintersecting subsets, potentially with different partitions for each agent; for Cowles, this would be renounced as tantamount to backsliding into arbitrary neutralization of full Walrasian interdependence, capitulating to the dreaded Marshallian tradition of partial equilibrium.[58] (The echoes of the stringent ethos of "identification" in structural econometrics were perhaps inevitable.) The entirely plausible cyborg alternative was often banished by fiat through postulation of initially opaque axioms about the cognitive horizons of the economic agent, most notably the widely deployed axiom of "independence of irrelevant alternatives" (Arrow, 1951a; Nash, 1950a; Savage, 1954). If conceptual relevance cannot be elevated to the a priori status of a cognitive universal once and for all, then it follows that probabilistic independence can never be a strict desideratum of rational allocation of

[57] This provides the explanation as to why the following historically innocent counterfactual could not have happened within the Cowles program, although it clearly did happen in the cyborg sciences outside of economics: "Imagine that modern decision theory began, not with perfect rationality and imperfect information, but the opposite" (Conlisk, 1996, p. 691). A more elaborate history than we have provided here would have to take into account the longer-term disciplinary functions of the image of agent as passive signal processor. One could do worse than to begin with Foucault: "From the master of discipline to him who is subjected to it the relation is one of signalization: it is a question not of understanding the injunction but of perceiving the signal and reacting to it immediately, according to a more or less artificial, prearranged code. Place the bodies in a little world of signals to each of which is attached a single, obligatory response" (1977, p. 166).

[58] The mathematical formalization of the intuitive idea of "information" as a partition imposed on a preexistent sample space, so redolent of the econometric approach to inference, is traced from Princeton to RAND in Chapter 6. There it is attributed in the first instance not to the ingenuity of economists, but rather instead first to controversies over the interpretation of game theory.

cognitive resources; but this was not the Cowles way of attacking the problem.

A third commitment followed rather directly from the second. Cowles would retain its unwavering adherence to constrained optimization as the very essence of economic rationality, now extended to such cognitive aspects as expectations and inductive inference. The endless paeans to Panglossian optimality were a direct consequence of socialist calculation controversy, but had next to nothing to do with cyborg models of feedback, accommodation under ill-defined objectives, and the evolutionary teleology of muddling through. Cowles researchers would have no truck with the characterization of the efficacy of an economic agent as possessing the capacity to be open-ended and flexible in the face of contingency, because that smacked of the historicism and institutionalism which they all abhorred. No matter where you found them, markets all worked alike, they believed; that was why someone from Mitteleuropa or Friesland could even aspire to plan an American economy. Moreover, the grail of "optimality" took on a life of its own. Once the Cowlesmen became themselves acquainted with the novel mathematical techniques of optimization with inequality constraints, measure-theoretic approaches to probability, fixed-point theorems, and limit sets, then the economic agent seemed to pick up these techniques with alarming cupidity. Indeed, it seemed that this notional agent veritably wallowed in concepts involving infinities that would flummox any self-respecting Turing machine. Even though much of the mathematics was cribbed from inspirational texts by von Neumann, the flagrant disregard for any cognitive prudence or computational relevance could not be attributed to the cyborg inspiration as much as to the Walrasian heritage. In sharp contrast with physics, the Walrasian organon had never definitively placed any bounds on the dimensions of the economic problem – not on the definition of the commodity space, or on the actual process of reconciliation of prices, or the nature of signals passed between the actors and the spectral auctioneer. It is probably just as well that the Walrasian agent had preferences but no detectable emotions, given that the world within which he had to operate was in equal parts forbiddingly complex and yet distressingly ontologically intangible.

Again, one should be careful in coming to understand the potential validity of the complaint that the Cowlesmen were busy building little mechanical versions of themselves (or their erstwhile econometrician selves). This has been a theme structuring many histories of the cyborg sciences, as well as various feminist commentaries on biases in science. The nature of the criticism proposed here is that the Cowlesmen did not *learn* anything from the economic agents that they had fashioned, and in this they diverged dramatically from experience in the parallel cyborg sciences. We might point to the field of artificial intelligence as a constructive counterexample. "If AI workers first aimed at mimicking what they considered cleverest in themselves, they managed to show the world that these activities – puzzle solving, game playing, theorem proving – are really pretty simple. . . . What are really hard [to program] are the things that humans do easily" (McCorduck, 1979, p. 337). The neoclassicals at Cowles never got to first base in this regard, because they never quite understood what it was about game playing or theorem proving that could be reduced to machine operations – or at least not in the time frame encompassed in this chapter.

The fourth commitment of Cowles was keenly raised by von Neumann in his letter to Koopmans reproduced in the previous section. Cyborgs were poised on the brink of a general theory of organization, but one that would explain the diversity of forms of organization and institutions within which all biological and cognitive entities would operate. Cowles, by contrast, had just come away from a battle with the Institutionalists at the NBER and cherished their Walrasian model as an eminently "institution-free" model of markets. Hence the very meaning of "information processing" could not be understood at Cowles in the same manner that it was coming to be construed in numerous cyborg precincts. Eventually cyborgs wanted to regard various multiple forms of markets as situated on an equal footing with other species of organizations, and thus subsume the entire set as individual instantiations of that larger genus "computer," whereas the Cowles neoclassicals stridently insisted on a unique generic "market" populated by homogeneous rational agents as unbounded information processors. The cyborgs wanted something like an ecosystem, whereas the Cowlesmen reveled in a monoculture. In the subsequent elaboration of the program, Walrasians would extend their portfolio to "institutions," but only as derived artifacts of the generic market, either as distorted protomarkets or else as stopgaps due to some postulated market imperfections. Herein lay the political origins of the Cowles doctrine of justifying government as physician to localized "market failures," confined in their consequences and amenable to diagnosis and treatment, a notion virtually unprecedented in the history of economic thought. From henceforth, the neoclassicals would hew tenaciously to their transhistorical and transcendental notions of an overarching "law of the market" to be touted in invidious contrast with what were perceived as parochial or historicist forms of explanation in the other social sciences. In this way, they would drastically misapprehend the cyborg imperative to uncover transhistorical laws of evolution or complexity while maintaining respect for the local, the diverse, the specific, and the adventitious.

The fifth and final commitment brings us full circle to the military. The postwar stress on the centrality of "uncertainty" as an irreducible fact of life, a veritable litany at Cowles Mark II, was predominantly a military imperative. "Uncertainty [is] the central fact that all command systems have to cope with" (van Creveld, 1985, p. 266); and a case can be made that the enhanced pace of tactics and logistics had brought this home to the C^3I structure of military hierarchy in World War II. The postwar military establishment sought to deal with this disturbing development by tightening up on the centralization of control of most aspects of the command structure, especially since the argument was frequently made that nuclear weapons had rendered a more decentralized and dispersed command structure obsolete. The promise of the cyborg sciences was that the computer would aid and assist in achieving centralized command and control, rendering armies mere extensions of the single panoptic intelligence located in HQ. On the face of it, this claim sounds a lot like the ambition of the Cowles market socialists; but, in fact, the respective visions of the taming of uncertainty were radically at odds with one another.

The first inclination of the military cyborgs was to banish uncertainty by removing the human being out of the command loop to the maximum possible extent. This approach, for instance, was the clear motivation of Wiener's seminal work on gun-aiming devices; it was equally the essence of the later cruise missile, which combined inertial guidance with computer terrain contour matching; and it was the ultimate agenda behind the military support of artificial intelligence and a means of producing "smart weapons." Of course, the complete elimination of the human presence from the command loop was a pipe dream, little better than mediocre social science fiction; but once this was adequately understood, the cyborgs rapidly accommodated to the new realities by reconstructing their computers as prostheses, the ultimate aids in "decision making under uncertainty" in a fluid and ill-defined fog of war. The Cowlesmen, by contrast, kept insisting that "uncertainty" was merely a narrow problem of inductive inference, something that market participants had effectively already resolved, having previously mastered them through maximum likelihood estimation algorithms of varying compositions. In place of Turing's deduction machine, they proposed to substitute an infallible "induction machine," something that not even the most visionary of computer engineers had yet to design. If some concession to realism was demanded, the Bayesian doctrine of the "money pump" was trundled out – namely, that agents who made mistakes in inductive inference would be "punished" in the marketplace in complex sequences of bets and would therefore "learn" to shape up and behave like the dependable statistical estimation packages that Cowles was convinced they already were.

The contrasting images of uncertainty could not have been more divergent. The military wanted enhanced centralization, communication, and control and was supporting the cyborg sciences in order to forge new techniques to achieve it. Cowles believed that the Walrasian model demonstrated that ideal centralized planning and decentralized market operation were really identical, both instances of the machinations of an intuitive statistician present in every grunt infantryman and weary supermarket cashier, such that elaborate schemes of communication and control were superfluous. Cyborgs imagined individuals as components (expendable or no) of a larger machinic infrastructure, whereas Cowles insisted upon the full "integrity" of the individual as something pitched less than a Turing machine but more than Jevons's logical piano, suspended in a Pareto-optimal stasis, dangling within an unspecified market "mechanism." Cowles was unabashedly "humanist" (though in a way both alien and incomprehensible to any standard referent of that term), whereas the cyborgs were rapidly turning posthumanist. Both had resort to the terminology of information processing, and both could make use of certain technical tools of optimization, such as linear programming or maximum likelihood algorithms; however, when push came to shove, the neoclassical economists rode the cyborg bandwagon for no longer than circumstances warranted.

However much their commitment to stand by Walrasian general equilibrium created all manner of problems for the Cowlesmen in coming into contact with the cyborg sciences, it is nonetheless crucial to understand the allure of the cyborg in the 1950s to appreciate why Cowles did not avoid them altogether. Most of the figures at Cowles were refugees thrice over: from various war-torn European nations, from the dislocations of the Great Depression and its aftermath, and from careers in the natural sciences. It was a foregone conclusion that each of them would be infatuated with the vision of an "institution-free" economics, a virtual reality extracted from the disappointments of their own histories. In operations research they discovered a full-blown virtual reality: war as a problem in logic; politics as a problem in logic; machines as the best embodiment of logic. In statistics they thought they had found a way of conjuring order out of chaos. And then they were already predisposed to believe in the self-sufficiency of the asocial generic individual, for how else had they themselves persevered? They were more than relieved to erase the burdens of history and start afresh, to impose "equilibrium" no matter how drastic the defalcations of the past. Manfred Clynes had once suggested that the first commandment of the cyborg creed was to "free themselves from their environment"; the market socialist planner dreamed of rising above all mundane determination in the name of the exercise of pure will, after the style of the military commander. There was much to admire in the brave new cyborg manifold for the Cowles economist; and admire it he did.

These five generalizations may seem to the reader to be painted with too coarse or broad a brush, but they could easily be rendered more detailed and concrete through individual biographies of the Cowlesmen after their stint in Cowles Mark II. To that end, we shall next opt to survey the careers of Koopmans, Marschak, Hurwicz, and Arrow – not that the exercise might qualify as serious biography, or that it might do their personal concerns justice, or to capture what it may have felt like to be them, but rather simply to indicate the paths that this Walrasian information processor could take once it had taken up residence in the vicinity of RAND. The anatomy of the neoclassical agent should be fleshed out with detailed consideration of the work of many more key figures associated in one way or another with Cowles and RAND, but that exercise will have to be postponed for a future comprehensive history of Cowles and RAND.[59]

Four Careers in Data Processing: Koopmans, Marschak, Hurwicz, Arrow

The career of Tjalling Koopmans has been accorded extended attention in this chapter, so it makes some sense to tie up loose ends and cast a glance at the aftermath of his all-important interactions with von Neumann and RAND. Subsequent to the publication of his *Three Essays* in 1957, Koopmans occupied the role of public spokesman for Cowles. At age forty-five when he moved to Yale, he briefly continued to proselytize for "activity analysis," asserting that it would substantially revolutionize economic theory. This prediction did not turn out quite as Koopmans had anticipated: linear programming did provide some important inspiration in the formalization of the Arrow-Debreu existence proofs (see Chapter 6), but after much promotion, the topic rapidly faded from the economics curriculum, only to flourish in operations research, software engineering,

[59] It would be significant for our thesis that even many of the "lesser" figures at Cowles also abandoned econometric empiricism for consideration of the agent as information processor, albeit temporarily to a greater or lesser degree, modulo individual careers. See, for instance, Clifford Hildreth (1953) or that of Andrew W. Marshall, discussed intermittently in this chapter. Likewise, there is the example of the ambiguous insider-outsider who could recognize which way the winds had shifted but bore sophisticated reservations about the treatment of agent as statistician. The most underappreciated example of this role was occupied with distinction by Nicholas Georgescu-Roegen; but one might also point to Roy Radner. Finally, there is the rare but telling example of the Cowles insider who decided that cyborg innovations really did justify the jettisoning of the Walrasian tradition. This latter persona, very important for our thesis, was occupied by Herbert Simon. His case will be taken up in Chapter 7.

and management curricula.[60] In the neoclassical theory community, attention returned rapidly to smooth production functions whose perceived virtue was that they rather more closely resembled standard-issue preference functions (Mirowski, 1989a, chap. 6). Attempts were made to turn linear programming into an all-purpose theory of the management of uncertainty (Charnes & Cooper, 1958; Markowitz, 1959), but these too were superseded within economics and business schools by other rival approaches. Consequently, Koopmans enjoyed one reputation in the OR community for activity analysis, but a rather different one in economics proper, where his own subsequent research tended to drift away from this emphasis and toward some rather more characteristic Walrasian concerns.[61] He never did any work on game theory, and the few comments he made on it were guardedly negative (1991, p. 176). Perhaps inevitably, his politics also veered in a more conservative direction.

Starting in 1960, Koopmans reoriented many of his publications to deal more directly with the theory of the individual Walrasian agent as a cognitive utility processor, still bereft of any computational considerations. These papers, frequently written with students, sought to characterize the problem of maximization of utility over an infinite time horizon, and to model something in this context he dubbed "impatience" (Koopmans, 1970; 1985). (If death lost its sting, would impatience be far behind? But one should never approach these exercises by searching for their literal meanings.) In the context of the 1960s and 1970s, these models were taken to have germane implications for a neoclassical theory of growth, which became another of Koopmans's specialties. All concessions to computational themes raised in his earlier correspondence with von Neumann evaporated from this work. That did not dictate that von Neumann had ceased to serve as his inspiration; indeed, if anything, Koopmans increased his rate of citation of his mentor. However, the erstwhile student now tended to upbraid his master for not being sufficiently neoclassical for his tastes.[62]

60 "One need only check where the important theoretical and applied papers in linear programming have been and are being published. Certainly not in economics journals. In fact, since the early involvement of economists with linear programming, there has been, with a few notable exceptions, little research by economists in this field. . . . it is a rare economics department that has a course in linear programming, or even a lecture or two on the subject" (Saul Gass in Schwartz, 1989, p. 148). See also Kendrick, 1995, p. 71; Lenstra et al., 1991; Keunne, 1967.

61 Koopmans bemoaned the increasingly obvious separation of the OR and economics communities in his talk "Some Early Origins of OR/MS" delivered to the ORSA/TIMS meeting November 8, 1977, but his analysis of the causes is rather superficial. Box 22 folder 445, TKPY; see also 1991, p. 186.

62 The freedom to reconstruct the master became more evident once he was safely in his grave. Here is a passage dating from 1964: "Paradoxically, von Neumann's paper shows

Koopmans's insistence upon the central role of "normal" preference orderings in any rigorous theory rendered him frequently in demand to pronounce upon the models of agents as statistical processors produced by others. In these academic arenas, the most noteworthy aspect of his comments was the firewall he built between his own disappointing experience as a statistical empiricist and his unwavering faith in the neoclassical agent as econometrician, combined with the conviction that increasingly elaborate formal models of preference orderings would eventually serve to dispel all pesky philosophical problems about the exact nature of probability and inductive inference. Problems of uncertainty would always be traced back to distinctly nonstrategic pathologies in a generic entity called "information":

> However, perhaps the most crucial kind of uncertainty, and certainly the one most susceptible to being diminished by social organization, arises from the lack of information on the part of any one decision-maker as to what other decision-makers are doing or deciding to do. It is a puzzling question why there are not more markets for future delivery through which the relevant information about concurrent decisions could circulate in an anonymous manner. It is also puzzling that where future markets do not exist, trade associations or other interested associations have stepped in to perform the same information-circulating function.[63]

It always seemed to perplex the mature Koopmans that the world did not simply acquiesce in the blinding power of serene omniscient rationality and omnipotence of "the" market. Questions of deception, dissimulation, and computation, so very important at RAND, were for him shoved to the margins of discourse, because they could, in principle, be "solved" by abstract market mechanisms. The question of how one might go about solving them, or else living in the material world in which their solution was a forlorn hope, was avoided by imagining the Walrasian system as an utterly ideal and "institution-free" mechanism safely ensconced in the millenarian future: the ultimate apotheosis of the socialist calculation

that for a piece of work to spark several new developments in economic theory, it is not necessary that it have any particular claim to realism in its portrayal of economic life. Actually, the paper is rather poor economics. I am not speaking merely of the assumption of an unchanging technology. . . . A more unusual defect is that consumption is not treated as in any way an end in itself" (Koopmans, 1970, p. 430).

63 From "Comments in Thursday afternoon session," Conference on Expectations, Uncertainty and Business Behavior, Pittsburgh, October 27–29, 1955, box 5, folder 81, TKPY. These comments were taken from a dismissive critique of a paper by Nicholas Georgescu-Roegen at this conference, itself a philosophically sophisticated survey of the problems of application of conventional axioms of probability theory to the cognitive phenomenon of uncertainty.

controversy. Koopmans testified to this belief in some comments upon Janos Kornai's book *Anti-Equilibrium*: "Dr. Kornai seems to identify the use of mathematical economics with the centralized computational planning solution. This does not seem entirely right to me. Mathematical economics also contains theorems that justify the market solutions under conditions of a sufficient range of perfect spot and future markets. These theorems establish the equivalence of perfect planning and perfect competition, although under conditions that neither planning nor competition has been able to attain."[64]

It has occurred to more than one of his co-workers that, in striving to describe the Walrasian cognitive agent apotheosized by Cowles, he was in some way seeking to describe himself. "Throughout his life, Koopmans never acted rashly and came as close to being a rational decision-maker as is humanly possible" (Beckmann, 1991, p. 266). The capstone of his career was the award of the Nobel Prize in 1975. Herbert Scarf reports: "Much of our conversation was taken up by Tjalling's distress about the fact that George Dantzig had not shared the prize. In a characteristic gesture, involving a fine blend of morality and precise computation, Tjalling told me that he had decided to devote one-third of his prize to the establishment of a fellowship in honor of Dantzig at IIASA [the Viennese think tank]. As we left the house for a press conference at Cowles, Tjalling said, with a certain shy amusement about what was awaiting him, 'Now I have become a public man.'"[65] Koopmans succumbed to cerebral stroke on February 26, 1985.

Jacob Marschak's conversion to the gospel of information processing was, if anything, more dramatic and thoroughgoing than that of Koopmans. If Koopmans had never been more than a reluctant applied econometrician, it was Marschak's empirical program for the "testing" of Walrasian theory that had been rudely relinquished in Cowles Mark II. Far from suffering discouragement at this repudiation, it was Marschak who enthusiastically embraced the economic agent as econometrician around 1950, and devoted the rest of his life to the explication of a Walrasian conception of the processing of information. Marschak, a scholar in his fifties by this time, was much more willing to engage with the cyborgs on their own turf than many a younger colleague, and, for this

[64] "Notes of a discussion talk on Kornai's book," box 18, folder 343, TKPY. Although undated, these notes are filed with reading notes of the book from the year 1968–69. Something about Kornai's work touched a chord in Koopmans, the faltering market socialist, and he became one of his major promoters in America.

[65] Scarf, 1997. This, along with Beckmann, 1991, is the only published biographical source on Koopmans.

reason, "on many occasions during the 1950s and 60s we heard economists question whether Marschak has actually left economics for other disciplines, such as psychology [or] information science" (McGuire & Radner, 1986, p. viii). Marschak embarked on extensive correspondence with such mavens of the "cognitive revolution" of the 1950s as George Miller and John McCarthy (boxes 94, 157, JMLA). In 1955 he convened a seminar on "The Formal Theory of Organization," which included in its number the cybernetician Ross Ashby as well as Howard Raiffa and Martin Shubik. At Yale, he taught a course called "The Economic Theory of Information and Organization," which used Ashby's *Introduction to Cybernetics* (1964) as a set text. He did leave Yale and Cowles in 1960 for UCLA, where he convened an interdisciplinary seminar in the mathematical social sciences, largely devoted to these issues. He maintained his ties to RAND until 1965, and engaged with a stellar sequence of collaborators, including Roy Radner, Gary Becker, and Arrow. He was elected president of the American Economics Association in 1977, but tragically died before assuming the office.

More vigorously than his fellow Cowlesmen, Marschak grasped at the lifeline of von Neumann–Morgenstern expected utility, only to acknowledge that empirical criticism by Maurice Allais and a handful of psychologists had rendered it an exceedingly slippery hawser. Marschak gained a firsthand appreciation for psychological experimentation at RAND and was one of the very first neoclassical economists to conduct systematic controlled experiments in decision theory in an attempt to transform utility theory into a viable simulacrum of cognitive science. Marschak innovated the position that, given Cowles's revised cognitive stance, empirical findings on the psychology of rational deliberation would have to be admitted into the theory as auxiliary hypotheses or special case adjustments, though not, of course, as grounds for rejection of utility theory (Marschak, 1974, 2:29). Early on, Marschak saw fit to criticize the Shannon conception of "information" as not sufficiently reconciled with the Walrasian approach to value theory – an incident described in detail in the next chapter. In these and other respects, Marschak was a one-man Maginot Line, pitted against the onslaught of the cyborgs, armed with little more than pluck and determination. Because neoclassical economists in the 1960s had grown rather complacent about the cyborg threat, of all the Cowlesmen he remains perhaps the most underestimated and least appreciated of the Mark II cohort.

The active engagement with nascent trends in cognitive science was not the only aspect of his work that rendered him less than popular with his neoclassical colleagues. More than any other Cowlesman, his published work is bedecked with open acknowledgments of the military inspiration of much of his research. "Organizations have often been compared with machines: armies and navies have been called 'fighting machines.' . . . Just as the technology of ordinary machines . . . is helped by the use of well-defined scientific concepts, so might the technology of human organizations. The modern naval vessel [is] a highly efficient inanimate machine. . . . But what about the 'animated' part of the ship, the group of men?" (1974, 2:63). Wearing your inspiration and patronage on your sleeve was deemed not so much wrong as unseemly, and it was something other Cowlesmen such as Arrow and Koopmans had instinctively learned to avoid. However, Marschak did come to share one reservation with other Cowlesmen about RAND and its military patrons, namely, the initial enthusiasm in OR for game theory. Interestingly, given the subsequent history of game theory, Marschak's primary complaint was that von Neumann's mathematical doctrine did not leave sufficient room for the diversity of human behavior, a diversity to which he believed that Walrasian theory subscribed. "Any application of the theory of games is based on the assumption that this symmetry of behavior norms is an actual fact. However, this need not always be the case. . . . It may be more useful to study my adversary, as I would study the weather or soil or any other uncertain natural phenomenon" (1974, 1:93). The response on his part was the construction of a "theory of teams," which absorbed the remainder of his research career.

Team theory has not subsequently enjoyed much favor in neoclassical economics, so perhaps it is best to delegate description of its objectives to Marschak's collaborator on the project, Roy Radner (1987, p. 349):

> In an economic or other organization, the members of the organization typically differ in (1) the actions or strategies available to them, (2) the information on which their activities can be based, and (3) their preferences among alternative outcomes. . . . Marschak recognized that the difficulty of determining a solution concept in the theory of games was related to differences of type 3. However, a model of an organization in which only differences of types 1 and 2 existed, which he called a team, presented no such difficulty of solution concept, and promised to provide a useful tool for analysis of efficient use of information in organizations. Such a model provided a framework for analyzing the problems of decentralization of information so central to both the theory of competition and the operation of a socialist economy.

Here once again was dished up the Mark II Mulligan stew: the Cowles fascination with socialist calculation à la Walras, the proposed conflation of economic planning and military command, the implacable insistence upon the validity of utility theory, the treatment of information processing as a special case of optimal allocation of commodities, and the not-so-subtle suppression of inconvenient cyborg themes. This was a watered-down version of the theory of games or, as some had it, a "simplification," an attempt to talk about C^3I in the context of something that looked very much like an army. "One needs information about the ever-varying situation . . . about the weather and the enemy dispositions in the various points of the sea, in the business of naval fighting" (Marschak, 1974, 2:65). However much oriented toward the elucidation of problems of information processing, its provenance as a defanged and desiccated game theory held few attractions for the bulk of the profession. With most strategic elements consciously deleted, and all the computational or interpretative aspects of information buried under the portmanteau of costs of conveyance, team theory found few adherents. It was, in the end, a theory of "organizations" more or less indistinguishable from the Walrasian system. The easy conflation of statistical estimation with machine calculation with cognition allowed this elision to proceed unchecked in Marschak's work. "In the theory of games a sequence of decision rules is called a 'strategy.' . . . In statistics it is called a 'statistical decision function.' Its name in the current literature of operations research is 'dynamic programming'" (1974, 2:121). Marschak, like many at Cowles, never believed in any intrinsic obstacles to translation.

Leonid Hurwicz gave the Cowles doctrine on information processing a somewhat different twist.[66] Hurwicz had initially come to the University of Chicago as a statistical scientist for the Institute of Meteorology in 1942 but rapidly migrated over to Cowles, where he first was employed as associate director of the "Price Control and Rationing" study under Yntema and later became research associate in 1944. Hurwicz was deputized by Koopmans as Cowles liaison with the Air Force on linear programming, in part due to his background in mathematical statistics, but his initial personal reaction was that "the linear programming model seemed totally unrelated to the kind of things that economists do." Hurwicz's point of departure, like most of the rest of Cowles, was the Lange-inspired Walrasian model of socialist planning; he had trouble seeing how linear programming or game theory would provide much in the way of tools to elucidate that problem. "Of course, we economists talked about particular mechanisms, such as central planning, competition, but we did not have a general concept of an economic mechanism. . . . Once you have this concept, but not before, you can

[66] Leonid Hurwicz: born 1917, in Moscow, Russia; LLM, University of Warsaw, 1938; emigration to the United States, 1940; Institute of Meteorology, University of Chicago, 1942–44; Cowles Commission, 1942–46, 1950–51; associate professor, Iowa State, 1946–49; professor, University of Illinois, 1949–51; RAND consultant, 1949–; professor of economics and statistics, University of Minnesota, 1951–99. Jacob Marschak attempted to have Chicago award him a Ph.D. on the strength of his published papers, but this was blocked by Yntema in 1947. See box 88, file: Hurwicz, JMLA. The best source on his career is the interview with him in Feiwel, 1987a.

formulate some axioms that it might or might not obey" (in Feiwel, 1987a, p. 271). Instead of following up on the statistical analogies of linear programming, Hurwicz took his cue from RAND and came to regard economies as "mechanisms" that displayed various properties that resembled computers. His long association with Kenneth Arrow biased this interpretation in two major directions: first, toward the use of some of the mathematics of game theory to help in the axiomatic formalization of Walrasian general equilibrium (p. 275) and consequently to explore possible models of dynamics of convergence to equilibrium; and, second, in his work on the design of incentives to produce optimal allocation mechanisms-institutions, all, it goes without saying, taking the Walrasian system as an adequate portrayal of market operation. In Hurwicz's mind, both projects were direct outgrowths of the original work on linear programming. Nevertheless, he encountered various indications early on that game theory per se would not provide an alternative framework for economics (1953).

The turning point in his career, now no longer merely a coincidence, was also located circa 1950. "My work in this area started around 1950–1 when I was still at the Cowles Commission. I was writing a more or less expository paper dealing with activity analysis. . . . when I used the word 'decentralization' I thought I should explain what I meant. . . . But then it struck me that I did not in fact know what we mean by decentralization. That was the beginning of many years of work trying to clarify the concept" (in Feiwel 1987a, pp. 271–72). The crux of the matter was the explication of the slippery word "information"; but not, as it turns out, in the way that Marschak or Hayek might do. He started out from much the same point of departure:

> The research outlined in this present note is focused on decision-making under uncertainty. The emphasis is, however, not so much on the criteria of optimality among alternative choices as on the *technology* of the processes whereby decisions are reached and choices are made. Under the conditions of 'rationality,' the final decision is preceded by certain operations which may, in general, be characterized as *information processing*. . . . When the information processing aspects of the problem are explicitly taken into account it is found that the concept of 'rational action' is modified.[67]

This inquiry was to be prosecuted by means of axiomatic models of convergence to Walrasian general equilibrium; but the objective of "global

[67] "Economic Decision-Making and Their Organizational Structure of Uncertainty" (1951?), file: Hurwicz, box 91, JMLA. This draft was envisioned to end with a section entitled "Relation to Cybernetics, Communication and Information Theory," which, however, is not present in the file.

stability" was called into question by counterexamples proposed by Herbert Scarf (1960) and David Gale (1963). Rather than give up altogether on the Walrasian project, or regard this as an insuperable empirical obstacle, Hurwicz then began to wonder if stability could be more concertedly "built in" to a neoclassical model, "in some sense 'design' the economic system so that it would have a more universal property of stability. . . . much of this work goes in the direction of designing a convergent *computational* system rather than designing a mechanism that could be applied in a real economy" (in Feiwel, 1987a, p. 262). At this juncture, the metaphor of the "program" came overwhelmingly to color his conception. Around 1960 Hurwicz attempted to define the economic environment as the Walrasian trinity of endowments, preferences, and technologies; the "mechanism" was recast as a "language," some response rules, and an outcome rule. "The interest was to isolate those processes whose *informational* requirements were no greater, and if anything less, than those of the perfect competitive process. . . . Our concept of informational decentralization takes into account the initial dispersion of information together with the limitations of communication. Both of the latter concepts can be rigorously defined" (Hurwicz, 1986a, pp. 301–2).

Although it might appear on its face that Hurwicz had come closer to an explicit computational approach to markets than his Cowles comrades, this was in fact not the case. Although the argument cannot be made here in detail, what is noteworthy about the evolution of Hurwicz's attempts to explicate his planning conception is the extent to which it progressively diverged from any commitment to an algorithmic or cognitive approach. In consort with the rest of his cohort at Cowles, "information" was treated strictly as a matter of finer partitions of a given state space of contingent commodities, after the manner of an individual Bayesian statistician; all practical strategic considerations were suppressed. The point appeared to be that convexity violations of the Walrasian model drastically increased informational requirements for the market, but only relative to the conventional Pareto optimum as benchmark. Inconveniently, the stability counterexamples of the 1960s called into question this approach asserting that Walrasian "price-taking" was really so transparently computationally efficient. Hurwicz's response from 1972 onward was to reinject some quasi-strategic elements in the format of having some agents willfully "misrepresent" their preferences in order to influence the resulting equilibrium (with stability issues now largely suppressed); the task of the planner now was reconceptualized as concocting "incentive compatible" mechanisms that would force agents to reveal their "true" preferences. What is fascinating for the historian is that, whereas the attempt to specify the informational requirements for the operation of the Walrasian marketplace was roundly ignored, the idea that "market failures" – read, divergences from the elusive Walrasian ideal – create insuperable problems for the planning of incentive structures proved a solid hit with the neoclassical profession, giving rise to a huge literature on incentive compatibility (Ledyard, 1987), to the outlandish extent of even being used to upbraid Hayek as having neglected the true lessons of Walrasian general equilibrium (Makowski & Ostroy, 1993). In Hurwicz's own words, "what economists should be able to do is figure out a system that works without shooting people. So that led me to the notion of incentive compatibility" (in Feiwel, 1987a, p. 273). The panacea, not unexpectedly, was found to be dictated by the Walrasian model. Market coordination would prove a snap if you just treated information like one more commodity; it would only be thwarted where pesky "imperfections" (which always boiled down to various nonconvexities of sets) supposedly paralyzed smooth communication. Although the objective was often portrayed as "self-reinforcing rules of communication," what strikes the outside observer is the extent to which issues of the specification of actual market coordination (not to mention appreciation for the implications of MOD) were persistently elided in favor of notional comparisons to the benchmark Pareto optimality of the posited Walrasian model. These supposed "mechanisms" bore no detectable relationship to specific market institutions or to well-defined algorithmic or cognitive science conceptions of information processing. It was yet another manifestation of the Cowles fascination with an "institution-free" market, as Hurwicz was the first to acknowledge: it was the formal investigation of "many constructs which according to my definition are mechanisms, although they may not resemble any mechanisms which exists or even one that I would seriously propose" (in Feiwel, 1987a, p. 270). Better to indulge in machine dreams than to construct real machines.

Whatever the successes or failures of the previous individual research programs, the one figure most personally responsible for the acceptance of the research agenda embodied in Cowles Mark II by the American economics profession was Kenneth Arrow.[68] As Arrow correctly diagnosed, Cowles in the 1940s "felt persecuted" (1995a). Marschak, Koopmans, and Hurwicz may each have been clever, but they possessed

[68] Kenneth Arrow (1921–): B.Sc. in mathematics, CCNY, 1940; M.A. in mathematics, Columbia, 1941; U.S. Army Air Corps, 1942–46; Ph.D. in economics, Columbia, 1951; research associate, Cowles Commission 1947–; assistant professor, University of Chicago, 1948–49; professor, Stanford, 1949–68; professor, Harvard, 1968–79; professor, Stanford, 1979–.

neither the temperament nor the *gravitas* to turn a breakaway sect into a stolid orthodoxy, even bolstered by the support of the military; but Arrow did. Adequately characterizing the means by which he worked this wondrous deed is not a simple matter; one might someday assay the full array of causes, ranging from his keen sensitivity to contemporary trends in the sciences to the capacity to make it seem that neoclassical economics possessed the wherewithal to illuminate moral issues, but predominantly it had to do with his unabashed capacity to cast himself as an unprepossessing spokesperson for reason itself. "An economist by training thinks of himself as the guardian of rationality, an ascriber of rationality to others, and the prescriber of rationality to the social world. It is that role that I will play" (Arrow, 1974b, p. 16).

Gaining the reputation as the embodiment of the voice of reason was no mean trick, given that he explicitly absolved himself from conforming to any of the broader notions of rationality prevalent in the culture. As late as 1996, he still stubbornly defended the canonical neoclassical agent as the embodiment of rationality, in the face of repeated criticism and disconfirmations.[69] He would unselfconsciously regale general audiences with the warning that, "the everyday usage of the term 'rationality' does not correspond to the economist's definition. . . . [it] is instead the complete exploitation of information, sound reasoning, and so forth" (Arrow, 1987, p. 206). But what precisely constituted the *complete exploitation* of sound reasoning? And how would we know when "information" was fully exploited, even if the silent caveat was appended that this was restricted to some narrow economic conception, especially when it was also acknowledged that, "it has proved difficult to frame a general theory of information as an economic commodity, because different kinds of information have no common unit that has yet been identified" (1984b, p. iii). Time and again he would insist the Cowles Walrasian model was the only rigorous framework within which to discuss legitimately cyborg themes like the nature of uncertainty or the functional prerequisites of social organization or even the nature of science, only to turn around and admit, "It is a matter of controversy how to represent the concept of uncertainty" (1984b, p. 198) or "I am not going to attempt a formal definition of an organization, which would probably be impossible" (p. 176), only to turn once more and write down a formal model. He seemed always to be prudently warning of some limitation or other of the

[69] See, for instance, Arrow's preface in Arrow et al., 1996, esp. p. iii): "Rationality is about choice. . . . Rationality of actions means roughly that the alternative choices can be ordered in preference, the ordering being independent of the particular opportunity set available. . . . rationality of knowledge means using the laws of conditional probability, where the conditioning is on all available information."

Walrasian model, even as he simultaneously expressed intolerance for any alternative. Even politics itself was to be approached as a rational discipline, only to turn around and admit this had no rational basis: "The only rational defense of what may be termed a liberal position . . . is that it is itself a value judgment" (1983a, p. 67). It exuded humility, it was formidable, and it worked.

Nevertheless, the key to Arrow's career is his essential similarity to the other Cowlesmen in background and preoccupations, and the consequent parallel course in career trajectory. Arrow began with an undergraduate fascination with mathematical logic, taking a course in the calculus of relations from Alfred Tarski the year he taught at City College of New York (Arrow, 1991a, p. 2). This course would provide one later link with RAND, because Olaf Helmer served as the translator of Tarski's logic textbook for which Arrow read the proofs. While the surviving course notes reveal exercises in abstract orderings that would recur in his later work on preferences, they also are bereft of any of the cyborg themes percolating in the 1930s and, especially, Gödel's proofs.[70] He then decided to embark on a career as a mathematical statistician, without forewarning that statistics had not yet been bracketed out as a separate academic discipline. He went to Columbia in 1940, where he came under the tutelage of Harold Hotelling. Although completing a masters in mathematical statistics, it was Hotelling who convinced him to make the conversion to mathematical economics.[71] Hotelling's influence on Arrow was "profound and indeed decisive," as he has testified on numerous occasions.[72] One

[70] See the notes and exercises for philosophy 246, box 28, KAPD. The lack of concern with paradoxes of self-reference, a phenomenon already alluded to in the previous paragraph, will later play a more significant role in our own subsequent narrative. Arrow's other major personal connection to RAND (independent of Cowles itself) was through Meyer Girschick, for whom Arrow's wife had previously worked as secretary and research assistant (Feiwel, 1987b, p. 22). It was Girschick who first invited Arrow to RAND.

[71] "Hotelling was giving a course in mathematical economics that really fascinated and impressed me. Statistics was still sort of a main goal, then Hotelling informed me that mathematics departments were very hostile to statistics; however, if I were to change my enrollment to the Economics department . . . he thought he could get some financial support for me. So I switched to economics – it was an economic motivation if you like" (Arrow, 1995a, p. 8).

[72] The quotation is from a letter to William Frazer Jr., November 21, 1978, in box 22, KAPD. This letter compares Hotelling to Abraham Wald, also on the Columbia faculty, stating "Wald was more 'modern'; he was fully embued with the Neyman-Pearson point of view and with the minimax point of view which was just developing." Arrow implies he preferred Hotelling's rather more old-fashioned approach, although one of his earliest papers makes use of Wald's sequential analysis, developed at the SRG. Arrow's *Collected Works* are dedicated "To the memory of Harold Hotelling." The implications of this deference for Arrow's approach to price theory are discussed in some detail in Hands & Mirowski, 1999.

aspect of this multifaceted heritage which we shall choose to highlight here had to do with the peculiar political slant that was imparted with respect to the Walrasian model:

> The other thing about Hotelling is that his work was in a kind of mildly socialist ideology. He did not present price theory so much as a *description* of the actual world, but rather as a kind of *ideal* for a good socialist society. Given my own political opinions at the time, that was just exactly the right congruence. I was enthralled by the idea that price theory was a poor description of the real world. . . . In a sense, Hotelling's views legitimized my study of price theory because it could be thought of as important, if not as description, then as prescription. (in Feiwel, 1987b, pp. 639–40)

Thus Arrow was exquisitely attuned to resonate with the party line at Cowles after he returned from his wartime stint in the Air Force as a captain in the Weather Service – something Hotelling must have appreciated, because he provided the recommendation for Arrow to move to Cowles in 1947.

Political orientation is not a minor consideration in this instance, for, as Christopher Bliss (in Feiwel, 1987a) has observed, all of Arrow's work has been informed by the vision of market socialism. It especially accounts for "his feeling for the importance of information and communication in economic systems. And this in turn is the view of the economic planner" (p. 302). It also accounts for the fact that Arrow, to a greater extent than his confreres, was eminently suited to help usher in the transition to Cowles Mark II. "Unconsciously, I very much resisted the topics that were being thrust on me by Marschak. What he wanted at that stage was model building as a basis for fitting. . . . When Koopmans shifted his interest from statistical methods to linear programming, then I felt much more comfortable" (in Feiwel, 1987b, p. 646). The time had come to shift the onus of statistical inference from the economist and onto the agent, a theme we have already broached. The motivation in this instance was that "planning" had become for Arrow synonymous with rational choice under uncertainty in the context of a Walrasian system.[73]

[73] Arrow acknowledges this in the preface to his *Collected Works*: "My ideal in those days was the development of economic planning, a task which I saw as synthesizing economic equilibrium theory, statistical methods, and criteria for social decision making. I was content to work on separate pieces of the task and not seek a premature synthesis" (1983a, p. vii). The fact that the projection of the econometrician onto the economic agent came first, and was inspired by reading von Neumann, is documented by one of Arrow's first Cowles Staff Papers dated March 11, 1947, entitled "Planning under Uncertainty: A Preliminary Report," CFYU. This paper reveals the extent to

The real breakthrough did not come at Cowles as much as it did at RAND. In Santa Monica, Arrow was in his element (Feiwel, 1987b, p. 648). "Operations research" had not been on the agenda when he had been in the Air Force Weather Office, but once at RAND, Arrow embraced it as the epitome of scientific planning expertise. He helped create an OR department at Stanford (Feiwel, 1987b, p. 36) right at the time Stanford was becoming the major West Coast outpost of military-academic research,[74] and maintained a consultancy at RAND throughout his career. Most notably, each of his premier economic contributions, from his "Impossibility Theorem" to his existence proof for general equilibrium, from his "learning by doing" paper and his explorations into the stability of the Walrasian equilibrium to his extensive commentaries on the economics of information, all had their genesis and motivation in some RAND research initiative.[75] All this and more was beholden to the military problem frame, but in contrast to Marschak, he managed to keep the military applications muted and separated out from the more abstract "economic" considerations. This was not done out of some duplicitous desire to mask the origins of his research, but rather for the more complex motivation that Arrow, like many of his Cowles confreres, was fundamentally skeptical of the tendencies of the cyborg sciences being incubated at RAND.

The major thrust of Arrow's rare published explicit commentaries on military topics is that problems arising in that context are qualitatively of a character no different than that encountered in the marketplace. "A very basic description of an organization is that it is composed of

which inductive inference was conflated with the playing of a game: "The concept of policy used here is clearly analogous to the von Neumann–Morgenstern definition of a 'strategy.' In effect, planning is considered as a game in which a (fictitious) opposing player is assigned the job of selecting which of the possible configurations of future events (more precisely, which of the possible joint probability distributions of future events) is actually to prevail" (p. 2).

74 The military background to the rise of Stanford's reputation is covered in Lowen, 1997. Interestingly enough, the Stanford economics department made a bid to incorporate RAND when it was seeking to break away from Douglas in 1948 (Collins, 1998, p. 294n).

75 The genesis of the impossibility theorem is discussed in the next section. The relationship of the existence proof of general equilibrium to game theory and RAND is covered in the next chapter. The origins of the "learning by doing" paper in questions of Air Force aircraft procurement has been very nicely covered in a paper by David Hounshell (in Hughes & Hughes, 2000). The inspiration for the stability work with Hurwicz is acknowledged in Feiwel, 1987a, pp. 198–99. The Cowles turn to an "economics of information" has been covered in this chapter, although some specific comments on Arrow's curious relationship to computers and computation will be discussed in the next chapter. I should acknowledge that Arrow does not agree with my assessment of the centrality of RAND in each and every instance. See Kenneth Arrow to Philip Mirowski, August 9, 2000.

a number of decision-makers, each of whom has access to some information. . . . These problems occur whatever the purpose of the organization. In many respects, the issues are sharper for military organizations. . . . But similar issues arise in economic organization also, though usually at a slower pace."[76] For instance, in a comment drenched in self-reference, Arrow once wrote, "The rapid growth of military R&D has led to a large-scale development of contractual relations between producers and a buyer of invention and research. The problems encountered in assuring efficiency here are the same as those which would be met if the government were to enter upon the financing of invention and research in civilian fields" (1962, p. 624). The whole mind-set of military uncertainty that had called forth the cyborg efforts was mistakenly being retailed as a thoroughgoing revision of the concept of information processing, in his view; the neoclassical market "mechanism" could handily provide an adequate comprehension of any such phenomenon. "Uncertainty usually creates a still more subtle problem in resource allocation; information becomes a commodity" (p. 614). Now, this "commodity" might possess some troublesome characteristics from the pure Walrasian vantage point: it might be "indivisible"; "costs of transmission of information create allocative difficulties which would be absent otherwise" (p. 616); property rights might present some difficulties, nonconvexities give rise to imperfect competition, and so on. The difficulties, while real, would not prove insuperable once the Walrasian agent was augmented with some statistical expertise, because "The rational theory of planning under uncertainty is identical with the foundations of statistical inference" (1951a, p. 88).

If military planning was nothing more than conventional optimization augmented by classical statistics, then there was little need for many of the profound innovations of American OR at RAND that ventured beyond the rather modest conceptions of British OR. Decision theory, for instance, need not be concerned with cognitive or experimental aspirations. "The modern theory of decision making under risk emerged from a logical analysis of games of chance rather than a psychological analysis of risk and value" (Tversky & Kahneman in Hogarth & Reder, 1987, p. 67). Computers in OR were treated as mere tools for calculating successive approximations to models without closed form-solutions (Arrow, 1984b, p. 63) rather than artifacts for illustrating abstract principles of organization or theories of automata. Because Cowles would never allow itself to be held to any standards of historical accuracy,

[76] Grant proposal "Information as an Economic Commodity" by Arrow, Mordechai Kurz, and Robert Aumann, September 6, 1985, pp. 32–33. A copy may be found in box 28, KAPD.

a just-so story would be told in which economics had always already been about "computers" *avant la lettre*, suppressing the concern with effective computation that was so prevalent at RAND and so absent in the Walrasian tradition.[77] And then there was the skeptical distance maintained from game theory, incongruous in someone who had made extensive use of the mathematics of game theory in his own theoretical work. These comments range from early intimations that the theory of games for situations greater than two persons is "still in a dubious state" (Arrow, 1951b, p. 19) to an invidious comparison of game theorists to "partial equilibrium theorists" in the early 1980s (in Feiwel, 1987a, p. 215). He even once had the imprudence to suggest that game theory might be less attractive than Walrasian general equilibrium because it dictated greater reserves of rationality on the part of its agents: "It may well be that you will never get a good theory. If you take formal game theory, it tends to have a lot of indeterminateness. It also makes great demands on rationality, well beyond those that neoclassical theory imposes" (Feiwel, 1987a, p. 241).

None of this should be taken to imply that Arrow did not progressively come to appreciate the significance of at least some cyborg themes over the course of his career. To his credit, he was among the first to direct economists' attention to the experimental results of Kahneman and Tversky (1982) calling into question the cognitive models of probability of real agents. He later tempered his position from a critic of Herbert Simon's notion of bounded rationality (Feiwel, 1987a, p. 231) to qualified support. After a career of repeatedly comparing the market mechanism to computing, the later Arrow conceded, "I would like to go into computing. I have not done anything in that area yet, in fact, I am not quite sure there is a field there" (in Feiwel, 1987a, p. 242).[78] He pursued this ambition by becoming the main economic liaison with the fledgling Santa Fe Institute, a notorious nest of cyborgs in the 1980s, serving to shape profoundly the character of the economics program there. In the final accounting of cyborg resistance and accommodation, Arrow's cumulative impact upon economics has yet to be gauged. Yet there is no question that his lifelong themes of command, control, communications, and information were bequeathed him by the military organization of scientific research growing out of World War II.

[77] Arrow, 1984b, p. 44: "The role of the price system as a computing device for achieving an economic optimum has been one of the main strands of economic theory since the days of Adam Smith."

[78] The relationship of these interests to his experience with the work of his student Alain Lewis is covered in the next chapter.

ON THE IMPOSSIBILITY OF A DEMOCRATIC COMPUTER

Can a computer have a politics? Can a politics be explicitly computational? We began this chapter with the question, and now we come full circle to encounter it once more in a very specific manifestation. Kenneth Arrow's first major contribution to the economics literature, which served both as his Ph.D. thesis and as one of Cowles's Mark II intellectual landmarks (as cited in its 1951 grant proposal quoted earlier), was the result in "social choice theory" now widely known as the Arrow Impossibility Theorem. The cyborg genealogy of this landmark product of Cowles and RAND warrants some closer attention for a variety of reasons. First, and most directly, it was a direct product of Cowles's participation in the socialist calculation controversy, although few have seen fit to situate it within that context. Second, it marks a defining moment when the metaphor of the machine is extended to politics in a formal manner, even though this was an old metaphorical theme in the literature of statecraft (Mayr, 1976); yet a curious set of conditions conjured it by a circuitous route, notably, by an intermediate comparison of politics with the Walrasian market mechanism. Third, while acknowledging the work was prompted by an explicit military problem generated at RAND, we shall offer a reading of the text that suggests it stands as a repudiation of a substantial portion of the RAND agenda in the late 1940s, and especially of von Neumann's game theory. Fourth, it will be salutary to plumb the conviction that this was a breakthrough in the *logic* of politics, on a par with that other great twentieth-century breakthrough: a proposition "which is to mathematical politics something like what Godel's impossibility theorem is to mathematical logic" (Samuelson, 1977, p. 935).

What was this theorem, which Arrow frowardly insisted on calling a "Possibility theorem" in his *Social Choice and Individual Values* (1951a)? Essentially, it was a statement that if one took the neoclassical characterization of individual preferences as inviolate, which Arrow glossed as his definition of "rationality," and one accepted the premise that a social ordering of a set of alternatives must represent the aggregation of those individual preferences, then the political procedure of majority ballot will generally produce social rankings of alternatives that violate that "social welfare function." In his own words, Arrow interpreted the result as implying, "If we exclude the possibility of interpersonal comparisons of utility, then the only methods of passing from individual tastes to social preferences which will be satisfactory and will be defined for a wide range of sets of individual orderings are either imposed or dictatorial. . . . the doctrine of voters' sovereignty is incompatible with that of collective rationality" (1951a, pp. 59–60).

Of course, such a bald statement demands the explication of many vague terms, such as preference, "imposition" of orderings, dictatorship, voter's sovereignty, and so forth. The imposing character of Arrow's argument derived in no small part from the rhetorical method he employed to make the case – namely, a set of axioms that were asserted to support these interpretations, leading inexorably to the theorem in question. The notion of a social welfare function had boasted a hallowed heritage in the history of neoclassical economics, but the treatment of individual preferences, social welfare, *and* voting outcomes as all instances of the same logic of orderings was pure Tarski (p. 13n). "The analogy between economic choice and political choice has been pointed out a number of times" (p. 5), but it took a certain hubris (and a belief that "Rationality is about choice") to expand a mere analogy into the conflation of the Walrasian model with the whole of politics; it might not have caught on if it had not already become stock-in-trade in a literature quite familiar to Cowles, namely, the "socialist calculation controversy." Revealing its sympathies, *Social Choice* cites Lange five times, and Hayek not at all. This was a bow to necessity rather than fashion, because the structure of the argument ruled out any considerations of process, "taste change," learning, and strategic communication a priori (p. 7). In other words, the text did not so much answer Hayek's concerns as shove them aside without comment. The argument was prosecuted not as a measured consideration of the pros and cons of the Walrasian market analogy for political equilibration, but rather as a consistency counterexample, such as one might find in a logic text or in a dissertation on metamathematics. That is, a certain set of hastily motivated "reasonable conditions" were posited that put restrictions on the individual and social welfare functions; "imposed" and "dictatorial" regimes were also defined as commensurate restrictions on welfare functions; and then a proof was offered showing the "conditions" would only exhibit logical consistency under dictatorial or imposed regimes.

There now exists a huge literature subjecting almost every technical aspect of the axioms and lemmas of this proof to criticism and scrutiny, and the size of the literature is a testament to the perceived significance of the result. Nevertheless, unless we succumb to the crudest sort of citation counts, it is difficult to escape the conclusion that its full historical and cultural significance has generally eluded the epigonai. Statements such as, "Aristotle must be turning over in his grave. The theory of democracy can never be the same" (Samuelson, 1977, p. 935) serve more to distract attention from the high-stakes game played by Walrasian economists. For anyone steeped in the socialist calculation controversies of the 1930s, it is hard to see it as anything other than a reprise of the Cowles theme that the Walrasian market is a computer *sans* commitment to any computational architecture or algorithmic specification; the novel departure came with the assertion that democratic voting is an *inferior* type of computer for calculating the welfare optima already putatively identified by the Walrasian computer. Because neither the process of market operation (recall Arrow's stability research lay in the future), nor the mechanics of individual preference calculation, nor that of voting procedures were accorded any explicit consideration, the exercise partakes of an air of virtual reality, in contrast to empirical enquiry, bearing all the earmarks of Cowles Mark II.

The interesting aspect of *Social Choice* lay rather in its relationship to the Lange political tradition of equation of centralized planning with Walrasian market operation. Although Arrow took pains to indicate that he favored democratic structures and individual freedom, the message of the proof itself belies these claims. However one evaluates the plausibility of the commensurability of centralized planning and the Walrasian model of perfect competition, there is no doubt that in Arrow's text they are treated as jointly superior to the ballot box. Voting is simply strangled communication through a degraded channel. As Jean-Pierre Dupuy so aptly put it, "the outcome of voting is simply to maximize entropy, that is to say, disorder" (1994, p. 174). This constitutes, in *sotto voce*, the answer to Hayek: the mechanical market and the centralized plan are more rational than the dead hand of tradition, more rationally justified than mere historical accident of past political decisions, even if those decisions were made under the imprimatur of democratic vote. Raw individual desire will not long be suppressed. Machines will rule, because the purest freedom is that solidly grounded in necessity. This becomes clear in the final chapter, where Arrow expresses his frustration with Kantian notions of the moral imperative, and in the process comes close to acknowledging an alternative conception of the very meaning of politics: "Voting, from this point of view, is not a device whereby each individual expresses his personal interests, but rather where each individual gives his opinion of the general will" (1951a, p. 85). This alternative vision rapidly disappears in the mist, however, since it cannot be conveniently expressed as a machine congenial to the Walrasian tradition.[79]

Arrow seemed incapable of grasping that politics was merely not a problem in "decidability," such as one might encounter with a Turing machine. True, it did sport superficial resemblance to a "measurement" or a "calculation"; but the whole atmosphere at RAND blocked serious

[79] Arrow does try and suggest it could be captured by a model of "the statistical problem of pooling the opinions of a group of experts," and significantly, in a footnote, admits "This analogy was pointed out to me by O. Helmer" (1951a, p. 85). Helmer went on to construct just such a parliament of experts later in his career. On this, see Rescher, 1997.

consideration of the way in which the ubiquitous computer metaphors were tendentious or inappropriate. Suppose, for the moment, that the purpose of a voting procedure really was the efficient extraction and collation of a set of fixed and independent "preferences" of a large populace. Then why bother with the whole awkward rigmarole of free elections? Why not take a page from the textbook of the statistician, and conduct a scientifically designed stratified random sample from the population? Or – more efficiently – maybe we should let people buy and sell the right to become incorporated into the sample in the first place? It would seem that this would be the obvious prescription to take away from reading *Social Choice*: it would be cheaper, it would directly deploy the technologies of the reduction of uncertainty so championed by Arrow, and, better yet, built-in randomization might actually "resolve" the vexed problem of inconsistency. The only minor glitch might come in getting a democratic populace to acquiesce to the panacea in the first place, because, as already noted, its notions of "rationality" might not jive with Arrow's.

Not only is *Social Choice* a riposte to Hayek; it is equally a remonstrance nailed to the portal of RAND. The conditions of the genesis of the text have often been related by Arrow; we quote a recent version, since it most closely addresses this theme.

> I spent the summer of 1949 as a consultant to the Rand Corporation. . . . There was a philosopher on the staff, named Olaf Helmer, whom I had met earlier through Tarski. . . . He was troubled by the application of game theory when the players were interpreted as nations. The meaning of utility or preference for an individual was clear enough, but what was meant by that for a collectivity of individuals? I assured him that economists had thought about the problem. . . . He asked me to write up an exposition. I started to do so and realized that the problem was the same I had already encountered. I knew already that majority voting would not aggregate to a social ordering but assumed that there must be alternatives. A few days of trying them made me suspect there was an impossibility result, and I found one very shortly. (Arrow, 1991a, pp. 3–4)[80]

In this extremely revealing passage, Arrow ties the military concern for C^3I and machine rationality to the planning orientation of the market

[80] A letter of Arrow to Marschak, August 24, 1948, suggests this may have been off by a year: "Of considerably more importance from the viewpoint of economics, I have shown that it is not possible to construct a social welfare aggregate based on individual preference functions which will satisfy certain natural conditions. The proof is actually extremely simple, and can, I believe, be easily expressed without the use of symbols (though personally, I find truth hard to understand in words). The first drafts of these results will be prepared as Rand memoranda" (box 154, JMLA).

socialists, and admits that the original purpose of the exercise was to evaluate the validity of game theory using neoclassical preferences for Cold War strategic gaming. Helmer was then in the throes of a large-scale evaluation of game theory as a strategic tool at RAND (see Chapter 6). The military had thought it was purchasing an algorithmic strategic rationality that would remove the lower echelons out of the command loop, but this seemed paradoxical when it came to democratic politics: what precisely were the objectives of Democratic Blue team versus the Dictatorial Red team? Did Blue stand to lose because it could not manage to get its strategic house in order, ironically fatally flawed because it was united under the banner of political incoherence?[81]

The Arrow Impossibility Theorem was a doctrine to dispel the paradox, but only at the cost of dispensing with the game-theoretic notion of strategic planning, and perhaps cyborg science *tout court*. First off, the theorem evaded the central C^3I question, because it had nothing substantive to say about the information-processing or feedback-control aspects of either democratic politics or market organization. The ironclad presumption of the truth of the Walrasian model preempted that line of analysis. And then, one major subtext of *Social Choice* was a negative evaluation of game theory (1951a, pp. 19, 70), mainly on the grounds that it clashed with the Walrasian approach to individual rationality. Most revealingly, the most important development in twentieth-century mathematical logic was dismissed in two sentences as irrelevant for the portrayal of politics and the market as computational devices: "From a logical point of view, some care has to be taken in defining the decision process since the choice of the decision process in any given case is made by a decision process. There is no deep circularity here, however" (1951a, p. 90).

So far, all that was being proffered to the military were negative results: RAND's enthusiasms were deemed barren. But wait: the apparent paradox of Blue team strategic vulnerability could be dispelled if the market and central planning were just two sides of the same optimal coin. The Pentagon can "impose" rational preferences about community political behavior toward the Red team (though it helps if they are kept secret from unwitting Blue team citizens), while the Walrasian market can give expression to the optimal rational desires of Blue team members, and there need be no inconsistency! (Consistency has always stood as Arrow's benchmark of rationality in preferences.) If either paragon of rationality

81 Arrow explicitly dealt with this possibility in a RAND research memo coauthored by Armen Alchian and William Capron in 1958, where the possibility of the Soviets winning the arms race was traced to their ability to draft scientific talent into military research as needed, whereas the United States relied on the "free choice" of scientists.

were to be resisted at the ballot box, then that would constitute prima facie proof that democratic voting is a degraded and undependable mechanism of the expression of rationality. The Cold War regime is described and justified, summed up in a few tidy axioms.[82]

I do not intend to suggest that Arrow was alone in expressing Cold War politics in his treatments of rationality, or was in any way unique. I should think the very same case could be made for the history of artificial intelligence. "For the past 30 years or so, computational theorizing about action has generally been conducted under the rubric of 'planning'" (Agre, 1997, p. 142). Machines most definitely have harbored some politics, and, until very recently, it was very hard to imagine any form of rationality (algorithmic or no) which was not centralized, hierarchical, and deeply fearful of loss of top-down control. This is the main reason why the plethora of fin-de-siècle effusions about the democratic promise of computers and the Internet is acutely embarrassing and acridly jejune. The extent to which the palpable planning bias in computation can be attributed to the very architecture of the first computers, and the extent to which it can be traced to the concerns of their military sponsors and users, is something that cannot be settled here. What is at issue is the fact that collateral disciplines like software engineering, artificial intelligence, and the like were able to make headway in explicating (at minimum) some limited aspects of human rationality because they were held to the requirement that their algorithms be effective: that the lines of code actually do something useful and, furthermore, the analyst had demonstrably constructed a model of how they were to be accomplished. The hallmark of the Cowles treatment of machine rationality was that it took the diametrically opposite approach.

This brings us, finally, to the "resemblance" of the Arrow theorem to Gödel's rather more fundamental impossibility result. There is a resemblance, but perhaps not the sort that Samuelson has suggested. Gödel, as is well known, suffered from mental instability and paranoia from the mid-1930s onward (Dawson, 1997, pp. 111ff.). Due to his disabilities, when he decided to apply for American citizenship after World War II, there was much trepidation amongst his friends concerning his possible behavior during the naturalization process. Oskar Morgenstern,

[82] "The purest exemplar of the value of authority is the military.... Under conditions of widely dispersed information and the need for speed in decisions, authoritative control at the tactical level is essential for success.... the aim of designing institutions for decision making should be to facilitate the flow of information to the greatest extent possible.... this involves the reduction of the volume of information while preserving as much of value as possible. To the extent that the reduction in volume is accomplished by reduction in the number of communication channels, we are led back to the superior efficacy of authority" (Arrow, 1974b, pp. 69–70).

who along with Albert Einstein was to serve as one of his two witnesses at the proceedings, reports that Gödel had taken the injunction to study the American system of government for the naturalization exam quite seriously, so much so that he confided in Morgenstern that, to his distress, he had discovered an inconsistency in the American Constitution.[83] Morgenstern, fearful that this would jeopardize the swearing-in ceremony, conspired with Einstein on the drive to the courthouse to distract Gödel's attention. Of course, the judge scheduled to administer the oath was acquainted with Einstein, and so Gödel was accorded special treatment when the appointed time arrived. The judge ushered them all into his chambers, began chatting with Einstein and Morgenstern, and as part of the process of making polite conversation, queried Gödel: "Do you think a dictatorship like that in Germany could ever arise in the United States?" Gödel, with all the tenacity of a logician, the fervor of a firsthand witness of Hitler's Anschluss, and the rationality of a paranoid, became animated, and launched into an elaborate disquisition on how the Constitution might indeed allow such a thing to happen, due to a subtle logical inconsistency.[84] The judge, wiser in the ways of man, quickly realized that something had gone awry, and thus quashed Gödel's explanation with an assurance that he needn't go into the problem, and proceeded to administer the citizenship oath.

We should ask ourselves, What is the difference between Gödel's inconsistency, and that putatively uncovered by Arrow's proof? More distressingly, what is the difference between the neoclassical economics profession and that prudent judge?

[83] This anecdote is reported in the Morgenstern diaries, entry dated December 7, 1947, OMPD. It is also discussed in Dawson, 1997, pp. 179–80.

[84] I confess with some trepidation to having written this before the presidential election debacle of 2000.

6 The Empire Strikes Back

> I'd like to think that computers are neutral, a tool like any other, a hammer that could build a house or smash a skull. But there is something in the system itself, in the formal logic of data and programs, that recreates the world in its own image. ... It is as if we took the game of chess and declared it the highest order of human existence.
>
> Ellen Ullman, *Close to the Machine*

PREVIEWS OF CUNNING ABSTRACTIONS

What did it mean to set out to construct a cyborg around 1950 in America? We are not talking engineering specs here, although it should go without saying that everything, hardware included, matters when it comes to easing into the American cybernetical sublime. It was a further testimonial to the planned science regime of World War II that the immediate postwar era found itself awash in auspicious gizmos. The transistor was invented in 1947 at Bell Labs; magnetic disk storage was implemented at the National Bureau of Standards in 1951; the first magnetic drum memory was installed in a computer for, the lineal predecessor of the National Security Agency in 1950 (Bamford, 1982, p. 99); magnetic core memories were innovated at Project Whirlwind at MIT in the early 1950s. In 1953, IBM built its first commercial electronic stored program computer (which immediately went to Los Alamos: Ceruzzi, 1998, p. 34), the IBM 701; RAND got its own JOHNNIAC machine with core memory in 1953 (Gruenberger, 1979). The Air Force persuaded IBM to redesign its new 704 computer to take advantage of transistors in 1959 (Ceruzzi, 1998, p. 70). Neither are we talking wetware, or at least not yet. The Watson-Crick model of DNA would burst upon the scene in April 1953; the PaJaMo experiment which factored into the understanding of messenger RNA happened in 1958.[1] The first transfer of a nucleus from a frog blastula into a second frog egg, which then subsequently developed into an adult frog, happened in the period 1951–52, the first pale intimation of the "selfish gene" as an information packet usurping control of a pliable cyborg husk (Kolata, 1998, chap. 3). Must this be a conventional history of technological innovations? No, the purpose of our interrogation is to inquire into what it was like to dream a zesty machine dream in the cat's cradle of von Neumann's accomplishments, soothed by the gentle lullaby of military largesse in the immediate Cold War era – whatever would greet the dreamer awakening to a new dawn of communications, control, and information, where the Mark of Gain was inscribed on every forehead?

Constructing a cyborg would have to be an exercise in making do. One would have had to face up to the fact immediately that von Neumann's legacy was ambiguous, at best. What the fledgling cyborg scientists at RAND had close to hand was, mostly, game theory and then, latterly, the computer. Von Neumann had bestowed his imprimatur on the development of game theory at RAND, to be sure, but then went on his merry way, pursuing the computer and related forays into what would subsequently become cognitive science and automata theory, and leaving the gamesters more or less to their own devices (Poundstone, 1992, p. 167). Furthermore, the notions of solutions to games that he had said he found most appealing were those that formalized the multiplicity of collaborative structures – this, incongruously, coming from the scientist second most closely identified with a hawkish stance toward the Soviet Union in the early 1950s. (Edward Teller would occupy first rank in that category for the next two decades.) The one true path to Cyberspace had not been decked out in Day-Glo colors; traffic signals emanating from the military were cryptic at best; and the omens were not uniformly propitious. Pileups and wrong turns occurred early on at RAND with clashes between the mathematical logicians and the bean counters; the former entrenched in their conviction that mathematical wizardry would eventually deliver to the military something it might someday want, whereas the latter sought a more direct method for administering what they regarded as a salutary dose of astringent rationality to a military mired in contradictory doctrines of strategy, tactics, logistics, and weapons procurement.

[1] These developments are not so distantly related as they might first appear. For instance, the role of Leo Szilard, the very same atomic scientist encountered in previous chapters, in the PaJaMo experiment is discussed in Judson, 1979, pp. 400ff. Moreover, the language of C^3I was never very far from the minds of the biologists, as documented in Kay, 1995; 2000. One can savor this in the popularized description given by François Jacob of the latter work in terms of two planes flying around with H-bombs, and a transmitter vainly attempting to signal "*don't drop don't drop don't drop*" (Judson, 1979, p. 421).

The glory of RAND in the 1950s – and the reason it would come to be remembered so wistfully in some quarters – was that the phalanx of talent gathered together there managed to conjure up so many momentous innovations out of such unpromising materials. In effect, American-style operations research and game theory were entertained as heuristic suggestions rather than fixed programs of academic dogma: the computer was the instrument that permitted marvelous latitude in interpretation and flexibility in negotiations over that vexed imperative of "rationality." Although the fruits of interdisciplinary camaraderie can always be overstated, it was the case that, at least in the 1950s, no single discipline gained an upper hand in defining the essence of a RAND analysis. Indeed, the glaring weakness of otherwise perceptive commentaries on the place of RAND in American strategic thought (e.g., Rapoport, 1964; Green, 1966; Kaplan, 1983; Herken, 1987; Hounshell, 1997b; Gray, 1997; Hughes, 1998) is that an appreciation of this important fact appears to have eluded them. They uniformly seemed to think that "the economists" hijacked RAND (and later the Department of Defense) through some nefarious species of technocratic imperative or ideological zealotry. It was undeniable that economists, after constituting a mere 3 percent of RAND research staff in 1949, swelled to a prodigious 35 percent of listed consultants by 1950 (Jardini, 1996, p. 37), and that this alone would go some distance in accounting for the profound reverse impact of RAND on the economics profession; but their participation proved negligible in the elaboration of game theory and the refurbishment of computational themes, and the local ecology on the ground in Santa Monica persisted in a much more varied and fluid state than any superficial observation that "game theory suffused every prescription at RAND." As we have already observed in Chapter 5, the Cowles economists were about as reticent when it came to various cyborg themes as was the randomly chosen rocket scientist at RAND; it is arguable that, on balance, they resisted cunning and clever innovations at RAND more than they participated in their elaboration. Moreover, the politics of the situation was substantially more complicated than that captured by conventional dichotomies of "left versus right." Hence the problem becomes, How can one understand the efflorescence of cyborgs at RAND, given the rather unpromising configuration of interests that were present at their creation?

The narrative advanced in this chapter is one in which many individual researchers at RAND in the 1950s, each in his own way, eventually paralleled the intellectual trajectory of von Neumann in the last decade of his life: namely, as they were forced by circumstances and events to become more "worldly," retracing his path away from game theory and toward something with a closer resemblance to a computer-inspired program of a generalized science of society and an abstract theory of automata. That is not to suggest they were engaged in conscious mimesis, nor that they were subject to direct guidance from von Neumann. If anything, von Neumann the man had receded into the status of a remote deity at RAND in the 1950s, whose infrequent utterances were treated as Delphic messages, candidates for perscrutation, but whose transcendental virtue had little practical impact upon RAND's quotidian problems with its major sponsor, the Air Force.

Far from being cheerful puppets of the military, the Punch and Judy show of the Cold War, RAND believed itself beset on all sides by hostile forces. "RAND had to be careful. It had powerful enemies in the Air Force, the Congress and in the universities" (Bellman, 1984, p. 157). It was commonplace to hear RANDites bemoan incidents where Air Force brass would tell them to their face that asking RAND about nuclear strategy was like asking people in the street about how to fight a nuclear war (Herken, 1987, p. 80). From the very start, Ed Paxson's attempt to produce something "useful" in the form of a "Strategic Bombing Assessment" was rejected with scorn by Air Force analysts. Retrospective accounts by insiders are replete with anecdotes of ways in which the Air Force would not only dismiss but actively undermine RAND's attempts to subject questions of strategy and logistics to rational assessment.[2] Scientists may presume it is their mandate to rationalize organizations, but the lesson learned with some pain at RAND was that, oftentimes, science alone was not sufficient to bring wisdom to the stubbornly irrational. Questions raised at RAND about such issues as the cost-effectiveness of jet airplanes, the wisdom of the Air Force monopoly on nuclear weapons delivery, and the overall utility of manned bombers did not endear the think tank to its patrons, causing relations with the Air Force to deteriorate irreparably by the early 1960s (Jardini, 1996, pp. 151, 154; Smith, 1966, p. 126). The degradation of the Air Force connection coincided in the minds of many with the drawing to a close of RAND's intellectual "golden age" (Herken, 1987, p. 98; Jardini, 1996, p. 146; Enthoven, 1995). To

[2] See, for instance, the interviews with Edward Quade, David Novick, and Robert Belzer, NSMR. Relationships with General Curtis LeMay were particularly prone to suspicion and distrust on both sides, signified by his quip that RAND stood for "Research And No Development." As analyst David Novick recounts, "just before Christmas in 1949 ... RAND people learned what the Air Force giveth, the Air Force can also take away. It turned out that between Washington and Dayton the cost of the turbo-prop bomber had doubled, and the cost of the pure jet had gone down 50%. As a result, when the RAND people came back to Santa Monica, they felt they had been had" (1988, p. 2). Ivan Getting reports that, "by 1950 General Vandenberg became unhappy with RAND. ... the RAND products were often not timely nor responsive to changing conditions." Getting was sent to read the riot act to RAND (1989, p. 240). Numerous clashes are recounted in Jardini, 1996, 1998.

outsiders, by contrast, this seemed to manifest the very apogee of RAND's influence and clout, just when Secretary of Defense Robert McNamara recruited his contingent of "Whiz Kids" from RAND to staff the Department of Defense in 1961 (Kaplan, 1983, pp. 252–53; Shapley, 1993, pp. 99–102). "The containment of emotion was the precise appeal to McNamara of the RAND strategists as staff advisors" (Shapley, 1993, p. 190).

This Machiavellian moment, in which no one could be altogether trusted, not even those picking up the tab, and wearing your emotions on your sleeve was fatal, is crucial for *verstehen* concerning the RAND mind-set in the 1950s, as well as for the genesis of various cyborg sciences. RAND's curious status suspended between neither/nor – neither a real corporation nor a philanthropic nonprofit; neither a real university nor a commercial R&D unit; neither fully organized along academic disciplinary lines nor truly free of disciplinary dogmas; an advocate for neither the physical sciences nor the social sciences; neither subject to the military command structure nor entirely free of its purview; neither consistently secretive nor fully open – rendered it a suitable incubator for monstrous hybrids. A world where everyone could potentially be an opponent in disguise, and all epic glories lay in the past, is a world primed to accept MOD as well as MAD. It is a world of office politics, Byzantine intrigues, mirrors within mirrors – eminently, a closed world of apocalyptic conflict. Whereas denizens of such millieux in the past often consumed themselves with fragmenting implacable religious cults, frenzied witch-hunts, or internecine dynastic feuds, conditions at RAND had injected a new, self-reflexive aspect into the situation: it was supposed to be the haven of steely-eyed technocrats with ice in their veins, pursuing the mandate to think through a real apocalyptic conflict, one fought with unimaginably devastating nuclear weapons.

These layers of looming apocalypse upon apocalypse, each as repressed and unintelligible as its stylobate, had the effect of escalating the drive for an abstract rationality well beyond that expounded in any previous social theory. It sought reassurances that could not rest satisfied with a mechanism for giving people whatever it was they thought they wanted, as with some pie-in-the-sky Pareto optimality; neither would it accept that the consequences of present actions could merely be extrapolated into the future using some mechanical inductive procedure. In an inherently unstable situation fraught with extremes, where meanings were elusive and one slip might mean catastrophic loss of everything that ever mattered for humanity, what was required was an ironclad standard of rationality imposed upon the threatening chaos by means of some external governor. The establishment of control was the essential precept in a nuclear world: passive resignation would never suffice; laissez-faire was out of the question. Each and every facet of the quandary nudged RAND further in the direction of a machine rationality.

The quest for a robust rationality suitable for the Cold War context began with game theory and OR, but it did not long remain tethered there. At RAND, game theory did not prove everything it had been cracked up to be,[3] and so therefore the cyborgs went to work transforming it into something just a bit more useful. Intensive scrutiny of the mathematics of games at RAND did not just lead to a reevaluation of von Neumann and Morgenstern's solution concepts – although that did happen, as we describe in the next section – but also more significantly to ontologically amorphous "games" mutating into all sorts of collateral procedures, protocols, tools, and inquiries. Likewise, OR was found to be incapable of answering the really pressing questions of the nuclear age, and so it was concertedly reprocessed into "systems analysis." Both of these trends drew their participants inexorably in a single direction: closer and closer to the computer. The multiform roles of the computer – icon, template, scratchpad, spectacle, blueprint, muse, tool, oracle, Baal – encouraged rationality to become more polymorphous and, it must be said, more perverse.

What had begun as the quest for unimpeachable certainty through extraction of the human out of the control loop was channeled instead toward cybernation: the melding of the human with the machine, and vice versa. For instance, one version of dissatisfaction with game theory led directly to the playing of war games, and gaming begat intensified simulation with machine prostheses, and machine prosthesis begat a novel field of research, that of engineering the man-machine interface.[4] Another,

[3] This, one of the most important facts about RAND, was first documented in detail by Jardini (1996) and is discussed in detail in our next section. On this, see also Nasar, 1998, p. 122. Numerous histories such as Hughes, 1998, p. 157, have stumbled over this point, and therefore are untrustworthy when it comes to understanding the history of OR and systems theory.

[4] "Well, first of all, the gaming from Kriegspiel to the war game exercise I suppose influenced research in the sense of people thinking about trying to simulate the environment in which these things might operate. Whether the simulation was like the psychologists' later Air Defense simulation, or whether that simulation was a map exercise on a map, or that simulation was a computer exercise in which you sort of formalized some assumptions about attrition and fighters attacking bombers or whatever – so that in all those cases, you had the sense, I suppose, of trying to simulate. . . . Lloyd Shapley was the chief Kriegspieler. . . . the impact of Kriegspiel was largely [on] those same people who were interested in the mathematical theory of games. . . . They along with the economists thought that they could use the theory of games to simulate conflict of all kinds, whether corporate or military, and where the war gamers, whether over the map or the people over in the simulated Air Defense Detection Center, were trying to do a simulation of a different type, not formalized in the mathematical computer type programs." Robert Specht interview, p. 22, June 26, 1989, NSMR.

related, set of qualms about the solutions to games led to experimentation with human subjects playing games, and this begat standardization of experimental protocols with the use of computers, which in turn begat machines and humans testing and probing each others' rationality, and this led to machines playing other machines for the purpose of illustrating the laws of rationality. Game theoretic strategies were deemed (reasonably enough) to be inadequate representations of what it meant to "communicate" through moves in the game, so this begat a renewed concern with codification of the algorithmic character of strategies; closer attentions to the deployment of systems of algorithms led in turn to the engineering of the human production of "programs," and program engineering was spun off as a novel professional activity (Baum, 1981).

The same sorts of things happened to OR at RAND. For instance, what had begun as the technique of linear programming, the optimization of fixed objectives over a given domain, proved fairly ineffectual when it came to efficiently ordering weapons whose technical specs were little more than the gleam in some colonel's eye, and so planning as optimization morphed into planning as the rational imagination of entirely fanciful scenarios programmed into a computer: the very essence of systems analysis. Imaginary scenarios on machines begat the idea of machines displaying imagination, and in conjunction with the man-machine interface in radar early warning systems, that inquiry begat artificial intelligence. The injunction to tame the unchecked proliferation of imaginary scenarios begat a renewed concern over what was widely accepted as not only credible but also believable, and this begat RAND's Logistics Simulation Laboratory (Shubik, 1975, chap. 5), and this segued in turn into the creation of computerized virtual realities.

The shifting emphases on technical and mathematical subjects within the nascent cyborg sciences is nicely captured in Table 6.1, compiled by Stephen Johnson, which documents the changing textbook pedagogy of "operations research" and its morph into "systems engineering" over the decades of the 1950s and 1960s. The quadrivium of statistics, linear programming, queuing theory, and game theory that had originally constituted the mathematical infrastructure of American OR in the immediate postwar era gave way to a broader-based set of mathematical techniques, most of which were more directly tethered to the computer. Cyborgs did not want to be fenced in by tradition: "the quarter-century old determinate techniques of game theory, programming, and operations research appear dated and overly restrictive upon the imagination" (Kuenne, 1967, p. 1400). Systems analysts were expected to be much more broadly based, familiar with Shannon's information theory, the rudiments of recursive function theory and/or Turing machines, formal logic, control theory, and some simulation algorithms, as well as basic physics and mathematical statistics.

Table 6.1. Subject Matter in Operations Research and Systems Engineering Textbooks

	Operations Research							SE & OR	Systems Engineering					
	1951 Morse	1954 McCloskey	1957 Churchman	1963 Goddard	1964 Stoller	1965 Enrick	1967 Ackoff	1960 Flagle	1957 Goode	1962 Hall	1965 Machol	1966 Porter	1967 Shinners	1967 Wymore
Probability and statistics	x	x	x	x	x	x	x	x	x	x	x		x	
Linear programming		x	x	x	x	x	x	x	x		x		x	
Queuing theory		x	x	x	x	x	x	x	x		x			
Game theory	x	x	x		x		x	x	x		x			
Network analysis						x	x	x		x				
Management		x	x			x	x			x	x		x	
R&D process	x		x				x	x	x	x	x			
Testing			x				x		x	x	x			
Matrix methods				x	x	x					x	x	x	
Information theory		x						x	x	x	x			
Computers		x						x	x		x			
Formal logic		x						x			x	x		x
Reliability and quality assurance								x			x		x	
Control theory								x	x		x	x	x	
Simulation								x	x		x		x	
Human factors								x	x		x		x	
Component hardware									x		x			

Source: Johnson, 1997, p. 912.

It is also noteworthy that systems theorists tended to regard themselves as encompassing theories of "human factors," which boiled down primarily to engineering approaches to social interactions, be they ergonomics, computerized experimental psychology, or various computer-inspired theories of organization and management.

As OR withdrew further into academic self-absorption and mathematical isolation, and systems analysis grew increasingly distant from academic economics, the emerging realignment of allegiances would dictate that systems theory would find a foothold in the postwar American renaissance of the business school, as well as in schools of engineering. AI, with its consanguinal relationship to systems theory, would find a home in the most amazingly diverse places, bringing as it did an ample military dowry, from departments of psychology to electrical engineering to schools of "computer science" to business schools; the only place it never really was welcome was in economics.

As with all really interesting generalizations, there were partial exceptions. A gaggle of scholars gathered around the charismatic figures of Kenneth Boulding and Herbert Kelman at the University of Michigan was one such anomaly. Boulding, increasingly dissatisfied with what he perceived as the narrow mind-set of economics and revulsed by the McCarthyite witch-hunts of the period 1949–54, had sought to give voice to his pacifism by instituting an interdisciplinary workshop on the subject of "conflict resolution" (Korman & Klapper, 1978). In Boulding's view, what was required was a theory of "general systems," which in practice was a mélange of now-familiar cyborg concerns such as Wiener's *Cybernetics*, Shannon's information theory, and von Neumann's game theory (Boulding, 1956, p. 153; Richardson, 1991). He managed to attract such diverse figures to his cause as the biologist Ludwig von Bertalanffy; the mathematical-biologist-turned-game-theorist Anatol Rapoport; and some students of Arthur Burks and John Holland, the preeminent scholars devoted to keeping the flame of von Neumann's theory of automata burning bright in the postwar period.[5] The Michigan Center for

[5] Biographical information on John Holland, and to a lesser extent Arthur Burks, can be found in Waldrop, 1992, pp. 151–97. Burks had worked with von Neumann on ENIAC, and was the editor of his *Theory of Self-Reproducing Automata* (1966). In the 1950s, Burks ran the "Communications Science" group at Michigan, where Holland was a student. Because the Burks group also had an interest in game theory, it kept in contact with the Conflict Resolution group, primarily through Anatol Rapoport and, later, Robert Axelrod. Some background on the influence of Boulding on the Michigan group can be found in Rapoport, 1997. The Rapoport connection is discussed further in Hammond, 1997. The Michigan BACH group (Burks, Axelrod, Cohen, Holland) provides the link between the Michigan peacenik's RAND and the automated Prisoner's Dilemma tournament discussed in Chapter 7; it also links up with the rise of one particular school of evolutionary

Conflict Resolution became known as the "peacenik's RAND," providing a counterweight to the perceived hawkish tendencies and military allegiances of the denizens of Santa Monica, all the while conducting their research within essentially the same idiom. For the nonce, one of the most noteworthy aspects of this anomalous antimilitary cyborg unit was that, the more Boulding grew enamored of systems theory, the less credibility he enjoyed as an economist within the American economics profession.

Finding themselves holding some lemons, a colloquy of cyborgs set about making lemonade with the assistance of heavy-duty military hardware. Finding themselves revived and refreshed by their newfound military drinking buddies, the neoclassical economists spurned the proffered lemonade in favor of the tried-and-true beverage, Chateau Walras 1874. The economists may have preached *de gustibus non est dispudandum* in their professional capacity, but brawls follow drinking bouts like mushrooms follow rainstorms. In the 1950s and 1960s, it seemed every time that someone at RAND proposed a novel approach to the vexed problem of rationality, there stood planted a neoclassical economist ready to throw cold water upon it. The cyborgs wanted to fire up their JOHNNIAC to plumb the depths of rational cathexis, whereas the neoclassicals scoffed such spelunking was superfluous, given the innate accessibility of the Utility Machine. The cyborgs sought to proselytize for systems analysis, only to run up against their local economists sneering at its ambitions (Hounshell, 1997a). The cyborgs flirted with some biological analogies, while the neoclassicals insisted that there was nothing there not already encompassed by their doctrine of constrained optimization.[6] Time and again, the neoclassical economists stubbornly reprised their role as wet blankets and stodgy naysayers, proud of their self-proclaimed status as guardians of the Good Old Time Rationality. They were ripe for a rumble, and so, eventually, the cyborgs struck back.

computation. Thus, it provides an interesting alternative genealogy for the Santa Fe Institute, as indicated by Waldrop.

6 For the sake of symmetry, all three instances in the text involve Armen Alchian, although numerous other economists with some connections to RAND could be assayed for their individual acts of resistance. Alchian will make a further appearance when we revisit the Prisoner's Dilemma.

Armen Alchian (1914–): B.A., Fresno State, 1936; Ph.D., Stanford University, 1944; Army Air Force, 1942–46; economist, RAND, 1947–64; professor of economics, UCLA, 1964–; member, Mont Pelerin Society. Alchian's thesis advisor was Allen Wallis. Alchian was exposed to cyborgs early in his career: "I learned about and bought the book [*TGEB*] before I even came to RAND – when it first came out I bought the book. And at RAND the first couple of years, von Neumann came out and gave a seminar and taught a course, and so I took the course. There were about six others – von Neumann, Morgenstern, Ken Arrow, Dresher . . . and here I was!" (interview by Mie Augier, February 28, 2000).

This chapter relates four instances when cyborgs, primarily but not exclusively emanating from RAND, sought to give the economists a taste of their own medicine: (1) reactions to the vicissitudes of game theory in the 1950s; (2) the tussle over Shannon information as an economic valuation principle; (3) a bizarre case of mathematical economics being "defended" against the depredations of RAND; and (4) some nascent indications that computers might not serve analytically to underwrite the neoclassical conception of rational choice. We choose to focus on these particular squabbles, drawn out of a substantially larger field of candidates, primarily because they highlight the problems of rationality, computability, and mathematical tractability that would come to haunt the orthodox economics profession at the end of the twentieth century. Each individual tiff or quarrel probably did not portend all that much in splendid isolation – and that may be one explanation why none of these controversies has yet to receive any attention in the historical literature – but when situated side by side, they reveal a certain pattern of repression and neglect that came to define the tone and tenor of orthodox economics in the United States. This complexion, like so much else, originated in the C^3I doctrines of the Cold War mobilization of scientific activity.

IT'S A WORLD EAT WORLD DOG: GAME THEORY AT RAND

Nietzsche once said that some people do not become thinkers because their memories are too refined. The economics profession has had more than its just share of thinkers in the postwar period. A fair number of those thinkers, especially the game theorists, are so impressed with their own powers of ratiocination that they have become convinced they can derive the full panoply of fundamental precepts of rationality from first principles, ex niholo and de novo. Wave after wave of aspiring code givers break rhythmically upon the shoals of practical wisdom, never once suspecting that they are all propagating out of a single source, RAND in the 1950s. They billow with pride over their acuity in detecting patterns, but seem incapable of seeing themselves as just another surge. They have congratulated themselves that, "the analysis is from a rational, rather than a psychological or sociological viewpoint" (Aumann, 1987, p. 460); but the historical vicissitudes of game theory can *only* be understood from those despised and spurned perspectives. The situation has degenerated in the past few years, with all manner of bogus "histories" being produced to whitewash recent events. For once rationality is irrevocably augmented by strategic considerations: what spirit could possibly move us to trust anything which the Strategically Rational Man says? To the Paradox of the Cretan Liar, we now add the Paradox of the Dog in the Manger.

Game Theory in the Doghouse; or, The Colonel's Dilemma

John Williams, and implicitly John von Neumann, had stocked the Military Evaluation Section at RAND with game theorists in order to give game theory a run for its money, and that is just what happened. Olaf Helmer convened a Conference on the Applications of Game Theory to Tactics at Chicago in March 1949; already by February 1949 Helmer could provide Oskar Morgenstern with twenty-seven separate research memoranda on game theory written at RAND about which he had been previously unaware.[7] Mathematicians most closely associated with this initial burst of activity were Williams and Helmer themselves, Lloyd Shapley, Oliver Gross, Melvin Dresher, Richard Bellman, Abraham Girschick, Merrill Flood, David Blackwell, Norman Dalkey, H. F. Bohnenblust, Robert Belzer, and J. C. C. McKinsey. Once subjected to a modicum of critical scrutiny, there seemed to be two initial reactions to game theory at RAND: first, reservations were voiced about von Neumann's own conception of the solution of a game; and, second, deep misgivings surfaced about whether the notion of "strategy" professed by game theory had any salience for military strategy or tactics. The two classes of objections were frequently not treated as separate, although they would become more distinguished by 1950, with the appearance of the rival Nash solution concept.

The objections to von Neumann's approach to solution concepts tended to play out along lines already broached in Chapter 3. Objections had already surfaced at the 1949 conference: "there is the disturbing likelihood that the military worth of an outcome to one opponent will be appraised differently by the two opponents, in which case we have a non-zero-sum game to deal with, which cannot be resolved by the von Neumann theory" (Helmer, 1949, p. 4). A convenient summary of the view from the vantage point of the RAND skeptics around 1950 was provided in the paper by J. C. C. McKinsey (1952a). He astutely began with the issue of information processing, accusing von Neumann of having prematurely made the leap from extensive to "normal" form games, thus suppressing most relevant aspects of cognitive interaction in a theory aimed at illumination of social structures. In such a world, "the players might as well pick their strategies

[7] See Olaf Helmer to Morgenstern, February 18, 1949, box 14, OMPD. Subjects included tank duels, silent duels, games with continuous payoffs, computation of strategies in games with convex payoffs, and strategic equivalence of games. It was not unusual for those interested in game theory in the 1950s to be uninformed about the extent of RAND's research in game theory, because the bulk of these early research memos were classified. Today, long after they might bear any military relevance, many remain classified, although RAND has very recently undertaken to open its archives to some historians. The description of the 1949 meeting can be found in Helmer, 1949.

[beforehand], and then leave to a computing machine the task of calculating the outcome" (p. 595). Yet, even in this most favorable case of an environment of mechanized play of a two-person, zero-sum game, finding minimax solutions rapidly becomes intractable when the number of possible strategies is only moderately large. He credited Olaf Helmer with raising the possibility that a player might not know which move gave rise to a subsequent opponent's move in a sequence of play, especially in situations of real conflict, and that these "pseudo-games" would be impervious to von Neumann's type of analysis. As for von Neumann's "characteristic functions" for games of three or more players, there was too wide a range of indeterminacy for the theory to be of any use, primarily due to the fact the theory tells us nothing about how agreements that constitute the coalitions are arrived at. Moreover, characteristic functions rule out the complications of the interpretation of play activity, a consideration that should only loom larger in importance in instances of multilateral interactions.

McKinsey was not alone in this assessment; he was merely the most precise and open about the nature of his objections. Another major critic was Richard Bellman.[8] "Thus it seems that, despite the great ingenuity that has been shown in the various attacks on the problem of general games, we have not yet arrived at a satisfactory definition of the solution of such games. It is rather likely, as has been suggested by Bellman, that it will be found necessary to distinguish many kinds of games, and define "solution" differently for different types; the theory of Nash of course represents a step in this direction" (McKinsey, 1952a, p. 610). Other RANDites also chimed in. "Nothing is said [in von Neumann & Morgenstern] about how coalitions are formed. . . . Whatever the process of their formation, it seems to admit of no realistic interpretation in typical economic situations, since for large n communication difficulties would prevent consideration of some (possibly optimal) coalitions" (Thrall et al., 1954, p. 17). Merrill Flood wrote: "The nonconstant sum case of the theory of games remains unsolved in the sense of von Neumann and Morgenstern. In recent years, some theoretical contributions have been made. . . . This effort has not yet been successful" (1958a, p. 5). The more that some mathematicians at

8 "Thus we have no unique way of determining optimal play for these realistic processes, even if we can decide on an appropriate criterion function. N-person games of any type, $N > 3$, are virtually impossible to cope with analytically" (Bellman et al., 1957, p. 480). It is an interesting sidelight that the two most vocal critics of game theory at RAND in the early 1950s were both subject to protracted loyalty investigations, McKinsey due to his homosexuality and Bellman due to his supposed links to Communists. McKinsey was forced to resign from RAND (Jardini, 1996, p. 96); Bellman survived his vetting but abandoned game theory. In the local ecology, this may have initially weakened the anti–game theory contingent.

RAND took a hard look at game theory, especially those with no history of being particularly beholden to von Neumann for their careers, the less they liked what they saw. It was not so much that they objected to the ambitions or scope of game theory: it was rather that the solution concepts seemed dubious.

Of course, by the early 1950s von Neumann had his own doubts about the universal applicability of the solution concept. As another early participant remembers:

> From the start of my interest in game theory, I found that the amount of information required for defining a game was unreasonable. Nevertheless, with cooperative games, one could safely abstract out information.... von Neumann and Morgenstern had made it clear, both in their book and in conversation, that they felt an attempt to develop a satisfactory dynamics was premature. (Shubik, 1997, p. 101)

Whereas a subset of the RAND mathematicians demurred on formal grounds, the receding prospects of practical application of game theory to military matters unsettled a larger contingent at RAND. On the face of it, this development was most unexpected. As Rufus Issacs (1975, p. 1), an innovator of the theory of differential games, put it: "Under the auspices of the US Air Force, RAND was concerned largely with military problems and, to us there, this syllogism seemed incontrovertible: Game theory is the analysis of conflict. Conflict analysis is the means of warfare planning. ... Therefore game theory is the means of warfare planning." He continued: "despite some excellent output, the fruits proved meager. The initial ardor slowly abated." It did indeed seem odd that a formalism apparently tailor-made for the military would fail so utterly to capture so many of its concerns. The reservations surfaced almost immediately at the 1949 Chicago conference. There Olaf Helmer strove to spell out the kind of scenario in which game-theoretic reasoning could be imposed upon a battlefield encounter.

> Suppose an objective defended by Red forces is to be taken by Blue, and the terrain conditions are such as to offer several methods of attack as well as of defense. Then the steps through which the Blue commander will have to go (though not necessarily explicitly) are the following: (i) he will have to use his imagination in order to obtain a complete list of tactics at the disposal of each side (these constitute 'pure strategies' in the sense of game theory); (ii) for each of the possible pairs of tactics that can be chosen by himself and the Red commander respectively, he must estimate the probable outcome (in terms of achievement of the objective, men and ammunition lost, etc.); (iii) for each possible outcome he must make an estimate of the relative military worth (this will establish the coefficients of the payoff matrix); (iv) he will have to make at least a

> partial estimate of the probabilities which his opponent will choose, or refrain from choosing, certain tactics at his disposal; this may be accomplished by estimating how the Red payoff matrix would appear to the Red commander and how, consequently, he would be inclined to behave; (v) he will have to select the best tactic. (Helmer, 1949, p. 6)

That is, unless Red commander overran the immobilized Blue forces while Blue commander was preoccupied somewhere between steps (iii) and (iv).

It didn't take long to come to the realization that anything less than theater-level combat commanders would never engage in something even remotely resembling this scenario, which unavoidably raised the nagging question of where or how the formalism would ever be brought to bear in the military. Certainly, the need for both sufficient time and a synoptic viewpoint would push the locus of calculation higher and higher up the chain of command, away from tactics and toward strategy; but this shift in itself was not necessarily objectionable, because it resonated with all the tendencies toward centralization of command and communication we have already mentioned. Possibly a further mechanization of the decision-making process was being called for; and that, conveniently, was also a burgeoning specialty at RAND. If this elaborate sequence of "estimates" was ever to be carried out, it was more likely to occur in a situation of nuclear brinkmanship rather than, say, a dogfight or a tank battle; this, yet again, had become a trademark preoccupation at RAND. Nevertheless, even in this best of all possible worlds, few analysts at RAND were persuaded that game theory was at all ideally suited to military applications; and, for their own part, the military officers were distinctly reserved in their enthusiasm.

At RAND, for all its public reputation as the House that Games Built, most of the systems analysts charged with interacting with the military were not at all reticent about making public evaluations concerning the irrelevance of game theory. Charles Hitch, director of the economics section, said in 1960, "Game theory is helpful conceptually, but so far as its theorems are concerned, it is still a long way from direct application to any complex problem of policy." In *Harper's* in the same year, he ventured that, "For our purposes, game theory has been quite disappointing" (in Poundstone, 1992, p. 168). Alex Mood, temporarily responsible for the game theory section, wrote that the theory of games "has developed a considerable body of clarifying ideas and a technique which can analyze quite simple economic and tactical problems. . . . these techniques are not even remotely capable, however, of dealing with complex military problems." Edwin Quade opined that, "Only an occasional problem associated with systems analysis or operations research is simple enough to solve by actually computing the game theory solution – and some of those are only marginally related to the real world."[9] When a certified RAND booster of game theory such as Melvin Dresher sought to codify the lessons of strategic doctrine deduced from game theory after two decades of effort, the best he could muster were inane platitudes such as "no soft spot in defenses," "weaker side bluffs," and "defender postpones, attacker acts early." Even Dresher, when challenged, had to admit that, although game theory could be found in some military textbooks, it had never been used in the field.[10]

Helmer, the RAND scholar most in charge of the initial game theory effort, went public with his disenchantment in the 1960s. In a 1963 article, he wrote: "both game theory and organization theory are in real trouble today. . . . the theory of games in my opinion has reached a state of near-stagnation with regard to its applicability to the real world" (pp. 245–46). Interestingly, Helmer felt impelled to insist this was equally the case in military and civilian contexts; furthermore, he explicitly pointed to the Nash solution concept as being unavailing in this regard. "[Nash] equilibrium point theory often suggests what appears to be an unreasonable solution, in the sense there may be a different outcome which is preferred by all players, and that this preferred solution can often be achieved by a rather minute and obvious amount of cooperation" (p. 249). The rather chastened hope he still held out for the project was recourse to greater psychological content and resort to "expert judgment."

The reactions of active military officers were more intriguing. Many who had come equipped with mathematical training would insist upon the myriad ways in which the formalism would be conducive to misappropriation and misuse: "game theory cannot correct a deficiency of intelligence" (in Mensch, 1966, p. 368). They were capable of making the usual criticisms from within the game-theoretic mind-set, such as ignorance of payoff matrices, the dubiousness of treating war as a zero-sum situation, and the essential intractability of solving any realistically large game; but some went further, questioning the very premise that a mere algorithm could serve to dispel the "fog of war" in any really important battle. They were cautious about the use of computers in battle, more because of their potential to separate the commander from tacit knowledge gained in combat than due to any residual Luddite tendencies. There was also a distinct bias toward self-fulfilling prophecies amongst

[9] All otherwise unattributed quotations are taken from Shubik, 1964, pp. 218–19.

[10] The first quotation is from Mensch, 1966, pp. 360–61; the paraphrase is from p. 385.

Melvin Dresher: born Krasenstadt Poland; B.S., Lehigh, 1933; Ph.D. in mathematics, Yale, 1937; instructor of mathematics, Michigan State College, 1938–41; statistician, War Production Board, 1941–44; mathematical physicist, NDRC, 1944–46; professor of mathematics, Catholic University, 1946–47; RAND 1948–.

game theorists that disturbed some of the military: "At a meeting of the AMS in the fall of 1948, Ed Paxson of RAND threw out a challenge to solve such games [as one-round duels]. Many accepted, and soon discovered the wisdom of a recipe formulated by Girschick and his colleagues at RAND: First guess the solution of a game, and then prove it" (Thomas in Mensch, 1966, p. 254). The conundrum, often encountered in applied mathematics, was how to reconcile the mathematicians' respect for formal rigor with their cavalier attitudes toward semantics and relevance. "The difficulty of making such important notions as deterrence precise, and subject to rational analysis, is a severe limitation to game theory in applications" (p. 260).

The military was forced to confront a much thornier problem in the evaluation of game theory than were the putatively cool and detached defense analysts at RAND. The enigma, almost a paradox, was, What should the military do, if it really believed game theory a serious contender as a rational theory of conflict? Certainly it would seem obvious to promote its development, but if its strictures concerning strategic behavior and rational response really rang true, then the military should find game theory applicable to the issue of the encouragement and deployment of game theory itself. To put the issue crudely: should game-theoretic advances and assessments be published or not? After all, the very term "think tank" was derived from World War II jargon which designated a secure room within which military strategies could be safely discussed (James Smith, 1991). The initial knee-jerk response was to treat game theory just like any other cutting-edge strategic weapon, as say, the hydrogen bomb: keep the maximum of details secret for as long as possible. RAND had already begun the practice of classifying much of the work of its mathematics section when it separated from Douglas Aircraft. But this was where a "true" game theory would turn out to be a qualitatively different appanage from nuclear weapons, because it was a self-reflexive doctrine.

The lessons of game theory up to that juncture tended to assume the format of injunctions that weaknesses of opponents should be exploited, and that one should obscure one's own rationality through bluff, evasion, deception, and randomization in any strategic situation, and that value was derived from exclusionary coalitions. Thus, simple restriction of access to the mysteries of game theory would never suffice: one would have to somehow create a closed world where game theory could be both "true" and "false" simultaneously, such that one could randomize one's moves in the larger game of the Cold War. The prognosis seemed to be that one should broadcast disinformation, or disseminate signals that one simultaneously both believed and disbelieved in the efficacy of the mathematics to dictate strategic behavior, such that the military would be seen to be both supporting and disparaging its development. The reflexivity could be taken to ever more refined levels, with the published literature assuming one position, while signaling to the cognoscenti that the reverse position might be found in the classified literature. There existed the distinct possibility that any attempt to keep game theory secret would inadvertently demonstrate its essential inapplicability to war or superpower conflict. One of the most frequently classified types of military information was precisely the range and character of strategies available to Blue Team in a given situation of potential conflict. But if the opponent was effectively kept in the dark concerning the contents of the relevant strategy set, and Red Team therefore had no solid idea about what game it was engaged in playing, then game theory as a formal structure was intrinsically irrelevant to the rivalry. Hence, what you chose to publish was as much a strategic move as the weapons you deployed.

Game theory did have a profound effect on RAND, but it could not be summed up as any simple catechism concerning its mathematical legitimacy. Rather, it fostered a particularly labyrinthine mind-set concerning the relationship of finely honed rationality to deception and illusion. Early on, RANDites realized that they were situated within a nexus of layers of deception – the United States had to deceive the Russians, to be sure, but the U.S. government would also have to deceive its own citizens; and RAND would have to sometimes deceive the populace and, more deviously, mislead its own patrons in the military. This issue explicitly was aired when discussions turned to how and what to publish in the "open" scientific literature: "Actions which our government may be forced to take in view of the world situation . . . may involve the necessity of some deception by us of our own population. This is of course a very touchy subject, but intuitively it seems a very important one and the incentive aspects of how to go about this are very fascinating."[11]

The house of mirrors would threaten to ensnare outside readers of the literature in the uberous double bind: if they believed in game theory (and who else would devote much time to this arcane literature?), then the realization that the military was cognizant of the strategic problems of reflexivity therein would render the published literature essentially untrustworthy; it was little better than white noise. But if one decided instead that game theory really did *not* manage to capture the true essence of strategic rationality – that there was something fundamentally suspect about the published literature – then an attempt to correct it through rational argumentation in the open literature would only painfully reveal

[11] Memo, L. J. Henderson to Franklin Collbohm, November 2, 1949: "Discussion concerning the McGraw Hill series and RAND Social Sciences," folder: History – Henderson, 1945, RAND Classified Library, quoted in Jardini, 1998.

that the critic naively misunderstood the role of game theory in a strategic environment. It was this paradoxical aspect, as much as any humanistic aversion to "thinking the unthinkable," that repulsed numerous mathematicians from game theory in the 1950s, with the exception of a few strategically placed individuals with high-level military clearances and an acquired taste for deception.

A telling illustration of this paradox appears in the work of Colonel Oliver Haywood, which, incidentally, is actually misleadingly cited in the most widely consulted guide to game theory in the later 1950s, the book by Luce and Raiffa (1957, p. 64); further, Haywood's paper is elsewhere called a "classic" (Coleman, 1982, p. 48). Haywood wrote a thesis for the Air War College in 1950 on game theory, which was incorporated into a RAND restricted research memorandum in 1951. In the memo, he provided a very jaundiced account of the potential applicability of game theory for military situations. He surveyed the problems that a zero-sum restriction would impose on the understanding of military outcomes and pointed out that a military commander could never appeal to randomization as a rationale for his behavior in combat. He noted that the aura of technocratic rationality surrounding game theory would prove a chimera: "There have been hopes that a mathematical theory employed in conjunction with a qualitative concept of worth will tend to eliminate bias and emotional influences from our decisions. These hopes appear largely unfounded" (Haywood, 1951, p. 67). Some appreciation of dynamics could not be brushed aside as a mere complication, because combatants tended to change the rules of engagement when they faced the prospect of utter defeat. Curiously enough, in this document Haywood tried to make the case that all these drawbacks should dictate that "the theory would appear to be of greater value for use by smaller military units than for making major decisions of national strategy" (p. 4), the diametrical opposite of the inference drawn above that only the pinnacle of military command would ever have the incentives and the ability to claim to make use of game theory.

The situation was more or less reversed when Haywood sought to publish an evaluation of game theory in the open literature. By 1954, no longer a colonel, but now a "civilian" manager at Sylvania's missile division, Haywood chose to analyze two battles from World War II and assert that the choices of the commanders could be explained in a post hoc fashion by game-theoretic considerations or, as he put it, "to demonstrate to the skeptical reader that actual and well-known battle decisions should be discussed in terms of game theory" (1954, p. 382). This version contains no hint of reservations about the implications of mixed strategies or the minimax solution, no acknowledgment of the divergence between opponents of knowledge of available strategies, and no warnings concerning the misleading pretensions of a rationalist picture of an emotionless conflict. Most notably, and in contrast with the previous RAND memo, there is no citation of the numerous classified RAND documents on game theory.

Today, reading these documents side by side, it is not so much that we might harbor the suspicion that there existed a superposition of Haywoods, a colonel and his civilian clone occupying alternative realities where every conviction placidly cohabited with its opposite in the MirrorWorld; but rather there were (at least) *two audiences for game theory*, and they were deemed to be worthy of being told different things. Therefore, when game theorists such as Luce and Raiffa assumed an air of po-faced innocence with statements like, "Presumably the recent war was an important contributing factor to the later rapid development [of game theory]" (1957, p. 3) – this in a text explicitly paid for by the ONR – a question arises as to the extent to which rationality was being parodied in the name of being explicated.[12]

The predicament of the postwar game theorist was further complicated by the reaction of the British OR contingent. Recall from Chapter 4 that they had never cottoned to the American style of operations research stemming from von Neumann's innovations; and now their fears had been exacerbated by what they perceived to be an unwise American escalation of nuclear weapons development, propelled largely (as they saw it) by strategic doctrines inspired by game theory. Whereas the older string-and-sealing-wax tradition of OR had helped win the war, the British progenitors now felt that something had gone very wrong with its American offshoot, and they focused upon game theory as the primary symptom of the disease. P. M. S. Blackett went public with the statement that "the theory of games has been almost wholly detrimental" (1961, p. 16); Robin Neild accused the Americans of imagining a world populated exclusively with hostile amoral gambling robots. Because these people were nominally allies in the Cold War, it became imperative for the Americans to open up a "third channel" to the European military establishment in order to bring it into line with the public relations face of the novel strategic approaches, without of course letting on the extent to which the original inventors of "operations research" were now being superceded

[12] It is only when one becomes aware of the tortured reasoning concerning what would qualify as legitimate military use of game theory that statements such as the following become comprehensible, although no less Jesuitical: "The basic developments in the theory of games were not motivated by defense considerations. The later defense considerations were influenced by the development of game theory. The military funding for further developments at RAND and elsewhere came after von Neumann's work on game theory. The work at RAND on duels, deterrence, and operational gaming were only part of the broad development of game theory" (Shubik, 1998, p. 43).

by experts in an improved American technics of rationality. One way this was facilitated was through the instrumentality of NATO, which through its Scientific Affairs Committee introduced game theory into the OR directorates of individual national military services in Europe, as well as recruiting various European academics into the fold. The problem of this European acquaintance with and conversion to American OR is explored in the next chapter.

RAND eventually dealt with the paradoxes and internal problems of game theory through an internal decision to redirect research efforts into the broad areas of systems analysis, wargaming, and computer engineering; a more immediate response was to adopt an external stance that distanced RAND from game theory, without going so far as to openly repudiate it. The trademark refrain from the mid-1950s onward was that the mathematics of game theory was a salutary bracing regimen (perhaps like daily piano practice, or Grape Nuts in the morning), and a wonderful inspirational Muse, but in their day-to-day affairs, smart defense analysts didn't actually make use of it. This doctrine was enshrined in RAND's own promotional publications – "In studies of policy analysis, it is not the theorems that are useful but rather the spirit of game theory and the way it focuses attention on conflict with a live dynamic, intelligent and reacting opponent" (RAND Corporation, 1963, pp. 26–27) – but it also appeared frequently in the comments of RAND employees. Edward Quade, for one, mused:

> Game theory turned out to be one of these things that helped you think about the problem and gives you maybe a way of talking about the problem, but you can't take one of these systems analysis problems and set it up in a game and solve the game. Only in the very simplest cases would you actually be able to get a game theory solution. It gives you a background of how to think about a problem. I find discussions like the discussion of the Prisoner's Dilemma – it's very helpful in illuminating your thought but you don't solve the problem that way, you know.[13]

James Digby suggested that game theory was "often more of a help as an attitude than in direct applications" (1989, p. 3). Albert Wohlstetter wrote a spirited defense of game theory in 1963 called "Sin and Games in America," where he excoriated various detractors, only to then admit, "I would say that game theory has been useful in some conceptual studies, of trivial use in the empirical analysis of tactics and, as a theory, hardly used at all in the complex empirical work on strategic alternatives" (in Shubik, 1964, p. 218).

[13] Edward Quade interview, pp. 33–34, NSMR.

The most successful popularizer of the RAND doctrine that game theory was chock full of perceptive insights about military and social organization, enjoying a vaticinal virtue completely divorced from its actual mathematical instantiations, was Thomas Schelling, whose book *The Strategy of Conflict* (1960) did more for the intellectual legitimacy of game theory than the entire stable of RAND mathematicians combined. Schelling worked the miracle of turning game theory into a sort of Kantian a priori: "the greatest contribution of game theory to social science in the end is simply the pay-off matrix itself that, like a truth table of a conservation principle or an equation, provides a way of organizing and structuring problems so that they can be analyzed fruitfully and communicated unambiguously" (in Mensch, 1966, p. 480). Schelling's book, which enjoyed a revival in the 1990s and plays a role in subsequent developments in economics, deserves our attention. For the moment, however, it should suffice to proffer the following broad generalizations about the career of game theory at RAND.

Von Neumann's theory of games had provided a focus and a rationale for the gathering of logicians and mathematicians in the early RAND Corporation, promising a rigorous basis for a new science of war. This promise was rudely dashed by some of the mathematicians and ridiculed by some of the military patrons. A few zealots like Herman Kahn (1962) who took the promissory note a bit too literally became an embarrassment at RAND, and were consequently ostracized. Nevertheless, there was a more Machiavellian reading of the situation, one in which the proscriptions of game theory applied to game theory would dictate a profoundly ambivalent public treatment of the intellectual content of game theory; in the panoptic House of Mirrors, only the most privileged of the inner circle would ever have access to the "real" intentions and bona fide evaluations (and classified models) of those charged with the development and application of game theory.[14]

A few mathematicians remained comfortably ensconced at ground zero in this closed MirrorWorld at RAND; for most of the other analysts at RAND, however, "games" had careened off into computer simulations, wargaming, construction of elaborate futuristic scenarios, systems

[14] "For years, it had been a tradition in game theory to publish only a fraction of what one had found, and then only after great delays, and not always what is most important. Many results were passed on by word of mouth, or remained hidden in ill-circulated research memoranda" (Aumann, 1987, p. 476). Far from sharing Aumann's nostalgia for those bygone days when a doughty band of stalwart men reasoned their way through intricate mathematical problems in splendid isolation, one is prompted to ask: How was it that game theorists could persist in their project in such unprepossessing circumstances? Who would support such indulgence? What rendered this a cumulative research program? The unspoken answer: the military.

engineering, laboratory experimentation, artificial intelligence, software engineering, and the like. The official line was broadcast that game theory had been a marvelous source of inspiration, but should never be taken too seriously as a mathematical description of rationality. The irony that the premier attempt at a rigorous mathematical codification of rationality, the very source of the cool, unemotional technocratic reputation at RAND, was now being downgraded to the hermeneutic status of an insightful parable, a suggestive metaphor, or convenient cognitive frame, seemed to have been lost on all those who espoused it at RAND. All the more telling, this sequence of events happened with the full awareness of all participants of what has come in retrospect to be regarded the signal mathematical development in game theory in the 1950s, the event most consequential for the subsequent history of *economics*, the invention of the "Nash equilibrium" concept.

Twilight of the Dogs: Nash Equilibrium

Some of the most bizarre and outlandish statements about the history of economics have recently been made about the role and accomplishments of the Nash equilibrium concept in game theory. Much of this cynosure is due to the fact that a broad regiment of economic theorists have recently sought to write von Neumann out of the history of game theory to the maximum extent possible, striving to supplant his legacy with the work of John Forbes Nash. Gerard Debreu, for instance, pronounced: "the six years that followed the first edition of the work of von Neumann and Morgenstern disappointed the expectations of the economists. Most of the research done on the theory of games at that time focused upon the MinMax theorem, which eventually moved off center stage, and now plays a minor supporting role. Only after 1950, and in particular after the publication of John Nash's one page article . . . did the theory start again to acquire the dominant features that characterize it today."[15] One would have thought rather that it was the economists who had disappointed von Neumann. And was all that activity at RAND – activity that Debreu had personally witnessed – just an insignificant detour?

[15] Debreu, 1983a, p. 6. See also the interview by Feiwel (1987a, p. 252): "The theory of games has not yielded exactly the results that von Neumann and Morgenstern expected. The main concept of the theory, the concept of a solution, which in modern terminology is a stable set, has not turned out to be fruitful. The most fruitful solution concept in game theory, the Nash equilibrium and the core, are not stressed in the book." It would have been difficult for von Neumann to do so, because they were both invented in reaction to his program. Revealingly, the other progenitor of the Arrow-Debreu model adopts essentially the same position: "Nash suddenly provided a framework to ask the right questions" (Kenneth Arrow in Nasar, 1998, p. 118).

Kenneth Binmore presents the Nash equilibrium as the "natural" generalization of von Neumann's minimax, and then asks, "Why did von Neumann and Morgenstern not formulate this extension themselves?" His answer was that they presciently foresaw the problem of the multiplicity of Nash solutions, and "therefore said nothing at all rather than saying something they perceived as unsatisfactory" (in Nash, 1996, p. xi). If Binmore had attended more closely to the text of *Theory of Games and Economic Behavior* (*TGEB*), he would have discovered that Von Neumann and Morgenstern positively embraced the prospect of a multiplicity of solution points. In Robert Aumann's historical exegesis of game theory, *TGEB* is lauded as an "outstanding event," but then treated as a mere transition between the earlier work of von Neumann (1928) and the subsequent "Nash program" (1951) of subsuming cooperative games within noncooperative models. It is praised in a mildly backhanded manner as breaking "fundamental new ground" in defining a cooperative game, the axiomatization of expected utility, and making "the first extensive applications of game theory, many to economics" (1987, p. 463). All of these claims require a fair supplement of interpretative license in order to accord them legitimacy, something Aumann apparently felt no compunction to supply. We have already noted in Chapter 3 that *TGEB* bore negligible economic content and was not deeply concerned with shoring up utility theory, for instance; the idea that it set out to define the subset of "cooperative games" is a post hoc construct. And then there is the crusade by numerous economists to fabricate a long and hallowed genealogy for the Nash solution concept entirely bypassing von Neumann, for, "as everybody knows, Cournot formulated the idea of a Nash equilibrium in 1830" (Binmore in Nash, 1996, p. xii). This bit of wishful thinking has been refuted by Robert Leonard (1994a).

At the end of the twentieth century, this quest to doctor the record with regard to Nash and his equilibrium has attained the status of a public relations campaign with the publication of a biography of Nash (Nasar, 1998) that became a bestseller (soon to be a major motion picture directed by Ron Howard and starring Russell Crowe! *Happy Days* meets *Gladiator*! The possibilities for revisionism are mind-boggling . . .) and a commissioned survey in the *Journal of Economic Literature* (Myerson, 1999). Although this is not a treatise on the contemporary sociology of the economics profession, a small caveat needs to be inserted here about this unusual effusion of infotainment concerning what must seem, to even the most avid follower of economics, a topic of rather arid compass appealing to only the most limited circle of cognoscenti. Incongruity turns into downright incredulity when one learns that John Nash refused to cooperate in any way with his biographer; and that the commissioned *JEL* survey was never vetted by any historian familiar with the relevant events, and consequently the only "history" to grace the text appears as a word in its title. The reader should approach these texts forewarned that John Nash and his equilibrium, through no effort of his own, have become the object of a vast ceremonial purification exercise.

Nasar, a journalist, admits that she was tipped off that there would be a juicy story here before Nash became newsworthy; only later was he awarded the Nobel Prize in Economics in 1994 jointly with John Harsanyi and Reinhard Selten. She was recruited as the public face of an extensive campaign of spin doctoring engineered behind the scenes by a few key figures, primarily because awarding the prize to Nash itself threatened to pry open a whole pallet of Pandora's boxes: problems ranging from the potential embarrassment of an award of a Nobel for the formalization of rationality to a mentally ill individual, to problems of justifying the Nash equilibrium as the central solution concept of choice of the fin-de-siècle economic orthodoxy in the era of its conceptual disarray, to a threatened uncovering of the extensive military involvement in modern orthodox microeconomics, to reopening the suppurating issue of the very legitimacy of an economics Nobel situated on a par with the Nobels for the natural sciences. Nasar does mention each of these issues but, like most modern journalists, sought to deal with the seamier side of science by repeating the technical opinions she was told by her behind-the-scenes informants (citing "confidential sources"), and then tarting up the story with lots of titillating sexual details, shameless appeals to sentimentality, and irresponsible hyperbole: "Nash's insight into the dynamics of human rivalry – his theory of rational conflict and cooperation – was to become one of the most influential ideas of the twentieth century, transforming the young science of economics the way that Mendel's ideas of genetic transmission, Darwin's model of natural selection, and Newton's celestial mechanics reshaped biology and physics in their day" (1998, p. 13). Writing serious history of economics is not easy when boilerplate hyperbole crowds out analysis.

For all its drawbacks, however, Nasar's work is redeemed by her reporting the best short description of the Nash equilibrium concept that I have ever read: according to Martin Shubik, someone who was very familiar with Nash during his Princeton days, "you can only understand the Nash equilibrium if you have met Nash. It's a game and its played alone" (in Nasar, 1998, p. 366). The relationship to his "beautiful mind" is a conundrum best left for the reader to reconcile.

No, there is something eerily hollow about the histories of the Nash solution concept on offer in the last half of the twentieth century, something that is beginning to be sensed by a few intrepid authors (e.g., Rizvi, 1994, 1997; Kirman, 1999; Hargreaves-Heap & Varoufakis, 1995, pp. 106–9). If Nash equilibrium really stands as the monumental philosopher's stone of game theory that some economics microtexts (Kreps, 1990; Binmore, 1992a) and recent infotainments assert it to be, then why was it so rapidly contemplated, and then roundly *rejected* at RAND in the 1950s? The mathematicians whom we cited in the previous section, men such as McKinsey, Bellman, Helmer, and Flood, all explicitly entertained the Nash equilibrium in the papers referenced, and did not regard it as appreciably changing their diagnosis of the crippling flaws of game theory. Melvin Dresher, as late as 1961, managed to write a RAND-endorsed textbook of game theory without once mentioning the Nash equilibrium concept. Luce and Raiffa (1957) did elevate the significance of Nash equilibria by structuring their textbook chapters as minimax, Nash equilibria, von Neumann stable sets, other refinements – only to conclude on the rather downbeat note that "a unified theory for all non-cooperative games does not seem possible" (p. 104) and "It is doubtful that a mathematician could be found today holding any hope for a moderately simple and general characterization of solutions" (p. 210). But, most significantly, *John von Neumann himself explicitly rejected the Nash equilibrium as a "reasonable" solution concept.*

This deeply inconvenient historical fact, pitting the mature polymath von Neumann against the perfervid callow John Nash, accounts for more of the purple prose and sophistical gloze which infect this literature than any other single consideration. For the longest time, this fact was common knowledge amongst the original game theory cognoscenti but hardly ever broached in print, and then dealt with only by euphemism and ambages.[16] Of course, this would have come as no surprise to anyone who might have talked at length with von Neumann about economics, or plumbed the published record, as we have tried to document in Chapter 3. But the relationship between the Walrasian orthodoxy and the Nash concept was so intimate and yet so conflicted that no one appeared interested in dredging up bad blood. But the revival of the Nash equilibrium in the 1980s and then the engineered Nobel in 1994 provoked a few witnesses to concede the existence of the problem.

For instance, Martin Shubik went on record stating that von Neumann "felt that it was premature to consider solutions which picked out a single point. . . . he particularly did not like the Nash solution and [felt] that a cooperative theory made more social sense" (1992, p. 155). Harold Kuhn,

[16] One example may be found in Aumann, 1987, p. 475, where he acknowledges that Oskar Morgenstern was never convinced of the validity of what he awkwardly dubs "the equivalence principle," by which he means the interminable exercises of the 1950s and 1960s to demonstrate how one contrived equilibrium concept after another would somehow "converge" to or subsume Walrasian general equilibrium. In this instance, as in others, Morgenstern was echoing the attitudes of von Neumann.

one of Nash equilibrium's main boosters, admitted in a Nobel seminar that, "Von Neumann and Morgenstern criticized it publicly on several occasions. In the case of the extensive form, von Neumann claimed it was impossible to give a practical geometric extensive form [for the concept]" (1995, p. 3). But the real revelation came in Nasar's chapter 11, where it was documented that Nash's concept was explicitly offered as a rival to von Neumann's, and that, as usual, von Neumann had gotten there first: "That's trivial, you know. That's just a fixed point theorem" (1998, p. 94).[17]

Von Neumann was not an active participant in the elaboration of game theory in the 1950s, as we have repeatedly stressed, but he was fully aware of Nash's work, and he did have some impact on attitudes at Princeton, the other major locus of game theory research in that era. The bald fact is that almost no one at the white-hot mathematical centers regarded the Nash equilibrium as the profound contribution which the revisionists would like to assert. Nasar admits this in a backhanded way: "Nobody at RAND solved any big new problems in the theory of noncooperative games" (1998, p. 120); Nash's "thesis had been greeted with a mixture of indifference and derision by pure mathematicians" (p. 124); "Nash's mentors at Carnegie and Princeton were vaguely disappointed in him" (p. 128). With some exceptions, Nash equilibria and the Nash program in game theory in fact garnered little further attention until the 1980s, when it took off as the supercharged engine of the revival of the game theory project within economics, as well as the instrumentality of a renewed attempt to mathematize the theory of evolution (Maynard Smith, 1982). It would seem, at minimum, getting the history of Nash equilibrium straight would be a necessary prerequisite for understanding the role and significance of game theory in fin-de-siècle economics.

Dogs and Monsters: The Nash Story

The history begins with John Forbes Nash himself.[18] Nash was a mathematics major at Carnegie Mellon when he took a single elective

[17] One should exercise caution with regard to the exact phrasing of von Neumann's response, because Nasar's footnote attributes this account third hand to Harold Kuhn; Kuhn, along with Tucker, was one of the prime movers behind the demotion of the von Neumann program in the history of game theory, and Nasar's prime informant. What is more significant is that Nash clearly got the message that von Neumann had considered his approach to equilibrium and rejected it. The hostility of the great man clearly troubled Nash long after the event; for instance, during his illness, Nash was heard to call himself "Johann von Nassau" (Nasar, 1998, p. 286).

[18] John Forbes Nash (1928–): M.A. in mathematics, Carnegie Mellon, 1948; Ph.D. in mathematics, Princeton, 1950; RAND consultant, 1950–54; professor of mathematics, MIT, 1951–59. The biographical data available are recounted in Nasar, 1994, 1998; Nash, 1994. His larger mathematical work is summarized in Kuhn, 1995.

course in international economics, by all accounts the only formal economics training he ever had. A term paper he wrote for the course provided the basis for his first published paper, "The Bargaining Problem" (Nash, 1950a). The significance of this paper, which claims to provide the single unique bargaining solution to a situation that takes the neoclassical characterization of preferences and endowments for granted, rests not so much in the seeming parsimony of its assumptions (Luce & Raiffa, 1957, pp. 124–34; Hargreaves-Heap & Varoufakis, 1995, pp. 118–28), but rather in its demonstration that Nash was reviving a cognitive variant of the neoclassical program that had, initially, no bearing upon or relationship to von Neumann's theory of games.[19] The most noteworthy aspect of the paper, beyond its jejune examples of Jack giving Bill a pen or toy or knife, is that the entire analytical superstructure is based on restrictions upon preference functions and neoclassical welfare definitions and had nothing whatsoever to do with strategic behavior. "A solution here means a determination of the amount of satisfaction each individual should expect to get from the situation" (Nash, 1996, p. 1).

There is no doubt that at this juncture Nash regarded himself as codifying a general principle of rationality from within the neoclassical tradition, as he then understood it; it bore no game-theoretic content or implications. His biographer Nasar admits that his abiding faith in the virtues of rationality was "extreme" (1998, p. 12), eliding the sad implications. A number of aspects of this version of "rationality" would have significant consequences down the line: for instance, it meant that "each [individual] can accurately compare his desires for various things, that they are equal in bargaining skill, and that each has full knowledge of the tastes and preferences of the other" (p. 1). Furthermore, there was the guileless suggestion that "rationality" would entail automatic agreement without any need for intersubjective interaction and that such a state would be unassailably acquiesced in by both parties as "fair": "one anticipation is especially distinguished; this is the anticipation of no cooperation between the borrowers" (p. 3).

When Nash arrived at Princeton to do graduate work in 1948, he was struck by the fact that his notions of rationality were divergent from those of von Neumann and others hard at work on the newfangled game theory. This sequence of events became obscured by the happenstance that Nash

[19] This is acknowledged indirectly in Nash, 1994; Kuhn, 1995, p. ii. That the bargaining paper shaped the later interest in games, and not vice versa, is acknowledged in Nasar, 1998, p. 91. In retrospect, it might appear that Nash's inquiry bore a close relationship to the work of Francis Edgeworth in his *Mathematical Psychics* (1881), which is often asserted in modern contexts; however, Nash had no familiarity with Edgeworth, and the problem of relating Edgeworth's objectives to those of modern game theory is questioned in Mirowski, 1994b.

showed the "Bargaining" paper to von Neumann and subsequently inserted some relatively mild critical remarks on game theory into the published version of 1950. This version complains that solutions to n-person games in *TGEB* are not determinate, and that, "It is our viewpoint that those n-person games should have values" (p. 3). Thus, what had begun as a restatement of neoclassical economics mutated into a critique and revision of von Neumann's game theory. In very short order, Nash produced his theory of "equilibrium points" (1950b), *not* to be confused with his previous "bargaining solution."

The circumstances surrounding this event are frequently neglected. The aspect of the proof which most immediately captivated later mathematical economists, the use of Kakutani's fixed-point theorem, was apparently due to David Gale, who first drafted and communicated the written-out solution concept to von Neumann, since he was listed on the manuscript as joint author.[20] In later proofs, Nash opted for the Brouwer theorem (1996, p. 24), a fixed-point theorem of lesser generality. The original existence theorem was swiftly communicated to the National Academy of Sciences by Solomon Lefschetz on November 16 with Nash as sole author, and it was deemed in due course that this work should constitute Nash's Ph.D. thesis, after only fourteen months of graduate study. Nash was worried that this work "deviated somewhat from the 'line' (as if of 'political party lines') of von Neumann and Morgenstern's book" (Nash, 1994, p. 2) and therefore, as a safety measure, began to prepare an alternative topic on real algebraic manifolds. "Many mathematicians, largely unschooled in game theory, consider Nash's work on noncooperative games as trivial in comparison with his later work in differential geometry and partial differential equations" (Kuhn, 1995, p. ii), so there were some grounds for trepidation. Nevertheless, Nash was persuaded to submit his twenty-seven-page thesis by Albert Tucker, and it was accepted; von Neumann did not serve as a reader.[21]

It all seemed a bit of a rush; perhaps it happened all too fast. Nash was widely regarded by his acquaintances as eccentric or worse; Lloyd Shapley,

[20] See box 32, VNLC: "Stable Points of N-Person Games," by J. F. Nash and David Gale, n.d., and the acknowledgment of receipt by von Neumann, November 5, 1949. Von Neumann wrote: "I had a short and interesting talk with Mr. Nash on this and connected subjects and I would be very glad if we, too, could get together sometime. The idea of discussing n-person games 'without cooperation' is an attractive one which had occurred to me too, but I gave it up in favor of the coalition-type procedure because I didn't see how to exploit it." It seems apparent that Gale assumed the responsibility of contact with von Neumann about the equilibrium concept; Nasar (1998, p. 95) glosses this as "Gale acted as Nash's agent." Gale had been working on OR at the Rad Lab during World War II.

[21] The readers were Albert Tucker and John Tukey. Nash acknowledged support from the Atomic Energy Commission.

a fellow student, remembers him as "obnoxious." People harbored the suspicion that there was something not entirely right with his mental equilibrium as early as his undergraduate years (Nasar, 1998, chap. 2); it only got worse with the passage of time. A common refrain was that "You couldn't engage him in a long conversation" (p. 72). All of this tended to get overlooked and forgiven at Princeton because of his phenomenal originality in mathematics; furthermore, anyone who spends enough time around mathematicians soon learns to make the crucial distinction between "logic" and "rationality." Nash spent the summer of 1950 at RAND because of his interest in game theory, and maintained a RAND affiliation until 1954. Things really began to degenerate when he was expelled from RAND as a security risk due to being arrested for homosexual solicitation. In the interim, he had accepted a prestigious position as Moore Instructor in Mathematics at MIT in 1951. His mathematical prowess was legendary, but by the mid-1950s things started going seriously wrong. In his own words:

> the staff at my university, the Massachusetts Institute of Technology, and later all of Boston were behaving strangely towards me. . . . I started to see crypto-communists everywhere. . . . I started to think I was a man of great religious importance, and to hear voices all the time. I began to hear something like telephone calls in my head, from people opposed to my ideas. . . . The delirium was like a dream from which I never seemed to awake.[22]

In the spring of 1959 Nash was committed to McLean Hospital in Belmont, Massachusetts, diagnosed as a paranoid schizophrenic. He has subsequently written:

> After spending 50 days under "observation" at the McLean Hospital, [I] traveled to Europe and attempted to gain status there as a refugee. I later spent times on the order of five to eight months in hospitals in New Jersey, always on an involuntary basis and always attempting a legal argument for release. . . . Thus time further passed. Then gradually I began to intellectually reject some of the delusionally influenced lines of thinking which had been characteristic of my orientation. This began, most recognizably, with the rejection of politically-oriented thinking as essentially a hopeless waste of intellectual effort. (Nash, 1994, p. 3)

The ensuing tale of Nash's collapsed career and unsavory behavior has been covered elsewhere (Nasar, 1998); almost everyone who has consulted him since has found him at some fundamental level uncommunicative.

This sad course of events in the career of an important scholar, should be treated with circumspection out of regard for the terrible tribulations

[22] See <www.groups.dcs.st-andre . . . story/ Mathematicians/ Nash.html>.

suffered by anyone struck down by a tragic devastating illness. It explains, for instance, the rather unusual preparations that had to be made in order to award Nash the Nobel Prize for economics in 1994, jointly with John Harsanyi and Reinhard Selten: the standard Nobel lecture, clearly out of the question, was replaced by a "seminar" conducted by the other two laureates and Harold Kuhn. It helps explain the curious eclipse of discussion of Nash equilibria in the 1960s. It also explains a problem that we, the inheritors of this tradition, have in understanding the character of the innovations promulgated by the Nash program; we are forced to depend upon layers of second- and thirdhand exegesis of self-appointed spokesmen because we cannot directly query the progenitor in any satisfactory sense. Nevertheless, we should not permit these unfortunate circumstances to obstruct a very basic question that must be asked about the Nash program: what kind of rationality is supposed to be expressed by Nash equilibria in game theory?

Kenneth Binmore has written, "An authoritative game theory book cannot possibly recommend a strategy pair as the solution to a game unless the pair is a Nash equilibrium" (in Nash, 1996, p. x). Perhaps; but the evasive wording here is itself significant. "Authoritative" texts did, in fact, violate this precept in the past, as we have documented; and what counts as "authoritative" in the present is at least in part due to the active interventions of Binmore as it is due to the putatively obvious virtues of the Nash concept. *Pace* Aumann, perhaps sociology and psychology have not been so successfully banished. One trembles at the ukase of "authority" here, if only because it is so antithetical to the spirit of what Nash himself was aiming at: a definition of rationality in game play so transparent and unassailable that everyone would voluntarily acknowledge its salience and conform to its dictates, entirely independent of any external interpersonal considerations. We shall dissect the Nash equilibrium in detail, but, for starters, it will suffice to define "Nash equilibrium" as the personal best strategic reply to an opponent (i.e., a constrained maximum) who is himself trying to discern your best strategic option and deploy his own best response, where the two desiderata coincide. It is clearly an attempt to extend constrained optimization of something which behaves very much like expected utility to contexts of interdependent payoffs. Nash himself described it as "an n-tuple such that each player's mixed strategy maximizes his payoff if the strategies of the others are held fixed. Thus each player's strategy is optimal against those of the others" (1996, p. 23). It all sounds so simple, except when you pick up diverse game theory texts and find they rarely tender the same explanation of the meaning of this "rationality" from one instance to the next.

In this volume, we shall treat the Nash equilibrium as the next logical extension of the Walrasian general equilibrium tradition into the Cold

War context; far from standing tall as some sort of culmination of von Neumann's project, perhaps subsuming it as a special case, it is instead the ultimate rebuke to his legacy. It was a gauntlet thrown down before the keeking juggernaut of an interpersonal theory of rational coordination, as Nash himself made abundantly clear. Could it nonetheless qualify as a cyborg innovation? This question of intricate subtlety, demands delicate discriminations. That Nash continually measured his status against von Neumann and Wiener gives some indication of where he conceived his intellectual sympathies lay (Nasar, 1998, p. 146). It is also significant that Nash worked his way through von Neumann's *Mathematical Foundations of Quantum Mechanics* as an undergraduate (p. 45) and harbored ambitions to alter quantum mechanics well into the period of his illness. "He considered 'thinking machines,' as he called them, superior in some ways to human beings" (p. 12). When it came to his version of game theory, the equilibrium concept did appear to embody a curious kind of virtual simulation. Nash also wrote on cyborg issues, such as parallel computation and machines playing other machines at games while at RAND, but that work has not enjoyed widespread dissemination.[23] It may also appear that the Nash program in economics has been increasingly shunted into cyborg channels in modern times, as we shall learn in the next chapter.

Nevertheless, the Nash equilibrium is ultimately disqualified from the cyborg program on at least four counts. First, the notion of rationality encapsulated in the Nash formalism contradicts the predilections of the cyborg sciences to construct that treacherous concept as less than omniscient and all-encompassing of all possible worlds, more situated on the level of interactive emergence, and closer to an entropic process of muddling through. In a phrase, cyborgs may be preternaturally skeptical, but Nash agents are inflicted with terminal paranoia. Second, and contrary to numerous assertions by game theorists, von Neumann's vision of the solution of a game was diametrically opposed to that of Nash at the very level of fundamental definitions. Third, while to all appearances striving to come to grips with problems of communication and the treatment of information in game theory, the Nash program should be recognized as serving to banish those phenomena from consideration instead of subjecting them to analysis. Fourth, the Nash program shares a fundamental attribute with its Walrasian cousin that it is generally noncomputable; and there are few sins more cardinal in the cyborg creed than a mathematical procedure being noneffective. Existence proofs, while nice, do not cut much ice with cyborgs. All in all, the Nash program exploded

[23] See Nash's paper on "Parallel Control," 1954, RAND RM-1361, which turns out to be a rather naive discussion of parallel computation of no technical competence; the papers on machine play of games are Nash, 1952; Nash & Thrall, 1952.

onto the scene not to praise cyborgs but to bury them. Not unexpectedly, the cyborgs decided to strike back.

How It Feels to Be Rational in the Sense of Nash: It's a Dog's Life

For all the reams of commentary on the decision-theoretic aspects of Nash equilibria (and it has blossomed well beyond the bounds of a narrow specialist avocation: see Poundstone, 1992; Skyrms, 1996; Dennett, 1995, chap. 9), it is profoundly disturbing to realize there exists no serious meditation on what it *feels* like to play a Nash equilibrium strategy in a game. Perhaps those mathematicians have themselves fallen prey to all that self-serving PR about the cold-blooded absence of emotions in game theory, the haughty pride in "thinking the unthinkable." It would be a dire mistake to let the mathematics obscure the very real emotional content of the Nash solution concept, for that would leave us bereft of an appreciation for the nature of its appeal in the postwar era.

Although it far outstrips our capacity, or even interest, to engage in psychological theorizing, a very stimulating comparison can be found between paranoia and the cognitive style of the masculine scientist in the work of Evelyn Fox Keller (1985). We quote some of her analysis at length, not so much to endorse it as to evoke some parallels with the Nash solution concept.

> The cognitive style of the paranoid . . . [is] grounded in the fear of being controlled by others rather than in apprehension about lack of self-control, in the fear of giving in to others rather than one's own unwelcome impulses, the attention of the paranoid is rigid, but it is not narrowly focused. Rather than ignore what does not fit, he or she must be alert to every possible clue. Nothing – no detail, however minor – eludes scrutiny. Everything *must* fit. The paranoid delusion suffers not from a lack of logic but from unreality. Indeed, its distortion derives, at least in part, from the very effort to make all the clues fit into a single interpretation. . . . For the paranoid, interpretation is determined primarily by subjective need – in particular, by the need to defend against the pervasive sense of threat to one's own autonomy. . . . the very fact of such vigilance – even while it sharpens some forms of perception and may be extremely useful for many kinds of scientific work – also works against all those affective and cognitive experiences that require receptivity, reciprocity, or simply a relaxed state of mind. The world of objects that emerges is a world that may be defined with extraordinary accuracy in many respects, but is one whose parameters are determined principally by the needs of the observer. (pp. 121–22)

This mortal fear of abject capitulation to others is the moral center of gravity of the Nash equilibrium, and its implacable commitment to the solitary self-sufficiency of the ego is the marginal supplement that renders the otherwise quite confused and contradictory textbook justifications of the Nash equilibria comprehensible. It is also the key to linking the Nash equilibrium to the Cold War. The Nash equilibrium stands as the mathematical expression par excellence of the very essence of the closed-world mentality of impervious rationality: "paranoids exhibit remarkable consistency and coherence in their belief systems. However, it is extremely difficult to convince such people that they have made errors in reasoning" (Einhorn in Cohen, 1981).

One of the rarest situations in the abstruse field of game theory is to find someone proffering a clear and straightforward explanation of when it is rational to play a Nash equilibrium. Bluntly: to what question is "Nash equilibrium" the answer? Nash, who personally initiated the distinction between "cooperative" and "noncooperative" games, sought to situate the latter in a pristine society-free state: "Our theory, in contradistinction [to von Neumann], is based on the *absence* of coalitions in that it is assumed each participant acts independently, without collaboration or communication with any of the others" (1996, p. 22). What it would mean to play a game without any open acknowledgment of an opponent whatsoever did seem to present a paradox;[24] for that reason, many who followed after Nash tended to seize upon the absence of *communication* as the defining characteristic of the Nash solution concept (McKinsey, 1952a, p. 608; Luce & Raiffa, 1957, p. 89). Perhaps the frustrating experience of trying to converse with Nash had something to do with this inference. This immediately raised the conundrum of what exactly it was that constituted "communication" in a preplay situation versus immersion in actual gameplay; the postwar preoccupation with C^3I thus further suggested that the Nash concept was really about the specification of "information" and its transmission in games (Kuhn, 1953; Myerson, 1999).

At first, this seemed to reinforce the cyborg credentials of game theory; but the dependence on "information" to formalize the motivation for the solution concept rapidly encountered formidable obstacles, especially with respect to the information-processing capacity presumed of the agent. Confusion was compounded over what an adequate information processor would look like, a theme explored in the next section of this chapter. Others sought to locate the desideratum in the complete and total absence of "binding agreements," evoking dependence upon some original Hobbesian state of nature.[25] Still others sought to ground the justification

[24] Unless, of course, the opponent was a machine. This reconstruction of the Nash justification is the topic of Chapter 7.

[25] See, for instance, the explicit statement of Harsanyi in 1982, p. 49: "Nash . . . defined *noncooperative* games as games permitting *neither* communication nor enforceable agreements. Later writers have found these definitions unsatisfactory. . . . It turns out that

in the presumed inviolate truth of the Walrasian economic actor: "The Nash equilibrium is the embodiment of the idea that the economic agents are rational; that they simultaneously act to maximize their utility; Nash equilibrium embodies the most important and fundamental idea in economics" (Aumann, 1985, pp. 43–44). This latter attempt at justification, while a reasonably accurate description of Nash's own personal route to the concept, was hardly calculated to win over the bulk of the prospective clientele for game theory.[26] In the 1960s, it didn't even get to first base with its target group, the economists. As game theory found its second wind in the 1980s and 1990s, Nash equilibria began to be justified in more abstruse cognitive terms, having to do with notions of "rationalizability" and the requirement of "common knowledge" (Hargreaves-Heap & Varoufakis, 1995) – something far beyond the wildest nightmares of a John Nash. The most common contemporary justification tends to combine an unwavering adherence to the maximization of utility with an attempt to enforce consistency of cognition with recourse to Bayesian decision theory, something Nash declined to do.

The Nash solution concept was not a drama scripted by Luigi Pirandello or a novel by Robert Musil; it was much closer to a novella by Thomas Pynchon. Just as von Neumann's minimax solution is best grasped as the psychology of the reluctant duelist (Ellsberg, 1956), the Nash solution is best glossed as the rationality of the paranoid. Nash appropriated the notion of a strategy as an algorithmic program and pushed it to the nth degree. In the grips of paranoia, the only way to elude the control of others is unwavering eternal vigilance and hyperactive simulation of the thought processes of the Other (Pierides, 1998). Not only must one monitor the relative "dominance" of one's own strategies, but vigilance demands the complete and total reconstruction of the thought processes of the Other – *without* communication, *without* interaction, *without* cooperation – so that one could internally reproduce (or *simulate*) the very intentionality of the opponent as a precondition for choosing the

enforceability or unenforcability of agreements is a much more important characteristic of a game than presence or absence of communication is." Why exactly that had to be so was never made clear by Harsanyi, although one might speculate that his involvement in modeling superpower nuclear proliferation treaties may have played a role.

26 Defenders of Nash sometimes adopted a certain intolerant hectoring tone concerning the previous justifications of Nash solutions adopted by their predecessors; for example, Aumann, 1987, p. 469; Myerson, 1999, p. 1080. The more careful the authors, the more prolix and convoluted become the definitions of rationality required to motivate the Nash equilibrium – for example, Hargreaves-Heap & Varoufakis, 1995, p. 53. This, more than anything else, accounts for the modern expressions of relief attendant upon the revival of Nash's unpublished "mass action" interpretation of the solution concept, and the "evolutionary" interpretation of equilibrium discussed in Chapter 7.

best response. An equilibrium point is attained when the solitary thinker has convinced himself that the infinite regress of simulation, dissimulation, and countersimulation has reached a fixed point, a situation where his simulation of the response of the Other coincides with the other's own understanding of his optimal choice. Everything must fit into a single interpretation, come hell or high water.

There may be a multiplicity of such possible interpretations, but that is no cause for undue alarm; all that really counts is that one such comprehensive account exists, that it can be demonstrated to the player's satisfaction that it is *consistent*. (When others resist the paranoid's construction of events, usually it only serves to reinforce his paranoia.) Patently, the fact that this happens within the confines of a single isolated consciousness suspends the maximization calculation outside the conventional bounds of time and space: instead it occurs within the closed world of closely scripted apocalyptic conflicts. Equilibration is not a process, isn't learned, and it is most assuredly asocial. The fact that the mere choice of strategies could itself be construed as an act of "communication" is irrelevant in this context; all play is unremittingly virtual. Appeals to learning or signaling or shared "common knowledge" would be most emphatically beside the point, because the Other has been rendered irrelevant.

This peculiar attitude of Nash is nicely captured in a story related by Martin Shubik (1992, p. 159). "Hausner, McCarthy, Nash, Shapley and I invented an elementary game called 'so long, sucker'[27] where it is necessary to form coalitions to win, but this alone is not sufficient. One also has to double-cross one's partner at some point. In playing this game we found that it was fraught with psychological tensions. On one occasion Nash double-crossed McCarthy, who was furious to the point that he used his few remaining moves to punish Nash. Nash objected and argued with McCarthy that he had no reason to be annoyed because a little easy calculation would have indicated to McCarthy that it would be in Nash's self-interest to double-cross him." The scene is shot through with pathos: Nash doggedly upbraiding others for some supposed deficiency in their full meed of rationality; yet, taken to the limit, his own construction of rationality would dictate that no one would ever voluntarily play this, or indeed any other game, with him.[28]

[27] Rather characteristically, Nash preferred to call this game "Fuck Your Buddy" (Nasar, 1998, p. 102). Nash himself invented a board game, as described in Milnor, 1995.

[28] This is not merely an observation of psychobiography, but also a formal consequence in the Nash program of "refinements" of equilibrium. As Abu Rizvi notes, "the common prior assumption together with the common knowledge assumption is actually inconsistent with differential information having any importance.... It is a small step from the impossibility of agreeing to disagree about an 'event' to showing that two risk averse agents will not want to bet with one another" (1994, p. 18).

This vignette of how it feels to be ensconced in Nash equilibrium can serve to clear away the decades of superfluous commentaries on the meaning and significance of the Nash program in game theory and prepare the way for an appreciation of the only historically faithful account of the Nash program, that of Christian Schmidt (1995a). Schmidt deftly explains why assertions that the Nash solution concept is merely a generalization of von Neumann's minimax are seriously misleading and fundamentally flawed. As we have indicated in Chapter 3, von Neumann sought to preserve the minimax solution throughout the various extensions and elaborations of games in *TGEB*. This commitment to minimax dictated that the two-person zero-sum game would be preserved as the essential invariant model of strategic interaction; three- and higher-person games would be resolved down into aggregations of two-person games through the intermediary of "dummy" players, while non-zero-sum games would be recast as their zero-sum counterparts through the analytical artifice of the fictitious player. For this reason, *TGEB* is populated with a whole raft of dummies, automata, and robots, as befits a cyborg text. These dummies form a prophylactic phalanx around the minimax concept, supposedly multiplying its power and extending its purview. They embodied von Neumann's ambitions for computers as prostheses, not as replacements for the brain. Unfortunately, the ploy did not work; the dummies did not turn out to be harmless or well regimented, in that they wreaked havoc with the originally proffered interpretations of the solution concept, and tended to subvert the ambition for a theory of institutions. Readers did not accept von Neumann's posited coalitions; nor did they endorse the idea that correlated payoffs could so effortlessly be neutralized.

Nash's innovation was to cast out the dummies once and for all by revising the very meaning of a game. Whereas the major fault lines for von Neumann ran between zero-sum and non-zero-sum and between two and three persons, these were downgraded to relatively minor distinctions for Nash. Instead, Nash invented a distinction that did not even exist in *TGEB*, that between cooperative and noncooperative games. Questions of the actual numbers of opponents (they're everywhere!) and the extent of their hostility (as expressed in conservation principles of joint valuation) are not matters of import for the paranoid mind-set. Rather, in Nash's scheme, von Neumann's minimax-cum-imputation values were to be relegated to the narrow cooperative category, as a prelude to being swallowed up by the "more general" noncooperative approach. This mantra became the hallmark of the Nash program: "One proceeds by constructing a model of the pre-play negotiation so that the steps of the negotiation become moves in a larger non-cooperative game [which will have an infinity of pure strategies] describing the total situation" (Nash, 1996, p. 31). The "game" would therefore need to expand exponentially to encompass everything that would potentially bear any possible relevance to strategic play in any conceivable scenario; it would be hard to discern where the game stopped and life took up the slack. But paranoia is marked by the belief that one can defend oneself against *every possible contingency*. The mathematical price of demonstrating that every game had a fixed value was the unloading of all analytical ambiguity onto the very definition of the game. Far from being daunted at a game without determinate bounds, Nash was already quite prepared to relinquish any fetters upon the a priori specification of the structure of the game, given that the entire action was already confined within the consciousness of the isolated strategic thinker. The game becomes whatever you think it is. He dispensed with dummies, exiled the automata, and rendered the opponent superfluous. This was solipsism with a vengeance.

There is some evidence that Nash himself was aware of the price to be paid. In his published work, and in most of the discussions of his solution concept in the first three decades following his first paper (1950a), this self-contained rationality was the only reigning interpretation of the equilibrium concept. In his unpublished Ph.D. thesis (1950), however, he acknowledged that the solipsistic interpretation required that "we need to assume the players know the full structure of the game in order to be able to deduce the prediction for themselves. It is quite strongly a rationalistic and idealizing interpretation" (1996, p. 33). In order to connect the Nash equilibrium more readily "with observable phenomena," he briefly proposed an alternative "mass action" interpretation as well. There, "It is unnecessary to assume that participants have full knowledge of the structure of the game, or the ability and inclination to go through any complex reasoning processes. But the participants are supposed to accumulate empirical information on the advantages of various pure strategies at their disposal" (p. 32). In a few paragraphs, Nash sketched the idea of populations of entities mechanically playing pure strategies, with some ill-specified, hill-climbing algorithm standing in for the process of "learning." He then asserted (and did not prove) that the distributions of entities playing various pure strategies in the population would converge to the mixed-strategy Nash equilibrium. "The populations need not be large if the assumptions still hold. There are situations in economics or international politics in which, effectively, a group of interests are involved in a non-cooperative game without being aware of it; the non-awareness helping to make the situation truly non-cooperative" (p. 33). In these few paragraphs were buried an incipient thermodynamical interpretation of games, more MOD than MAD. Nothing more was ever heard from Nash on the mass-action interpretation from thenceforward.

This brief attempt at putting ourselves in the shoes of a Nash player may assist in understanding the ways that Schmidt indicates that the Nash program in game theory diverged dramatically from that of von Neumann. First, Nash did not approach the technique of axiomatization in the spirit of von Neumann. Chapter 3 demonstrated how von Neumann's disaffection from the Hilbert program led him to search for an alternative grounding for the certainty provided by mathematics; the terminus of that quest was not game theory but the theory of automata and the computer. Axiomatization was a means to an end but no longer a desideratum, much less an ultimatum. For Nash, by contrast, the act of axiomatization was the paradigm of making everything "fit" the scheme of rationality (1996, p. 42); consistency would invariably trump tractability and practicality. Here, of course, resides the paranoid core of the Cold War fascination with formal axiomatics[29] that so pervaded the social sciences in the postwar period.

There is no evidence that Nash ever betrayed any appreciation for the logical significance of the work of Gödel, or Turing, or indeed any other philosopher. Second, in contrast with von Neumann, Nash understood strategy to be bound up with the process of the formulation of speculative models of the expectations of the opponent, and this would subsequently encourage an approach to strategy as a subset of statistical inference. Von Neumann had very little patience with subjectivist approaches to probability, and as for the brain, he repeatedly warned that no one knew enough yet to describe its operation. Game theory was not a substitute for psychology in his books.[30] Third, von Neumann was never very interested in attempts to subject game-theoretic predictions to experimental protocols, as befitting his belief that games described social, but not individual, regularities (Poundstone, 1992, p. 117). Perhaps unexpectedly, the pure mathematician Nash did become briefly embroiled in the earliest attempts at RAND to provide game theory with an experimental complement. This incident, fraught with consequence for the subsequent path of economics, is limned in the next section.

Although it is not our ambition to trace out all the implications of the Nash program in game theory, or to track all the twists and turns in its subsequent "refinements" and revisions in the postwar period, we do need

[29] Marvin Minsky, who knew Nash at Princeton, is reported to have said: "We shared a similarly cynical view of the world. We'd think of a mathematical reason for why something was the way that it was. We thought of radical, mathematical solutions to social problems. . . . If there was a problem we were good at finding a really ridiculously extreme solution" (in Nasar, 1998, p. 143).

[30] This fact alone explains why the account in Myerson, 1999 of how von Neumann "failed" has absolutely no relationship to the historical record. The idea that Nash possessed acute insights into human psychology and social institutions that von Neumann somehow missed would be ludicrous, were it not persistently found in conjunction with the obvious motivation to rewrite von Neumann out of the history.

to do something to counteract the opinion prevalent at the turn of the century that Nash succeeded where von Neumann "failed," at least when it came to the project of revamping economics. True enough, Nash did provide an existence proof for his equilibrium concept, whereas von Neumann was not able to do the same for his stable set. Also, to his credit, Nash managed to appeal to just those aspects of the formalism that were guaranteed to endear his approach to the doctrinal predispositions of neoclassical economists – hyperindividualism, nonaccessible utility functions, constrained maximization, and cognition as a species of statistical inference. Nevertheless, at this late date, it is more than a little strained to assert that the Nash program was a smashing success and that von Neumann's final analytical program of automata theory was not. Many of the invidious comparisons that are made to von Neumann's discredit accomplish their transvaluation of values by arbitrarily restricting the accounting to *TGEB* versus the ensuing modern economic orthodoxy, while ignoring von Neumann's subsequent intellectual innovations, not to mention heavy-duty repression of the sequence of intellectual embarrassments that the Nash equilibrium concept has suffered in the interim. Some may be contemptuous of appeals to comprehensive intellectual genealogies, believing all's fair in love and war, but our intention here is to simply provide a historical account of how the game has been played and not pretend that some Platonist essence of distilled rationality has been titrated out of the Cold War.

To that end, we pose the question, Did Nash succeed in formalizing a general characterization of "rationality"? I doubt even the most avid partisans of fin-de-siècle game theory would go that far.[31] But we are free to venture further and ask whether Nash really succeeded in banishing the "dummies" and the "fictitious players" from game theory, and therefore absolving it from dependence on the rational prosthetics pioneered by von Neumann. Initially this may have seemed to have been the case, but in the fullness of time we can come to appreciate the extent to which this belief

[31] "Which solution concept is 'right'? None of them; they are indicators, not predictions" (Aumann, 1987, p. 464). "Unfortunately there may be many equilibria that satisfy the condition of mutually consistent expectations.... As soon as the extensive form is examined, many new problems with the concept of the noncooperative solution concept appear. These problems are so critical that it can be argued that the context-free, rationalistic, high-computation, high-information, common knowledge assumptions behind the noncooperative equilibrium solutions must be replaced, or at least reinforced, by more behavioristic 'mass particle' models" (Shubik, 1998, pp. 38–39). The various conceptual obstacles that confront Nash equilibria as a paradigm of a rational solution concept, ranging from nonuniqueness to the folk theorem, are conveniently covered in Rizvi, 1994. The newer misnamed "evolutionary game theory" approaches to the mass action gambit can be found in Samuelson, 1997; Weibull, 1995.

was sorely mistaken. This argument has two components. The first is that, as the computational and cognitive demands of the Nash solution became increasingly apparent, a realization emanating from von Neumann's own computational legacy, one subset of game theorists began to appeal to automata in order to justify the "rationality" of various refinements of Nash equilibria. Instead of renunciation and excommunication, one set of "dummies" was just traded in for another in game theory. The irony of this development is explored in detail in the next chapter. The second component is the observation that Nash's alternative "mass action" justification for his noncooperative approach betokens a different, but comparable, retreat from the expulsion of the dummies. Here, instead of a human player facing down an automata across the tapis, each and every player is now demoted to the status of an automata in order that rationality finally "make sense" in repeated play. Machine dreams are not so effortlessly repudiated as perhaps first thought; sometimes the sensation of waking from a troubling dream itself constitutes a dream, albeit one with that unmistakably delicious sensation of reality, of sanity, of safety.

Unleashing the Dogs of Wargaming

Nash's innovation occurred at an interesting juncture in the evolution of RAND's attitudes toward game theory. In the early 1950s, a substantial contingent of analysts had become disillusioned with its promise to underwrite a "science of war"; they were casting about for something more substantial to supply to the Air Force (Jardini, 1996, p. 110). While some historians have characterized the shift as a reorientation toward cost-benefit analyses as providing the central organizing principle around which the diverse researchers could rally, this interpretation depends too heavily upon the elevation of Albert Wohlstetter's bomber basing study as the quintessential RAND product in this period (Hounshell, 1997b. p. 245). "Systems analysis" admittedly came to constitute the public face of the think tank's expertise, but other initiatives were simultaneously in the offing at RAND, ones more immediately relevant to the nascent cyborg sciences. There was, for instance, a cadre that argued that game theory had even tarnished the systems analyses that were being produced at RAND:

> It is clear that the real-world systems for which RAND is attempting to provide analytic counterparts are systems of organized human action, our own and those of a potential enemy. The quality and quantity of the various combinations of material resources at disposal and of the military strategies with which they are employed could only provide system solutions were one to suppose that the human effort involved in their exploitation is strictly determined by the potentialities of material and strategy, and that human action automatically and invariably realizes

> these potentialities. This assumption is not only not made in current RAND systems formulations, it is emphatically rejected.[32]

The solution proposed was not recourse to the established social sciences for sustenance but, rather, to embark upon a novel program of research into the man-machine interactions implied in modern weapons systems. A group working under the direction of John L. Kennedy, which included Robert Chapman and Allen Newell, was assigned to study the McChord Field Air Defense Direction Center in Tacoma, Washington, to explore the problems and prospects for the "man-machine interface."[33] The idea was essentially to remedy the presumption of strategic omniscience in game theory with empirical research into the way that human beings would react to threats, crises, and attacks under varying levels of stress, differing procedures for learning, and alternative management structures. The RAND innovation was to employ machines to put the "social" back in social science, initially through simulated mockups of radar screen displays, but quickly through the interposition of actual electronic computers between the subjects and their information displays. The Systems Research Laboratory was set up at RAND in 1952 to bring a simulacra of Tacoma to Santa Monica, and to build one of the first experimental laboratories in management science.

In the escalation of the Cold War, world events conspired to render this machine-based social science a more credible option. In September 1949 the Soviet Union exploded its first test nuclear device, shocking the United States out of its complacency about its relative strategic superiority. The Air Force initiated a crash program in early warning systems, one that through a sequence of coincidences deflected the MIT Whirlwind computer project away from flight simulation and air traffic control and toward the far and away more lucrative air defense clientele (Edwards, 1996, pp. 90-101). The project was transformed into the Semi-Automatic

[32] Herbert Goldhamer, "Human Factors in Systems Analysis," April 1950, RAND D-745, quoted in Jardini, 1996, p. 102.

[33] Robert Specht interview, p. 13, NSMR: "A group of psychologists in the Mathematics Division, headed by John Kennedy, wanted to study how people and machines work together under stress – how people set up their own informal patterns of information and direction. The psychologists and a group of computer experts simulated an Air Defense Direction Center where information on an enemy raid comes in from radars, has to be analyzed and evaluated, and then orders are sent to a fighter group. Initially they used college students as guinea pig crews. The Air Force was impressed by the productivity of these student crews during simulated air raids and saw that this simulation would be a useful training tool. The work grew in size – first into the System Development Division of RAND, formed in 1954, and then into the System Development Corporation, spun off in 1957." On the importance of these initial simulation exercises, see Edwards, 1996, p. 123; McCorduck, 1979, pp. 120ff.; Chapman et al., 1958; Capshew, 1999.

Ground Environment, or SAGE, which was the test-bed for a whole array of technological innovations, including magnetic core memory, video displays, the first effective algebraic computer language, synchronous parallel logic, analog-to-digital conversions, digital transmission over telephone lines, duplexing, real-time control, and, not the least, simulation of attack scenarios on the very same equipment that would be used in the eventuality of real nuclear war.

SAGE was a massive undertaking; for instance, it made IBM the premier computer manufacturer of the postwar period (Flamm, 1988, pp. 87–89). The software requirements were equally daunting, and after some initial forays at MIT's Lincoln Labs, RAND was given the job of programming SAGE. Consequently, RAND became one of the premier software engineering centers in the world, eventually spinning off its software division as the Systems Development Corporation in 1957 (Baum, 1981). Some have asserted that the rise of NATO and the outbreak of the Korean War served to displace RAND's concern over nuclear confrontation in the direction of tactical and logistical systems, and that this implied a downgrading of game theory in favor of more conventional economics (Hounshell, 1997b, p. 256). The actual situation was much more complex, with the Economics department under Charles Hitch opting for a cost-benefit view of the world, but other units taking their cues from developments in computation and elsewhere to produce more hybrid social sciences.[34]

Simulation of combat behavior and systems logistics was not the only response to the perceived drawbacks of game theory at RAND. There were also close ties to the burgeoning practice of Monte Carlo simulation pioneered by von Neumann and Stanislaw Ulam at Los Alamos. Extracted from its original context of the design of atomic bombs, Monte Carlo turned out to be a generic technique that could be used to "play" a model in situations where direct experimentation was inaccessible (Galison, 1997, p. 756). With Monte Carlo, simulation was layered upon simulation: the computer generated "pseudorandom" numbers, and then reversed the more standard procedure of solving stochastic problems by reducing them to differential equations, instead evaluating deterministic integrals

[34] The fragmentation into numerous departments and research units such as SRL and SDC was a symptom of the lack of unified approach to defense research at RAND. For instance, David Novick split off the Cost Analysis Department from the Economics Department in 1950; Stephen Enke founded the Logistics Department in 1953. All were nominally "economists," although there were already disputes over what that meant. Even Hitch's trademark doctrine of "suboptimization" (1953) should be understood in retrospect as a retreat from Cowles's Walrasian doctrine of full interdependence in the direction of an older, Marshallian version of OR – that is, some backpedaling from a hardcore scientism so that analysts could deploy some tractable quantification.

by sampling techniques characteristic of probability models. There was also a connection to mathematical logic and its paradoxes, as Ulam himself noted: "Metamathematics introduces a class of games – 'solitaires' – to be played with symbols according to given rules. One sense of Gödel's theorem is that some properties of these games can be ascertained only by playing them" (1952, p. 266).

The systems analysts at RAND, many of whom were erstwhile logicians, had confronted various intractable modeling problems right off the bat, such as the mathematical problem of estimation of success in hitting targets from a B-36 gun turret. Von Neumann suggested early on that these problems could be attacked by means of the Monte Carlo method.[35] The ambidextrous character of the technique would encourage inferences to slide back and forth in both directions: you could "game" a simulation in order to obtain results resembling those more conventionally derived from experiments; or you could "experiment" with ways in which a game could be simulated in order to gain insight about regularities denied to more conventional analytic tools. George Brown and von Neumann (1950; Brown, 1951) had pointed the way to the latter by describing a setup where two fictitious players, programmed to behave according to certain statistical algorithms, would play an iterated game, generating observations on distributions of possible relevant solution concepts.[36]

All of these developments, in conjunction with the emerging doctrine of C^3I as the essence of military coordination, are essential for an under-

[35] Edward Quade interview, p. 10, NSMR. It is no accident that von Neumann would raise an analogy between games and Monte Carlo: "the evaluation of a multidimensional integral inside a unit cube . . . is a one person, non–zero sum von Neumann formal game for mixed strategies" (Galison, 1997, p. 764). The spread of the Monte Carlo method explains the fact that RAND's all-time best-selling publication was the book *A Million Random Digits* . . . (1955) which was used for simulation in the days before cheap computerized random number generators. It is a characteristic cyborg irony that RAND's most popular publication was a book of utter gibberish. For more on the book, consult <www.rand.org/publications/classics/randomdigits>.

[36] The circumstances leading up to what would be von Neumann's last published papers on game theory are briefly described in the George Brown interview, March 15, 1973, p. 96, SCOP. George W. Brown is another major cyborg figure who has so far eluded attention. Brown was responsible as a student for drafting the extensive form graphs in *TGEB*, trained in mathematical statistics at Princeton, and then moved back and forth between OR and early computer development in such a way that maintained contact with von Neumann for the last decade of his career. After a stint at RAND as administrator in charge of numerical analysis, Jacob Marschak recruited him to join the business school at UCLA and start up one of the nation's first university computer centers. Brown was somewhat rueful about how universities got their computers: "Look into how universities financed their participation with computers and you will discover that they sold their souls to Defense Department bookkeeping" (Brown interview, p. 66, SCOP).

standing of what happened to game theory, first at RAND and then, with a pronounced lag, in economics. Economists had been insisting that the scientific approach to experimentation was impossible in their discipline for well over a century; indeed, the promise of econometrics in the first half of the twentieth century was that it would serve as a surrogate for experimentation. This fundamental prejudice was circumvented at RAND through the instrumentality of the computer, which fostered an environment where distinctions between experiment, simulation, and prosthesis could be temporarily blurred, all in the interests of getting beyond the tiresome insistence of neoclassical economists that individual rationality was virtually pervasive and faithfully captured by their formalisms. Game theory did not long remain the province of a few cloistered mathematicians, but got repackaged and reprocessed into a number of protocols and practices and devices that the military found more useful, such as Monte Carlo simulation packages, air defense training programs, software modules, war games, and weapons R&D projections evaluations. Wartime science policy had irreversibly changed attitudes; as Peter Galison has written, "It is impossible to separate simulation from World War II" (1997, p. 775).

The computer permitted the analysts to exploit common aspects of all these endeavors, be it computerized war games with human and machine opponents, Monte Carlo simulation of learning curves in weapons development, or the study of human behavior in electronically mediated command and control situations. Treating humans as information processors was not only an obvious projection of the predicament of their being forced to interact in close quarters with computers, but it was also a response to the perceived weaknesses of game theory as any sort of valid theory of strategic rationality. The beauty of the computer was that it was itself ambivalent with regard to the specific approach to psychology that it represented and might well be retailed to the particular client: it could appear either fully behaviorist, hewing to a pristine mathematical input-output characterization, or appear fully functionalist, teeming with internal process descriptions of cybernetic goal states (Edwards, 1996, p. 184). Due to this protean character, it was not immediately necessary to specify exactly where the machine left off and the human agent began.

One of the most important persons at RAND to prosecute these cyborg initiatives was Merrill Flood. In a more just world, Flood would surely deserve a more exalted status in the pantheon of economic thought. As the person who coined the term "software," he foresaw the implications of the computer for social theory with much greater clarity than many of its more avid boosters and, as a mathematician who was not afraid of psychology, he pioneered multiple strands of empirical research through the erasure of the man-machine interface. It was Flood who was tapped by *Econometrica* to review that revolutionary manifesto *Cybernetics*, and he made it plain that it was a challenge to business as usual in economics:

> This reviewer will hazard the guess that the most likely future connection between cybernetics and problems of economics will in some fashion be concerned with individual differences. Wiener has touched on this point indirectly in his economic argument when he employs such terms as: nonsocial animals, knaves, fools, altruistic, and selfish. Some of these differences might conceivably be treated as variables in a utility function, but it seems more reasonable to give them a primary role in the economic model so they cannot be overlooked so easily. (1951b, p. 478)

Some modern economists are vaguely familiar with Flood as the inventor (jointly with Melvin Dresher) of the "Prisoner's Dilemma" game, arguably the most important individual game scenario in the entire history of game theory; but very few have yet seen fit to credit him with the rise of experimental economics.[37] Flood was a Princeton mathematical statistician who read *TGEB* in manuscript and spent the war working on aerial bombing strategies for the OSRD. This curriculum vitae was thought to render him an ideal candidate to research game theory for RAND, where he worked from 1949 to 1952. In contrast to some of the other mathematicians there, Flood became rapidly disaffected from the game theory research program, because: (1) so-called irrational behavior was more commonplace in real-life situations than the game theorists would care to admit; (2) the individualist character of rationality had been exaggerated to an untenable degree; and (3) the Nash bargaining solution and the Nash equilibrium concept were deeply unsatisfying and implausible.[38] He also was none too impressed with the Cowles contingent at RAND. His response to these reservations was to construct various games that would drive some of these skeptical points home and to conduct numerous experiments, both formal and informal, in order to

[37] This amnesia is somewhat mitigated by comments such as those of Martin Shubik in Weintraub, 1992, p. 249. However, he makes some distinctions about the orientations of various players, which ended up much less sharp in practice. "The people at RAND more or less split in two – Merrill Flood and the other people involved in the Thrall book and the more social-scientist-oriented. Thus one had Helmer, Dalkey, Shapley, Paxon [*sic*], Goldhamer, and Speier in one form or the other involved with COW (Cold War Game), and Flood, Shapley and the others noted, and Thrall considering small, more or less formal, experimental games. Earlier still was the group at Princeton – Shapley, Nash, myself, McCarthy, Hausner – 1949–52 – we were not directly considering experimental games – but were considering *playable* games to illustrate paradoxes and cute aspects of game theory."

[38] Evidence for (1) can be found in Flood, 1958a, p. 20; for (2) in 1958a, p. 17, and in Thrall, Coombs, & Davis, 1954, p. 140; the suspicions about Nash are covered in Poundstone, 1992, p. 129.

isolate the phenomena that had escaped the notice of game theorists and neoclassical economists. Although many of these experiments remain buried as unpublished RAND reports, a few did see the light of day, and their heritage continues to shape the subsequent evolution of economics.

A telling example of how Flood served as the counterweight to various attempts of Cowles economists to domesticate the cyborg initiatives is provided by an unpublished transcript of a conference on the Theory of Organization held in September 1951.[39] Leonid Hurwicz opened the discussion by describing the recent controversy between Cowles and the Institutionalist economists; he anticipated that Cowles's work on the theory of organizations would alleviate the controversy. For Hurwicz, organization theory meant asking how an optimum would be obtained. "Even Robinson Crusoe would have a limited time to compute an optimum program." The answer would be found in the design of an "optimum flow of information." He dabbled in computer analogies, but only in the standard Cowles modality of imagining the right utility computer for the right job. "It is similar to the problem of choosing between two computing machines when the type of problems to be solved is not completely known." No interest was evinced in the actual theory of computation: for instance, the pivotal idea of a universal Turing machine was not even mentioned in this context. David Gale, also a participant, seemed to think that the theory of organization was primarily about imposing restrictions on individual preference functions so that one could aggregate up to community welfare functions. Herbert Simon suggested that a choice had to be made between the alternatives of optimization versus an empirically informed description of behavior. Flood's contribution was by far the most spirited in the report:

> A mathematical theory of Organization does not yet exist. The attempts to mathematize the works of Simon and Bernard have not yet succeeded. . . . Koopman's [*sic*] transportation problem [:] simple application, but when attempts are made to apply it, it is discovered an inadequate list of factors have been considered. . . . The objective of the theory of organization is to describe a wide range of phenomena precisely. A good theory of organization should be equally applicable to corporations, organisms, physical parts of the body or of animals, as well as to such things as family structure. Much stress placed on the need for versatility. Attempts at RAND have been made along "Robotology" lines. Robotology is not to be specifically animate or inanimate. Tests are

[39] "Decision-Making and the Theory of Organization," sponsored by the Econometric Society and the Institute of Mathematical Statistics, report by T. M. Within, September 6, 1951, copy in box 22, OMPD. All quotations in this paragraph are taken from this document.

> being made of humans, of physical objects, of nerve networks, etc. Householder described the nerve network of the RAND TOAD, a mechanical contraption built at RAND that reacts in a predictable way to certain stimuli. Kleene put the nerve network of the Toad into mathematical form, and constructed existence proofs that other networks could similarly be put into mathematical form. Robot Sociology is being studied. Bales, at Harvard has constructed a set of postulates about leadership etc. Robots may be constructed that behave according to Bales' postulates. A Monte Carlo process is used. Synthetic equivalents to human beings can be built and observed.

Flood himself became a wizard of Applied Robotology, innovating ways in which its principles could be rendered perspicuous for social scientists still mired in their precomputational preoccupations. His crusade for Robotology lasted well after he departed from RAND (e.g., Flood, 1978). In a retrospective address on his decades in OR and computational models, he mused:

> Guided missiles, moon-crawlers, and various other cybernetic systems that now exist, provide us with relatively simple examples of inductive machines, and it is in this direction that I predict the next major breakthrough in the inevitable sequence of intelligent machines. . . . the future of intelligent machines depends vitally upon progress in developing valid and practical theories of social relationships. Fortunately, intelligent machines are already helping us in the scientific work to understand social processes and to develop proper theories of economics, psychology, sociology and international relations. (Flood, 1963, pp. 223, 226)

A recurrent theme of Flood's experiments in the 1950s was to expose the way that social relationships mattered for economic outcomes. In one of his "experiments," he offered a RAND secretary a choice: either accept a fixed cash prize (say, $10) or opt for the case where she and another secretary would get a larger sum (say, $15) on the condition that the duo could agree how to split the money between themselves and tell Flood their rationale. One Nash solution was that the duo would split the marginal difference of the latter prize (i.e., $12.50 and $2.50), but Flood found instead that the secretaries appealed to fair division of the total amount (i.e., $7.50 each). Another "experiment" involved a modified Dutch auction for a baby-sitting job amongst his three teenage children. Instead of forming a coalition to prevent downward bidding pressure on the price of the job, as would be predicted by the von Neumann stable set, the three stubbornly tried to undercut each other, with one eventually gaining the job for a pittance.

But Flood was just as capable of running high-tech experiments as these low-tech exercises. In a paper that bemoaned the lack of interest of game theorists in the phenomenon of learning, Flood created an artificial player that he called a "stat-rat," which was in fact a computer algorithm constructed to simulate the Bush-Mosteller (1951) Markov learning model, itself constructed to explain the ubiquitous "Skinner box" behaviorist experiments of that day. Using Monte Carlo simulation techniques, he first explored whether the stat-rat could learn to play a credible game of Morra, first against a randomized opponent, but then against actual human beings recruited from the RAND researchers (Alex Mood and Robert Belzer). He concluded that the "stat-rat usually learns a good strategy when a constant mixed-strategy is played against him" but that the game theorists at RAND could usually defeat the stat-rat easily (in Thrall et al., 1954, p. 156). (Score one for humans at RAND!)

Flood's psychological experiments provoked at least one Cowlesman to defend neoclassical preference theory from his depredations. Kenneth Arrow took exception to the idea that Flood's experiments called into question the existence of stable consistent orderings, primarily on the grounds that the preferences elicited did not correspond to the "true" but distal desires. Arrow also raised the specter of the "complexity of choice," without realizing that this sword might also damage its swordsman. Nevertheless, these sorts of considerations provide the backdrop for Arrow's subsequent work on choice functions, taken up later in the chapter. Whatever may be judged the import of the insights provided by these diverse exercises, far and away the most famous product of Flood's short stint at RAND was the Prisoner's Dilemma game, invented and deployed in an experimental setting by Flood and Dresher in January 1950.[40] Flood, with his Princeton background and RAND location, was one of the first to hear of Nash's newly proposed solution concept for his "noncooperative games," and felt convinced that there was something dubious about both the "rationalistic" and "mass action" interpretations. Who else but an Applied Robotologist would be affronted by the Nash program to banish all dummies from von Neumann's game theory? He therefore

[40] In one of the ubiquitous injustices of scientific misattribution of credit, Flood (and Dresher) were long denied credit for their invention and promulgation of the Prisoner's Dilemma. The game was devised and tested in January 1950, and its fame spread rapidly throughout the closed community of game theorists. The experiment was first described in RAND RM-798 (Flood, 1952), and first published in Flood, 1958a, by which time it had already become a topic of extensive commentary in Luce & Raiffa, 1957 and elsewhere. Luce and Raiffa misattributed the game to A. W. Tucker, who had only served to popularize it amongst psychologists, thus setting in train a long series of apocryphal histories. Flood actually wrote to Raiffa protesting the misattribution, but Raiffa persists in denying him priority in Weintraub, 1992, p. 173. We have already had occasion to caution the reader about various efforts of Tucker and Kuhn to revise various aspects of the historical narrative related herein. This injustice has been rectified in the popular sphere by Poundstone, 1992, but somewhat mitigated by the account in Nasar, 1998, p. 118.

Table 6.2. Flood and Dresher's Original Prisoner's Dilemma Payoff Matrix, *cents*

	JW Defect	JW Cooperate
AA Cooperate	−1, 2	1/2, 1
AA Defect	0, 1/2	1, −1

collaborated with Dresher to concoct a payoff matrix where the Nash equilibrium would not coincide with the more standard equilibria that might maximize joint or "cooperative" payoffs. They were so impatient to see how things would turn out that they recruited RAND analysts John Williams (JW) and Armen Alchian (AA) to play 100 repetitions of the game in normal form with the payoff structure (AA, JW), reproduced here in Table 6.2.

Suppose Alchian and Williams both chose "cooperation"; then Alchian would receive half a cent, while Williams would win a cent (Flood had them playing for pennies). The point of the payoff matrix was that if either player examined this cooperation point, they would realize that they personally could do better by shifting to noncooperation or "defection"; the Nash solution concept insisted both would necessarily comprehend this about the other, and therefore the Nash equilibrium would be for both to defect. Alchian would have to settle for zilch, and Williams for half a cent, if Nash had fully captured the notion of strategic rationality. Flood, characteristically, was not happy with this construction of rationality, and decided to plumb the *social* aspects of coordination and defection through repeated play of the game, having both Alchian and Williams record their motivations and reactions in each round (reported in Flood, 1958a, pp. 24–26, and in Poundstone, 1992, pp. 108–16).

Although both RAND researchers were familiar with von Neumann's game theory, Flood reports that they were not aware of the Nash solution concept at that juncture. Both players, aware of each other's identities, were enjoined not to make any efforts at direct communication or side payments. The prescribed Nash equilibrium turned out not to be the preferred or dominant modality of play, with Alchian choosing cooperation 68 times out of 100, and Williams choosing cooperation 78 times. In their recorded comments, Williams began by expecting both to cooperate to maximize their earnings, whereas Alchian expected defection (recall Alchian was the neoclassical economist, Williams the mathematician). Williams also believed that, because the payoffs were skewed in his favor, he had effective control of the game, and although both had equivalent access to the same payoff matrix, his play was devoted to getting Alchian to learn that fact. Their comments were studded with non-Nashian reflections such as: "I'm completely confused. Is he trying to convey information to me?" and "This is like toilet-training a child – you have to be very patient." Flood concluded, "there was no tendency to seek as the final solution either the Nash equilibrium point or the [split the difference] point x = y = 0.6 which was available to them" (1958a, p. 16).

Nash, who was immediately apprized of these results, of course did not agree with their interpretation. His response was a litany of the Duhem-style auxiliary hypotheses, which were more revealing of his own approach to game theory than of any behaviorist concern with gauging the extent of the psychological accuracy of the exercise:

> The flaw in this experiment as a test of equilibrium point theory is that the experiment really amounts to having the players play one large multimove game. On cannot just as well think of the thing as a sequence of independent games as one can in the zero-sum cases. There is too much interaction. . . . Viewing it as a multimove game a strategy is a complete program of action, including reactions to what the other player has done. . . . Since 100 trials are so long that the Hangman's Paradox cannot possibly be well reasoned through on it, it's fairly clear that one should expect an approximation to this behavior which is most appropriate for indeterminate end games with a little flurry of aggressiveness at the end and perhaps a few sallies, to test the opponent's mettle during the game. It is really striking, however, how inefficient AA and JW were in obtaining rewards. One would have thought them more rational. If this experiment were conducted with various different players rotating the competition and with *no information given to a player of what choices the others have been making until the end* of all the trials, then the experimental results would have been quite different, for this modification would remove the interaction between the trials. (in Flood, 1958a)

Here we observe a number of attributes of what was characterized in the previous section as the paranoid approach to play. When confronted with disconfirmations, the first inclination of the paranoid is always to enlarge the conspiracy: that is, expand the definition of the "game" to encompass considerations previously external to the formal specification of the game. There is nothing absolutely fallacious about this predisposition, except for the fact that, once begun, it is very difficult to justify any preset boundary conditions whatsoever for the game, and, consequently, everything becomes potentially a candidate for a solution. Second, in a move incompatible with the first, Nash entertains the possibility of interaction not as symptomatic of imperfect noncooperation (i.e., deficient total paranoid control), but rather as itself a rational response. Suppose that one did not possess a complete and total plan of action, but only a provisional algorithm, conditional upon the actions of the opponent, due to computational or cognitive limitations. Nash then tries to discount this as a little bit of "sparring" and "testing of mettle" on the road to quick convergence to his Nash equilibrium; but this route is blocked by the intransigence of Alchian and Williams: too bad they weren't "more rational." Third, we are briefly ushered into the vestibule of the "mass action" interpretation of equilibrium, which of course is diametrically opposed to every other consideration raised hitherto by Nash. His comment starkly reveals the purpose of the mass-action scenario was primarily posited to neutralize all social interaction and all interpretation, rendering the player a windowless monad by fiat. Flood would have been least swayed by this third consideration, because his point was that this paranoid isolation was precisely the polar opposite of what goes on in rational social interaction, and therefore game theory had yet to earn its spurs as social theory. Even some open-minded game theorists came to acknowledge this: "one of the most important lessons that game theory has to teach is that a concept of individual rationality does not generalize in any unique or natural way to group or social rationality" (Shubik, 1975, p. 24). It is little wonder that Flood wrote: "Dr. Dresher and I were glad to receive these comments, and to include them here, even though we would not change our interpretation of the experiment along the lines indicated by Dr. Nash" (1958a, p. 16).

Walking the Dog: War Games

Most of the defense analysis community did not waste much time bemoaning the infelicities and drawbacks of game theory in general and the Nash equilibrium in particular, although it does seem the Air Force did fund some follow-up experiments at Ohio State University prompted by Flood's initial findings (Scodel et al., 1959; Minas et al., 1960; Poundstone, 1992, p. 173). For the most part (and to the consternation of modern game theorists), Nash solutions just faded from the radar screens in the 1960s. To the extent game theory was kept on life-support in America, it was at least in part due to pacifist critics of game theory at Michigan such as Rapoport (1964), thinking that it still constituted a major inspiration for strategic thought at RAND and in the military and therefore required further experimental and theoretical critique. Here, paradoxically, the inside-outside distinction described as the "Colonel's Dilemma" also played an inadvertent role in keeping Nash equilibria alive. Meanwhile, a substantial cadre at RAND had decided instead that war games as computer simulations were the real wave of the future. Ed Paxson himself had drifted in that direction, beginning with his Aerial Combat Research Room, constructed to permit the simulation of various aerial combat maneuvers (Jardini, 1996, p. 53). As Paxson became a major proponent of systems analysis, he estimated that by 1963 about a quarter of all organizations engaged in military decision analysis were making use of wargaming and simulation techniques (1963, p. 4). He also noted the increasing prevalence of "man-machine games": "A symbiosis between man and machine is indicated."

Wargaming is a very old tradition; it is often asserted that chess itself began as a simulation or training for war. Most histories trace documented practices of wargaming to the Prussian tradition of *Kriegsspiel* at the beginning of the nineteenth century (Hausrath, 1971, p. 5). Wargaming was integral to the military heritage of Germany, and much of what would later be discussed under the supposedly novel rubric of game theory could be found to be anticipated in the voluminous officer training manuals of the German empire. Indeed, an echo of the postwar dissatisfaction with the Nash solution concept could be heard in disparaging comments made by British authorities about the numbing effects of *Kriegsspiel*.[41] To a lesser extent *Kriegsspiel* had taken root in American military pedagogy, being taught (for instance) at the Naval War College since 1889. Yet, wargaming was not just used for teaching purposes; Douglas Aircraft, the progenitor of RAND, had a "Threat Analysis Model" in 1945, which it used in a supplementary way to boost demand for its aircraft (Allen, 1987, p. 134). Ultimately, it was American-style operations research that took wargaming out of the realm of the sandboards and the tin soldiers and ushered it into the twentieth century as a research tool and, beyond, into something less distinguishable from real battles.

The Operations Research Office (ORO) at Johns Hopkins was one of the first centers of wargaming as organizational analysis, although most of the details remain unpublished (Hausrath, 1971, p. 192). From the late 1940s onward, ORO developed an air-defense simulation scenario in order to analyze various combinations of simulated bomber, missile, and land attacks; this is often celebrated as the site of the first extensive use of computer simulation techniques in defense analysis, and the first computerized war game (Allen, 1987, p. 133). A small glimpse of this activity surfaced in the unclassified literature in the form of George Gamow's tank game, sometimes called "Tin Soldier" (Page, 1952–53; Gamow, 1970, pp. 171–72). The game, pallidly tame by modern standards, holds little interest as a set of rules per se; its importance was rather that Gamow's experience at Los Alamos and his central location in Hopkins's OR unit fostered a

[41] "Such a lack of imagination was common in the German army; and a constant playing of war games [against] a mental replica of itself almost certainly made it worse. In the case of 'free' war games the problem was aggravated by uniform solutions imposed by General Staff umpires, trained on the principle that the army should 'think with one mind.'" Quoted in Allen, 1987, p. 122.

situation where his little game became a laboratory prototype for the computerization of wargaming in general, evolving into the histrionically dubbed "Maximum Complexity Computer Battle" by 1954. The objective of the exercise was to extricate gaming from the grip of the colonels and make it the preserve of the operations researcher; games suitably simulated would reveal unanticipated intricacies in strategy. To this end, Gamow sought the cooperation of von Neumann. Gamow sent von Neumann a précis of his tank game in March 1952. He wrote: "In connection with my work here [at ORO] I have developed a tactical tank battle game which we play here with real tanks (from the 5&10 cents store). The rules are included, and it is a lot of fun! But the actual purpose of the game is to put it on the IBM. I visualize something in the way of the Montecarlo method. . . . All of this has great possibilities, except that I am worried about the size of the sample needed here. . . . What is your opinion? Or is it too much for any machine?" [42]

What is significant is that such a simple game – far more rudimentary than either chess or poker – could have been potentially approached through the analytical techniques and axiomatic attitudes of *TGEB*; but, instead, von Neumann endorsed the computational-simulation approach, writing that, "your plan to investigate it by playing random games with the help of a fast machine excites me, too." He offered to facilitate Gamow's access to various computing machines and encouraged him to consult in person about some of the fine points of Monte Carlo simulations.

Although von Neumann had little further opportunity to have input into the burgeoning field of war games, it would be premature to dismiss this as an irrelevant or insignificant offshoot of the "more legitimate" tradition of mathematical game theory. In any resource terms one might care to index, be they dollars or man-hours or military interest, wargaming and simulation far outstripped the efforts devoted to analytical game theory in the decades from the mid-1950s to the mid-1970s, and even perhaps beyond. The Johns Hopkins's ORO developed a whole array of games, from ZIGSPIEL (Zone of the Interior Ground Defense Game) to LOGSIM-W (Army Logistics Inventory control), while RAND produced its own diversified portfolio from Cold War to MONOPOLOGS; all were disseminated throughout the military command structure by the late 1950s (Hausrath, 1971, pp. 192-93). In 1958 the Naval War College brought the Naval Electronic Warfare System online, which had cost more than $10 million (Allen, 1987, p. 135). SIMNET, a tank team interactive trainer produced in the 1980s, cost more than a billion dollars (Office of

[42] Gamow to von Neumann, March 11, 1952, box 22, VNLC. Von Neumann's response, quoted in this paragraph, March 17, is in box 3, folder 14.

Technology Assessment, 1995, p. 15). The first technological forays into "virtual reality" dated from the late 1960s and were so dauntingly expensive that they were jointly funded by the Air Force, the ONR, and the CIA (National Research Council, 1997). In 1961 Robert McNamara ordered the Joint Chiefs of Staff to create a Studies, Analysis and Gaming Agency (or SAGA) in the basement of the Pentagon; not only the budgets but also the records of the elaborate scripts and actual gaming exercise transcripts are treated as classified (Allen, 1987, pp. 28–29).

Wargaming also recruited quite a few of its most eminent practitioners from the ranks of the game theorists. Martin Shubik served for years as one of the main links between the game-theoretic and gaming communities (Shubik, 1975). Andrew Marshall started out at the Cowles Commission at Chicago working on Klein's U.S. model (1950); he was one of the very first economists recruited to RAND to work on game theory; but like so many others, he rapidly gravitated toward wargaming and other forms of strategic analysis at RAND. Marshall, initially at RAND and later as head of the Pentagon's Office of Net Assessment, became one of the foremost promoters and funders of military gaming in the 1970s and 1980s.[43] Other illustrious game theorists who crossed over to become major figures in the wargaming community included Richard Bellman, Daniel Ellsberg, and Thomas Schelling.

Yet the extent of the impact of wargaming on game theory ventured well beyond questions of funding and personnel. The effusion of gaming activity and research had the effect of pushing the whole complex family of "games" in directions that it might not ordinarily have gone, had it been left exclusively to the coterie of pure mathematicians. Whereas the purist mathematicians had seemed relatively preoccupied with certain limited classes of abstractions, such as existence, closed-form solution concepts, and reconciliation with utilitarian-based mechanisms of inductive inference, the war gamers accelerated the exploration of various cyborg trends which we have foreshadowed in previous sections of this chapter. The war gamers proudly regarded themselves as taking a cue from von Neumann: "the political game is somewhat similar to the use of Monte Carlo methods, whereby machine simulation takes over the job of a purely mathematical derivation of the results" (Herbert Goldhamer & Hans Speier in Shubik, 1964, p. 268).

War games, for instance, tended to focus attention on the open-closed distinction, a theme that had been a cyborg preoccupation dating from the

[43] See Allen, 1987, p. 148. Andrew W. Marshall is another major cyborg figure who has attracted little scholarly attention. From sources such as Kaplan, 1983 and Adams, 1998, and Schwartz, 1995, he would appear to be one of the foremost actors in the intersection of strategic thought and computerized OR. We encounter him again in Chapter 7 as a guru of twenty-first-century cyberwar.

founders' fascination with thermodynamics. In their view, mathematical game theory had only dealt with "rigid" games, namely, those with externally fixed and inviolable rules that putatively covered all possible outcomes. War gamers, by contrast, imagined a two-dimensional taxonomy between rigid and "free" games (usually governed by the uncodified judgments of an umpire) superposed upon a distinction between "open" and "closed" games (the latter restricting player knowledge to his own relatively isolated situation). Nash theory represented the extreme locus of rigidity and closure; playable war games tended to be situated in an orthogonal quadrant. "The closed game is complex and formal; the open game simple and informal" (Hausrath, 1971, p. 125). The "open" game was an acknowledgment that there would always be some relevant social structures remaining outside the purview of the specification of the game, in rough parallel with the precept that there would always be numbers pitched beyond the computational capacity of the Turing machine, and that one intrinsic objective of gaming was to get the participants to think more carefully about the place of those structures in strategic action.

A trend of even greater consequence was that wargaming inadvertently fostered the deconstruction of the unquestioned integral identity of the player of a game. The marriage of the computer and wargaming initially sped up game play, and, of course, the whole point of the exercise was to "put yourself in someone else's shoes," to simulate the actions of another. The acceleration of play in combination with use of actual machine interfaces to be used in real war (war room displays, radar screens, interactive maps, communications devices), raised the possibility that humans facing each other across the Blue versus Red tapis would supply only a low-quality learning experience, compared with wargaming on the C^3I devices themselves. Indeed, simulation and communication aspects were themselves becoming blurred together. "In a different interpretation, the command and control system is considered to *be* a simulator, not just a means of communication with one" (Office of Technology Assessment, 1995a, p. 13). The drawbacks of human gaming were brought home by observations made on low-tech war games of the 1950s: real human beings seemed (irrationally?) extremely averse to exercising the nuclear option – bluntly, dropping the bomb – even in a gaming environment; and, moreover, it had proved nearly impossible to find anyone to play a convincing "Red" opponent (Allen, 1987, pp. 40–41). The solution, as Allen put it, was that "RAND went completely automatic" (p. 328). First, RAND developed an automatic "Ivan"; then found it had to also produce a simulated "Sam"; and soon, the machines were furiously playing each other, slaloming down the slippery slope to Armageddon. This progressive dissolution of the human subject as archtypical player, and his displacement by machines, is the topic of the bulk of the next chapter.

Wargaming and simulation stimulated each other to ever dizzying heights of speculation and innovation. For instance, the SDC spin-off from RAND produced a simulation called "Leviathan" in the late 1950s (Dyson, 1997, p. 178). The gist of the exercise was to "think the unthinkable" about the strategic consequences of the automation of nuclear war. Leviathan consisted of a mock-up of a semiautomated defense system (patterned upon SAGE), which grew too complicated for any analytical mathematical model to comprehend – note the implied criticism of game theory! Things would get more nonlinear if one included the possibility that the Russians also built a computerized defense system of similar capabilities: what would one computer system "think" about the other? Would they opt to communicate directly with one another? This eventuality became so mesmerizing that it even served as the premise for the 1960s movie *Colossus: The Forbin Project*. But wargaming did not just provide nightmares for defense analysts and movie plots for Hollywood: it played a major role in the next technological watershed of the computer, the spread of the personal computer. Conventional histories often point to spreadsheets as the "killer app" that broke open the business market for personal computers, but it is less frequently noted that the "killer app" that brought Apple computers, with their GUI interface, kinesthetic mouse, and color graphics into the home market were interactive war games. In 1977 the Apple II was mostly "suitable for fast-action interactive games, one of the few things that all agreed PCs were good for" (Ceruzzi, 1998, p. 264).

Wargaming and its technological automation had further profound implications for game theory and social theory. The increasingly insular community of mathematical game theorists may have convinced itself periodically that it had "solved" the problem of information in the playing of games – perhaps through some Bayesian device, or an especially perspicuous partition of the total state space – but the war gamers knew better. Reading Clausewitz had taught them to respect the irreducible fact of the "fog of war," and they maintained a healthy appreciation for its implications throughout their endeavors. Simulations, amongst other potential purposes, were run to give players a taste of how they and their opponents would react to the chaotic conditions of battle. The operations researchers were also aware that if you gave most people too much information, this could be as damaging and disruptive as giving them too little (Shubik, 1975, p. 23). Hence, the war gamers bore little patience for the attitude that "the more information the better," which had become prevalent amongst the mathematical decision theorists and the neoclassical economists. Their dissatisfaction with more conventional images of rationality led instead in a number of different directions. One such possible research program was to inquire directly into the nature of the various cognitive limitations of individual humans, which would come into play in situations of threats and fast-moving developments. These inquiries would become the direct forerunners of cognitive science and theories of "bounded rationality."

Alternatively, one might seek out ways in which automation might mitigate the problems of cognitive weakness; and this was one of the main inducements for the military support for the nascent field of artificial intelligence. As automation spread, a third inquiry became increasingly salient. All of these proliferating simulation and augmentation programs would repeatedly bump up against ceilings of machine capacities and computational limitations. Indeed, compared with radiation cross sections and bomb explosion hydrodynamics, the other early computer application that most doggedly pushed the envelope of computation technologies involved wargaming and simulation.[44] The recurrence of bottlenecks of computation raised the possibility that some games were resistant to effective computation in principle. Michael Rabin initiated this research program in 1957, when he asked whether a machine player of certain classes of formal games would be capable in principle of calculation of a winning strategy. This inquiry, so fraught with significance for modern economics, will be covered later in this chapter.

The 1960s and 1970s were the decades of war games, and *not* mathematical game theory per se. Game theory, and especially noncooperative game theory, had been dragooned to provide some formal rigor and direction for wargaming (Thomas & Deemer, 1957); but in practice, it was wargaming that ultimately recast and reconfigured game theory. This had not yet become evident to the denizens of the think tanks circa 1960; but in the meantime, RAND was suffering through one of the more dire threats to its continued existence (Jardini, 1996, pp. 143ff.). In an attempt to assert greater control over RAND, the Air Force commissioned a Strategic Offensive Force Study (SOFS) and an Air Battle Analysis, the former dealing with force structures and weapons selection, the latter with wargaming and simulation. In a manner never before attempted, the Air Force sought to insinuate its control over the timing, participation, and content of this study, to the extent that, incredibly, "Many RAND people felt the organization was being invaded by the military" (Smith, 1966,

[44] See Quade interview, p. 40, NSMR. Incidentally, Quade also indicates the connection between wargaming and bounded rationality, when he insists on p. 32 that optimization was the hallmark of operations research, but Herbert Simon's satisficing approach was characteristic of systems analysis. Simon's relationship to OR and computation at RAND is considered in Chapter 7.

p. 171). In retrospect, the results were disastrous from both the perspective of the Air Force and that of RAND. The Air Force took one look at the preliminary SOFS report, "didn't like it one damned bit," and promptly suppressed it (p. 148). The pressure precipitated open warfare in the Strategic Objectives Committee within RAND – which included Andrew Marshall, Thomas Schelling, Albert Wohlstetter, and Herbert Goldhamer – and, in Jardini's words, "the disintegration of RAND's strategic analysis community can be dated to the Strategic Objectives Committee meetings of 1959" (p. 146).

Within this context one must come to situate the most important work on game theory from this period, Thomas Schelling's *Strategy of Conflict* (1960).[45] It was the virtue of Schelling's work to make it appear as though one could still reconcile the war gamers and the mathematical game theorists, the pro- and anti-Nash contingents, the hawks and the doves, the cold-blooded scientists and the intuitive strategists, and the Air Force and RAND. For the modern-day economic theorist, Schelling's book is remembered for such ideas as "focal points," "precommitment," and the coordination of rational expectations. In the context of the late 1950s, however, Schelling was the one high-profile defense analyst who made a plausible case that noncooperative game theory had something tangible to offer the military, but at the cost of stressing the form without the substance: he repeatedly admitted he was retailing game theory without the mathematics. Although he might at some points appear to be criticizing Nash theory (1960, app. B), in general he was promoting much of the thrust of the Nash program: for instance, the definition of the game in that book ballooned to encompass "the aesthetic properties, the historical properties, the legal and moral properties, the cultural properties" (p. 113). In another instance, he would simply deny Flood's thesis that the Prisoner's Dilemma was a critique of Nash: "If game theory can discover the existence of prisoner's dilemma situations, it is not a 'limitation' of the theory that it recognizes as 'rational' an inefficient pair

[45] Thomas Schelling (1921–): B.A., Berkeley, 1943; Ph.D. in economics, Harvard, 1951; U.S. Bureau of Budget, 1943–46; Economic Cooperation Administration, 1948–50; Office of the President, 1951–53; associate professor of economics, Yale, 1953–58; RAND consultant, 1956–68; professor, Harvard, 1958–90; professor, University of Maryland, 1990–. Biographical information can be found in Swedberg, 1990. There he describes the genesis of his book: "Most of the time I was in government I was dealing with negotiations. These were international negotiations, where you used promises of aid, threats of withholding aid, and things of that sort. So when I went back to academic life in 1953 at Yale . . . I soon perceived that military strategy was like a form of diplomacy and therefore was a good field of application for me. So I read memoirs of World War II and so on, and started to write what I thought was a book" (p. 188). Modern impressions of the significance of Schelling can be perused in McPherson, 1984; Myerson, 1999.

of choices. It is a limitation of the situation" (in Mensch, 1966, p. 478). He disparaged the relevance of explicit communication, after the manner of Nash, only to treat all play as surreptitious communication (1960, p. 105). If everything a player did could potentially constitute an act of communication, then nothing would effectively be distinguished as communication, and the paranoid doctrine would be triumphally reinstated. Tacit knowledge, tacit bargaining, and limited war were all different facets of the same phenomenon (p. 53). Rejecting the previous Air Force doctrine of massive nuclear retaliation, he rendered the concept of "limited war" palatable, and the pursuit of "arms control" just another strategic move in the game of threats and counterthreats. The choice of certain moves such as punishments were not only the mechanical deployal of "strategies," but in (Nash!) games they were themselves attempts to communicate with the otherwise blinkered opponent. His was the doctrine of "limited reprisals" and "graduated escalation," similar in practice to Herman Kahn's proposals for "Tit-for-Tat Capability," itself derived from observations of the Prisoner's Dilemma (Kaplan, 1983, p. 332). Whereas some cyborgs had been insisting that irrationality tended to predominate in really vicious games, Schelling offered the soothing prescription that threats (virtual or no) could serve to focus the mind and inspire strategically (i.e., game theoretically) correct choices. While this could be demonstrated mathematically, he suggested there was no need, because it was also homespun common sense.[46] If one still stubbornly remained a skeptic, then a salutary dose of wargaming experience might bring the handpicked politician or recalcitrant colonel around to this rational point of view.

Schelling did not placate all his colleagues at RAND: for instance, Karl Deutch parodied his theory by asserting, "The theory of deterrence . . . first proposes we should frustrate our opponents by frightening them very badly and that we should then rely on their cool-headed rationality for our survival." Even his most-quoted example of two people managing to coordinate a previously unanticipated meeting in New York City at Grand Central Station at noon did not demonstrate the power of tacit knowledge and focal points to solve bargaining games, as he maintained (1960, p. 56); rather, in the eyes of many RAND strategists, it merely illustrated the

[46] [His topic] "falls within the theory of games, but within the part of game theory in which the least satisfactory progress has been made, the situation in which there is common interest as well as conflict between adversaries: negotiations, war and threats of war, criminal deterrence, tacit bargaining, extortion. The philosophy of the book is that in the strategy of conflict there are enlightening similarities between, say, maneuvering in limited war and jockeying in a traffic jam, between deterring the Russians and deterring our own children." (Schelling, 1960, p. v).

poverty of game theory in confronting the "frame problem" in the definition of rationality.[47] For many outside RAND, Schelling seemed to exemplify the paradox that we must learn to be irrational in order to become truly rational (Green, 1966, pp. 144–45). For some analysts within RAND, by contrast, his book was a symptom of the urgent need to either reduce their overwhelming dependence upon the Air Force for their daily bread, or else to abandon RAND altogether for more placid academic pastures (Jardini, 1996, p. 263).

Schelling's appeal to game-theory-without-game-theory, to Nash equilibrium without all the mathematical fuss, communication-without-communication, and rationality-without-rationality did occupy a special pride of place in the RAND consultant's portfolio for the decade of the 1960s. Schelling and his colleagues could retail this doctrine-without-a-center to their increasingly disgruntled military sponsors through the instrumentality of war games, although it is noteworthy that these were primarily exercises that did *not* depend in any critical way upon the computer. Schelling himself was one of the foremost experts to conduct high-level SAGA exercises in the basement of the Pentagon from 1958 onward (Allen, 1987, p. 151). Nevertheless, this turned out to be only a temporary expedient. As wargaming increasingly became defined on the computer and underwent automation over the decade, Schelling first abandoned game theory, and then the defense analysis community altogether. He did, however, turn his wargaming experience to good account when he turned his attention to issues such as racial discrimination and urban economics: his simple simulation exercises, really cellular automata minus the theory of computation, would later garner him praise as a precursor of cyborg artificial economies.[48]

[47] In a footnote on page 55, Schelling admits that asking the question of people in New Haven may have had something to do with the level of agreement in their answers to his hypothetical predicament. As anyone who has ever lived there for any length of time knows, the North Shore railway is the only halfway convenient way to escape town into New York, and therefore everyone that one encounters at Yale has most likely spent numerous dreary hours in transit at Grand Central Station. If the question had been asked in Princeton, the answers may have been wildly different. The point of this observation is that rationality cannot be treated as context-independent, but instead rests upon elaborate layers of experience and local knowledge, which constitutes one of the major obstacles to the ambitions of game theory (and artificial intelligence, and cognitive science . . .). As John McCarthy glosses what is called "the frame problem" in AI, "the frame problem is not having to mention all the things that didn't change when a particular action occurs" (in Shasha & Lazere, 1995, p. 33).

[48] "The first concerted attempts to apply, in effect, agent-based computer modeling to social science explicitly are Thomas Schelling's" (Epstein & Axtell, 1996, p. 3). This is egregiously garbled history, especially coming from authors with prior RAND affiliations.

THE HIGH COST OF INFORMATION IN POSTWAR NEOCLASSICAL THEORY

> Information or intelligence has been a crucial part of all warfare, but as a means to an end. In the Cold War, it became an end in itself.
>
> Paul Ceruzzi, *A History of Modern Computing*

The Nash equilibrium in game theory may not have seemed all that descriptive of normal human behavior, or especially useful to the military in its quest for C^3I, but that doesn't mean it wasn't important for the subsequent development of neoclassical economics in the United States. Indeed, one can discern a direct filiation from the Nash equilibrium to each of the other major topics covered in this chapter: the debut of the information concept in neoclassical economics, the Arrow-Debreu model of Walrasian general equilibrium, and the issue of the computability of the entire neoclassical program. The pivotal character of the Nash formalism in this history does not derive from any immediate unqualified adoption of game theory by the Cowles economists – that didn't happen – but, curiously, from the skeptical evaluation of the Nash approach at RAND, which was then perceived to have all manner of profound implications for the neoclassical program. Neoclassical economists with RAND affiliations were slowly introduced to the significance of issues of information, formalization, and computability because they were recruited to help evaluate the Nash formalism, and not because they perceived the Nash equilibrium as the obvious telos of neoclassical rationality, as some would have it in retrospect (Myerson, 1999). The neoclassical fallout from the Nash program was not a fluke, because as we argued in the previous section, Nash's own inspiration derived from a reconciliation with his own prior encounter with the neoclassical model; RAND just conveniently provided the arena in which Cowles economists were brought into rapid and intimate contact with the game-theoretic innovation, and with the jaundiced reactions to it. In particular, as Ceruzzi so astutely noticed, the paranoid fascination with complete and total intelligence started to become an end in itself.

It is truly astounding just how prevalent is the conviction in the fin-de-siècle economics profession that there exists an "economics of information," cheek by jowl with the absence of any agreement as to precisely what it encompasses (Lamberton, 1998; Stiglitz, 2000). Because one cannot invest much credence in any particular volume that happens to have "information" in its title, it might be more prudent to ask, When and how did information become such an overriding concern for neoclassical economists? The answer has been tentatively broached above in Chapter 5: It started in the 1930s with the socialist calculation controversy. But that controversy had little to do with the subsequent format assumed by the slippery concept of "information": for that we have to thank the cyborg sciences, and primarily the computer, cryptography, and game theory; for the Cold War above all brought information to the forefront of theoretical concern. One could search in vain the texts of the first two generations of mathematical neoclassical theorists for anything approaching a coherent conception of information as an entity or as a process;[49] yet within another generation it had emerged triumphant from a "slum dwelling" (Stigler, 1961) into the limelight. The impetus came from a small number of developments: briefly, Shannon's wartime formalization of the information concept patterned upon thermodynamics; the drive in the immediate postwar social sciences to treat cognition as intuitive statistics; the attempt to either demolish or defend the Nash equilibrium on the grounds of its presumed information requirements; and, last but hardly least, the computer itself as a concrete model of an information processor. It is indicative that the closer one got to RAND, the more likely it was that the analyst in question would find these issues compelling.

Hayek had proposed that "the market" be conceptualized as a gigantic distributed information conveyance device; but, as usual, it took a more substantial metaphor to get this conviction off the ground. Initially, it became fashionable in the postwar period to conceive of the market system as a sort of gigantic virtual telephone exchange, or perhaps simply the juice that kept the relays clacking, shuttling signals hither and yon, market bells and whistles for the perplexed yet needy consumer. Once again, Kenneth Arrow strikingly managed to give voice to this curious amalgam of econometrics, telephone switches, Cold War readiness, virtual dynamics, and agents as information processors:

> The information structure of individual economic agents powerfully conditions the possibilities of allocating risk bearing through the market. By information structure here I mean not merely the state of knowledge existing at any moment of time but the possibility of acquiring relevant information in the future. We speak of the latter in communication terminology as the possession of an *information channel* and the information to be received as signals from the rest of the world. Thus the possibility of using the price system to allocate uncertainty, to insure against risks, is limited by the structure of the information channels in existence. . . . The transformation of probabilities is precisely what constitutes the acquisition of information. (1974a, pp. 37–38)

The extent to which the neoclassical agent actually suffered from the trademark 1950s malaise of sitting stranded by the telephone will not

[49] There are, of course, interesting exceptions, among whom Oskar Morgenstern must surely number as one of the more significant. On Morgenstern, see Mirowski, 1992; Innocenti, 1995. An attempt to sketch in this lineage is Lamberton, 1996.

tempt us to linger long in this chapter; we are more inclined at this juncture to probe the ways in which cyborgs proved unwilling to concede control over this telephone metaphor to the neoclassical economists outright, and subsequently sought to assert their own versions of information processing in its stead, especially originating in various conceptions of cryptanalysis coming out of World War II. Cowles had to bear the brunt of this contention, and that goes some distance in explaining why a formal "economics of information" first surfaced at Cowles.

Bannin' Shannon

References made in passing to "communication terminology" in the early postwar American context would invariably mean just one thing to your average engineer or natural scientist: Claude Shannon's "information theory." Whereas the Cowles economists may not initially have enjoyed intimate familiarity with the charms of information treated after the manner of thermodynamics, events rapidly conspired to bring them up to speed. The encounters were many and varied, but we focus on a single incident, that of a paper by J. L. Kelly Jr. entitled "A New Interpretation of Information Rate" (1956). This incident is worth recounting, not only because it briefly pitted Cowles against Bell Labs, but because, more than in any other instance, key Cowles protagonists kept revisiting it again and again decades after the fact, reassuring each other just once more for the road that "Shannon information theory has no relevance for economics."[50] The proleptic desire to give that damned spot one more quick scour may have been indicative of a deeper and more pervasive tarnish on those sleek neoclassical information channels, something industrial-strength repetition just couldn't reach. Maybe "information economics" requires something more than lustration, perhaps something closer to a Truth and Reconciliation Commission (Boyle, 1996).

J. L. Kelly Jr. was an acoustics researcher and computer scientist at Bell Labs. Perhaps his best remembered work was on speech synthesis, although he is also credited with the design of an echo canceler on telephone lines, and composition of one of the very first software compilers for computers (Millman, 1984, pp. 111–17). Kelly's interest in communications derived in the first instance from his telephone work but was also due to the fact that "Acoustics researchers often suffered a wearisome cycle of interaction with the computer: acoustic recording, analogue-to-digital conversion, computer processing, digital-to-analogue conversion, and finally audio playback" (p. 370). The need for repetitive

[50] Examples can be found in Arrow, 1984b; McGuire, 1986; Dasgupta & David, 1994, p. 493; Marschak, 1973; R. Elias in Machlup & Mansfield, 1983, p. 501; Arrow, 1996; and elsewhere.

signal conversions, especially under the limited technological conditions of the 1950s, brought home the conviction that communication bottlenecks were the bane of computation, even in cases where actual transmission was not the crux of the problem or the existence of efficient codes at issue (1956, p. 917). The search for efficient communication as a generic abstraction came to constitute a virtue in its own right, and the attractions of both automation of the process and calculation of its theoretical limitations would have been readily transparent to an engineer such as Kelly. The connection was reinforced by the fact that Kelly knew Shannon personally at Bell Labs; indeed there is some evidence that Shannon may have encouraged Kelly in his extension of information theory to the economic sphere.[51]

Kelly penned his "New Interpretation" to demonstrate that Shannon information theory could be used to describe the economic situation of a gambler awaiting news of the outcomes of baseball games on a communication channel, although incidental comments at the beginning and end of his paper reveal much more vaunting ambitions, something approaching an alternative to the neoclassical theory of value in economics. "The utility theory of von Neumann . . . would depend on things external to the system and not on the probabilities which describe the system, so that its average value could not be identified with the [information] rate as defined by Shannon" (p. 918). The idea was that, if calculated optima were intended as context- and content-free, and one were committed to the idea of markets as information processors, then one should go the full nine yards and not make empty appeals to a construct like "utility," which merely reprocessed context as idiosyncratic inaccessible preferences. Nonetheless, Kelly was forced to work out his economic analogue of Shannon's information within the ambit of a very limited situation, for the reason that he could only derive the key expression [Σ p_i log p_i] under inordinately restrictive circumstances.

Kelly imagined a gambler sitting by the phone (a noiseless binary line), waiting to hear which team had won a baseball game before the local bookies were clued in to the scores. In the absence of noise, the gambler should supposedly bet all his money on the team identified on the line as victorious. In an absolutely indispensable and equally absolutely unjustified assumption, Kelly posited that the winning bet would invariably double the gambler's money (just as two punch cards would supposedly contain twice the information of one card). Kelly next considers a noisy channel and posits that the gambler should opt to

[51] Kelly (1956, p. 926) thanks Shannon for his assistance. This would support our thesis in Chapter 2 that Shannon was not so loath to extend information theory outside the narrow ambit of communications engineering as he has been frequently made out in retrospect.

maximize the rate of growth of his money. Predicated on prior knowledge of the probabilities of winning and losing teams, Kelly carried out the optimization exercise and arrived at the Shannon entropy measure. The analogy was really quite simple. Restricting all logarithms to base 2, in a noiseless channel $H(X) = \log 2 = 1$, which just restated the assumption that correct bets produced 100 percent augmentations of stakes. With the presence of noise – read, "uncertainty" – the reliability of the bets would be reduced by the amount of "equivocation" hampering the transmission rate $H(X) - H(X|Y)$,[52] diminishing the growth rate of wagers proportionately. If any one of the following presumptions were inapplicable – value is adequately measured in money; the probability of a correct bet is identical with the probability of a correct transmission; the net return on a successful bet always equals the magnitude of the bet – then all connection to Shannon's information theory would be severed.

As a descriptive economic theory, this little parable would hardly attract a moment's notice; but, in a sense, that was beside the point. At RAND, Richard Bellman rapidly picked up on Kelly's exercise as a salutary instance of his own preferred technique of dynamic programming (Bellman & Kalaba, 1957), precipitously raising the credibility stakes of the approach. John McCarthy (1956) then jumped in with his own proposal for an abstract measure of information. All sorts of economists were similarly busily occupied, comparing market operations to telephone transmissions in the 1950s, and they, too, were keen to reduce all information processing to the manipulation of a few known probabilities. Now, who could legitimately claim that they possessed the more plausible framework for treating information as the manipulation of some prior probabilities, the johnny-come-lately neoclassical economists or Claude Shannon and his doughty band of cyborgs?

Cowles was not slow to rise to the bait. The first to snap at the interlopers was Jacob Marschak in 1959 (in Marschak, 1974, 2:91–117). He had already demonstrated some familiarity with Shannon's work as early as 1954, but only to dismiss its relevance for economics peremptorily (1974, 2:47). Kelly's manifesto dictated that information theory would require more elaborate refutation, which he undertook to provide in the 1959 piece. Marschak began by contrasting the approach of the engineer to that of the economist, the former hewing to "purely physical criteria," whereas the latter was concerned with "human needs and tastes, profits and utilities, and this makes him a nuisance," presumably to all those shortsighted engineers at Bell Labs. He then proceeded to launch into an illustration from World War II–vintage OR, suggesting that the nagging

[52] For the derivation of the mutual information measure and its relationship to the transmission rate, see van der Lubbe, 1997, p. 118.

of economists had been critical in keeping the larger objectives of the war in the driver's seat, as against the blinkered prognostications of the engineers. (Clearly this was a prescient example of other later economists' forays into virtual history.) Perhaps the engineers could treat the amount of information as a parameter for their own purposes, Marschak chided, but the economist could not, since he was concerned with the *value* of information. Here Marschak's argument went slightly awry, for although he was accusing the engineers of rendering information as too thinglike, he simultaneously insisted the economist would wish to argue in terms of a demand for and supply of information – that is, to treat it as a free-standing *commodity*. To some extent this was a function of his elision between a set of messages and the channel that existed to convey them: the first of a whole sequence of perilous ellipses. Once one was brought to admit demand price to have a bearing on the value of information, Marschak wrote, even if one permitted the entropy index as a cost measure, entropy could no longer be treated as a surrogate for the economic valuation of information.

Next he homed in on Kelly's attempt to equate the value of information with Shannon's entropy measure. Marschak indicated contra Kelly that he had to insist upon the validity of the expected utility concept of von Neumann, thus strategically insinuating he was recapturing the Neumannian mantle from Kelly and Shannon. Next he sought to turn the tables on Kelly, asserting that Kelly himself had not managed to avoid appeal to a criterion function but had only spirited in one that had been left implicit. The quest to institute some index of channel capacity as a measure of expected payoff or objective valuation of information was misguided, in Marschak's opinion. Nevertheless, in the interest of finding some middle ground, he did allow that Shannon's theory could perform some limited economic service when subordinated to the neoclassical theory of demand, because "Kelly's quantity is related to the demand price" (p. 114). This pioneered a line that subsequently became hallowed dogma at Cowles, namely that welfare economics was intimately bound up with the economics of information: "we can regard the macro-economics of information as an extension of the theory of welfare economics, or public policy. It would attempt to characterize a socially optimal allocation of channels, given the distribution of tastes and beliefs" (1974, 2:126). Kenneth Arrow, as we have already witnessed, was even more explicit about this construction of the role of information within orthodox economics.[53]

[53] "I had the idea of showing the power of sequential decision procedures in a context suggested by Shannon's measure of the cost of communication. . . . The dialogue could be regarded as a special case of the Lange-Lerner process of achieving an optimal resource allocation" (Arrow, 1984b, p. 262).

Marschak devoted a fair proportion of the rest of his career struggling to confine Shannon's information measure within some sort of supply-demand framework, which meant in practice either tacking perilously between incorporation of the formalism within first the demand and later the supply function (1973, p. 264); or conversely excoriating the formalism for not sufficiently taking into account that the noise or "errors" themselves were not adequately reduced to some commensurate valuation principle (1971, p. 212). This erratic performance was spurred on by the fact that Marschak could never be held to hew consistently to a single construction of that nebulous term "information": sometimes it would denote rates of transfer of signals, sometimes amounts of a commodity, sometimes an actual semantic conception, sometimes a decision technology within an individual, and sometimes it stood for the entire transpersonal system of conveyance of messages within which choice must be exercised. Marschak found himself torn between the analytical construction of Shannon and that of David Blackwell (described below), conceding the virtues of both but never entirely satisfied with either. The spectacle of his gradual weaning from the Neyman-Pearson framework, the tradition within which he had conducted his erstwhile econometric exercises, in the direction of something more closely resembling a Bayesian framework, tended to further militate against according any stable meaning of the term "information."

Toward the end of his life, he tended to jumble these repeated forays into information economics together as special cases of the economics of a hulking bit of hardware: "The economist can conclude: communicating and deciding belong to a common theory of optimal symbol processing. . . . Inquiring, remembering, communicating, deciding – the whole information system, to be acquired and used by a firm or a public agency, determines the expected benefit to the user, given his prior probabilities and the benefits he attaches to each pair of benefit-relevant action and event" (1971, p. 215). If you encountered trouble pinning down exactly what it was that information economics was all about, when in doubt you could always gesture sagely in the general direction of the computer. All qualms were promptly dispelled by the mesmerizing banks of blinking lights, at least in the 1960s.

Kenneth Arrow became equally embroiled in the upbraiding of Kelly and the domestication of Shannon's theory of information. He wrote an entire paper that took issue with Marschak's method of subsuming the entropy measure: "Marschak refers to the value of information as the demand price. That is, a channel will be worth acquiring if the value exceeds the cost. This cannot be the case" (1984b, p. 112). Instead, Arrow assumed the position that there was one version of neoclassical theory where the value of information would coincide with the definition of the amount of information, namely, where the utility function was defined as separable and logarithmic in "states of nature." However, because restriction of the functional form of the utility function stood as the paradigm of ad hocery and illegitimate model building at Cowles, the upshot of the exercise in his view was to discredit Shannon information as an economic entity. This was made explicit elsewhere: "the well-known Shannon measure which has been so useful in communications engineering is not in general appropriate for economic analysis because it gives no weight to the value of information" (1984b, p. 138). One might then be permitted to wonder why such an inauspicious and ill-suited concept had to be repeatedly consigned to the grave, over the extended space of decades. The answer, never actually allowed to surface in Arrow's work, was that Shannon information required repeated internment because it resembled Arrow's own approach to an uncanny extent. The ensemble of channels, signals, and probabilities of messages characterized Arrow's construction of the information economy, almost as much as it did Shannon's telephone wires. "By 'information,' I mean any observation which effectively changes probabilities according to the principles of conditional probability" (1984b, p. 199). But cognition as intuitive statistics never completely quelled the irritation provoked by the Shannon concept; as late as 1990, Arrow was insisting that, "Information, and the lack of it, and differences in information, play an absolutely key role in the way the economic system operates" (quoted in Lamberton, 1998, p. 225). Unfortunately, "there is no general way of defining units of information" (Arrow, 1996, p. 120); and worse, it was still unclear if neoclassical theory had much cogent to say about the elusive concept.

Arrow, to an even greater extent than Marschak, was willing to entertain the idea that information was a generic thing conveyed by means of channels that exhibited a separate integrity, defined almost entirely by its stochastic characterization. It was conjured out of a pre-Adamite state by the brute reality of uncertainty. "Uncertainty usually creates a still more subtle problem in resource allocation; information becomes a commodity" (1962, p. 614). For Arrow, information was the antidote for uncertainty, the emollient for equivocacy, the incoming electron beam collapsing waves to precise particles. It could flow after the manner of Shannon; but where he would depart from the engineering portrayal was that the volumetric dimension was left vague; nonetheless, the proviso was appended that economic valuation had to be attached to the flow. Valuation considerations hinted at a Gresham's Law for the information economy: "the aim of designing institutions for making decisions should be to facilitate the flow of information to the greatest extent possible.... This involves the reduction of the value of information while preserving as much of value as possible. To the extent that the reduction of volume is accomplished by the reduction in the number of communication channels, we are led back to the superior efficiency of authority" (1974a, p. 70). Here we can observe the confluence of Shannon information theory and the socialist calculation controversy, all expressed in an idiom which was arguably antithetical to the gist of both, namely, the Walrasian conception of the economic actor.

This cavalier usage of the terminology of "information" was extremely hidebound and parochial, as noted, among others, by Fritz Machlup (in Machlup & Mansfield, 1983, p. 649). Metaphoric invention may frequently slip the surly bonds of earth when it comes to contemplation of information; but only a decision theorist of neoclassical pedigree could find the analogy of mineral extraction and signal extraction compelling, in pursuit of an explanation of why "research is a form of production" (Arrow, 1984b, p. 141). The production metaphor was more than a *jeu d'esprit*, however, because it was part and parcel of Arrow's lifelong project to shift information away from cognitive questions and process orientations in the interests of extricating it from the troubled "demand side" of the neoclassical equation and safely confining it to the putatively safer "supply side." If information could be swiftly rendered thinglike, then one could proceed with confidence to delineate its supposedly special characteristics: indivisible, difficult to appropriate, perhaps even subject to increasing returns in production (1984b, p. 142; 1996, p. 120). A commodity so slippery and intangible might not have any effectively functioning market to allocate it: this would then connect up with Arrow's favorite Walrasian theme of "market failure": "Once information has been obtained, it can be transferred cheaply. . . . As a result, it is difficult to make information into property. If information is not property, the incentives to create it will be lacking" (Arrow, 1996, p. 125).

Much of this would then be put to use at Cowles (and, later, elsewhere) to argue that the stripped down Walrasian model was not an altogether adequate characterization of "reality"; and that the welfare implications of that obstreperous commodity "information" would more than warrant various forms of statist intervention in the economy, due to the "distortions" it induced. Information might "flow" in Arrow's world; but it was so viscous, that more often than not it just gummed up the works. Hidden among all this verbiage was the infrequent admission that these prognostications had nothing whatsoever to do with the cognitive act of information processing (1984b, p. 200).

Breaking the Code

Tracking the vicissitudes of "information" with gun and camera through the career of any single neoclassical economist, even one so important as

Arrow or Marschak, cannot begin to make sense of the convoluted history of contrapuntal cyborg initiatives and neoclassical responses in the last half of the twentieth century. For instance, it is fascinating the ways in which one vague and ill-specified concept – such as "transactions costs" (Klaes, 1998) – served mutually to buttress and propagate another concept such as "information" over the very same postwar time frame in the writings of many of the same protagonists of our present narrative, such as Marschak, Alchian, and Arrow. (We specifically return to the saga of Arrow's own struggle with information processors later in the chapter.) Once neoclassicals started to entertain the idea of information as a thing, they were brought up sharply against the cold steel of paradox: they certainly did not want anyone to calculate the quantum of information contained in any of their own articles (Dorfman, 1960)! Or, more disturbing, perhaps neoclassical economics itself was evidence for the pervasive market failures so trumpeted at Cowles. Or maybe information was something that the great unwashed had to contend with in their daily lives, while mathematical economists (conveniently?) dealt in something more refined and less common? Or else one might have taken the position of the Chicago school, that all this folderol about information was a tempest in a teapot. If information was a commodity, then it was no different from guns and butter, and there was no call for any amendment of the neoclassical model. Leonard Savage once tendered an especially clear summary of this position to the premier convocation of cyborgs in the 1950s, the Macy Conferences on Cybernetics.[54]

Indeed, the cautionary lesson for outsiders is that you cannot understand modern economics simply by confining your attention to self-identified economists; it is always far more fruitful to keep your eyes on the military and the operations researchers. In this particular instance, the neoclassical formalization of information has been inseparable from the history of the development of the computer, primarily due to a sequence of alternative mathematical treatments of information emanating out of cryptanalysis. In Chapter 2, we indicated that Shannon's information theory had its genesis in his work on cryptography, even though its published incarnation was primarily addressed to communications engineers. There existed another approach to ciphers and codes and duplicitous transmissions in the immediate postwar era, a research

[54] *Savage:* In the last analysis . . . the *value* of information is its cash value. . . .
McCulloch: I am afraid "value" in your sense would turn out to be a multidimensional affair, with very little chance of it being simplified to a measure.
Savage: No, it is simply one-dimensional. . . . This value is von Neumann Morgenstern expected utility.
McCulloch: Very familiar and very illusory.
Pitts: The El Dorado. (in von Foerster, 1952, pp. 217–18)

tradition unrelated to Shannon's entropy, one closer to game theory in inspiration, which was destined to echo down the corridors of economic orthodoxy. For lack of a better term, one could call it "surveillance in state space."

Perhaps a more promising approach to understand the curious way in which "information" entered the world view of the postwar neoclassical economist was to see the neoclassical project as hamstrung between a pillar and a post or, less figuratively, between a transmitter and an antenna or, more precisely, between Bell Labs and RAND. Shannon's theory at Bell Labs had lent "information" a distinct thinglike cachet, an ontological warrant divorced from any human or social activity, as has been repeatedly noted by commentators. But game theorists at RAND in the early 1950s had undergone their own initiation into the mysteries of "information" by a different route, primarily through the early work of Bohnenblust et al. (1949) and David Blackwell (1951, 1953), and codified in the 1953 paper by Harold Kuhn (reprinted in 1997). The Bohnenblust paper attributes the original idea to discussions taking place at RAND in 1949 about the need to investigate "reconnaissance games," or games where one player makes certain moves in order to find out intelligence about the other player.[55] This latter model of information had been heavily informed by the prior formalisms of game matrices and strategic payoffs, as opposed to Shannon's dependence on thermodynamic entropy. This mathematical formalism began by imagining a full enumeration of "states of the world" (discrete for mathematical tractability) and potential acts by an individual in each of those various states. "Utility" would then be conceptualized as a matrix of "payoffs," sporting one discrete entry for every act occurring in every possible state of the world. The problem was then expostulated that the player was "uncertain" about which state of the world actually had obtained; but she was fortunate enough to receive "signals" (from where? from whom?) concerning the identity of the realized state of the world. These signals might bear a stochastic relationship to the probabilities of occurrence of different states of the world; both sets of probabilities were known to the player, and this could be expressed in a second matrix. A third matrix, known as a "decision rule" or a strategy, would relate choice of acts to signals received. Multiplying the two matrices, we would discover: [acts × states] [states × signals] = [acts × signals]. Consequently, the "payoff" from each possible method of resolution of "uncertainty" – or, as some would

[55] Bohnenblust et al. (1949) credits von Neumann with the original idea, but this seems rather unlikely for a number of reasons. For instance, the game setup is extremely awkward, with the second player assumed neither to engage in countermeasures nor himself to be engaged in reconnaissance. Again we encounter a closed world where MAD reigns but the possibility of MOD is ruled out of bounds.

put it, the acquisition of information – would become blindingly transparent to the agent; in practice, she would seek to maximize utility subject to choice of decision technology, given a full specification of states of the world and available signals.

The initial objective of the researchers at RAND had been to extend the notion of a "game" to conditions of uncertainty about other players, but that specific interpretation did not long hold sway. The problem was that the setup did not lend itself to a *strategic* setting, in which duplicity and deception were paramount. For Kuhn, this implied that one had to relinquish the von Neumann approach altogether in favor of the Nash equilibrium. By contrast, David Blackwell sought to reinterpret this "game against nature" setup as a full formalization of the problem of the extraction of information in the scientific context of "experimentation"; in other words, it was to do yeoman service as the statistician's version of "information processing" circa 1950. To this day, explications of the Blackwell model often are motivated in terms of "uncertainty" about the weather or some other naturalistic and nonintentional stochastic process (Rubinstein, 1998a, p. 42). It had little or nothing to do with the computer but did share one ambition with Shannon information theory: Blackwell sought to characterize when one intermediate configuration of [states × signals] would be uniformly "more informative" than another, abstracting away from all considerations of valuation or the specific decision rules deployed. The relationship to cryptanalysis was always quite near the surface of this research program, with Blackwell explaining that if situation (matrix) P was "less informative" than situation Q, then that meant that signals from P could always be "garbled" in such a way that they would be indistinguishable from those emanating from Q, but that the reverse would not be true.[56]

Arrow was one of the major collaborators with Blackwell when they were both at RAND in the late 1940s; previously familiar with Abraham Wald's statistical decision theory, an old OR standby, he would readily appreciate the attractions of the Blackwell approach; moreover, Marschak and Arrow were both acquainted with Shannon's theories at RAND over the same time frame. It was therefore quite a "natural" move that the major protagonists of Cowles Mark II would see fit to take the measure of these two alternative models of statistical "signal extraction" as potential templates for the new model Walrasian agent as miniature prototype econometrician. Moreover,

[56] The hope that this would provide an unambiguous measure of the "amount of information" for economic purposes was dashed, however; for this ordering there was no real-valued function f such that it would always be the case that f(P) > f(Q) iff signal structure P was less informative than signal structure Q. In this respect, Blackwell's ordering somewhat resembles the idea of strategic dominance in game theory. See McGuire, 1986, p. 109.

it should be remembered that general enthusiasm for "information theory" reached its peak in most of the human sciences in the mid-1950s (Newell & Simon, 1972, p. 880; Garner, 1962). The puzzle that the Cowles experience at RAND thrust upon the vanguard was that neither of these two alternatives completely satisfied their prior requirement that the Walrasian model emerge unscathed. Shannon's "information" was pleasingly thinglike and, therefore, evoked the treatment of communication as just another economic commodity, but the experience with Kelly revealed that it was a potential Trojan Horse: the entire neoclassical theory of value might just come crashing down around them if entropy were allowed through the front gates.[57] Shannon's theory worked in terms of "symbols," and, although he had forsworn "meaning," even this limited linguistical construction threatened to revive all manner of suppressed issues of cognition and semantics that the neoclassicals would just as soon pass up. Shannon entropy had also become the province of the computer and telephone engineers; they seemed none too impressed with an insistence upon a pervasive market "optimality" that somehow magically obtained and for which no one was prepared to offer a plausible process algorithm. The point of Shannon's theory, after all, was that perfect transmission, while conceivable, would never be obtained in practice. These considerations obliged the Cowlesmen, however reluctantly, to denounce Shannon's information theory loudly and often.

Blackwell's cyryptanalytic construction, by contrast, initially promised a much more salutary fit. Probabilities were applied to "states" rather than symbols, and this resonated with the Debreuvian reinterpretation of the Walrasian model as encompassing "uncertainty" by simply extending the definition of the "commodity" to encompass variant states of the world. It also resonated with the trend toward appropriation of mathematical formalisms from game theory to shore up the Walrasian tradition (discussed further in the next section). The cognitive capacities of the agent were summarily repressed, with the question of decoding reduced to "choice" over a fixed "decision rule" about how to respond to disembodied signals. These decision technologies themselves sported a gratifying resemblance to commodities: this was an alternative portrayal of the market as disseminating "knowledge." From a different vantage point, the comparison of the rational economic agent to the scientist conducting "experiments" also undoubtedly appealed to Arrow's political sensibilities. Koopmans thought that Hayek's manifesto could best be understood in terms of Fisher's "sufficient statistics" (1991, pp. 22–23n). The character

[57] The tensions between the neoclassical model and entropy are discussed in some detail in Mirowski, 1988; 1989a, pp. 386–95. The problems of accommodation to entropy in economics were first raised in a perspicuous manner by Georgescu-Roegen, 1971, 1975.

of the communication involved was breathtakingly vague, even in comparison with the ontologically sparse Shannon theory: no specification whatsoever of the source of the signals, the nature of the alphabet, the possible motives or cognitive capacities of the sender, the cognitive potential of the receiver to learn, the source and nature of noise, . . . the list goes on and on. It resembled nothing so much as the jury-rigged bottle tapping out a meaningless telegraphic message from no one to nowhere in the landmark 1959 nuclear holocaust movie *On the Beach*. The fundamental emptiness of the model, precisely the quality that rendered it unsuitable as a serious description of scientific induction, or indeed in any branch of the nascent cognitive sciences, was the very aspect that rendered it so very attractive to the Cowlesmen. Here, it seemed, was an exemplary mechanical econometrician, reading off the implications from a communications device (possibly provided by the market), merrily maximizing utility in much the same manner as it did in those hallowed pretelephonic (and pre-Neumann) steam engine days of the 1890s.

The perceived neoclassical edge that Blackwell enjoyed over Shannon when it came to information concepts would have profound consequences for the subsequent evolution of the economics discipline in the period from the 1960s to the 1980s. Although it is hardly ever mentioned, Blackwell's cryptanalytic device is the lineal predecessor of a large proportion of postwar neo-Walrasian and game-theoretic work on the "economics of information"; anyone who starts off by modeling differences in "information" by writing "People in London have a different information partition from those in Milan" is deploying pseudocryptanalysis without being altogether aware of it.[58] Furthermore, an OR-flavored tradition of "decision theory" that metamorphosed into a subset of analytic

[58] The quotation is from Hammond, 1990, p. 3. Incidentally, in this paper not only does Hammond not mention Bohnenblust or Blackwell in his potted "history," but he instead attempts to attribute the "idea of representing information by means of a partition of possible states of the world into events" to von Neumann, possibly to lend it greater credibility. While von Neumann did indeed provide one axiomatization of utility, and Blackwell did come up with his model in a game-theoretic context, there is no shred of evidence that von Neumann would ever have endorsed such an insipid concept of information as a formalization of modern notions of information processing.

When they are not trying to pin the Blackwell formalism on von Neumann, neoclassicals are busy banishing the origins of the approach back to the murky mists of the seventeenth century: "the fundamental conceptual too we shall use is the state of the world. Leibniz first initiated the idea; it has since been refined by Kripke, Savage, Harsanyi and Aumann" (Geanakoplos, 1992, pp. 56–57). This genealogy is more interesting for whom it manages to vault over than for those whom it graces with laurels.

Other clear examples of the Blackwell influence in modern neoclassicism can be found in Hirshleifer & Riley, 1992, chap. 5; Lipman, 1995; Plott & Sunder, 1988; Sunder, 1992; Geanakoplos, 1992; Rubinstein, 1998a, chap. 3; Arrow, 1996.

philosophy also traces its antecedents to these military innovations: here the illustrious cast includes Alfred Tarski, Willard Quine, Saul Kripke, Patrick Suppes, Donald Davidson, and David Lewis.[59] One might then think that so influential a formalization of information or knowledge would have enjoyed a more robust reputation in the annals of cyborg fame. Nevertheless, regardless of its apparent longevity, the Blackwell model did not ultimately prove the neoclassical philosopher's stone, for a number of reasons, which only came to be understood over the longer term. (We shall decline any speculation as what it did for analytic philosophers.) In another of those pesky postwar cyborg ironies, the neoclassical economists inadvertently came to appreciate the vast Saharas of their own ignorance, the poverty of their own cherished trove of information, by pretending that every economic agent, in effect, knew everything.

The turnabout came in the interplay between the vague specification of the definition of "the state of the world" and the injunction emanating from game theory that agents should be endowed with strategic intelligence. We have previously mentioned that Blackwell's portrayal of information was antistrategic. It was a symptom of this inherent contradiction that Blackwell's initial separation between "states" and "acts" was itself poorly delineated: for how was it possible to provide a full specification of states of the world independent of the actions taken to conform to them (Rubinstein, 1998a, p. 42)? And worse, should knowledge and beliefs of others be incorporated into the definition of a state? Once one started down that road, Bedlam seemed to beckon:

> A state of the world is very detailed. It specifies the physical universe, past, present and future; it describes what every agent knows, and what every agent knows about what every agent knows, and so on; it describes what every agent does, and what every agent thinks about what every agent does, and what every agent thinks about what every agent thinks about what every agent thinks, and so on; it specifies the utility to every agent of every action, not only of those that are taken in that state of nature, but also those that hypothetically might have been taken, and it specifies what everybody thinks about the utility to everybody else of every possible action, and so on; it specifies not only what agents know, but what probability they assign to every event, and so on. Let [omega] be the set of all possible worlds, defined in this all-embracing sense. We

[59] "My theory of convention had its source in the theory of games. . . . Coordination games have been studied by Thomas Schelling, and it is he who supplied me with the makings of an answer to Quine and White" (David Lewis, 1986, p. 3). Because the voluminous literature on "common knowledge" discussed in Chapter 7 acknowledges its debt to Lewis, we here discover the subterranean connections to RAND laid bare. The debt owed by American analytical philosophy to RAND and OR is explored in Elliott, 2000. McCumber, 2001, points to the larger Cold War context.

> model limited knowledge by analogy with a far-off observer who from his distance cannot quite distinguish some objects from others. (Geanokoplos, 1992, p. 57)

This conjured the Cyborg quandary, already encountered by Wiener during the war: what happens when the world to be described was not passive and inert, but potentially malevolent and deceptive, actively simulating us as we strive to subject it to our wills? The neoclassical project had initially set out to concoct a science of economics by treating everyone as passive and deterministic pleasure machines, with sharp separation of motive from action, thus finessing all questions of will, intentionality, and interpretation (Mirowski, 1989a). State-space formalisms were the paradigm of that world view, incorporating conservation laws into their very specification. Thermodynamics had asserted the centrality of randomness and indeterminism for the operation of steam engines; but this was deemed by the economists as not requiring any revision of their fundamental premises or mechanical metaphors. By World War II, Wiener and others had realized that the mere introduction of statistics was not enough: they believed human beings were a different sort of machine, one that exhibited self-reference, duplicity, interactive emergence, and possibly even novelty. In the terminology of Chapter 2, humans were MOD and not just MAD.

This realization had prompted the development of game theory in wartime, but then rapidly left it behind as insufficiently appreciative of interactive nuance. The cyborgs then hoped that their insight would be expressed through the alternative instrumentality of the computer. When Bohnenblust et al. reintroduced the state-space approach, ostensibly to augment game theory, they were (perhaps unwittingly) repudiating MOD in favor of a world strictly MAD: hence Blackwell's reinterpretation in terms of "experiments" and games against nature. The opponent (Nature) was a pushover, an empty theater of action *sans* motive, an emitter of signals without intent. In order to navigate their MAD MAD MAD world, however, agents had to supposedly "know" everything already, as described in the preceding quotation, and thus transcend the capacities of any abstract computer. This could hardly be promoted as a promising framework for understanding communication, information acquisition, and learning. Worse, these superagents, however computationally endowed, would be stupider than most eight-year-olds, because their inferences would be vulnerable to corruption by the slightest hint of strategic deception or interaction. MOD would not be so effortlessly banished. Any information imparted to manipulate the mechanical induction procedures of the agent to their own detriment and to the benefit of the opponent could not itself be expressed as a partition of states

(Rubinstein, 1998a, pp. 53–56). This, in turn, raised the issue, What if one of the state spaces in the vast permutation of possibilities encompassed the proposition that the state-space portrayal of the world was itself a false representation of knowledge? People in the 1990s sometimes contemptuously disparaged cultural relativism as a self-refuting doctrine; but postmodernists have had no ironclad monopoly on self-refuting portrayals of knowledge in the postwar period.

The myriad contradictions of the Blackwell approach to information slowly played themselves out in neoclassical economics and game theory. A full enumeration of these events would trace the variant constructions of "information economics" through the three postwar American schools of orthodoxy. At Chicago, Stigler (1961) first attempted to restrict "information" to refer to distributions of Marshallian price statistics and nothing else; but this construction rapidly gave way to the rise of "rational expectations" theory (Sent, 1998). At Cowles, the generalized Blackwell approach was first used to discuss various market imperfections in an Arrow-Debreu framework, but then morphed into the doctrine that agents could infer everything they wanted to know from knowledge of the Arrow-Debreu model plus observed price movements (Radner, 1989); this was a major inspiration of the rational expectations approach and, of course, a major repudiation of the trademark Cowles political message. The infamous "no-trade theorems" lay in wait at the terminus of this program. At MIT, Paul Samuelson (1966; Bernstein, 1992) responded to the complaint that price movements were stochastic whereas the neoclassical models were deterministic by proposing that "properly anticipated prices" would fluctuate randomly due to the fact of having incorporated all relevant "information." This in turn led Joseph Stiglitz (1993, 2000) and others to explore paradoxes of treating information as a commodity in a partial equilibrium framework. It will be necessary to make note that *none* of these developments made the slightest use of computational theory during their chequered histories. Once these and other trends devolved to their bitter and unpalatable economic conclusions, most theorists in the neoclassical tradition by the 1990s had become fed up to the gills with information, perfect or imperfect, symmetric or asymmetric, Blackwell or Shannon.

Rather than sketch a history of information internal to economics, it might be better to close this impressionistic survey with the view from someone situated outside of economics – that is, from a more concertedly cyborg vantage point. Economists were not the only ones interested in information as a "thing"; the spread of the computer was forcing reconceptualization of bits as commodities in many walks of life. Battles were being fought in the courts and in the workplace over the progressive commodification of information (Branscomb, 1994). Pundits were dismissing the obvious exodus of manufacturing industry to offshore low-wage sites by extolling the dawn of a new "information economy" (Porat, 1977). Something new was clearly happening, and the neoclassicals may have felt that they were in the *avant garde*, contemplating the future with the aid of their newfound Blackwell formalism. Nevertheless, in stark contrast with Shannon entropy, information was most decidedly not a "thing" in this model. Attempts to impose an ordering on the domain of [signal × states] matrices that would provide an unambiguous index of "informativeness" generally failed. There was no dependable method to translate "informativeness" into "benefit" (Marschak, 1971, p. 207). One direct implication was that Arrow's persistent commodification of information in his Walrasian model, and in his policy discourse, was groundless: there was no justification for it in the Blackwell tradition. This explains why partisans of this school often had to offer up some rather reckless rodomontade, whenever anyone ventured to ask the naive question, What is that thing called information in your mathematics?

> Economic agents have different "amounts" of information. Speaking about amounts of information suggests we have an economically relevant measure of the quantity or value of information. That is not in fact the case. . . . However, in the special case in which one person knows everything that another knows and more, we can say the former has more information. That suffices for my discussions of asymmetrically located information. (Spence, 1974, p. 58)

Frequently, the mathematics, far from instilling a little conceptual clarity, has fostered the practice of freely moving between the Shannon "thinglike" and Blackwell "ordering" conceptions of information, often within the ambit of a single model. The utter incoherence of a half century of information economics has been deftly dissected by James Boyle in his *Shamans, Software and Spleens*:

> Perfect information is a defining conceptual element of the analytical structure [of neoclassical economics] used to analyze markets driven by the absence of information in which information itself is a commodity. I have even offered an analogy. Imagine a theology that postulates ubiquitous God-given manna in its vision of a heavenly city, but otherwise assumes that virtue and hard work are both maximized under conditions of scarcity. Now use that theology to provide the basic theoretical structure for a practical discussion of the ethics of food shortages. (1996, pp. 35–36)

The second drawback of the Blackwell model in the context of the other sciences was that some respected voices at RAND tended to disparage the matrix decomposition as the wrong way to approach the social issues which one might think would be the province of economics. Not unexpectedly, given his interest in Shannon entropy and his skepticism about game theory, Richard Bellman was among those who questioned the cogency of the Blackwell framework. In common with control theory, it presupposed,

> albeit tacitly, that we have cause and effect under control, that we know both the objective and the duration of the control process. As a matter of fact, also implicit is the assumption that one knows what to observe and that the state variables can be measured with arbitrary accuracy. In the real world, none of these assumptions are uniformly valid. . . . we don't know how to describe the complex systems of society involving people, we don't understand cause and effect, which is to say the consequences of decisions, and we don't even know how to make our objectives reasonably precise. None of the requirements for classical science are met. (1984, p. 183)

For Bellman, the answer was to pursue a postclassical science, and not pine after a social physics dangling tantalizingly just beyond our reach. Bellman opted for information processors which could be computed, preserved in metal and bits, rather than those which stopped at existence proofs.

The third drawback of the Blackwell formalism in the context of contemporary science was that adherence to a frequentist conception of probability, the very hallmark of an objective science in the 1950s, was very difficult to reconcile with the idea that agents somehow came equipped with differential partitions of their otherwise commonly experienced state space. Where the probabilities of the "states" came from was an increasingly nagging problem, not adequately addressed by Neyman-Pearson calculations of Type I and Type II error. As neoclassical theoreticians, and indeed Blackwell himself, moved ever closer to subjectivist notions of probability, the cryptanalytic framework was rendered less and less relevant. Hence, economists increasingly "went Bayesian" while physical scientists predominantly remained frequentist. The paradox of elevating a limited statistical method of induction to pride of place as the alpha and omega of information processing had festered at the heart of the Cowles project from the war onward, and these patent divergences only made things worse: if Arrow et al. really believed that cutting-edge econometrics was incapable of settling the issue of the truth or falsity of neoclassical price theory, whyever would it solve any of the economic agent's problems of inductive inference in the marketplace?

The history of the postwar mangle of information theory by economists still awaits its chronicler; we cannot hope to provide a comprehensive survey here. Our limited purpose in highlighting the Shannon and Blackwell formalisms is to demonstrate that economists did not come to analyze information because they suddenly noticed – mirable dictu! – *knowledge* played an inescapable role in market operation, or *understanding* was indispensable to market functioning, or *information* loomed large as an object of market exchange. If empirical imperative had been the supposed trigger, then where had all those neoclassicals been for over a century? Had they all been massively misled by training their blinkered attention on poor stranded Robinson Crusoe? No tangible empirical discovery could possibly account for the novel fascination for information beginning in the 1950s on the part of the Cowlesmen, which then spread in traceable ways to the rest of the profession. Instead, the causes were very simple: there was Hayek, the socialist calculation controversy, the computer, and then the mandate of the cyborg sciences to explore the various aspects of C^3I for the military.

The Cowlesmen sought to retreat to a portrayal of the agent as a little econometrician, under the imprimatur of von Neumann–Morgenstern expected utility, but then found themselves suspended between three versions of agent as information processor, *no one of which they had proposed themselves*. There was the newfangled "information theory" of the communications engineers, brought to their attention by Kelly. There was the rapidly expanding cadre of "decision theorists" nurtured under the sign of OR and a few analytic philosophers, and formulated for a neoclassical context by Bohnenblust and Blackwell. And, finally, there was the real thing, the explicit computational models of mind as symbol processors, including the military-induced field of "artificial intelligence," to which Herbert Simon was a major contributor. Simon took it upon himself to remind the Cowlesmen from time to time that there really was a third alternative, however much they remained deaf to his entreaties: this is the topic of Chapter 7.

In this sense, orthodox economics owed its postwar shape and fin-de-siècle fascination with "information" to the cyborgs and to the military – its fascination, and its incoherence. While the original Cowlesmen may have known from whence their inspiration came, their students, their student's students and their epigones did not. This was the source of one of the most crippling aspects of postwar neoclassicism, namely, the utter absence of any consistency and coherence in assertions to model "information" or cognitive processing. It is not hyperbole to note that for the longest time anyone could write essentially anything he pleased about "information" in neoclassical journals; models were as idiosyncratic as the preferences which Walrasians are pledged to honor unto their last. The term "information" was not used with much of any mathematical consistency from one paper to the next; its referent was more insubstantial than Hegel's *Geist*.[60] The exile of the history of thought from economics departments, combined with general innocence of the history of science, made it possible for each successive generation of tyro scientists to believe that it would be the first, the blessed, the one to finally divine the secrets of information in the marketplace. It would comprise the intrepid cryptanalysts, cracking the secret codes of capitalism. This, too, can all be traced back to those original encounters with the cyborgs in the 1950s.

RIGOR MORTIS IN THE FIRST CASUALTY OF WAR

The American economics profession only adopted widespread uniform standards of mathematical training and research after World War II. This fact, perhaps prosaic and unprepossessing in itself, assumes heightened significance when firmly situated within the histories of mathematics, the natural sciences, and the contending schools of economics in that era. For the heritage of the cyborgs was not limited to questions of the contested meanings of communication, control, information, and the "correct" approach to the computer; it extended to the very evaluation of the promise and prospects of mathematics for fields which had yet to experience the bracing regimen of formalization. As we argued in Chapter 3, von Neumann himself had been inducted as a lieutenant in Hilbert's formalist program, only to switch metamathematical allegiances upon grasping the import of Gödel's theorems. His program of automata and his alliances with the American military were intended to infuse mathematics with a new and improved raison d'être, one grounded in the projects and inspirations of the applied sciences. This may very well have been the strategy of von Neumann and his confederates, but it was most definitely not the game plan for the preponderance of the postwar American mathematics community. Mathematicians had recoiled from the implications of Gödel's, and by implication von Neumann's, work, but in a different direction: one, for lack of a better label, we shall crudely refer to as "Bourbakism."

> For a few decades in the late thirties, forties and early fifties, the predominant view in American mathematical circles was the same as Bourbaki's: mathematics is an autonomous abstract subject, with no need of any input from the real world, with its own criteria of depth and beauty, and with an internal compass for guiding future growth. . . . Most

[60] Not everyone was as candid as Kreps (1990, p. 578n): "The terms of information economics, such as moral hazard, adverse selection, hidden action, hidden information, signalling, screening and so on, are used somewhat differently by different authors, so you must keep your eyes open when you see any of these terms in a book or article. . . . As a consumer of the literature, you should pay less attention to these labels and more to the 'rules of the game' – who knows what when, who does what when."

> of the creators of modern mathematics – certainly Gauss, Reimann, Poincaré, Hilbert, Hadamard, Birkhoff, Weyl, Wiener, von Neumann – would have regarded this view as utterly wrongheaded. (Lax, 1989, pp. 455–56)

The across-the-board postwar withdrawal of American mathematicians to their lairs deep within their citadel of "rigor," untroubled by the demands and perplexities of application, away from the messy realities of cross-disciplinary collaboration encountered in the late war effort, would normally be of little practical consequence for our narrative, except for the fact that this occurred precisely at the same instant that American economists were being drawn in the direction of greater mathematical training, pedagogy, and sophistication – that is, right when Cowles was being subsidized by RAND and the military to develop mathematical economics. The economists may not have fully realized it, but they were at the mercy of crosswinds howling out of mathematics and logic, not to mention the backwash from the Air Force and the squall of the Red Scare. The turbulence would trap more than one economist in a career downdraft.

The operant question after the Hotelling-Schultz debacle of the 1930s was not *whether* economics should be mathematized; rather, it was *how* this should be accomplished, and to what ends. It would have been a relatively straightforward process to import mathematics and mathematicians into economics if the procedure was to conform to a paradigmatic instance of the leavening benefits of formalization in some already existing science; but this shortcut was obstructed by sharply divergent images of the promise and prospects of mathematics prevalent in the postwar context. One could, of course, simply mimic specific mathematical models lifted wholesale from some existing natural science, which usually meant physics, and many did; but this did not automatically carry the cultural gravitas and authority that it might have done in the later nineteenth century (Mirowski, 1989a). After Gödel, it was an open question whether the use of just any mathematics of whatever provenance would guarantee consistency or comprehensive enumeration of logical consequences, much less something as forbidding as impersonal truth and transparent communication.

Outsiders to the cloisters of mathematicians often have the mistaken impression that once one establishes one's bona fides as a trained mathematician, then after that almost anything goes, and thenceforward all mathematical discourse somehow occupies the same glorious plane of legitimacy. Nothing could be more removed from the actual historical record. Some feisty postwar neoclassicals like Paul Samuelson tended to insinuate that the major conflict was between the mathematical illiterates and luddites in the economics profession and those blessed with mathematical talent and training, but this question of premature career obsolescence was a mere sideshow, and not at all the main event of the decade of the 1950s. The real action involved decisions about the sorts of mathematics to be used, the formats of legitimation they would be held to meet, the reference groups that would define the standards for superior mathematical expression, as well as the hierarchy of division of labor that the development of mathematical research would entail. Although a comprehensive survey would acknowledge that any given individual economist could occupy a philosophical point in a space of prodigious combinatorial proportions, when it came to the use of mathematics there did exist a small number of relatively discrete positions that came to characterize the bulk of the avant-garde of the economics profession.

One influential position, centered at Cowles, tended to look to the American mathematics profession to set the tone and tenor of a successful project of formalization: this ultimately constituted the influence of Bourbakism within economics. A second position, held by a substantial subset of analysts at RAND, looked instead to von Neumann's program of computation and automata to provide a foundation for a new generation of mathematized disciplines.[61] Because Cowles and RAND had become yoked together in the 1950s, just when Cowles Mark II had repudiated its earlier research mandate, it was inevitable that these divergent visions of mathematization would clash. Because historians have avoided treating the mathematization of economics as the expression of a struggle over potential options for legitimation, they have tended to overlook the outbreak of a rash of very serious controversies over the format of formalization in economics in the 1950s. A comprehensive history of the mathematization of the economics profession is desperately needed, because the issues have been brutally repressed in the interim; here we can only point to one or two incidents in order to evoke the 1950s atmosphere in which fundamentalist faith in mathematization came back to torment its staunchest advocates.

One cannot adequately comprehend Cowles Mark II without giving some attention to the metamathematical movement called "Bourbakism" in the immediate postwar period.[62] This initially Francophone movement sought to unify and codify the vast bulk of mathematics predicated

[61] Other possible approaches to the role of mathematics were hinted at in Chapter 4, associated with the Chicago school and the MIT approach. Because of our concentration on the advent of cyborgs at Cowles and RAND, we unfortunately cannot explore these variant constructions here. See, however, Hands & Mirowski, 1999.

[62] This all-too-brief summary is based on the more elaborate and careful description in Weintraub & Mirowski, 1994. The best technical description is Corry, 1992, 1996, 1997a. An interesting recent attempt to place Bourbaki in its social context is Aubin, 1997.

upon the conviction that there were a very few mathematical "structures" from which could be derived the whole body of rigorous mathematical analysis. Although it swept the postwar American mathematics profession, this approach was very "French," as one of its most perceptive critics noted:

> [The advent of Bourbakism exacerbated the contrast] between the "top down" approach to knowledge and the various "bottom up" or self-organizing approaches. The former tend to be built around one key principle or *structure*, that is, around a tool. And they rightly feel free to modify, narrow down and clean up their own scope by excluding everything that fails to fit. The latter tend to organize themselves around a class of *problems*. . . . The top down approach becomes typical of most parts of mathematics, after they have become mature and fully self-referential, and it finds its over-fulfillment and destructive caricature in Bourbaki. For Bourbaki, the fields to encourage were few in number, and the fields to discourage or suppress were many. They went so far to exclude (in fact, though perhaps not in law) most of hard classical analysis. Also unworthy was most of sloppy science. (Mandelbrot, 1989, pp. 10–11)

Bourbakists tended to favor set theory and topology to express their structuralist tendencies, and to disdain any appeal to paradigmatic applications in order to justify their own taxonomies and classifications of the mathematics they thought warranted rigorous treatment. Another skeptical mathematician recalls the mood of the 1950s: "A distinguished mathematician . . . pointedly remarked to me in 1955 that any existence theorem for partial differential equations which had been proved without using a topological fixed point theorem should be dismissed as applied mathematics" (Rota, 1997, p. 51). Heuristic concessions to the reader were considered bad form amongst the Bourbaki; it is not even clear this mathematics was written to be *read*.[63] But more to the point, Bourbakism made it *seem* as though the foundations of mathematics could be shorn up against the corrosive influence of Gödel's theorems; although in retrospect we can see this consisted more of smoke and mirrors than of actual accomplishment.[64]

[63] "It seemed very clear that no one was obliged to read Bourbaki. . . . a bible in mathematics is not like a bible in other subjects. It's a very well arranged cemetery with a beautiful array of tombstones" (Guedj, 1985, p. 20).

[64] The case against the foundational ambitions has been made with great verve by Leo Corry (1992, 1996). Outsiders should realize that the ambitions of Bourbaki are now treated in many quarters as sheer delusions: "the identification of mathematics with the axiomatic method for the presentation of mathematics was not yet [in the 1940s] thought to be a preposterous misunderstanding (only analytic philosophers pull such goofs today)" (Rota, 1997, p. 15). He neglects to include their cousins, the neoclassical economists.

The Bourbaki strain was introduced into Cowles in 1949 direct from France in the person of Gerard Debreu (Weintraub & Mirowski, 1994); but the susceptibility to the doctrine preceded his appearance, largely due to the installation of Tjalling Koopmans to the position of research director in 1948. Koopmans's relations with Marshall Stone in the Chicago mathematics department, another Bourbaki booster, helped prepare the way. Whatever the specifics of the matchup, the marriage of Cowles Mark II and Bourbaki was a match made in heaven. Everything about the philosophy resonated with Koopmans's and Arrow's predilections, from its "top-down" planner's orientation expressed through mathematics to its renunciation of grounding the success of economics in its empirical instantiations. Both Arrow and Koopmans were inclined to conflate mathematical rigor with the search for rationality and truth (Arrow, 1951b, p. 130; Simon 1991a, pp. 106–7); and this was the ne plus ultra of Bourbakism. The hyperelitism that was the hallmark of Bourbaki could easily be adapted to the regime of newly imposed security classifications, which hemmed them all round. The sloppy science promulgated down the Midway could be upbraided with a brisk dose of topology. The Cowlesmen especially grew enamored of the idea that Walrasian general equilibrium could serve as the "mother structure" for all of rigorous economics, a position adumbrated in Debreu's *Theory of Value* (1959) and Koopmans's 1957 *Three Essays on the State of Economic Science* (reprint 1991). Last but hardly least, the Bourbakist feint of behaving for all the world as though the nastier implications of the theorems of Gödel and Turing had effortlessly been circumvented through redoubled axiomatization would shape Cowles's attitudes toward computation for decades to come. In practice, Bourbaki would become a charm to ward off cyborgs.

The Opportunity Costs of Formalism

What has gotten lost in the effusions of self-congratulation of American economists about their sagacious embrace of formal methods and Bourbakist rigor is an awareness of the extent to which the novel aesthetic concerning mathematics after circa 1950 was resisted, not by the mathematically illiterate but rather by a broad swathe of the (admittedly small) community of mathematical economics (Weintraub & Gayer, 2000, 2001). Bluntly, very few economists thought the project of Arrow, Koopmans, Debreu, et al. to "save" neoclassicism by rendering it impregnably rigorous and rarifiedly abstract was such a keen idea. These practitioners refused to accept the insinuation that their only choice was to embrace or abjure all mathematical expression *tout court* on Cowles's terms; in their view, Cowles was using mathematics in a defensive and retrograde fashion, given trends in research in other sciences. What has often been portrayed as ineffectual dissent over the legitimacy of all mathematical expression, a futile resistance against the all-engulfing tide of history, was in fact at least in part an argument over conflicting images of good science. This difference of opinion over the future of mathematical economics flared up in the early 1950s in swift reaction to the New Order at Cowles Mark II, both within the Econometrics Society and at RAND.

The first flare-up was provoked by that perennial maverick and frequent stand-in for von Neumann in the postwar economics profession, Oskar Morgenstern.[65] Morgenstern had come to resent the antiempirical and Bourbakist turn taken at Cowles, and decided to do something to prevent the Cowles dominance of logistical organization of the Econometrics Society and the house journal *Econometrica* from becoming a vehicle for the spreading of their new attitudes amidst the profession. His chosen angle of attack in 1953 was to circulate a letter suggesting that it be made an ironclad prerequisite of admission as a Fellow of the Econometrics Society that the candidate have actually done some econometric work: "they must have been in one way or another in actual contact with data they have explored and exploited, for which purposes they may have even developed new methods." This was a direct slap at Koopmans and Cowles, and was understood as such by all parties involved. Koopmans was consequently directed to poll the other Fellows of the society on Morgenstern's proposal, only to discover to his dismay that support for Cowles within the society was actually in the minority. Koopmans himself argued against the proposal, as did Marschak, who had resort to the curious riposte that, "If [Morgenstern's] suggestions were adopted the following men could never aspire to the fellowship: John v. Neumann, Vilfredo Pareto, and Leon Walras." [66] The only allies of Cowles were the mathematician Griffith Evans, and economists Luigi Amoroso and René Roy. Agreeing with Morgenstern were Jan Ackerman, Oskar Anderson, R. C. Geary, P. C. Mahalanobis, Erich Schneider, and one of the original founders of the Society, Charles Roos. The proposal was only subsequently defeated by some subtle procedural maneuvers by Koopmans; but it did serve notice that there festered some very sharp differences amongst the

[65] The evidence for the following paragraph comes from an Econometric Society memorandum, September 18, 1953, distributed to all Fellows of the Society. A copy can be found in box 6, TKPY.

[66] Considering the Cowles's members reflex contempt for history, Marschak probably did not realize that he was mistaken about Pareto, who had indeed performed a few empirical exercises. Where he passed from ignorance to duplicity, however, came with his mention of von Neumann. He must have known that von Neumann would never have endorsed the Bourbakist turn at Cowles, or countenanced the idea that a mathematician could make substantial contributions to science entirely divorced from a grounding in the practical concerns of the subject matter.

membership as to the role and functions of mathematics in economic research, and the breadth of disquiet at the attempt by Cowles to set the terms under which such methodological questions might be discussed.

The second incident occurred as a "symposium" on the use of mathematics in economics in the *Review of Economics and Statistics* of November 1954, with backlashes echoing for another year. In what one participant gleefully called a "slugfest," a relatively obscure economist, after penning a two-page plea for discussion of the limitations of some of the practices of postwar mathematical economists, was subjected to what by any estimation must appear the overkill of nine different economists piling on abuse and scorn. The identity of the economists tapped to discipline the poor sacrificial lamb named David Novick gives an even better idea of the imbalance of this supposed "debate": Lawrence Klein, James Duesenberry, John Chipman, Jan Tinbergen, David Champernowne, Robert Solow, Robert Dorfman, Tjalling Koopmans, and Paul Samuelson. The sheer disproportion of it all did not go unnoticed: "I am puzzled as to how one should interpret this phenomenon. Is it a bold denial of the law of the conservation of momentum? – for a glancing blow from an unchalked cue has set some of the biggest balls in mathematical economics rolling. Or is it a manifestation of collective guilt?" (Stigler, 1955, p. 299).

For some, this incident marks the first time American neoclassicals felt sufficiently confident to flaunt their triumph over rival schools of economic theory; Koopmans in his *Three Essays* portrayed it as a turning point in economic methodology, seeing that, "in a somewhat organized response to a passionate and heroic revival of the challenge by David Novick, the discussion has reasonably well clarified the general issues involved" (1991, p. 172); but, in fact, a little bit of scrutiny reveals that there was something altogether different going on. The first clue is that of the nine respondents, six had some affiliation with RAND as of the date of the symposium, as did Novick's lone delayed defender (Enke, 1954). The second, more significant clue, nowhere stated or cited in the symposium or elsewhere, is that Novick was also a resident economist at RAND at the time. At the height of the Cold War, nothing – philosophy, politics, even mathematics – nothing was adequately understood on the basis of first impressions and open public discourse, and here was another confirmation of that principle. What at first glance might seem an insignificant flare-up of arcane methodological dissent in economics was in fact a mere symptom of a higher-stakes battle going on at RAND in the mid-1950s. The subtext of the debate, and the reason it belongs in this chapter, is that RAND had paid for Cowles to help develop a mathematical social science, but now some at RAND were having second thoughts about what had been purchased.

Rectifying the injustice begins with restoring the figure of rebuke, David Novick, to his rightful role as a significant protagonist in the dispute.[67] Novick was a product of the Columbia school of statistical economics in the 1930s who went on to a career in government service and, in a wartime capacity, as an economic planner with the War Production Board. In many ways he was the antithesis of the median Cowles economist: where they treated "planning" as an abstract theoretical proposition and "uncertainty" as a subset of mechanical statistical inference, Novick had been engaged in nitty-gritty disputes over tariff barriers, property taxes, gold standard fluctuations, and the like for over two decades. Much of his time was taken up with the statistics newly generated for the purposes of government management and regulation, which had instilled in him a healthy respect for the problems of constructing useful data, an appreciation he retained for the rest of his career. Novick was recruited by Charles Hitch to come to RAND in 1949, just as it was gearing up its economics program. The 1949 summer conference at RAND was intended to acclimatize the new recruits to their RAND assignments, and, initially, Novick was dazzled by the likes of Kenneth Arrow and Paul Samuelson. The recruits were nominally encouraged to pursue their individual research interests, but some unsubtle guidance was provided as to the Air Force's desires in the matter.[68] Novick, due to his earlier work on industrial planning, was tapped to do studies of vulnerabilities of particular industries to nuclear attack.

Later in 1949 RAND found itself stymied by the Air Force actively scuppering the results of weapons studies it was inclined to dislike by altering the previously supplied underlying cost data; RAND decided to protect itself from further cynical manipulation by instituting its own department of military cost analysis, and Novick was appointed its first

[67] Most of the information in the following paragraph comes from transcripts of interviews conducted with Novick, February 24 and June 30, 1988; copies can be found in NSMR. David Novick (1906–91): Ph.D., Columbia; U.S. Tariff Commission, 1934–38; National Security Resources Board, 1947–49; University of Puerto Rico, 1947–49; RAND, 1949–71.

[68] *Collins:* Was there some sense that these papers would have some possible relationship to the RAND enterprise?

Novick: Right. By this time they had already been talking about bombardment. They had not yet gotten to air defense. But strategic bombardment was the theme song of the day. . . .

Collins: But was there a shared sense that economists could be helpful in thinking through some of these issues related to bombardment?

Novick: Well, actually, Ed Paxson had already identified this. Because Paxson took bombardment into the concept of benefits. And benefits meant, what did your bombs do?

(Novick interview, by Martin Collins, the NSMR interviewer, February 24, 1988, pp. 12–13, NSMR)

head (Novick, 1988). What had begun as a simple proposition of pricing massive lumbering hardware like the B-36 bomber rapidly ran into the obstacle that the Air Force itself did not really know how much it cost to operate any particular airplane or weapons system; so RAND became embroiled in the protocols of virtual costing of weapons systems that it had not proposed and did not control or – worse – did not, strictly speaking, actually exist. Actual planes would exhibit widely variable costs configurations depending on a staggering array of variables; but because the military procurement thrived on accelerated technical change, cost horizons would of necessity refer to technologies that had yet to exit the drawing boards. What the neoclassical economist would treat as the most dependable single bit of information in the marketplace, the price, was a very insubstantial and intangible entity in the world of military procurement; so RAND nurtured another competence, which it could eventually sell to the military and its linked industries: cost estimates. By 1953 only RAND could in the United States claim the capacity to produce rationalized sets of cost estimates on weapons systems that had up to that instant been nothing more than the machine dreams of some colonels: for instance, Novick's unit provided the first cost estimates of the Atlas intercontinental ballistic missile to von Neumann's Teapot Committee at Palm Springs on Thanksgiving 1953.

Novick's notions of how to derive costs rapidly diverged from those favored by the neoclassical theorists at RAND. In his view, one did not begin by presuming the agents in the market already "knew" the range of possibilities; nor did one start by presuming that technologies must conform to one or another abstract production function amenable to econometric estimation. Some empirical regularities had to be built into the theory of costing, but these ideas had nothing to do with any "marginal" conditions and were rather situated somewhere between formal mathematics and the raw data: "The life cycle, the distinction between investment and recurring costs, these things are automatic to anyone who's ever been in the cost business. . . . if you go out with a stop watch and a tape measure and do a cost based on that attempt at precision, you are not costing reality." One did not passively extract "information" of quantifiable precision; nor did one passively depend on "the market" to settle the matter: one directly intervened in the process. This is nicely illustrated by one of Novick's anecdotes:

> *Collins:* How did you go about providing a cost estimate for these things where the technology was unproven?
>
> *Novick:* Not only unproven, undeveloped.
>
> *Collins:* . . . How does an economist, someone who does cost analysis, grapple with that kind of situation?

Novick: You go to San Diego and get hold of a Belgian refugee by the name of Bossert. . . . you learn for the first time about a thing called liquid oxygen, which is the fuel that is going to propel these things, and you learn about the delicate balance between fuel and throw weight and vehicle carrier weight, and you learn it's a tough job. You then go to Huntsville, where they have already built a Redstone. . . .

Collins: What did these interviews tell you? They told you the complexity of the technology. What did that enable you to do in terms of the costing activity?

Novick: Dream and I mean that literally. Because once I learned about liquid oxygen, I first went to Union Carbide, which was regarded as *the* source. I felt they were too smug and too uncooperative to be of any use to me. I had learned there were three or four smaller outfits. . . . I chose Air Products. When I got to Air Products, they saw this as an opportunity to sell their products, so they essentially educated me.[69]

Still later, it became obvious that what clients wanted was not for RAND to perform one fixed set of estimates for weapons systems with configurations dictated from elsewhere, or simply provide bland impartial measurements of how well existing weapons worked; they importuned RAND to help them to play with various virtual realities themselves, trying out alternative configurations and variant cost structures. Thus Novick's cost analysis department joined forces with RAND's computer sciences unit, and developed something very like the spreadsheet programs that subsequently became one of the first "killer apps" of the later generation personal computers. As cost analysis went virtual and became computerized, its distance from American OR and especially systems analysis grew less and less distinct. Novick realized that his earlier experiences with the War Production Board could be put to good use in convincing the military to operate more like a planned economy; but that would entail using RAND's expertise not just as a source of statistics or projections of virtual weapons profiles and costs, but in having RAND become engaged in all levels of the conceptualization of strategy.[70] These ambitions were realized to a large extent when Robert McNamara introduced "Program Budgeting" into the Department of Defense in

[69] Novick interview, by Martin Collins, June 20, 1988, pp. 22–23, NSMR.

[70] "I reasoned that, if USAF had a good expense accounting system at the base and command levels, the numbers for operating costs would fall out automatically. . . . it seemed that earlier work I had done in costs at the Tariff Commission and on the War Production Board's Controlled Materials Plan . . . might be useful. CMP was a budgeting system, planning system, or programming system to manage the nation's resources for war" (Novick, 1988, pp. 8–9).

1961 (Shapley, 1993); and the expert who had written the textbook was Novick.

While Novick's cost analysis department was busy congealing virtual weapons out of real interventions in the innovation process, and generating computer simulations of costs out of time-consuming immersion in the details of weaponry, the Mathematics and Economics sections of RAND were busy antiseptically thinking their way through the nuclear unthinkable, and conjuring the pure laws of planning out of their traditions of Walrasian general equilibrium and game theory. Novick maintains that he enjoyed good relations with them, though primarily because of his own rather catholic notions of tolerance.[71] But this outward show of camaraderie must be qualified by the tempestuous context of RAND in the early 1950s. Following upon the relatively disastrous failure of Ed Paxson's early "Strategic Bombing Systems Analysis" and the Air Defense Study of 1951 (Jardini, 1996), RAND's Mathematics and Economics departments undertook a large-scale review and reconsideration of the prospects and pretensions of systems analysis.

In the early 1950s six economists (Armen Alchian, G. Bodenhorn, Steven Enke, Charles Hitch, Jack Hirshleifer, and Andrew Marshall) subjected the early techniques of systems analysis to scathing critique and eventually sought to replace it with a theoretical "economics of innovation" (Alchian et al., 1951; Hounshell, 2000). Their major objection to the systems analyses then being carried out at RAND was that "specialized knowledge and extensive computations are insufficient to indicate which systems are better or worse than others." As neoclassical theorists, they were convinced that they understood the problem of choice better than someone like Novick or Paxson, even if the range of alternatives were so ill-specified as to border on the ethereal. The critics sought to fortify the scientific credentials of systems analysis by restricting the operational objectives to concrete measurable criteria, and then ranking alternatives according to the only index they felt could summarize the complicated trade-offs – namely, costs provided by the marketplace. The analyst should not violate his neutrality by attempting to intervene in the process of concretizing the as-yet unrealized weapons system; instead, he should rely on

[71] *Collins:* "What was the relationship between your division and the social sciences? There must have been some fairly close points of interaction.

Novick: Only because I was interested in what they were doing. They weren't interested in what I was doing at all. But I thought they were doing a great job, and I would continuously tell them so, so that I was welcome, not sought out but welcome. Furthermore, I'd spent a lot of time in middle Europe, and most of them came from middle Europe.

(Novick interview, June 20, 1988, p. 50, NSMR)

an abstract general theory of research and development to be provided by the economic theorist. The critics reveled in the hard-nosed operationalist rhetoric broadcast throughout economics by Paul Samuelson in the 1930s (Mirowski, 1998c) and pioneered the cold-blooded deployal of maximization, which later became the basis for every RAND stereotype from Dr. Strangelove onward: "Given (1) the pattern of Research and Development expenditure among weapons in the current Air Force program, and (2) the implied dates at which they will probably become available for operational use, what time phasing and selection of weapons (and quantities) will yield a specified time pattern of kill potential at minimum cost?"[72]

The demands of von Neumann's "Teapot Committee" to accelerate the intercontinental ballistic missile program in 1953, vesting control for its development with the main proponents of the efficacy of systems engineering, the firm of Ramo-Wooldridge and General Bernard Schriever's Western Development Division, brought this internal RAND controversy to a head.[73] Armen Alchian in particular went on the offensive in 1953, attacking Edward Quade's systems analysis course and Novick's cost analysis section. Alchian, pursuing his analogy of neoclassical maximization with biological "evolution" first proposed in 1950, opined that multiple configurations of generic weapons systems should be pursued simultaneously by the government contracts through relatively unstructured initiatives proposed and controlled by private contractors, with something very much like the market left to "select" the optimal choice. Kenneth Arrow, somewhat less inclined than Alchian to challenge the defense contractors and indulge a political distaste for planning, recast the problem of systems analysis in an unpublished 1955 RAND memo as yet another instance of his favored trope of costly search under uncertainty.[74] Whereas Alchian was convinced there was no economic

[72] Armen Alchian, "Phasing Problems II," RAND D-1136, January 8, 1952, quoted in Hounshell, 1997a, p. 10.

[73] On Ramo-Wooldridge and Schriever, see Hughes, 1998, chap. 3; Getting, 1989; Hounshell, 1997a. The history of the Teapot Committee is covered in Neufeld, 1990. Ramo-Wooldridge's entire role as a contractor in the production of the ICBM was predicated upon its expertise in systems engineering and systems analysis, completely removed from any aspect of hardware engineering.

[74] Kenneth Arrow, "Economic Aspects of Military Research and Development," August 30, 1955, RAND D-3142. The crux of this argument became the basis for Arrow's classic article "Economic Welfare and the Allocation of Resources for Invention" (1962), only now shorn of most of its crucial military content and context. Arrow's later work on "learning-by-doing" was itself a denatured and unacknowledged version of the widely used "learning curves" found throughout the wartime aircraft engineering profession, first developed by Theodore Wright of Cornell Aeronautics. Learning curves and product life cycles were everyday tools of Novick's Cost Analysis section.

problem that systems analysis had any business or credentials to solve – perhaps the evolution of private-sector R&D would itself inevitably bequeath the military an ICBM – Arrow felt that his usual panoply of rigidities and market failures dictated that there was some role for government subvention of those perpetually underfunded aerospace firms. But between those two positions, there was still substantial agreement that the kinds of hands-on planning and quantification and prediction practiced by Novick and the systems analysts had trespassed the limits of science and social planning that their tradition had pledged to defend. Thus, systems analysis was deemed antithetical to the Cowles doctrine of the economic agent as statistician, information processor, and utility computer.

Strangely enough, in 1954 this dispute came to be played out upon the rather unlikely battleground of the correct or defensible mathematization of political economy. Novick, understandably, felt that all the hard-nosed rhetoric about costs and options was wildly overstated, and that the pretensions of the neoclassicals to understand production processes, much less R&D, were more than one sandwich short of a picnic. Rather grandiose projects like ICBMs simply would not happen unless someone in charge made them happen according to fairly well understood planned sequences, in his view; one did not wait for the various contending factions to work out their needs and desires beforehand but instead anticipated and even conjured the goals that the project might meet, out of imagination and more than a little cajolery of the clients involved. In a word, the economists counseled patience, passivity, and letting the market decide, whereas the systems analysts believed they were actively pioneering the rational practice of the building of weapons systems. Yet, systems analysts like Novick suspected that the economists made up for their lack of operational specificity in their catechism of market superiority with a spurious precision of analytical, primarily mathematical, technique. The contrast came down to what both sides meant by being "quantifiable." The neoclassicals often engaged in operationalist rhetoric of "kill ratios" or welfare indices, but in the final analysis they revealed they respected mathematical pyrotechnics far more than demonstrable familiarity with heat resistance or thrust capabilities measured on the shop floor. This attitude, of course, was one of the main hallmarks of Cowles Mark II. Novick, by contrast, knew when he was dealing with squidgy numbers, but drew the line when it came to those objective functions so beloved by the neoclassicals. As he reported in retrospect: "If you were going to ask me to quantify things, I would not endanger my reputation by trying to do nonquantifiable things. . . . what was involved for the most part were social values. . . . In other words, when would people surrender? What was the maximum fire you could put to their heels? Things of that kind."[75]

Here, in all its complicated complicity, is the background to Novick's two-page *cri de coeur* published in the November 1954 *Review of Economics and Statistics*. His complaint, philosophically unsophisticated but nonetheless heartfelt, was that logical consistency should never be confused with relevance or salience, and that algebraic expression does not imply empirical quantitative availability. "The current use of mathematics in social science is largely a form of intellectual shorthand and in no way demonstrates that the methods heretofore so successful in the physical sciences have suddenly become adaptable to the social sciences. . . . the theory may be a most interesting one, susceptible to 'toy' proofs, but not at all adaptable to the facts of the real world" (1954, p. 357). Although the language might today appear homespun and the sentiments philistine, with a modicum of distance we can appreciate that the indictment of the practices of mathematical economists was essentially the same as that proffered by John von Neumann, and described in Chapter 3. Economics mimics physics but, for various structural reasons, does not behave like physics. The legitimacy of mathematical expression should therefore derive from its solid grounding in the applied phenomenon and not from any magical potency of a sacerdotal language or platonic rationality. Cutting-edge techniques may serve equally to disguise conceptual bankruptcy as to demonstrate originality. Of course, Novick's own right to be heard was severely compromised by his admission that, "those of us who have only a limited training and a still more limited experience in mathematics are too often cowed by the symbols and are afraid to challenge them lest we be embarrassed by showing our ignorance"; but his sociological point nevertheless emerged loud and clear that Cowles's embrace of mathematics (really, Bourbakism) was willfully obscurantist and exclusionary.

The sheer brio with which many of the respondents lit into Novick makes for amusing reading a half century later. Lawrence Klein, speaking for the erstwhile econometric wing of Cowles, was possibly more profound than he realized when he suggested, "Perhaps we would not have come upon the 'fundamental equation of value theory' (the Slutsky equation) without the help of mathematics" (Novick, 1954, p. 360). After sneering at Novick, James Duesenberry did admit that, "In the final triumph of complete generalization no observation whatsoever will be inconsistent with maximization theory. This seems to be a fruitless game" (p. 362). John Chipman sought to absolve mathematical economics of its sins by insisting, "The lack of empirical content in our field is not peculiar to

[75] Novick interview, June 20, 1988, p. 45, NSMR.

mathematical economics. It is characteristic of the whole history of the subject" (p. 363). Jan Tinbergen, as would be expected, was the other staunch defender of econometrics, although he conceded, "Being myself a mathematician of only modest knowledge, I often experience considerable difficulties when reading Cowles Commission stuff" (p. 367). Robert Solow's screed must be read to be believed. It flits from disparaging Leontief's input-output analysis ("one of the few cases where . . . the fact-to-theory ratio is if anything too high") to noblesse oblige ("I have often my self thought it rather dense of foreigners not to speak English, so I can understand his point of view"), to a version of sociobiology *avant la lettre* ("Why is economic theory becoming more not less mathematical? . . . As a good Darwinian, I believe this is no accident") (p. 373). But one must bypass the pleasures of lighthearted banter and masculine ridicule to finally encounter at the real target of Novick's distress, the designated hitter of Cowles, Tjalling Koopmans.

Koopmans, more than any other participant in the "symposium," understood that the real complaint was not so much directed at "mathematical tools of long standing in economics, such as diagrammatic analysis or simple calculus"; rather, it was the appropriateness of "matrix algebra, set theory, difference equations, stochastic processes, statistical inference, and the axiomatic method, which are now the issue" (p. 377). Undaunted, Koopmans suggested that disquiet over "increasing tendencies to formalism" also had swept the physics profession with the advent of quantum mechanics, but with time "the clamor has abated" and everyone had come to voluntarily accept the fruitfulness of high-powered mathematics. (Of course, some students had abandoned physics in the interim; but few would be privy to actual biographies of our protagonists – not to mention any awareness of the relative *unimportance* of axiomatization in physics.) His presumption that the history of physics provided the best guidance to the future of research practices in economics was, interestingly enough, the occasion of his only oblique reference to the Cold War milieu that saturated the background to this discussion: "In fact, the headstart of physics over the political and social arts and sciences has since become the major threat to contemporary civilization." Next, he chastised Novick for his naiveté in thinking that anyone believes that mathematics was being used to browbeat or mesmerize an unsophisticated audience. Notwithstanding that claim, he then embarked upon a paean to mathematics as "the realm of ideas par excellence" and proceeded to contradict Novick's own primary thesis by insisting, "The appropriateness of mathematical reasoning in economics is not dependent upon how firmly or shakily the premises are established" (p. 378). His prime exhibit for this claim was the brand-new publication by his Cowles confederates Kenneth Arrow and Gerard Debreu on the existence theorem for Walrasian general equilibrium (1954). "Is there any topic more basic in contemporary economic theory?" he pleaded.

But who, in the final analysis, would serve as judge and jury as to which of these arcane exercises would eventually be deemed basic or worthwhile? Here Cold War discretion repressed the vital issues of science funding and science management, which have been the subject of much of this volume; instead, without once explicitly mentioning Bourbaki, the axiomatic approach was repackaged and retailed as a user-friendly doctrine for economists as familiar and unthreatening as the age-old division of labor:

> It seems impractical, for the few people who have made the investment of time and effort necessary to understand and use these mathematical and statistical tools, to try and meet Dr. Novick's request for a verbal restatement. . . . In some cases originators of mathematical economic theories provide such statements. In other cases, they seem to be restrained by a curious preference, found in even more pronounced form with pure mathematicians, for letting the bare logic of their problem stand out alone, while considering any discussion of motivations, relevance or subsequent objectives as an undesirable admixture of inevitably somewhat subjective elements. Whether this attitude reflects true wisdom or false shame, it will in some cases be more efficient for other minds with different motivations to act as intermediaries. . . . For his immediate progress [the mathematical economist] is often more dependent on communication with mathematicians than economists. (Koopmans, 1954, p. 379)

In effect, Cowles had already thrown in its lot with the "mathematicians"; the literary economists would have to search out their own allies and make their own separate peace. Koopmans had previously revealed his contempt for the Institutionalists in the "Measurement without Theory" controversy in 1947; he was now serving notice that Cowles regarded itself as unencumbered by much of any demands for accessibility emanating from the economics profession at large. But in the context of the dispute between the systems analysts and the economists at RAND, there was another, albeit almost invisible, line drawn in the sand. Koopmans and some of the economists in residence at RAND were quietly declaring their independence from requirements that they produce analyses that were deemed "useful" by their primary patron, the Air Force, and, by implication, the cadre of systems analysts. The direction in which they were inclined to go was indicated in an internal Chicago memo by Koopmans:

> When a research contract between RAND and the Cowles Commission was taken on in 1949 it constituted exactly the mixture of theory and application on which I wanted to work. More recently RAND has exerted influence to narrow down the work to problems of transportation

> and location. While I am still interested in facilitating and stimulating work of Cowles Commission staff members in this area, my own interests are shifting to problems more central to economics in which the interaction between large groups of decision makers is studied as a process determining states of equilibrium or change.[76]

It appears that this brace of disputes at RAND, and to a lesser extent contretemps within the community of mathematical economists, acted to push a few major players at Cowles closer to Bourbakist orientations and to have the Cowles program coalesce around an anti–von Neumann alliance. The timing of these disputes is not immaterial: the year 1954 marks the publication of the centerpiece of the postwar neoclassical orthodoxy, the Arrow-Debreu model of general economic equilibrium. It may seem counterintuitive to claim (as we did in Chapter 5) that the Cold War experiences of a few key figures were responsible not only for the postwar orthodoxy of the neoclassical economic agent as information processor, and for events at RAND to turn them against von Neumann, but to venture further and assert that the exemplar of neoclassical orthodoxy, the very style of mathematics and the understanding of its role in intellectual inquiry, was directly linked to the immersion of the neoclassical program into the big chill of the superpower standoff.

For those who dislike consolidated explanations, any air of incongruity is dispelled by the realization that the operation of "influence" is never solely that of passive acceptance or imitation. The Cowlesmen may have basked in the genius of the comet von Neumann as it streaked through Chicago and Santa Monica; but that didn't mean that they had to genuflect toward his "third period" enthusiasms, nor conform to his vision of a less-than-celestial brand of mathematical rigor. Ultimately, the hostility of von Neumann to the neoclassical program took its toll, both directly through his comments and indirectly through his progeny; yet his stature and influence were such that the mathematical neoclassicals had no choice but to define themselves against his legacy. There is no better example of how love turns into hate and imitation breeds contempt than the Arrow-Debreu model.

Fixing the Point

The only substantial historical account of the genesis of the Arrow-Debreu model is that provided by Weintraub (1985, chap. 6). The story therein starts with Karl Menger's Mathematical Colloquium in Vienna in the 1930s – among whose participants were numbered Kurt Gödel and Alfred Tarski – and the papers on the equations of equilibrium by Kurt

[76] Tjalling Koopmans, "Replies to Questionnaire of Self-Study Committee," January 11, 1954, box 17, folder 312, TKPY.

Schlesinger (1933) and Abraham Wald (1936). The narrative then hops to von Neumann's 1937 paper "On an Economic Equation System and a Generalization of the Brouwer Fixed Point Theorem," which had been publicly presented as early as 1932 to a seminar at Princeton, and in 1936 to Menger's Colloquium. Weintraub calls this paper "the single most important article in mathematical economics. . . . [It] contains the first use in economics of certain now common tools: explicit duality arguments, explicit fixed-point techniques for an existence proof, and convexity arguments" (Weintraub, 1985, pp. 74–75). Menger's colloquium was disrupted and scattered by the *Anschluss*, and the center of gravity subsequently shifted to America. Von Neumann and Morgenstern's *TGEB* provides the next landmark, primarily because it acknowledges two alternative methods of constructing the minimax proof, one citing Kakutani's 1941 simplification of the fixed-point theorem, and the second relying upon a nontopological approach predicated upon the Hahn-Banach theorem, derived from a paper by Jean Ville. Finally, there is the work done at RAND: primarily, John Nash's 1950 "Equilibrium Points" paper, also using the Kakutani fixed-point theorem (noted in the previous section as due to an initial suggestion by Gale). Weintraub then quotes an extended passage from a personal letter from Arrow tying together all these threads:

> According to my recollection, someone at RAND prepared an English translation of the [Wald] *Ergebnisse* papers to be used by Samuelson and Solow in their projected book (sponsored by RAND), which emerged years later in collaboration with Dorfman. I read the translations and somehow derived the conviction that Wald was using a disguised fixed-point argument (this was after seeing Nash's papers). In the Fall of 1951 I thought about this combination of ideas and quickly saw that a competitive equilibrium could be described as a [Nash] equilibrium point of a suitably defined game by adding some artificial players who chose prices and others who chose marginal utilities of income for the individuals. The Koopmans paper then played an essential role in showing that convexity and compactness conditions could be assumed with no loss of generality, so that the Nash theorem could be applied. Some correspondence revealed that Debreu in Chicago . . . was working along some very similar lines. . . . We then combined forces and produced our joint paper. (1985, p. 104)

Accepting Weintraub's account, and realizing that the existence proof was brought to fruition precisely in the midst of the controversy raging at RAND over the correct approach to mathematical economics, we can begin to contemplate the significance of each of the nodes of this particular narrative. In Vienna, there were actually two different streams of inquiry: one, running from Schlesinger to von Neumann (1937) and onward into activity analysis, was concerned with feasible and maximal growth rates dictated by technological relations in a dynamic circular-flow system; and the other, the question of existence of equilibrium in an avowedly static Walrasian system, found solely in Wald. The tension in the narrative arises because if the purpose were simply to track the evolution of "tools," and in particular the fixed-point argument, then von Neumann would absorb the lion's share of attention, which was undeniably his due; but if the telos of the narrative resides instead in an inquiry into the cogency and consistency of the Walrasian model, then instead Wald becomes the major progenitor. The revealing aspect of Arrow's account is that the way to bring the two separate strains of fixed-point theorems and Walras together was to *leave von Neumann out:* and the catalyst of this realization was RAND. Cowles in general and Arrow in particular were not enamored of von Neumann's conceptualization of the equilibrium of a game (Arrow, 1951b); but Nash provided an alternative that was much more solidly rooted in the Walrasian tradition of the constrained optimization of utility. Von Neumann embraced the multiplicity of social conventions, which would downgrade the salience of the question of existence, not to mention anything so pretentious as the "fundamental welfare theorem"; but Nash sought the unique paranoid solution of strategic rationality. Hence the burst of hothouse attention accorded to Nash equilibrium at RAND in the early 1950s, along with all the Cold War considerations that had brought that situation about, precipitated out the central doctrine of the postwar neoclassical orthodoxy, the Arrow-Debreu model. In a historical generalization that takes on greater urgency in the fin-de-siècle world of economics, game theory begat the iconic model of postwar American Walrasian orthodoxy, and not the reverse. In this sense even the Arrow-Debreu model bears the mark of the cyborg on its forehead.

The path to the Arrow-Debreu model reveals the extent to which game-theoretic innovations first dominated but subsequently became repressed in the mathematical elaboration of the model. First came the political interest in welfare economics on the part of both Arrow and Debreu in earlier writings. Next came Debreu's 1952 paper in which he defined a game in a more abstract modality than Nash, bringing to bear an improved version of the Kakutani fixed-point theorem due to Eilenberg and Montgomery.[77] This paper has a "historical note" which acknowledges von Neumann's priority in the fixed point theorem, but reserves primary credit for the "notion of equilibrium point first formalized by J. F. Nash."

[77] See the reprint in Debreu, 1983b, chap. 2. There the paper is listed as having been written under a RAND contract, while Saunders Maç Lane and André Weil are thanked for their help. Both Weil and Eilenberg were erstwhile members of Bourbaki.

Next came the joint Arrow-Debreu paper, written under ONR contract and first read at the Econometric Society meetings in Chicago in December 1952, for which the earlier Debreu fixed-point proof constitutes the "heart of the existence proof" (Weintraub, 1985, p. 104). The Walrasian problematic is ushered into the realm of formalization of games by means of some tinkering with definitions: "A central concept is that of an abstract economy, a generalization of the concept of a game" (in Debreu, 1983c, p. 69).

What is especially noteworthy is Arrow and Debreu find themselves impelled to resort to an old trick of von Neumann, namely, to introduce a *dummy* or, in this case, what they call a "fictitious participant who chooses prices." But instead of being dragooned into reducing more complex solution concepts to simpler ones, the purpose of the dummy here was to circumvent the awkward situation where the garden-variety "agents" got to choose quantities of purchases but not prices. As Werner Hildenbrand felicitously puts it (in Debreu, 1983b, p. 18), the fictitious agent "whose role is to choose the price system" reduces the problem of existence of Walrasian equilibrium to a problem of existence of a Nash equilibrium. The dummy here is the alter ego of Arrow and Debreu themselves, "choosing" the prices as planner which would exactly mimic the "perfect" market.[78] In the "Historical Note" to this paper, the lineage is traced from Cassel to Schlesinger to Wald: von Neumann is now nowhere in evidence. By the time we get to Debreu's 1959 *Theory of Value*, the last vestiges of von Neumann are swept away, for the 1954 proof technique is entirely replaced by the "excess demand approach," which represses any explicit reference to game theory, and conveniently exiles the dummy from the land of Walras, without seriously addressing any of the problems concerning who or what sets the prices (p. 89, n4). The entire model has been rendered subordinate to the fixed-point theorem, credited now only to Brouwer and Kakutani. Von Neumann does not even make it into the text as progenitor of game theory or rate an entry in the index.

Observing how von Neumann was repudiated even in the arcane region of proof techniques is equally poignant. In his 1937 expanding economy

[78] This is rendered especially explicit in an unpublished Cowles Commission staff paper by Kenneth Arrow, March 11, 1947, entitled "Planning under Uncertainty: A Preliminary Report," CFYU. Here the transition from agent as econometrician to game theoretic "dummy" as social planner to the fixed-point manifestation of the existence of Walrasian equilibrium is rendered especially transparent: "The concept of policy used here is clearly analogous to the von Neumann-Morgenstern definition of 'strategy.' In effect, planning is considered as a game in which a (fictitious) opposing player is assigned the job of selecting which of the possible configurations of future events (more precisely, which of the possible joint probability distributions of future events) is actually to prevail. (See Marschak's review of von Neumann and Morgenstern, pp. 109–10)."

model, von Neumann pioneered the first use in economics of the Brouwer fixed-point theorem explicitly in the context of a nonconstructive proof: basically, he showed the negation of his theorem would lead to a contradiction. In that period of his life, making such a logical point devoid of any algorithmic content was sufficient – existence just suggested some state of affairs was conceivably possible. By the time of *TGEB*, however, he had changed his mind about the wellsprings of useful mathematics, and the document is consequently a transitional affair, as we argued in Chapter 3. The axiomatizations play little substantive role, whereas the proofs of the minimax theorem are still nonconstructive; but the way forward toward an algorithmic approach appears in section 16 on "Linearity and Convexity." This program was essentially carried forward by George Dantzig, who provided the first constructive proof of the minimax in 1956, as well as in the late paper by Brown and von Neumann (1950) on automated game play. So by the 1950s, at least for von Neumann, the fixed-point theorem had been downgraded in significance in favor of constructive proofs for what he considered to be the central theorems of game theory. Indeed, the contrast in *TGEB* section 17.8 between the "indirect" and "direct" methods of proof of the minimax later became for von Neumann one of the main reasons to privilege the minimax over solutions such as Nash's equilibrium point: it was susceptible to constructive proof, whereas the Nash equilibrium was not. Still later, constructive algorithms eclipsed the importance of games altogether.

We can now appreciate from the documentation presented here that the Cowlesmen were traveling in the diametrically opposite direction by the 1950s. Whereas neoclassical economics had a lineage rooted in mechanics and therefore constructive models, the lesson derived by Arrow, Debreu, and Nash from Bourbaki was that questions of existence of equilibrium were really just demonstrations of the logical consistency of the model; there was no pressing commitment to models as a calculative device that mimicked reality. They all enthusiastically embraced fixed-point theorems in all their nonconstructive glory as *defining* the essence of equilibrium, to the neglect of whether and how it came about. In this sense they did finally cut themselves free from their origins in classical mechanics, which may go some distance in explaining how, in their own estimation, the history of their own economic tradition ceased to matter for their project. This embrace of nonconstructive proof techniques would constitute one of the major barriers to reconciliation with cyborgs, as we shall discover in the next section. It may also contribute to an explanation of von Neumann's disdain for Nash's solution concept as "trivial": after all, he had deployed the Brouwer theorem in economics more than a decade before and had subsequently decided that it was a dead end.

Pace Koopmans, mathematics is not some empty infrastructure of logical inference that merely passes along any premises, as if on a conveyer belt, which are openly and transparently inserted at the beginning. Specific mathematical techniques bear all sorts of unstated and partially obscured metaphorical baggage, traditional connotations and accretions accumulated over long histories of usage. What, then, were some of the half-buried implications of this signal resort to Brouwer's and Kakutani's fixed-point theorems? To begin, we can reproduce Koopmans's own gloss on the Cowles interpretation of the theorems:

> A simple prototype of this class of theorems says, roughly, that if one is given a continuous mapping (however distorting) of the points of a circular disk into points of another circular disk of the same radius, and if thereafter the second disk is placed in any position on top of the first, then there will be at least one point which is found directly under (coincides with) its image. Later versions allow greater generality in the shape and dimensionality of the set of points considered, and allow each point of the first set to be mapped not merely into a single point, but into some subset of points of the (identically shaped) second set, with appropriate generalization of the concept of continuity. The statement then becomes that, upon putting the two sets together, at least one point of the first set coincides with some point of its image subset in the second set. In the application to competitive equilibrium, the mapping projects the bundle of choices made by the various market participants, now to be regarded as one single point, into a certain set of choice bundles. This set contains for each participant all those choices that maximize his goal (profit, satisfaction), in comparison with all the other choices that remain available to him, given the way in which the choices made by other participants affect his budget restraint. If each participant finds that his choice maximizes his particular goal within the subset of choices remaining available to him, then he has no incentive to change his choice, and a competitive equilibrium therefore exists. (1991, pp. 57–58)

Of course, there persists the question of what the demonstration of the existence of such a possibility achieves for the postwar economist. Assertions that this was how a "real science" like physics did things would simply not wash, as Koopmans the lapsed physicist acknowledged in his *Three Essays* (1991, p. 58). Working physicists almost never attended to existence theorems. Fixed-point theorems at Cowles tended to telegraph a different set of commitments and expectations. First of all, they were impossible to separate from the larger trend toward escalated standards of acceptable mathematical proof. As Rota stated, any existence proofs concerning partial differential equations not using fixed-point theorems in the 1950s were just deemed awkward applied trash. The mere fact that economists could actually refer to a proof technique published little more than a decade previous would broadcast the subliminal message that economists could cut the mustard; they were as au courant as those newfangled TV consoles appearing in those newly constructed recreation rooms. But this resort to topology was equally consciously calculated to conjure the rarified atmosphere of Bourbaki, and the standards of mathematical rigor increasingly prevalent in mathematics departments. Even though von Neumann had been the first to import such proof techniques into economics, and his intervention in Cowles is what had effectively first brought them to many economists' attention, his was not the legacy Cowlesmen wanted to genuflect before (Debreu, 1959, p. ix).

Topological arguments were *global* arguments; and this also sported certain connotations in the 1950s. For instance, it encouraged an ambition toward "generality," which might not be fully motivated from within the tradition of economic theory alone. (Recall: Walras's book may have been "pure," but that was not necessarily a synonym for generality.) It imagined the theorist looking down on the economy from a detached god's-eye view and, by proving a general equilibrium was a Pareto optimum, pronouncing it good. This enjoyed a certain resonance with the Cowles understanding of the terminology of "planning" and market socialism, albeit one that drove their colleagues in the Chicago economics department to distraction. It also implied, without ever having to proffer any justification, that there was really only one generic sort of market, at least at that dizzying level of abstraction.

For the nonce, however, there was one further subtle implication of the resort to fixed-point arguments. Fixed-point theorems seemed to hold out the prospect that individual calculations of self-interested advantage need not succumb to problems of infinite regress or paradoxes of self-refutation in the aggregate. The worry that unchecked competition would dissolve into Pyrrhic victory for the market – perhaps in a Hobbesean downward spiral of cutthroat competition, or maybe the economy as one giant supermonopoly, or possibly in one final orgy of "creative destruction"– was one that dogged economics throughout the first half century. The Kakutani theorem provided the soothing prescription that there was at least one configuration of prices and quantities where everyone was doing their individual best, given that everyone else was doing their level best. The attendant logic of "price-taking behavior" and their agnostic cognitive stance rendered this interpretation a bit strained in the Arrow-Debreu framework, although the Nash interpretation rendered the nature of the problem a bit clearer: the strategic paranoid knew they were doing their personal best, given the opponents were doing their best, given that the paranoid knew that the opponents were doing their best, given that the opponents knew that the paranoid knew they were doing their best . . . ad infinitum. The prime attraction of the fixed-point argument, and the reason it came to be the proof technique of choice throughout the economic orthodoxy of the second half of the twentieth century, was that it appeared to circumvent the paradoxes of self-reference that had haunted mathematical arguments in logic and the social sciences since the beginning of the century.

The allure of the fixed-point theorem was such that it was used not once, but thrice at Cowles in 1954, each time in order to disarm the objection that prior knowledge of the existence of a deterministic equilibrium would be a self-falsifying proposition, and therefore permanently rule out of bounds any ambition that the social sciences could partake of the same formal structure enjoyed by the natural sciences.[79] The first and most high-profile employment was in the construction of the Arrow-Debreu model, with the reassuring moral that "competition" might lead to "efficiency"; but the other two uses were commensurately consequential for the neoclassical orthodoxy: they were the paper by Grunberg and Modigliani (1954) on the "Predictability of Social Events" and the paper by Herbert Simon (1954) on the "Possibility of Election Predictions." One troubling aspect of the Cowles shift to portraying the economic agent as little econometric information processor was the age-old complaint that successful predictions, if made public, would be self-refuting, because they would prompt destabilizing feedback. In one sense, this was precisely the challenge to which both cybernetics and game theory had arisen to confront; but for Cowles, the more immediate worry was that this objection would obviate the cogency of the neoclassical equilibrium theory. Again under the inspiration of Nash, Modigliani brought the fixed-point theorem to bear upon a simple supply-and-demand model in order to argue that *in principle* correct predictions need not derange the model, because there exists at least one point where all prophecies could be self-fulfilling. Similarly, Simon used a fixed-point argument to insist that correct predictions are possible *in principle* even if predictions sway voting behavior. The importance of both these papers is that they served as the immediate impetus for the 1961 paper by Muth on the concept of "rational expectations"; and, as this paper is treated as the font of the spread of the rational expectations theory throughout the profession in the 1970s, we again detect the cyborg genesis of yet another tenet of the modern orthodoxy.

[79] The first writer to notice this motive for the use of fixed-point theorems was Hands (1990). It will be relevant for cultural historians to come to realize that appeal to fixed-point arguments spread throughout the human sciences favorably inclined toward "structuralist" explanations from the 1960s onward: this even extends to the seemingly insular community of French structuralists (see Dupuy, 1996). The subterranean linkages of structuralism and game theory are further explored in Leonard, 1997.

However versatile the fixed-point theorem has proved in buttressing the axiomatic rigor of the neoclassical project, and however much it has drawn much of its potency from being inextricably entangled with cyborg themes, it must be noted that, in the final analysis, it shared all the weaknesses of the Bourbakist project itself. Bourbakism only made it seem as though the foundations of mathematics had been shorn up on formalist principles: Gödel's theorem and other deep paradoxes of metamathematics were not so much confronted as simply bypassed and ignored. The whole of rigorous mathematics was consequently not derived from a very few "mother structures"; indeed, those maternal structures rapidly got abandoned in the process of producing new mathematics. Likewise, the ubiquitous fixed-point theorems did little or nothing actually to confront paradoxes of self-reference in economics; they just made it seem as though the Walrasian project had been anchored to some rigorous foundations. The whole of economics was consequently never derived from the "mother structure" of Walrasian general equilibrium; indeed, the Sonnenschein-Mantel-Debreu results of the 1970s finally drove that point home (Sonnenschein, 1972; Rizvi, 1998). Public predictions could backfire, leading to paralysis and indecision, as the notorious "no trade" theorems would demonstrate (Milgrom & Stokey, 1982).

All those arguments *in principle* never once confronted the real problem that the raft of claims to generality had foundered upon the fact that both the Arrow-Debreu model and the portrait of agent as econometrician never were committed to any acceptable scenario of how the market actually worked. Thoroughgoing interdependence of agents and markets was never fully entertained, if only because formal reference of the market system to itself by itself was banished by fiat instead of reasoned argument. If the dummy (or the "auctioneer" or the omnisciently rational agent or Bayesian predictor or . . .) really were a price computer, then the very first mandate should have been to research the conditions under which the computer did not halt. Instead, the Cold War mind-set provoked economists and decision theorists to focus their attention on whether the process of reasoning was "globally complete and consistent," leaving no chink in the armor through which the wily opponent might infiltrate. The twentieth-century yearning for an ironclad guarantee of the complete absence of contradiction in the marketplace (or perhaps more threatening: between the ears of the rational individual) was a mirage, little better than the nineteenth-century yearning for a perpetual motion machine. It was a machine dream; and dreams can sometimes be salutary, so long as they are never confused with conscious reality. As von Neumann had insisted, after Gödel everything would have to be different. There was only one proverb to broadcast to those fearing internal contradictions in their systems: provided there were no contradictions, then absence of contradiction would, of necessity, be undecidable.

DOES THE RATIONAL ACTOR COMPUTE?

Given that the cyborgs and their fleeting fascination with game theory in the 1950s were more or less responsible for the sudden infatuation with Walrasian existence proofs at Cowles, it might then come as no stunning epiphany to appreciate that the cyborgs were also first off the mark to ask the next obvious question: how likely was it that the rational economic actor could actually carry out the sorts of computations which were implied and imagined by the game theorists and, consequently, the Cowles neoclassicals? In other words, just how powerful were all those little "utility computers" that the Cowlesmen thought inhabited the marketplace? Because almost the entire discussion tended to take place outside of the conventional precincts of economics proper, this section can only offer a preliminary sketch of a history that, like so much else in this volume, awaits a more seriously detailed narrative. A major theme of this section is that while the questions were often asked in remote and often exotic locations, and were often couched in unfamiliar idioms of recursive function theory, one should not therefore infer that many key Cowlesmen were blissfully unaware of the challenge they posed; indeed, on the contrary, this is yet another installment in the saga of cyborg raids upon the neoclassical citadel.

The story begins back at Princeton's mathematics department. There, David Gale and F. M. Stewart, following up on Harold Kuhn's (1953) attempt to introduce some concept of "information" more explicitly into the theory of games, decided to extend his formalism to the case of an infinity of possible moves for each player; but for tractability, they restricted their inspection to two-person, zero-sum games. The intention of their exercise was to show that, although von Neumann had proved that finite games of that class were strictly determined, this attribution would not hold for infinite games (1953, p. 253). They noted that their proof might lead to certain paradoxical results, which in turn might call into question the entire program of treating information as 'atomic' and cumulative, at least in the context of game theory: for instance, a subgame of a strictly determined game need not itself be strictly determined (p. 264). Apparently, this paper set more than a few people in the mathematics community thinking that game theory might fall prey to the same sorts of incompleteness considerations that had already beset formal logic and theories of computation. One such individual was Frank Louis Wolf.[80] In

[80] Frank Louis Wolf (1924–): Ph.D. in mathematics, University of Minnesota, 1955; professor, St. Cloud State College, 1949–67; professor, Carlton College, 1967–. Unfortunately, I have been unable to discover who at Minnesota might have put him onto what

his 1955 thesis, he began with a subset of Gale-Stewart type games and then clarified the issue of whether a game could be considered "solvable" by defining solvability as the specification of a mechanical procedure for playing the game in an optimal fashion. In a direct extrapolation of Gödel's theorem, he suggested that there was no effective constructive method that would uniformly serve to find optimal strategies for all constructively defined games (p. 37).

By all accounts quite independently of Wolf, someone else at Princeton also saw the implications of the Gale-Stewart exercise; but this was someone not so easily overlooked as the relatively isolated lone Wolf. Michael O. Rabin was a student of Alonzo Church at Princeton from 1953 to 1957; he would later become the Thomas J. Watson Professor of Computer Science at Harvard.[81] In what will surely rank as a classic in the annals of cyborg history, Rabin published a paper in 1957 that calmly impugned many of the conceptual pretensions of game theory, as well as the drive on the part of Harold Kuhn and others to introduce considerations of C^3I into the Nash tradition. Rabin signaled that he was not at all interested in calling into question the "realism" of game-theoretical accounts of human interaction: "It is quite obvious that not all games which are considered in the theory of games can actually be played by human beings" (p. 147). Instead, "The question arises as to what extent the existence of winning strategies makes a win-lose game trivial to play: Is it always possible to build a computer which will play the game and consistently win?" The surprising answer was no – all the more surprising because it was not couched in the supposed generality of the Nash approach, but rather confined to the rather simpler and more computationally accessible program of von Neumann, where the two-person, zero-sum paradigm was still predominant.

Beginning with a Gale-Stewart specification, Rabin made an observation that would often be overlooked in the subsequent annals of game theory: "there would be no sense in worrying about the effective computability of a strategy when the rules of the game in which it applied were not effective." If games were to be played by machines, then all the relevant aspects of the game had better be reducible to lines of code; sloppiness on the part of relentlessly rigorous mathematicians about what actually constituted the rules and structure of the game (usually by conflating the "normal" and extensive form specifications) and the exact

must have been at the time a very unusual topic for anyone in mathematics outside of Princeton. His unpublished 1955 thesis is available from University Microfilms.

81 Michael O. Rabin (1931–): M.Sc. in mathematics, Hebrew University, 1953; Ph.D. in mathematics, Princeton, 1956. The best biographical source on Rabin is Shasha & Lazere, 1995, pt. 2, chap. 2. All unattributed quotations in the text are from this source.

situations to which a player's parcel of "information" referred would be the bane of cogent thought in game theory for decades to come. Because Rabin restricted himself to two-person games with invariant alternating moves, he was able to confine an effective specification of the game to decision processes which would: (a) tell at any given position whose turn it was to play; (b) tell whether a given move was in conformity to the rules of the game; and (c) tell whether play terminates at a given position and which player won.

Once Rabin had platted the foundations of an effectively playable game, the rest of his paper was devoted to constructing a counterexample that would demonstrate that, for this particular game, existence of a winning strategy could be proved, but it could equally well be proved that there was no machine-computable strategy for playing the game. This counterexample was clearly aimed at the nonconstructive existence proofs, such as the ubiquitous fixed-point theorem exercises, which were sweeping the enthusiasts for Nash equilibria at Princeton in that era; but it did not have much of an immediate impact on economists. Rabin ventured even further with his second theorem, suggesting that any computable strategy that player two might use could never be a winning procedure in repeated play, because the first player could discover it after a finite set of plays (Gödel and universal Turing machines as cryptanalysis once again!) and defeat the opponent as long as he persisted in his strategy. Rabin concluded: "An intelligent player will of course change his strategy when he sees that he is consistently losing. But Theorem 2 proves that it is not advisable to use a computer instead of player II, because after a finite number of plays it will always lose (when talking about computers we exclude the possibility that they contain some device for *mixing* strategies)" (1957, pp. 153–54).

A sounding of the subsequent game-theoretic literature reveals that Rabin's warnings fell on deaf ears in both the economics and defense analysis community. Major surveys (e.g., Luce & Raiffa, 1957; Shubik, Brewer, & Savage, 1972; Aumann, 1987) neglected to cite it. Even as late as 1996, the *Handbook of Computational Economics* somehow manages to overlook it. One might venture that the reason it seemed to have very little salience for the economists who were so clearly its intended target was that the specific counterexample constructed by Rabin relied on such an obscure condition for the player to claim victory that it may have appeared on its face to have very little relevance for any economic problem.[82] Perhaps

[82] The game had player α choose integer i while player β had to choose integer j, knowing i; finally, player α got to pick integer k, knowing both i and j. If a given function $G(k) = i + j$, then player α would win; otherwise, his opponent was the victor. The perplexity came in the extremely obscure specification of the function G, which involved the enumeration of a recursively enumerable set and its complement. J. P. Jones (1982) alleviated this

another factor was that the economists had already more or less given up on von Neumann's canonical approach to game theory, with its elevation of the two-person zero-sum game and the minimax as the central solution concepts; it would take quite some time for it to dawn upon the cognoscenti that the vaunted "generality" of the Nash equilibrium would not extend much beyond the actual algorithmic calculation of finite solutions in noncooperative games with two players, because in that eventuality, "even if the input data are rational, the set of Nash equilibria with a given support need no longer be a convex, or even connected set" (McKelvey & McLennan, 1996, p. 89). A third possible reason for ignoring Rabin's game is that all games in the Gale-Stewart tradition sported some essential infinity in their specification: in Rabin's game, it was located in the convoluted specification of the victory condition. It may have been that any economist or defense analyst who took the time to understand the paper might just have discounted its importance, based on some unspoken belief that all economic games were intrinsically finite in extent. In any case, the noncomputability of game strategies, with its implied flaw at the heart of the image of economic rationality, did not show up on the radar screens of neoclassicals for at least another three decades.

Rabin's point may have eluded the economists, but there remains the tantalizing thesis that it could nevertheless have played an important role in the subsequent development of the theory of computation. Rabin, as we learn from Shasha and Lazere (1995), went on to become one of the major pioneers of von Neumann's theory of automata and computation; among his more important contributions was the formalization of the theory of nondeterministic finite state automata (Rabin & Scott, 1959). Unlike the original Turing machine, which passes through exactly the same sequence of states every time it is fed the identical input, the nondeterministic machine actually makes random moves at certain specified junctures. Rabin later explained this in terms of the economically evocative language of "choice":

> We postulated that when the machine is in a particular state and is reading a particular input, it will have a menu of possible new states into which it can enter. Now there is not a unique computation on a given input, because there are many paths. Consider a menu in a fancy French restaurant.... You are in the start state for selecting a meal. In the first instance, you select the hors d'oeuvres, and then you select the soup, and so on. This is not a random process. We are not tossing a coin. But it is nondeterministic because you have choices as to the next state (i.e., menu

problem in Rabin's proof by replacing G with a Diophantine equation – something more transparently algebraic, yet hardly a quotidian concept in the economist's toolkit. The proof was rendered more user-friendly for economists in Velupillai 1997.

> item selection) that you move into. The choices are the states. Each of these sequence of choices represents a possible computation and results in either satisfaction or dissatisfaction with the meal. Acceptance or rejection. (in Shasha and Lazere, 1995, p. 74)

Rabin, as a good cyborg, did not presume a single choice would always lead to the "same" outcome – consistency being the bugaboo of small minds – but appears to have been influenced by the issue raised at the end of his 1957 paper: namely, a computer might evade the paradox of dependence upon an inferior computational strategy if it had the ability to *randomize* – a Neumannian theme if there ever was one. In computational terms, choice can easily lead to ambiguity, and so Rabin and Scott defined a nondeterministic machine as "accepting a sequence" if at least one of the possible computations would reach an accepting (read: winning) state. Von Neumann had originally imagined multiple virtual plays of a strategy as a species of quantum mechanical wave packet, collapsing to some summary statistic of winnings; now Rabin extended the analogy, imagining the repeated calculations of a nondeterministic machine collapsing to some central tendency of computational output. A certain proportion of "acceptable" answers might be good enough. This instantiation of a "guessing machine" might seem rather fanciful, except for the fact that it led to one of the most important concepts in the theory of computation, the notion of a hierarchy of computational machines arrayed in order of their computational power and complexity.

The basic theory of computational complexity, discussed in Chapter 2, is a direct outgrowth of Rabin's work on automata. Rabin and Scott explicitly sought to explore and taxonomize the properties of machines "less powerful" than a Turing machine, primarily those equipped with finite memory capacities and finite internal states that limit what the machine can do. Augmenting those devices with a "guessing module" raised the issue in stark immediacy of whether and under what circumstances a mechanical randomizer might circumvent problems of undecidability and unpredictability of the halting problem. Augmenting rudimentary automata with the guessing module held out the promise of cheap fortification of computational power while remaining restricted to a small number of internal states; it looked like a reprise of von Neumann's trademark thesis that rationality might be fortified by randomness, perhaps taking the Monte Carlo method to a novel transcendental level. However, Rabin and Scott proved the rather unexpected proposition that any problem solved with a nondeterministic machine could also be solved by a deterministic finite-state automata, although they might differ in resource usage, given the scarce resource was time or memory storage.

The now standard hierarchy (in order of increasing computational capacity) of types of automata, from finite, pushdown, and linear-bounded to Turing machine, as well as their correlation with what is now known as the "Chomsky hierarchy" of language recognition devices – regular, context-free, context-sensitive, and recursively enumerable strings (Badii & Politi, 1997, pp. 166–67) – is an expression of this original insight into the stratified power and complexity of various abstract computational devices. So, too, is the idea that Turing machines can be augmented with multiple tapes and "oracles" that give random answers to noncomputable functions; and the option of nondeterministic Turing machines in turn led to the now burgeoning field of classification of the time- and space-complexity characterizations of individual algorithms.[83] Perhaps this ricochet of inspiration from thermodynamics to quantum mechanics to economics to computation does not strain credulity when we come to view Rabin's work as unified by a single overriding concern: "a general investigation of the minimum amount of work needed to perform a given computational task, i.e., the inherent difficulty of that task" (Shasha & Lazere, 1995, p. 80) – an economic conception of the world par excellence, one might opine, until one learns that Rabin's most recent claim to fame is the theory and implementation of encryption systems to enforce computer system security (Kolata, 2001). The economy of information has never ventured very far from military and espionage concerns.

For a myriad of reasons, the computational limitations of strategic thought did not make much in the way of inroads into the mathematical elaboration of game theory in the 1950s through the 1980s; but the situation with regard to the fundamental primitive of the Walrasian tradition, the utility or "preference" function, was a different story. From the mid-1950s onward, Herbert Simon began to complain of the computational implausibility of the neoclassical portrayal of "rational choice"; but because his own relationship to the theory of computation bears more than a modicum of ambivalence, we postpone consideration of his theme of "bounded rationality" to the next chapter. Some philosophers with close associations with decision theory also began to wonder if "our machines are *rational agents* in the sense in which that term is used in inductive logic and economic theory" (Putnam, 1967, p. 409; also 1960). Computer scientists such as Richard Karp were chiming in by the 1970s: "economics traditionally

[83] The work of Stephen Cook and Richard Karp on the notion of NP-completeness is presented in a nontechnical manner in Shasha & Lazere, 1995, pt. 2, chap. 6, and in Association of Computing Machinery, 1987. The standard pedagogical introduction is Garey & Johnson, 1979. Parenthetically, "random" Turing machines are not machines that grind away randomly, but rather machines that output the answer "yes" if half of all halting configurations are yes, outputs "no" if none of the halting configurations say yes, and doesn't halt under any other circumstances. See Cutland, 1980, p. 167.

assumes that the agents within an economy have universal computing power and instantaneous knowledge of what's going on throughout the rest of the economy. Computer scientists deny that an algorithm can have infinite computing power" (in Association of Computing Machinery, 1987, p. 464). But general equilibrium theorists, being good Bourbakists, had long ago given up on philosophers as having anything germane to contribute to economics; they also shunned computer science as too "applied." Instead, one must attend to the tortured history of the attempt to "purify" the preference concept of any residual taint of psychological commitment within the neoclassical camp to witness the subtle incursion of computational themes. The reason this tradition is relevant to our present concerns is because just as the concept of "choice" was putatively being emptied of all mental referent, the cyborgs began to press home the indictment that the strategic duplicity that so exercised both cybernetics and game theory was not necessarily germane to the fundamental computational critique of economics; instead, even in the simplest case of individual "choice" over a fixed and given domain of alternatives, the neoclassicals were imagining that their little utility computers were doing something that was, quite literally, impossible. And (as if the reader had not already tired of the prolepsis), the prime stage setting of the major encounter between the cyborgs and the neoclassicals was RAND.

This chapter of the cyborg incursion begins (for once) outside the immediate ambit of Cowles, in the precincts of that third school of American neoclassicism, MIT. In 1938 Paul Samuelson asserted that he would avoid the dreaded curse of utilitarianism altogether and derive all of the fundamental content of demand theory from his "operational" axioms of choice, which became known as the theory of "revealed preference." He framed these assumptions in a quasi-algorithmic manner by imagining them as pairwise comparisons between discrete commodity bundles, combined with price considerations, concatenated together to form preference profiles. Although we cannot pursue here the vicissitudes of the changing meanings of "revealed preference" in the hands of Samuelson and others,[84] it will suffice to say that Samuelson's program did not end up providing an alternative to the more conventional specification of individual utility within the context of neoclassical price theory, in part because it was later proved to be isomorphic to the standard integrability conditions in 1950, and in part because the MIT school never betrayed any interest in computational considerations. What it did achieve, nevertheless, was to provoke reconsideration of the theory of choice on

[84] But see Hands & Mirowski, 1999. The standard reference, still unsurpassed, on the history of Samuelson's theory of "revealed preference" is Wong, 1978. Samuelson, 1998, provides a response to his critics.

the part of someone much more inclined to regard the issue as a subset of the abstract "logic of choice."

Kenneth Arrow, as we have already indicated, displayed a penchant throughout his life to equate economic rationality with formal logic. His undergraduate years were shaped decisively by his courses in mathematical logic and philosophy of mathematics and, in particular, the tutelage of Alfred Tarski.[85] Possibly due to their timing, however, it appears he did not become acquainted with the implications of the undecidability theorems at that juncture; and his training preceded the development of the computer by something less than a decade. These historical accidents, we would argue, came to color his reactions to Samuelson's initiative to recast demand theory in a more austere yet impregnable mold. In Arrow's opinion, the trend to dispense with utility in favor of more abstract preference orderings had already swept the avant-garde of the profession with the work of J. R. Hicks and R. G. Allen and his own mentor Harold Hotelling in the 1930s. Nevertheless, something about the state of the discussion of Samuelson's revealed preference in the 1950s captured his attention and provided the impetus for his proposal of the formalism of "choice functions" in 1959.[86] The paper proposed that the choice function formalism constituted the general case, of which both revealed preference and conventional demand functions were more narrow special cases. Furthermore, it demonstrated that Samuelson's "weak axiom of revealed preference" was tantamount to an ordering derivable from a standard choice function as long as the choice function was defined over a finite set. In retrospect, one might have thought that, at that late date, either the necessity of finite domains, or the drive to guarantee both completeness and consistency, would have set all sorts of alarm bells ringing about uncomputable numbers and undecidable propositions; but perhaps hindsight really is too effortless a font of counterfactuals. In any event, Arrow gave no sign that he saw any potholes marring the path to smooth preferences of the generically rational agent.

Arrow may not have heeded the warning signs, but others did. One of the vigilant was a political scientist, Gerald Kramer. A small but stalwart band of political scientists in the 1960s had been grappling with the inter-

[85] See the lecture notes collected in box 28, KAPD, especially those on the "Nature of Mathematics"; those from Philosophy 12R, "The Consistency of a Mathematical System"; and Philosophy 246, "The Philosophy of Mathematics," fall 1939. It is especially noteworthy that the middle lectures do not deal anywhere with the recent theorems of Kurt Gödel.

[86] See the brief preface (1984a, p. 100) to the reprint of Arrow, 1959: "The ideas were indeed related to Paul Samuelson's concept of revealed preference, but unlike that work mine took an abstract view of the domains of choice instead of confining them to budget sets. . . . it represents a systematic comparison of alternative rationality concepts; an ordering is a consistency relation among choices from pairwise sets, and it is compared with other kinds of consistency relations."

pretation of Arrow's so-called impossibility theorem, described in Chapter 5; one way to defang the supposed "paradox" therein was to challenge Arrow's own account of rationality on its own terms; and that is what Kramer (1967) proceeded to do. For his allies, Kramer recruited the writings of Herbert Simon, and more to the point, Michael Rabin. The importance of this relatively out-of-the-way document is that it set the pattern for most of the noncomputability proofs for neoclassical choice functions that would subsequently follow. The way the paper works is straightforward. Define the environment for the agent as a set of alternatives that are denumerably infinite. Posit a collection of binary comparisons defined over the original alternatives after the manner of Arrow et al.; this results in a power set over the alternatives. Then posit the existence of a decision or choice function whose domain is the previously mentioned power set, and which assigns a direction of preference to each comparison; Arrow had shown this is isomorphic to the standard neoclassical preference or utility function. Finally, presuming that "a decision-maker will be considered as some sort of finite information-processing device, or *automaton*," and accessing results in Rabin and Scott, Kramer demonstrated that these axioms led to a contradiction. In a marginally more intuitive gloss, infinite sets concocted from infinite sets and then subjected to attempts to sort the results into infinite categories of discrimination will generally fail the computability test.

Kramer's paper, it must be insisted, did not appear like some bolt from the blue. He had begun this work at MIT and had revised the paper while a visitor at Cowles, albeit after Cowles had moved to Yale. The paper itself, while published in a relatively inaccessible annual, was later made available to the economics community as a Cowles reprint (no. 274). There certainly existed a constituency within the political science community in the 1960s that was more predisposed to believe that any conceptual paradoxes should not be laid at the doorstep of "democracy" but rather delivered to where they rightly belonged, namely, the nave of neoclassical economics. Yet, on the other hand, Kramer did not expend much effort to spell out the implications that he felt should be drawn from his demonstrable contradiction. But whatever the configuration of causes, this rather more damning "impossibility proof" did not receive a fraction of the attention of Arrow's similarly named result; and as for Arrow himself, what was required was a more personal messenger bringing the cyborg update. Hermes did finally arrive in the person of Alain Lewis.[87]

[87] Alain Lewis (1947–): B.A., in philosophy, economics, and statistics, George Washington University, 1969; Ph.D. in applied mathematics, Harvard, 1979; Lawrence Livermore Labs, 1978–79; RAND, 1979–82; University of Singapore, 1981–83; Cornell, 1983–87; University of California, Irvine, 1987–. The primary sources of biographical information are personal correspondence with Philip Mirowski and the file of correspondence in box 23, KAPD.

Soliloquy

Here, at the pivotal point of our narrative, I feel impelled to pause for a brief soliloquy. I have rarely addressed you directly, dear reader, but here abides a conundrum so deadening and dismal that I am driven to break the unwritten rules of academic narrative. The problem I face is that I cannot resist mentioning some background facts that will render me even more vulnerable to ad hominem calumny than anything I may have written previously; and yet, to repress it would force me to engage in the very same Cold War duplicity that we have observed in others in the course of this narrative. The problem I confront is that I must reveal that our next protagonist, Alain Lewis, suffered severe spells of mental illness during the course of his career. I do not do this to pander to the fin-de-siècle frenzy for personal tragedy and scandal, which seems to wash over our hoi polloi in their ever more clamberous search for stimulation and gossip. Nor do I do it to impugn the intellectual character of the personalities discussed. Rather, it is precisely because mental instability has been so very prevalent amongst many truly epochal thinkers in this narrative, and that it has been used elsewhere time and again to discount certain ideas or dismiss certain figures, that I feel it has been elevated to the plane of a structural problem in my chronicle. I am not the first to notice this. Gian-Carlo Rota has written, "It cannot be a complete coincidence that several outstanding logicians of the twentieth century found shelter in asylums at some time in their lives" (1997, p. 4). Rota was referring to Ernst Zermelo, Kurt Gödel, and Emil Post; but I don't think the coincidences stop there. Even the normally unflappable von Neumann spent his last bedridden days raving with fear and despair, welcoming a deathbed conversion to a theological doctrine that he had contemptuously ridiculed in his salad days; Alan Turing committed suicide. We have already described Nash's predicament. Even Gerald Kramer suffered from severe depression toward the end of his days. Anyone who has worked around mathematicians knows they include some of the more eccentric characters to be found anywhere in human society; but here the obvious disconnect between the Platonic Ideal of calm omniscient calculation and the lives filled with frenzy, folderol, and menace looms as more than background noise, threatening more cognitive dissonance than one can ignore when it comes to people claiming to theorize the very pith and moment of human rationality. The question simply cannot be avoided any longer: why didn't theorists of rational inference appear more rational?

The folk wisdom that hyperrationality can easily turn into its opposite is a platitude that hardly merits repetition in this context. So is the lesson of Catch-22 that military rationality is a species of logic more alien than any sheltered civilian can imagine. Rather, I should like to suggest that the Cold War took what would have been, in calmer and less doom-laden circumstances, a mildly paradoxical situation, and pushed it to baroque extremes where such contradictions themselves came to assume the reputation of a higher, more ethereal plane of rationality. What did this hyperrationality look like? For instance, defense experts like Thomas Schelling were telling their clients that it was "sensible" to risk all of life on earth to gain a few years' temporary political advantage over an opponent, that it was possible to frighten them silly to make them more "rational." Were these experts doing to their clients what they wanted their clients to do to the Soviets? Or: Certain members of Cowles were spinning tales of imaginary "incentive compatible mechanisms" that would purportedly induce the full and free disclosure of information in the marketplace, all the while repressing the nature of their own funding mandates and, sometimes, the very content of some of their own publications. Other economists deadpanned that governments could do nothing right, all the while depending unashamedly upon direct military funding to underwrite their very existence.

Everyone and their neoclassical brother in the 1950s was praising mathematics to the skies as enforcing the definition of rationality as consistency in preferences, while at the same instant nearly everyone involved knew that hardly anyone in their acquaintance was behaving in a fully consistent manner. Dropped into the midst of such sophistry, a John Nash or a Gerald Kramer or an Alain Lewis was not merely a stranger in a strange land; he was a prophet of a new mode of existence. The abnormally skewed perceptions and insistence upon following conventional ideas of rationality into nastier culs-de-sac could appear to more pedestrian souls as extraordinary insight into the predicaments of Cold War man. The very aspects of the Cold War that rendered the mathematical codification of rationality as the ultima Thule of intellectual distinction – and, as such, constitute the motor behind much of our narrative – also dictated that the metallic tincture of madness would glint behind almost every formal argument in this period.[88]

[88] While we have no pretensions to interweave the images of popular culture with academic accounts of the mandarins of mathematical rationality, unlike, say, the masterful Edwards, 1996, it is impossible to resist the temptation to remind the reader who has viewed Stanley Kubrick's classic film *Dr. Strangelove* that the inspiration for the eponymous character was a convex combination of Herman Kahn, Henry Kissinger, and John von Neumann.

But there is one more dark undercurrent to this account about which we must not be coy, dear reader. Precisely because mental instability has been so rife in this narrative, with numerous protagonists perched precariously near their own personal deep end, the role of their colleagues and their science managers looms much larger than it might otherwise have done. In the face of this surfeit of unconventional ingenuity – the gift of the mad – when it comes to formal argumentation, it would make all the difference in the world exactly who were deemed sufficiently "brilliant" to warrant having their numerous gaffes and transgressions overlooked and forgiven, and those for whom the exhausting effort of forbearance and accommodation was deemed just an undue imposition. These friends and factotums quite literally tipped the balance between incarceration and indulgence. The handlers and managers held the lives of these fragile souls in their hands, deciding what to reveal and what to drape with the veil of privacy. They placed themselves strategically between sender and receiver, dear reader, like Shannon's original model of an encryption device; and, by so doing, in the final analysis it was they who controlled the very meaning of rationality. To put it bluntly, in some situations some such Dionysian soul might eventually rate the services of a journalist expending effort to burnish their legacy, or a respected graybeard to stage-manage their public appearances, whereas another such soul under only slightly differing circumstances would be consigned to the depths of obscurity or the dungeons of New Bedlam, even though the actions cited as virtues in the former case would be equally cited as vices in the latter. One will be praised for his beautiful mind, whereas the other will be disparaged for his crackpot ideas. In the case of mathematicians, it will all resolve down to the opinions of a handful of strategically located gatekeepers. Anyone who seeks to reconfigure this framework of absolution and censure will himself transgress upon an elaborate web of pacts and alliances.[89] *For my own part, I regard this as confirmation of the overwhelming role that social context played even in that most austere and abstract of the sciences during the most frigid stretches of the Cold War.*

* * * * *

[89] As an example, see the review by Joan Didion of Sylvia Nasar's biography of Nash in the *New York Review of Books*, April 23, 1998, and the letter to the editor by Peter Lax in the October 8, 1998 issue. Didion's reply (p. 58) is germane to this issue: "As Mr. and Mrs. Lax of course understand, the piece in question was not an 'attempt to link Ted Kaczynski and John Nash.' It was instead an attempt to suggest that much current discussion of the mystery of human behavior has been reduced by politicization to a factitiously moralized rhetoric, a point that would seem supported by this letter."

With that interlude, we may now return to the career of Alain Lewis. As a student at Harvard in the 1970s, Lewis was guided by Kenneth Arrow into areas of fleeting interest, such as the application of nonstandard analysis instead of measure theory to the characterization of "atomless agents" in game theory and general equilibrium analysis, and further refinements of his own impossibility theorem. Upon completion of his thesis, Arrow ushered him into the netherworld of classified research, first at Lawrence Livermore and then at RAND.[90] During his stint at RAND, sometime in 1980, Lewis set out on a research project of his own, delving into the implications of computability for Arrow-style economic analysis. The novel departure was less disconnected from his previous experiences than the topics might initially suggest. This was explained in a draft press release on Lewis's research:

> The research program is important, says Lewis, because it is the first significant application of recently developed techniques in recursion theory to two mathematical topics: N-Person von Neumann Games and Weak Combinatorial Versions of the Axiom of Choice. . . . Some of the models he is working with are so complex that they defy an answer. In one case, says Lewis, "God forbid there should be a real life response to the theoretical model." Lewis is referring to nuclear war where, despite the best intentions, strategists and other players cannot risk a true scenario. Thus Lewis' theoretical structure devises a multi-person game that represents levels of conflict in a nuclear war and provides optimal strategies for such an event. "My work assesses the usefulness of these theoretical models and can apply to military science and political science. . . . If the models are too complex, they are no good. I must figure out how large a machine I would need to simulate the actions of a scenario, and I critique the components. While I'm simulating actions, I'm dealing with reality and the games must be realistic," says Lewis.[91]

To Arrow's credit, he initially encouraged Lewis in his exploration of the computability of the Walrasian model, via the same path Arrow had originally taken in his own quest, namely, via game theory. Whatever the initial impetus, it seems that, quite early on, Lewis came to appreciate Kramer's point that strategic considerations found in game theory and in

[90] See Lewis, "Vita," October 1987, box 23, KAPD. Classified topics included antiterrorist measures in systems design for nuclear plants; command and control networks for the strategic nuclear force; and computer-assisted battle management by means of simulation. Lewis characterized his objective in one of his published works as "the task of constructing models of complex military socio-political phenomena, typified in C^3 + I systems as found in theories of command and control" (Lewis, 1985c, p. 211). As can be observed, Lewis spent his early years much closer to the wargaming side of the cyborg sciences.

[91] "Mathematician Becomes Master of the Game," draft press release, January 16, 1984, box 23, KAPD. The piece ends with Lewis quoting Rabin, 1957.

the various treatments of how the neoclassical agent deals with uncertainty only tended to confuse the issue of understanding the abstract nature of the agent as information processor, if the root problem was the non-effective specification of the act of choice itself, shorn of all these superimposed complexities. Thus Lewis was drawn directly to Arrow's formalism of choice functions; and once there, he proceeded to recast Kramer's proof in terms more conformable to the standards of mathematical rigor of his day, and to remove numerous mathematical infelicities that might have stood in the way of making the transition to the mathematical tools of choice taken for granted amongst the next generation of Cowles theorists.[92] The argument ran up against the stubborn opposition of some mathematical economists at high-profile economics journals and some logic specialists at applied mathematics journals; however, it finally appeared (Lewis, 1985a), and is now considered in some circles as an underground classic in the theory of computational economics.

Kramer had shown that there was no finite automata that could make the choices that Arrow said choice functions of rational agents were capable of doing. Lewis realized that the unaddressed issue really was one of hierarchies of computational power, and that the benefit of the doubt should be accorded to Arrow by pitching the question at the level of maximum capacity, namely, the Turing machine. He did not make the claim that rational agents *were* Turing machines; only that appealing to computational viability meant that whatever was being asserted about agents' ratiocination abilities had better be programmable on a Turing machine, or else by Church's Thesis it did not meet anyone's criteria of a "calculation." "It is obvious that any choice function C that is not *at least computationally viable* is in a very strong sense *economically irrational.* Unless the computation is trivial, i.e., involves no computation at all, and thus is both complete and accurate without any use of the Turing machine, the choices prescribed by a *computationally nonviable* choice function can

[92] We briefly indicate some of the more striking innovations in this footnote for those interested in following up on his proof techniques. First, he makes a rather critical distinction between a computable *representation* of the choice problem, and a computable *realization* of a rational choice process. The structure of the paper is an argument that, for any reasonable choice of domain, a nontrivial recursive rational choice has an unsolvable graph and cannot be recursively realized. This is accomplished without recourse to a diagonalization argument, a proof technique pioneered by Cantor that is rarely persuasive in applied contexts. Lewis was aware that neoclassicals would not acquiesce in Kramer's restriction of the functional domain to the natural numbers, so Lewis instead worked in terms of the recursive reals. However, it must be said that Lewis's proof style does not aim for elegance; rather, it seeks to clear away all the irrelevant objections that were being proposed by mathematical economists who regarded the proof as destructive.

only be implemented by computational procedures that do one of two things: either (a) the computation does not halt and fails to converge, or (b) the computation halts at a non-optimal choice. . . . whatever purpose there can be to a mathematical theory of representable choice functions is surely contradicted in such circumstances" (Lewis, 1985a, pp. 45–46). The crux of the issue was the very existence of a "logic of rational choice"; and Lewis demonstrated that for any neoclassically relevant domain of alternatives, Arrow-style nontrivial rational choice had to traverse an unsolvable graph and therefore stranded beyond recursive realization.

Kramer the political scientist had not been willing to follow this conclusion to its bitter end: if there was no point to nonviable preference representations of rationality, then there was certainly no point to stacking up cathedrals in the air concerning aggregate demand, market equilibrium, incentive-compatible institutional mechanisms, Pareto optima, and every other Walrasian notion beloved at Cowles.

> The theorem of representable choice functions in the neoclassical setting, and thus consequently the theory of neoclassical demand correspondences and the theory of SDF-derived social welfare functions, when defined on families of compact subsets of [the reals], presumes the possibility of a mathematical correspondence that, even in principle, cannot be performed or realized in effectively computable terms under the weakest, and therefore best, possible circumstances of recursive approximation. As we have mentioned previously, and do not mind stating yet once more, this appears to have serious consequences for the foundations of neoclassical mathematical economics. (Lewis, 1985a, p. 68)

Lewis then proceeded to follow through on this prognosis, and explicate the ways in which this flaw would exfoliate throughout the analytical structure of neoclassical economics. First, he elaborated upon his findings on uncomputable choice in the format of a book manuscript aimed at a larger audience; but for reasons we have been unable to discover, it was never published. This revision did reveal, however, that Lewis was taking up the anti-Bourbaki cudgels that had been dropped since Novick:

> Abstraction for the sake of sophistication of technique whose utility does not extend beyond the development of that technique has had a pernicious influence in many other areas of mathematical discipline, and it seems that in the post World War II period of development, the Bourbaki school of thought for the foundations of mathematics and its supporters has been most culpable in this proliferation, of what we call for want of a better term, "technicism." To inquire of a model expressed very technically within a subject of mathematics whether it is constructive or recursively realizable brings the affliction of technicism down to earth, so to speak, in that we inquire thusly as to the *real or effective content* of such models. It is the author's opinion that only a model of social

> phenomena that possesses effective content can serve as a meaningful guideline for positive-normative prescriptions.[93]

Next, he sought to show that issues of nonviable choice would be found at every level of the Walrasian program. Of course, if preferences or utility functions were frequently computationally nonviable, then the very project of making further inferences from them about aggregate welfare, market operation, and the like would have been rendered procedurally groundless; but Lewis did not pursue that line of attack, presuming that the reader was capable of drawing that conclusion on their own. Instead, he opted to examine various landmark arguments in the history of neoclassicism, and subject them to the same scrutiny regarding computationally effective specification. Beginning with his patron Arrow's "impossibility theorem," Lewis (1988) cast the computational critique as "strengthening" the original argument against voting. Then Lewis (1991) subjected the standard notion of the convergence of the game-theoretic solution concept of the "core" to Walrasian general equilibrium to the cold stare of the computer and pronounced that it, too, was a pipe dream. "Edgeworth would have understood that . . . a precondition for the contract curve to 'narrow' is that there be a sympathy with each other's interests. As we add more players to the markets, the complexity of the core conditions become synonymous with verifying whether there is sufficient sympathy with not only the old interests of the players of the market but now those of the new players, mixed, permuted, coalesced and decoalesced" (p. 277). A more straightforward example of combinatorial explosion of intractability could not be imagined. Lewis (1992a,b) demonstrated that if demand correspondences derived from choice functions were not computable on the domain of the recursive real numbers, then it would not be possible to realize recursively the outcome of a Walrasian general equilibrium system. "Do non-trivial demand correspondences really exist in any meaningful (i.e., effectively) constructive sense? Within the confines of Church's Thesis . . . the answer seems to be No" (1992b, p. 220). These papers also extended noncomputability results to Nash equilibria in game theory, using various complexity notions to rank degrees of unsolvability in Gale-Stewart games and noncooperative games. The literature of "Hurwicz allocation mechanisms" also suffers from a computable vantage point (1992b, p. 225). Clearly, by the time of the last paper, Lewis had heard every possible objection to his theorems, including the last refuge of the scoundrel, that

[93] Lewis, 1986, p. 134; manuscript copy in possession of the author. Immediately after this passage Lewis suggests that Gödel, Church, or Turing would not have "sneered at Walras' concern over the solvability of his model." This, of course, was a backhanded reference to the genealogy of progress which Arrow and Debreu had constructed for their own achievement.

economic agents need not be able actually to do the mathematics in order to act *as if* they were neoclassically rational. His frustration with economists surfaces in an entertaining fashion in the last paper:

> We would also like to take issue with the analogy of dogs catching frisbees and solutions to differential equations being necessary for the performance of these acts with the computability of demand correspondences. It does not seem obvious to us that the construction of a demand correspondence is equivalent to the solution of a differential equation . . . the construction of a demand correspondence is really the *cognitive resolution* of the allocation of bounded resources of energies between two or more competing alternative uses. . . . Viewed in this light, it seems fairly obvious that some effective means must exist to provide the resolution of alternatives in a uniform way over finite subsets of the budget set when required in order for demand correspondences to exist in any constructive sense of the word. (1992a, pp. 143–44)

Although Lewis was trying to play the theory game in a responsible fashion, and by the early 1990s all the relevant Cowles theorists were aware of his work, it is noteworthy that every single paper in economics he ever managed to get published appeared in the same journal; worse, in the view of the elite of the mathematical neoclassicals, it was an obscure backwater. This problem of limited exposure was both exacerbated and possibly influenced by evidence that Lewis's mental balance was progressively destabilized over this period. But through it all, he never presented himself as a wrecker of the neoclassical program. He consistently praised the progenitors of the neoclassical program such as Walras and Edgeworth. His desire to be regarded as someone who had taken all the professions of adherence to rigor regardless of their social consequences to heart is best exemplified by a letter he wrote to Gerard Debreu:

> If you will allow me to take the opportunity of this letter to express the wish that my results not be interpreted as any form of nihilism for the mathematical groundwork laid by Arrow and yourself in neoclassical theory, I should be most pleased. I am a child of neoclassical mathematical economics and had it not been for the rigorous model-theoretic frameworks I read and re-read in my early graduate school days, no serious inquiry into the effectiveness of these models could have ever begun. . . . I believe my results have the following meaning by way of analogy with models of Peano Arithmetic. It is known that the ordinary operations of addition and multiplication are recursive, and we see this as evidence daily in hand-calculators and elsewhere. On the other hand, *these models pay for their recursiveness with a form of incompleteness* – e.g., there are theorems that are *true* in Peano Arithmetic, but not *provable* within Peano Arithmetic. Here I am of course referring to the work of [*sic*] C. Gödel. . . . Now, if we deal with only *totally finite* models

> of Walrasian general equilibrium, with not only a finite set of agents, but also a finite set of alternatives for each sort of agent, then most assuredly these models will be recursively realizable, since everything is bounded by the totality of the components of the structure. . . .
>
> In exact analogy to the nonstandard models of arithmetic, the continuous models of Walrasian general equilibrium pay for the use of continuity, and the "smooth" concepts formulated therein, with a certain noneffectiveness, that can be made precise recursion-theoretically, in the realization of the prescriptions of the consequences of such models. By the way, if ever you are interested in an analysis of the effective computability of rational expectations models that are all the rage in some circles, it would not be hard to make the case that such models are irrational computationally. . . . When I first obtained the result for choice functions, I thought my next task would be the reformulation of *The Theory of Value* in the framework of recursive analysis. I now have second thoughts about the use of recursive analysis, but I still feel that a reformulation of the foundations of neoclassical mathematical economics in terms that are purely combinatorial in nature – i.e., totally finite models, would be a useful exercise model-theoretically. If successful, then one could "add on" more structure to just the point where effectiveness goes away from the models. Thus we ourselves could effectively locate the point of demarcation between those models that are realizable recursively and those which are not.[94]

Someday, Lewis's papers may come to be regarded as landmarks in the history of computational economics, and perhaps even decisive nails driven into the coffin of the Cowles project.[95] Whatever tomorrow might bring, the record of the past is there for us to discover in the Arrow papers. Therein we find that Arrow was in close and repeated contact with Lewis throughout the decade of the 1980s, and that Lewis doggedly tried to get Arrow to acknowledge the seriousness of the computational challenge to his own lifelong theme of economic agent as information processor. Here, finally, we reprise the theme of this chapter: cyborgs would not just sit still and let the Cowlesmen misrepresent the cut and thrust of the cyborg sciences. It is most illuminating to track Arrow's written statements about

[94] Alain Lewis to Gerard Debreu, December 12, 1985, box 23, KAPD. With hindsight, we know that the quest for this sharp boundary between recursiveness and nonrecursiveness is itself undecidable.

[95] There is already a reaction that seeks to soften the blow struck by Lewis's papers, centered primarily at the University of Minnesota. See Richter & Wong, 1998, 1999a, 1999b, 1999c. These authors, however, shift the goalposts in order to render the Walrasian project more computable. For instance, they simply ignore any decidability questions with respect to Arrovian choice functions; and they "do not shrink from using classical nonconstructive methods in proving the existence of digital algorithms." Nevertheless, they do concede the central point that "Brouwer's theorem fails for computable functions" (1999a, p. 4).

information and computation and juxtapose them to what we can learn about Lewis's interventions in his own understanding of the issues involved. To wit: what did Arrow know, and when did he know it?

It seems fairly apparent that early on, Arrow had no trepidation that paradoxes in logic might have some bearing on his economic concerns. The closest he ever approached the problem was at the end of his *Social Choice and Individual Values* (1951a, p. 90): "From a logical point of view, some care has to be taken in defining the decision process since the choice of decision process in any given case is made by a decision process. There is no deep circularity here, however." As described earlier, Arrow in the 1950s and 1960s tended to conceptualize information as a *thing*, and as such more or less as unproblematic as any other commodity definition within the Walrasian framework. This coincided with his repeated statement that Arrow-Debreu general equilibrium demonstrated that the individual economic agent had very few cognitive demands made upon him in a competitive economy, given that all he had to do was know his own preferences, and the perfect market would do the rest. This, of course, was the conventional Cowles riposte to Hayek. By the 1970s, however, largely through his interaction with Kahneman and Tversky, he revealed that there might be some psychological evidence that human beings faced some "limits as information processors" but that the market could even obviate this problem, because one could just "substitute" other factors to offset the unfortunate limitations (Arrow, 1974a, p. 39). In this period Arrow indulged in maximum confusion of "learning" with "purchasing" something, to the extent of encouraging all sorts of research into education as signaling and screening. At this stage, Arrow began trying to read Herbert Simon as some minor variant of neoclassicism, purportedly arguing that hierarchy and bureaucracy were merely evidence of some lingering "market failures" or nonconvexities (p. 64). People might appear limited in their capacities simply because an optimization process had encouraged them to rise to their own level of excellence. "The aim of designing institutions for making decisions should be to facilitate the flow of information to the greatest extent possible. . . . this involves the reduction of the volume of information while preserving as much of the value as possible. To the extent that the reduction of volume is accompanied by reduction in the number of communication channels, we are led back to the superior efficacy of authority. . . . The purest exemplar of the value of authority is the military" (pp. 70, 69).

By the mid-1980s there was some change to Arrow's tune; one could attribute this to the repeated interactions with Lewis. At a conference on psychology at Chicago in October 1985, he acknowledged, "The main implication of this extensive examination of the use of the rationality concept in economic analysis is an extremely severe strain on information-gathering and computing abilities. . . . The next step in analysis, I would conjecture, is a more consistent assumption of computability in the formulation of economic hypotheses" (Arrow, 1987, pp. 213–14). While not exactly giving up his older mantra that information requirements were "low" in competitive environments (p. 207), there was now a sense that his favored notions of rationality could lead to paradoxes, such as the no-trade theorem or the requirement of complete common knowledge. What he did *not* do, however, is explicitly cite Lewis, nor did he exactly admit that the problem went all the way down to his choice function formalism or, indeed, his broad-church Bourbakism. Instead, he gave indications in interviews and other venues that he was "working on communication and computing" (Feiwel, 1987a, p. 242). Some idea of what this meant in the mid-1980s can be gleaned from an NSF grant proposal he composed with Mordechai Kurz and Robert Aumann in this period on "Information as an Economic Commodity."[96] This document presages the trend toward having finite automata play games, which we cover in the next chapter, and it seeks to treat computational limitations as themselves the resultant of an optimization process: "we hope to obtain a constructive theory of memory selection in strategic contexts." But while others rapidly came round to that research program, Arrow was still restless, looking for other, perhaps better ways to reconfigure the economic agent as information processor. The one path he did not choose to explore was Lewis's suggestion to produce a truly computationally effective version of his trademark Arrow-Debreu model. Although he never published anything on the issue, in correspondence with Lewis from 1986 onward he tended to reject most of the implications of Lewis's critique. For instance, he wrote Lewis on July 21:

> [In the example, max xy subject to x + y = A,] the demand function for each good is, of course, simply A/2. This would appear to be as computable as any function one could imagine, short of the identity. If the only problem is that the function is defined over a set which is itself too large to describe by a Turing machine, then I must wonder whether the right question has been asked. . . . To compute *equilibrium* one does need in principle the whole correspondence. But if one takes the algorithms actually used (Scarf's or homotopy algorithms), the demand needs to be computed only for a finite (or in the limit, denumerable) set of parameter values. It is true that the sense in which these algorithms yield approximations to equilibria is less certain than one would like. . . . But the claim the excess demands are not computable is a much

[96] Arrow, Kurz, & Aumann, "Information as an Economic Commodity," submitted September 6, 1985, box 28, KAPD. It is noteworthy the extent this draft makes reference to military matters, and the extent to which they were deleted for submission to the NSF.

> profounder question for economics than the claim that equilibria are not computable. The former challenges economic theory itself; if we assume that human beings have calculating capacities not exceeding those of Turing machines, then the non-computability of optimal demands is a serious challenge to the theory that individuals choose demands optimally. The non-computability of equilibria is merely a statement about how well economists can use theory to predict; this is methodologically serious but is not *ipso facto* a challenge to the validity of economic theory.[97]

In this response we catch a fleeting glimpse of the reprise of the attitudes of Cowles Mark II – here, the complaint that you can't actually calculate a Walrasian equilibrium is just brushed off as a problem in decision theory for the economist. Nonetheless, the notion that the global pretensions of general equilibrium, here conflated with economic theory *tout court*, were being seriously challenged was beginning to bite, in part because of other negative developments in the interim, such as the Sonnenschein-Mantel-Debreu results on the lack of any useful restrictions imposed by Walrasian theory upon excess demand functions. This, I suggest, provides one motive for why Arrow was predisposed to participate in the fledgling Santa Fe Institute when it undertook to develop an economics program in the later 1980s. With the repeated incursions being made by the cyborgs into the normally self-confident neoclassical program, *what was called for was a new incarnation of Cowles*: that is, a reconstitution of the original cadre of neoclassicals, perhaps replenished with new blood (from willing natural scientists), to regroup and reinvigorate the economists for their confrontation with the cyborg challenge. What better place to do it than the new mecca of the cross-pollination of computation with the natural sciences? Lest the reader thinks this reading of motives too harsh or instrumental, I point to the fact that it was Arrow himself who first made the comparison of Santa Fe with Cowles (Waldrop, 1992, p. 327).

And so the long march of the cyborgs continues unbowed. Arrow was never fully won over to a full-fledged computational economics; but, then, the experience with modern natural scientists has nonetheless altered his

[97] Arrow to Alain Lewis, July 21, 1986, box 23, KAPD. Arrow persisted in his insistence that the theory of recursive functions was practically irrelevant to economics in a series of letters and conversations into early 1987. For instance, on February 23, 1987, he wrote: "I say that if the domain of the identity function is some computationally convenient approximation to the reals, then computing the identity function and even characterizing its entire graph should be regarded as trivial in a suitable definition. Evidently, and you are perfectly convincing on this point, classical recursion theory does not lead to that result. Then I say that, important as recursion theory is from a foundational viewpoint, it is inadequate as either a guide to computability in a very meaningful sense or as a test whether the computational task is too difficult to ascribe to human beings."

outlook to a certain extent. In another bouquet of unintended consequences, after the computational critique, economic rationality has gone searching for a new imprimatur. This is best illustrated by some programmatic statements made of late by Arrow in public contexts of "science in the next millennium" (Arrow, 1995b):

> The foundations of economic analysis since the 1870s have been the rationality of individual behavior and the coordination of individual decisions through prices and the markets. There has already been a steady erosion of these viewpoints, particularly with regard to the coordination function. Now the rationality of individual behavior is also coming under attack. What is still lacking is an overall coherent point of view in which individual analysis can be embedded and which can serve as a basis for new studies. What I foresee is a gradual systematization of dynamic adjustment patterns both at the level of individual behavior and at the level of interactions and transactions among economic agents. Indeed, the distinction between these levels may well be blurred and reclassified. In the course of this development, the very notion of what constitutes an economic theory may change. For a century, some economists have maintained that biological evolution is a more appropriate paradigm for economics than equilibrium models analogous to mechanics.... Methodology will also change. Formal theory-building, with assumptions and logical inferences, will never disappear, but it will be increasingly supplemented by simulation approaches.

7 Core Wars

INHUMAN, ALL TOO INHUMAN

> "Freedom of the will" – that is the expression for the complex state of delight of the person exercising volition, who commands and at the same time identifies himself with the executor of the order – who, as such, enjoys also the triumph over obstacles, but thinks within himself that it is really his will itself that overcame them. In this way the person exercising volition adds the feelings of delight of his successful executive instruments, the useful "under-wills" or under-souls – indeed our body is but a social structure composed of many souls – to his feelings of delight as commander. *L'effet c'est moi.* What happens here is what happens in every well-constructed and happy commonwealth; namely, the governing class identifies itself with the successes of the commonwealth.
>
> Friedrich Nietzsche, *Beyond Good and Evil*

Once the nascent postwar neoclassical orthodoxy had divaricated outward from Cowles and Chicago and MIT to the rest of the nation and beyond, the garden-variety negative reaction to these doctrines was that they were too "methodologically individualist," too solipsistic, or, if you happened to be put off from the rhetoric surrounding the formalism, too "selfish." On any given Saturday night, so the scuttlebutt went, it would invariably be the neoclassical economist who would be the last to offer to pay for a round of drinks, and the first to insist that everyone get separate checks. Many postwar economists wore these epithets as badges of honor, testimony to their thick skins and their tough-minded attitudes toward the bitter truth. They had learned a thing or two from RAND about "thinking the unthinkable." Whether you were a hardened Thatcherite crowing, "There is no such thing as society" to a crowd of cowed Labourites, or Kenneth Arrow intoning, "let each individual decide what is essential to his or her self-constitution and act accordingly . . .

who is to decide what is essential to the constitution of the self other than the self?" (1997, p. 761), they were echoing a long and hallowed Western tradition of appeal to the inviolability of the individual self as the alpha and omega of all "social" theory. "Free to be you and me!" – after all, wasn't that the rallying cry of both the hawks and doves in the Cold War?

The perpetual battle between the oversocialized individual and the overindividuated society has preoccupied far too many scholars for too long for there to be any realistic prospect of bringing the dispute to any near-term resolution. One more additional economic treatise eviscerating the possibility of altruism will surely propel even the most rationalist of moral philosophers in search of blessed immersion in a long night of reruns of the *X-Files*. Thankfully, no rehash of the tired question of the inevitability of selfishness occupies our agenda; rather, we shall simply take note of the fact that the most commonly retailed reason amongst the American economic elect for the rejection of von Neumann's program for economics was that it was *not sufficiently individualist.* Somehow, first in his favorite game-theoretic solution concept and then in his insistence upon the centrality of organizations and automata for the future of economics, the opinion gradually grew entrenched amongst economists that he had not adequately grounded the stability of law-governed structures of society in the rational Self. Yet, repetition of ceremonial exorcisms of von Neumann along these lines would eventually echo as hollow cant, once one realized the One True Faith had itself come to uncannily resemble the thing it had so long professed to despise.

It is rather my intention to suggest that something rather paradoxical has been happening to this Self, this base camp of behaviorism, this propugnaculum of steadfastness, this command post of the soul; and moreover, to predict that, as the neoclassicals tenaciously persist in their stalwart defense of this citadel of mastery and control in economics, they may soon get the shock of their careers when they find it deserted and hollow. Indeed, the purpose of this chapter is to document the extent to which a stealthy retreat from the Self has already occurred in some of the more avant-garde quarters of the economics profession. Whether the neoclassicals are conscious of this fact, much less come to the realization that John von Neumann has been the patron saint (or Maxwellian Demon?) of this farewell to arms, is something we shall have to leave for others to judge. For instance, Kenneth Arrow himself did yeoman service as an exemplar of a Cold War intellectual who has inadvertently promoted the dissolution of the economic agent for more than two decades, all the while loudly proclaiming from the rooftops, there is no self but the self, and Pareto (and/or Nash?) is its prophet.

Who or what is this vaunted individual self that neoclassical economists have pledged their troth to preserve and protect?[1] There are so many different identities and metaphors dangling from this sacral scarecrow of a self that it is difficult to know where to begin. In the interests of clarifying the recent natural history of the Self, we might subject its personality profile to a diagnostic checklist, although subcomponent individuals will resist any overarching regimentation. First off, there is the all-important image of the individual as the conceptual analogue for the atom in a natural-science inspired theory of society. The fact that the analogy runs afoul of all sorts of infelicities in the actual elaboration is not of immediate relevance here.[2] Next, there subsists that imposing fact of one's own physical animality, the incredible anisotropic kinesthesis of feeling oneself an organism. Deny that, cackles the unrepentant realist, slamming a fist on the table so hard it bleeds. The first two commonplace incarnations of selfhood imperceptibly blend into a third, the feeling of being the only real center of causal agency in a world where all other activity may or may not be properly understood as the resultant of similar kinds of agency, animate or inanimate. Anthropomorphism constitutes the house doctrine of this version of the self. The fourth Self, often confused with the third but effectively separate, is the notion of the individual as the sole locus of consciousness and the seat of intentionality. Guns don't kill people; people kill people, or so the slogan goes.[3]

Collectivities can't think; and therefore they possess the same ontological status as the tooth fairy, full stop. Fine discriminations concerning the stability of selfhood are frequently overridden by appeal to a fifth

[1] "Politicizing activities is no greater guarantee of preserving individuation than commodifying them" (Arrow, 1997, p. 764). This, of course, is the very same Arrow who believes his "impossibility theorem" depoliticized politics. Others know better: "The history of psychology . . . is intrinsically linked to the history of government. . . . economics, in the form of a model of economic rationality and rational choice, and psychology, in the form of a model of the psychological individual, have provided the basis for similar attempts at unification of life conduct around a single model of appropriate subjectivity" (Rose, 1996, pp. 11, 28). It seems "modern individuals are not merely 'free to choose,' but obliged to be free, to understand and enact their lives in terms of choice" (Rose, 1999, p. 87).

[2] The belief in the individual as the atom of social action has always enjoyed an uneasy existence in the history of economics. For instance, the physical field theory, which was the inspiration for utility theory, actually was thought at the end of the nineteenth century to dispense with many forms of atomism (Mirowski, 1989a, pp. 56–57). Hence the rather curious status of atomist claims in Cohen, 1994; Redman, 1998; Georgescu-Roegen, 1971.

[3] After I had written this, I discovered to my delight that Bruno Latour had subjected this very mantra of the NRA to his actant-network analysis (1999, chap. 6).

notion of the individual as the only temporal invariant in an otherwise chaotic field of flux. Days may come and days may go; but only the Self can string them together – or maybe not, said David Hume on one of his bad days (1966, p. 256). Immanual Kant, jolted from his slumber by Hume, proposed a sixth version of the Self: the moral agent worthy of autonomy and freedom. Here the individual is analytically isolable from his surroundings precisely because freedom is possible in principle, if not always actually experienced or enjoyed. The authentic self could learn to become what it already was in embryo, namely, a self-actualized individual. And then there is a seventh Self: the individual as the only reliable storehouse of memory in a world of relentless degradation and noise. This is Schrödinger's self, encountered in Chapter 2. An eighth Self is frequently conflated with the seventh: the individual self is here defined by the boundaries of consistency in theories of rational choice. A house divided against itself is a prescription for disaster, or so say the acolytes of this catechism. There subsist many more versions of the individual Self than I have listed here, at least in Western thought; and in the spirit of individualism, the reader should certainly feel free to augment the list to her own satisfaction.

Individual selves mill about the intellectual landscape in untold numbers; but individual*ism* is an eminently nineteenth-century achievement, as are "society," social*ism*, and "social science." This is important to keep in mind when one sets out to discuss the putative methodological individualism of a nineteenth-century doctrine like neoclassical economics. The conviction of rock-bottom foundations to be found anchoring the Self, such as that guilelessly expressed by a William Stanley Jevons, was much easier to retail in *fin de dix-neuvième siècle* Britain than might now be possible in end-of-millennium cyberspace: "Human nature is one of the last things which can be called 'pliable.' Granite rocks may be more easily moulded than the poor savages which hide among them" (1890, p. 290). It is not that one cannot still find the self-made economic man grimly sallying forth on a daily basis to do battle with the ghostly apparitions of social forces and collective consciences, kicking a rock now and then in honor of Dr. Johnson. Indeed, at this late hour these kinesthetic encounters remain as Sissiphysian as they are tedious. Rather, the reservations proffered here are of a different stripe. It could be paraphrased as: too many kicks will turn any boulder to rubble, given enough time and repetition. No matter how much Ayn Rand protested it just ain't so, there is just no escaping that appeals to the vaunted solidity of the self are notorious for their lack of specificity in the twentieth century; this would have become elevated to the status of a cliché, were it not for the fact that most economists and even some philosophers have grown so blinkered in their cultural outlook that they somehow have come to mistake insularity for consensus.[4] Selves are not what they used to be: things fall apart; the center does not hold. Part of the change can be attributed to the dawn of a modicum of self-awareness on the part of the Self: we are told by some modern prophets to approach the Self "not as a centralized and all-powerful entity, but as a society of ideas that include both our image of what the mind is and our ideas about what it ought to be. . . . The idea of a single, central Self doesn't explain anything" (Minsky, 1985, pp. 40, 50). Of course, the fact it never did adequately explain anything does not rule out that it nonetheless served a purpose or two; but that raises the possibility of Self-deception, a specter that haunts the byways of the theory of rational choice and the boulevards of psychology alike (Dupuy, 1998).

It would seem that the primary reason why the Self has experienced a certain deliquescence in the larger culture is that the cyborg sciences have served to undermine each and every definition of the individual Self enumerated here. In a phrase, methodological individualism is being slowly and inexorably displaced by methodological cyborgism. The physically intact and cognitively integrated seat of autonomy – the cohesive locus of responsibility – is rapidly giving way to the heterogeneous and distributed jumble of prostheses, genes, hybrids, hierarchies, and parallel processors. Indeed, the granddaddy of cybernetics, Norbert Wiener, feared the encroachment of the natural sciences upon the social sciences precisely because he saw the writing on the wall: "The problem of cybernetics, from Wiener's point of view, is that it annihilates the individual as locus of control" (Hayles, 1990b, p. 224). But it didn't stop there; perversely, cybernetics and its progeny have acted to undermine the Self in the name of the triumph of the individual will. Donna Haraway has insisted that "the cyborg is also the awful apocalyptic *telos* of the West's escalating domination of abstract individualism, an ultimate self untied at last from all dependency, a man in space." This is one of the acute insights of science studies scholars such as Haraway, Evelyn Fox Keller, Lily Kay, Ian Hacking, Andrew Pickering, Katherine Hayles, and Paul Edwards, and a major reason why the analytical category of "cyborg science" stands for much more than just an arbitrary taxonomic device, or a fashionable phrase.

This saga of the development of a world-historical theme into its antithesis may smack of the Hegelian dialectic – a *genre* that has never entirely lost its appeal, no matter how despised and disparaged – but it is

[4] Some of the better discussions of the fate of the self in the twentieth century are Ewing, 1990; Bergmann, 1977; Rose, 1996, 1999; Hayles, 1999; Zizek, 1999. For some recent attempts of economists to be a bit more open to nominal interdisciplinary influences, but failing miserably when it comes to the self, see Frank, 1988; Rabin, 1998; Elster, 1998.

also something else; it illustrates the ease with which machines undergo transubstantiation into thoughts, as well as vice versa. The dissolution of the Self was a direct consequence of the cyborg sciences' faithful adherence to the precepts of individualism in all their motley manifestations; thus the technologies and theories that corrode the unified Self were first promoted as pure expressions of the drive to reduce all phenomena to their individualist components. The twin talismans of the cyborg sciences, the computer and the gene, were initially thought to help clarify what it meant to be a self in good standing: the former conjuring the cognitive self-sufficiency of the mind, and the latter the barcode to be read at Nature's checkout counter. But these technologies refused to stay securely confined within their conceptual boxes; and what followed was a riot of ontological promiscuity. "Late twentieth century machines have made thoroughly ambiguous the difference between the natural and the artificial, mind and body, self-development and external design. . . . Our machines are disturbingly lively, and we ourselves frighteningly inert. . . . The cyborg is a kind of disassembled and reassembled postmodern collective and personal self" (Haraway, 1991, pp. 152, 163).

One need not consort with wicked postmodern pundits to discern the disturbing outlines of the mix-and-match cyborg; one need only attend to the nightly news. When Deep Blue beat Gary Kasparov at chess, news anchors shook their heads in sorrow over the coming ascendancy of the Machine. When Dolly the sheep posed prettily in her paddock for a perplexed world, it didn't take a genius to see that humans would soon be next. When a computer scientist had a silicon chip implanted in his arm so he could feel just that much closer to the electronic devices in his lab, most sensed that this was a vague premonition of the shape of things to come. When Donald McCloskey had a sex change operation and wrote a real-time diary about the experience for all to empathize on the Internet, then economists got a premonition of where rational choice theory Chicago-style was really headed. What has been missing so far in these blastoplasts from the past is measured consideration of the extent to which everyone (or at least those not sequestered from all electromagnetic emanations in the past decade) can comprehend that the Self has been beset by serious encroachments from all sides, and thus can have a visceral appreciation of the ways in which the late-modern ascent of Info-Capitalism and its sidekick, the Human Genome Project, have not underwritten the catechism of "from each according to their innate talents, to each according to their scarce individuality." People have not congealed into the one big undifferentiated "Borg" of pop science fiction, or at least not yet; but the comfortable intellectual justifications for the primacy of the Self in social theory have long since lost their unquestioned assent.

Invasion of the Body Snatchers

Let us just quickly review the above checklist of the sources of the Self, and how they have fared in the age of methodological cyborgism. Starting from the top: The hoary chestnut of the individual as the metaphorical atom in a social physics is so hopelessly out of date that only the sorry state of science education in elementary and secondary schools might help explain its dogged persistence. Over the course of the twentieth century, physics has repeatedly announced it had arrived at the ultimate fundamental constituents of the universe, only to have those tiny building blocks become resolved into ever smaller components. The grail of a unified ylem, be it particle or field, always has and always will lie just beyond our powers of resolution (Lindley, 1993); it has been an ontological imperative which has repeatedly encountered disappointment in its realization; and therefore a crudely physics-inspired methodological individualism is a will-o'-the-wisp, and cannot be said to have dictated some parallel mandate in the social sciences.

More likely, the pivotal conception of the Self in social theory derives not from some nuanced understanding of the history of science, but rather the brute immediate physicality of our being. My somatic self ends at my skin; there just seems no way around this fact. Yet, here is where both genetics and the computer as prosthesis undermines such deeply held intuitions. Rightly or wrongly, the advent of DNA as code has encouraged a "beanbag" conception of the identity of the organism (Keller, 1995; Kay, 1997a; Maynard Smith, 1988). This has run riot in the year of the public announcement of the "completion" of the mapping of the human genome. If I am supposedly the sum total of my genes, as we are reminded by every news anchor, when do I suddenly stop being me, after some base pairs are swapped or replaced? This turns out not to be an idle speculation; the status of transgenic organisms, especially where human genes are introduced into other species, is one of the most explosive latent controversies in the vast minefield of bioengineering (Weiss, 1998). And then there is xenotransplantation, the genetic manipulation of animal parts for transplantation into humans (Butler, 1998). At what point will the transgenic mice and pigs and sheep (or birds or abalone – patents pending!) be seen as a new caste of slaves, like some sort of twisted reprise of an H. G. Wells novel? How much longer will the Thirteenth Amendment block the American patenting of humanoids? The Trojan pig is coming to a pen near you. Such mix-and-match Chimeras are only the most literal instantiations of a whole array of hybrids being forged in the cyborg workshop. It is now commonplace to speak of DNA as a literal computer: that is, computations can be performed using the biochemistry of DNA replication in the lab.[5] This convergence of the two trademark cyborg technologies in the vital machine, once only a dream, is now a reality. Once DNA can be manipulated as a computer, then the cross-engineering of cells and microelectronic silicon chips cannot be far behind. Indeed, start-up firms already seek to meld combinatorial biochemistry with silicon-based electronics (Alpers, 1994). While this may carry the notion of computer as prosthesis to exaggerated lengths, for our purposes it mainly serves to project the implications of von Neumann's ambitions for computational prostheses in stark contrast with Jevons's fusty quaint appeal to the rock-hard character of human nature. You *can* change the nature of those savages, Stanley! Just ask Monsanto and DARPA and Applied Genetic Technologies!

How about the Self as center of causal agency? With some trepidation, I would petition my patient reader to consult the *Manual of Mental Disorders* – known in the psychiatric trade as *DSM-III* – under the heading of the disorder known as "anosognosia." This syndrome, which is estimated to inflict somewhere between 30 and 70 percent of the population at some time or other in their life-span, is defined as, "an alteration in the perception and experience of the self . . . manifested by a sense of self-estrangement or unreality. . . . the individual may feel 'mechanical' or as though in a dream. . . . people may be perceived as dead or mechanical" (Spitzer, 1979, p. 259). Machine dreams are a telltale symptom of anosognosia. Whatever the actual incidence of anosognosia in the general populace, I think the reader may concede it was the syndrome most characteristic of quotidian experience in the Cold War era and, more pertinently, of many of the major protagonists covered in this volume. What the phenomenon of anosognosia suggests is that the experience of the self and its efficacy is not a dependable invariant but can be warped and distorted and molded in certain distinct ways; and when instruments like the atomic bomb and the computer impinge on the consciousness of large proportions of the population simultaneously and in a correlated fashion (say, at Alamogordo, or in the Cuban Missile Crisis), the putatively imperturbable conception of the self may be shaken to its very core.

But let us suppose, dear reader, you are one of those thoroughly modern breed of skeptics who thinks that Freud was a fraud and psychiatry a pseudoscience constructed primarily to render analysis interminable and reduce the poor dupe of a patient to a human money pump. For you, the mind is nothing more than a bubbling brew of catecholamines, serotonin, cholinergics, and dopamines. Anosognosia may not phase you, although

[5] A wonderful web site full of how-to information on DNA computation is Lila Kari's home page at ⟨www.csd.uwo.ca/faculty/lila⟩.

you may be surprised to realize you are equally vulnerable in your belief in the inviolate self. For the problem of causal agency still haunts the microtubules of neurochemical brain, as much as the social psychology of the anosognosiac. Depression, anxiety, dementia, and schizophrenia all impinge in differing ways on the ability of the agent to conceptualize himself as a coherent self, one capable of exercising causal agency in the world. But even if hobbled, at least prior to the second half of the twentieth century, you could still maintain that the uneasy amalgam of behaviors really belonged to you. However, with the help of the nouveau cyborg science of psychopharmacology, the clinical psychiatrist can now raise and lower the brain levels of the four previously mentioned chemicals, and indeed many others, with all the panache of adjusting the pneumatic dials on a machine (Healy, 1998). Now, who or what is the *cause* (Aristotelian or otherwise) of the activities of the liberally plumbed and pinguefied self? If we really "listened to Prozac," would it tell us that it had finally liberated the "true self," or merely added emotional plastic surgery to the already formidable armamentarium of the prosthetic surgeon?

But, surely, many reading these lines will hasten to object that none of this whatsoever threatens the Self as the seat of human consciousness and intentionality. Yet this is easily the single-most salient consequence of the development of the computer in the postwar period. From the 1940s onward, the foremost popular fear of the computer was that it would degrade human intentionality by constituting a rival intentionality, one largely indistinguishable from the original, but lacking personhood. There was nothing quite so comparable to that little frisson of grammatical epiphany when confronted with machines who think. Such computer anxiety may appear painfully naive today, given the rudimentary character of actual machines in the 1960s and 1970s (Turkle, 1984); but the repeated controversies over whether machines can think should signal that the challenge of computer consciousness has only become exacerbated as computers have become more ubiquitous. In our first chapter, Daniel Dennett was quoted on the cyborg dictum that "we" would have to relinquish our quaint affection for all things human as a prerequisite for understanding consciousness. It may not be necessary to follow Dennett's lead very far in this respect in order to appreciate the impending dissolution of the idea that the self is the *fons et origo* of intentionality.

Ever since the illustrious David Hume entertained the notion that the Self was no temporal invariant, need we reiterate that this apprehension has only grown more acute in the interim? The more that biological evolution becomes the locution of choice in scientific explanation, the more temporal invariance of the self will come to seem old-fashioned, or quaint, almost nostalgic. One of the piquant aspects of the recent fascination with cloning is the question of the extent to which one would or would not feel a fundamental bond of identity with one's genetically engineered *Doppelganger*. It is a good bet you really will not enjoy the prospect of immortality in any palpable sense if the Self that lives on becomes a stranger – although the temptation for the rich to find out anyway will surely be irresistible.

But how about Kant's riposte about the moral autonomy and freedom of the Self that comes to learn its own Selfhood? Here is where the fallout from the cyborg sciences becomes positively lethal. The more that decision theorists and philosophers jump on the cognitive science bandwagon, and fall all over themselves to portray the rational agent as an automaton of various levels of computational capacity coupled to an arbitrary aggregation of genetic material, then the less likely it becomes that the Self will earn its spurs by learning to exercise its inviolable Selfhood. Instead, we can only be confronted with a heterogeneous jumble of prostheses, organs, and replacement modules, each of which can be augmented or disposed of as circumstances warrant. When comprehensive adjustment of the organism can potentially take place in roughly the same time frame as the alteration of its environment, then there is no telos of development, no terminus of the self. Everything from liposuction to *lapsus memoriae* militates against it. And as for moral responsibility, there are few things less "responsible" for their actions than a computer. Don't blame the mainframe or the program, blame the programmer; garbage in, garbage out. Our culture is replete with little lessons that when agents become machines, then they must be permanently exiled from the Garden of Eden. One of the deep homilies of Stanley Kubrick's *2001* is that when HAL the Computer transgresses his role in the master-servant scenario, he is summarily decorticated and prevented from partaking in the transcendental communion with a higher intelligence at journey's end.

The Self as the well-apportioned storehouse of memory has been especially vulnerable to the onslaught of the cyborg sciences. One of the primary components of the von Neumann architecture for the computer, and the single most astringent constraint in its early development was the contrivance of the organ of "memory." This limitation was rather quickly projected upon the architecture of human cognition as well, with one of the key controversies that ushered in the advent of cognitive science being the dispute over the limitations of human memory.[6] The parallel discussion of the hierarchy of automata of increasingly greater com-

[6] The paper by George Miller (1956) on "magic number seven" is frequently cited as one of the founding texts of cognitive science (Baars, 1986). The rival traditions of conceptualizing memory as a storehouse versus a correspondence are discussed in Koriat & Goldsmith, 1996. Other approaches to human memory and its disorders are surveyed in Schacter, 1996.

putational capacities discussed in Chapters 2 and 6 was also predicated upon specifications of various mechanisms of memory. Memory and its attributes had become one of the pivotal theoretical terms in the cyborg sciences; but all this newfound concern served to call attention to the profound weaknesses and instabilities of human memory, and thus, inadvertently, call into question the use of memory as *the* definitive characteristic of the self. For instance, it has been recently suggested that the rise of the "sciences of memory" made possible the discovery of "multiple personality syndrome" (Hacking, 1995); and nothing more graphically calls into question the very existence of an integral Self than the spectacle of multiple warring personalities frantically fighting for control over a single human body.

We conclude our checklist with that perennial favorite of the neoclassical economist, the definition of the Self as consistency in rational choice. If, given a fixed set of relative prices, you opted to choose A over B over C; and then subsequently, under the same set of circumstances, you chose C over A, or perhaps B over A, then you just did not make the grade as someone it was worth theorizing about, much less gracing with the honorific of "rational." Without the peremptory conflation of consistency of preference orderings with coherence of the Self, there was no "there" there for the neoclassical model to grab hold of; at best, from Pareto onward, identifying intransitivity as the utter disarray of the rational self implied that one might have to defer to the tenets of (gasp!) sociology in order to make any sense of this residuum of behavior. Much of the pronounced neoclassical hostility toward psychology over the past century can be understood as instinctive rejection of any rival notion of selfhood. Paul Samuelson's program of "revealed preference" theory (and thus Arrow's choice function framework) was an operationalist restatement of the dictum that in order to qualify as a neoclassical self in good standing, one must pass the test of perfect self-consistency.

One could easily maintain that this portrait of the self has been repeatedly disconfirmed in empirical tests, ranging from the "Allais paradox" to the phenomenon of "preference reversals" (Camerer in Kagel & Roth, 1995); but we shall opt not to take a position on the thesis that neoclassical rationality can actually be definitively refuted in any experimental setting (Cohen, 1981). Rather, we shall accuse the neoclassicals themselves of internal inconsistency. In the previous chapter, we have already outlined how this catechism fell afoul of the cyborg sciences. To reprise Chapter 6, Gödel's theorem states that formal systems of the power of arithmetic cannot be guaranteed both complete and consistent, and this has all sorts of implications for the theory of computation. But the portrayal of the economic agent has persistently been revamped in the likeness of a utility computer in the postwar era. Consequently, we observed in Chapter 6 that the construct of the "rational choice function" can be shown to disintegrate on contact with the abstract limits imposed by the possibilities of computation; the fact that the consistency of the self is unattainable for the situations imagined by the economist has been couched as a subject of mathematical proof by Lewis, Nachbar, Prasad, Richter, Wong, and a host of others, and not simply contingent empirical disconfirmation. This all amounts to the thesis that if the Self depends crucially on some form of computational consistency for its economic integrity, then someone needs to break out the Superglue.

If a long-overdue calm reassessment of the Culture Wars should ever materialize, we would eventually come to realize that it was not those wily irresponsible postmodernists who wrought the most havoc within the house that solid bourgeois virtues built with their "decentering of the self" and their fragmentations of the body (Amariglio & Ruccio, forthcoming; Rose, 1996). Rather, postmodernism was itself an effluvium of the intellectual innovations bubbling to the surface from the cyborg sciences, that is, originating in the *natural* sciences and their collateral pursuits (Forman, 1997). Methodological cyborgism is the natural and normal progeny of methodological individualism, sharing with it all the endearing qualities of scientism, hypertrophic mathematical formalism, monistic ambitions, and catchy slogans. Wherever the computer has cast its allure, there be cyborgs. Indeed, some of the most poignant evidence of this trend is the parade of intellectuals – Daniel Dennett here again springs to mind, but also sociobiologists like John Maynard Smith, or Deirdre McCloskey or Kenneth Arrow – setting themselves up as defenders of the Good Old Time Religion, all the while sowing the seeds of postmodern fragmentation in their wake. It is our present task in this chapter to outline how the very same dynamic has expressed itself in the elaboration of recent economic theories.

I Wanna Be a Unique Individual, Just Like You

The treatment of the self in economics has languished as a curiously neglected topic. One might search the pages of *Economics and Philosophy* or the *Journal of Economic Methodology*, or sample the nascent specialty of "behavioral economics" with its ubiquitous discussions of psychology and economic behavior and never once catch a fleeting glimpse of what might, with some justice, be called the central problem of the orthodox world picture of the neoclassical economist. Only very recently has the issue surfaced in a few philosophically informed discussions of recent economic theory (Kirman, 1992; Janssen, 1993, 1998; Sent, 1998, pp. 104–6). There it has slowly dawned upon the perceptive few – perhaps significantly, almost exclusively Europeans – that the neoclassical championing of the primacy of the individual has been more or less a sham throughout the history of mathematical neoclassical economics. The irony has been that, whereas the accompanying textual commentary will inevitably praise the individual self as the font of all analytical determinacy and the market as the only device that accords the idiosyncratic needs of the self their due meed, when one stares long and hard at the mathematics, all this vaunted respect for the essentiality of diverse individuality flies right out the window. In the Marshallian case, the contradiction is readily apparent, with the "representative agent" really little better than the "group mind" or *Volkerpsychologie* of the Germanic tradition so despised by tough-minded British thinkers. Yet other, more intricate models also commit similar solecisms.

Close readings of the annals of the mathematically oxymoronic individuals in the history of neoclassical economics will have to await a more propitious moment; in the meantime, we shall inadequately gesticulate toward where one would start to uncover the basic contradictions. The lineup would begin with Francis Edgeworth, who in his quest to argue that increased numbers of bargainers would render the utilitarian equilibrium unique, was forced to imagine each new entrant as a clone (in all relevant respects) of an already existing transactor (Mirowski, 1994b). In the Chicago school of neoclassicism, Nobelists George Stigler and Gary Becker felt impelled to argue that "tastes neither change capriciously nor differ importantly between people" (1977, p. 76) on conceptual, not empirical, grounds. One might suspect it was the hard-core Walrasians who could have been expected to stand up for the sanctity of individual difference; but there, also, one would be sadly disappointed. The chill wind of mendacity was already wafting about Cowles Mark II in its infancy, when it was acknowledged that everyone had to be treated as "self-sufficient" if the newly produced existence proof really were to guarantee equilibrium of an "exchange economy" (Rizvi, 1991): otherwise, irreducible differences between workers and capitalists might become apparent in the mathematics. The disingenuous stance toward the individual in the theory was further exacerbated by the tendency to treat firms and individuals as imperfectly ontologically differentiated (Ellerman, 1980) in Walrasian general equilibrium, a practice that would shortly lead to an effulgence of impostor individuals conflated with firms, clubs, teams, families, and even nation-states. Things came to such a pass that there was a phase in which it became fashionable in the neoclassical literature to decompose the rational agent into multiple warring selves, in order to purportedly patch up gaps in the utilitarian organon (Thaler & Shefrin, 1981; Schelling, 1980, 1984).

The dispensability of the individual was pushed to extremes by Robert Aumann's (1964) attempt to "generalize" the Arrow-Debreu model by stocking it with "atomless agents" – in effect, agents so "small" that their preferences taken individually didn't matter for price formation; not insignificantly, the implication was that they were all economically indistinguishable. This mathematical idiom not insignificantly became the stock-in-trade of high theory of the Cowles Foundation after the move to New Haven (Cowles Foundation, 1991). Yet the antiindividualism of the Walrasian tradition became painfully palpable in the 1970s, with the elaboration of the Sonnenschein-Mantel-Debreu theorems (Rizvi, 1998). Once the bad news about the inability of the unrestricted Walrasian model to say much of anything about uniqueness or stability of equilibrium became well known, the preferred escape route was to arbitrarily posit that everyone was identical: that was the one situation in which all the sought-after properties of equilibrium would obtain.[7]

The litany of spurious individualism continues with the accession of Nash game theory to pride of place as the centerpiece of orthodox microeconomics in the 1980s. The perverse misrepresentation of the independent experience of other individuals began with Nash himself, as we saw in Chapter 6: strategic rationality for Nash was captured formally by the ability of the gamester to reconstruct the entire thought processes of all rivals within the space of their own skulls, which was about as close to the concept of identity of individuals that one might get and still have an opponent different enough from oneself to make the game worth playing. But even this rudimentary level of individual difference proved refractory in much the same manner and for much the same reasons it had done in the Walrasian tradition. Nash equilibria, just like Walrasian general equilibrium, could be shown to exist under fairly general conditions; but uniqueness and stability were proving elusive. The redoubtable Aumann, along with other Nash enthusiasts such as Harsanyi, advocated a position which would "save" the Nash equilibrium by making the players ever more alike. This began with an argument by Aumann (1976) that statistical induction, in conjunction with common knowledge, would render it impossible for two rational agents to agree to disagree. In their Bayesian portrayal of the agent, "rationality" purportedly dictated that all individuals should possess the same prior probabilities – surely a travesty of the subjectivist position. Harsanyi seemed to think the shock of homogenization could be softened by having everyone come equipped with a fixed repertoire of "types" that constituted identities they could assume. Further scrutiny of the game-theoretic setup

[7] Professor Rizvi reminds me that, under general conditions, positing identity was not enough: if all agents are identical and possess homothetic preferences, then the exchange economy as a whole obeys WARP in the aggregate – but then there really is just one individual in the economy, writ large.

also led many to concede that every player had to possess the same stock of "common knowledge," including the game structure, the nature of the rationality of their opponents, their posterior beliefs on strategy choice, and any other seemingly irrelevant consideration that might constitute a "focal point" of attention: by this time it was beginning to seem that the game was not worth the candle, because everyone would only be furiously trying to outwit themselves. The mad rush in the 1970s and 1980s to provide a "justification" of the Nash equilibrium concept collapsed in a frenzy of counterexamples, as it became evident to all and sundry that "it turns out to be difficult to regard the Nash equilibrium concept as a consequence of a theory of individual rationality" (Janssen, 1998, p. 3).

I do not mean to propose here the patently untenable assertion that no neoclassical economist of consequence has ever produced a model populated with richly differentiated individuals, much less to inquire in detail how it was that such a surfeit of neoclassical economists could so proudly preach methodological individualism while managing to observe it only in the breach. Rather, our more targeted task is to focus upon the grand contradiction embedded in the fact that the rejection of von Neumann's vision for economics by everyone from Nash to Koopmans to Simon to Aumann to Myerson to Binmore was predicated upon a defense of the sanctity of a rational self that was perilously absent within their own tradition; and the more they protested von Neumann's corporatist inclinations, the worse the contradictions grew. Von Neumann's conception of social theory was simply more prescient than any of the numerous individualist attempts to resist it.

Curiously enough, defending the primacy of the self was much easier in that bygone era when neoclassical economics had proclaimed pristine agnosticism about the nature of the mind. When Milton Friedman opined assumptions don't matter, or Gary Becker insisted that no commitments concerning rationality were required, or Paul Samuelson made a show of abjuring utility for brute behavior, then there was quite explicitly no place left to go when searching for the elusive self. But once the cognitive (half-)turn was made at Cowles, once rationality started to get conflated with econometric inference, once hostages were surrendered to the utility computer, once the economy was recast as the coding and transmission of messages, then attention could not help but be shifted from selfless outcomes to processes. This was the kiss of death for the Self in economics, because now there was a name, an address, a sender, a receiver, and maybe even a telephone number for the agent; now one could come calling and find there was no one at home. Worse, suppose you happened to be one of the few who took von Neumann's strictures to heart: say you came to believe that computers really were the physical embodiment of individual algorithmic rationality, and thus it became imperative to learn something about the Code of the Cyborg in order to retrofit rational economic man. The ultimate irony is that, with the best of intentions, you would find yourself inadvertently participating willy-nilly in the wholesale deconstruction of the Self in economics.

The remainder of this chapter relates the narrative of two or three "individuals" (really, a dynamic duo and a gaggle of game theorists – even in historical narratives, individualism is a pain to maintain) who did just that. They illustrate just how far the cognitive turn in economics has proved a debacle for the deliberative Self. Herbert Simon was one of the original members of Cowles Mark II who defected early on from Koopmans's program of cyborg accommodation in the direction of something he dubbed (in conjunction with Allen Newell) as "bounded rationality" and "artificial intelligence." We argue that Simon has essentially abandoned the old-fashioned self in favor of a program of constructing simulacra of people in explicit reaction to John von Neumann, thereby promoting an uneasy alternative accommodation with computational themes. Then we shall return once more to the microeconomic orthodoxy of the 1980s, namely, Nash equilibrium and its serried ranks of defenders. Game theory may have gone to sleep in the 1960s in economics, but it nonetheless did find a home in operations research and computer science, especially in Israel. In that exotic clime we will observe Robert Aumann disseminating little machines playing games throughout the landscape, almost brazenly ignoring the way in which his virtual agents tended to undermine the coherent economic self. In the final section, we shall encounter a mathematician-turned-economist Kenneth Binmore (in conjunction with a phalanx of fellow game theorists) proposing the intriguing scenario of von Neumann–style automata playing each other in an economic version of core wars. William Gibson couldn't have spun better science fiction in any of his neuromances: what better icon of cybotage than two computers caught up in an endless feedback loop trying to outsmart one other?

HERBERT SIMON: SIMULACRA VERSUS AUTOMATA

> Herb Simon once said to me: Don't waste your time commenting on someone else's history. Wait until you write your own.
>
> Edward Feigenbaum, "What Hath Simon Wrought?"

From Ratio to Imbroglio

This quotation captures, for me, the discreet charm of that major protagonist in our cyborg narrative, Herbert Simon. For those willing to make the effort to peruse his far-flung writings in economics, psychology, operations research, management science, statistics, artificial intelligence, politics, science policy, and especially his extremely artfully constructed autobiography, which manages simultaneously to seem guilelessly confessional and yet hinges on the trope of comparing his own life to a rat running a maze (1991a, p. 86), they will discover that he has in effect pioneered many of the pivotal theses that have structured this book. Simon first observed that, "physicists and engineers had little to do with the invention of the digital computer – that the real inventor was the economist Adam Smith" (1982, 1:381). I wish I could claim priority, but I must confess it was Simon who first wrote: "Perhaps the most important question of all about the computer is what it has done and will do to man's view of himself and his place in the universe" (1982, 2:198). He was one of the first (in 1978!) to suggest to economists in a public lecture that their previous definition of economics as "the allocations of scarce resources to competing ends" was sadly out of date (1982, 2:444). He has been steadfast as an unapologetic prophet of the ontological indeterminacy of the Cyborg: "Here is the computer. Here's this great invention, here's a new kind of organism in the world. . . . Let's have a science of them in the same way we have a science of *Homo sapiens*" (in Baars, 1986, p. 367). Simon cautioned that one must approach the innovations of operations research, game theory, cybernetics, and computers as all of a piece in any understanding of the rise of American mathematical economics (Simon, 1991a, p. 107). He also took a political concern over "planning" and transmuted it into a uniform psychological environment for decision making in everything from the individual consciousness to the large multiperson organization.[8] Thus in Simon's work we observe most clearly how it came to be that "planning" was rendered innocuous for the American postwar context. It was once again Simon who first noticed, "the computer is the perfect tool for simulation because it is the perfect bureaucracy" (in Crowther-Heyck, 1999, p. 429).

Simon has also provided some of the best clues for the cyborg historian; for instance, he has modestly insisted that one should credit much of his subsequent recognition within economics (such as it is) to connections forged in the early days of RAND and Cowles (1991a, p. 115; Baars, 1986, p. 380). He significantly has pointed us in the direction of Merrill Flood, Robot Sociologist, as the person who initially summoned him to RAND in 1952 as a summer consultant; and he cheerfully concedes

[8] This point is made in Crowther-Heyck, 1999, p. 190. The cyborg enthusiasm for a flat ontology was the subject of a letter from Ross Ashby to Simon, August 23, 1953: "It is my firm belief that the principles of 'organization' are fundamentally the same, whether the organization be of nerve cells in the brain, of persons in society, of parts in a machine, or of workers in a factory" (quoted in Crowther-Heyck, 1999, p. 315).

that RAND awoke him to the possibility that machines can think. Simon was the only insider to Cowles Mark II to blow the whistle on the rather strained attitude toward mathematics and the virtues of rigor prevalent there (1991a, pp. 104–6), as well as the only member to cheerfully describe its politics. "I decided that I had been a closet engineer since the beginning of my career" (p. 109). And, of course, he has been one of the most disarming critics of the postwar orthodoxy: "How any grown-up, bright human being can go satisfied with the neoclassical theory is kind of hard to understand" (in Baars, 1986, p. 379). At first glance, the work you hold in your hands may seem little more than a series of footnotes to Simon's pronouncements. Simon, true to form, has written his own history.

And yet . . . And yet . . . the maxim imparted to Feigenbaum also encapsulates the characteristic Cowles contempt for history, and it should serve as a warning that Simon has been every bit as ruthless in constructing his own persona as, say, Nietzsche was in another era. Simon has been many things to many people, but he is no historian. All his prodigious autobiographical epopee have been tuned to conjure a world as much as they have been plucked to record it.[9] For all the worldly success, the Nobel Prize, the transdisciplinary passports, all the exhilaration of attaining the status of Cyborg's Cyborg, Simon has had quite a bit to be circumspect about; for Herbert Simon has also played the part of the consummate Cold War intellectual.[10] Simon has demonstrated all the talent of the successful espionage agent: he can *pass* under almost any circumstances. It is no accident he is a specialist in *intelligence* and the sciences of the artificial. It is appropriate that he has moved from one discipline to another, never calling any one home for long. The reader of his complete oeuvre cannot but help notice that he is entirely capable of telling widely diverse audiences in different disciplines what he thinks they would like to hear. In the final analysis, and in the interest of maintaining multiple aliases, Simon resists calling things by their real names: "We must either

[9] This is intended as a compliment rather than a criticism. The behavior of the Nietzschean "Strong Methodologist" who does not legislate method for others, but instead cobbles together a novel scientific identity out of a *bricolage* of far-flung materials is discussed in Mirowski, 1994c, pp. 53–54.

[10] There is a sense in which Simon's military ties permitted him to break new ground as the first social scientist in such strategic advisory bodies as the National Academy of Sciences, PSAC, and COSPUP (Cohen, 1994, pp. 165, 171). His early RAND connections paved the way for him to catch the eye of Bernard Berelson and Rowan Gaither at the Ford Foundation, who in turn supplied substantial subsidies for both his research into AI and the founding of GSIA at Carnegie Mellon. It was as an AI expert that he infiltrated the bureaucratic preserves of the natural scientists, thus opening doors for his Cowles comrades into the science policy elite.

get rid of the connotations of the term, or stop calling computers 'machine'" (1982, 2:199). In an interview with *Omni* magazine (1994), the interlocutor put to him the deep question: "Could a computer have come up with your theory of bounded rationality?" to which Simon responded (testily): "In principle, yes. If you ask me if I know how to write that program this month, the answer is no." Or, yet, even better: "There does not have to be a real thing for every noun – take *philosophy*" (in Baumgartner & Payr, 1995, p. 234).

I believe that Herbert Simon is one of the most egregiously misunderstood figures in the history of modern economics. Much of this, as we shall observe, can be attributed to the obtuseness of the neoclassical economics profession when confronted with a master polymath – some, but by no means all. Another moiety must be laid at Simon's own door. The very same Simon who is so quick to disenfranchise the individual economic agent and insist on the primacy of the social organization generally presents his own history as though it were the saga of the Promethean Self, abstracted from detailed social context (Sent, forthcoming). The result is that there are some awkward blank pages, and some torn-away sheets, in the ledger of his accomplishments.

The first glaring elision that jumps out at the historian is the deafening silence on the overwhelming role of the military in Simon's career. It is now widely conceded that "artificial intelligence" would never have had such a stellar run for the money were it not for the military (Edwards, 1996; Guice, 1998; Sent, 1998); no one has yet observed that "bounded rationality" equally owes its inspiration to the military. But if that is so, what can it mean for Simon to opine: "I don't know how [questions of war and peace] could be approached in a scientific way with our present scientific knowledge" (1983, p. 103)? A second elision is his reticence to provide context concerning his own prodigious role in shaping government policy toward the social sciences, and therefore the very milieu in which social science was recast in the postwar period (Cohen, 1994, pp. 165, 171). Simon is as much the inheritor of the legacy of Warren Weaver as he is of Norbert Wiener; but both are judiciously downplayed. As a historian of artificial intelligence has written, Newell and Simon's "means-end analysis is just Wiener's feedback principle carried to a higher level of abstraction" (Crevier, 1993, p. 53), yet Simon sought to elevate Ross Ashby's importance over Wiener. Symmetrically, Simon's theory of organizations is just Weaver's theme of "complexity" also carried to a higher level; he himself prefers to trace the theme to Chester Barnard. Hence, just as he models problem solvers as independent of their context, Simon seems unwilling to regard his research, or indeed the entire postwar social structure of science, as the conscious product of an environment of science managers.[11] A third lacuna concerns the observation that Simon gathers no moss: he repeatedly tends to exit a discipline for greener pastures just as his work begins to enjoy a modicum of familiarity in its erstwhile setting. The more sedentary denizens of the recently evacuated field often respond to his brief sojourn, not with hosannas of happiness, but more often than not with carping resentment and hostility.[12] We are not concerned in deciding whether intermittent outbreaks of ingratitude are just sour grapes or something a bit more yeasty – the question of Simon's impact across the academic board is actually quite a difficult one. Rather, our own disquiet arises from an apparent inability of Simon to confront the potential reasons why he has been able to enjoy the peripatetic gypsy life in the hidebound groves of academe without suffering the consequences. Our contention is that much of his freedom must be understood as an artifact of the advent of the computer, and consequently owes much to his deeply ambivalent relationship to John von Neumann.

How Economic Rationality Found a Bound

Although von Neumann's abortive intervention in the postwar construction of mathematical economics did not initially transform the content (though it did profoundly affect the form, as outlined in Chapter 6) of the neoclassical orthodoxy, it did have a more profound effect on at least one Cowles researcher, namely, Herbert Simon. In the first instance, were it not for Cowles, Simon might not have been later accorded much credibility as an economist, for his early career of the 1930s was squarely

[11] "The question of context [of problem solving] was just not asked" (Allen Newell in Baumgartner & Payr, 1995, p. 151).

[12] We shall make do with one representative complaint from psychology and one from AI, although examples could be plucked from any other field in which Simon has visited: "Everybody in psychology knows he's great, but nobody's ever read him. I have never taken him seriously, and I can tell you why. When you have a man who sits there and looks you straight in the eye and says the brain is basically a very simple organ and we know all about it already, I no longer take him seriously. . . . Simon is not a human psychologist; he is a machine psychologist. . . . That is not cognitive science, but abject cognitive scientism" (Walter Weimer in Baars, 1986, pp. 307–8).

John McCarthy has written: "In my opinion, [Simon's] GPS was unsuccessful as a general problem solver because problems don't take this form in general and because most of the knowledge needed for problem solving and achieving goals is simply not representable in the form of rules for transforming expressions" (in Association of Computing Machinery, 1987, p. 259). It may also help to note that Simon was essentially exiled from the economics department at Carnegie Mellon into the psychology department in 1970. See Simon, 1991a, pp. 250ff.

An assessment from within RAND comes from the George Brown interview, March 15, 1973, SCOP: "the problem is not how to do better with a computer what people do pretty well; the problem is how man plus computer can do better than the man without the computer."

situated in political science, in that subfield concerned with bureaucracies, municipal government, and the codification of the principles of administration. Simon was nevertheless an enthusiastic comrade-in-arms of the small band of theorists seeking to advance the mathematization of the social sciences in the interwar era and, as such, was ushered into the inner circle of Cowles in the early 1940s, during his stint at the Illinois Institute of Technology (1942–49). Through Cowles he came in contact with John von Neumann[13] and thence to RAND; and it was these two encounters which brought him in contact with his life's thematic of bounded rationality.

The senses in which Simon's concept of the boundedness of rationality derived from a rejection of the neoclassical mathematization of economics in conjunction with a reaction against game theory and the automata program of von Neumann have only recently begun to be explored. It appears that initially Simon believed he had found a kindred spirit in von Neumann's project to develop a mathematical theory of social organization, for from his earliest work on *Administrative Behavior* (1947), he felt confident that "organizations can expand human rationality, a view quite opposed to popular folklore in our society" (1991a, p. 72). He recalls that *TGEB* "hit me like a bombshell" ("Personal Memories"). He aired this initial opinion in his glowing review of *Theory of Games*: "it leads to most important sociological results – notably to a demonstration that in games with more than two persons coalitions (organizations of two or more persons who co-ordinate their behavior) will in general appear. . . . many of the research problems which appear to social scientists to be significant lend themselves rather directly to translation into the theory of games and, hence, to rigorous treatment" (1945, p. 560). On further reconsideration of *TGEB*, however, he rapidly reversed his initial favorable impressions, expressing qualms about the solution concepts as incapable of saying anything about observable behavior, and doubting that they approximated any actual social forms of organization, either. He tried to initiate a dialogue with von Neumann and Morgenstern on these issues, but was accorded peremptory treatment by both. The state of his attitude toward game theory soon after von Neumann's lectures to Cowles is captured in a long letter he wrote to Oskar Morgenstern in August 1945:

> This is a long-delayed reply to your letter concerning my review of the *Theory of Games*. Since you wrote I have had the opportunity to discuss your book with professor von Neumann when he was in Chicago. . . .

[13] See, in particular, Simon, 1991a, p. 108; 1992b; and, most importantly, his unpublished "Personal Memories of Contacts with John von Neumann," December 20, 1991, included in Herbert Simon & Mie Augier, *Models of Herbert Simon* (forthcoming).

> In Chapter III . . . this analysis of "randomization" as a defense, and the defensive interpretation of bluffing in poker may be a part of the truth, but certainly not the whole truth. In military tactics, for example, the attempt to secure surprise certainly evidences a more aggressive tactic than what you call a "good strategy." Hence your definitions eliminate from consideration any games where players deliberately seek to "find out" the opponent. . . . I recognize that abandoning the present definition of "good strategy" would undermine the idea of the "value of the game," and hence the entire theory of coalitions. In spite of the productivity of the concept for the formal theory, I am convinced it will have to be given up or greatly modified in the further development of the theory for application to social phenomena. . . .
>
> Being a social scientist rather than a mathematician, I am not quite willing for the formal theory to lead the facts around by the nose to quite this extent [with regard to the definition of the solution concept of the stable set – P.M.] – although I recognize that similar methods have been very fruitful both in relativity theory and in quantum mechanics. I have further difficulty in pinning down the operational meaning of the term "solution." . . . I might sum up these objections by saying that I am not at all clear as to the behavioristic implications of your concepts of "solution" and "stability," and that I feel the only safe approach to a definition of stability is an explicitly dynamic one.[14]

Over time, he began to make increasingly harsh comments about the structural weaknesses of von Neumann's game theory as any sort of codification of rationality, as highlighted in his Nobel lecture: "Game theory addresses itself to the 'outguessing' problem. . . . To my mind, the main product of the very elegant apparatus of game theory has been to demonstrate quite clearly that it is virtually impossible to define an unambiguous criterion of rationality for this class of situations" (1982, 2:487–88).

But this attempt in the later 1940s to draw von Neumann out on the implications of his construct of game theory would have been unavailing, because as we have suggested in Chapter 3, von Neumann had already more or less abandoned game theory himself in favor of automata

[14] Herbert Simon to Oskar Morgenstern, August 20, 1945, box 32. VNLC. Morgenstern appended a handwritten note to "Johnny" in German: "I have not yet written to Simon. I have also not given much thought to his points, but I'm not very impressed. However, it is nice that he is so busy scrutinizing the theory."

Parenthetically, the evidence in this letter apparently clashes with the assertions made by Simon in "Personal Memories" that his first substantive encounter with von Neumann was at the 1950 Econometrics Society Meeting in Boston, and that "None of my interactions with the authors of *The Theory of Games* were with him [i.e., von Neumann]." Indeed, it is quite likely that Simon attended von Neumann's lectures on games at Cowles in 1945.

theory in the era in which Simon sought contact. Simon was nevertheless induced to confront the shift in von Neumann's program by some incidents that were crucial for the genesis of his own program (Sent, 1997b, pp. 34–36; Simon, 1988; "Personal Memories"). In the first, Simon was dragooned as discussant of von Neumann's 1950 address to the Econometrics Society, because "all of the people I approached for this task declined, perhaps intimidated by von Neumann's colossal reputation" ("Personal Memories"). In his address, entitled "The Theory of Automata," von Neumann elaborated on his third-phase trademark topic, the disanalogies between the digital computer and the brain. Simon, with characteristic feistiness, opted to challenge the mathematical demigod and insist upon some positive analogies between computer software and the hierarchical organization of human thought, predicated upon his own study of bureaucratic administration. The anticipation of later themes is apparent from the abstract of the discussion prepared by Simon:

> The interest of social scientists in the recent rapid development of automata has two bases: they are interested in powerful computing devices that would enable them to handle complex systems of equations; and they are interested in the suggestive analogies between automata on the one hand, and organisms and social systems on the other. With respect to the second interest, Mr. von Neumann's strictures on the limitations of the automaton-organism analogy are well taken. The analogies are fruitful in exposing some of the very general characteristics that such systems have in common – for example, characteristics centering around the notions of communication and servomechanism behavior. But it is dangerous to carry such analogies in detail – e.g., to compare specific structures in a computer with specific neural structures.
>
> An important aspect of the theory of automata needs to be further explored. Rationality in the organisms often exhibits an hierarchical character. A frame of reference is established, and the individual or social system behaves rationally in that frame. But rationality may also be applied in establishing that frame, and there is usually a whole hierarchy of such frames. Some automata exhibit the rudiments of such a hierarchy, but it may be conjectured that the greater flexibility of organismic behavior is somehow connected with the more elaborate development of hierarchical procedures.[15]

[15] Memo to Participants in Harvard and Berkeley Meetings of the Econometric Society, September 19, 1950, in Simon & Augier, *Models of Herbert Simon* (forthcoming). In "Personal Memories" Simon writes: "I have no recollection that I ever had a serious conversation with von Neumann about the similarities and differences of our views about the analogy between brains and computers."

And once again, at a 1952 RAND seminar, von Neumann expressed skepticism about the quality of chess-playing programs written for computers, touching on the ways both game theory and computer architectures encountered obstacles in capturing high-quality strategic play. Simon again sought to contradict the master, and in his own words, this stance set him on course for the symbol processing approach to artificial intelligence and his initial brace of papers announcing the birth of the concept of "bounded rationality."[16] "I proceeded, rather soon thereafter and in direct response to the lecture, to sketch out a chess-playing program based on Shannon's earlier proposal and augmenting the heuristics he had discussed" ("Personal Memories").

The key to attaining an understanding of bounded rationality is to apprehend that its genesis was essentially simultaneous with Simon's induction into the world of artificial intelligence, and that it was intended from the beginning not only to stake out his independent position in opposition to game theory and neoclassical economics but also simultaneously to construct an alternative to von Neumann's nascent theory of automata.[17] All of these aspects were laid out in the watershed 1955 *Quarterly Journal of Economics* paper (Simon's most frequently cited paper by economists), although due to some unfortunate historical accidents, such as the elimination of an appendix on chess-paying programs in the published *QJE* version, as well as the publication of the complimentary portion of the paper on the "environment of choice" in a psychology journal (1956b), few if any economists were readily apprised of this fact. The connections between bounded rationality, artificial intelligence, psychological experimentation, and von Neumann have been obscured because they were originally forged at the Systems Research Laboratory (SRL), a subset of RAND which did not itself produce any high-impact academic research. However, some historians are nonetheless coming to regard this unit as one of the defining moments in the early construction of cyborg science (McCorduck, 1979, pp. 124ff.; Edwards, 1996, pp. 122–25). As Allen Newell reminisced: "I have considered myself as a

[16] "My brash youthful reaction to the lecture was that the job couldn't be quite as hard as [von Neumann] suggested, and I resolved to try to do something about it. That soon led to my partnership with Allen Newell and Cliff Shaw, and nothing in my life has been the same since" (Simon, 1988, p. 10).

[17] While some of Simon's later themes concerning rationality can be found in his earlier *Administrative Behavior* (1947), especially his fascination with hierarchies, I believe Simon would concede that these ideas would neither have gained the attention nor drawn sustained intellectual succor merely from the field of the political study of bureaucracies. But I would venture further to suggest that without the computer, and a fortiori, without von Neumann, the program of bounded rationality would have had no independent substance.

physicist for a long time, but at RAND I worked on experiments in organizational theory" (in Baumgartner & Payr, 1995, p. 148).

The SRL can be summarized as starting with the techniques of wargaming, crossing them with the managerial task of organizing the command and control structure of the newly installed early-warning radar systems, recognizing that the computer would play a central role in the new configuration, and proposing to engineer and to improve the man-machine interface through the technique of simulation. The three psychologists plus Allen Newell who initially composed the SRL found it quite easy to slide from approaching the computer as an instrument to treating the air defense detection center itself as a computer: "A system's environment can be inferred from, and its actions controlled by, information – a vital commodity to any organization" (Chapman et al., 1958, p. 251). By 1951 they had built a mockup of the McChord Field Air Defense Detection Center in Santa Monica and, using IBM calculators, had simulated attacks appear on radar screen displays.

When Simon was hired in as a consultant on the project in 1952, it was as if he had been struck down on the road to Damascus, and many distinct theoretical entities coalesced into a single phenomenon: "But that air defense lab was a real eye-opener. They had this marvelous device there for simulating maps on old tabulating machines. Here you were, using this thing not to print out statistics, but to print out a picture, which the map was. Suddenly it was obvious that you didn't need to be limited to computing numbers" (Simon in McCorduck, 1979, p. 125). The liberation from Cowles's prior commitment to cognition as intuitive statistics in favor of pursuit of organizations as symbol processors, via the novel technique of simulation, courtesy of the military and IBM, was immediately enthralling. Simon claims in retrospect an immediate rapport was struck with Newell, and the two began to collaborate on the problem of producing a chess-playing machine. The RAND connection permitted Simon to fortify his challenge to Johnny von Neumann using the eponymous Johnniac at RAND. "Perhaps my most important encounter with von Neumann was not with the man but with the computer he designed for Rand. . . . All of our early artificial intelligence work was done on Johnniac, and the details of construction of our first IPL's (information processing languages) were considerably influenced by the architecture of Johnniac and its assembly language (called Easy Fox)" ("Personal Memories"). Enlisting the RAND programmer J. C. Shaw, Newell and Simon completed the first artificial intelligence program, the Logic Theorist, by 1956. This program, and the chess-playing algorithms that followed, were based on various heuristics, which became the first instantiation of Simon's later trademark doctrines of "satisficing" and "bounded rationality."

For our present purposes, it is not quite as important to enumerate the various ways Simon's project of the boundedness of rationality diverged from the neoclassical orthodoxy as to highlight the ways in which it should be set in opposition to von Neumann's automata – or, as he maintains, "von Neumann's work on "neural" automata had little influence on our own research" ("Personal Memories"); and here this notion of "computer simulation" is crucial. Von Neumann was always a proponent of the logical possibilities for innovative reasoning opened up by the computer, and, consequently, it was Alan Turing's notion of the Turing machine which became pivotal in his elaboration of the prospective theory of automata. By contrast, the tradition of formal logic never held much allure for Simon, although this fact may itself be occluded by the happenstance that Logic Theorist was an algorithm dedicated to proving thirty-eight of the first fifty-two propositions in Chapter 2 of Russell and Whitehead's *Principia Mathematica*.[18] Simon shrewdly understood that, strategically, it would take the demonstration of a machine performing manipulations of formal logic in order to make people entertain the notion that a machine could think. Nevertheless, Simon's philosophical position has consistently maintained that formal logic does not empirically describe how humans think, and therefore he has always been much more drawn to Alan Turing's other epochal creation, that of the "Turing Test."

The Turing Test had been proposed in 1950 to provide a quick and dirty definition of machine intelligence. As we saw in Chapter 2, Turing imagined what he called an "imitation game," where an average subject is allowed to pose questions to a human and a computer, suitably disguised and dissembled, with the objective of identifying the machine. Turing proposed that if the machine could fool the target subject with a frequency approaching that which a man could fool the same target subject by pretending to be a woman, then that, for all practical purposes, would constitute evidence of machine intelligence. Furthermore, Turing predicted that by the turn of the millennium, machines would be capable of passing this test more than 70 percent of the time. We must forgo extended meditation upon the rich overtones of wartime deception, gender identification, the intentionality of dissimulation, and game theory here to focus on the pivotal point: Turing set the dominant tone

[18] See Russell & Whitehead, 1910. This incident is described in Crevier, 1993, pp. 44–48; O'Leary, 1991. For an insightful history of automated theorem proving, see MacKenzie, 1996. Simon's own relationship to logic is exemplified by the comment in his autobiography: "logic is not to be confused with human thinking. For the logician, inference has objective formal standards of validity that can only exist in Plato's heaven of ideas and not in human heads" (1991a, p. 192).

for decades of conflating machine intelligence with *simulation*. Simon has been known on occasion to call the Turing Test "the method of science" (in Baumgartner & Payr, 1995, p. 241). This was not simply a harmless turn of phrase: Simon expanded this thematic into a full-blown alternative to von Neumann's automata.

Whatever the other progenitors of the field of artificial intelligence may have believed (Crevier, 1993; Button et al., 1995), Simon has consistently hewn to the precept that if a machine can simulate the behavior of human problem solvers in well-specified situations, and can reproduce their observable outputs, then the machine is thinking. As he puts it in his own inimitable fashion, "I am still accused of positivism, as though that were some kind of felony" (1991a, p. 270). Thus it is an egregious error to write off Simon's bounded rationality as merely the shopworn complaint of the naive empiricist that "people really don't think like that," perhaps pleading for a more nuanced input from psychology into economics (although he sometimes does personally come across this way). Simulation has permitted Simon to propose an altogether different style of theorizing, one where mimicry displaces the previous dictum that one must grasp the structure of explanation through mathematical intuition or axiomatic display. As his collaborator Newell comments: "In Herb Simon's and my stuff . . . [the] concern for artificial intelligence is always right there with the concern for cognitive psychology. Large parts of the rest of the AI field believe that is exactly the wrong way to look at it; you ought to know whether you are being a psychologist or an engineer. Herb and I have always taken the view that maximal confusion between those is the way to make progress" (in Crevier, 1993, p. 258). He has been an articulate advocate of "theories of the middle range," by which he means that he neither wants to resolve individual cognitive processes down to their neurophysiological components, nor does he aim to model the global general cognitive architecture. He is quite happy to continue to "simulate human behavior over a significant range of tasks, but do not pretend to model the whole mind and its control structure" (Simon, 1991a, p. 328). He is the prophet of the construction of simulacra as the prime analytical technique of the "sciences of the artificial."

In this conviction that simulation is a legitimate research activity, he has been quite happy to claim Turing (but not von Neumann) as a precursor:

> It has been argued that a computer simulation of thinking is no more thinking than a simulation of digestion is digestion. The analogy is false. . . . The materials of digestion are chemical substances, which are not replicated in a computer simulation. The materials of thought are symbols. . . . Turing was perhaps the first to have this insight in cleat form, more than forty years ago. (Simon, 1996, p. 82)

Perhaps this brief and somewhat superficial characterization of Simon's professional quest can help us begin to understand the sheer boorishness of the perennial challenge of the orthodox economist that Simon provide him with *the* model of bounded rationality, the plaint that there is no *single* procedure that is warranted by the concept of satisficing, the lament that the revision of aspiration levels in the course of satisficing can itself account for no *equilibrium* position, the insistence that limitations of computational capacity represent just another category of scarcity over which optimization can be deployed. It is akin to insisting that a Zen master produce his Ten Commandments inscribed in stone on the spot; or, perhaps more to the point, importuning Bill Gates to explain the indispensability of a college education in our modern economy; as such, it is symptomatic of an obtuse category mistake.

Contrary to popular opinion, Simon has not been engaged in proposing a model of rationality as consistency for use in economics; nor has he simply been occupied by importing some intriguing observations from empirical psychology, some exotic quirks of our biological makeup that explain our divergence from full-blooded rationality, after the manner of "behavioral economics." He has been *simulating* the operation of a number of problem-solving tasks as though they were the manipulation of symbols on something very nearly approximating a serial von Neumann architecture.[19] Each simulation module is relatively self-contained; the success of the simulation calibrated by comparison to experimental evidence gathered primarily from protocols where subjects "think aloud" while they attack a problem.[20] The proof of the pudding is in the eating; or, as they so trenchantly put it, "a good information processing theory of

[19] The von Neumann architecture consists of four principal components: a memory unit, a control unit, an arithmetic-logic device and input-output devices. Programs are stored in the finite memory unit, along with data in process. The control unit executes the program according to a sequential clock cycle. The restriction to sequential operation has been dispensed with in other computer architectures, called "connectionist" systems. One reason why Simon's work is sometimes regarded as outmoded in the AI community is that his notion of symbol processing remains wedded to the sequential architecture, whereas the bulk of the profession has shifted its allegiances to a greater or lesser extent to connectionist architectures. Simon has maintained, however, that, "As an overall model for human thinking, the von Neumann architecture is, in my view, considerably closer to the mark than is a connectionist system" ("Personal Memories").

[20] This raises the interesting issue of Simon's hostility to statistical procedures as a legitimate form of empirical inquiry in his work on cognitive information processing. "We never use grouped data to test the theory if we can help it. The models describe individuals, so the hard part is to say with precision what is common to all information processors. With this approach it does not seem natural to assume that human behavior is fundamentally stochastic, its regularities showing up only with averaging" (Newell & Simon, 1972, p. 10).

a good human chess player can play good chess; a good theory of how humans create novels will create novels" (Newell & Simon, 1972, p. 11). The chess-playing program does not even attempt to compose novels; nor does the novel composer play chess. Neither has a prayer of matching the ability of a three-year-old to negotiate her way through a room strewn with toys and furniture, or to intuit grammar from mere eavesdropping upon human conversations.

It becomes apparent that this extreme modularity and specificity of each of the simulation exercises has in fact facilitated a situation in which computers can produce "results" that are locally impressive to the untutored layperson, without actually having to solve many of the knottiest problems of the nature and operation of generalized intelligence in all its versatile splendor. Indeed, one observes the logical extension of Simon's simulacra in the development of "expert systems" which mimic the stylized responses of a suitably chosen set of narrowly defined experts: the ultimate Turing Test. But other than a technology slated to displace some high-priced human labor with machines, something of practical interest to the military and corporate sponsors of AI, what is the ultimate goal of Simon's rather motley portfolio of simulation programs? Is Simon just himself following his own dictates, "satisficing" with respect to his version of artificial intelligence just far enough to placate his sponsors and produce something that can readily be sold to his (gullible? instrumentalist?) public? Simon adamantly insists this is not the case. Indeed, he is unusual in that he hotly denies that AI oversold its own promise and therefore consequently suffered in credibility.[21] Rather, here is where his personal version of a theory of "complexity" and his appeal to evolutionary theory enter to provide intellectual credentials for his program.

Simon harkens back to his early experience as a theorist of bureaucracy in order to argue that hierarchies and chains of command really do embody principles of human intelligence. Not only is it his credo that "nature loves hierarchies" but that it especially loves hierarchies that come in a particular format: namely, loosely coupled trees of relatively self-contained modules.[22] Simon says Nature loves hierarchies because

[21] *Omni:* Maybe its that AI, unlike genetics, has had a disappointing record in the last few years.
Simon: That was a myth that Mr. Dreyfus began promulgating back in the Seventies.
Omni: He's not the only one.
Simon: He started it. There is a mythical history of AI in which an initial burst of activity was followed by a long period of disappointment. (Simon, 1994)

[22] I believe this is most clearly explained in Simon, 1973, although it is a major theme of his more widely available lectures on *The Sciences of the Artificial* (1981). For those familiar with Simon's earlier work at Cowles on econometric identification, they will recognize his

everything cannot consistently depend upon everything else; like a good corporate CEO, Nature knows how to delegate authority, automate the accounts and concentrate her attention upon the really important issues. All systems – inanimate, cognitive, social – are divided into low-level modules that deal with high-frequency changes in the environment, middle-management modules that deal with moderate frequency types of coordination between tasks, and a few modules (maybe just one) that deal with very low frequency changes in the overall homeostatic behavior of the system; it is a corporate organization chart. "Motions of the system determined by the high frequency modes . . . will not be involved in the interactions of those subsystems. Moreover, these motions will be so rapid that the corresponding subsystems will appear always to be in equilibrium. . . . The middle band of frequencies . . . will determine the observable dynamics of the system under study – the dynamics of interaction of the major subsystems. As we have seen, these dynamics will be nearly independent of the detail of the internal structure of the subsystems, which will never be observed far from equilibrium" (Simon, 1973, pp. 10–11).

In order to develop a *theory* of hierarchy, it is only necessary to describe the middle layer of modules, since those are the only system dynamics accessible to observation from outside the system. The very high frequency base modules appear inert, largely decoupled from the operation of the system as a whole, performing their repetitive tasks; while the CEO lowest-frequency, highest-position module will appear as either purely invariant or else never fully law-governed, due to the time scales on which it influences the dynamics of the system. The lesson to be drawn from this portrait of distributed delegated authority is that only theories of the "middle range" are conceptually efficacious for limited organisms such as ourselves, and furthermore, middle-range modules in a hierarchy can be efficiently understood one at a time, because their loose coupling (or "near decomposability") dictates that their aggregation does not appreciably alter their functions and operation within the hierarchy. "In a world that is nearly empty, in which not everything is closely connected to everything else, in which problems can be decomposed into their components – in such a world, [bounded rationality] gets us by" (Simon, 1983, pp. 22–23). This ontology of the modularization of reality underwrites Simon's research credo that he need not take either the neurophysiology of the

description of hierarchy as a projection of his conditions for causal identifiability in statistical estimation of linear systems. This, of course, suggests that his general characterization of hierarchy is in fact inapplicable in cases of nonlinear interaction or when stochastic terms are not themselves linear and separable. The pervasiveness of this theme of modularity is adumbrated in Sent, forthcoming.

brain or the global cognitive architecture of the mind into account; he can just proceed sequentially studying individual midrange cognitive modules devoted to highly delimited tasks of problem solving. It also provides reassurance that once he has successfully simulated a cognitive module, he rests assured he has more or less permanently added another discrete building block to the ultimate goal of the cognitive architecture of the brain. "I do not believe that more specific programs for particular task domains or ranges of such domains will be displaced when the "right" general architecture comes along. They will simply become essential components in the larger system" (1991a, p. 328).[23]

At that juncture, Simon (1983) sought to connect his program to his own relatively idiosyncratic understanding of biological evolution. He has asserted on any number of occasions that the loose coupling he deems the defining characteristic of hierarchies actually accelerates the processes of natural selection and therefore is favored in a Darwinian sense by nature. "Hence, almost all the very large systems will have hierarchical organization. And this is what we do, in fact, observe in nature" (Simon, 1973, p. 8). Furthermore, Simon displayed a distinct tendency to conflate the notion of the "complexity" of a system with the extent of its hierarchical organization, correlating the level of complexity with the "peakedness" of the organization chart and the relatively limited "span" of connectivity at the higher levels of control relative to that at lower levels (1981, p. 198). Because nature purportedly loves hierarchies, this is asserted to be evidence that all evolution moves in the direction of increased complexity.[24] On some occasions, this has encouraged him to venture to suggest that the modular trial-and-error architecture of many of his individual cognitive simulation programs merely reproduces the trial-and-error architecture of Darwinian evolution, and that this attests to their legitimacy (e.g., 1983, p. 40).

This subversive project to irreversibly blur the boundaries between machine psychology and human psychology, between the biological and the conceptual, all through the instrumentality of computer simulation is the very essence of the cyborg imperative to erase the boundary between the Natural and the Artificial. It is the New Age version of an age-old longing for our oneness with the cosmos. "What the computer and the

[23] Simon's student Edward Feigenbaum subsequently became the major promoter of the version of AI known as "expert systems," taking this logic to its ultimate instrumentalist conclusions.

[24] Simon does aver that this scheme "assumes no teleological mechanism. . . . Direction is provided to the scheme by the stability of complex forms, once these come into existence" (1981, p. 203). Interestingly enough, however, "the evolution of complex systems from simple elements implies nothing, one way or the other, about the change in entropy of the entire system" (p. 204).

progress of AI challenge is an ethic that rests on man's apartness from the rest of nature" (1982, 2:200). Equally transgressive is the procedure of decomposing rationality into numerous cognitive modules which may (or may not) be in procedural conflict with one another from a detached system-level perspective; this found its apotheosis in unruly computer programs, such as Oliver Selfridge's "Pandemonium," and may help explain some of Simon's hostility toward connectionism. This flirtation with rationality as inconsistency is redolent of the primal thermodynamic inspiration of the cyborg sciences, but in Simon's simulacra it still diverges significantly from von Neumann's conception. Simon has on various occasions acknowledged this fact:

> I don't think von [Neumann] did have a very deep appreciation, *there's no reason why he should have had*, for what we now call artificial intelligence. He obviously did have very deep insights into the logical structure of the computer and the whole notion of putting the program inside the computer in terms of storage of programs. . . . But on artificial intelligence he was always full of warnings. We shouldn't think that this imitates the human being. I heard him do this twice in particular.[25] (emphasis added)

> Out of [automata capable of producing automata, von Neumann] hoped would come a way of generating complexity from simplicity – of creating creativity. . . . his focus was still that of a mathematician, more upon questions of logical possibility than on concrete empirical schemes to emulate human thinking. . . . We can think of von Neumann's work on the computer and the brain as a return to the preoccupations of his earliest years with the foundations of logic and mathematics. (1992b, p. 575)

These eulogies might at first glance appear incongruous, because Simon's prosecution of the program of research simulacra would appear on its surface to share so much with von Neumann's innovation of a novel theory of automata. In the first instance, both programs depend crucially on the computer as a research tool, but also on the computer as a metaphor for their respective theoretical ambitions for a mathematical social science. Both programs worked in terms of sequential computational architectures. Both saw themselves as providing an ur-theory of organization, both Natural and Social, at its most abstract level. Both bore little patience with the neoclassical presumption of economic agents as homogeneous globules of desire spinning about on their celestial axes with infinite precision. Both wanted to shift the cognitive center of gravity

[25] From "Conversation with Herbert A. Simon at Carnegie Mellon University on 4/9/75," by Pamela McCorduck, pp. 20–21, box 40, HSCM (emphasis added).

of that nebulous agent toward reconceptualization as an information processor, with inherent cognitive limitations giving rise to social arrangements of varying degrees of complexity to cope with the environment. Both were searching for a novel alliance with evolutionary biology in order to underwrite their general theory of a computational approach to the origins of order.

And yet these two theories of automata and simulacra ended up diametrically opposed, and opposed they remain; and the nature of their dissonance brings us to the crux of the dispute over the future elaboration of the notion of "bounded rationality."

Corps Wars: Simulacra versus Automata

The goal of contrasting von Neumann's program of automata theory with Simon's procedure of simulacra is not solely to understand better their respective motivations (and to demonstrate once again that the shade of von Neumann stalks the dreams of the fin-de-siècle economist), but more to the point, to compare and contrast the two major paths of accommodation that modern economics faces in response to the spread of the computer. The inconvenient fact that neoclassicals are increasingly being forced to confront toward the end of the century is: the more you know about computation, the more untenable the entire Walrasian project becomes. Chapter 6 has sketched the outlines of this case. Whatever it is that neo-Walrasian economists apparently want agents to do to "process information," it seems both machines and people prove equally obstreperous and unable to bend to their wishes. There will always be those who, unrecusant and unrepentant, refuse to abandon the Good Ship Walras (or Nash) as it slips beneath the waves;[26] but time and tide is on the side of the cyborgs. Hence once jammed in their lifeboats, it seems likely surviving mathematical economists will find themselves either plotting a course toward the USS *Simon* or else toward the flotilla *von Neumann*. Perhaps these preliminaries are redundant; the precocious, not needing a weatherman to know which way the wind blows, have already clambered aboard either the simulacrum of a dreadnought or the armada of automata; but then again, perhaps in their haste, they don't know just what they've gotten into.

What they will find when they arrive can be described in two successive stages: once as novel naval engagement and then once again as tragedy. From a distance, the Good Ship Simon has initially looked vastly more inviting to the orthodoxy since it is festooned with mottos seemingly

[26] The paper by Rust (1997) is a veritable catalog of various ways to deny the distressing content of computational theory. We provide our own catalog of alternative responses in Chapter 8.

unthreatening to neoclassicals: "Procedural rationality is the rationality of a person for whom computation is the scarce resource" (1982, 1:470): sounds just like Lionel Robbins all over again. Simon himself surely bears some responsibility for fostering the impression in many hard-core neoclassicals that they can discern in Simon one of their own, however much he inconveniently tends to deny it on occasion. His version of "bounded rationality" works in terms of individuals and their psychologies, and builds up social phenomena from them as mere aggregations of loosely coupled components. His advocacy of simulation, again contrary to initial expectations, resonates with a certain inbred distaste inherent in the neoclassical profession for actually having to account on a case-by-case basis for the real actions of the diversity of real people. This constitutes the hollow methodological individualism identified at the outset of this chapter. Simon further insists that "thinking" is just physical symbol manipulation, and it takes little imagination to appreciate that for certain audiences this can sound an awful lot like the manipulation of prices as information in the mind of the economic agent. Simon, after all, has been striving all along to come up with a new definition of "intelligence," and that is something that can be easily mistaken for a species of optimizing "rationality" by a neoclassical theorist cast adrift sufficiently long in the Sargasso Sea of economics: the awkward terminology of "bounded" rationality renders it all but inevitable.

The entire program would have been better served by being retailed under a more truthful banner like "modular intelligence" or "subroutine behaviorism"; but not only do such gonfalons lack the requisite 'tude for the nineties, but they also do not sufficiently allow for the recurrent gaps between what Simon does and what Simon Says. What the program essentially amounts to is a low-impact treatment of computation: precisely because it makes few formal computational demands, it apparently need not necessarily render obsolete the vast bulk of the orthodox training in economics.

By contrast, the von Neumann armada can seem gunmetal gray, sitting distressingly low in the water but dispersed from here to eternity, and more than a little forbidding to human-centered concerns. It is not centrally engaged with psychology or neurophysiology, and therefore is not really occupied in building up the architecture of societies from individual beliefs and intentions. It is not even concerned with getting any one agent "right" in any particular simulation, because the individual agent will not bulk large in the Neumannesque scheme of things. Rather, it is the root theory of abstract information processors and their interactions, ranging all the way from the genome to the market. The operation of computation is Topic One on the research agenda, with humans treated as the environment within which the computers play out their imperatives.

Casting causal entities as machines is standard modus operandi; erasing boundaries between humans and machines hardly raises an eyebrow. Because the theory of automata is primarily about the reflexivity of self-reproduction and the generation of novelty, some rigorous codification of "rationality" is difficult to portray as the ultimate objective of the exercise. It is not especially user-friendly, in that it makes no palpable immediate reference to human purposes and intentions. It veers toward abstract theories of computation and logic, themselves limited and bounded in various ways, but gives few hostages to methods of simulation.

As the previous sections have suggested, the counterpoint between the two programs was and is no accident. From a more nuanced vantage point, one can regard Herbert Simon's simulacra as a "humanist" and antifoundationalist version of von Neumann's elevation of the computer as the central subject of the cyborg sciences. It displaces von Neumann's automaton at stage center with an abstract human of middling capacities; a figure with few, if any, noteworthy qualities, say, more like Josef K than Seymour Cray or Monsieur le K(apital); someone who just wants to be a good cog in a bureaucracy, but slips up from time to time in ways he himself cannot discern; the ideal blank slate upon which to project any of a motley array of disciplinary peccadilloes (Crowther-Heyck, 1999, p. 195). The blandness of Simon's Prometheus may harbor a cunning unbecoming to such a protagonist, however. For in diametrical opposition to von Neumann, and contrary to all superficial appearances, Simon aims at sanctioning an autonomy of the social sciences from the natural sciences.[27] True, the social sciences stand in need of a bracing shot of mathematical rigor, or so Simon says, but on their own terms, and not those of the physicist – or even the biologist. In his famous lectures on *The Sciences of the Artificial* he deploys the natural-artificial dichotomy in order to conflate the contrast between the "pregiven/constructed" with that of the nonteleological/goal-directed, in order to assert the fundamental methodological poverty of the natural sciences in handling the second terms in each of those distinctions. "The central task of a natural science is to . . . show that complexity, correctly viewed, is only a mask for simplicity" (1981, p. 3); but the methodological sin of the natural sciences is to make that leap prematurely, and without justification. If there really do exist phenomena of daunting complexity, and a case can be made that they tend to fall into the purview of the sciences of the social, then his preferred method of simulation is deemed the correct protopadeutic (pp. 18–21). Moreover, if the social sciences, and in particular neoclassical economics, need to abjure their unseemly weakness for the bald imitation of physics

[27] This idea was first suggested to me by David Walton, and is the subject of his forthcoming Ph.D. thesis.

because they suffer from *simplicitas praecox*, then the doctrine of the "sciences of the artificial" sanctions their recourse to alternative (and unorthodox) research procedures and therapy. In a deft move, Simon's critique of neoclassicism dovetails with his postulation of the necessity of a separate and distinct mathematical social science. Neoclassicals, flailing about for a lifesaver but oblivious to the spiky critique, are therefore bound to end up wondering whether their rescue by impalement was a fate worse than the original capsizing of the Good Ship Walras.

Simon's previous pleas for a simulation-based alternative science of the artificial have repeatedly fallen on deaf ears in the economics profession, ranging from the abortive program of the "behavioral theory of the firm" to his program for a new institutional economics of organizations (subsequently domesticated as Oliver Williamson's "transactions costs") to his explanations for skew distributions in firm size and income distribution. When push comes to shove, orthodox economists couldn't shake the feeling that simulation a la Simon never amounts to much as science. Now that "bounded rationality" is all the rage, can economists conveniently ignore all that attendant baggage that comes with (as they reckon) taking Simon on board? I believe the response must operate on at least two levels: one that grapples with Simon's own insistence on the validity of his entire research agenda as a whole; and a second that seeks to comprehend the orthodox inclination to deep-six Simon and maintain appearances that no one ever was forced to walk the plank. I shall concentrate in the next subsection on the cases of Ariel Rubinstein and John Conlisk, because they conveniently serve as "representative agents" with regard to the issues I wish to highlight. Their claim on our attention derives from the fact both recently have written surveys arguing the case for reconsideration of something they call "bounded rationality" by the economic orthodoxy.

How Not to Make Friends and Influence Cyborgs

Ariel Rubinstein is someone squarely situated in the white-hot center of the orthodox economics profession: a well-known defender of the Nash program in game theory, and an adapter of computer models to characterize agents as part of a quest to update the older Walrasian framework, which he professes to accept at some level.[28] His role in the cyborgization of Nash equilibrium is covered in the next section. John Conlisk is a bit

[28] Ariel Rubinstein (1951–): Ph.D., Hebrew University, 1979; professor, Hebrew University, 1981–90; professor, Tel Aviv University, 1990–; professor, Princeton University, 1991–. For more on Rubinstein and his background, see Rubinstein, 1993b. There he states, "The issue of interpreting economic theory is, in my opinion, the most serious problem facing economic theorists at the moment" (p. 81).

harder to characterize: situated a further distance from the neoclassical orthodoxy, perhaps more skeptical about its past, yet part of the whole culture of decision theory and operations research that has thrived in the military milieu of southern California. Both have engaged in advocacy of bounded rationality; but both also approach Herbert Simon as a kind of *idiot savant*, unselfconsciously proposing the brilliant idea of bounded rationality without really doing much to bring it to fruition. Conlisk penned an entire survey article for economists on "bounded rationality" without once explaining Simon's role in AI or mentioning simulation; the closest he comes to a characteristic Simonized theme is to briefly entertain the notion that utility maximization is actually conceptually and empirically empty; but sensing that such sentiments start one skidding down the slippery slope to Perdition, he quickly backpedals, writing, "Discipline comes from good scientific practice, not embrace of any particular approach" (1996, p. 685). The meaning and reference of the term "science" is presumed never to have been altered over the past century or more, yet the content of "economics" has; the economic orthodoxy was never actually mistaken, although something has gone awry; nothing is ever new under the sun; but nonetheless for some mysterious reason, the core content of economic "rationality" must change. This is just one instance of Conlisk haplessly tying himself up in knots.

Rubinstein looks rather similar in overall approach. His credo is unapologetic aggressive Bourbakism: "the focus is not on the substantive conclusions derived from the models but on the tools themselves." Facility with producing mathematical models is the alpha and omega of economic theory; why some models draw attention of the cognoscenti in one era and languish unappreciated in another is treated as simply ineffable, eluding mere mortals. The only empirical psychology seriously considered in both texts is some findings of the statistical decision theorists Kahneman and Tversky, and in any event, neither economist can really be long distracted by such minutiae. The glittering prize for which they strain and yearn is to somehow catch the fancy of that small coterie of self-identified orthodox economic theorists: "The crowning point of making microeconomic models is the discovery of simple and striking connections (and assertions) that initially appear remote. . . . microeconomists are not prophets or consultants; neither are they educators of market agents. . . . Those models do not pretend to predict how people apply the values of truth to statements of a natural language, or to provide instructions for their use; neither do they attempt to establish foundations for teaching 'correct thinking'" (1996, p. 191). We could be back at Koopmans's *Three Essays* in 1957 – almost.

Rubinstein's rectitude, such as it is, resides not in his exertion of any effort to try and understand what Simon is all about, or to provide a workable definition of bounded rationality for the orthodox (for as we shall see, there is no such philosopher's stone), but rather to confront Simon with some lectures on the topic of "bounded rationality" in manuscript and then to reproduce Simon's reactions in truncated form. Given Simon's half century of open opposition to game theory, one can only marvel at Rubinstein's disingenuousness in apparently expecting some form or another of imprimatur from Simon. Nevertheless, Simon's rejoinder is entirely consistent with the program of simulacra that we have outlined here:

> You are very generous in crediting me with a major role in calling the attention of the economics profession to the need to introduce limits on human knowledge and computational ability into their models of rationality. . . . But you seem to think that little has happened beyond the issuance of a manifesto, in the best tradition of a Mexican revolution. And you mainly propose more model building as the way to progress. You show no awareness of the vast amount of research (apart from the work of Tversky) that has been done (and mostly published in psychological and artificial intelligence journals) since the 1950s to provide empirical evidence about the phenomena of human decision making. . . . Nor can it be objected that bodies of facts are useless without theoretical analysis, because most of these facts have now been embedded in (explained by?) fully formal theories that take the shape of computer programs. . . . For mathematicians, the unhappy detail is that these equations are almost never solvable in closed form, but must be explored with the help of simulation.[29]

Theory, theory, who's got the theory? If we are to play the theory game, it would help if we could get straight on the rules. Sometimes the economic orthodoxy likes to assert that the standards are all self-imposed, tacit, and entirely self-referential (Samuelson's "applause of our peers"); but in fact, there are certifiable guidelines. The secret to achieving fame amongst

[29] Simon to Rubinstein, December 2, 1996, excerpted in Rubinstein, 1998a, p. 189. While Rubinstein does a good job of capturing most of the content of Simon's objections, he does omit a few things from the original correspondence. First, Simon situates Rubinstein's work as part of a "growing inclination" by game theorists to talk about bounded rationality, a trend he views with tempered enthusiasm. Second, Rubinstein omits a paragraph concerning Simon's attempt to simulate scientific discovery itself, thus repressing the acknowledgment that some of the battle is being fought out on the terrain of the history and philosophy of science, a ground Rubinstein studiously avoids. Third, he omits a brief disparagement of econometrics. Fourth, he omits what I think to be the key phrase of the letter: "And even if that condition [having good empirical knowledge of which procedural considerations to incorporate into economics] is met, I will place my bets in theories of the form of symbolic difference equations (computer programs) over other mathematical formulations, and on simulation as a necessary substitute for most theorem-proving." In other words, Simon rejects Rubinstein's entire project.

the economic orthodoxy is to begin with a textbook neoclassical agent in his standard problem setting, be it a Marshallian partial equilibrium, full Walrasian interdependence, or Nash game-theoretic equilibrium, and to tweak the model with some limited aspect of an information-processing characterization that seems to "resolve" some isolated prior problem with the orthodox tradition. Conlisk is quite straightforward in his eagerness to conform to this genre: "Why not condemn problem solving which leads to systematic error? The answer is simple: Deliberation cost" (1996, p. 671). Trumping this, Rubinstein masters this protocol in his lectures: altering the way agents partition information spaces to "dissolve" the no-trades paradox; arbitrarily limiting the memory capacity of the agent to "reduce" the multiplicity of Nash equilibrium or "solve" the Prisoner's Dilemma, positing a player as encompassing "multiple selves" to get round the unsatisfying and implausible interpretations of mixed strategies, and so on.

No specific "tweak" can solve each and every neoclassical infelicity; so no collection of mathematical "tweaks" is intended as a benchmark characterization of the one true model of the agent. But more significantly, the appeal to computer metaphors is both relentless and simultaneously consciously superficial, in that appeals to the established consequences of theory of computation must never be allowed to impugn the validity of the canonical characterization of the neoclassical agent introduced at the outset. You can play with what an agent is mathematically defined to be in the model; but you must never never mention what attributes of the agent are definitively ruled out of bounds by computational considerations. In other words, the threatened invalidating aspects of the literatures cited in Chapter 6 are to be brutally repressed.

The reason Rubinstein is tropismatically drawn to "bounded rationality" is that he thinks it sanctions appeal to the outward trappings of computation without ever engaging with the actual process. Once again we encounter the fundamental trope of the Cowles Commission. The threatened deconstruction of neoclassical rationality by mathematical theories of computation is ruled inadmissible by fiat. He writes, explicitly referring to the Turing machine, "I doubt that the nonexistence of a machine that is always "rational" validates the bounded rationality perspective" (1998a, p. 184). Hence the objective in Rubinstein's theory game is not to build himself an agent, or even to render neoclassicism computationally effective from the ground up; it is rather to conjure surprise and wonder (shades of Adam Smith!) amongst the closed circle of mathematical elect at his own ingenuity in mimicry of mathematical artifacts (from finite automata to perceptrons) which are related to the computer. But the source of his edification derives from virtuosity in proof technique, not in constructing a cognitive science. (In the next section, however, we encounter game theorists who sought a different sort of rapprochement with computer science.)

There are rules in Simon's theory game, and as we have seen, they are altogether different, having to do with everything from abandoning the neoclassical preferences problematic to asserting the legitimacy of simulation as a scientific technique to the privilege of intermediate levels of hierarchy in scientific explanation. However, there remains just enough family resemblance to help explain the unrequited attraction on Rubinstein's part. Rubinstein is concocting little modular mathematical simulations of agents to solve Rubinstein's (and the economics profession's) problems, whereas Simon constructs modular problem-solving algorithms to solve his agents' problems. Rubinstein will never be troubled to reconcile all his individual discrete "models" into a single cognitive agent; Simon absolves himself of a unified theory of the cognitive actor. In Rubinstein's case the reconciliation would be unattainable by definition, because agents often are saddled with unmotivated contradictory computational restrictions (e.g., being able to formulate elaborate game-theoretic strategies while being unable to count – see 1998a, p. 167) within the ambit of a single model. For a long time Simon thought he could treat learning as a "second-order effect" (Newell & Simon, 1972, p. 7); Rubenstein thinks it is possible to blithely discuss "knowledge" with no commitment as to how that state was arrived at (1998a, p. 61). Rubenstein would have his agents access various automata without any commitment to the theory of computation; and Simon has maintained a deeply ambivalent relationship to computational theory throughout his career.

This shared matrix of reticence and resemblance, so fundamental to understanding the discreet charm of Simon for neoclassical economists, is brought into focus when one scans Simon's writings for explicit discussions of mathematical computational theory – recursive function theory, Turing machines, and computational complexity. The citations are few and far between, but they do indicate a rather stern dismissal of their practical relevance:

> Early in its history, computer science became concerned with the decidability problem. . . . It gradually dawned on computer scientists, however, that the decidability question was not usually the right question to ask about an algorithm or a problem domain. (So great was the fascination with automata theory and the prestige of the Gödel theorem that the dawning took several decades.) (1982, 2:466–67)

There is an element of truth in this characterization: for many of the main protagonists of the artificial intelligence movement did make their break with tradition by renouncing the importance of formal computational theory and the theory of automata.[30] But von Neumann never once conceded that AI was asking the "right questions," and even in modern AI, the debate continues over the correct role of logic in modeling (MacKenzie, 1996). Further, there persists some controversy in the computer science profession over the extent of the relevance of the Universal Turing machine to their various concerns. Thus, Simon could continue, "The important questions for computing were the probabilities that answers could be produced in a reasonable time." However, his simulacrum tradition made no analytical progress in that area: modern conventional measures of computational complexity such as NP-completeness of calculations are all based upon the Turing machine as the benchmark computational model.

Be that as it may, Simon was well aware that he was taking the position that real people don't regularly use formal logic in problem solving, and that this dictum sanctioned his trumpeting the unimportance of formal theories of computation. Rubinstein, who harbors no such ontological commitments, is bereft of any plausible justification for his quarantine of computational theory. For someone who has written, "Economic models are viewed as being analogous to models in mathematical logic" (1998a, p. 191), it would seem the height of irresponsibility to neglect one of the most important developments in mathematical logic in the twentieth century, namely, Gödel's incompleteness theorems and Turing's uncomputable numbers. Far from being an abstract critique, this has immediate and direct relevance for the orthodox economist's ambitions for the program of bounded rationality.

Both Rubinstein and Conlisk seem to know that there is something very redolent of a logical paradox in the way that neoclassicals want to make use of bounded rationality; it redounds to Conlisk's credit that he is a bit more forthright about it than Rubinstein (1998a, p. 98; Conlisk, 1996, p. 687). Suppose we think the reason that bounded rationality exists in the world is because somebody somewhere (the agent herself, natural selection as blind watchmaker, the economist) is really optimizing. In the conventional case, full and total optimization is costly, due to costs of deliberation, costs of cognition, costs of information, or (insert your own favorite justification here). So in response, the economist posits a metaoptimization, in order to calculate the "optimal" amount of irrationality for the agent. The whiff of impropriety comes with the idea that

[30] Of course, this may have had something to do with the preferences of the military in supporting "problem solving" over formal models of automata and self-organization. The military, and particularly DARPA, was far and away the primary sponsor of AI in its early years. On the possible biases that may have skewed the history of computer science, see Guice, 1998.

metaoptimization can somehow achieve a determinacy that lower-level bounded optimization could not, and that small divergences from neoclassical rationality conform to some well-behaved distance metric. Plainly: why isn't the metaoptimization also bounded in some manner? Is this to be explained by some meta-metaoptimization? A common rebuttal to the charge of "infinite regress" is to deploy the same sorts of fixed-point theorems used in the original neo-Walrasian existence proofs at Cowles (Lipman, 1991); but this entirely misses the point and is unavailing in any event. There is a logical paradox of self-reference lurking within, which can exfoliate throughout the whole axiomatic system, in exactly the same way Gödel numbers served dual roles as reference tags for statements and statements in their own right within the system. A severely rigorous logical theory would confront the implications of the implied paradox, namely, that no optimization can be both complete and consistent.

Conlisk is certainly correct in observing that every single neoclassical theorist who has glimpsed the paradox has effectively peremptorily truncated the infinite regress at the first metalevel – that is, the superoptimizer is deemed to encounter no trouble deciding to opt for boundedly rational procedures – with no plausible justification other than convenience. Rubinstein certainly goes for this gloze. But the nasty implication of this jury-rigged repair scheme is that there can be no unified theory of bounded rationality in the sense in which economists wish it so. There is absolutely no reason to believe that there is a unique fixed procedure of being boundedly rational, which ever could be identified by metaoptimization.

As the neoclassicals had always feared, there is always more than one way to be "irrational," if that means divergence from standard optimization; a metric for their comparison seems a forlorn hope. But worse, if one does not truncate the regress, considerations from the theory of computation comprehensively seal off any such escape route through searching for the preferred "equilibrium" amount of irrationality, selecting one such optimum from the potential continuum of possibilities. No procedure for such a selection can be guaranteed to halt at the correct answer: this is simply the "halting problem." The fixed-point theorems of Lipman and others are deeply flawed because predicated upon nonconstructive mathematics; there is no effective algorithm for choosing the correct "amount" of bounded rationality. Indeed, because of the initial analytical move made to render optimization itself subject to optimization, we have the standard setup for Cantor's diagonal argument, the most commonplace proof procedure for demonstrating the existence of noncomputable numbers. "Bounded rationality" has only one consequence for those unwilling or unable to fully take on board Simon's entire program of consciously constructing simulacra. The bottom line is that neoclassical economists must of necessity renounce formal computational theory: they have no choice in the matter.

And in having no choice, perhaps they finally can come to empathize with the qualms of all those who have had the definition of welfare as unfettered choice thrust upon them over the past century against their own better judgment.

SHOWDOWN AT THE OR CORRAL

There is more than one way to demolish the Self. Herbert Simon initiated his own process of deconstruction through the twin techniques of decomposing the self into little hierarchies of problem-solving modules, in passing blurring the boundaries between organism and environment, and by pursuing the method of simulation, which effaced the external integrity of the functioning self. But few took him all that seriously in economics until the very end of the century. In the interim, the true-blue defenders of the Self in economics were more likely to be found pursuing the logic of game theory into some of its darker recesses, working on what they considered to be the true and only doctrine of self-interest. The thesis of this chapter is that one's initial position as critic or defender made no difference to the ultimate fate of the Self: wherever cyborgs intruded, then the days of the integrated self were numbered. No matter how much the economists strained to reprimand and challenge the legacy of von Neumann, their speculative forays ended up in complicity with his research program, witting or no. In this section, we briefly survey how this happened in game theory, especially when it came to mechanizing the agents that played the perilous games associated with the Cold War. The last section concludes with the specific case of Kenneth Binmore, as someone appearing to embrace the cyborg at the end of the century. To do so, we must take up the history of game theory from where it was left off in Chapter 6.

Where Software Dares

When the history of economics in the last quarter of the twentieth century finally comes to be written, the most important and incongruous phenomenon that will cry out for explanation will be the fall and rise of game theory, from disparaged minor adjunct of operations research and decision theory to glistening centerpiece of orthodox neoclassical microeconomics. Sometime in the 1970s game theory started to venture out of its gated ghetto in the military barracks and in RAND and turn up in all sorts of new venues. John Harsanyi appeared bent on turning Nash noncooperative theory into some grand philosophical dogma aimed at reviving literal utilitarianism; William Hamilton and John Maynard Smith started subjecting poor unsuspecting animals to its rigors; Reinhard Selten began doing gaming experiments with his students; David Kreps started promoting various Nash equilibria as expressing various commonsense economic notions; a small cadre of Europeans sought to explore the ways in which game-theoretic solutions could be construed as dovetailing with Walrasian themes; and Robert Aumann was everywhere at once. The explosion of pages devoted to game theory in microeconomic texts from, say, the first edition of Varian (1978) to Kreps (1990), was astounding in speed and scope; and it was commonly said that the unprecedented rapid propulsion of the *Journal of Economic Theory* (begun in 1969) to the front ranks among the cognoscenti was due to its catholic tastes and openness to the new idiom of modeling. A full-scale accounting for the change would require a comprehensive survey of the intellectual crosscurrents in neoclassical economics; but, as we have repeatedly insisted, this is not a comprehensive history of postwar microeconomics. Instead, we aim to call attention to the ways in which cyborg themes infused the game-theoretic instauration, simultaneously facilitating and undermining the new model orthodoxy.

We must begin with scrutiny of that shunned sibling of game theory, war games, for its ultimate feedback upon economics. Gaming has always been composed of three collateral branches during the Cold War, namely, military exercises, computer war games, and academic game theory; and in the 1960s computer gaming had come into its own, far surpassing game theory in usefulness and sophistication. Some economists at RAND, such as Alan Enthoven,[31] found the computerized technology repugnant (Allen, 1987, p. 137); others insisted that war games have "little to do with game theory except for being equally cute and equally subject to abuse" (Schelling, 1964, p. 1082); but it was hard not to notice that the colonels had grown fonder of their little Panopticons of Bits than anything evoked by the formal theorems about game theory that their mathematicians were busy producing. Many analysts at RAND were quite happy to respond to market signals and give the colonels what they wanted, and so all sorts of computer games full of sound and fury were conceived at RAND.

One early software product was STROP (Strategic Operational Planning), which was proudly touted as evaluating the outcome of a

[31] Alain Enthoven (1930–): B.A., Stanford, 1952; Ph.D. in economics, MIT, 1956; RAND, 1956–60; deputy assistant secretary of defense, 1961–65; assistant secretary of defense for systems analysis, 1965–69; vice-president, Litton Industries, 1969–73; professor, Stanford, 1973–. Further information is available in Hughes, 1998. Enthoven is one of the economists we have to thank for the modern innovation of the "health maintenance organization."

nuclear exchange in about one-fiftieth of a second; finally, nuclear war could be simulated in a time frame shorter than the real thing. The accompanying documentation admitted that it was a substitute for what the people in the Mathematics section at RAND had been retailing all along: "The knottiest problem in analyzing a central nuclear war is the payoff and [win] criterion. Central nuclear war is not only highly nonzero sum but the outcomes can include cases that are catastrophic to one or both sides. The theory of nonzero sum games is in an unsatisfactory state, and methods of dealing with catastrophic payoffs are extremely elementary" (in Allen, 1987, p. 146). The war gamers proffered a resolution to this problem by providing programmed schedules of potential scales of destruction and, in the new improved STROP-II, even included pecuniary budgets for weapons systems: war simulations never ventured very far from economics, and vice versa. Other incarnations of war games were even more imaginative (Enke, 1958b; Geisler, 1959). RAND's Systems Research Laboratory, the site of Herbert Simon's fateful encounter with Allen Newell, was itself a bustling testbed of wargaming. The simulation of an air defense center was, in the first instance, a war game; the fact it became a training device and a structure for cognitive experimentation merely demonstrated the effusive creativity that came from crossing the computer with the war game. Simon's subsequent forays into AI and chess-playing programs would themselves reverberate back into wargaming at RAND, as we shall soon observe.

One pivotal figure in the history of wargaming is Andrew Walter Marshall, the very same Cowles economist who had moved to RAND to do game theory in the early 1950s and quickly crossed the line to Monte Carlo simulation, producing scenarios of a possible Soviet version of a Pearl Harbor–style attack. Marshall subsequently became the head of the Pentagon's Office of Net Assessment, where he was known as the major promoter and watchdog of war games (Allen, 1987, p. 149) and "one of the most prominent strategic thinkers of the age" (Adams, 1998, p. 255). Dissatisfaction with the level of sophistication of gaming prompted Marshall to make the unorthodox move of commissioning a fully automated war game from RAND in the 1970s, and to pioneer the adaptation of commercial war games for personal computers to automate cost-effective training exercises in the military in the 1980s. All those scruffy hackers in the 1960s playing rudimentary versions of "Star Trek" and Maze Wars late into the night at the "computer center" find their apotheosis in the image of corporals playing commercial video games in the 1990s.[32] It would be a short extrapolation from artificial players in war

[32] Point your Web browser at ⟨http://138.156.15.33/doom/doom.html⟩ to access the Marine Doom home page. The Marine version has downloadable software to modify the

games to the idea of "information warfare" or cyberwar, the new buzzword of the post-1989 military, a blue-sky concept credited to Marshall (Adams, 1998, p. 56).

Once you had created "artificial autonomous agents" who could plausibly play war games, then the possibility that they could be "turned" to mount an attack on the very computer system and the infrastructure of C^3I upon which the military had come to depend was an obvious extrapolation of existing trends. Viruses, computer worms, logic bombs, and sniffers (Lohr, 1996; Boulanger, 1998) are merely the visible tip of a submerged teeming mass of "autonomous artificial agents" in cyberspace unleashed by the military and their subcontractors. (The media tendency to blame every outbreak of cybotage and computer mayhem on testosterone-crazed teen cybernerds itself spoke volumes about the extent of persistent self-censorship amongst intrepid American journalists.)[33] Information, which was generally supposed to make us stronger and more self-reliant, now served only to ratchet up our paranoia. "Information and communications technologies will change how conventional battles are conducted. Instead of bombing factories, the aim may now be to penetrate information networks. As more and more of the economy of any country is embodied in its information systems, that country will be more vulnerable to disruption" (Marshall in Schwartz, 1995). Indeed, the U.S. Department of Defense reported an estimated 250,000 attacks on its computer systems in 1996 (Boulanger, 1998, p. 106). What had started out as simulation rapidly escalated into a new kind of reality, one distinctly redolent of cyborg preoccupations (Cronin & Crawford, 1999); RAND once more was the epicenter (Khalilzad & White, 1999).

In plumping for this novel notion of automated cyberwar, the visionaries at RAND derived a modicum of their inspiration, not from the neoclassical economists and their baroque notions of information, but

ubiquitous teen program to include tactical fire, the fog of war, and friendly fire. When it comes to fun with your computer, commerce and combat only differ by five characters. In the new leaner, meaner army, the synergy between previous military research and innovative consumer products is now encouraged. On this, see National Research Council, 1997 and Lenoir, 2000. "Maze Wars" at MIT in the early 1970s had a filial relation to SimNet, built for the Defense Department (Dertouzos, 1997, p. 93).

[33] One way to counteract this tendency is to look more carefully for where these creepy-crawlies first surfaced: "Several variants of a virus created as an act of terrorism were detected spreading among IBM PCs in Israel in early 1988" (Moravec, 1988, p. 128). Another is to track what happens to the supposed culprits. In 2000, the press made much of the apprehension of the Filipino student putatively responsible for the "Love Bug"; but almost no one reported that he was immediately released due to (1) lack of evidence and (2) the convenient fact that there was no law against such activities in the Philippines.

rather from the AI tradition, and especially from Herbert Simon. The guru of RAND's automated war games, Paul Davis, put it like this:

> To gain control over the enormous number of variables in a war game it would be necessary to *automate* the game . . . which involves replacing all human players of a traditional military-political war game with automatons called "agents" . . . achieving military realism implies: (1) discarding optimizing models based on simple-minded images of war; (2) dealing with political-military constraints and command-control problems; (3) relegating to secondary status quantitative criteria for action . . . and (4) paying attention to cybernetic phenomena in which decisions are not really made consciously at the top at all . . . all this having to do with the *management of complexity*. . . . As a whole, the war plan is "almost decomposable" in the sense of Simon. (Davis et al., 1983, pp. 8, 11, 19).

The beauty of this quotation is that it illustrates the way in which the computer tends to foster the existence of what Paul Edwards has called the "closed world": as an artifact of the algorithmic imperative, the computer game comes to swallow up the players, incorporating them as cyborgs, just another part of the software. Most really trenchant theories in the history of Western social thought have sought to turn humans into automata of one stripe or another (e.g., Mayr, 1976): but it was at RAND for the first time in history that the players of formal games on computers were themselves rendered as formal automata, not just out of convenience, but as an integral part of building the theory of social processes that became embodied in the game. This recourse to Robot Sociology would have profound consequences for game theory, as well as the subsequent understanding of the Self in orthodox economic theory.

It was not only RAND that "went completely automatic" in the dawning era of cybernation. The Russians had conceived of an enthusiasm for operations research and cybernetics by the early 1960s, and it appears that they were somewhat ahead of the West in explorations of automation of rudimentary war games. A glimpse of these developments can be gleaned from a memo written by Tjalling Koopmans after a visit to Russia in May 1965.[34] He reports that a conversation with I. M. Gel'fand had alerted him to the existence of "an extensive literature in this field

[34] "Work by Gel'fand and Tsetlin on games with automata" by Tjalling Koopmans, n.d. [but internal evidence suggests May–Oct. 1965], TKPY. It is significant for our narrative that the distribution list of this memo includes Robert Aumann, Anatol Rapoport, Thomas Schelling, Herbert Simon, and Albert Tucker. Clearly, someday the potted history of so-called evolutionary game theory will need to be rewritten by an historian familiar with the Russian sources.

in Russia, which consists mostly of rather terse reports on analytical or simulation results, and of a few more explicit expository papers," primarily concerned with the play of games by various classes of automata. Although Koopmans said nothing about the military applications, it is significant that he signaled to his recipients that one motive behind this work was the ambition of "constructing a theory of games in which the players are extremely limited, in memory, observation, and analytic powers – the opposite extreme to von Neumann–Morgenstern game theory. . . . one of the motivations arose from looking at the operation of the nervous system as an interaction of its cells. While the 'behavior' of each cell is presumably simple, the performance of the entire nervous system is of a much higher order of complexity and effectiveness. I also have an impression that work along these lines has a potential bearing on economics as well. Not all of us are perfect calculators in our economic decisions."

Anyone seeking to confront game theory in the 1960s couldn't avoid some sort of rendezvous with the computer. Kenneth Boulding's Center for Peace Studies and Conflict Resolution and the affiliated BACH group at the University of Michigan thought they saw a way to turn the computer into a weapon against the game theory they felt was muddling the minds of the generals, but doing so in a manner consonant with the precepts held dear at the "peacenik's RAND." The major figures at Peace Studies were Boulding and Anatol Rapoport,[35] with Rapoport providing the bridge to the BACH group, consisting of Arthur Burks (the von Neumann protégé and editor of the *Theory of Automata*), Robert Axelrod, Michael Cohen, and John Holland (the progenitor of genetic algorithms). Boulding notoriously attempted to ally "peace research" with General Systems Theory (Hammond, 1997; Rapoport, 1997), thus jeopardizing his credibility within economics; whereas others at Michigan opted for a more technical route to cross-fertilize research traditions. One recurring theme of the Michigan group was its attempt to bleach out the

[35] Anatol Rapoport (1911–): born in Russia, emigrated to the United States in 1922; Ph.D. in mathematics, University of Chicago, 1941; Air Force, 1941–46; assistant professor of mathematical biology, University of Chicago, 1947–54; Mental Health Research Institute, University of Michigan, 1955–1969; University of Toronto, 1970–76; director, Institut für Hohere Studien, Vienna, founded by Paul Lazarsfeld and Oskar Morgenstern, 1980–84. Rapoport has published an autobiography in German, *Geweissheiten und Zweifel* (1994), translated as Rapoport, 2000. The primary source in English on his biography is Hammond, 1997. His career is one of the more underresearched in the history of economics and the cyborg sciences. In particular, the contemporary blindness to his role, as well as that of the Michigan group in general, in the rise of experimental economics is deplorable, as is the neglect of his connection of early work on neural networks to game-theoretic themes.

military stain on much of von Neumann's legacy. Rapoport, for one, reports that one of the turning points of his life consisted of a conversation with von Neumann on possible sources of support for research on mathematical biology. When he learned that von Neumann was advocating a preventative atomic war against the USSR, he temporarily felt his faith in science shaken (Hammond, 1997, p. 190). Rapoport's accomplishments in challenging the military definition of the cyborg was sometimes acknowledged, albeit in code, by the Dark Side (Aumann, 1962; Schelling, 1964). In many respects, the Michigan anti-RAND would become as important for economics as its dark *Doppelganger*, although this has totally escaped scholarly notice.

Although "experimental game theory" may have begun life at RAND, it grew to adolescence at Michigan under the tutelage of Rapoport. By the 1960s Rapoport was acknowledged as the premier expert on empirical investigation into the legitimacy and prevalence of game-theoretic solution concepts, as well as a respected linguist. He coauthored the first review of the experimental gaming literature (Rapoport & Orwant, 1962); wrote a number of popularizations for *Scientific American*, and loomed large in the subsequent gaming literature in social psychology (Colman, 1982). In the 1960s Rapoport had decided that Flood and Dresher's "Prisoner's Dilemma" scenario was an excellent way to prosecute an attack on the pretensions of Nash to have understood the psychology of strategic rationality (Rapoport & Chammah, 1965). The "inferior" equilibrium identified by Nash was viewed at Michigan as a symptom of the incoherence of formal game theory; both Rapoport and the BACH group believed what should replace it was some form of computer simulation-cum-evolutionary theory. "I'd always been tremendously bothered by the Prisoner's Dilemma, says [John] Holland. It was just one of those things I didn't *like*" (in Waldrop, 1992, p. 264). The project of synthesis fell to Holland's colleague Robert Axelrod, a political scientist with a computational background. Axelrod took Burks's and Holland's fascination with von Neumann's automata, and combined it with the idea of the Prisoner's Dilemma game as a literal tournament, or perhaps a simulation of an ongoing ecology, run entirely on a computer. It was already obvious that the game-theoretic notion of a "strategy" resembled the software engineer's notion of a "program"; all one had to do then was conflate the program with an agent and have the agents all play each other in the computer.

Axelrod's rather cathectic inspiration was to extend experimental game theory to encompass computers: he solicited programs to play the iterated Prisoner's Dilemma from experts in various fields and pit them against one another in a round-robin tournament of two hundred moves each. Because they were run on the same computer, all other aspects of the game would be effectively standardized, and a uniform score could be attributed to each program. The structured repetition would allow the various programs to engage in signaling, learning, or any of the other "interpersonal" aspects of game play (as long as they were algorithmic) resisted by Nash but touted by scholars in other fields. It is interesting to note that, out of the fourteen programs submitted for the first tournament, few from RAND – with the partial exception of Martin Shubik and Gordon Tullock – entered the contest.[36] In the first round, the economists' programs were soundly drubbed by that submitted by a mathematical psychologist – not altogether coincidentally, the very same Anatol Rapoport. Rapoport's program was not only the canniest; it was also the shortest in terms of lines of code, heaping indignity upon defeat. His algorithm was called TIT FOR TAT; on the first move, it would unconditionally play "cooperate"; on each subsequent move it would do exactly what its opponent had done in the previous round. It was "nice," yet "punished" bad behavior; and it was transparently simple. The result could easily have been an artifact of the small numbers of programs and the lack of focus upon representative opponents, so Axelrod decided to run the tournament a second time, only now presenting all comers with all the details from the first round. In the sequel, sixty-two programs were entered; and TIT FOR TAT won once again hands down. All of this was reported in Axelrod's high-visibility book, *The Evolution of Cooperation* (1984).

Although some commentators try to conjure an atmosphere of suspense by suggesting Axelrod's results were "surprising" and "unexpected," I rather think the evidence militates against this view. First, there is the fact that Rapoport clearly was involved in the early stages of the research program at Michigan, although he had moved to Toronto (in protest against the Vietnam War) by the time of the actual tournament. Rapoport had himself conducted numerous human experiments with games of all stripes and had a fair idea of what would succeed in the highly stylized iterated case. Second, the BACH group had already had some interaction with the application of game theory and simulation to evolutionary biology and knew that the conventional game theorists had not. The fact that formal notions of "rationality" in game theory had grown so prodigiously complicated as to become unwieldy was also something about which they were already well aware. Third, Axelrod possessed an agenda which he wanted to promote well before the tournament was staged: "*The Evolution of Cooperation*, with its focus on the Prisoner's Dilemma, was written during the Cold War. Indeed, one of its primary motivations was to help promote cooperation between the two

[36] The names are reproduced in Axelrod, 1984, p. 193. The only economists to enter were Shubik, Nicholas Tideman, Paula Chieruzzi, James Friedman, and Gordon Tullock.

sides of a bipolar world" (Axelrod, 1997, p. xi). Axelrod clearly had wanted to find that nice guys don't finish last, and that Blue team and Red team could manage to attain peaceful coexistence. This helps explain the rather tendentious layer of interpretation with which the tournaments were served up:

> If a nice strategy, such as TIT FOR TAT, does eventually come to be adopted by virtually everyone, then individuals using this nice strategy can afford to be generous. . . . Cooperation can begin with small clusters. It can thrive with rules that are nice, provocable, and somewhat forgiving. And once established in a population, individuals using such discriminating strategies can protect themselves from invasion. The overall level of cooperation goes up and not down. (1984, p. 177)

Of course, there were no real humans anywhere in Axelrod's world; instead, it was populated entirely by little automata. Yet this simple observation, so frequently made in order to dismiss the tournament, would come to be seen as the point of the whole exercise. What Axelrod and the war gamers had both achieved was to mix simulation and computers as prostheses to "solve" a problem in rationality more conventionally attacked by formal proof techniques. It would serve to prod the economists ever so slightly out of their complacent slumber,[37] forcing them to reconsider the internal workings of their own little "agents" and worry a tiny bit more about whether their preferred species of rationality might itself be an impossible dream.

Have Your Agent Call My Agent

The fractal history of game theory in the Cold War cannot be adequately understood from within the vantage point of any single discipline; and most certainly not from a location within the relatively parochial confines of economics. There is no period where this is more true than the crucial decades from 1955 to 1975, when support for its mathematical development had began to wane at RAND, until the renaissance of interest within economics in the 1970s. Game theory tended to eke out a peripatetic existence in this interval, depending upon the kindness and largesse of a few related cyborg sciences to keep it alive. This was the era of the confident preeminence of the Self, yet curiously enough, game theory was experiencing difficulty attaching itself to these Individuals that bestrode the world like a colossus. But before we come full circle to recount the saga of diehard machines dueling to the death in the 1980s, it might be prudent to survey a few of the niches within which game theory

[37] Which is not to suggest the economists would acquiesce in Axelrod's interpretation of his results. See Nachbar, 1992; Linster, 1994; Milgrom, 1984; Binmore, 1998a.

subsisted during the interregnum. The primary refuges were operations research and computer science.[38]

As we noted in Chapter 4, game theory tended to be treated as standard-issue equipment in the tool kit of the operations researcher in the postwar period, even though it rarely found much practical use in the actual solution of managerial or military problems in the field. Because operations research as an academic discipline tended over time to retreat ever further into mathematical obscurity, this did create a sheltered niche within which game theory could thrive in a parasitic sort of fashion. Nevertheless, suffering much the same fate as linear programming, the faltering of interest in RAND and elsewhere in the United States did not bode well for the future vibrancy and intellectual development of the mathematical technique. Thus it is noteworthy that game theory tended to garner recruits primarily through the extensive spread of operations research in this period, and not generally from within the economics profession. Far from being drawn to game theory by its wonderfully insightful intricate portrayal of strategic rationality and rich appreciation for the ambiguity of social experience, game theory instead tended to draft recruits by following the barrel of a gun or, more correctly, by hitching a ride on the missile warheads of the inexorable postwar spread of atomic weapons.

If we consider its provenance, it should come as no surprise that OR was diffused outward in the Cold War along the gradients of strategic military alliances and treaties. As America set about reorienting the vectors of power and influence in the postwar world, it would find itself having to coordinate with its allies multilateral military exercises, intelligence, communications, and the whole bundle of advanced technologies which it had innovated in World War II. Given the initial imbalance of power, this was not a simple matter of give-and-take; America bestowed and the junior partners in the alliance received; American military doctrines, and hence American-style OR, were deemed unquestionably superior; the only problem was how to bring the military cadres of the allies up to speed so that tactical and strategic plans could mesh with some tolerable degree of consilience. This concern grew especially acute when it came to the use of atomic weapons; and especially delicate when the ally in question was putatively conforming to the American doctrine of nonproliferation, but was surreptitiously running an undisclosed program of nuclear weapons development and acquisition. To deal with this problem, the American military began to consciously

[38] The third niche, and one that deserves significantly more attention than we can spare it here, is the burgeoning area of sociobiology and its attendant parallel field of animal ethology. On this, see Haraway, 1981–82; Sismondo, 1997.

promote the training in computer-based C^3I amongst the allies (Ceruzzi, 1998, p. 105), to be sure, but also propagated the collateral trappings of American military science, which meant OR and its concomitant, game theory. If the allies would never fully bend to the will of America, perhaps at least they could be prevented from inadvertently setting off World War III.

Hence from roughly 1952 onward the history of game theory goes global and is inextricably intertwined with Cold War diplomatic and military history. Histories now trace the intricate considerations that lay behind the formation and stabilization of the North Atlantic Treaty Organization (NATO), the most detailed being that of Tractenberg (1999). The complex interplay of overtures toward European integration, American desires to exert military control, and the effective delegation to SACEUR of authority to initiate nuclear combat proved one of the thorniest issues of the 1960s. As Robert McNamara insisted,

> There must not be competing and conflicting strategies in the conduct of nuclear war. We are convinced that a general nuclear war target system is indivisible and if nuclear war should occur, our best hope lies in conducting a centrally controlled campaign against all the enemy's vital nuclear capabilities. Doing this means carefully choosing targets, preplanning strikes, coordinating attacks, and assessing results. . . . These call, in our view, for a greater degree of Alliance participation in formulating nuclear policies and consulting on the appropriate occasions for using nuclear weapons. Beyond this, it is essential that we centralize the decision to use our nuclear weapons to the greatest extent possible. We would find it intolerable to contemplate having only a part of the strategic force launched in isolation from our main striking part.[39]

The Americans had been trying to get NATO to foster an internal capacity for C^3I and operations since 1952, but for various reasons they had not been satisfied with the progress along these lines. NATO formed a Science Directorate in the mid-1950s to fund collaborative research (Williams, 1997) and convened a conference in Paris to proselytize for OR amongst NATO member countries in April 1957 (Davies & Verhulst, 1958); but the view from Washington was that parochial academic schools and bureaucratic rivalries were obstructing the spread of American doctrines. The U.S. Air Force had awarded more than one hundred research contracts through the European Office of the Air R&D Command by 1957, but the situation demanded something with greater coherence and staying power. What was needed was something like RAND

[39] McNamara remarks, NATO ministerial meeting, May 5, 1962 (quoted in Tractenberg, 1999, p. 316).

for Europe; but political considerations dictated that it could not be sited in England, France, or especially Germany, and the recruitment of ideal personnel was proving refractory.

This situation provides the background for the formation of the Center for Operations Research and Economics (CORE) at the Catholic University of Louvain in 1965, conveniently located near the NATO directorate in Brussels, under the charismatic leadership of Jacques Drèze.[40] CORE was first inspired by the Carnegie-TIMS connection to provide modern American OR training and research in English; the Ford Foundation offered substantial research funding in 1966 to expand activities and recruitment. It is no exaggeration to state that CORE became the first respected center of American-style game theory and Cowles-inflected microeconomics on the European continent. Early research topics included linear programming, computational algorithms, the Slutsky equations, Bayesian decision theory, and the newfangled treatment of the continuum of agents. Early American visitors included Jack Hirschleifer, Merton Miller, Tjalling Koopmans, David Gale, Robert Aumann, and Bruce Miller. Once institutionalized, regular ties could be forged to RAND and the Cowles figures so they might provide the mentoring that might not be found within their own national boundaries. "I remember vividly and most gratefully how during a summer workshop at RAND [Aumann] spent long hours explaining to me all the basics and tricks of the trade" (J. F. Mehrtens in Cornet & Tulkens, 1989, p. 19). Regular exchanges with the Israeli OR community began as early as 1968.

A rather more significant instance of the military-inspired spread of OR can be discovered in the new postwar entity of Israel. The culturally attuned historian of economics cannot fail to notice the overwhelming dominance of Israeli names in the late twentieth-century development of game theory but, equally, cannot but wonder at the deafening silence that surrounds this phenomenon. The prominence of game theory can only be traced to two broad causes: the decision of the Israeli Defense Force (IDF) to pour vast resources into an indigenous capacity to develop an indigenous strategic doctrine; and a geopolitical situation that, if anything, was even more suited in its apocalyptic tenor to game-theoretic reasoning along the lines of the Nash theory than even the superpower confrontation between Russia and the United States. These should be juxtaposed with the fact that Israel opted to undertake an "opaque" or unacknowledged

[40] Jacques Drèze (1929–): Ph.D., Columbia, 1958; NATO consultant, 1955–56; assistant professor, Carnegie Tech, 1957–58; professor, Catholic University of Louvain, 1958–64, 1968–; University of Chicago, 1964–68. The information in this paragraph comes from the author's interview with Drèze, May 1999, and research reports in CORE.

program of the development of nuclear weapons ever since the Suez crisis, and hence has been engaged in a military posture of secrecy and deception with regard to its nominal allies (not to mention its hostile neighbors) unparalleled anywhere in the world.[41] As one historian of the covert weapons program has written, "This kind of planning was unique to Israel, as few nations have military contingency plans aimed at preventing apocalypse" (Cohen, 1998, p. 14). Indeed, the unique experience of Israel in never having formally acknowledged nor denied having a substantial offensive nuclear capability stands as one of the very few concrete instances one can point to in the Cold War of a whole nation actually living a mixed strategy, in the game-theoretic sense.

The proclivities of the cyborg sciences would themselves appear to have been tailor-made to suit Israeli circumstances, and vice versa. Few countries on earth have harbored such a fervent propensity to believe that science could compensate for the palpable national deficiencies in resources, geographical situation, historical blessings, or internal homogeneity. As early as 1948 the Israeli military instituted the "Hemed" or Science Corps; by the 1960s the military was running huge black operations off budget to pay for the notorious weapons facility at Dimona. Israel also enjoyed one of the first JOHNNIAC-class computers built anywhere outside of the United States: the WEIZAC.[42] More germane to our present concerns, there was no private computer industry in Israel in the 1960s, and yet it was well known by that era that Israel ranked with the United States and Japan as world leaders in software design and production (Hersh, 1991, p. 137). It is intermittently admitted in the open press that the more recent Israeli computer sector, or "Silicon Wadi," owes its continued vitality almost entirely to the military sector (Steinberg, 1998). This hothouse computer development has had extensive consequences for Israeli military doctrine and strategic theory – and, it follows, for our cyborg narrative.

After Israel went on nuclear alert during the Yom Kippur War, it fell to Andrew Marshall (the very same) to try and start a dialogue with Israel on the possibility of at least coordinating basic issues of nuclear target selection with the United States. The American government has been absorbed ever since with the drastic implications of lack of strategic coordination with the government of Israel, both in the Middle East and

[41] The Israeli weapons program is described in Hersh, 1991; Cohen, 1998; and Schmemann, 1998. Israeli strategic military doctrine is described in Levite, 1989. To my knowledge, there is no history of Israeli game theory in English. However, this may have something to do with the fact that "Israel is the only Western democracy that has a military censor who oversees every publication dealing with security issues" (Cohen, 1998, p. 345).

[42] See George Brown interview, March 15, 1973, SCOP.

elsewhere; and one language in which this found expression was the idiom of game theory. In some quarters, the experience of the IDF is treated as a major source of data on the impact of modern C^3I on what are regarded as the "information pathologies" of warfare (van Creveld, 1985). Let one nuclear land mine be tripped on the Golan Heights in the early 1980s, or have Israel launch a single nuclear counterattack on Baghdad during the Gulf War, and every military war plan of the superpowers (and probably everything else) would not be worth the paper it was written on. Israel has been ground zero of strategic thought for almost every military theorist of the postwar era.

Consequently, the Israeli military establishment has maintained a strong cadre of research game theorists and operations researchers since the later 1950s. The leading light in this community for four decades has been Robert Aumann.[43] Aumann has stood as the major intellectual exponent of the Nash approach to game theory in the postwar period; not unexpectedly, he has also occupied a singular position as an expert consultant to both the IDF and the American military (e.g., Aumann & Maschler, 1995, p. xii) from at least the 1960s onward. He has acknowledged that his career orientation was a product of the military: "The Naval electronics problem was my entrance into decision theory and from there into game theory" (1997, p. 15). While building an Israeli center of game-theoretic research, he maintained a substantial presence in the American OR community, not the least through his close connections to Kenneth Arrow at Stanford. This dual status – dual citizenship, dual allegiance – has been instrumental in constructing a bridge between Israeli military doctrines and modern American OR and strategic doctrines during the critical period of the "gray" weapons program and the posture of "nuclear opacity" toward the United States in order to drive hard bargains for supplies of conventional weaponry. "Until about 1966 there was little systematic effort to define the political and strategic objectives of the nuclear project" (Cohen, 1998, p. 235); and therefore the earliest uses of game theory supposedly involved bureaucratic questions of weapons choice and cost-benefit analysis. This function was neatly

[43] Robert John Aumann (1930–): born Frankfurt, Germany; dual citizen, United States and Israel; B.S., CCNY, 1950; Ph.D. in mathematics, MIT, 1955; research assistant, Princeton, 1954–56; professor of mathematics, Hebrew University, 1956–; Center for Rationality and Interactive Decision Theory, Hebrew University; visitor, RAND, 1963, 1968. Aumann had early in his career encountered John Nash at MIT (Nasar, 1998, p. 140). The only source of biographical information is the brief entry in Hart & Neyman, 1995, pp. 20–28. This otherwise colorless curriculum vitae offers one shred of personal insight (p. 20): it seems Aumann is known simultaneously as Bob, Johnny, and Yisrael, and "His wife Esther tells the story of the difficulties she once had while issuing passports for their children; the clerk could not understand why each had a different father!"

illustrated by one of Aumann's early position papers on the bearing of game theory on military doctrine:

> We wish here to outline a game-theoretic approach to some of the strategic problems that are inherent in the cold war situation. . . . It is *not* meant to answer questions of over-all political or military policy or military policy. Questions like "Shall we share atomic armaments with NATO?" or "Shall we defend Guantanamo if attacked?" cannot be answered with this approach, nor can it shed any appreciable light on these questions. We will go further and say that in our opinion mathematical game theory is inherently unsuited to deal with such problems. The questions that we will try to analyze here are basically quantitative questions that arise in the context of a particular complex of weapons systems. For example, our approach is designed to shed light on questions like: "For a given budget, what should be the balance between Minuteman and Polaris missiles?" or "In designing a new weapons system, what balance should we strike between the effectiveness of the individual weapons and the quantity of weapons produced?"[44]

The useful ambiguity of whom the "we" (America? Israel?) references reveals one of the roles which Aumann was well suited to play in this larger strategic context; another, the ability to mediate between critics of game theory at RAND (covered in Chapter 6) and the new potential clientele for game-theoretic analyses in the IDF and NATO, is also nicely illustrated in this document. But as the relative positions of Israel and the superpowers shifted after 1966, and the intellectual credentials of game theory reached a nadir and then revived dramatically, Aumann proved exceedingly nimble in forging new alliances to buttress the claims of the program, to such an extent that he came to be regarded throughout the world as one of the major interpreters of the goals and aims of game theory. For instance, once a strategic rationale was required for Israel's nuclear capabilities, Aumann rapidly reversed course and asserted the capacity of game theory to discuss the nuclear standoff in the later 1960s (Aumann & Maschler, 1995). Yet, in a reprise of the colonel's dilemma discussed in Chapter 6, there persisted the strategic question of what one tells the man in the street versus what one tells the prime minister: "Sometimes when people interview me for newspapers in Israel, they ask questions like, can game theory predict whether the Oslo agreement will work or whether Saddam Hussein will stay in power. I always say: Those situations are not sufficiently structured for me to give a useful answer"

[44] Robert Aumann, "Game Theoretic Approach to Deterrence," box 10, OMPD. Although undated, internal evidence suggests the paper dates from the very early 1960s. The ambiguous referent of the "we" is particularly poignant from a game-theoretic perspective of the supposed integrity of the individual player.

(1997, p. 11). Game theory thus assumed a flexibility in Israel that had somehow initially eluded it in the United States. This explains the rather curious phenomenon that in the period circa 1955–75, almost all the signal breakthrough developments in game theory (such as they were) originated outside the original American centers of game theory at RAND and Princeton.

Throughout his long career, Aumann maintained a ruthlessly instrumentalist attitude toward the justification of game theory, combined with a thoroughly unshakable allegiance to neoclassical economic theory: a combination rather rare in the United States in the period before 1980, but one that took root and flourished in Israeli soil, largely on the strength of his advocacy.[45] Curiously enough for a trained mathematician, Aumann staunchly defended the principle that utility maximization was a non-negotiable precept of "rationality": it was "the underlying postulate that pulls together most of economic theory; it is the major component of a certain way of thinking, with many important and familiar implications which have been part of economics for decades and even centuries" (1985, p. 35). Bald disregard for intellectual history could be shrugged off with nary a qualm by many a mathematician; but what was more striking was that Aumann openly denied that science was driven by pursuit of "the truth" (1985, pp. 31–34; 1997, p. 8). Instead, the more success that Aumann enjoyed with his clientele, the more he was willing to assert that the primary criterion of success is success.

Although utility maximization (and thus the Nash program) was deemed the very essence of rationality, Aumann rarely went out on a limb to defend any particular strand of game theory. Instead, he would aver, "Game theoretic solution concepts should be understood in terms of their applications, and should be judged by the quantity and quality of those applications" (1985, p. 65). But because so many of those applications lay inaccessible, buried in classified and unpublished research reports, or shared in the oral tradition amongst the elect, their quantity and quality sported a certain ineluctable insubstantial air.[46] The benefits of game theory did not appear to be something that would thrive in bright sunlight or open air: "experimental work in game theory . . . is a contradiction in

[45] These guiding principles are evident throughout his career, beginning with his review of Rapoport, 1962, and continuing in his influential overview articles "What Is Game Theory Trying to Accomplish?" (1985) and his *New Palgrave* entry (1987), as well as his revealing interview (1997). The latter adumbrates his position that science is not about truth but about usefulness; again, the "we" for whom use is judged is conveniently left vague.

[46] "The world will not long support us on our say-so alone. We must be doing something right, otherwise we wouldn't find ourselves in this beautiful place today" (Aumann, 1985, p. 37). The style of rhetoric here is really quite breathtaking, given that this published paper acknowledged support by ONR.

terms, and it is no wonder that so many studies yield results 'contrary to the conclusions of game theory'" (1962, p. 678). Instead, one had to vaguely gesture in the direction of shadowy virtual actors lurking in restricted areas off-limits to the civilian populace. For Aumann, game theory had managed to transcend the mundane demands imposed upon neoclassical economics:

> The general aims of social physics are similar in spirit to those of natural science. Not so with game theory; this is a *normative* theory. It gives advice, describes the behavior (in conflict-cooperative situations) of theoretical beings called "rational persons," suggests arbitration schemes, defines "fair outcomes," and so on. It does *not* purport to describe actual behavior of any kind, this being notoriously irrational. The general aims of game theory are comparable to those of certain parts of applied statistics, such as quality control, or to those of operations research. (1962, p. 676)

It is those operations researchers, those elusive rational (espionage) agents, that insist we should devote more attention to Aumann. If game theory didn't apply to flesh-and-blood selves in the Era of Individualism, then just who or what did it describe? Much later, when game theory had spread throughout the economics profession as microeconomic orthodoxy, and not only the empirical but the internal contradictions of various game-theoretic propositions had become common knowledge, the instrumentalist defense took on a much more aggressive edge: "I don't believe in justifications; it's not religion and it's not law; we are not being accused of anything, so we don't have to justify ourselves" (1997, p. 31). Yet this defense was not pure bluster; for in the interim Aumann had in effect discovered a solution to the conundrum of the Self: he had helped to *construct* a new world populated by a new kind of agent of indeterminate ontological pedigree. First in Israel, and then around the world, economists were introduced to a novel idea – cyborgs were the beings that frolicked in the Nash netherworlds of apocalyptic conflict: they were the natural, laudable, excellent, and compelling paragons of rationality that constituted the soul and subject matter of game theory. Game theory had found its true home in cyberspace. If our protagonists were not so contemptuous of history, and therefore of the heritage of von Neumann, they might have applauded this trajectory as the Return of the Native.

What game theorists did from the 1970s onward was to find and cultivate new alliances beyond the immediate defense analysis community; and here, as elsewhere, Aumann was in the forefront of reconnaissance. If game theory was running into brick walls trying to come up with the ultimate formal definition of the rationality of the human Self, then the time had come to recast it as the herald of the Posthuman and the framework for a Grand Unification Theory of all of science. Aumann was one of the first to announce: "Game theory is a sort of umbrella or 'unified field' theory for the rational side of social science, where 'social' is interpreted broadly to include human as well as non-human players (computers, animals, plants)" (1987, p. 460). The hierarchical ordering of the "non-humans" was significant. While game theory had found a berth amongst the animals since the early 1970s,[47] they nevertheless were not the sorts of neighbors and clientele to replace the military as supporters of mathematical research. The computer community was a different matter. There was a thriving and vibrant computer industry putting on weight cheek by jowl with the operations researchers in the Israeli military incubator; and, moreover, Israel housed one of the premier centers of research on the theory of computation and automata, thanks to pioneers like Michael Rabin. The consanguinal relations of game theory and computation, never long suppressed, were destined to once more come to the fore. The way the Nonhumans made their entrance was through a two-phased approach: first, the theory of automata was accessed for what it might do to help the beleaguered game theorists; and then, second, the phalanx of computer programmers became intrigued with what game theory might do for them. First, little cyberentities were proposed as the algorithmic paragons of Nashionality; and then, when that grew stale, the Artificial Intelligentsia intervened with the idea of appealing to Nash game theory for help in constructing an artificial society of artificial agents in cyberspace. Robert Aumann was poised there waiting strategically situated to nurture both phases.

This is not the appropriate place to enumerate and explain the numerous conceptual problems that the Nash program encountered in game theory and economics in the 1970s and 1980s.[48] Suffice it to say for present purposes that the problem of the so-called folk theorem in repeated games, the endless sequence of "refinements" to Nash equilibrium to try and whittle down the daunting plurality of possible points; and the thorny issue of "common knowledge of rationality" requirements for equilibrium to make sense played havoc with any assertion that game theory had successfully codified the meaning of that notoriously slippery term "rationality." Indeed, it had occurred to more than one observer that various Nash refinements sought to "use *irrationality* to arrive at a strong form of rationality" (Aumann, 1987, p. 478), a paradox only for those fainthearted whose vision of game theory was

47 See note 38. The classic locus is Maynard Smith, 1982.

48 Luckily, that task has been insightfully pioneered by Abu Rizvi (1994, 1997). A textbook primer that faces up to these problems in an unflinching manner is Hargreaves-Heap & Varoufakis, 1995.

couched in outmoded notions of "truth." In the late 1970s, Aumann decided to venture further down this path by exploring the extent to which "bounded rationality" might further serve formally to buttress the Nash program and game theory in general. As we have now become accustomed to expect, this did *not* mean Herbert Simon's version of "bounded rationality." In the 1981 survey where Aumann first signaled his intentions, Simon did not even rate a mention, and the concept is instead credited to Roy Radner (p. 21); by his 1986 Nancy Schwartz lecture, the Nobelist is peremptorily dismissed: "Much of Simon's work was conceptual rather than formal. . . . Particular components of Simon's ideas, such as satisficing, were formalized by several workers, but never led to any extensive theory, and indeed did not appear to have any significant implications" (p. 5). No, for Aumann, the only legitimate concept of bounded rationality would have to take its cue more directly from the computer, which in the early 1980s, meant a restriction on memory capacity which implicitly demarcated the hierarchy of automata of various computational powers (as explained in Chapter 2). Hence, in 1981 Aumann began to explore various ways in which arbitrary memory limitations might be used to "solve" proliferating anomalies in repeated games, such as "explaining" the emergence of cooperation in the repeated Prisoner's Dilemma. At this juncture, there was no explicit recourse to the theory of computation in the published work; however, he was clearly pointing his students to look in that direction, including Abraham Neyman, Bezalel Peleg, and Ariel Rubenstein.

Thus it ensued that, first in Israel and later elsewhere, "the mainstream of the newer work in bounded rationality [became] the theoretical work that has been done in the last four or five years on automata and Turing machines playing repeated games" (Aumann, 1986, p. 17). By most accounts, the landmark paper was Neyman (1985), which managed to derive a cooperative Nash equilibrium out of the repeated Prisoner's Dilemma game by restricting the fractious machines to using mixtures of pure strategies, each of which can be programmed on a finite automaton with an exogenously fixed number of states. "This is reminiscent of the work of Axelrod, who required the entrants in his experiment to write the strategies in a fortran program not exceeding a stated limit in length" (Aumann, 1986, p. 17). Ben-Porath (1986) then argued that there was no advantage to having a "bigger" automaton play a repeated zero-sum game; so it became apparent that the fascination with automata was all tangled up with concern over providing justification for Nash solutions to noncooperative games. This was followed with a paper by Rubenstein (1986), which sought to render the mental "size" of the player automaton endogenous by attaching a "cost" to memory. In short order, the most prestigious economic theory journals were whirring and humming with the wistful sounds of finite automata – most of them patterned upon Moore (1956) machines – busily playing games with all manner of their cyborg brethren.[49] All of these cyborgs and their engineers could congratulate themselves that they had gotten a grip on the notoriously thorny problem of the "complexity" of rational strategic action – couldn't they?

Actually, this was only the first rumblings of the coming-out party of cyborgs in game theory. For instance, the game theorists were not generally accessing the terminology of "complexity" in the same way it was coming to be used by the computer scientists in that era. First off, the decision theorists wanted "complexity" to be limited to the number of states that could be accessed by the finite automaton, which essentially boiled down to the number of strategies available to the machine. In the 1980s game theorists hadn't yet begun to worry about such issues as time and space complexity and computational intractability, primarily because they hadn't graduated yet to full Turing machines. (This will be the primary topic of the next section.) So the meaning of "complexity" was still roughly where Warren Weaver had left it a generation before. Second, the immediate reason the Israeli game theorists remained enamored with the finite automata was the initial problem situation that had been bequeathed them by Aumann and others: namely, "saving" Nash rationality by "bounding" some arbitrary aspect of algorithmic calculation. The little finite automata they favored were stopgaps to plug a gaping hole in the economists' research; they were not (yet) full-blooded supercomputers dueling to the death, unlike their cousins running the fully automated war games in the basement of the Pentagon and the depths of Dimona. Yet, as so often happens, the strain to plug one leak exacerbated pressures which allowed others to appear. Here, recourse to finite automata raised to consciousness the much more perilous issue of the nature of the Self playing these games.

Thus, once the contrived pretense of humans "choosing" the various machines to play games in their stead was summarily dropped, then the real issues of conflating players with machines had to come to the fore. For instance, Aumann began to cite the work of Alain Lewis as relevant

[49] This literature is just too large to adequately survey here. Some highlights include Ben-Porath, 1990; Kalai & Stanford, 1988; Zemel, 1989; Lehrer, 1988; and Megiddo & Widgerson, 1986. A very nice survey, from which we derive our generalizations in the text, is Kalai 1990. Another overview is provided by Holm, 1992. As mentioned earlier, the previous Russian literature is a vast terra incognita, although perhaps not so much to some of our protagonists.

A quick reading of Moore, 1956, reveals the extent of the military-cyborg inspiration of this literature: "one or more copies of some secret device are captured or stolen from the enemy in wartime. The experimenter's job is to determine in detail what the device does and how it works. . . . the device being experimented on may explode" (p. 131).

to issues being broached by automata playing games; but he then found himself unable to confront the fact that Lewis's work was explicitly geared to demonstrate the utter incoherence of the portrayal of rationality in the game-theoretic literature. Patently, Aumann had admitted that the Nash program was predicated upon acceptance of the neoclassical characterization of the individual agent; but if "rational choice theory" a la Arrow was already demonstrated to be computationally intractable, then what difference did it make whether the artificially hobbled machine could or could not manage a few strategies more or less in a repeated game? The entire project of finite automata "saving" neoclassical theory and the Nash program was thoroughly riven with contradictions, perplexities that eventually had to break out in the open literature, and therefore did – although not, it seems, in Israel.

However, even as the tortured recourse to automata for justifying noncooperative game theory was running its course, a far more portentous reverse salient made its appearance in the mid-1980s. It seems that a number of computer scientists and programmers, themselves disappointed by various perceived failures of artificial intelligence and computer architectures in that era (Crevier, 1993; Anderson & Rosenfeld, 1998), began to turn to game theory for inspiration. Many, though not all, were trained in Silicon Wadi, and had begun to make their names in the corporate-military milieu of software engineering: Yoav Shoham, Nimrod Megiddo, Yoram Moses, Jeffrey Rosenschein, Yishay Mor, Nir Vulkan, and Bernardo Huberman. The connections between Aumann's Center for Rationality and Interactive Decision Theory and Stanford were conveniently mirrored in connections between Silicon Wadi and Silicon Valley, including computer scientists Joseph Halpern and Daphne Koller at Stanford. Thus it transpired that under the auspices of Xerox Parc and IBM, feelers were extended from the computer science community to many of the Israeli game theorists, and various interdisciplinary conferences ensued, such as the TARK (Theoretical Aspects of Reasoning about Knowledge) confabs, which began in 1986. The computer scientists just mentioned began to publish papers in game theory journals. The motives for this meeting of minds ranged from superficial attractions of economic metaphors for those seeking to break away from the socialistic biases of GOFAI (Good Old-Fashioned AI) and toward notions of distributed intelligence,[50] to

[50] As Michael Wellman put it, "Researchers adopting this approach would like to harness the awesome power of mechanisms like the Invisible Hand and the Blind Watchmaker for the task of distributed computation" (1994, p. 336). Parenthetically, this review is a perceptive critique of the program proposed in Huberman, 1988. Wellman continues, "While not technically a public good . . . information is a notoriously difficult commodity to produce and distribute efficiently by a market mechanism" (p. 345).

problems of message passing and coordination amongst networked computers (Fagin et al., 1996), to more focused projects seeking to construct artificial software "agents" who might interact in a rational manner over the most important computer development since time-sharing, namely, the Internet. But whatever their idiosyncratic personal motivations, the computer scientists could not help but bequeath a certain gravitas and legitimacy upon game theory, merely by taking it seriously as a resource for their innovations in computer programming. The incipient alliances being forged between some orthodox microeconomists and some computer scientists is the phenomenon most freighted with consequence for economics in the twenty-first century. We can tentatively identify two different coalitions, which have begun to coalesce around divergent images of the computer.

The first example of the alliance between economics and computer programmers is exemplified by the work of Bernardo Huberman (1988, 1998; Huberman & Hogg, 1995). Huberman would like to enlist economists to help revise computational architectures away from the conventional von Neumann approach. Huberman's essential dependence upon the neoclassical tradition can be gleaned from the following quotation: "a competitive market has three main components: agents, resources and preferences. . . . The new perspective here is to view the programs themselves making choices as agents. Their preferences are dictated by computational needs for resources to complete their tasks. . . . Associated with a choice there is a perceived payoff, which is the analog of the utility function in economics" (1998, p. 1170). Now, computer programmers have always had to take the allocation of computational resources into account – CPU, memory space, pipeline bus, and so forth – but tended to solve them after the fashion of planned economies, with algorithmic rules for allocation. Huberman and his associates became enamored of the idea that there must be a better way to structure the allocation, based on free-market principles. But just as generations of economists have never quite grasped everything that might be implied by their progenitors' advocacy of "conservation laws" or energy potentials as the paradigm of individual cognition, Huberman and his collaborators have had some trouble pinning down what is meant by a "market." Many honorific citations of Adam Smith and Friedrich Hayek do not in and of themselves give practical guidance to coding practice and software engineering. This difficulty in "using" economics made itself manifest in such AI programs as "Spawn" and "Enterprise," where it was unclear that all the extra coding of "prices," "markets," and "auctions" did actually constitute an efficient improvement upon the daunting problem of software engineering for parallel distributed environments. Was economics really performing any valuable function as a guide in these situations? "Developing and evaluating a variety of auction and price mechanisms that are particularly well suited to these computational tasks is an interesting open problem. This fits nicely with economic theory, where there have always been questions of how the Walrasian auctioneer adjusts prices to equate supply and demand" (Huberman & Hogg, 1995, p. 145). If Huberman had enjoyed an earlier acquaintance with the work of Lewis or the Sonnenschein-Mantel-Debreu results, he might have come to realize that his second sentence actually contradicted his first. The Walrasian tradition, in positing a single generic "market," has unfortunately had relatively little to say about the sought-after diversity of "market mechanisms," even though its postwar history owed some inspiration to the computer, as described in Chapter 5.

It was certainly not at all unreasonable of Huberman to expect that economics might have potentially provided some guidance in recourse to different market formats in order to alleviate various allocation problems; indeed, this will be the vision of economics to which we will pledge our own allegiance in the next chapter. Rather, it was prior historical trends in economics described in this volume, in conjunction with trends in the development of the computer, that have thrown up daunting obstacles in the way of an economics that can inform computer architectures. In the first instance, while interest in massively parallel distributed architectures has been avid and sustained, actual success in hardware engineering of such systems has proved less than stunning, with heavily subsidized firms like Thinking Machines going bankrupt in the 1990s. It has proved possible to simulate many forms of parallel distributed processing on more conventional sequential architectures, but this has not yet produced breakthroughs in computer design. The design community has instead been captivated by a different kind of distributed processing, which has been provoked by the phenomenal growth and expansion of the Internet at the end of the century. It has been the interaction between the World Wide Web and the post-1975 orthodoxy of the economics profession, which has revolved around Nash game theory (and *not* a concerted theory of diverse market institutions), that focused attention once more on the "agent" rather than the market, and gave rise to the most powerful alliance between orthodox economists and computer scientists.

The wave of the future in economics and computers – the very apotheosis of the vectors of influence covered in this chapter – is the construction of "autonomous artificial agents" (AAAs) to populate the Internet markets of the future (Knapik & Johnson, 1998; Jennings & Wooldridge, 1998; Huhns & Singh, 1998; Vulkan, 1999; Holland & Miller, 1991; Binmore & Vulkan, 1999). Game theorists in economics have had very little cogent to say about the institutional operation of diverse markets, but they have been hard at work constructing little automata to play games since the 1980s, especially in Israel. Hence it comes as no surprise to observe that many of the original computer pioneers of the construction of AAAs have ties to the Israeli Silicon Wadi and military OR establishments. It is but a minor leap from "worms," "sniffers," and logic bombs to software entities that "observe" your every preference and peccadillo as you surf the Web, and then roam the newly commercialized Internet in search of that special rare pinot noir or especially wicked Glenn Branca symphony that will make your own heart glad, driving a hard bargain with its AAA counterpart at megaglobalstore.com to get it at a favorable price. The military-issue cyborg thus undergoes post–Cold War conversion to the less prepossessing spy in the house of love. Every venture capitalist west of Yerevan is absolutely salivating over the prospect of automated purchases on a worldwide scale, and therefore development of AAAs proceeds apace wherever high-tech dreams dwell.

Our current concern is not to prognosticate about the future of computers; it is to understand the complex interplay of these developments with the elaboration of economic theory. Thus, at this critical juncture we observe the game theorists reaping the benefits of their earlier alliance with computer scientists. From the TARK conferences onward, it seems the economists have managed to convince a subset of the computer scientists that, whatever else one might say about it, "Game theory is the right tool in the right place for the design of automated interactions. Game theory tools have been primarily applied to human behavior, but in many ways they fall short: humans do not always appear to be rational beings, nor do they necessarily have consistent preferences over alternatives. Automated agents can exhibit predictability, consistency, narrowness of purpose . . . and an explicit measurement of utility. . . . Even the notion of a 'strategy' . . . takes on a clear and unambiguous meaning when it becomes simply a program put into a computer" (Rosenschein & Zlotkin, 1994, p. 6). In other words, Nash's machine dreams, if not made flesh, have at minimum been endowed with a virtual sort of reality. And, as the philosopher has maintained, if it is real in its effects, then believe it, it's real.

One should not overstate either the prevalence of AAAs or the extent of the sway that game theorists exert over their development. The game-theoretic approach to agent architectures is just one among many approaches currently vying to set the standards for Internet commerce, and furthermore, "We do not yet know how to build a generalized agent, and specialized agents are extremely limited. . . . Today agents are capable of performing a combination of low-level clerical tasks on a user's behalf, but users, for the most part, are not yet able to safely assume that these tasks will be done correctly" (Knapik & Johnson, 1998, p. 197). Nevertheless, the mere existence of the alliance of AAAs and orthodox game theorists evokes the future prospects for economics as a cyborg science more dramatically than anything else we have previously encountered in this narrative. What could be a more "literal" rendition of a cyborg than your own personal software *Doppelganger* roaming cyberspace looking for that special purchase, eluding rival zoids and outwitting the cyber-merchants with their powerful servers and superior resources? But more piquant, what if it raises the possibility that making cyborgs not just the ideal inspiration but the literal subject matter of economics dangles the prospect of permitting orthodox economists to transcend all the crippling criticisms of their program, which we have enumerated throughout this volume? Some especially foresighted economists have already glimpsed this Promised Land: "Fortunately, operating within a computer context torpedoes most of the difficulties that arise when trying to model players as people" (Binmore & Vulkan, 1999, p. 4). Or: "game theory (and mechanism design) seems much more suitable for automated agents than it is for humans" (Vulkan, 1999, p. F69). Cyborgs to the rescue?

Only someone innocent of history could regard this as a reasonable prospect. The fundamental message of this volume has been that the cyborg sciences have shaped most of the major developments in postwar economics, but that in doing so, they irrevocably transform the intellectual content of economics in ways unforeseen and ill-understood by those seeking to co-opt it to the orthodox enterprise. Cyborgs may be recruited to prop up neoclassical economics, but they will not be enslaved to the neoclassical view of the world.[51] They are bigger, and maybe less integrated, than that.

SEND IN THE CLONES

> "According to nature" you want to live? . . . Imagine a being like nature, wasteful beyond measure, indifferent beyond measure, without purposes and consideration, without mercy and justice, fertile and desolate and uncertain at the same time; imagine indifference itself as a power – how *could* you live according to this indifference?
>
> Friedrich Nietzsche, *Beyond Good and Evil*

As Nietzsche understood so well, a captivating narrative can't get off the ground without packing a moral. Thus far in this narrative, I have been more or less portraying my protagonists as sleepwalkers, dreaming their machine dreams about the economy almost oblivious to the cyborgs that jostle them hither and yon. Desperate to endow economics with scientific status, they seem unconcerned with the changes going on all about them

[51] But I find that my confident statements must even here be qualified. "It is important to identify the intended unit of autonomy. . . . An agent that appears autonomous to other agents may in fact be a slave" (Huhns & Singh, 1998, p. 3).

in science. Desperate for rigor, they skirt the most pressing logical paradoxes. Desperate for science to give their lives meaning and significance, they are revulsed by what modern science has done to meaning and the self. Desperate for philosophical succor, they redouble their efforts to concoct mathematical models. Desperate to paint the market as Natural, they conjure up sciences of the artificial. Desperate to assert the primacy of individual will over social determination, they end up effacing their own individuality. A narrative crafted from such protagonists is bound to be dispiriting, and not a little downbeat.

But suppose, just for a lark, we could conjure up an altogether different style of protagonist to nudge us onward toward our denouement. This would ideally be someone poised to intervene in economics with eyes wide open to the various crosscurrents out of the sciences buffeting the discipline, someone who had the courage to face up squarely to what the cyborg sciences were doing to the cultural images of the self, and to boldly go where other theorists feared to tread. This protagonist would not stoop to disparage philosophy;[52] nor, indeed, would he flinch at the prospect of starting a very long book on game theory with a quotation from Nietzsche, nor fear quoting him repeatedly on the lack of coherence of the ego so frequently identified as the "cause" of thought (1998b, p. 287). He would partake of Nietzsche's laughter by calling the book *Just Playing*. Moreover, he would realize that the predicament of the Self at the turn of the millennium poised a major challenge to the coherence of any social science.

This character, should he materialize, would be a person who had conceived of an interest in the foundational conundrums of the neoclassical economist toward the end of the century: someone (at least apparently) not beholden to the military but instead truly swept up in the philosophical problems of rationality for their own sake, at least as they bore upon the quandaries of the social sciences. Suppose further that this person was a trained mathematician; but, incongruously, someone who possessed a broad and deep background in literature and philosophy, but who also was confident enough in his education to wear his learning lightly. This would entail, for instance, an ability to write three graceful sentences in a row without the intervention of a miracle or a ghostwriter. As if this weren't already too much to ask, let us suppose further that this spectral special somebody actually had assumed a skeptical stance toward the Bourbakism that had swept the mathematics profession in the postwar period, realizing that hiding your head in the axiomatic sand only made things worse when it came to the physical and social sciences. He would

[52] He describes himself: "His ambition is to be taken seriously as a philosopher by the philosophy profession" (Kenneth Binmore in Ben-Ner & Putterman, 1998, p. xxvii).

preach that, "the rush to formalize so that a theorem can be proved is all too likely to result in important issues being overlooked when foundational issues are on the table. Once one has closed one's mind by adopting an inadequate mathematical formalism, it becomes hard even to express the questions that need to be asked if the inadequacy of the formalism is to be exposed" (1999b, p. 130). Rather, he would seek some justification of his preferred conceptions of rationality in algorithms grounded in the procedural problems of quotidian economic existence.

But because we are indulging ourselves in our own little reverie, let's not just stop there. Let us venture to posit an intelligence both more consistent and more ruthless than any we have yet encountered amongst our prior parade of game theorists. Just for a lark, suppose this thinker fully appreciated that the telos of the postwar neoclassical tradition was to portray people as computers: "we come equipped with algorithms that operate as though they were employing the principles of revealed preference theory to deduce preference-belief models from consistent sets of social behavior. The process being proposed is recursive. . . . we have large brains as a result of an arms race within our species aimed at building bigger and better computing machines for the purpose of outwitting each other" (1998b, pp. 194, 212). Yet, with a tough-mindedness characteristic of Nietzsche, he would realize that what was true for "us" must indeed be also true of himself: that he, too, must needs be indistinguishable from a robot: the inevitable fate of all Posthuman Selves. As a robot, he cannot claim to be an intelligence *sui generis*, but rather a mouthpiece attached to a program; and that consequently the process of intellectual discourse, if truly rational, is studded with deception, feints, emotional outbursts, and all the evasive gambits expected of a strategic automaton. Furthermore, in an era when biology has displaced physics as the cultural icon of choice, it is precisely where computers become conflated with genes that this reflexivity becomes most relevant:

> Like everyone else, social Darwinists are just mouthpieces for the memes that have successfully replicated themselves into our heads. These memes seek to replicate themselves into other heads, because they wouldn't have survived if they didn't have this property. . . . In seeking to persuade others, social Darwinists are constrained by the neo-darwinian meme to making pragmatic appeals to the self-interest of our fellow citizens. But traditionalists tell themselves and others much more exciting stories in which our mundane arguments are trumped by appeals to Moral Intuition or Practical Reason or some other modern successor to Mumbo-Jumbo and the other gods who served as sources of authority for our ancestors. Sometimes, as when traditionalists invoke the Sanctity of Life as a reason for inflicting suffering on babies who have yet to be conceived, it is tempting to fight fire with fire by inventing new sources

> of authority, like Humanism or Gaia or Science – but the meme that is pushing this pen wants so much to be replicated that it won't even let me appeal to Intellectual Honesty in arguing against the creation of such new graven images. (1998b, p. 180)

In such an anosognosic world, memes would constitute the programs that motivate all us lumbering robots; we cheerfully try and replicate the memes that move us by insinuating them into the central processors of other robots, perhaps through technologies of writing and speaking and interdisciplinary research at university Research Centres. However, should those Others prove a shade recalcitrant, perhaps blinded by some outdated graven images or obsolete processors, then, the blessings of modern Science will facilitate the construction of a master race of Superrobots, or *Übermaschinen*, to produce some especially devious offspring – let's call them "autonomous artificial agents" – who more readily encapsulate their memes, causing them to replicate even more efficiently, spreading them to the four corners of cyberspace.

This wizard will not be afraid to set the world aspin around an axis bold as love, though on occasion he can effortlessly display a harry-potterish sort of charm. "Nature red in tooth and claw also operates in the world of ideas" (1990, p. 66). He will finish up his book claiming to pronounce upon morality and politics and the good life – "an attempt to vindicate the basic intuitions of Rawls's *Theory of Justice* using the naturalistic approach of David Hume in place of the so-called rationalism of Immanuel Kant" (1999b, p. 136) – by reiterating that the citizens of his commonwealth are just a loose jumble of genes and memes and justice algorithms poised in Nash equilibrium (1998b, p. 513). Nobody here but us knowbots.

Surely such a person, such a fearless Nietzschean iconoclast, should he exist, would boldly set out to explore and expound upon the virtues of the cyborg sciences in clarifying the crisis of individual rationality in the postmodern age. . . . Wouldn't he?

Binmore and the Selfish Meme

There is good news and bad news. The good news is that such a person did indeed surface in the economics profession in the 1980s, and his name is Kenneth Binmore, the author in the foregoing phantom citations.[53] The bad news is that, through a convoluted sequence of events, it devolved to Binmore to assume the mantle of the primary spokesman for the

[53] Kenneth Binmore (1940–): Ph.D. in mathematics, University of London, 1964; professor of mathematics, London School of Economics, 1969–88; University of Michigan, 1988–94; University College London 1994–; director, Centre for Economic Learning and Social Evolution. Biographical information can be found in Binmore, 1999b.

anticyborg defense of Nash game theory in the 1990s and, in the process, thus proceed to deliquesce further the neoclassical rational individual agent in the interests of saving him from a fate worse than death – namely, a thoroughly implausible godlike omniscience. Henceforth all his formidable formal skill and rhetorical finesse was bent to the purpose of – portraying every economic agent as an identical computer! And as we have come to expect, this was proposed in the interests of rendering economics a "real" science: "unless and until real advances are made in game theory, the social sciences are doomed to remain but a poor relation to the physical sciences" (Binmore, 1990, p. 6).

Binmore merits our attention because he has trenchantly signaled his being patently aware of the cyborg juggernauts that stand arrayed against his crusade. We have already encountered Binmore in Chapter 3 disparaging von Neumann and Morgenstern's *Theory of Games* as something that should not be read by impressionable young minds (1992a, p. xxix). In one of his most revealing philosophical texts, his *Essays on the Foundations of Game Theory* (1990) – the other being *Just Playing* (1998b) – he takes pains to differentiate his position from that of Herbert Simon: "while Simon's observations are relevant, they are not in a form which makes them applicable in game theory" (p. 21). He has been known to upbraid Robert Axelrod for his tit-for-tat tournament and not according sufficient respect to the orthodox game-theoretic tradition (1998a). Indeed, in Binmore's texts one finds the primordial results of Gödel acknowledged and cited as relevant to economics in a straightforward manner, something that we have demonstrated in previous chapters as rather scarce on the ground in the precincts of orthodox neoclassical economic theory. He also comes the closest of any late-twentieth-century orthodox economist to realizing the profound continuity between the computer and the game-theoretic understanding of rationality, pausing at the brink of the Cold War paranoid conflation of "programs" and "strategies": "Universal Turing machines are particularly relevant for game theory because of the traditional implicit assumption that perfectly rational players can duplicate the reasoning processes of the opponent" (1990, p. 173). Nevertheless, for Binmore these are all treated as inconvenient obstacles to be surmounted, and not resources to be exploited or pointers to more subterranean trends altering the very definition of fin-de-siècle "science."

This intentional stance eventually led Binmore to propose in his *Essays* the scenario of what we have been calling in this chapter "core wars": that is, the lurid spectacle of two Turing machines pitted against one another in various games, deploying digital duels as a device for explicating the meaning of economic rationality. There are multiple ironies lurking in

Binmore's machinic morality play, but we shall dwell on only two: first, the fact that however far removed Binmore might initially appear to be from the milieu of the military-academic complex, he could not circumvent its pervasive influence, because his core wars scenario had already been fully anticipated by developments in wargaming and cyberwar preparedness described earlier; and, second, the greater the degree of his recourse to the cyborg sciences in the name of defending Nash game theory, the more that any justification of the neoclassical approach to human ratiocination lost its way.

We need to cast a fleeting glance at Binmore himself before we become better acquainted with his machines. Binmore has managed to stand out from run-of-the-mill game theorists because he has displayed a profound respect for the problem of what it could mean to "know" something and, by inference, what it might mean to be human. One suspects that this appreciation was provoked by the whole debacle over the meaning of "common knowledge" precipitated by Robert Aumann and others in the 1970s (1998b, p. 520); but whatever its cause, it has induced a heightened sensitivity to the problem of context in strategic inference. "In chess, for example, it simply does not make sense, given the environment in which it is normally played, to attribute bad play by an opponent to a sequence of uncorrelated random errors in implementing the results of a flawless thinking process" (1990, pp. 154–55). The concessions he appears willing to grant to the players of his abstract games, however, he has proved unwilling to extend to his intellectual rivals. If it is the case, as he concedes, that the game theorist "needs to be something of a propagandist as well as a theoretician" (p. 19) – and this is the mendacity of the memes broached earlier – then doesn't that suggest we the audience need to situate the game theorist within his own social context in order to evaluate our responses to his arguments and blandishments? If Binmore insists that von Neumann and Simon and Axelrod have been committing numerous analytical "errors" in playing the game properly,[54] then wouldn't it make sense for us, the reader over his shoulder, to situate them within their own contexts, if only in order to begin to gauge our own reactions to the accusation that they have hijacked commonsense notions of rationality? Can there really be only one successful mode of persuasion? If there is a pattern to their divergence from Binmore's account, then perhaps that fact alone calls into question the full adequacy of Nash's assertion that all rational players must accept the same algorithmic version of rationality?

[54] For instance, he has complained that, "the circle-squarers of social science bombard me with a constant stream of new fallacies" (Binmore, 1999b, p. 131).

Binmore's treatment of von Neumann exemplifies this paradox. On numerous occasions he has found it nearly impossible to restrain his impatience with von Neumann. One wonders if there was a lapse of memory concerning Nash's own history behind his rash accusation that "Von Neumann and Morgenstern's *Theory of Games and Economic Behavior* is more than a little schizophrenic" (1996, p. ix). No doubt, there were differences in concept and motivation between the two coauthors, as there were problems in the extension of the two-person, zero-sum framework to more elaborate models; but Chapter 3 has argued that more context, rather than some amateur folk psychology, is the prerequisite to understanding why the book turned out the way it did. In any event, von Neumann abandoned the game-theoretic account of rationality fairly quickly in favor of the automata theory account as a response to problems he perceived in its elaboration. At no time could the indictment stick that von Neumann was engaged in a "mindless attempt to apply the maximin criterion to all decision problems" (1996, p. xi); if any stratagem deserves the sobriquet "mindless," it would instead be the first commandment of decision theory that one resort to the optimization of utility in order to escape the morass of human psychology – a stratagem repeatedly endorsed by Binmore himself (1990, p. 14; 1998b, p. 513; 1999b, p. 131). Unable to rationalize to himself how the inventor of game theory could have passed up the juicy chance to elevate the Nash equilibrium as the central dogma of the game-theoretic formalism, Binmore has ventured so far as to endow von Neumann with clairvoyance with respect to the subsequent problem of equilibrium selection: "For a two-person zero-sum game, the answer to the equilibrium selection problem is irrelevant, because all the Nash equilibria in such a game are equally satisfactory. I think that von Neumann and Morgenstern saw that the same is not true in general and therefore said nothing at all rather than saying something they perceived as unsatisfactory" (1996, p. xi). The fact that *TGEB* is studded with comments about the novel conception of a collection of nonunique strategies as a solution concept should have squelched that interpretation, much less the evidence presented by Nasar (1998) that von Neumann unilaterally opposed Nash's approach. There may be a grain of truth to the complaint that "von Neumann and Morgenstern's approach left them with little to say" (Binmore, 1996, p. xiii) about the structure of cooperation; but any impatience might be tempered by the fact that Nash's approach left game theorists utterly bereft of any wisdom to impart concerning avoidance of paradoxes of rationality.

Binmore's disdain for context also colors his understanding of theorists whose ideas he favors, such as Nash himself. In a truly extraordinary passage, he seeks to exonerate Nash of any of the debacles that beset game theory in its heyday:

> [Nash wrote] in an abstract, laconic style in which only the considerations immediately relevant to the theorem to be proved are not suppressed. His paper therefore allowed economists, not only to appreciate the immensely wide range of practical applications of the idea of Nash equilibrium, it also freed them of the need they had previously perceived to tie down the dynamics of the relevant equilibrating process before being able to talk about the equilibrium to which it will converge in the long run. In retrospect, one can see that this freedom turned into license with the flourishing of the industry of refinements of Nash equilibrium, as game theorists in the 80s vainly sought to solve the equilibrium selection problem by inventing more and more elaborate definitions of a hyper-rational player. (1996, p. xii)

Nash's obvious penchant for an equilibrium being always already present derived directly from his conception of strategy as the complete reconstruction of the opponent within the confines of the brain of the player: anything less would be an expression of weakness and vulnerability. Nash adopted the idiom of neoclassical utility theory because it fit in so well with this conception: neoclassical economics had no worked-out doctrine of dynamics, *pace* Binmore, and failed to come up with anything more convincing in the era immediately after Nash's publications. Nash did not *free* the orthodox economist of any such fetters: rather, Nash voluntarily submitted to them himself. And as for the wonderful range of practical applications of Nash equilibria, they seemed to elude most economists, as well as operations researchers, until well into the 1970s. Finally, the burgeoning bevy of game theorists did imagine increasingly baroque versions of hyperrationality in the 1980s to "save" Nash; but this was due to their own tenuous grasp of the implications of half-buried computer metaphors that populated their models.

Be that as it may, Binmore proposed that endless dreary misunderstandings of the nature of rationality in game theory should be dispelled by more concerted recourse to the computer. Arguments over the incredible requirement of "common knowledge" had conjured the hope among the cognoscenti that introducing a little bit of uncertainty through the instrumentality of computational limitations – the trope of automata playing games in the previous section – would miraculously render seemingly irrational choices rational, rescuing Nash from irrelevance. Binmore correctly tagged this move as a "halfway" measure (1990, p. 155); and, in a similar way, he correctly diagnosed Simon as avoiding complete confrontation with the theory of computation (p. 182n). Games were best understood formally as programs run on a variety of architectures: "a *rational decision process* will be understood . . . to refer to the *entire* reasoning activity that intervenes between the receipt of a decision stimulus and the ultimate decision. . . . Such an approach forces rational behavior to be thought of as essentially algorithmic. This makes it natural to seek to model a rational player as a suitably programmed computing machine" (p. 153). Binmore submitted that one should start with an abstract machine of maximum complexity, namely, the Turing machine, and that this would immediately raise the issue of Gödel's incompleteness and Turing noncomputability.

> In brief, the notion of rationality used by economists assumes that agents can decide the logically undecidable. This doesn't matter much until one gets embroiled in the details of reasoning chains of the form, "If I think that he thinks that I think . . ." But when one does, one is led to precisely the sort of self-reference that Gödel used. (Binmore, 1999b, p. 132)

Never mind that Michael Rabin, J. P. Jones, Alain Lewis, Peter Albin, and a whole host of others had made the point before; it was Binmore who first managed to crack the consciousness of the orthodox economics profession about the dire consequences of identifying *Homo economicus* with a full-fledged computer in the context of game theory.[55] Nash's precept that a rational player should be able to reconstruct totally the thought processes of an opponent finds its "natural" expression in the formalism of a universal Turing machine accessing the Gödel number of its opponent and then simulating its entire process of strategic reasoning. Initially, this presumes a model of knowledge based on what Binmore calls a "closed universe" (and we called in Chapter 1, following Paul Edwards, a "closed world"), one in which "all the possibilities can be exhaustively enumerated in advance, and all the implications of all the possibilities explored in detail so that they can be neatly labeled and placed in their proper pigeonholes" (1990, p. 119). But once formalized with a computer, this closure proves elusive.

> To know a state includes knowing, not only everything there is to know about the state of the physical world, but also everything there is to know about everybody's *state of mind*, including their knowledge and beliefs. The self-reference implicit in such an interpretation brings Gödel's theorem to mind. Recall that this says that any sufficiently complex formal deductive system cannot be complete unless it is inconsistent. That is to say, in the world of theorem-proving, the "open universe" is a *necessary* fact of life with which one has to learn to live. One is therefore perhaps entitled to be suspicious of theories of knowledge in which this fact of life is somehow evaded. (p. 120)

[55] This happened first in his essay in *Economics and Philosophy* (1987–88), which was revised and expanded in his 1990 book. He does acknowledge, however, that "I have found few economists willing to believe that the work of Gödel or Turing might be relevant to their subject" (1999b, p. 132).

It might initially appear that a computer of maximum capacity, that is, a Turing machine, could circumvent this problem. Yet this was precisely the doctrine that Turing had confuted by the "invention" of his machine. Binmore continues:

> Suppose that the play of the game is prefixed by an exchange of the players' Gödel numbers. . . . A perfectly rational machine ought presumably to be able to predict the behavior of opposing machines perfectly, since it will be familiar with every detail of their design. And a universal Turing machine *can* do this. What it *cannot* do is to predict its opponents' behavior perfectly *and* simultaneously participate in the action of the game. It is in this sense that the claim that perfect rationality is an unattainable ideal is to be understood. . . . None of this is at all profound. Mathematically, all that is involved is a trivial adaptation of the standard argument for the halting problem for Turing machines. (pp. 173, 176)

One might perhaps be forgiven for thinking this would put the kibosh on the entire project of Nash equilibrium in game theory, at least in its neoclassical incarnation. But here is where Binmore's Nietzschean ambition confounded all conventional expectation. In his now-classic paper, he held out the promise of transcending this seemingly insuperable obstacle to game theory by shifting the entire logic of "explanation" from what he (uncharacteristically awkwardly) calls an *eductive* framework to an *evolutive* framework (1990, pp. 187ff.). The former is essentially another name for the quest for algorithmic rationality described earlier, the complete and consistent calculation of the strategies of the opponent, which has been demonstrated to be logically unattainable. The latter regards players as "simple stimulus-response machines whose behavior *has the appearance* of having adapted to the behavior of other machines because ill-adapted machines have been weeded out by some form of evolutionary competition" (p. 187; emphasis added). It might then have transpired that all he had accomplished was to reprise the literature on finite automata playing games surveyed in the previous section, where unmodeled "metaplayers" choose simple automata to play their games for them, an impression seemingly encouraged by Binmore in his citation of Rubinstein, Neyman, Megiddo, and others. But, instead, Binmore tipped his hand by suggesting that the "infinite regress" problem of the supposedly rational choice of boundedly rational automata can be short-circuited by a direct appeal to *evolution*. Because this paper heralded the mass exodus of economists into the promised land of "evolutionary game theory" in the 1990s in another apparent cyborg accommodation, it may prove worthwhile to pursue this argument in slightly more detail.

Perhaps inspired by Nietzsche's definition of "truth" as a "mobile army of metaphors," Binmore insisted, "meta-players can be seen as a metaphor for an evolutionary process. The question of the complexity of the decision-making process attributed to the meta-players then ceases to be an issue since it is unloaded onto the environment" (1990, pp. 190–91). In other words, the debilitating paradoxes of rationality could be *unloaded* onto biology, or at least a simulacrum of biology. It was at this precise location that Dawkins's (1976) "memes" – ideas that behave like genes, or better yet, software – made their debut in Binmore's oeuvre. This move was made easier by the preexistent use of Nash equilibria in behavioral ethology by John Maynard Smith and others; one would not expect your average economist to make a distinction between molecular biology and ethology – after all, to an outsider, isn't it all just biology? Binmore sought to portray this new-model evolutionary theory as more modest than its eductive predecessor: whereas game theorists had been searching for a "master-program" to play all possible games, evolution would only exercise selection upon computer program surrogates in repeated play of a single specific game. The scenario had some undeniable sci-fi overtones: swarms of machines itching to initiate games in a meme soup, randomly encountering one another like carp in a tidepool, mostly evoking total walleyed incomprehension, but eventually connecting with another machine that recognized the same game format; repeatedly making the same game moves irrespective of anything like a competent partner on the other end of the modem; toting up a "score" for itself; and then (something? someone?) swooping down like an angel of death and removing low-scoring machines from the soup. (Where *do* cyborgs come from, mommy? And where do they end up?) What did this have to do with the good old-fashioned Nash equilibrium?

The magic that Binmore wrought upon the computational critique is that he made it seem as though Nature would get herself back to the Nash equilibrium, which had been peremptorily banished from Eden by the Turing machine. In the course of his core wars, Binmore proposed that the inaccessible master program would arise out of the meme soup as a process of self-organization! "Think of a *large* population of hosts (the hardware), each of which may be infected with a master program (the software), with successful master programs sometimes displacing the less successful" (1990, p. 193). Because machines are randomly matched, many of the problems of the Nash program that travel under the rubric of the "folk theorem" are rendered irrelevant. Because the machines play a large statistical sample of the population over time, it is asserted that many of the problems of "common knowledge" are similarly dissolved. Multiple equilibria are literally bypassed. If one master program wins out over the long haul, then it has a high probability of *playing itself* – that is, it has attained the vaunted status of John Nash playing a fully rational and therefore fully identical player. The Nash equilibrium reappears, a cosmology congealed out of a chaos: only individualism gets lost in the process.

Binmore was quite careful to point out that this collapse of diversity to uniformity is neither necessary nor guaranteed under many conditions. Certain subsets of programs may persist under quasi-symbiotic conditions in the meme soup. However, Nash was still claimed to be rescued by a further recourse to John Maynard Smith and his own special revision of the noncooperative solution concept called the "evolutionary stable state" (Maynard Smith, 1982; Mailath, 1998; Samuelson, 1997). The entanglement of the literatures of Nash game theory and evolutionary stable strategies in fact has much less to do with anything a biologist would recognize as evolution than it does with the ongoing fascination of economists with machines; but even a brief survey of this literature would embroil us too deeply in contemporary controversies. In any event, Binmore himself subsequently became disillusioned with aspects of this research program.[56] The major point to be made here is that Binmore represents a further chapter in the dialectic of cyborg resistance and accommodation.

It would appear that Binmore's fundamental ambition is to defend the neoclassical project to the hilt, or, as he puts it, "Find the boundaries up to which neoclassical theory works."[57] The real conundrum is to come to understand precisely how the fin-de-siècle fascination with replicator dynamics in repeated games is supposed to achieve that result. What precisely is the interpretation of replicator dynamics which Binmore propounds? Does the meme soup exist in the bowl of the individual cranium? He is well aware that a wide range of "choice models" has been concocted to result in some favored version of replicator dynamics; but the whole point of his program was to dispense with this "eductive" approach. Furthermore, Binmore concedes that experimental evidence only supports this notion when the tasks posed to the subjects are exceedingly "simple." In any event, the "learning" interpretation of evolutionary games has itself been subjected to devastating computational critique (Nachbar, 1997). Alternatively, perhaps Binmore really takes the

[56] "Our study of evolution in repeated games taught us that the evolutionary stability of Maynard Smith and Price is an inadequate tool for studying the problems that arise in games with whole continua of Nash equilibria. We dealt with this problem by proposing a refinement of evolutionary stability, but we were satisfied neither with our own refinement nor with the refinements proposed by others" (Binmore, 1999b, p. 133).

[57] The following paragraph is based upon a seminar on evolutionary game theory given by Binmore at Notre Dame in October 1999.

automaton scenario exceedingly seriously. Here, individual people are thought to play games rather mechanically, with their cognitive abilities falling well short of the power of a Turing machine; it falls to Nature to cull the stupid strategies and remove the "irrational" behaviors. But this interpretation is equally untenable, since the replicator dynamics that guarantee convergence to Nash equilibrium bear essentially no economic justification and have little resemblance to biological reproduction. "I am not appealing to some vague analogy between economic processes and biological evolution" (1999b, p. 133). Most game theorists of whom I am aware stop well short of really positing "sudden death" for their little inept economic agents, much less advocating parthenogenesis as a winning economic strategy. Because Binmore rejects both the devil and the deep blue sea, he periodically is left straining for some third interpretation, one that he has designated on occasion as "non-cognitive learning." Although he never manages to give it a solid characterization, it appears to involve some sort of process of imitation, which somehow manages to evade any mental or computational characterization, yet nonetheless does possess some sort of integrity across individuals. It is a weirdly disembodied notion of learning; but to enjoy out-of-body experiences, you have to have a body in the first place, and this is an area where neoclassical theory has not enjoyed any notable successes. This, I believe, goes some distance in explaining the fascination that memes exert over Binmore.

It is a symptom of how far he has ventured down the cyborg path that Binmore could only entertain this curious "non-cognitive learning" in the company of a motley of sociobiologists and evolutionary epistemologists at his Centre for Evolutionary Rationality; but when it came to long-term funding, he found himself drawn to the real action amongst the hybrids of posthumanist rationality, namely, the burgeoning field of AAAs (Binmore & Vulkan, 1999). For, as even he had to admit, "Fortunately, operating within a computer context torpedoes most of the difficulties that arise when trying to model players as people" (p. 4).

Whether any of this scenario could have been played out before the last decade of the twentieth century is doubtful. Much of this had already been pioneered amongst the war gamers and computer scientists working under the direction and guidance of their military patrons. The ontological dissolution of the agent and the spread of the Internet are the preconditions for this novel defense of neoclassical rationality. "Carl Hewitt recently remarked that the question *what is an agent?* is embarrassing for the agent-based computing community in just the same way the question *what is intelligence?* is embarrassing for the mainstream AI community" (Wooldridge & Jennings, 1995, p. 1). The entire "evolutionary game" literature is but a pale reflection of themes pioneered by computers constructed to outwit one another in the Cold

War context and the next generation of cyberwar technologies spawned by their proliferation.

We leave this closed world of apocalyptic blipkrieg with one last observation on Binmore. It seems that he, also, has been motivated to somehow "defend the Self" from its depredations at the end of our century: "It is true that *homo economicus* is not a carbon copy of *homo sapiens*. But the discrepancies quoted by critics usually involve deviations from rationality that cost very little, or else occur only rarely" (1998b, p. 13n). But the more he took the cyborg sciences seriously, the more he reduced the Self to rubble. Take his argument about the emergence of machines playing their identical selves as the outcome of the process of evolution. How was this an improvement over the endless rubbishing of individual difference, which we have noted throughout the history of neoclassical economics? Or, take instead the concept of an evolutionary stable strategy. More than one commentator noted that if Nash equilibria embody rationality, then it is the entire population, and not the individual agents, that can be graced with the honorific of rationality in those models. But if it is the population that is rational, we are returned to the group mind of sociology, the anathema that neoclassical economics had set itself against. Combine that with the frequent observation that evolution has no telos or final objective, and it begins to appear that Nash equilibrium turns out to be a rather empty prize. Indeed, all the gene and meme talk by Binmore itself presages the ultimate dissolution of *Homo economicus* into a ragbag of bits and bobs derived from elsewhere, a picaro of subroutines, a *bricolage* of algorithms picked up along the way. There seems to be very little reason left for the Self to stand up and proudly exclaim "I gotta be me!"

8 Machines Who Think versus Machines That Sell

> We may trust "mechanical" means of calculating or counting more than our memories. Why? – Need it be like this? I may have miscounted, but the machine, once constructed by us in such-and-such a way, cannot have miscounted. Must I adopt this point of view? – "Well, experience has taught us that calculating by machine is more trustworthy than by memory. It has taught us that our life goes smoother when we calculate by machines." But must smoothness necessarily be our ideal (must it be our ideal to have everything wrapped in cellophane)?
>
> Ludwig Wittgenstein, *Remarks on the Foundations of Mathematics*

Once upon a time, a small cadre of dreamers came to share an aspiration to render the operations of the economy manifest and comprehensible by comparing its configuration to that of rational mechanics. It was a simple and appealing vision of continuous motions in a closed world of commodity space, uniformly propelled toward an equilibrium of forces; the forces were the wants and desires of individual selves. Each and every agent was portrayed as a pinball wizard, deaf, dumb, and blind to everyone else. Not everyone who sought to comprehend and control the economy harbored this particular vision; nor was the portrayal uniformly dispersed throughout the diverse cultures of the world; but the more people were progressively trained in the natural sciences, the more this dream came to seem like second nature. After a while, it no longer qualified as a dream, having graduated to a commonplace manner of speech. Economics was therefore recast in something tangible as the theory of a particularly simple kind of machine. This is very nicely illustrated by two exhibits on the same floor of the National Science Museum in London: in the center of the floor is A. W. Phillips's contraption of Perspex hydraulic pipes, with different colored liquids sloshing around a "national economy"; and off in a side gallery is Charles Babbage's "difference engine," a mill for dividing the labor of producing numbers. It transpired that the nineteenth-century neoclassical theorists of the so-called Marginalist revolution fostered a more abstract and mathematical version of the machine dream, although certain individuals such as William Stanley Jevons and Irving Fisher were not adverse to actually erecting their vision in ivory and steel and wood as well.

In this volume, it has been argued that something subtly profound and irreversible has happened to machine dreams sometime in the middle of the twentieth century. Instigated by John von Neumann, and lavishly encouraged by the American military, the new generation of machine dreamers were weaned away from their classical mechanics and made their acquaintance with a newer species of machine, the computer. Subsequently, the protagonists in the economists' dramas tended to look less like pinball wizards, and came increasingly to resemble Duke Nukem instead. There were many twists and turns in how this transubstantiation was wrought, from socialist turncoats to Twisted Metal 7; but one recurrent theme of this account has been the persistent tension between an unwillingness on the part of economists to relinquish their prior fascination with classical mechanics, and the imperative to come to terms with the newer computer. After all, aren't computers still made of metal and mineral and polystyrene, still subject to the same old rules of equilibria of classical forces as well? Must we perforce leave our familiar old rational mechanics behind? Gears grind and circuits flash, tradition and innovation clash, then become indistinct as software materializes out of hardware.

Indeed, one might suggest that by the end of the century, the embrace of the sharp-edged computer by the machine dreamers has nowhere yet been altogether wholehearted, indulgent, or complete. Rather, wave upon wave of computer metaphors keep welling up out of cybernetics, operations research, computer science, artificial intelligence, cognitive science, software engineering, and artificial life and washing up and over the economics profession with varying periodicities, amplitudes, and phase shifts. The situation has been exacerbated by the historical fact that "the computer" refers to no particular stable entity over our time frame. What had started out as a souped-up calculator-cum-tabulator grew under military imperatives to something closer to a real-time command-and-control device, complimentary to the discipline of operations research. Yet, there was simultaneously the mitigating circumstance that the business world pressed its own agenda upon the computer, and therefore we observe the machine being reconfigured as a search-and-sort symbol processor, monitoring time-sharing, surreptitiously collecting reams of data and supporting Web commerce. Under the imperative of mass commercialization, the PC familiar to almost everyone has come to resemble the offspring of some mutant miscegenation between the typewriter and the television. And in the not-so-distant future, the computer threatens to become really indistinguishable from the biological organism, with DNA performing various of the computational functions identified here, as well as many others more closely related to physiological processes. The biggest obstacle to answering the question – What is the impact of the computer upon economics? – is that the computer has not sat still long enough for us to draw a bead on the culprit.

Nevertheless, the sheer shape-shifting character of the New Model Machine has not altogether hampered our historical investigations. The saga of the computer has unexpectedly provided us with the scaffolding for an account of the constitution of the postwar economic orthodoxy. It is true, however, that we have only managed to cover a small subset of all the ways in which the computer made itself felt within the postwar neoclassical economics profession. For the most part, we have hewn to the heuristic to keep our attention steadily riveted upon the ways in which the computer has recast the economic agent by enhancing its cyborg quotient, which has meant in practice paying close attention to the ways in which neoclassical microeconomics and its mathematical practice in America have been transformed from an equilibrium metaphor to a command-control-communications-information orientation, the C^3I paradigm. A plethora of other manifestations of computer influence on postwar economics has suffered untimely neglect in the present account: one might cite the all-too-obvious alliance between econometrics, Keynesianism, and the spread of the computer in the immediate postwar period, rendering tractable all manner of national models and elaborate statistical calculations previously deemed inaccessible, and therefore enhancing the legitimacy of a separate "macroeconomics"; likewise, the field of financial economics found its footing in the computer-enabled ability to manipulate vast reams of real-time data and use them to construct new synthetic financial instruments. The efflorescence of experimental economics starting in the 1970s could never have happened without the large-scale computerization of experimental protocols and the attendant standardization of practices and data-collection capabilities, which in turn made it available for export to a broad array of aspiring laboratories. Cliometrics would never have displaced economic history without the computer; nor would econometrics have become a specialty in its own right. It is perhaps less frequently realized that almost all the earliest neoclassical discussions of evolutionary economics were produced at RAND, and thus de facto bore some relationship to the computer. The impact of Soviet cybernetics upon Soviet-era Marxism is a vast terra incognita. Regarded prosaically as a technology, the computer conjured up all sorts of novel activities and functions that could be brought fruitfully under the ambit of economic expertise.

In this our final chapter, however, we shall regretfully pass all those suggestive observations by, in the interests of a reconsideration of an even bigger question, namely, why it is woefully insufficient to treat the computer merely as a technology, just another gadget in that bodacious "box of tools," which the notoriously all-thumbs economists love to evoke. Economists, at least when they are not dreaming, still think that they live in a world in which inanimate objects are readily and obediently bent to the purposes of their makers; but their history discloses a different situation: their tools are amazingly lively, whereas their profiles of the human actors are distressingly reactive, if not downright inert. With increased recourse to the computer as an amazingly flexible and adaptable prosthesis, the Promethean device amplified feedback upon the very definition of what it means to be rational, not to mention what it means to be human. With increased dependence on the computer to carry out all manner of economic activities, it has and will redound back upon the very meaning and referent of the term "economic" as well.

WHERE IS THE COMPUTER TAKING US?

The core doctrines of the orthodoxy of neoclassical economics in the second half of the twentieth century were never quite as stable as they have appeared to those credulous souls gleaning their economics from textbooks (or, sad to say, from most standard works in the history of economics). The dominance of the Cowles version of Walrasian general equilibrium, busily promoting an agent who looked like nothing so much as a structural econometrician of the 1960s, gave way to the "rational expectations" approximation to information processing in the 1970s (Sent, 1998); and this, in turn, gave way to a "strategic revolution" in the 1980s, consisting primarily of dispensation with passive optimization for the rigors of the hermeneutics of suspicion after the manner of Nash game theory; by the 1990s, econometrics became increasingly displaced by experimental economics as the empirical procedure of choice by the avant-garde; and dissatisfaction with much of the accelerated obsolescence sweeping economic theory induced the appearance of a *nouvelle vague* associated with the Santa Fe Institute (see Mirowski, 1996; Arthur, Durlauf, & Lane, 1997) and often retailed under the rubric of "computational economics." One can imagine many alternative ways to account for these shifts in enthusiasm amongst the cognoscenti: some sought to preserve a maximum of continuity inherent in doctrines from one shift to the next, insisting upon some untouched hard core of neoclassical commitments; some simply reveled in the pop-cultural ethos of new toys for new boys, seeing each new mathematical artifact as inherently progressive; others greeted each new development as finally promising release from the conceptual tangle that had strangled the previous neoclassical tradition.

While perfectly comprehensible as the kind of spin that accompanies any promotional campaign in a consumer culture, these accounts are all irredeemably shortsighted.

This book takes a different tack. It seeks to frame each of these transformations as halting, incomplete accommodations to a larger complex of cyborg innovations, extending well into the next millennium. To frame the thesis with maximum irony, the serried ranks of orthodoxy of microeconomics have been imperfectly shadowing the trajectory of John von Neumann's own ideas about the most promising prospects for the development of formal economics, from their early fascination with fixed points and the linear expanding economy model, through game theory (and the red herring of expected utility), and finally (as we have been foreshadowing in Chapter 7) coming to invest its greatest hopes in the theory of automata. This scientific titan who could only spare a vanishing fraction of his intellectual efforts upon a science he regarded as pitifully weak and underdeveloped has somehow ended up as the single most important figure in the history of twentieth-century economics. This mathematician who held neoclassical theory in utter contempt throughout his own lifetime has nonetheless so bewitched the neoclassical economists that they find themselves dreaming many of his formal models, and imperiously claiming them for their own. This polymath who prognosticated that, "science and technology would shift from a past emphasis on subjects of motion, force and energy to a future emphasis on subjects of communications, organization, programming and control," was spot on the money. The days of neoclassical economics as protoenergetics (Mirowski, 1989a) are indeed numbered. Any lingering resemblances should be chalked up to nostalgia, not nomology.

But I should open myself to a well-deserved charge of inconsistency if I opted to frame this narrative in such a purely personalized manner. It was not the *person* of John von Neumann who was capable of mesmerizing generations of neoclassical economists to mince about like marionettes. He may at some points have resembled Darth Vader, but he could never have been mistaken for being Gepetto. (The intervention in helping bring together Cowles with the military and RAND, however, stands out in our history as a singular exception.) Everything written in the previous chapter about the deliquescence of individual selves as self-sufficient protagonists in our postmodern world would contradict that plot line. Rather, the computer (or, more correctly, the computer-plus-human cyborg) has stalked the dreams of each succeeding generation of economic sleepwalkers; and the computer continues to exercise its dormative sway over economists, otherwise swelled with drowsy pride over their personal innovative accomplishments. Without that protean machine, it would have been highly unlikely that the history of neoclassical economic theory would ever have taken the course that it did after World War II. In the absence of that thing-without-an-essence, cyborgs would not have infiltrated economics (and the other social sciences) in successive waves. Without the computer, it would still be obligatory to bend a knee to the mantra that economics really was about "the allocation of scarce resources to given ends" and not, as it now stands, obsessed with the conceptualization of the economic entity as an information processor.

Previous chapters have been almost entirely concerned with relating the narrative history of this epoch-making departure; now the opportunity has arrived in this final chapter to face forward briefly into the next century and speculate about the ways in which this overall trend might be extrapolated. The most clipped and concise way to accomplish this is to ask, What does it mean to claim there either now exists or shortly will materialize a coherent "computational economics"? The query is neither as straightforward nor as technocratically routine as it might initially seem. After all, given the attitudes documented herein, isn't the median attitude of the median economist: von Neumann is dead; long live von Neumann?

Admittedly there are now published many journals with titles like *Computational Economics* and *Netnomics*, and there are courses with these titles offered at many universities. There is a Society for Computational Economics, and a Society for Social Interaction, Economics and Computation, each with its own annual conclaves, awards, honorific offices, and all the other trappings of academic professionalism.[1] There also exists a substantial array of survey articles and books from many different perspectives.[2] There subsist the conferences and summer schools associated with the Santa Fe Institute, which is blessed with a cyborg history and relationship to economics that holds its own fascination, but which we must regrettably pass by in this venue.[3] Yet, amid this veritable beehive of cyberactivity, what seems altogether notable by its absence is any comprehensive map or outline of what "computational economics" now is or could ultimately aspire to become. Indeed, in my experience, many of those engaged in activities falling under this rubric rather enjoy their (sometimes undeserved) reputation as dweeby "nerds" who can be distinguished from the more general populace by their high-tech obsessions. The last thing

1 I should admit that I am a member of both organizations and have attended their meetings. The web site for SIEC can be found at ⟨www.econ.unian.it/dipartimento/siec⟩. A guide to various research groups roughly falling under these rubrics is maintained by Leigh Tesfatsion at the Web site ⟨www.econ.iastate.edu/tesfatsi/allist.htm⟩.

2 See, for instance, Bona & Santos, 1997; Judd, 1997, 1998; Velupillai, 2000; Amman, Kendrick, & Rust, 1996.

3 Some background on the history of the Santa Fe Institute can be gleaned from Waldrop, 1992; Helmreich, 1995; Bass, 1999; and the back issues of the *Bulletin* of the institute.

they would ever be caught dead doing would be dabbling in something as un-geek and un-chic as history or philosophy. Yet, one shouldn't write this off simply to bland self-confidence or technohubris: there persists the problem that the "computer" as an artifact has been changing so rapidly, and the dot-com financial mania has so abruptly transformed the commercial landscape, that to engage in free-form speculation about the impact of the computer on economics is to compound the ineluctable with the ineffable.

Thus, it falls to the present author to proffer some suggestions about what might just be the once and future impact of the computer upon how we think about the economy. Think of it, if you will, as what a cyborg does after a long hard day of information processing – time to go into sleep mode and access a few machine dreams.

FIVE ALTERNATIVE SCENARIOS FOR THE FUTURE OF COMPUTATIONAL ECONOMICS

Just as all really good science fiction is rarely more than a tentative and inadequately disguised extrapolation of what is recognizably current experience, prognostications about the future of economics should possess a firm basis in what is already recognizably ensconced on the horizon. William Gibson's *Neuromancer*, far from describing an alien posthuman world, bore the unmistakable shock of recognition to such a degree in 1984 that it has lately begun to curl with age, if only around the edges. To aspire to a similar degree of prognosis, I shall proceed to describe four different versions of computational economics that are firmly grounded in the existing academic literature, and then give some reasons for thinking that they have not yet been adequately thought through, much less provided with fully fledged coherent justifications. The fifth version of computational economics will prove an epsilon more insubstantial, but will nevertheless be firmly grounded in the previous narrative, because it most closely resembles the ambitions of John von Neumann for economics at the end of his career. Once delineated and distinguished, it will be left for you, the reader, to assess the odds and place your bets on the next millennium's Cyborg Hambletonian.

Judd's Revenge

A peculiar breed of economists is afoot in America. They think the economics they learned back in graduate school really is a Theory of Everything and thus can compensate for any deficiency of concrete knowledge of history, other disciplines, other people, or indeed anything else about the world they putatively live in. For them, the computer, like everything from falling birthrates to the fall of the Wall, stands as just one more confirmation of their poverty-stricken world view:

> Being economists, we believe the evolution of practice in economics will follow the laws of economics and their implications for the allocation of scarce resources. . . . In the recent past, the theorem-proving mode of theoretical analysis was the efficient method; computers were far less powerful and computational methods far less efficient. That is all changing rapidly. In many cases the cost of computation is dropping rapidly relative to the human cost of theorem-proving. . . . The clear implication of standard economic theory is that the computational modes of theoretical analysis will become more common, dominating theorem-proving in many cases. (Judd, 1997, p. 939)

In this particular mind-set, the computer isn't really capable of transforming anything, since it is just a glorified calculator. Ever cheaper number crunchers will only diminish the number of theorems proved in the *AER* and *JET* at the margin. Who ever thought that a mere monkey wrench could derail the market in its ongoing revelation of the Rational? We will call this blinkered vision of the future "Judd's Revenge," named after Kenneth Judd.[4]

This version of the future of computational economics will prove most congenial for a neoclassical economist to imagine, mainly because it combines adherence to the absolute minimum of concessions to the transformative power of the computer upon economic thought, with the maximum commitment to the maxim that tools, and not ideas, are the coin of the realm in economics. In brief, this position is built around the premise that *whatever* sanctioned neoclassical model one might choose to work with, its prognostications will be marginally improved by rendering it more explicit as a computer program, and then having the computer help the analyst calculate specific numerical results of interest dictated by the prior model, be they "equilibria" or boundary conditions or time paths of state variables. One must be very careful *not* to automatically equate the activities of this school with the building and calibration of so-called "computable general equilibrium models," if only because justifications of these latter models tend to migrate rather indiscriminately between the first three categories of computational economics herein enumerated. (The content-tool distinction turns out to be hard to maintain in practice, even in this instance.) Nevertheless, in rhetoric if not in technical practice, a computable Walrasian general equilibrium model exemplifies the style of theorizing most commonly favored by this cadre. The standard rationale of Judd's Revenge is that most formal neoclassical models are either too large (in principle), too nonlinear, or too "complex" to be solved analytically. The computer is conveniently

[4] Kenneth Judd: Ph.D., Wisconsin, 1980; professor, Northwestern, 1981–86; Hoover Institution, 1986–; coeditor, RAND *Journal of Economics*, 1988–95.

trundled onto the stage as the spiffy tool that will help us to do more of what it was that economists were doing at Cowles in the 1960s and 1970s – mostly optimization of one flavor or another – but not really to do it any differently. The computer as glossed here is closer kin to the improved buggy whip and not the horseless carriage.[5] This is economics as if John von Neumann (and most of the events recounted in this book) had never happened. The premier exponent of this version is Kenneth Judd (1997, 1998), although it is actually quite commonplace.

One can readily appreciate the attractions of such a position for those whose primary concern is to maintain the appearance of continuity in economic theory at any cost. The problem persists, however, that this version of computational economics must repress so much that has happened, that it is not at all clear why anyone should think they could cobble together some sort of interesting intellectual career out of such paltry materials. Not only are most of the postwar changes that have happened in the microeconomic orthodoxy rendered so bland and featureless as to be invisible, with any residual idea of cumulative progress left high and dry; but, furthermore, the only source of intellectual excitement available to this sort of economist is the thrill of coming up with a niftier (faster, cheaper, cleaner) algorithm than their peers. This, it should go without saying, is software engineering pure and simple, and *not* what most economists think of as doing economics. Other sorts of ancillary functions in the history of economics – and here one thinks of national income accounts generation, or enterprise accounting, or the development of econometric software – have rapidly been relegated to subordinate positions, and farmed out as distinct separate professional identities. Advocates such as Judd have never seen fit to provide an explanation why one should not, at minimum, expect the same to happen with the objects of their enthusiasm. In other words, why won't microserfs in their little cubicles be churning out these algorithms on an industrial scale sometime soon? It is difficult to regard this version of computational economics as little more than a briefly passing phase.

[5] There is also a literature in the same spirit as this work that imagines all manner of complex scenarios for "optimal" pricing schemes for Internet usage, marginal pricing and Lindahl pricing for routers and TCP protocols, and all manner of other "solutions" to problems that have yet to arise on the Web. From the vantage point of the present narrative, these have been little more than the wet dreams of neoclassical economists stymied by their painful irrelevance when it comes to almost everything that has happened since the first demonstration of ARPANET in 1972 (Hafner & Lyon, 1996; Abbate, 1999). If we actually believed neoclassical models, then the Internet should have never even come into existence.

Lewis Redux

Kenneth Judd betrays no outward sign of being aware that the theory of computation could be used to make theoretical criticisms of the practices and concepts of neoclassical microeconomics. Others (in increasing order of skepticism), from John Rust to Marcel Richter to Kumaraswamy Velupillai, are vastly more sensitive to this possibility. This awareness, as we observed in Chapter 6, dates back to Rabin's (1957) critique of game theory, but reached a new level of sophistication with the work of Alain Lewis. At various junctures, Kenneth Arrow has sounded as though he endorsed this version of computational economics: "The next step in analysis . . . is a more consistent assumption of computability in the formulation of economic hypotheses. This is likely to have its own difficulties because, of course, not everything is computable, and there will be in this sense an inherently unpredictable element in rational behavior" (1986, p. S398). The touchstone of this version of computational economics is the thesis that the nonrecursiveness of most neoclassical models does stand as a fundamental flaw in their construction, and that the flaw is serious enough to warrant reconstruction of the Walrasian approach from the ground up. Lewis's letter to Debreu, quoted in Chapter 6, remains the most elegant summary of this position:

> The cost of such an effectively realizable model is the poverty of combinatorial mathematics in its ability to express relative orders of magnitude between those sorts of variables, that, for the sake of paradigm, one would like to assume to be continuous. In exact analogy to the nonstandard models of arithmetic, the continuous models of Walrasian general equilibrium pay for the use of continuity, and the "smooth" concepts formulated therein, with a certain non-effectiveness, that can be made precise recursion-theoretically, in the realization of the prescriptions of the consequences of such models. By the way, if ever you are interested in an analysis of the effective computability of the rational expectations models that are all the rage in some circles, it would not be hard to make the case that such models are irrational computationally. . . . When I first obtained the result for choice functions, I thought my next task would be the reformulation of *The Theory of Value* in the framework of recursive analysis. I now have second thoughts about the use of recursive analysis, but still feel that a reformulation of the foundations of neoclassical mathematical economics in terms that are purely combinatorial in nature – i.e., totally finite models, would be a useful exercise model-theoretically. If successful, then one could "add on" more structure to just the point where the effectiveness goes away from the models. Thus we ourselves could effectively locate the point of

> demarcation between those models that are realizable recursively and those which are not.[6]

Despite some isolated attempts to recast the Walrasian system (or individual versions of Nash games) in recursive format, by and large they have not been accorded much in the way of serious attention within the economics profession; moreover, Lewis's suggestion of isolating the exact threshold where one passes over from recursivity to nonrecursiveness has attracted no interest whatsoever. In all fairness, it is doubtful that any such boundary would itself be computable. The situation at the end of the 1990s inherent in this version of computational economics is much more curious than all that.

First, what we tend to observe is that most of those in the forefront of this movement – one thinks of Arrow, for instance – never actually delve very deeply into the uncomputable bedrock of rational choice theory. While they might acknowledge some undecidability results at a relatively higher level of the theory – say, at the level of collective choice and the aggregation of preferences – they never concede that there would be deep impossibilities for anyone to even possess the capacity for constructing their own neoclassical preference function. Instead, orthodox figures such as Arrow or Marcel Richter or John Rust tend to speculate in vague ways how some future developments – maybe technological, maybe evolutionary – elsewhere in the sciences will someday break the dreaded deadlock of noncomputability for orthodox economists. Beyond the rather ineffectual expedient of wishing for pie in the sky, this literature is itself deceptive, for as we noted in Chapter 2, Turing noncomputability is a *logical* proposition and not predicated upon the unavailability of some scarce resource, and therefore it is not subject to be offset by any future *technological* developments, no matter how unforeseeable or unanticipated. In a phrase, someday in an advanced technological future it is conceivable that many computational problems will become less intractable, but it is far less likely that anything demonstrably formally uncomputable will be rendered computable. The careless conflation of *intractability* (NP-complete, NP hard problems) with *noncomputability* (undecidability on a Turing machine) under some generic rubric of computational complexity is one of the very bad habits prevalent in this literature.

Second, it seems that another bad habit of this literature is to refer blithely to Herbert Scarf's (1973) algorithm for the computation of Walrasian general equilibria as if it had already successfully carried out

[6] Alain Lewis to Gerard Debreu, December 12, 1985, box 23, KAPD.

the project outlined by Lewis. Here the word "computable" in the phrase "computable general equilibria" has fostered all manner of unfortunate confusions. What Scarf actually did was to devise an algorithm for *approximation* of a fixed point of a continuous mapping of a sphere S_{n+} onto itself under certain limited circumstances. Appeal is then commonly made to Uzawa's (1988) result on the equivalence between the theorem of existence of equilibrium prices for an Arrow-Debreu economy and Brouwer's fixed-point theorem on a vector field of excess demands. There are numerous slippages between this limited mathematical result and the empirical uses to which any such algorithm is put. First off, there is the insuperable problem that the Sonnenschein-Mantel-Debreu results[7] suggest that the fully general Walrasian system imposes almost no restrictions on observable excess demand functions (other than "Walras's Law" and homogeneity of degree zero of demand functions), so the fact you can construct *some* correspondence between a given set of data and some arbitrary Walrasian model chosen from a vast array of candidates is not too surprising. But more to the point, Scarf's algorithm does not *confront* the pervasiveness of noncomputability of the Walrasian model so much as simply work around it. Suppose the fixed point we search for turns out to be a nonrecursive real number – that is, it falls afoul of Lewis's proof. We might, for instance, try and find an approximate recursive real to the original point and construct an exact algorithm to calculate it in a guaranteed fashion; or, alternatively, we might restrict the vector field of excess demands to be defined over the computable reals, and then use an algorithm to determine an exact solution. Scarf's algorithm does neither: instead, it restricts the admissible excess demand functions a priori to one class of functions for which there are exact algorithms to compute equilibrium solutions. It is noteworthy that the meaning of "computable" is treated cavalierly: for Scarf only makes use of classical analysis and nowhere accesses the theory of computation as understood in computer science. Hence, it was possible for Richter and Wong (1999a) to prove that there exist economies for which there is no applicable Scarf algorithm. Indeed, "if Scarf's approximation technique were truly computable, it would yield a computable fixed point. However, Scarf uses a process that is non-computable in general" (p. 10).

Scarf, like Cowles in general, opted to deal with fundamental paradoxes of Bourbakism by simply looking away. In saying this, there is no attempt to indict him for duplicity: indeed, he was just importing some standard OR programming techniques back into neoclassical economics: he in effect was just applying a simplex algorithm to a grid. The lightly disguised

[7] See Sonnenschein, 1972, 1973; Mantel, 1976; and Debreu, 1974. A very good nontechnical summary of this literature is Rizvi, 1998.

dependence upon OR was another Cowles predilection that had paid off handsomely in the past.

Simulatin' Simon

The two previous scenarios are relatively easy to summarize because they take it for granted that nothing substantive in the neoclassical orthodoxy need be relinquished in order for economists to come to a satisfactory accommodation with the computer. In a sense, they assume the position: if classical mechanics was good enough for dear old Dad, therefore it's good enough for me. By contrast, from here on out, the world should be sharply divided into those who still take nineteenth-century physics for their exemplar of a supremely successful scientific method, and those who believe that profound changes in the relative standings of the natural sciences in the later twenteith century make it imperative to look to biology – and, in particular, the theory of evolution – for some cues as to how to come to a rapprochement with the computer. All of our subsequent protagonists do this to a greater or lesser degree. Our latter three versions of computational economics all find it salutary to make reference in one way or another to evolution as part of their program of reconciliation, although it may appear that each individual case may hew to a version of evolution that would be regarded as idiosyncratic and unreliable by their economic competitors, not to mention real biologists.

For many economists, the fugleman figure of Herbert Simon best represents the progressive future of computational economics. As we observed in the previous chapter, Simon's quest is to avoid explicit consideration of the formal theory of computation and instead to build computer simulations of economic and mental phenomena, largely avoiding prior neoclassical models. It cannot be chalked up to convenience or incapacity on his part, because it is the outcome of a principled stance predicated upon his belief in bounded rationality: there is only a certain "middle range" of observed phenomena of a particular complexity that it is even possible for us mere mortals to comprehend; and since reality is modular, we might as well simulate these accessible discrete subsystems. Because the computer is first and foremost a symbol processor, in his estimation, Simon believes the computer constitutes the paradigmatic simulation machine, both capturing our own limitations and showing us a more efficient means for doing what we have done in a less than stellar manner all along. Some day, these modular algorithms may be loosely coupled together to form a massive theory of the economy (or the brain) on some giant megacomputer; but in the interim, this simulation activity really is the best we can do, and thus is an end in itself. This humble prescription dovetails with Simon's own conception of biological evolution as piecemeal engineering, because he thinks that is what happens in Nature, as well as in the world of intellectual discourse.

A moment's meditation will reveal just how influential Simon's vision has been for computational economics (even though individual economists may be unfamiliar with the exact philosophical underpinnings). Computer simulations compose the bulk of all articles making reference to computation appearing in postwar economics journals. Simon's own "behavioral theory of the firm" (Cyert & March, 1963) was an early attempt to link computer simulations to empirical study of firm activities. When Richard Nelson and Sidney Winter (1982) heralded their novel "evolutionary economics," it consisted primarily of computer simulations of firms not so very far removed from those of Simon.[8] Whenever economists have made reference to cellular automata exercises, as in the case of Schelling (1969), they have in fact been engaging in wholesale simulation, not addressing the formal theory of automata. Simulation is a technique with a long established history in operations research and organizational studies (Prietula, Corley, & Gassar, 1998). The popularity of the Santa Fe Institute with its advocacy of the nascent field of artificial life has only enhanced the credibility of various economic simulation exercises emanating from that quarter.[9] Even the staid Brookings Institution felt it had to jump aboard the simulation bandwagon or else be left behind in the slums of cyberspace (Epstein & Axtell, 1996).

Computers do tend to foster a certain cultural atmosphere where simulations become much more common and therefore tend to appear more unexceptional, a case made by thinkers from Baudrillard (1994) to Edwards (1996) to Webster (1995). But once one gets over the frisson of

[8] For an early critique of the rather cavalier treatment of simulation in Nelson & Winter, 1982, see Mirowski, 1988, chap. 9. The early neoclassical attempt to appropriate "evolutionary" language to defend the program should be situated within the context of the early dispute between Cowles and Chicago over simulation versus "as if" theorizing. This accounts for the fact that most of the "seminal" attempts to recruit evolution began at RAND: Alchian, 1950; the early work of Nelson and Winter; Hirshleifer, 1977; and Houthakker, 1956.

[9] A very readable and entertaining popular introduction to the mysteries of artificial life is Levy, 1992; the book on Santa Fe that captured the attention of economists was Waldrop, 1992. The literature of Artificial Life has unearthed all the half-buried controversies over the nature of simulation and its place in science; for some sources on this mare's nest of conundra, see Boden, 1996; Helmreich, 1995; Langton, 1994, 1995; Langton et al., 1992. Some examples of the impact of artificial life on economics can be sampled in Lane, 1993; Friedman & Rust, 1993; Arthur, 1993; Palmer et al., 1994. One advantage that economics has over evolutionary biology in this regard is that it is entirely plausible to question whether little simulated organisms are really "alive," whatever that means; but it is less likely to question whether "artificial economies" are real when they are turned into technologies for conducting real on-line markets.

unreality, there still remains the nagging problem of the evaluation of the efficacy and quality of simulation in the conduct of scientific inquiry. Does the computer simulation differ in some substantial way from the process of mathematical abstraction when one constructs a "model" of a phenomenon? For all Simon's evocative appeal to his prospective "sciences of the artificial," he does display a predisposition to conflate the two distinct activities in order to justify computer simulation. An interesting alternative to Simon's own justification for the benefits of simulation can be found in Galison (1996). In that paper, he argues that early computer simulations in the construction of the atomic and hydrogen bombs were first motivated as bearing a close family resemblance to actual experiments, but experiments where controls could be more thoroughly imposed. Over time, however, neither the mathematicians nor the bench experimentalists fully accorded computer simulations complete legitimacy within their own traditions, regarding them as too farfetched, so a third category of specialization grew up, with its own lore and its own expertise, which served as a species of "trading zone" (in Galison's terminology) that mediated some research interactions of mathematical theorists and particle experimentalists.

Whatever the relevance and aptness of Galison's story for physics, it does appear dubious when applied to economics. First, contemporary advocates of economic simulation don't seem to mediate much of any interaction between theorists and empiricists, at least in orthodox precincts. Not only is the division of labor substantially less pronounced in economics than in physics, but the lines of communication between diverse specialists are more sharply attenuated.[10] Furthermore, simulation practitioners in economics have a lot to learn when it comes to protocols and commitments to reporting the range of simulations conducted, as well as following standard procedures innovated in physics and elsewhere for evaluating the robustness and uncertainty attached to any given simulation. Forty years on, the first complaint of a member of the audience for an economic simulation is, Why didn't you report that variant run of your simulations? Where is the sensitivity analysis? How many simulations does it take to make an argument? One often finds that such specific criticisms are often parried by loose instrumentalist notions, such as, "we interpret the question, can you explain it? as asking, can you grow it?" (Epstein & Axtell, 1996, p. 177). On those grounds, there would never have been any pressing societal need for molecular biology, much less athlete's foot ointment.

I would like to suggest a different set of criteria for the evaluation of simulations in economics. To set the record straight, computer simulations

[10] This point is documented in Mirowski, 1994d, and Mirowski & Sklivas, 1991.

never can and will never be banished from economics. Simulations will always be an important accessory to the spread of the cyborg sciences. Indeed, as we have argued, they are an indispensable component of the program, since they serve to blur the boundaries between digital and worldly phenomena. However, simulations will only be dependably productive in economics when they have been developed and transformed from the status of *representations* to the status of *technologies*. We have already witnessed this sequence of events more than once in this volume. For instance, early simulations of the staffing of air defense stations were transformed into training protocols for enlisted recruits; and then they became test-beds upon which further technological scenarios could be played out in preparation for choice of prospective R&D trajectories. Simulated control exercises became templates for further automation of actual command and control functions. They also provided inspiration of an entirely new class of technical developments later dubbed "artificial intelligence" in Simon's own research. Or again, as we observed in the previous chapter, simulated play of games by automata transmute into the design of "autonomous artificial agents" concocted to conduct real transactions over the Internet. The lesson of these examples is that simulations become fruitful when they are developed and criticized to the point that they can become attached in a subordinate capacity to some other activity, which is not itself a simulation. To put it bluntly, outside of some future prospective consumer markets for Sony Playstations and Nintendo virtual reality boxes, most simulations do not stand on their own as intellectual exercises or compelling catechesis, or at least not without a well-developed history of separate professional specialization in computer simulation, such as that found in particle physics. Here Simon and his minions have regrettably stopped well short of realizing the full potential of computational economics.[11]

Dennett's Dangerous Idea

The most common popular conception of computers at the end of the century (as Turing so accurately predicted) is of a machine who thinks. Because the computer so readily trespasses upon the self-image of man as the thinking animal, it has become equally commonplace to believe that the mind is nothing more than a machine; that is, it operates like the computer. Such deep convictions cannot be adequately excavated and

[11] This observation extends to Simon's recent project of mechanizing the very process of scientific research itself through simulation of various landmark "discoveries," described in Langley et al., 1987; Simon, 1991b. This project is subjected to critical scrutiny in Sent, 2001.

dissected in this venue;[12] but one must concede that this ubiquitous package of cybernetic preconceptions has profound implications for what a "computational economics" may come to signify in the near future. Although Herbert Simon is considered one of the progenitors of the field of artificial intelligence, it is of utmost importance to understand that he neither promotes a global unified computational model of the mind nor regards the neoclassical economic model as a serious or fit candidate for such a mental model. Others, of course, have rushed in to fill this particular void. A whole raft of self-styled "theorists" – although scant few empirical cognitive scientists among them – proclaims that it is possible to access some algorithms from artificial intelligence, combine them with a particularly tendentious understanding of the theory of evolution, and arrive at a grand Theory of Everything, all to the ultimate purpose of maintaining that all human endeavor is constrained maximization "all the way down." One infallible earmark of this predilection is an unaccountable enthusiasm for the writings of Richard Dawkins. The theory of rational choice (perhaps simple optimization, perhaps game theory) is unselfconsciously treated as the very paradigm of information processing for biological organisms and machines; consequently both computers and humans are just a meme's way of making another meme. Although one can find this hyperphysical sociobiologized version of economics in the works of a broad range of economists from Jack Hirshleifer (1977, 1978) to Kenneth Binmore (1998b), perhaps the most comprehensive statement of the approach can be found in the popular book by the philosopher Daniel Dennett, *Darwin's Dangerous Idea* (1995).

Dennett may not himself have become embroiled in much in the way of explicit economic prognostications, but that does not preclude him from retailing a certain specific conception of the economic as Natural common sense. "So there turn out to be general principles of practical reasoning (including, in more modern dress, *cost-benefit analysis*), that can be relied upon to impose themselves on all life forms anywhere" (1995, p. 132). "The perspective of game-playing is ubiquitous in adaptationism, where mathematical *game theory* has played a growing role" (p. 252). "Replay the tape a thousand times, and Good Tricks will be found again and again, by one lineage or another" (p. 308). However, once one takes the death-defying leap with Dennett and stipulates that evolution is everywhere and always algorithmic, and that memes can freely soar beyond the surly bounds of bodies, then one can readily extrapolate that the designated task of the economist in the bungee-jump is to explore how the neoclassical

[12] Some of the better meditations upon this incredibly contentious issue are Sherry Turkle in Sheehan & Sosna, 1991; Born, 1987; Button et al., 1995; Agre, 1997; Crevier, 1993; Penrose, 1989; Searle, 1992; Shanker, 1995; Collins & Kusch, 1998.

instantiation of rational economic man "solves" all the various optimization problems that confront him in everyday economic experience. Neoclassical economics is cozily reabsorbed into a Unified Science that would warm the cockles of a Viennese logical positivist.

The reader may be tempted to enter a demurrer: isn't this version of computational economics really just the same as our second option? Indeed not: the drama is in the details. In Lewis Redux, the analyst sets out from a standard neoclassical model and subjects it to an "effectiveness" audit using the theory of computation. Here, by contrast, the analyst starts out with what she considers to be a plausible characterization of the cognitive states of the agent, usually co-opted from some recent enthusiasm in a trendy corner of current artificial intelligence, and rejoices to find that neoclassical results can be obtained from a machinelike elaboration of agent states, perhaps with a dollop of "evolution" thrown into the pot. The literature on finite automata playing repeated games was one manifestation of this trend in economics (Kalai, 1990); the recent enthusiasm about what has been dubbed "evolutionary game theory" (Mailath, 1998; Samuelson, 1997; Weibull, 1997) is another. Much of what has come to be called "behavioral economics," in either its experimental (Camerer, 1997) or analytical (Rubinstein, 1998) variants, also qualifies. It is fast becoming the technique of choice at the economics program at the Santa Fe Institute (Lane, 1993; Casti, 1997a; Helmreich, 1995). More baroque manifestations opt for appropriation of some new strain of artificial intelligence in order to construct economic models of agents, be it genetic algorithms (Marimon, McGrattan, & Sargent, 1990; Arifovic, 1994; Arifovic & Eaton, 1995), neural nets, or fuzzy sets. The key to understanding this literature is to note that once "algorithmic reasoning" attains the enviable state of ontological promiscuity, then any arbitrary configuration of computers is presumed fair game for economic appropriation, as long as they arrive eventually at what is deemed to be the "right" answer. The distinctive move within this tradition is to make numerous references to an agent's mental operations as being roughly similar to some aspect of what computers are thought to do, but simultaneously to studiously avoid making reference to any computational theories: computers computers everywhere, but never a stop to think. The rationale behind this awkward configuration of discourse should by now have become abundantly apparent: no one here wants to openly confront the noncomputability of basic neoclassical concepts.[13]

[13] Dennett (1995, chap. 15) again provides an instructive parallel. He finds that he must strenuously deny any relevance of Gödel's Theorem to either Artificial Intelligence or Darwinian evolution, if only to maintain his Panglossian adherence to Universal Optimization. Some philosopher of economics might someday find it an entertaining exercise

There are one or two reasons, over and above crude recourse to bait-and-switch tactics, for thinking that this brand of computational economics probably does not possess a bright future or real staying power. One drawback is that the ambivalence on the part of most economists in forging a mutual alliance with artificial intelligence in the era of its chastened retreat from prior extremes of shameless hubris and undelivered-upon promises has been palpable. AI has apparently lost the knack of blinding people with science, at least for now. Economists have not historically been notably willing to ally themselves with crippled or flagging research programs in the natural sciences; they have been predictable dedicated followers of fashion. This must be compounded with the fact that prominent figures in artificial intelligence, such as Simon and Minsky, have not been all that favorably inclined toward neoclassical economics.

One other drawback is that most orthodox economists' level of faithfulness to the formal requirements of a theory of evolution is downright louche. For instance, as argued in the preceding chapter, to what economic phenomenon does the indispensable "replicator dynamics" refer in evolutionary game theory? In exercises with genetic algorithms such as that found in Sargent, 1993, do the individual strings of code refer to different ideas in the mind of a single agent, or is the pruning and winnowing and recombination happening in some kind of "group mind"? (Neither seems entirely correct.) Is anyone really willing to attest to the existence of any specific "meme," so that we would know one when and if we saw it? And then there is the dour observation that exercises in slavish imitation of AI and ALife have been produced at Santa Fe and elsewhere for more than two decades, and nothing much has come of them. But the bedrock objection can be stated with brutal simplicity: how likely is it that any economists will *ever* make any real or lasting contribution to cognitive science? And let's be clear about this: we are talking here about people trained in the standard graduate economics curriculum.[14] I have repeatedly posed this question to all types of audiences, running the gamut of all proportions of neoclassical skeptics and true believers, and not once have I ever encountered someone who was willing to testify in favor of the brave prospect of economists as the budding cognitive scientists of tomorrow. As they say in the math biz, QED.

to compare Dennett's position on "the intentional stance" with Milton Friedman's doctrine of "as if" maximization, if only to come to see that pop philosophy often only recapitulates pop economics with some long and variable lags.

14 Some whose grasp on history is less than secure might be tempted to counter with the example of Herbert Simon; but they neglect the fact that his Ph.D. was in political science. It might also be interesting to consider the extent to which "behavioral" experimental economics is disparaged or simply ignored by psychologists and cognitive scientists alike.

Vending von Neumann

There remains one final possibility, albeit one for which there is very little tangible evidence in the modern economics literature,[15] that economic research could attempt to conform more closely to von Neumann's original vision for a computational economics. The outlines of this automata-based theory were first broached in Chapter 3. Von Neumann pioneered (but did not fully achieve) a *logical* theory of automata as abstract information processing entities exhibiting self-regulation in interaction with their environment, a mathematical theory framed expressly to address the following questions:

Von Neumann's Theory of Automata

1. What are the necessary prerequisites for self-regulation of an automaton? [A: The von Neumann architecture for the sequential digital computer.]
2. What are the formal prerequisites for self-reconstruction of an abstract automaton of the same given level of complexity? [A: Von Neumann's theory of cellular automata, reliability, and probabilistic automata.]
3. Does a universal automaton exist that can construct any other given automaton? [A: Yes, Turing's notion of a universal computer.]
4. Are there regularities to formal methods of resistance to noise (entropic degradation) in the process of self-replication of automata?
5. Under what conditions can it be demonstrated that an automaton can produce another automaton of greater computational complexity than itself? In other words, what are the abstract preconditions for the possibility of evolution?

As one can observe from the absence of complete answers to some of these questions, von Neumann did not manage to bequeath us a comprehensive and fully articulated theory of automata. For instance, he himself was not able to provide a single general abstract index of compu-

[15] Perhaps the single-most visible exception to this generalization is the work of Peter Albin (1998) and Duncan Foley's introduction to that volume. Readers of this work will realize that I profoundly sympathize with almost all of their motivations, but that I differ with them about where and how one should introduce the theory of automata and complexity considerations into economics. Their predilection to regard this as an *epistemic* question for cognitive models of the agents, and not as a question of the evolution of market institutions, insufficiently differentiates their research from that found subsumed under option 4.

tational complexity; while some advances (described in Chapter 2) have been made in the interim, there is still no uniform agreement as to the "correct" measure of complexity (cf. Cowan et al., 1994). Furthermore, some aspects of his own answers would seem to us today to be unacceptably arbitrary. For instance, many modern computer scientists do not now believe that the best way to approach questions of evolution is through a sequential von Neumann architecture, opting instead to explore distributed connectionist architectures. Other researchers have sought to extend complexity hierarchies to such architectures (Greenlaw, Hoover, & Ruzzo, 1995). Still others speculate on the existence of computational capacities "beyond" the Turing machine (Casti, 1997b). Nevertheless, the broad outlines of the theory of automata persist and remain clear: computer science (from Rabin & Scott, 1959, onward) has been structured around a hierarchy of progressively more powerful computational automata to undergird "a machine theory of mathematics, rather than a mathematical theory of machines" (Mahoney, 1997, p. 628). Computational capacity has been arrayed along the so-called Chomsky hierarchy of language recognition: finite automata, pushdown automata, linear bounded memory automata, and, at the top of the hierarchy, the Turing machine.

The problem facing economists seeking to come to grips with the theory of computation has been to work out the relationship of this doctrine to a viable economic theory. Due to parochial training or stubbornness or historical amnesia, they have been oblivious to the possibility that John von Neumann did not anticipate that his theory be subordinated to explication of the psychological capacities of the rational economic agent, or even to be appended to Nash noncooperative game theory to shore up the salience of the solution concept. This would misconstrue the whole cyborg fascination with *prosthesis*. Instead of economic theorists preening and strutting their own mathematical prowess, what is desperately needed in economics is a "machine theory of mathematics," or at least an answer to the question, *Why is the economy quantitative?* especially given the irreducibly diverse and limited mathematical competence in any population. As we have argued throughout this volume, von Neumann consistently maintained that his theory of automata should be deployed to assist in the explanation of *social institutions*. If social relations could be conceptualized as algorithmic in some salient aspects, then it would stand to reason that institutions should occupy the same ontological plane as computers: namely, as prostheses to aid and augment the pursuit of rational economic behavior. The theory of automata would further instruct us, however, that these were prostheses of an entirely different character than hammers, and more Promethean than fire: they have the capacity to *reconstruct themselves* and to *evolve*. In the same sense that there could exist a formal theory of evolution abstracted from its biological constituent components (DNA, RNA, etc.), there could likewise exist a formal theory of institutions abstracted from their fundamental intentional constituents (namely, the psychological makeup of their participants).

Thus, von Neumann sought to distill out of the formal logic of evolution a theory of change and growth of sweeping generality and broad applicability. At base, very simple microlevel rule structures interact in mechanical, and possibly even random manners. Diversity of microlevel entities stands as a desideratum in order for this interaction to produce something other than meaningless stasis. Out of their interactions arise higher-level regularities generating behaviors more complex than anything observed at lower microlevels. The index of "complexity" is here explicitly linked to the information-processing capacities formally demonstrable at each level of the macrostructure; for our purposes, and although von Neumann did not propose it, this means the Chomsky hierarchy. Von Neumann justified the central dependence upon a computational metaphor to structure his theory of evolution because, "of all automata of high complexity, computing machines are the ones we have the best chance of understanding. In the case of computing machines the complications can be very high, and yet they pertain to an object which is primarily mathematical and we can understand better than most natural objects" (1966, p. 32). It might be prudent to realize that, instead of repeating the dreary proleptic Western mistake of idolizing the latest greatest technological manifestation of the Natural Machine as the ultimate paradigm of all law-governed rationality for all of human history, computers can better be acknowledged as transient metaphorical spurs to our ongoing understanding of the worlds that we ourselves have made.[16] Because the computer refuses to hold still, so too will the evolving theory of automata resist coming to premature rest.

So this clarifies von Neumann's projected role for the computer within the confines of a modern formal social theory; but it still does not illuminate the question of to which *entities* the overworked term "institutions" should refer in this brand of economics. What is it that economics should be *about*? Note well that whatever constitutes the vehicle of economic activity, they should be entities that can grow, reproduce, and evolve. Unfortunately, this was something concerning which von Neumann left little or no guidance; in each of the first four computational options presented in this section, the answer has always been presumed to

[16] The problem of the Western penchant for approaches to conflating Nature and Society is discussed in greater detail in Mirowski, 1994a, 1994c, as well as in Pickering, 1992; Latour, 1993, 1999; Robertson et al., 1996; and de Marchi, 1993.

be the rational economic agent herself, something that his heritage of skepticism would tend to rule out of consideration. It thus falls to the present author to suggest that the appropriate way to round out von Neumann's vision for economics is to construe *markets* (and, at least provisionally, not memes, not brains, not conventions, not technologies, not firms, and not states) as formal automata. In other words, the logical apotheosis of all the various cyborg incursions into economics recounted in this book resides in a formal institutional economics that portrays markets as evolving computational entities.

The alien cyborg character of this research program may take some getting used to for the average economist. Comparisons of markets to machines are actually thick on the ground in the history of economics, but the notion that one should acknowledge this construct as providing a heuristic for mathematical formalization seems mostly repressed or absent.[17] Markets do indeed resemble computers, in that they take various quantitative and symbolic information as inputs, and produce prices and other symbolic information as outputs. In the market automata approach, the point of departure should be that there exists no single generic "market" in any economy, but rather an array of various market algorithms differentiated along many alternative attributes – imagine, if you will, a posted-offer market, a double-auction market, a clearinghouse market, a sealed-bid market, a personalized direct-allocation device – and, moreover, each is distinguished and subdivided further according to the types of bids, asks, and messages accepted, the methods by which transactors are identified and queued, the protocols by which contracts are closed and recorded, and so on. In the market automata approach, it is deemed possible (to a first approximation) to code these algorithmic aspects of the particular market forms in such a manner that they can be classified as automata of standard computational capacities and complexity classes. The initial objective of this exercise is to array the capacities of individual market types in achieving any one component of a vector of possible goals or end states, ultimately to acknowledge that market automata are *plural* rather than singular because of the fact that no single market algorithm is best configured to attain all (or even most) of the posited goals. This stands in stark contrast to the neoclassical approach, which has the cognitive agent commensurate and mediate a

[17] "The economic system can be viewed as a gigantic computing machine which tirelessly grinds out the solutions of an unending stream of quantitative problems" (Leontief, 1966, p. 237), harkening back to the socialist calculation controversy discussed in Chapter 5. Some rare forays into an explicit automata conception of markets include (Miller, 1986, 1996; Domowitz & Wang, 1994). One might also cite Commons, 1934, as a forerunner to this approach.

range of diverse goals through the instrumentality of a single unique allocation device called "the market."[18] Diversity in markets is the watchword of the von Neumann automata approach, for both theoretical and empirical reasons. The bedrock empirical premise is that markets are and always have been structurally and functionally diverse in their manifestations in the world; only the neoclassical tradition and the field theory metaphor found itself driven to imagine a single generic market present throughout human history. The guiding theoretical watchword is that there can be no evolution without variability of the entities deemed to undergo descent with modification.

Once one can manage to entertain the gestalt switch from a single omnipotent market to a plurality of markets of varying capacities and complexities, then attention would immediately turn to the project of taxonomizing and organizing the categories. It gives one pause to come to realize how very little attention has been accorded to the taxons of market forms in the history of economic thought. Under the impetus of some recent developments recounted in the next section, there has begun to appear some groundbreaking work in differentiation of market algorithms, sometimes under the finance rubric of "market microstructure theory,"[19] and in other instances under the banners of experimental economics or evolutionary economics. It is thought-provoking to read there the nascent beginnings of attempts to produce "family trees" of various formats of interest, such as that reproduced as Figure 8.1. Although its author does not make anything of it, the resemblance of the diagram in Figure 8.1 to a phylogenetic tree is undeniable: a device commonly used to represent descent with modification in evolutionary biology. This diagram does not adequately capture any such phylogeny – indeed, its orthodox provenance has left the tradition that spawned it bereft of any means of judging whether one abstract market form could descend from another, much less the historical curiosity to inquire about whether it actually occurred or not.

My contention is that the market automata approach does provide the wherewithal to prosecute this inquiry. Once the particular algorithm that characterizes a particular market format is suitably identified and represented as a specific automata, then it becomes feasible to bring von

[18] In a slogan, neoclassical economics is like a promoter of Esperanto, whereas the von Neumann approach is very like a Chomskean linguist. Both believe in some fundamental unity inherent in the diversity of spoken languages; only the former simply thinks it can be imposed from above, whereas the latter seeks to distill the posited unity mathematically out of the phenomenological diversity. Of course, we don't presume to pronounce upon the success of either linguistic program here.

[19] See, for instance, O'Hara, 1995; Miller, 1990; Domowitz, 1990, 1993b; Clemons & Weber, 1996; Spulber, 1999; and Steven Wunsch's "Auction Countdown" at ⟨www.azx.com/pub⟩.

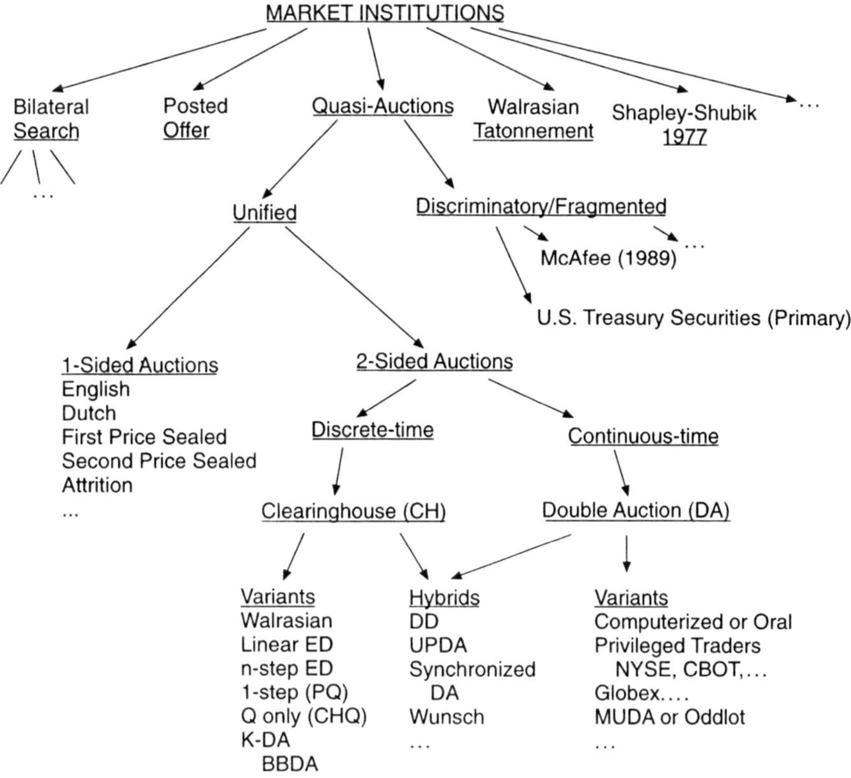

Figure 8.1. Market taxonomy. *Source*: Friedman & Rust, 1993, p. 8.

Neumann's full-fledged project back into economics. The agenda would look something like this. Starting from scrutiny of the algorithmic representation, the analyst would enquire whether and under what conditions the algorithm halts. (Markets do sometimes spin out of control, failing to produce plausible prices.) This would encompass questions concerning whether the algorithm arrives at a "correct" or appropriate response to its inputs. Is the primary desideratum to "clear" a market in a certain time frame, or is it simply to provide a public order book in which information about outstanding bids and orders is freely available? Or, alternatively, is it predicated upon a simple quantifiable halting condition, such as the exhaustion of arbitrage possibilities within a given time frame? Is it configured to produce prices of a certain stochastic characterization? Some would insist instead upon the attainment of some posited welfare criterion. The mix of objectives will be geographically and temporally variable: the very hallmark of an evolutionary process.

Next, the Neumannian analyst would gauge the basic computational capacity of the specific market format relative to its identified objective or objectives. Is the market a simple finite automaton, or perhaps something classed as more powerful, approaching the power of a Turing machine? If it attains such power, can it be further classified according to the computational complexity of the inputs it is prepared to handle? Then one might proceed to compare and contrast market automata of the same complexity class according to their computational "efficiency" by invoking standard measures of time or space requirements (Taylor, 1998). Once the process of categorization is accomplished, the way is then prepared to tackle von Neumann's ultimate question: namely, under what circumstances could a market of a posited level of complexity give rise to another market format of equal or greater complexity? In other words, in what formal sense is market evolution even possible?

It may be here, at the portrait of one specific market automata giving rise to another, that economic intuition may falter, or perhaps sheer incredulity at the prospect of posthuman cyborgs undergoing parthenogenesis across the cyberscape may stultify analysis. For what could it mean for a market automaton to "reproduce"? This is where the abstract essence of the computational approach comes into play. Concrete market institutions spread in an extensive manner by simple replication of their rules, say, at a different geographical location. This would not qualify as von Neumann reproduction, because it was not the market algorithm itself that was responsible for producing the duplicate. Rather, market automata "reproduce" in this technical sense when they are able to simulate the abstract operation of other markets as a subset of their own operation – that is, they can perform all the calculations and operations of the other market.

An intuitive understanding of this process of simulation as assimilation can be gleaned from a familiar market for financial derivatives. When agents trade in a futures market for grain contracts, they expect the collection of their activities to simulate the (future) outputs of a different distinct market, namely, the spot market for the actual grain. It becomes pertinent to note that frequently the spot market (say, an English auction) does operate according to a certifiably different algorithm than does the futures market (say, a double auction or dealer market); here, in the language of von Neumann, one automaton is "reproducing" an altogether different automaton. It may also be germane to note that the automaton may be simulating the activity of another market automaton of the same type, or, even more intriguingly, an automaton of higher complexity is simulating a market of lower complexity, as may be the case with the futures emulation of the spot market for grain. This very self-referential aspect of market automata suggests the relevance of machine logic for market operations, conjuring the possibility of a hierarchy of computational complexity, and opening up the prospect of a theory of evolutionary change. For while it may be formally possible for a market automaton of higher complexity to emulate a market of lower complexity, it is not in general possible for the reverse to take place.

Already at this stage we begin to glimpse how a "machine theory of mathematics" can begin to reorient the practice of economics. Turing machines may ideally possess an infinite "tape" or memory, but they are restricted to a finite number of internal states as well as a finite alphabet. The motivation underpinning this requirement is that we are enjoined to adhere to a constructivist approach to mathematics, the very antithesis of the Bourbakist approach favored at Cowles and discussed in Chapter 6. For technical reasons, it follows that our automata must be restricted to a discrete alphabet or, when calculating, restrict its operations to the set of countable (natural) numbers. Far from being an awkward nuisance, this constitutes the first sharp empirical prediction of the market theory of automata. An empirical regularity concerning the history of markets is that prices have always and everywhere been expressed as rational numbers – that is, as ratios of natural numbers – and, further, have been denominated in monetary units that are discrete and possess an arbitrary lower bound. This empirical regularity is not simply due to "convenience" or some virtual cost of calculation; it is a direct consequence of the algorithmic character of markets. The penchant of the Cowles economist for the real orthant is entirely a figment of the imagination, nurtured by Bourbakist ambitions for spurious generality. Markets and their participants don't care if some economists are talented enough to pine after the Fields medal; and it's a good thing, too. Markets must operate in an environment of limited and fallible calculators; they will only persevere if they prosecute constructive mathematics.

In the theory of market automata, many economic terms will of necessity undergo profound redefinition and revalorization. Henceforth "the market" no longer refers to a featureless flat environment within which agents operate; rather, there is posited an ecosystem of multiform diversity of agents and cultures in which markets ply their trade of calculation and evolve. Moreover, perhaps for the first time in the history of economics, a theory of markets exists that actually displays all three basic components of a modern theory of evolution: a mechanism for inheritance; explicit sources of variation; and one or more processes of selection.[20] In barest outline:

[20] See, for instance, Sober, 1984; Levins & Lewontin, 1985. The self-described "evolutionary economics" of Nelson & Winter, 1982, lacks one or more of the components, as first pointed out by Mirowski, 1988. "Routines" neither qualify as mechanisms of inheritance

1. Market automata "reproduce" as algorithms by the process of emulation and subsumption of algorithms of other market automata. They are physically spread and replicated by the initiatives of humans propagating their structures at various spatiotemporal locations for diverse commodities.
2. Market automata are irreducibly diverse due to their differing computational capacities, and to the diversity of the environments – namely, the goals and objectives of humans – in which they subsist. Sources of variation can be traced to the vagaries of recombination – glitches in one market automata emulating an automata of a different type – and a rare kind of mutation, where humans consciously tinker with existing market rules to produce new variants.[21]
3. Market automata are "selected" by their environments – ultimately by their human participants – according to their differential capacity to "reproduce" in the von Neumann sense, which means performing calculations that dependably halt and displaying the capacity to emulate other relevant market calculations emanating from other market automata. The theory presumes that the process of selection is biased in the direction of enhanced computational complexity, although here, as in biology, the jury is still out on this thesis. Because there is no unique function or purpose across the board which a market may be said to exist "for" in this schema, there is no privileged direction to evolution in this economics.

This sketch of the von Neumann alternative for computational economics has occupied slightly greater space than the prior alternatives, if only because of its relative absence in the postwar economics literature. This ephemeral quality might be understood in two ways. In the first, one might opine that its relative underdevelopment was deserved, citing all the rationales tendered when one tends to ignore inchoate research programs: it is vague, poorly linked to empirical inquiry, lacks resonance with the

nor exhibit intrinsic variability. Furthermore, their appeal to Lamarckian "theory" cannot belie the fact that if the entity subject to selection displays less inertia than the environment, then there is nothing for "selection" to operate upon. Winter (1987) confuses his own theory's nonconformity with the standard theory of evolution with the absence of an analogue of sexual reproduction in the economy. The former need have nothing whatsoever to do with the latter.

[21] An example of the former would be futures markets not only emulating spot markets, but introducing financial considerations and goals into the simulated operation. An example of the latter might be the proposal of a "second-price auction" by changing one of the rules of a "first-price auction" to award the contract to the second-highest bidder.

concerns of modern social theorists, is outlandish in its metaphorical inspiration, sports a dispiriting air about the human prospect, and so on. The reason that this book is written as *history* is to preclude all such bromides. Rather, its purpose has been to argue that von Neumann's machinic hum has haunted the American economics profession in the postwar period, and that proponents of the neoclassical orthodoxy have repeatedly conceived and shaped their doctrines in reaction against the cyborg imperatives that they think they have detected in von Neumann's work. The automata approach is not the rodomontade of a few marginalized souls easily drowned out by the strident sermons of the neoclassical elect; it is the characteristic allocution of cyborg culture, found throughout fin-de-siècle science. The remainder of this chapter reveals that, if one listens carefully enough, it's basso profundo can be heard echoing throughout the long corridors of some modern developments within the orthodox economics profession. Markets as automata are coming to a workstation near you.

In my experience even sympathetic audiences have difficulty seeing how von Neumann's automata theory is already present in the enthusiasms of modern orthodox economics. In lieu of retailing more lists, the next section briefly outlines a sequence of developments in "experimental economics" and its offshoots, which reveal a gaggle of economists speaking cyborg without being altogether aware of it. The purpose of this one last cybertale is to suggest that potential alliances can be built between von Neumann's program and some subsets of modern economics, were the protagonists so inclined.

THE HAYEK HYPOTHESIS AND EXPERIMENTAL ECONOMICS

The folkways of cyborgs often elude the commonplace understandings of men. We saw in Chapter 6 that one version of the experimental approach to economics was pioneered at RAND by Merrill Flood and his Robot Sociology. Indeed, one could easily make the case, although we shall regrettably decline to do so here, that von Neumann was also indirectly responsible for the genesis of yet another modern thema: namely, what passes as "experimental economics" in our turn-of-the-century economics profession.[22] This is not the way the story is told amongst the experimental

[22] This comment is prompted by the thesis found in Goldstein & Hogarth, 1997, that von Neumann provided the main impetus for the rise of experimental decision theory in the immediate postwar period, and some comments about RAND by Martin Shubik in Smith, 1992. We might go even beyond this and point out that the post-1970 successes of experimental economics were heavily dependent upon the incorporation of computer protocols both to render results fungible between laboratories, and for what Goldstein calls the "gambler metaphor," or the treatment of every psychological question as though

economists and, in particular, by their premier spokesman, Vernon Smith.[23] Our aim in the present venue is not to challenge Smith's history head on, but instead to recount how it was that Robot Sociology was eventually readmitted into good graces in experimental economics after a protracted period of neglect, and how this has had far-reaching consequences for the market automata approach to computational economics.

Smith tells us that he was not originally sold on the virtues of laboratory experimentation in economics, but instead was put off by a classroom exercise in a simulated market conducted by Edward Chamberlin while he was a graduate student at Harvard. Chamberlin had assigned costs and reservation prices on cards to individual subjects and had them mill about the room making trades, only to post the results triumphantly on the blackboard, revealing that the students had not converged to predicted Marshallian price and quantity (see Chamberlin, 1948). Something about the exercise nagged away at Smith, who was convinced that the deck had (literally) been stacked against neoclassical theory, or at least in favor of Chamberlin's theory of monopolistic competition.[24] He pondered how it might be possible to reduce the number of intramarginal trades and induce more price stability in the outcome. From our present vantage point, the relevant lesson to glean is that the Harvard experience had planted the suspicion in Smith that all markets were not all alike, and that it would be prudent to pay closer attention to the specific rule structures that made up the market experiment.

it were a response to a game by an isolated asocial subject. Other evidence for the automata inspiration of economic experiment can be found in Hoggatt, 1959, 1969; Hoggatt, Brandstatter, & Blatman, 1978. Parenthetically, we have in this yet another illustration of the doctrine of unintended consequences: von Neumann was personally never enamored with experimentation as a means of constructing social theory.

[23] Vernon Smith (1927–): B.S. in electrical engineering, Cal Tech, 1949; M.A., Kansas, 1952; Ph.D. in economics, Harvard, 1955; assistant professor, Purdue, 1955–67; consultant, RAND, 1957–59; professor, University of Massachusetts, 1968–75; University of Arizona, 1975–. For his version of the history, see Smith, 1991a, 1992, as well as Roth, 1993. The history of experiment in economics is one of the most sadly neglected areas in the entire corpus of the history of science. Smith himself is aware of the subterranean links of his program to the earlier Cowles fascination with an operations research approach to markets: "The formal study of information systems in resource allocation theory (Hurwicz, 1960) and the laboratory experimental study of resource allocation under alternative forms of market organization had coincident beginnings and, in important respects, have undergone similar, if mostly independent, intellectual developments" (Smith, 1991a, p. 254). This connection is explored in a forthcoming Notre Dame thesis by Kyu Sang Lee.

[24] E-mail from Vernon Smith to Philip Mirowski, November 8, 1999: "I expected to confirm Chamberlin's disconfirmation, thereby getting a stronger, less trivial result. . . . It 'disconfirmed' Chamberlin's 'disconfirmation' – you have that right, but I didn't believe the result, thought it was an accident. But the things that I thought that might have made it an accident were not borne out in subsequent experiments."

In the late 1950s, Smith ran a number of what he originally dubbed "market simulations" in his economics classes, all with the purpose of confirming Chamberlin's disconfirmation. He lit upon a structure based on then current New York Stock Exchange rules, essentially a double-sided sequential auction with a price improvement rule, which did manage to converge to the predicted Marshallian price and quantity in the majority of cases; and, moreover, Smith asserted that its operation resulted in greater "efficiency" of outcomes than that of other formats, in the importantly limited Marshallian sense of realized producer plus consumer surplus (crudely, the area between the left-hand blade-plus-handle of the supply and demand "scissors"). Smith had appreciated early on that experimental design bore more than a passing resemblance to computer programming: "To write the instructions for an experimental task is to define a trading institution in all its mechanical detail" (1991a, p. 109). He parlayed this insight into an actual computer code when he hooked up with Arlington Williams around 1977 to computerize his market experiments. The results then proceeded to display a machinelike dependability in two senses: the protocol could be programmed onto a computer to automate many aspects of the experiment and export it to other laboratories; and, when conducted with real money, the theory would be corroborated whatever the target population – experienced, inexperienced, economics majors, comp lit majors, math wizards, numerically challenged, whatever. Experiments with other rule structures did not turn out nearly so reliably or display such efficiency (Plott & Smith, 1978; Smith, 1991a, pp. 107ff.). Nevertheless, first Smith and then others were so impressed with the robust character of the double-auction experimental protocol that they sought to subject it to all sorts of "stress tests,"[25] and its persistent vaunted reliability prompted Smith to proclaim in the 1980s that this was one of the first and most solid lessons of the burgeoning field of experimental economics.

To that end, in 1982 Smith announced something he called "the Hayek Hypothesis": [experimental protocols of] "Strict privacy [regarding agents' valuation and cost data] together with the trading rules of a market institution are sufficient to produce competitive market outcomes at or near 100% efficiency" (in Smith, 1991a, p. 223). Elsewhere (1991b) he has drawn out what he considers to be the principal morals of the hypothesis: markets "work" because of an "economic rationality" attributed to "institutions," economic rationality is not reducible to individual rationality, economic rationality is neither cognitively nor computationally intensive, and therefore experiments that test economic rationality on an

[25] See Davis & Williams, 1991; van Boening & Wilcox, 1996; Davis & Holt, 1993.

individual level are nugatory. "What is imperfectly understood is the precise manner in which institutions serve as social tools that reinforce, even induce individual rationality" (1991b, p. 881).

Because this assertion that "experiments have shown" that markets work has become hallowed folklore amongst many economists in the absence of any criticism or philosophical reflection, it might behoove us to clarify briefly what the Hayek Hypothesis does and does not state. First off, there is almost no textual support in Hayek for much of what is being asserted by Smith about his experiments, beyond some generic political endorsements that the market does indeed "work," and that social and legal institutions are necessary prerequisites for its reliable operation. Hayek displayed progressively diminishing interest in defending the neoclassical research program over the course of his career and, in any event, would not have endorsed the explicit Marshallian devices (such as consumer's surplus as welfare index) that are so central to Smith's case. Of course, Smith may just be evoking Hayek in a merely ceremonial manner. Nevertheless, harkening back to Chapter 5, there is something apposite, something exquisitely *right* about taking Hayek's name in vain in this context. For Smith's proclamation is merely the latest continuation of the socialist calculation controversy by other means, those means still largely dictated by the same set of scientific influences bearing down upon Hayek in the 1950s. Fifty years on, the underlying motivation is still: where is the best location to project the metaphor of the computer, and what does that imply for the possibility of planning the operation of the economy? Only now, it would appear that Smith believes the best way to defend what he conceives to be the pith of the neoclassical program from the depredations of its detractors – either those who either bemoan the cognitive limitations of the real-world economic agent, or else disparage the computational limitations to the operation of the Land of Cockaigne ensconced in Walrasian theory – is to imagine the market institution more explicitly as a computer, but the same sort of computer that runs the market experiments in the lab.

Here, however, the neoclassical program comes acropper of the muted appeal to cyborgs by Smith and his minions. For the "defense" of free markets, if that is in fact what is intended, is much more narrowly based and strewn with paradoxes than Smith has yet openly conceded. The qualms begin with an insistence upon some precision concerning the theory that it is claimed is being supported. We have argued in this volume and elsewhere (Hands & Mirowski, 1998) that wholesale rejection of the Marshallian program provided the impetus for the Walrasian neoclassical ascendancy in the postwar United States; nothing revulsed that earlier generation of mathematical economists more than (in their view) the putatively sloppy way in which Marshallians made all sorts of outlandish claims for the significance of their partial equilibrium exercises; and, worse, the way the Marshallians wore their cavalier disregard for obvious objections to their confident welfare evaluations as badges of honor. It is not at all clear that Smith's championing of that very same Marshallian position amounts to something other than beating an unseemly retreat away from everything that had fostered the conditions for the neoclassical orthodoxy to displace its rivals in the first place. Smith sometimes suggests his generalizations about DAs have been experimentally extended to "Walrasian" multimarket situations; but I believe this obscures very real incompatibilities between the Marshallian and Walrasian traditions, as well as imposes a bland Whig interpretation on his own very evident hostility to the Walrasian tradition in his early work. To put this somewhat differently: it is a rare neoclassical economist these days who would recommend a controversial policy change on the basis of some expected improvement in Marshallian consumer surplus. The aura of orthodoxy of Smith's "defense" of neoclassical economics actually draws succor from confusing the Marshallian theory in introductory economics textbooks with what a graduate-trained economist is supposed to believe at this late hour.

There is an even deeper sense in which this defense is incoherent. In the laboratory, Smith takes great precautions to render his subjects as close to neoclassical agents as inhumanly possible: reservation prices and costs are imposed; controls are exerted so that the subjects both know and concentrate upon the data that the experimenter deems relevant; "strict privacy" means that social interaction and communicative interchange is reduced to the stark minimum; if the rules are violated the data are discarded; subjects are schooled in the protocols of machinelike behavior by being run through preparatory conditioning sessions on the work-stations. Once you understand everything that is involved with the imposition of Smith's experimental protocols, it seems more than a little disingenuous then to bemoan that it is "imperfectly understood" the ways in which these experimental markets "induce individual rationality." These experiments don't test neoclassical theory (as Smith sometimes will admit); they try and impose mechanical rationality upon subjects who might not otherwise display it.[26] They try, but never entirely succeed, because full

[26] Smith disputes this characterization in his e-mail to the author, November 8, 1999: "people come off the blocks as better do it yourself machines, than when aided by machines in DA trading. You have it turned upside down. . . . There have been far more oral DAs run than computerized. This error comes from only working with the published sources. Oral DA is not news, and is unpublishable."

neoclassical rationality is unattainable on a Turing machine.[27] Yet the bold assertion that the market works takes it for granted that the full panoply of decision-theoretic algorithms are firmly in place. If they are not really all there, if people are just a little inconsistent or confused or fickle, then most of the criteria used by Smith to evaluate equilibrium and efficiency in his markets are left without foundations. Hence, his statements about the absence of any need for individual rationality to make neoclassical stories work are simply groundless. One can't help but suspect that the existence of the computer in the lab has served to distract attention from paradoxes arising from the imperfect cognitive instantiation of the computer in the head.

Next, even if we ignore the previous caveats and concede that every empirical generalization that Smith reports for his lab subjects holds true (a tall order), it would still be the case that his Hayek Hypothesis is caught in a cyborg conundrum. For his vaunted demonstration of economic rationality at the machine-institution level only holds true for a very restricted version of the double-auction procedure – an institutional protocol with extremely circumscribed prevalence in the world outside the laboratory, however popular it may be within its confines. Smith's program seems to be trapped in a rhetorical bind: experiments suggest that "markets" really are diverse and different, but the need to preserve his bona fides dictates the generic mantra that "The Market" works goes unchallenged. It seems he can't decide if he wants to identify markets with institutions as normally understood by economists, or with their older supra-institutional status. Hence, what can only charitably be described as a local regularity is being blown all out of proportion into a lawlike generalization. The lack of self-awareness of this problem can be thrown into sharp relief by considering what is perhaps the second most popular market format amongst the experimental economists after the double auction, namely, the so-called posted-offer market.

The posted-offer market (unlike the double auction) is familiar to most readers as what they encounter every week at their local grocery store. Sellers post a fixed price for each item, and buyers can take it or leave it.

[27] The situation is even more complicated than the fact that the Lewis noncomputability results are effectively avoided through the presumed "unimportant" easing of the cognitive demands placed on the experimental subjects, such as restriction to whole numbers or simple fractions, limitations upon the number of commodities dealt with simultaneously, blocking out "extraneous" considerations. In an underappreciated article, Williams (1980) admits that there are measurable differences between market experiments run on "oral-auditory" versus "written-visual" computerized protocols. This important cognitive regularity has disappeared from the experimentalists' tradition, as has the dirty secret that "outrageous errors are more common in computerized markets" (Charles Holt in Kagel & Roth, 1995, p. 369).

If Piggly Wiggly won't provide a good price, perhaps the Tesco or the Carrefour down the lane will. Although prevalent throughout the developed world, experimentalists can barely repress their disdain for the posted-offer market. In their view it is "inefficient" relative to the double auction, it outputs prices that are "too high" for competitive equilibrium, it inordinately slows convergence to competitive equilibrium, and if it has a competitive equilibrium, it unfortunately does not coincide with the Nash equilibrium (Smith, 1991a, p. 300). Less frequently admitted is the fact that in the laboratory, playing the role of the "buyer" in a posted-offer setting is dead boring: so much so that, early on, most experimentalists decided to save some money by zeroing out the human buyers in their posted-offer experiments (Kagel & Roth, 1995, p. 381; Davis & Holt, 1993, p. 177). That is, they replaced all the buyers with computer programs! Sometimes they tell their experimental subjects they are playing simulated buyers, and sometimes not. The point here is not that there is something illegitimate about introducing simulations into experimental situations – this is indeed the trademark cyborg innovation – but rather that the Hayek Hypothesis is a house built on sand. If the posted offer is so inferior, so boring, so utterly trivial, and yet so ubiquitous, shouldn't those interested in singing hosannas to The Market buckle down and explain its prevalence? Furthermore, shouldn't any comparisons of the efficacies of different small-m market formats be expressly carried out in the two-by-two experimental design of human double auctions with or without computer participants versus human posted offer with or without computer participants? Although all the cells of that research design have yet to be sufficiently filled in, it was a foregone conclusion that the Hayek Hypothesis would eventually summon up a revival of Robot Sociology; and that is precisely what happened.[28]

GODE AND SUNDER GO ROBOSHOPPING

One research project highlighted by Vernon Smith's championing of his Hayek Hypothesis was to explain the observed superior efficacy of the double auction in the vast array of market experiments, auction and otherwise, that had been performed over more than two decades. Smith, as we saw, was predisposed to attribute their clockwork efficacy to the market institution as such, and not to concede much to the cognitive abilities of the experimental subjects. But if economists were avoiding

[28] Eugene Ionesco is reported to have once said that the ideal staging of his "farces tragiques" would involve the actors really turning into literal puppets at the end of the performance, with their arms and legs flung down to the floor. He sought "not to hide the strings, but make them even more visible." One thinks in this regard in particular of M. Smith in *La cantartrice chauve*.

cognitive science like castor oil, how would this question ever get resolved? It would turn out to be a stroke of genius to realize that the practical means to separate out the two classes of causal factors and subject them to individual test had already been pioneered within the experimental tradition and, in particular, by the use of simulated buyers to run posted-offer experiments. The experimental economists who first came to this realization were Dhananjay Gode and Shyam Sunder; and for this reason alone their 1993 paper, "Allocative Efficiency of Markets with Zero-Intelligence Traders: Markets as a Partial Substitute for Individual Rationality," will certainly go down in the annals of cyborg economics as a late-twentieth-century classic.

The paper was perhaps all the more effective because Gode and Sunder (henceforth G&S) were transparently not wholehearted converts to the cyborg cause. Beginning with the very title ("Partial Substitute"), their paper is studded with all sorts of qualifications and reassurances customized for the neoclassical audience of the *Journal of Political Economy,* amounting to assurances that their findings would not derange the stolid complacency of the neoclassical economist. Indeed, a number of those assertions made therein are demonstrably false but ultimately irrelevant to their argument, serving merely as window dressing to lull the reader into a state of placid calm, as a prelude to ushering them into the topsy-turvy world of cyborg roboshopping.[29] For what G&S came to appreciate was that Vernon Smith's experimental protocols were really attempts to discipline his subjects to act like machines, but that there would always be some residuum that would escape comprehensive control – "It is not possible to control the trading behavior of individuals" (1993, p. 120). So why not take the logic of Smith's method to the limit, and have robots – or, more correctly, algorithms – conduct the trades in the double-auction experimental setup? Other than exhilarating in the sheer cyborg delight in blurring the boundaries between man and machine, their exercise had the compelling virtue of more directly addressing Smith's

[29] Or else simply get it past the referees at the *JPE*. The most egregious example of their method of ambagious justification is their assertion that their results support Gary Becker's 1962 assertion that one could still derive neoclassical demand curves from random choices on the part of "irrational" agents. First off, Becker's assertion is itself wildly overstated, as demonstrated by Hildenbrand, 1994, pp. 34–35. Secondly, G&S don't "derive" any demand curve whatsoever, but simply impose it a priori, after the manner of Vernon Smith. Third, both Becker and G&S present themselves as "saving" Marshall: Becker attempts it by suggesting demand curves exist independent of the cognitive states of agents while ignoring how The Market works its magic, whereas G&S show that one particular set of market rules concerning process produce the same regularities predicted by Marshallian theory with minimal restrictions on agent cognition. Far from supporting one another, the two approaches are chalk and cheese.

contention that it was the market "institution" that was primarily responsible for the vaunted efficiency of double-auction outputs.

To that end, G&S concocted what they called "Zero-Intelligence Agents" to participate in double auctions; for convenience, in this volume we will call them "zoids." These agents, stripped of their agency, were just bits of software, whose sole purpose in ALife was to submit messages to the software used in market experiments that normally came from human subjects. What a zoid does, in its own inimitable fashion, is come up with a bid price or ask price for a single unit of a nondescript unspecific "commodity" (what else would a zoid want to buy, anyway?). Because it is a little challenged in the intelligence department, the zoid draws this bid or ask randomly from a uniform distribution of prices, which are bounded for the buyer zoid by the numbers 1 and its preassigned reservation price, and for the seller zoid by the preassigned cost and an arbitrary number bound.[30] Because the prior imposition of reservation prices and cost assignments are standard experimental protocol from Smith onward, in this respect, at least, zoids are being treated symmetrically with human subjects in experimental economics. The difference comes in what happens next. G&S then further differentiate two classes of zoids, which we will dub "A-zoids" and "B-zoids." A-zoids (or affectless zoids) have no further function in ALife other than to mechanically spit out bids or asks, with absolutely no further consequences for their own behavior. B-zoids (budgeted zoids), by contrast, come equipped with the further capacity to stay within their prior specified budget constraint; this is accomplished essentially by shrinking the support of their probability distribution as a consequence of their actually being chosen within the auction setup to complete a transaction.[31] It's not much of a brain, but

[30] The precise manner in which the support of the probability density is defined makes a big difference for the outcome of the exercise, as does the exact sequence of steps in the recognition and processing of individual bids and asks. For instance, since zoids are so zoidlike, the market program itself "chooses" the lowest of all zoid bids and the highest of all zoid asks in a given interval, rather than the zoids themselves obeying the improvement rule. G&S have not always been forthcoming in sharing the details of their computer codes for their zoids; but Shyam Sunder has made one version of the zoids available on the Web at his ⟨http://zi.gsia.cmu.edu/sunder⟩. Matthew Weagle has coded his own version of their exercise and consequently discusses many of the fine points of zoid construction in Weagle, 1996.

[31] The claim that B-zoids have "budgets" is one of the least examined aspects of G&S's exercises. In fact, what they have are shrinking probability supports and a no-loss constraint – not a preset budget that restricts the very opportunity to engage in further profitable trades. As explained in Mirowski, 1989a, neoclassicals are notoriously lax when it comes to invariance principles. For more on this, see Weagle, 1996. For these and other reasons, the description of G&S's results in the text should not be taken as endorsement of all their claims.

computational theory would alert us to recognize that augmentation of a memory capacity, however rudimentary, betokens a serious promotion up the scale of complexity. Once the marketplace is populated with the zoids, then G&S had the inspired idea to let the zoids trade amongst themselves on a computerized double auction and have human subjects trade on the exact same computerized double-auction market, and then compare the outputs.

The results for A-zoids were essentially what you would expect: brain-dead and affectless, A-zoids provide yet another illustration of the old computer adage, "garbage in, garbage out." Double auctions with A-zoids did not converge to predicted price and quantity but only emit noise. The shock came with the B-zoids. In that setup, price and quantity *did* converge to Marshallian predictions; more unexpectedly, measures of Marshallian efficiency also approached the high levels observed in human experiments! In some setups, the human outputs and B-zoid outputs were more or less indistinguishable in broad statistical terms. Realizing that this verged on the assertion that there is no detectable difference between zoids and humans, G&S recoiled in horror from the implications of their experiments: "Note that we do not wish to argue that human traders behaved like [zoids] in the market. They obviously did not. Human markets exhibit a pattern of lower efficiency in the first period, followed by higher efficiency in later periods" (Gode & Sunder, 1993, p. 133 & n). No elaborate quantitative statistical evidence was provided to back up this claim, however.[32] The terror before which G&S stood transfixed was not that economists would begin to confuse human beings with their bits of code (was it?), but rather that it was all too easy to draw some devastating morals from their exercise about the utter futility of more than a century of neoclassical economic theory. While in 1993 they tended to echo Vernon Smith's assertion that the double-auction framework had managed to induce "aggregate rationality not only from individual rationality but also from individual irrationality" (p. 136), the prognosis could just as easily have been that "aggregate rationality" had no relationship to anything that the neoclassicals had been trumpeting as economic rationality for all these years. If anything, they backpedaled more awkwardly over time as the bad

[32] Vernon Smith also disputes this interpretation of the G&S results: "We ran some experiments in which one human traded in an environment of robots, and the human beats the pants off the robot. The idea that robots can do as well as humans (except in providing expert assistance to check mechanical errors) is still (circa 2000) a pipe dream whether we are talking about markets or natural language. This is the fundamental flaw in game theory: Nash and subgame perfect equilibrium are for robots who alone can implement dominance and backward induction." E-mail to author, Nov. 8, 1999. The unpublished experiment that Smith mentions does not really address the issue of efficiency raised in the text, however.

news began to sink in: "We assume simple trader behavior for tractability, not to challenge or criticize utility maximization" (1997, p. 604). This even extended to the prospect that the rational economic man threatened to disappear into a welter of cyborg machines and institutions: "Market rules are a consequence of individual rationality because they evolve out of individual choices over time" (1997, p. 606). And yes, Virginia, economic valuations can all be resolved into embodied labor time, because some human was required to initiate all economic activity.

If Vernon Smith lifted the lid just a crack, G&S have pried the top off Pandora's Box, and no amount of posturing will disguise the fact that cyborgs and zoids and all manner of other hybrids now prowl the marketplace. Machines of middling capacities can produce market regularities, something rendered salient through agent simulations. The question now should be, What is the real lasting significance of the demonstrated result that the specific market format identified as the most efficient, the most efficacious, the most powerful embodiment of the neoclassical ideal of the Market, produces its hallowed results in experimental settings with severely impaired robots? Whatever could the overworked term "rationality" mean in this most artificial of environments? We might venture the observation that, whereas G&S have demonstrated that they possess prodigious instincts for concocting the most edifying thought experiments for provoking wonder and delighting the mind, their capacities for drawing out the implications of those exercises for the practice of economics nowhere rises to the same level of mastery. For those of us who still believe in the virtues of the division of labor, this should not cause anyone to lose any sleep. What is important to take into account is that their ceremonial homage to the neoclassical program has stood as an insuperable impediment to construction of a satisfactory theoretical account of this very intriguing finding. Zoids do not and cannot tell us anything about human cognitive capacities; nor do they illuminate awkward neoclassical notions of "institutions." Zoids are automata, interacting through algorithms running on other automata. The ghost of von Neumann has been whispering through G&S, whether they were aware of it or not.

The attempt to elevate the double-auction setup as the best instantiation of the generic neoclassical Market commits the fallacy of a logical conflation of the part with the whole; the attempt to isolate realization of Marshallian consumer-producer surplus as the *ne plus ultra* of all market activity only misrepresents the telos of the neoclassical model. These are incidental mistakes, artifacts of their origins in Smith's experimental economics, and once they are dispensed with, it becomes possible to recast G&S's research program as fitting snugly within the framework of markets as evolving automata. For once we reorient attention to the fact that markets are modular plural automata, a thesis already implicit in the experimental economics literature, then it follows we can come to entertain the notion that there are multiple, possibly incommensurate, quantitative objectives, within which individual market formats exist in order to satisfy. The masterful stroke of separating out the computational capacities of markets from the cognitive attributes of their human participants, so crucial to the procedures of G&S, only serves to open up the possibility from within the orthodox neoclassical literature that the market algorithms themselves can be sorted, graded, and analytically distinguished by their computational regularities. Something approaching this option clearly has already occurred to G&S, since in their next major paper (1997), they do precisely that. There they seek to attribute the capture of what they call "allocative efficiency" – really, Marshallian surplus – by B-zoids in their roboshopping activities to different configurations of market rule structures.

One endearing attribute of the G&S research program is that it encourages novel perspectives on age-old economic questions by making use of the essential modularity of programming in the von Neumann architecture.[33] Was it the people, or was it the market that accounted for the regularities of the experimentalists' double auction? Answer: Introduce zoids, and conceptually separate out pure market effects. But the double auction is itself an agglomeration of a number of individual component rules in the form of algorithmic specifications. What exactly is it about the double auction that interacts so efficiently with zoids? Answer: Break the market software down further into relatively separable subroutines, let the zoids interact on each, and factor out their distinct effects. But then, what are the relevant building blocks that comprise the double auction, and into which the rules should be decomposed? Answer: Other similarly coded market formats. Significantly, G&S concede that the latter two questions cannot be conveniently asked within any cognitive framework, and therefore (unwittingly?) find themselves venturing further down the road toward von Neumann's version of market automata.[34] In their 1997

[33] If the reader interprets this statement as suggesting that the market automata approach might encounter conceptual obstacles if situated within a connectionist setting, then they would probably be on the right track.

[34] "Allocative efficiency should not be confused with 'information efficiency' discussed in the accounting and finance literature" (Gode & Sunder, 1997, p. 603); "if the trading rules are smart, the traders need not be" (p. 623). Actually, there lurks a paradox in the G&S approach: if they are really serious about abjuring any dependence upon cognitive attributes of the traders, then it follows that their favorite Marshallian measure of "efficiency" loses all rationale. After all, does a zoid really "care" if it managed a trade at less than maximum consumer surplus? While this festers as an inconsistency in the G&S approach, it has no bearing upon von Neumann market automata, which have no commitment to any such measure of market efficiency.

paper, they posit a sequence of nested subroutines – a "Null market" that randomly allocates zoid asks to zoid buyers; a sealed-bid auction, which is comprised of the Null plus a price priority and binding contract rules; the preceding format augmented with a call provision that aggregates bids and asks over a fixed time frame; a double auction with one round; and a multiple-round double auction. What turns out to be fascinating about this exercise is that G&S had to decide where the boundaries of the various subroutines-rules should be located, and which subsets of rules could be abstracted away from for the purposes of their inquiry. Nothing in the neoclassical microtheory tradition would begin to inform their choices in this regard; and therefore, much of what they did decide was driven more by the exigencies of the experimental software than by any empirical description of any real-life market "in the wild" (as experimentalists are wont to say). Nevertheless, resorting to both simulation and some probabilistic analysis, G&S claim to demonstrate that the sealed bid algorithm results in a gain of Marshallian efficiency over the Null market; the call provision itself can build in a modicum of efficiency; and the repeated double auction routine guarantees even a higher level of efficiency. As they summarize:

> Efficiency is lower if (1) traders indulge in unprofitable trades; (2) traders fail to negotiate profitable trades, and (3) extramarginal traders displace intramarginal traders. Voluntary exchanges among traders who have the judgment to avoid losses eliminate the first source of inefficiency. Multiple rounds of bids and asks . . . reduce inefficiency due to the second source . . . the binding contract rule and the price priority rule . . . discriminates against extramarginal bidders because their redemption values are low, and their lower bids are given lower priority. (1997, pp. 622–23)

It will prove imperative to recast their summary with an enhanced modicum of precision and specificity. What G&S have shown is that there exists an implicit hierarchy of market rule structures that can be ranked according to *one and only one* (and a rather tendentious one at that) quantitative index of market "success"; furthermore, resorting to simulations with zoids, one can attribute relative proportions of improvement in performance in this single index to sequential augmentation of the market algorithm with more powerful rules, primarily involving the enforcement of consequences of zoid losses through memory calls, the consolidation and amplification of price comparisons, and imposition of price improvement and priority rules as a prerequisite of participation. (Given some rather astringent restrictions they felt they were impelled to make upon the versions of market algorithms considered, G&S had very little to say about market rules and their impact on quantity variables.) Hence, theirs is ultimately an extremely limited exercise, but no less instructive because of that fact. For, witting or no, what G&S have achieved in this paper is to pioneer an entirely novel approach to the kinds of phenomena that economics might aspire to explain.

If different market formats (or, using Neumannesque language, different market automata) such as the random allocator (Null), the sealed bid, and the double auction can be effectively ranked according to their efficacy in achieving some specific quantitative indicator of success, then would it not be possible likewise to conduct such a comparative exercise with respect to other quantitative indicators? There is no Commandment from On High to be held hostage to Marshallian notions of surplus; therefore, it becomes conceivable instead to resort to other bench marks. Is the objective to "clear" the market in a timely fashion? Or does it simply serve as a public order archive where outstanding bids and offers are smoothly and accurately transmitted to all participants? Is the prevalence of the particular market format predicated upon some other straightforward halting condition, such as the exhaustion of targeted arbitrage possibilities within a specified time frame? Or is it perhaps geared to produce a somewhat more complicated state of affairs, ranging from imposition of some cultural notion of "orderly" price dynamics (no quick drastic price movements, low volatility, low autocorrelation), to maintenance of a situation where firms might more readily acclimatize (if not more directly control) their client base to changes in product life cycles, technological developments, or even scheduled style changes? Indeed, we might subject each market automata to an effectiveness audit, running down the prospective list of all the things we think it is that markets do, and evaluating their computational capacities to attain each objective. Each of these proposed aspects of markets, and many others, have been discussed at one time or another in the history of postwar economics; the problem encountered most frequently was that, in each case, they were considered in splendid isolation, bereft of the insight that different market formats might exist to facilitate differing objectives, and thus blocked from consideration of the thesis that no "generic" market exists that could attain them all equally or adequately. The beauty of G&S's dependence on zoids to provide a base line from which to measure and highlight the effects of differences in market formats is that we can effortlessly dispense with the orthodox presumption that the raison d'être of markets stand or fall on what their idealized unified individual "feels" or "thinks" about them.

The research program herein envisioned, which takes its inspiration from von Neumann, is both extensive and ambitious; if anything, it threatens to appear infinite in all directions, lacking any concrete specificity or paradigm instances of the kinds of analysis that might be conducted under its imprimatur. To counter that hasty impression and to cement the case that modern economists have been propagating cyborg economics without being aware of it, I provide an exercise in the appendix to this chapter that explicitly links the G&S (1997) exercise to the treatment of markets as automata. G&S have shown that, in the presence of zoids and relative to their Marshallian conception of "welfare," the Null market is inferior to the sealed-bid market, which in turn is inferior to the double auction. Hence they propose a nested hierarchy of market algorithms of greater "power" to achieve their specified objective. What I and my coauthor Koye Somefun (1998) have argued is that the abstract G&S characterizations of each of those three formats can be restated in the idiom of the theory of automata and ranked along the Chomsky computational hierarchy. In the appendix, we simplify the exercise to portray both the sealed bid and double auction as pushdown automata with multiple one-way read-only heads; each market is imagined to be an abstract language recognizer that accepts certain stylized messages in the form of bids and asks. The appendix demonstrates that the G&S style two-head pushdown automaton can accept messages characteristic of the sealed bid, but *not* the double auction; whereas the G&S style four-head pushdown automaton can accept messages from both the sealed bid and double auction. Because the sealed bid can be encoded on an automata of lesser computational capacity than the double auction, we interpret this as suggesting the double auction can emulate the sealed bid, but not vice versa. In other words, we have demonstrated the existence of a *computational hierarchy* of market automata that maps directly into G&S's *welfare hierarchy* of market formats. It would thus seem fairly uncontroversial to conclude that the reason the double auction is more efficacious *in this instance* is that it exhibits greater computational power than the sealed bid. Because it can handle greater complexity in the differentiation and coordination of extramarginal traders, it can regularly realize greater Marshallian surplus. It is this general class of comparative argument, although not an endorsement of its idiosyncratic welfare commitments, that we anticipate will become increasingly familiar in a future computational economics hewing to von Neumann's tradition.

This brings us back once more to questions of evolution, so central to von Neumann's vision, and so utterly absent in G&S. The breathtaking lack of curiosity about computation so characteristic of American economics infects their inability not only to detect the mapping of their exercise into computational theory but also to ask what it means to maintain that there exist whole classes of markets that rate as inferior by their own standards. The siren song of the Hayek Hypothesis has done a disservice: it has suppressed contemplation of the conundrum presented to this research community by the proliferation of market forms in the world that do not appear to make the grade, at least by their own criteria. The notion of a nested hierarchy of market forms bears no theoretical salience whatsoever for G&S (and apparently for Vernon Smith); and they don't know what to make of its looming presence in their own work. In the von Neumann program, by contrast, its significance is elevated to prime importance. The fact that a market of lesser computational capacity can be subsumed within the code of another, more powerful market automata as a subroutine is the most important clue that *markets evolve* through emulation and accretion of algorithms. Markets that may appear to be rudimentary and of low computational capacity relative to a given quantitative objective manage to persist because they continue to occupy an effective niche in the larger economic ecology, be it through having proved itself robust to a local combination of diverse environmental demands (i.e., plans and objectives of humans), or else through symbiosis with market forms higher up the hierarchy. The arrow of time is oriented in the direction of the appearance of ever more computationally powerful market automata, but this says very little about the relative prevalence of earlier and simpler market formats over time. Indeed, one might expect to observe some restricted class of specialized markets – and here financial markets spring to mind – outstrip the larger population of market automata in terms of computational improvement and innovation. Automata in some environments undergo accelerated evolution. There may be many sensible explanations of this phenomenon, but none of them will dictate that the great bulk of markets for consumer goods is therefore obsolete or fatally flawed. The great delusion of the neoclassical project from the time of Walras has been to conflate the complexity of financial markets with the complexity of markets as a whole.

CONTINGENCY, IRONY, AND COMPUTATION

I am repeatedly taken aback at the capacity of the narrative recounted in this volume to provoke the most overwrought reactions: outrage, perfervid denial, disgust, but most frequently, the desperate question – are the cyborg sciences a good thing or a bad thing? Is economics finally ascending to the Valhalla of "hard science," or is it skidding wildly down the slippery slope to intellectual decrepitude? Do cyborgs herald a bright future of liberation, or are they just the next phalanx of dull foot soldiers of entrenched power? Aren't machines always co-opted for apologetics? For a long time, I used to plead agnosticism: isn't that your job to judge, dear reader? Of course, it is not as though I had studiously avoided all philosophical investigations in the course of my work. In my previous writings I have tried to explain that recurrent aspirations to the status of science in the modern world are fundamentally underdetermined, and fraught with all sorts of historical contingencies and strange ironies that are rapidly repressed upon the triumph of a research program over its rivals. Appeals to richly evocative metaphors or vague references to grand schemes of conceptual unification have played at least as important a part as bald appropriation of mathematical formalisms or conformity to putative rules of "scientific method"; and the reasons why a particular version of economics gains adherents and accrues societal support may have little or nothing to do with the immediate motivations of the thinkers who are identified as its progenitors. But I find time and again that it is the more expansive notions of the motive forces of *science* that languish unheeded. The historical accident of a great instauration in American science funding and organization from private to military dominance can have profound repercussions upon the very content of selected sciences for decades to come, long after their funding imperatives recede in perceived importance. Some protagonists whose encounter with economics was only fleeting may eventually have greater consequence for its future than those who devote their entire lives to it. Programs which proudly sport unshakable intellectual confidence and self-assurance can crumble overnight, given the right set of circumstances; other programs can blithely limp along in conceptual disarray for generations. All of these things and more are illustrated by events recounted herein.

Nevertheless, the sampling of raw emotions encountered on the road has convinced me that a few parting thoughts on the possible significance of this history are in order, especially given the fact that this history hits home in the way a narrative safely ensconced a century or more in the past cannot. What can it mean for our ideas of economic value to have once again – that is, after the last major shift from substance to field theories of value in the nineteenth century (recounted in Mirowski, 1989a) – undergone utter transubstantiation, this time from the previous neoclassical field definition of economics as "the allocation of scarce resources to given ends" to the cyborg definition of the economy as a giant information processor? What will it mean to experience the formal definition of *value* in economics suffering further dematerialization and abstraction relative to the early physicalist and materialist predispositions of classical political economy? Postwar economists have expressed nothing but disdain for value theory; has it finally given up the ghost?

There are some indications that these sorts of questions are already getting an airing in the popular culture. When the Mitchell Kapoors of the world intone, "It's not a problem of the have and have nots; it is the divide between the know and know nots," they may just be doing something more than vying for laurels as the emperor's new dry cleaner, although irony cannot be altogether ruled out. We have not yet come anywhere near to dispensing with manufacturing and raw materials extraction so we can spend our salad days communing with knowbots, Jeff Bezos notwithstanding. People have not left their lumbering meat machines behind for effortless frolic in cyberspace, nor will they do so anytime soon; but they do presently find themselves increasingly dealing with novel conceptual images and role models of what it means to earn a living and generate economic value in our brave new millennium. Economists, contrary to their own oral tradition, have not managed to relinquish value theory and say good-bye to all that. As this book argues, their notions of value have been repeatedly revised by the cyborg sciences. Because half-interred conceptions of value tend to lie at the root of what most people believe about justice, politics, community, and responsibility, the impact of the history of science upon the history of economic value reverberates far beyond the closed community of professional economists. Hot-button issues like the superiority of market organization, the inviolability of individual freedom, the legitimacy of affluence, the efficacy of democracy, and the grounding of the social in the natural all cannot escape being transformed by such revisions in the deep structure of valuation. But there persists much uncertainty about just how far down the fault lines run. Is it in fact the hidden agenda of this volume to intone, "The handmill gives us society with the feudal lord; the steam engine, society with the industrial capitalist; the computer, society with the military analyst"?

Although the role of the American military is one of the major overlooked factors in the stabilization of the American economic orthodoxy in the postwar period, and goes a long way to explain the fin-de-siècle fascination with game theory, one should not draw stark morals from that fact. It does not follow, for instance, that economics at Cowles harbored inherently sinister connotations; nor does it follow that the C^3I orientation that informs so much of modern economics is unavoidably inimicable to the freedom of citizens of a democratic society.[35] The American military, as we have tried to stress in this volume, was never a monolithic taskmaster in its goals and intentions; and the scientists who accepted its largesse were often themselves consciously at odds with it. Yet, acknowledging the excess parameters of freedom within which they moved, it did make a difference that it was the military (especially the Air Force and the ONR), and not, say, the U.S. Commerce Department, or (after 1945) the Rockefeller Foundation, or even the Catholic Church, that acted as the executive science manager in charge of economics in the

[35] For instance, the fact that the CIA opted to promote Abstract Expressionism as America's answer to Soviet Socialist Realism (Saunders, 1999) in the Cold War battle for hearts and minds need not detract one whit from my conviction that Jackson Pollack was one of the most imaginative painters of the twentieth century.

postwar period. The beneficiaries may protest that they were left alone to do their research as they wished with little or no interference; but we should take their testimony with the same grain of salt that we should imbibe when a child of the affluent suburbs testifies that there was no trace of racial prejudice in postwar America when he was growing up.

Likewise, one should not presume that, because the computer has been so central to the story related herein, there has been something pernicious or perverse about the fact that our own self-understanding has been informed, if not driven, by what is, after all, just a machine. For a person of a certain temperament, the mere suggestion that human inquiry might be strongly conditioned by something beyond its proper subject matter and its nominal objectives is a symptom of the decline of the West, or at least the breakdown of civilized discourse. Economists especially like to imagine themselves as rational self-made men (with a few self-made women thrown in for good measure), self-sufficient and transparently self-conscious and completely aware of every "choice" they have ever made in their careers, agents in intimate control of their every motive; no self-delusions allowed here! The merest whisper of the urgings of a collective unconscious risks toppling them over into a paroxysm of disdain for the "oversocialized individual" of the "soft sciences." To those imperious souls, I might just interject that there is nothing demonstrably fallacious about using machines to pursue our own self-understanding, even if it be in a less than fully self-conscious manner, because resort to metonymy, metaphor, analogy, and all the other props of reasoning are just what one would expect from limited cognizers and easily distracted *bricoleurs* such as ourselves. The "boundedness" of our rationality is expressed not by some imaginary divergence from optimality, but through the means and devices we access to think clearly. Far from regarding rampant computer metaphors as a debased pollution of purity of economic thought, I would suggest that they unpretentiously reprise a set of intellectual practices that have been solidly dominant in Western social thought ever since its inception. Machines, it seems, are good to think with.

There may be those who feel that the appropriate moral to draw from our little history of modern economic thought is to reject science and machines altogether as inspirations for economics. To them I would put the question, Where will you turn at this late date for your economics innocent of all scientism? Where will you find your economic Erewhon bereft of all machines? Scholars are free to seek their inspiration wherever they will, but as a historian, I think it would be unconscionable not to point out that every single school of economics that has ever mustered even a sparse modicum of support and something beyond a tiny coterie of developers in the West has done so by accessing direct inspiration from the natural sciences of their own era and, in particular, from machines. The challenge for those possessing the courage to face up to that fact is to understand the specific ways in which fastening upon the computer, instead of the steam engine or the mechanical clock or the telephone, has reconfigured our options for the development of social theory in the immediate past and the prospective future.

One of the direct consequences of this philosophical stance sketched in this chapter is that much of the *Methodenstreit* over the meaning of the computer will come to a head soon in the impending battle to decide (as has happened roughly once a century) where it is that valuation congeals for the purposes of a science of economics: will it be portrayed as happening within the recesses of the idealized computer situated between the ears of the representative agent? Or will it be staged as the intermediate output of a population of automata called "markets" scattered throughout the landscape of diverse and cognitively opaque human beings? Many will undoubtedly find the former option sweetly resonant with their own humanist commitments, whereas the latter would be deemed so "posthuman" as to be drably dispiriting and dehumanizing. I would implore these interlocutors to pause for a minute and reconsider this position. Is it really the case that further conflation of our somatic selves with machines is really the last best hope for a humanist economics? A humanist social science? Haven't we already caught a glimmer of where that road leads in Chapter 7? Once the computer metaphor goes cognitive, it seems to be as acidly corrosive to the constitution of the individual human being as anything nurtured in the wildest polymorphous fantasies of the postmodernists – selfish genes, devious memes, downloadable souls, other humans as expendable carbon copies of ourselves. The hot deliquescence of the *Homo economicus* is the dirty little secret of fin-de-siècle neoclassical economics, one that only becomes increasingly apparent with every subsequent game-theoretic model. Nothing seems poised to reverse the neoclassical hollowing out of human beings into hulking mechanical shells: not experimental economics, not evolutionary game theory, not Herbert Simon, not Robert Frank, not Amartya Sen, not the Santa Fe Institute, nothing.

Until someone comes along with a better suggestion, it would appear that the alternative program of treating markets as machines, while provisionally reserving for their human environment all the privileges of consciousness, intentionality, nonalgorithmic thought, and the entire gamut of diversity of cognitive capacities and idiosyncratic behavior is, ironically, the program more faithful to the humanist impulse. The quest to elevate humanity to deserve the vaunted honorific of "rationality" by painting humans as prodigious machines would seem so neurotically misplaced as to be scandalous, were it not so taken for granted at the end of the millennium. The arrogation of the epithet "rational" on the part of neoclassical economics has been a defalcation on a grand scale: as we have argued, they have altogether overlooked the prevalence of price as *ratio* as one of the prime bits of evidence that market operations have been restricted to a limited subset of mathematics – eminently computable mathematics – so that humans can feel free to impose any interpretation or construction they wish on economic events. Humans need not be instinctive mathematicians or intuitive statisticians in order to realize our cognitive potential; at least in the economic sphere, we can let the machines do it for us. As Wittgenstein said, life goes smoother when we calculate by machines. Perhaps the time has arrived to acknowledge a bit more liveliness on the part of our machines, so that we can ourselves appear a bit less mechanical. And once that tentative concession is made, then the real gestalt switch might click in: instead of repeatedly confusing ourselves with machines that represent for us the icon of rationality, we might come to see ourselves as organisms evolving in tandem with our machines.

As tends to be the case for most abstract philosophical discussions of rationality, however, we rapidly run the risk of becoming too arid, too idealist, too solipsistic. The theses suggested in this chapter in no way need depend for their salience upon ideas and abstractions bloodlessly dueling in cyberspace. For cyborg science in general, and the program of regarding markets as automata in particular, both possess a solid basis in material developments in the modern economy and society. In a phrase, the next generation of automated war games has already arrived, and not just on the Sony playstations found everywhere toys are sold. For the advent of the Internet in the 1990s has only accelerated developments pioneered at RAND in the 1960s and the National Association of Food Chains in the 1970s (Brown, 1997) and the SABRE airline reservation system and other lineal descendants of the SAGE air defense system. The automation of communications, command, control, and information processing attendant upon the military uses of the computer have come home to roost in the marketplace. Once upon a time the human being was thought to be expendable, extractable "out of the loop" in the military chain of command; now our crafty programmers have come to pursue the very same ambition for markets. The enthusiasm seemed to start with the "dot-coms," the retail firms promising to dispense with bricks and mortar and yet encompass the wired world into a single client base. But there is not much to keep a hacker occupied in servicing a bank of network servers in some warehouse in Cupertino or Seattle; and there is much more money to be saved in automating the human side of the activities of retailing and shopping. So in the later 1990s we embarked on a world of increasingly automated markets, first in finance and then elsewhere; and as we observed in Chapter 7, AI found itself a new lease on life in building roboshoppers, "automated agents," and a whole range of wormware and shopbots who seek unctuously to enhance your shopping experience, making you feel more like a pampered "individual," while they surreptitiously amass information about your every virtual move. Double agents indeed.

The military and the marketplace intersect once again in our post–Cold War world in the palpable automation of electronic markets. The question broached here is whether this trend reinforces the orthodox approach to economics, or whether it gives succor to von Neumann's market automata alternative. Interestingly, there has appeared the article by Nir Vulkan (1999), which is the first to my knowledge to argue the former thesis. In a delicious bit of spin, Vulkan backhandedly admits that neoclassical theory never actually fit those bumptious human beings very well, but now finds its appropriate apotheosis in application to artificial agents on the Internet. "Game theory (and mechanism design) seems much more suitable for automated agents than it is for humans" (p. F69). Moreover, three generations of neoclassical failure to "extract" well-behaved utility functions from obstreperous humans may now be at an end, thanks to those wily WebBots: "The burden of constructing preferences and utility functions from past behaviour and by asking the right questions, will therefore fall on the shoulders of those who design [automated] agents" (p. F70). How convenient that prior generations of economists went to all that trouble to construct mechanical models of *Homo economicus* without any legitimate referent for their labors, now that roboshoppers have come along to fulfill that calling! And it goes without saying that we must rewrite history in order to legitimate the triumph: "It is an interesting fact that AI and economics have had many overlapping interests over the years. John von Neumann's pioneering work had laid the foundations for modern AI as well as modern game theory" (p. F88). The Ministry of Truth couldn't have done better.

Although the reader must judge for herself, I think Vulkan's narrative is implausible in the extreme. Experts in AI who take it as their province to help construct these artificial agents are nowhere near as sanguine about the promise of game theory: "Despite the mathematical elegance of game theory, game theoretic models suffer from restrictive assumptions that limit their applicability to realistic problems" (Jennings et al., 1998, p. 22). And as for the stealthy elicitation of neoclassical preference functions by observant roboshoppers, it would seem that the Lewis proofs of the essential noncomputability of preference functions should scare off any reasonably sophisticated computer scientist. Of course, AI specialists were already aware of this: "first order logic is not even *decidable*, and modal extensions to it (including representations of belief, desire) tend to be *highly* undecidable" (Wooldridge & Jennings, 1995, p. 12). For those who would rather have empirical proof, I would suggest consulting the best instances of recent experimental economics (e.g., Sippel, 1997), where it is demonstrated that careful real-time elicitation of choices regularly violates the axioms of revealed preference. Thus the impending alliance of autonomous artificial agents with a cognitive version of neoclassical economics seems little more than a pipe dream.

It would appear much more plausible that as the initial hype over automated agents dies away, AI will retreat to what it has done best in the past: the development of programs that assist human beings in doing the things that they have themselves initiated. And it is here that the von Neumann market automata program will find support. For as programmers are hired to automate markets, they will be brought up abruptly against the fact that small differences in market algorithms make big differences in market outcomes; and, furthermore, that one cannot load all the computational capacity onto the artificial agents and none onto the environment within which they are presumed to operate. (Software engineers have a cute name for this – the "ontology problem.") When they turn to reference sources like the "market microstructure" literature for help, they end up concluding that it has little to offer that addresses their specific concerns.[36] In setting out to engineer markets, researchers at the MIT Media Lab, for instance, found themselves running their own versions of experiments and positing their own principles of cyborg economics (Guttman & Maes, 1998). At such junctures, with some experience it will become apparent that an approach that treats markets as automata that can be sorted by formal objectives and computational complexity is much more useful than the whimsical prospect of an utterly obedient cyberslave. Indeed, the real epiphany will dawn when experts in cyborg science come to realize that, once markets grow more automated, it will be difficult for their human users to tell the difference.

APPENDIX 8.1

Double Auction and Sealed Bid Encoded onto Automata

In this appendix we show that the algorithm underlying a sealed bid is of lower computational complexity than the double auction (DA), thus formally illustrating the claims made in this chapter concerning the relationship of the theory of automata to the work of Gode and Sunder (1997). This result is obtained by showing that a sealed-bid auction can be encoded onto an automata of less computational capacity than the DA.

[36] Even Vulkan (1999, p. F69) has to admit: "it is disappointing how little work has been carried out by economists in response to these developments."

The class of machines we access is that of a push-down automata with multiple one-way read-only heads that move from left to right scanning the input tape. We might regard such an automata as a language acceptor where a DA automata and sealed-bid automata are automata constructed so that they only accept the language underlying a DA and sealed-bid market, respectively. "Language" in this context just refers to the characteristics of the algorithm representing a market institution. The language underlying the sealed-bid algorithm (with accumulation rule) is any sequence of bids preceded by an ask (the minimum ask) where there is a designated bid that is at least as large in magnitude as all other bids and which is at least as large as the minimum ask. Similarly, we can describe the language underlying a DA market institution as any sequence of bids and asks (with at least one active bid and ask) where there is a designated bid and ask. The designated bid is at least as large in magnitude as the designated ask, the designated bid is at least as large as all other bids, and the designated ask is at least as small as all other asks. We use the notion of a language to abstract out the essential computational features of a certain market algorithm. The introduction of this form of abstraction allows us to more rigorously compare and contrast the essential features of different market institutions. Moreover, it enables the analyst to categorize markets with respect to their computational capacity, where the notion of computational capacity is very precisely defined.

Before we can move to the formal analysis it is necessary to introduce some notions taken from the theory of computation (see Lewis & Papadimitriou, 1981, p. 29). We start with the notion of an alphabet, which is a finite set of symbols. An obvious example of an alphabet is the Roman alphabet $\{a, b, \ldots, z\}$. We follow the convention of having Σ denote an alphabet. A string over an alphabet is a finite sequence of symbols from the alphabet. A string may have no symbols, in this case it is referred to as the empty string and is denoted by *e*. The set of all strings – including the empty string – over an alphabet Σ is denoted by Σ^*.

In this appendix we use the alphabet $\Sigma = \{I, a, b\}$ to represent orders. The string S represents an individual order where $S = xI^m$ for $x \in \{a, b\}$ and I^m abbreviates that S contains mI's. The length of S minus 1 gives the size of the submitted order and $|S| = m + 1$ denotes the length of S. We write $S^{(i)}$ to emphasize that S is the ith submitted order and we write $S^{(i)}(a)$ $(S^{(i)}(b))$ to emphasize that the ith order is an ask (bid). The string π represents the concatenation of all orders submitted in a certain trading period; i.e., $\pi = S^{(1)} \ldots S^{(n)}$. π^i denotes the tail of π, which are all orders following the ith order. $S^{(j)} \subseteq \pi^i$ means that $S^{(j)}$ is an order in π^i, so that by definition of π^i we have $S^{(i)} \not\subset \pi^i$. A multiple-headed push-down automata scans the input tape on which π is encoded. The heads are one-way, which means that once a head scans up till π^i it can no longer scan any $S^{(j)}$ for $j \leq i$. In addition to the multiple heads the push-down automata possesses restricted storage facility in the form of a stack of arbitrary size. The stack is organized according to the last-in, first-out principle. This means that the symbols stored last onto the stack are deleted first. More formally we can describe a push-down automaton with multiple heads as follows.

Definition

A k-headed push-down automaton (henceforth k-PDA) is a sextuple $M = (K, \Sigma, \Gamma, \Delta, s, F)$, where K is a finite set of states; Σ is an alphabet (the input symbols); Γ is an alphabet (the stack symbols); $s \in K$ is the initial state; $F \subseteq K$ is the set of final states; Δ, the transition relation, is a mapping from $Kx(\Sigma \cup \{e\})^k x\Gamma$ to finite subsets of $Kx\Gamma^*$, where $(\Sigma \cup \{e\})^k$ abbreviates $(\Sigma \cup \{e\})x \ldots x(\Sigma \cup \{e\})$ the k symbols (possibly the empty symbol/string) read by the k-heads.

Let M be a k-DPA, p and q be two states in M, u an input symbol, γ, β stack symbols and $((p, u, e, \ldots, e, \beta), (q, \gamma)) \in \Delta$. Then M, whenever it is in state p with β on top of the stack, may read u from the input tape with head I, read nothing with the other heads, replace β by γ on top of the stack, and enter state q. A symbol is deleted whenever it is popped from the stack and a symbol is stored whenever it is pushed onto the stack. M for example pops (deletes) β from the stack with the transition $((p, u, b, \ldots, I, \beta), (q, e))$ and pushes (stored) β with the transition $((p, u, b, \ldots, I, e), (q, \beta))$. M is said to accept a string $\pi \in \Sigma^*$ if and only if (s, π, e) yields (p, e, e) for some state $p \in F$ and a sequence of transitions. The language accepted by M, denoted L(M), is the set of all strings accepted by M.

With this notation in hand, we can more explicitly define the languages underlying a DA market and sealed-bid market institution. Let L(DA) and L(SB) denote the languages underlying a DA and sealed bid, respectively. Then we can define L(DA) and L(SB) as follows.

Definition

$$L(DA) = \{\pi = S^{(1)} \ldots S^{(n)} : (\text{for } S^{(i)}(a), S^{(j)}(b) \subset \pi)[|S^{(j)}(b)| \geq |S^{(i)}(a)|,$$
$$|S^{(j)}(b)| \geq |S^{(m)}(b)|, |S^{(i)}(a)| \leq |S^{(m)}(a)| \text{ for } 1 \leq m \leq n]\}$$

$$L(SB) = \{\pi = S^{(1)} \ldots S^{(n)} : (\text{for } S^{(j)}(b) \subset \pi)[|S^{(j)}(b)| \geq |S| \text{ for } \forall\, S \subset \pi]$$
$$\text{and } S^{(1)}(a), S^{(m)}(b) \text{ for } 1 < m \leq n$$

An automaton M that accepts L(SB) has the ability to scan the input tape π for the largest bid, henceforth denoted by $S^{(j)}(b)$, and determines if $S^{(j)}(b)$

$\geq S^{(1)}(a)$ where only the first order should be an ask. M should reach a final state only if $S^{(j)}(b)$ is identified and it is verified that $S^{(j)}(b) \geq S^{(1)}(a)$ holds. Similarly an automaton M that accepts L(AD) has the ability to scan the input tape π for the largest bid $S^{(j)}(b)$ and the lowest ask, henceforth denoted by $S^{(i)}(a)$. Moreover, M checks if $S^{(j)}(b) \geq S^{(i)}(a)$ holds. M should reach a final state only if $S^{(j)}(b)$ and $S^{(i)}(a)$ are identified and it is verified that $S^{(j)}(b) \geq S^{(i)}(a)$ holds. An essential feature of any machine accepting either L(DA) or L(SB) is that it can compare the length of a particular string $S^{(j)}$ with an arbitrary number of strings in π^j. A machine accepting L(DA) requires the additional feature of doing this for an additional string $S^{(i)}$ and the machine also needs to compare two designated strings $S^{(i)}$ and $S^{(j)}$ with each other. In the following we will make use of these observations to show that a 1-PDA cannot accept either a DA or sealed-bid market and that both a 2- and 3-PDA cannot accept a DA market. In the next appendix we show that a 2-PDA and 4-PDA can accept the language underlying a sealed-bid and DA market respectively. This completes the proof that the algorithm underlying DA market is of a higher computational capacity than the algorithm underlying the sealed bid auction.

Lemma I

Let M denote a k-PDA that needs to compare the length of $n \geq 1$ substrings in π^i with $|S^{(i)}|$, where $S^{(i)} \subset \pi$ and $k \leq n$. Then M can only perform this task if $S^{(i)}$ is stored onto the stack at some point of the computation.

Proof (Proof by contradiction)

To compare $S^{(i)}$ with $S^{(m)} \subset \pi^i$ requires at least two heads whenever $S^{(i)}$ is not stored onto the stack: one to scan $S^{(i)}$ and one to scan $S^{(m)}$. Thus to compare $S^{(i)}$ with n substrings of π^i requires at least n + 1 heads. n heads are needed to scan $S^{(i)}$ n-times and the first time the length of $S^{(i)}$ is compared with a string in π^i an additional (n + 1)th head scans the other string simultaneously. But by assumption there are only $k \leq n$ heads available. QED.

The implication of lemma I is that every k-PDA that successfully performs the task of comparing the length of $S^{(i)}$ with an arbitrary number of substrings of π^i will always have to store $S^{(i)}$ onto the stack. If the machine does not always manage to store $S^{(i)}$ onto the stack, it might have too few heads to perform the task. Therefore we can henceforth assume that $S^{(i)}$ is always stored onto the stack.

Lemma II

A 1-PDA cannot perform the task of comparing the length of $S^{(i)}$ with an arbitrary number of substrings of π^i.

Proof

Pushing (storing) $S^{(i)}$ onto the stack enables the machine to compare $|S^{(i)}|$ with the length substring of π^i denoted by $S^{(m)}$. To compare $|S^{(i)}|$ with $|S^{(m)}|$ requires popping (deleting) parts of $S^{(i)}$ at the same time $S^{(m)}$ is scanned. After $S^{(m)}$ is scanned π^m remains where $\pi^m \subset \pi^i$. Thus the machine cannot recover $S^{(i)}$ because it is not a substring of π^m. QED.

Lemma III

Let M denote a k-PA that compares the length of an arbitrary number of substrings in π^i with a string $S^{(i)} \subset \pi$. Then M should have at least two heads to perform this task.

Proof

It follows from lemma I that 1-PA cannot compare the length of an arbitrary number of substrings. Thus it suffices to show that a 2-PA can perform this task. Without loss of generality we can assume that $S^{(i)}$ is pushed onto the stack and both heads are positioned, so that π^{m-1} remains for both heads, where $S^{(m)}$ is the next string which length machine M compares with $|S^{(i)}|$. Furthermore we can assume without loss of generality that head 1 scans $S^{(m)}$. While scanning $S^{(m)}$ symbols of $S^{(i)}$ are popped (deleted) until either head 1 completed scanning $S^{(m)}$ or $S^{(i)}$ is completely popped (deleted) from the stack. This procedure enables M to determine which string is longer. Next M restores $S^{(i)}$ by having head 2 push $S^{(m)}$ onto the stack after which head 1 scans the remainder of $S^{(m)}$ while for every scanned symbol a symbol is popped from the stack. QED.

Observe that the procedure described in lemma III is the only procedure that can successfully compare the length of $S^{(i)}$ with that of $S^{(m)}$ using only two heads. Lemma I already establishes that $S^{(i)}$ has to be stored onto the stack to permit an arbitrary number of comparisons with a k-PDA. Furthermore comparing $S^{(i)}$ and $S^{(m)}$ implies popping $S^{(i)}$ from the stack. Thus any successful strategy – that only uses two heads – needs to be able to restore $S^{(i)}$ onto the stack using the head that was not used to scan $S^{(m)}$ the first time. The only possible way to extract this information from π^i is described by the procedure in lemma III.

Lemma IV

Let M denote a 2-PA; then M cannot perform the task of comparing both $|S^{(i)}|$ and $|S^{(j)}|$ with an arbitrary number of strings in π^i and π^j, respectively.

Proof

From lemma III it follows that at least two heads are needed to restore a string onto the stack. This implies that M should simultaneously compare $|S^{(i)}|$ and $|S^{(j)}|$ with strings in π^i and π^j, respectively. Furthermore it follows from lemma I that $S^{(i)}$ and $S^{(j)}$ have to be stored onto the stack to allow an arbitrary number of comparisons with other strings. But to compare $|S^{(i)}|$ and $|S^{(j)}|$ simultaneously implies that sometime during the computation the string stored first – let's say $S^{(i)}$ – has to be compared with a string currently read from the input tape, which implies that the string stored last – $S^{(j)}$ in this case – has to be deleted. After deleting this $S^{(j)}$ it is impossible to restore it onto the stack because the heads are occupied comparing and restoring $S^{(i)}$. QED.

Proposition I

Both the languages underlying sealed-bid and DA market cannot be accepted by a 1-PDA.

Proof

From L(SB) and L(DA) it follows that any machine M accepting L(SB) or L(DA) should have the capacity to compare the length of a string $S^{(i)}$ with an arbitrary number of strings in π^i. It follows directly from lemma II that a 1-PDA cannot perform this task. QED.

Proposition II

The languages underlying a DA market cannot be accepted by k-PDA for $k \leq 3$.

Proof

Let M be a 3-PDA that accepts L(DA). From lemma III it follows that at least two heads are needed to compare $|S^{(i)}|$ with the length of an arbitrary number of strings in π^i. This means that M cannot first scan π for $S^{(i)}(a)$ (the lowest ask) and then scan π for $S^{(j)}(b)$ (the highest bid) because this would require at least four heads. Thus M has to scan π simultaneously for $S^{(i)}(a)$ and $S^{(j)}(b)$. It follows from lemma I that both intermediate values of $S^{(i)}(a)$ and $S^{(j)}(b)$ have to be stored onto the stack. Furthermore it follows from lemma IV that two heads are not sufficient to scan π for $S^{(i)}(a)$ and $S^{(j)}(b)$ and store the intermediate values. Therefore we can assume that all three heads are occupied discovering $S^{(i)}(a)$ and $S^{(j)}(b)$. Once M completes this task both $S^{(i)}(a)$ and $S^{(j)}(b)$ are stored onto the stack. But now M cannot check if $|S^{(i)}(a)| \geq |S^{(j)}(b)|$ so that any 3-PDA fails to accept L(DA). QED.

APPENDIX 8.2

Sealed-Bid Auction with Accumulation Rule

In this appendix we give the pseudo code for a 2-PDA that accepts a sealed-bid auction (with accumulation rule). For simplicity we assume that the input string π already has the correct format – that is, the first order is the minimum ask; all following orders are bids in the format $S = xI^m$ for $x \in \{a, b\}$. The five steps are as follows:

I. Push the minimum ask $S^{(1)}(a)$ onto the stack, using head I.

II. Compare stored ask with the next bid of the submitted bids, using head I.

If ask > bid then recover ask by adding bid to remainder of ask, using head II.

Otherwise, push bid onto stack, using head II.

III. Repeat step II until ask ≤ bid or end of input tape is reached. When the latter happens the automaton terminates without reaching a final state.

IV. Compare stored bid with bid currently scanned by head I.

If stored bid ≥ scanned bid recover stored bid by adding scanned bid to remainder of stored bid, using head II.

Otherwise, pop remainder of stored bid and push currently scanned bid onto stack, using head II.

V. Repeat step IV until the end of the input tape is reached, after which the machine reaches a final state.

To construct a 4-PDA that accepts a DA is not fundamentally different from the foregoing machine. Such a machine first determines the lowest ask using a similar approach. Next this machine is used as subroutine to determine the maximum bid and whether it exceeds the minimum ask.

Envoi

Try to imagine the virtual worlds that will be made possible by the power of a shared parallel computer. Imagine a world that has the complexity and subtlety of an aircraft simulation, the accessibility of a video game, the economic importance of the stock market, and the sensory richness of the flight simulator, all of this with the vividness of computer-generated Hollywood special effects. This may be the kind of world in which your children meet their friends and earn their living. . . . Whatever you imagine virtual worlds will be like, or whatever I imagine, is likely to be wrong.

Daniel W. Hillis, "What Is Massively Parallel Computing?"

References

ARCHIVES

Archives, Center for Operations Research, Economics and Econometrics, Université Catholique de Louvain, Louvain-la-neuve, Belgium
Kenneth Arrow Papers, Perkins Library, Duke University
Computer Oral History Project, Smithsonian Museum of American History, Washington
Cowles Foundation Archives, Yale University
Nicholas Georgescu-Roegen Papers, Perkins Library, Duke University
Harold Hotelling Papers, Columbia University Library Manuscripts Collection
IEEE-Rutgers Center for the History of Electrical Engineering
Radiation Lab Interviews: web archive <http://199.172.136.1/history_c..._projects/rad_lab>
Tjalling Koopmans Papers, Sterling Library, Yale University
Jacob Marschak Papers, Young Library, University of California, Los Angeles
Oskar Morgenstern Diaries, Perkins Library, Duke University
Oskar Morgenstern Papers, Perkins Library, Duke University
National Air and Space Museum, RAND Oral History Project, Washington, D.C.
National Security Archives, Gelman Library, George Washington University <www.seas.gwu.edu/nsarchive/nsa>
Michael Polanyi Papers, Regenstein Library, University of Chicago
Rockefeller Foundation Archives, Sleepy Hollow, New York
Herbert Simon Papers, Carnegie Mellon University Archives
Jan Tinbergen Papers, Erasmus University, Rotterdam
John von Neumann Papers, Library of Congress
Warren Weaver interviews, Oral History Collection, Columbia University
Norbert Wiener Papers, MIT Archives
Edwin Bidwell Wilson Collection, Harvard University Archives

THESES

Alexander, Jennifer. 1996. "The Meanings of Efficiency." Ph.D. diss. University of Washington.

Bailey, Joseph. 1998. "Intermediation and Electronic Markets: Aggregation and Pricing in Internet Commerce." Ph.D. diss. Massachusetts Institute of Technology.

Bowles, Mark. 1999. "Crisis in the Information Age." Ph.D. diss. Case Western Reserve University.

Collins, Martin. 1998. "Planning for Modern War: RAND and the Air Force, 1945–50." Ph.D. diss. University of Maryland.

Crowther-Heyck, Hunter. 1999. "Herbert Simon, Organization Man." Ph.D. diss. Johns Hopkins University.

Gilles, Donald. 1953. "Some Theorems on N-Person Games." Ph.D. diss. Princeton University.

Hammond, Debora. 1997. "Toward a Science of Synthesis: The Heritage of General Systems Theory." Ph.D. diss. University of California–Berkeley.

Hay, David. 1994. "Bomber Businessmen: The Army Air Forces and the Rise of Statistical Control." Ph.D. diss. University of Notre Dame.

Helmreich, Stefan. 1995. "Anthropology Inside and Outside the Looking-Glass Worlds of Artificial Life." Ph.D. diss. Stanford University.

Jardini, David. 1996. "Out of the Blue Yonder." Ph.D. diss. Carnegie-Mellon University.

Johnson, Stephen. 1997. "Insuring the Future: The Development and Diffusion of Systems Management in the American and European Space Programs." Ph.D. diss. University of Minnesota.

Klaes, Matthias. 1998. "The Emergence of Transactions Costs in Economics." Ph.D. diss. University of Edinburgh.

Nanus, Burton. 1959. "The Evolution of the Office of Naval Research." Master's thesis. Massachusetts Institute of Technology.

Nash, John. 1950. "Non-Cooperative Games." Ph.D. diss. Princeton University.

Rau, Eric. 1999. "Combat Scientists: Operations Research in the United States during World War II." Ph.D. diss. University of Pennsylvania.

Rius, Andres. 1999. "On the Rationality of Democratic Macroeconomic Policies." Ph.D. diss. University of Notre Dame.

Solovey, Mark. 1996. "The Politics of Intellectual Identity and American Social Science, 1945–70." Ph.D. diss. University of Wisconsin.

Wolf, Frank Louis. 1955. "Machine Computability in Game Theory." Ph.D. diss. University of Minnesota.

SECONDARY SOURCES

Abbate, Janet. 1999. *Inventing the Internet.* Cambridge: MIT Press.

Abreu, D., & Rubinstein, A. 1988. "The Structure of Nash Equilibrium in Repeated Games with Finite Automata." *Econometrica*, 56:1259–81.

Ackoff, Russell, ed. 1961. *Progress in Operations Research.* New York: Wiley.

1978. "Operations Research." *International Encyclopaedia of Statistics*, pp. 667–71. New York: Free Press.

Adam, Alison. 1998. *Artificial Knowing.* London: Routledge.

Adams, James. 1998. *The Next World War.* New York: Simon & Schuster.

Adelman, L. 1994. "Molecular Computation of Solutions to Combinatorial Problems." *Science*, 266:1021–24.

Agar, Jon, & Balmer, Brian. 1998. "British Scientists and the Cold War." *Historical Studies in the Physical and Biological Sciences*, 28:2:209–49.

Agre, Philip. 1993. "Interview with Allen Newell." *Artificial Intelligence*, 59:415–49.

1997. *Computation and Human Experience.* Cambridge: Cambridge University Press.

1998. "Yesterday's Tomorrow." *Times Literary Supplement*, July 3, pp. 2–3.

Agre, Philip, & Rotenberg, M., eds. 1998. *Technology and Privacy: The New Landscape.* Cambridge: MIT Press.

Albers, D., Alexanderson, G., & Reid, C., eds. 1990. *More Mathematical People.* Boston: Harcourt Brace.

Albin, Peter. 1980. "The Complexity of Social Groups and Social Systems by Graph Structures." *Mathematical Social Sciences*, 1:101–29.

1998. *Barriers and Bounds to Rationality.* Princeton: Princeton University Press.

Albin, Peter, & Foley, D. 1992. "Decentralized, Dispersed Exchange without an Auctioneer." *Journal of Economic Behavior and Organization*, 18:27–51.

Albin, Peter, & Gottinger, H. 1983. "Structure and Complexity in Economic and Social Systems." *Mathematical Social Sciences*, 5:253–68.

Alchian, Armen. 1950. "Uncertainty, Evolution and Economic Theory." *Journal of Political Economy*, 58:211–22.

1963. "Reliability of Progress Curves in Airframe Production." *Econometrica*, 31:679–92.

Alchian, Armen, Arrow, Kenneth, & Capron, William. 1958. "An Economic Analysis of the Market for Scientists and Engineers." RAND RM-2190-RC. RAND Corporation, Santa Monica, Calif.

Alchian, Armen, Bodenhorn, G., Enke, S., Hitch, C., Hirshleifer, J., & Marshall, A. 1951. "What Is the Best System?" RAND D-860. RAND Corporation, Santa Monica, Calif.

Allen, Beth. 1990. "Information as an Economic Commodity." *American Economic Review, Papers and Proceedings*, 80:268–73.

Allen, Thomas. 1987. *War Games.* New York: McGraw-Hill.

Alpers, Joseph. 1994. "Putting Genes on a Chip." *Science*, 264:1400.

Alt, James, Levi, Margaret, & Ostrom, Elinor, eds. 1999. *Competition and Cooperation.* New York: Russell Sage.

Amariglio, Jack, & Ruccio, David. Forthcoming. *Economic Fragments.* London: Routledge.

Amihud, Y., & Mendelson, H. 1989. "The Effects of Computer-Based Trading on Volatility and Liquidity." In H. Lucas & R. Schwartz, eds., *The Challenge of Information Technology for Securities Markets*, pp. 59–85. Homewood, Ill.: Dow Jones Irwin.

Amman, Hans, Kendrick, David, & Rust, John, eds. 1996. *Handbook of Computational Economics.* Vol. 1. Amsterdam: Elsevier.

Anderlini, Luca. 1990. "Some Notes on Church's Thesis and the Theory of Games." *Theory and Decision*, 29:15–52.

Anderson, James, & Rosenfeld, Edward, eds. 1998. *Talking Nets.* Cambridge: MIT Press.

Anderson, P., Arrow, K., & Pines, D., eds. 1988. *The Economy as an Evolving Complex System.* Redwood City, Calif.: Addison-Wesley.

Anderson, T., et al., eds. 1957. *Selected Papers in Statistics and Probability by Abraham Wald.* Stanford: Stanford University Press.

Ando, Albert. 1979. "On the Contributions of Herbert A. Simon to Economics." *Scandinavian Journal of Economics*, 81:83–93.

Andreoni, James, & Miller, John. 1995. "Auctions with Artificial Adaptive Agents." *Games and Economic Behavior*, 10:39–64.

Anon. 1997. "Intelligent Agents Roboshop." *Economist*, June 14, p. 72.

Arbib, Michael. 1969. *Theories of Abstract Automata.* Englewood, Cliffs, N.J.: Prentice-Hall.

Arifovic, Jasimina. 1994. "Genetic Algorithms, Learning and the Cobweb." *Journal of Economic Dynamics and Control*, 18:3–28.

Arifovic, Jasimina, & Eaton, Curtis. 1995. "Coordination via Genetic Learning." *Computational Economics*, 8:181–203.

Arquilla, John, & Ronfeldt, David. 1993. "Cyberwar Is Coming!" *Comparative Strategy*, 12:141–65.

1996. *The Advent of Netwar.* Santa Monica, Calif.: RAND.

Arrow, Kenneth. 1951a. *Social Choice and Individual Values.* New York: Wiley.

1951b. "Mathematical Models in the Social Sciences." In H. Lasswell & D. Lerner, eds., *The Policy Sciences in the US*, pp. 129–54. Stanford: Stanford University Press.

1959. "Rational Choice Functions and Orderings." *Economica*, 73:292–308.

1962. "Economic Welfare and the Allocation of Resources for Invention." In Richard Nelson, ed., *The Rate and Direction of Inventive* Activity, pp. 609–25. Princeton: Princeton University Press.

1969. "The Organization of Economic Activity: Issues Pertinent to the Choice of Market versus Nonmarket Allocation." In *The Analysis and Evaluation of Public Expenditures*, 1:47–66. Joint Economic Committee. Washington, D.C.: GPO.

1974a. *The Limits of Organization.* New York: Norton.

1974b. *Essays on the Theory of Risk Bearing.* Amsterdam: North Holland.

1978. "Meyer Girschick." In *International Encyclopedia of Statistics*, pp. 398–99. New York: Free Press.

1979. "The Economics of Information." In Michael Dertouzos & Joel Moses, eds., *The Computer Age: A Twenty year View*, pp. 306–17. Cambridge: MIT Press.

1983a. "Social Choice and Justice." In *Collected Papers*, 1:306–17. Cambridge: Harvard University Press.

1983b. "Behavior under Uncertainty and Its Implication for Policy." In B. Stigum & F. Wenstop, eds., *The Foundations of Utility and Risk Theory*, pp. 20–32. Boston: Reidel.

1983c. *General Equilibrium.* Vol. 2 of his *Collected Papers.* Cambridge: Harvard University Press.

1984a. *Individual Choice under Certainty and Uncertainty*. Vol. 3 of his *Collected Papers.* Cambridge: Harvard University Press.

1984b. *The Economics of Information*. Vol. 4 of his *Collected Papers.* Cambridge: Harvard University Press.

1986. "Biography of Jacob Marschak." In McGuire & Radner, 1986, pp. 19–56.

1987. "Rationality of Self and Others in an Economic System." In Hogarth & Reder, 1987, pp. 201–16.

1990. "Interview." In Swedberg, 1990, pp. 133–51.

1991a. "The Origins of the Impossibility Theorem." In Lenstra et al., 1991, pp. 1–4.

1991b. "Cowles in the History of Economic Thought." In Cowles Commission, 1991.

1994. "Beyond General Equilibrium." In Cowan et al., 1994, pp. 419–50.

1995a. "Interview." <http://woodrow.mpls.frb.fed.us/pubs/region/int9512.html>.

1995b. "Viewpoints: The Future." *Science*, 267 (March 17):1617–18.

1996. "The Economics of Information: An Exposition." *Empirica*, 23:119–28.

1997. "Invaluable Goods." *Journal of Economic Literature*, 35:757–65.

2000. "Increasing Returns: Historiographic Issues and Path Dependence." *European Journal of the History of Economic Thought*, 7:171–80.

Arrow, Kenneth, Columbatto, E., Perlman, M., & Schmidt, C., eds. 1996. *The Rational Foundations of Economic Behavior.* London: St. Martin's Press.

Arrow, Kenneth, & Debreu, Gerard. 1954. "Existence of an Equilibrium for a Competitive Economy." *Econometrica*, 20:265–90.

Arrow, Kenneth, Harris, L., & Marschak, J. 1951. "Optimal Inventory Policy." *Econometrica*, 19:250–72.

Arrow, Kenneth, & Hurwicz, Leonid. 1962. "Decentralization and Computation in Resource Allocation." In R. Pfouts, ed., *Essays in Economics and Econometrics: Essays in Honor of Harold Hotelling*, pp. 34–106. Chapel Hill: University of North Carolina Press.

1977. *Studies in the Resource Allocation Process.* Cambridge: Cambridge University Press.

Arrow, Kenneth, & Karlin, S. 1958. *Studies in the Mathematical Theory of Inventory and Production.* Stanford: Stanford University Press.

Arrow, Kenneth, Mnookin, R., Lee, R., Tversky, A., & Wilson, R. eds. 1995. *Barriers to Conflict Resolution.* New York: Norton.

Arthur, Brian. 1993. "On Designing Economic Agents That Behave like Human Agents." *Journal of Evolutionary Economics*, 3:1–22.

Arthur, Brian, Durlauf, Steve, & Lane, David, eds. 1997. *The Economy as an Evolving Complex System II.* Reading, Mass.: Addison-Wesley.

Asaro, Peter. 1997. "The Evolution of Biological Computers, 1943–1963." Unpublished manuscript, Department of Philosophy, University of Illinois.

1998. "Design for a Mind: The Mechanistic Philosophy of W. Ross Ashby." Unpublished manuscript, Department of Philosophy, University of Illinois.

Ashby, W. Ross. 1964. *An Introduction to Cybernetics.* London: University Paperbacks.

Ashworth, William. 1994. "The Calculating Eye." *British Journal for the History of Science*, 27:409–41.

1996. "Memory, Efficiency and Symbolic Analysis." *Isis*, 87:629–53.

Aspray, William. 1985. "The Scientific Conceptualization of Information: A Survey." *Annals of the History of Computing*, 7:117–29.

1990. *John von Neumann and the Origins of Modern Computing*. Cambridge: MIT Press.

Association of Computing Machinery. 1987. *ACM Turing Award Lectures: The First 20 Years.* Reading, Mass.: Addison-Wesley.

Aubin, David. 1997. "The Withering Immortality of Nicholas Bourbaki." *Science in Context*, 10:297–342.

Aumann, Robert. 1960. "A Characterization of Game Structures of Perfect Information." *Bulletin of the Research Council of Israel*, 9F:43–44.

1962. "The Game of Politics." *World Politics*, 14:675–86.

1964. "Markets with a Continuum of Traders." *Econometrica*, 32:39–50.

1974. "Subjectivity and Correlation in Randomized Strategies." *Journal of Mathematical Economics*, 1:67–96.

1976. "Agreeing to Disagree." *Annals of Statistics*, 4:1236–39.

1981. "Survey of Repeated Games." In Martin Shubik, ed., *Essays in Game Theory and Economics in Honor of Oskar Morgenstern*, pp. 11–42. Mannheim: Bibliographisches Institut.

1985. "What Is Game Theory Trying to Accomplish?" In K. Arrow & S. Honkapohja, eds., *Frontiers of Economics*, pp. 28–76. Oxford: Basil Blackwell.

1986. "Rationality and Bounded Rationality." Nancy L. Schwartz Lecture, Kellogg School of Management, Northwestern University.

1987. "Game Theory." In J. Eatwell et al., eds., *New Palgrave*, pp. 460–80. London: Macmillan.

1989. *Lectures on Game Theory*. Boulder, Colo.: Westview.

1992a. "Irrationality in Game Theory." In P. Dasgupta, D. Gale, O. Hart, & E. Maskin, eds., *Economic Analysis of Markets and Games*, pp. 214–27. Cambridge: MIT Press.

1992b. "Perspectives on Bounded Rationality." In Yoram Moses, ed., *Theoretical Aspects of Reasoning about Knowledge*. San Mateo, Calif.: Morgan Kaufman.

1995. "Backward Induction and Common Knowledge of Rationality." *Games and Economic Behavior*, 8:6–19.

1997. "On the State of the Art in Game Theory." In W. Guth et al., eds., *Understanding Strategic Interaction: Essays in Honor of Reinhard Selten*, pp. 8–34. Berlin: Springer.

Aumann, Robert, & Brandenberger, Adam. 1995. "Epistemic Conditions for Nash Equilibrium." *Econometrica*, 63:1161–80.

Aumann, Robert, & Kruskal, Joseph. "The Coefficients in an Allocation Problem." *Naval Research Logistics Quarterly*, 5:111–23.

1959. "Assigning Quantitative Values to Qualitative Factors in the Naval Electronics Problem." *Naval Research Logistics Quarterly*, 6:1–16.

Aumann, Robert, & Maschler, Michael. 1985. "Game Theoretic Analysis of a Bankruptcy Problem from the Talmud." *Journal of Economic Theory*, 36:195–213.

1995. *Repeated Games with Incomplete Information.* Cambridge: MIT Press.
Aumann, Robert, & Sorin, Sylvain. 1989. "Cooperation and Bounded Recall." *Games and Economic Behavior*, 1:5–39.
Axelrod, Robert. 1984. *The Evolution of Cooperation.* New York: Basic.
1997. *The Complexity of Cooperation: Agent Based Models.* Princeton: Princeton University Press.
1998. Agent-Based Models of Conflict and Cooperation Web Site. <http://pcps.physics.lsa.umich.edu/Software/ComplexCoop.html>.
Baars, Bernard. 1986. *The Cognitive Revolution in Psychology.* New York: Guilford Press.
Baas, N. 1994. "Emergence, Hierarchies and Hyperstructures." In Langton, 1994.
Babbage, Charles. 1989a. *The Collected Works of Charles Babbage.* New York: New York University Press.
1989b. *On the Economy of Machinery and Manufactures.* 1832. Vol. 8 of *Collected Works.* New York: New York University Press.
1994. *Passages from the Life of a Philosopher.* New Brunswick: Rutgers University Press.
Babe, Robert. 1994. "Information as an Economic Commodity." In Babe, ed., *Information and Communication in Economics*, pp. 41–69. Boston: Kluwer.
Backhouse, Roger, et al., eds. 1998. *Economics and Methodology: Crossing Boundaries.* London: Macmillan.
Badii, R., & Politi, A. 1997. *Complexity: Hierarchical Structures and Scaling in Physics.* Cambridge: Cambridge University Press.
Baggott, Jim. 1992. *The Meaning of Quantum Theory.* Oxford: Oxford University Press.
Baker, Oliver. 1999. "Schrödinger's Cash Register." *Science News*, 156.22:344–46.
Bamford, James. 1982. *The Puzzle Palace.* Boston: Houghton Mifflin.
Barber, William. 1985. *From New Era to New Deal.* Cambridge: Cambridge University Press.
1996. *Designs within Disorder.* Cambridge: Cambridge University Press.
Barricelli, Nils. 1962. "Numerical Testing of Evolution Theories, Parts I & II." *Acta Biotheoretica*, 16:69–98, 99–126.
Barwise, Jon, & Etchemendy, John. 1993. *Turing's World 3.0.* Stanford: CSLI.
Bass, Thomas. 1999. *The Predictors*. New York: Holt.
Basu, K. 1990. "On the Non-Existence of a Rationality Definition for Extensive Games." *International Journal of Game Theory*, 19:33–44.
Baudrillard, Jean. 1994. *Simulacra and Simulation.* Ann Arbor: University of Michigan Press.
Baum, Claude. 1981. *The System Builders.* Santa Monica, Calif.: SDC.
Baumgartner, Peter, & Payr, Sabine, eds. 1995. *Speaking Minds: Interviews with Twenty Eminent Cognitive Scientists.* Princeton: Princeton University Press.
Baumol, William. 2000. "What Marshall Didn't Know: On the 20th Century's Contributions to Economics." *Quarterly Journal of Economics*, 140:1–42.
Bausor, Randall. 1995. "Liapounov Techniques in Economic Dynamics and Classical Thermodynamics." In Rima, 1995, pp. 396–405.
Baxter, James. 1946. *Scientists against Time.* Boston: Little Brown.

Beatty, John. 1995. "The Evolutionary Contingency Thesis." In G. Wolters & J. Lennox, eds., *Concepts, Theories and Rationality in the Biological Sciences*, pp. 45–82. Pittsburgh: University of Pittsburgh Press.

Becker, Gary. 1962. "Irrational Behavior and Economic Theory." *Journal of Political Economy*, 70:1–13.

Beckmann, Martin. 1991. "Tjalling C. Koopmans." In Shils, 1991, pp. 253 66.

Begehr, H., et al., eds. 1998. *Mathematics in Berlin.* Berlin: Birkhauser.

Beller, Mara. 1999. *Quantum Dialogues.* Chicago: University of Chicago Press.

Bellin, David, & Chapman, Gary. 1987. *Computers in Battle.* New York: Harcourt Brace Jovanovich.

Bellman, Richard. 1953. "On a New Iterative Algorithm for Finding the Solutions of Games." RAND P-473. RAND Corporation, Santa Monica, Calif.

1957. *Dynamic Programming.* Princeton: Princeton University Press.

1963. "Dynamic Programming, Intelligent Machines and Self-Organizing Systems." In *Proceedings of the Symposium on the Mathematical Theory of Automata*, pp. 1–11. Brooklyn: Polytechnic Press.

1984. *Eye of the Hurricane.* Singapore: World Scientific.

Bellman, Richard, et al. 1957. "On the Construction of a Multistage Multiperson Business Game." *Journal of the Operations Research Society of America*, 5:469.

Bellman, Richard, & Kalaba, R. 1957. "On the Role of Dynamic Programming in Statistical Communication Theory." *IRE Transactions on Information Theory*, IT-3:197–203.

Benaroch, Michael. 1996. "Artificial Intelligence and Economics: Truth or Dare?" *Journal of Economic Dynamics and Control*, 20:601–5.

Bender, Thomas, & Schorske, Carl., eds. 1997. *American Academic Culture in Transformation.* Princeton: Princeton University Press.

Beniger, James. 1986. *The Control Revolution.* Cambridge: Harvard University Press.

Ben-Ner, A., & Putterman, L., eds. 1998. *Economics, Values and Organization.* Cambridge: Cambridge University Press.

Bennett, Charles, & Landauer, Rolf. 1985. "The Fundamental Limits to Computation." *Scientific American*, 253 (July):48–56.

Bennett, Edward, Degan, James, & Spiegel, Joseph, eds. 1964. *Military Information Systems.* New York: Praeger.

Ben-Porath, Eichanan. 1986. "Repeated Games with Finite Automata." IMSSS, Stanford University.

1990. "The Complexity of Computing a Best-Response Automaton in Repeated Games with Mixed Strategies." *Games and Economic Behavior*, 2:1–12.

1993. "Repeated Games with Finite Automata." *Journal of Economic Theory*, 59:17–32.

Bergmann, Frijthof. 1977. *On Being Free.* Notre Dame: University of Notre Dame Press.

Berlinski, David. 1975. *On Systems Analysis.* Cambridge: MIT Press.

Bernal, John David. 1949. *The Freedom of Necessity.* London: Routledge Kegan Paul.

Bernstein, Michael. 1998. "Professional Authority and Professional Knowledge." *Poetics Today*, 19:375–99.

Bernstein, Nina. 1997. "On Line, High-Tech Sleuths Find Private Facts." *New York Times*, September 17, pp. A1, A12.

Bernstein, Peter. 1992. *Capital Ideas*. New York: Free Press.

Binmore, Ken. 1981. *Foundations of Analysis*. Cambridge: Cambridge University Press.

1987–88. "Modeling Rational Players I & II." *Economics and Philosophy*, 3:179–214; 4:9–55.

1990. *Essays on the Foundations of Game Theory*. Oxford: Basil Blackwell.

1992a. *Fun and Games*. Lexington, Mass.: Heath.

1992b. "Evolutionary Stability in Repeated Games Played by Finite Automata." *Journal of Economic Theory*, 57:278–305.

1996. Introduction to Nash, 1996.

1997. "Rationality and Backward Induction." *Journal of Economic Methodology*, 4:23–41.

1998a. Review of *Complexity of Cooperation*, by Robert Axelrod. *Journal of Artificial Societies and Social Situations*. <http://www.soc.surrey.ac.uk/JASSS/1/1/review.html>.

1998b. *Just Playing*. Cambridge: MIT Press.

1999a. "Why Experiment in Economics?" *Economic Journal*, 109:F16–24.

1999b. "Goats Wool." In Heertje, 1999, pp. 119–39.

Binmore, Ken, & Shin, H. 1992. "Algorithmic Knowledge and Game Theory." In C. Bicchieri & M. Chihara, eds., *Knowledge, Belief and Strategic Interaction*, pp. 141–54. Cambridge: Cambridge University Press.

Binmore, Ken, & Vulkan, Nir. 1999. "Applying Game Theory to Automated Negotiation." *Netnomics*, 1:1–9.

Black, Fischer. 1971. "Towards a Fully Automated Exchange." *Financial Analysts Journal*, 27 (July):29–35; (November):25–28, 86–87.

1986. "Noise" *Journal of Finance*, 41:529–43.

Blackett, Patrick 1961. "Critique of Some Contemporary Defense Thinking." *Encounter*, 16:9–17.

1962. *Studies of War: Nuclear and Conventional*. New York: Hill & Wang.

Blackwell, D. 1949. "Noisy Duel, One Bullet Each, Arbitrary Non-Monotone Accuracy." RAND D-442. RAND Corporation, Santa Monica, Calif.

1951. "Comparison of Experiments." In J. Neyman, ed., *Proceedings of the Second Berkeley Symposium*.

1953. "Equivalent Comparisons of Experiments." *Annals of Mathematical Statistics*, 24:265–72.

Blackwell, D., & Girschick, M. 1954. *Theory of Games and Statistical Decisions*. New York: Wiley.

Bloomfield, Brian, ed. 1987. *The Question of Artificial Intelligence*. London: Croom Helm.

Bloor, David. 1973. "Wittgenstein and Mannheim on the Sociology of Mathematics." *Studies in the History and Philosophy of Science*, 4:173–91.

Blumenthal, Marjory. 1998. "Federal Government Initiative from the Foundation of the Information Technology Revolution." *American Economic Review, Papers and Proceedings*, 88.2:34–39.

Bode, Hendrik. 1960. "Feedback – the History of an Idea." In *Proceedings of the Symposium on Active Networks*. New York: Polytechnic Press.

Boden, Margaret. 1981. *Minds and Mechanisms*. Ithaca: Cornell University Press.

ed. 1996. *The Philosophy of Artificial Life*. Oxford: Oxford University Press.

Bohnenblust, H., Dresher, M., Girshick, M., Harris, T., Helmer, O., McKinsey, J., Shapley, L., & Snow, R. 1949. *Mathematical Theory of Zero-Sum Two-Person Games with a Finite Number or a Continuum of Strategies*. Santa Monica, Calif.: RAND.

Bohnenblust, H., Shapley, L., & Sherman, S. 1949. "Reconnaissance in Game Theory." RM-208. RAND Corporation, Santa Monica, Calif.

Boltzmann, Ludwig. 1974. *Theoretical Physics and Philosophical Problems*. Ed. Brian McGuinness. Dordrecht: Reidel.

Bona, Jerry, & Santos, Manuel. 1997. "On the Role of Computation in Economic Theory." *Journal of Economic Theory*, 72:241–81.

Boole, George. 1954. *An Investigation of the Laws of Thought*. New York: Dover.

Born, Rainer, ed. 1987. *Artificial Intelligence*. New York: St. Martin's.

Bosch-Domenech, Antoni, & Sunder, Shyam. 1999. "Tracking the Invisible Hand." <http://zi.gsia.cmu.edu/sunder>.

Boulanger, A. 1998. "Catapults and Grappling Hooks: The Tools and Techniques of Information Warfare." *IBM Systems Journal*, 37:106–13.

Boulding, Kenneth. 1956. *The Image*. Ann Arbor: University of Michigan Press.

1966. "The Economics of Knowledge and the Knowledge of Economics." *American Economic Review*, 56:1–13.

Boumans, Marcel. 1998. "Lucas and Artificial Worlds." In John Davis, ed., *The New Economics and Its History*, pp. 63–88. Supplement to vol. 29 of *History of Political Economy*. Durham: Duke University Press.

Bowden, B., ed. 1953. *Faster than Thought*. London: Pitman.

Bowker, Geof. 1993. "How to Be Universal: Some Cybernetic Strategies." *Social Studies of Science*, 23:107–27.

1994. Review of *Toward a History of Game Theory*, by E. R. Weintraub. *British Journal for the History of Science*, 27:239–40.

Boyle, James. 1996. *Shamans, Software and Spleens*. Cambridge: Harvard University Press.

Brams, Steven, & Kigour, D. 1988. *Game Theory and National Security*. Oxford: Blackwell.

Brands, H. W. 1993. *The Devil We Knew*. New York: Oxford University Press.

Branscomb, Anne. 1994. *Who Owns Information?* New York: Basic.

Brazer, Marjorie. 1982. "The Economics Department of the University of Michigan." In Saul Hymans, ed., *Economics and the World Around It*, pp. 133–275. Ann Arbor: University of Michigan Press.

Brenner, Alfred. 1996. "The Computing Revolution and the Physics Community." *Physics Today*, 49.10 (October):24–39.

Brentjes, Sonja. 1994. "Linear Optimization." In Grattan-Guinness, 1994a, pp. 828–36.

Brewer, Gary, & Shubik, Martin. 1979. *The War Game*. Cambridge: Harvard University Press.

Bridges, Donald. 1991. "The Constructive Inequivalence of Various Notions of Preference Ordering." *Mathematical Social Sciences*, 21:169–76.

1992. "The Construction of a Continuous Demand Function for Uniformly Rotund Preferences." *Journal of Mathematical Economics*, 21:217–27.

Bridgman, Percy. 1927. *The Logic of Modern Physics*. New York: Macmillan.

Brillouin, Leon. 1951. "Maxwell's Daemon Cannot Operate: Information and Entropy I." *Journal of Applied Physics*, 22:334–57.

1953. "Negentropy Principle of Information." *Journal of Applied Physics*, 24:1152–63.

1962. *Science and Information Theory*. 2nd ed. New York: Academic Press.

Brodie, Bernard. 1949. "Strategy as a Science." *World Politics*, 1:467–88.

Bronowski, Jacob. 1973. *The Ascent of Man*. London: BBC.

Brooks, D., & Wiley, E. 1988. *Evolution as Entropy*. 2nd ed. Chicago: University of Chicago Press.

Brown, E. Cary, & Solow, Robert, eds. 1983. *Paul Samuelson and Modern Economic Theory*. New York: McGraw-Hill.

Brown, George. 1951. "Iterative Solutions of Games by Fictitious Play." In Koopmans, 1951.

Brown, George, & von Neumann, John. 1950. "Solutions of Games by Differential Equations." In Kuhn & Tucker, 1953, pp. 73–79.

Brown, L., Pais, A., & Pippard, B., eds. 1995. *Twentieth Century Physics*. Vol. 3. Philadelphia: American Institute of Physics.

Brown, Michael. 1992. *Flying Blind: The Politics of the US Strategic Bomber Program*. Ithaca: Cornell University Press.

Brown, Stephen. 1997. *Revolution at the Checkout Counter*. Cambridge: Harvard University Press.

Brush, Stephen. 1978. *The Temperature of History*. New York: Burt Franklin.

1983. *Statistical Physics and the Atomic Theory of Matter*. Princeton: Princeton University Press.

Brynjolfsson, Erik, & Hitt, Lorin. 2000. "Beyond Computation: Information Technology, Organizational Transformation." *Journal of Economic Perspectives*, 14:23–48.

Brynjolfsson, Erik, & Kahin, Brian, eds. 2000. *Understanding the Digital Economy*. Cambridge: MIT Press.

Buchan, Glenn. 1996. "Information War and the Air Force: Wave of the Future? Current Fad?" RAND Issue paper 149. <www.jya.com/1p149.html>.

Buck, George, & Hunka, Stephen. 1999. "W. S. Jevons, Allan Marquand and the Origins of Digital Computing." *Annals of the History of Computing*, 21:21–26.

Buck, Peter. 1985. "Adjusting to Military Life: The Social Sciences Go to War." In Smith, 1985, pp. 203–52.

Buckland, Michael. 1991. "Information as Thing." *Journal of the American Society for Information Science*, 42:351–60.

Buckley, Walter, ed. 1968. *Modern Systems Research for the Behavioral Scientist*. Chicago: Aldine.

Budieri, Robert. 1996. *The Invention That Changed the World*. New York: Simon & Schuster.

Bud-Friedman, Lisa, ed. 1994. *Information Acumen*. London: Routledge.

Burgos, Glen. 1993. "Manufacturing Certainty: Testing and Program Management for the F-4 Phantom." *Social Studies of Science*, 23:265–300.

Burke, Colin. 1994. *Information and Secrecy: Vannevar Bush, Ultra, and the Other Memex.* Metuchen, N.J.: Scarecrow Press.

Burks, Arthur. 1970. *Essays on Cellular Automata.* Urbana: University of Illinois Press.

1986. *Robots and Free Minds.* Ann Arbor: University of Michigan College of LSA.

Burns, Arthur. 1951. "Mitchell on What Happens during Business Cycles." With comment by Jacob Marschak and reply by Burns. In *Conference on Business Cycles*, pp. 3–33. New York: NBER.

Bush, R., & Mosteller, C. 1951. "A Mathematical Model for Simple Learning." *Psychological Review*, 58:313–23.

Bush, Vannevar. 1945. *Science: The Endless Frontier.* Washington, D.C.: GPO.

1970. *Pieces of the Action.* New York: Morrow.

Butler, Declan. 1998. "Xenotransplantation." *Nature*, 391 (January 22):309.

Button, Graham, Coulter, Jeff, Lee, John, & Sharrock, Wes. 1995. *Computers, Minds and Conduct.* Cambridge: Polity Press.

Caldwell, Bruce. 1988. "Hayek's Transformation." *History of Political Economy*, 20:513–41.

1997a. "Hayek and Socialism." *Journal of Economic Literature*, 35:1856–90.

ed. 1997b. *Socialism and War.* Vol. 10 of the *Collected Works of Friedrich Hayek.* Chicago: University of Chicago Press.

Callebaut, Werner, ed. 1993. *Taking the Naturalistic Turn.* Chicago: University of Chicago Press.

Callon, Michel, ed. 1998. *The Laws of the Markets.* Oxford: Blackwell.

Camerer, Colin. 1997. "Progress in Behavioral Game Theory." *Journal of Economic Perspectives*, 11:167–88.

Campbell, R., & Sowden, L., eds. 1985. *Paradoxes of Rationality and Cooperation.* Vancouver: University of British Columbia Press.

Campbell-Kelley, Martin, & Aspray, William. 1996. *Computer.* New York: Basic.

Candeal, J., Miguel, J., Indurain, E., & Mehta, G. 2001. "Utility and Entropy." *Economic Theory*, 17:233–38.

Canning, D. 1992. "Rationality, Computability and Nash Equilibrium." *Econometrica*, 60:877–88.

Cannon, Walter. 1939. *The Wisdom of the Body.* New York: Norton.

Capshew, James. 1999. *Psychologists on the March.* Cambridge: Cambridge University Press.

Carpenter, W. B. 1875a. "On the Doctrine of Human Automatism." *Contemporary Review*, 25:399–416, 940–62.

1975b. *The Doctrine of Human Automatism.* London: Collins.

Carrithers, M., Collins, S., & Lukes, S., eds. 1985. *The Category of the Person.* Cambridge: Cambridge University Press.

Cashen, Walter. 1955. "War Games and Operations Research." *Philosophy of Science*, 22:309–20.

Cason, Timothy, & Friedman, Daniel. 1997. "Price Formation in Single Call Markets." *Econometrica*, 65:311–45.

Casti, John. 1997a. *Would-Be Worlds.* Cambridge: Wiley.

1997b. "Computing the Uncomputable." *Complexity*, 2.3:7–12.

Caywood, T., & Thomas, C. 1955. "Applications of Game Theory to Fighter vs. Bomber Combat." *Journal of the Operations Research Society of America*, 3:402–11.

Cerf, Vinton. 1973. "Parry Encounters the Doctor." *Datamation*, 19 (July):62–63.

Ceruzzi, Paul. 1989. *Beyond the Limits.* Cambridge: MIT Press.

1998. *A History of Modern Computing.* Cambridge: MIT Press.

Chaloupek, Gunther. 1990. "The Austrian Debate in Economic Calculation in a Socialist Society." *History of Political Economy*, 22:659–75.

Chamah, Brigitte. 1999. "The Emergence of Cognitive Science in France." *Social Studies of Science*, 29:643–84.

Chamberlin, Edward. 1948. "An Experimental Imperfect Market." *Journal of Political Economy*, 56:95–108.

Channell, David. 1991. *The Vital Machine*. Cambridge: Oxford University Press.

Chapman, Robert, Kennedy, John, Newell, Allen, & Biel, William. 1958. "The System Research Laboratory's Air Defense Experiments." *Management Science*, 5:250–69.

Charnes, Abraham, & Cooper, William. 1958. "The Theory of Search: Optimum Distribution of Search Effort." *Management Science*, 5:44–50.

Christ, Carl. 1952. "History of the Cowles Commission, 1932–52." In *Economic Theory and Measurement: A 20 Year Research Report*, pp. 3–65. Chicago: Cowles Commission.

1994. "The Cowles Commission's Contributions to Econometrics at Chicago, 1939–55." *Journal of Economic Literature*, 32:30–59.

Church, Alonzo. 1940. "On the Concept of a Random Sequence." *Bulletin of the American Mathematical Society*, 46:130–35.

Churchland, Patricia, & Sejnowski, Terrence. 1992. *The Computational Brain.* Cambridge: MIT Press.

Churchland, Paul. 1979. *Scientific Realism and Plasticity of Mind.* Cambridge: Cambridge University Press.

Clancey, William, Smoliar, S., & Stefnik, S., eds. 1994. *Contemplating Minds.* Cambridge: MIT Press.

Clark, Ronald. 1962. *The Rise of the Boffins.* London: Phoenix.

Clarkson, G. 1962. *Portfolio Selection: A Simulation*. Englewood Cliffs, N.J.: Prentice-Hall.

Clemons, Eric, & Weber, Bruce. 1996. "Competition between Stock Exchanges and New Trading Systems." Working paper, New York University.

Cliff, Dave, & Bruten, Janet. 1997. "Minimal Intelligence Agents for Bargaining Behaviors in Market-Based Environments." HP Labs Technical Report, HPL-97-91. Palo Alto, Calif.

Clynes, Manfred, & Kline, Nathan. 1995. "Cyborgs and Space." Reprinted in Gray, 1995, pp. 29–33.

Cohen, Avner. 1998. *Israel and the Bomb.* New York: Columbia University Press.

Cohen, I. B. 1994. *Interactions*. Cambridge: MIT Press.

Cohen, Kalman, et al. 1983. "A Simulation Model of Stock Exchange Trading." *Simulation*, 41.11 (November):181–91.

Cohen, Kalman, & Schwartz, Robert. 1989. "An Electronic Call Market." In Henry Lucas & Robert Schwartz, eds., *The Challenge of Information Technology for Securities Market*, pp. 15–58. New York: Dow Jones Irwin.

Cohen, L. J. 1981. "Can Human Irrationality Be Experimentally Determined?" *Behavioral and Brain Sciences*, 4:317–70.

Colander, D., & Landreth, H. 1996. *The Coming of Keynesianism to America.* Cheltenham: Elgar.

Coleman, Andrew. 1982. *Game Theory and Experimental Games.* Oxford: Pergamon Press.

Collins, Harry. 1985. *Changing Order.* London: Sage.

1990. *Artificial Experts.* Cambridge: MIT Press.

1991. "Comment." In Neil de Marchi & Mark Blaug, eds., *Appraising Economic Theories*, pp. 227–31. Aldershot: Elgar.

Collins, Harry, & Kusch, Martin. 1998. *The Shape of Actions: What Humans and Machines Can Do.* Cambridge: MIT Press.

Commons, John R. 1934. *Institutional Economics.* Cambridge: Macmillan.

Conlisk, John. 1996. "Why Bounded Rationality?" *Journal of Economic Literature*, 34:669–700.

Cooper, W., et al. 1958. "Economics and OR: A Symposium." *Review of Economics and Statistics*, 40:195–229.

Copeland, B., & Proudfoot, D. 1996. "On Alan Turing's Anticipations of Connectionism." *Synthese*, 108:361–77.

Cornet, Bernard, & Tulkins, Henry, eds. 1989. *Contributions to Operations Research and Economics.* Cambridge: MIT Press.

Corry, Leo. 1992. "Nicholas Bourbaki and the Concept of Mathematical Structure." *Synthese*, 22:315–48.

1996. *Modern Algebra and the Rise of Mathematical Structures.* Basel: Birkhauser.

1997a. "The Origins of Eternal Truth in Modern Mathematics: Hilbert to Bourbaki." *Science in Context*, 10:253–96.

1997b. "David Hilbert and the Axiomatization of Physics." *Archive for the History of the Exact Sciences*, 51:83–198.

1999a. "David Hilbert between Mechanical and Electromechanical Reductionism." *Archive for the History of the Exact Sciences*, 53:489–527.

1999b. "Hilbert and Physics." In Jeremy Gray, ed., *The Symbolic Universe*, pp. 145–88. Oxford: Oxford University Press.

Cottrell, Allin, & Cockshott, W. P. 1993. "Calculation, Complexity and Planning: The Socialist Calculation Debate Once Again." *Review of Political Economy*, 5:73–112.

Cover, T., & Thomas, J. 1991. *Elements of Information Theory.* Cambridge: Wiley.

Cowan, G., Pines, D., & Meltzer, D., eds. 1994. *Complexity: Metaphors, Models and Reality.* Reading, Mass.: Addison-Wesley.

Cowen, Tyler. 1993. "The Scope and Limits of Preference Sovereignty." *Economics and Philosophy*, 9:253–69.

Cowles Commission. 1951. *Rational Decision Making and Economic Behavior.* 19th Annual Report. Chicago: Cowles Commission.

Cowles Foundation. 1964. *Report of Research Activities.* July 1961–June 1964. New Haven: Cowles Foundation.

1971. *Report of Research Activities.* July 1967–June 1970. New Haven: Cowles Foundation.
1974. *Report of Research Activities.* July 1970–June 1973. New Haven: Cowles Foundation.
1991. *Cowles' Fiftieth Anniversary.* New Haven: Cowles Foundation.
Craver, Earlene. 1986a. "The Emigration of the Austrian Economists." *History of Political Economy*, 18:1–32.
1986b. "Patronage and the Direction of Research in Economics." *Minerva*, 18:1–32.
Crevier, Daniel. 1993. *AI: The Tumultuous History of the Search for Artificial Intelligence.* Cambridge: Basic Books.
Cronin, Blaise, & Crawford, Holly. 1999. "Information Warfare: Its Applications in Military and Civilian Contexts." *Information Society*, 15:257–63.
Cronon, William, ed. 1995. *Uncommon Ground.* New York: Norton.
Cross, John. 1961. "On Professor Schelling's Strategy of Conflict." *Naval Research Logistics Quarterly*, 8:421–26.
Crowther, J., & Whiddington, R. 1948. *Science at War.* London: His Majesty's Stationery Office.
Crutchfield, James. 1994a. "Is Anything Ever New? Considering Emergence." In Cowan et al., 1994, pp. 515–38.
1994b. "Observing Complexity and the Complexity of Observation." In H. Atmanspacher & G. Dalenoort, eds., *Inside vs. Outside*, pp. 235–72. Berlin: Springer.
1994c. "The Catculi of Emergence." *Physica D*, 75:11–54.
Curtiss, John. 1949. "Some Recent Trends in Applied Mathematics." *American Scientist*, 37:587–88.
Cushing, James. 1994. *Quantum Mechanics: Historical Contingency and the Copenhagen Hegemony.* Chicago: University of Chicago Press.
Cutland, Nigel. 1980. *Computability: An Introduction to Recursive Function Theory.* Cambridge: Cambridge University Press.
Cyert, Richard, & March, James. 1963. *A Behavioral Theory of the Firm.* Englewood Cliffs, N.J.: Prentice-Hall.
Dantzig, George. 1951. "A Proof of the Equivalence of the Programming Problem and the Game Problem." In Koopmans, 1951.
1956. "Constructive Proof of the Min-Max Theorem." *Pacific Journal of Mathematics*, 6:25–33.
1963. *Linear Programming and Extensions.* Princeton: Princeton University Press.
1987. "The Origin of the Simplex Method." Technical Report SOL 87-5. Stanford Optimization Laboratory, Stanford.
1991. "Linear Programming." In Lenstra et al., 1991, pp. 19–31.
Danziger, Kurt. 1990. *Constructing the Subject.* Cambridge: Cambridge University Press.
1992. "The Project of an Experimental Social Psychology." *Science in Context*, 5:309–28.
2000. "Making Social Psychology Experimental." *Journal of the History of the Behavioral Sciences*, 36:329–47.

D'Arms, J., Batterman, R., & Gorny, K. 1998. "Game Theoretic Explanations and the Evolution of Justice." *Philosophy of Science*, 65:76–102.

Darnell, Adrian, ed. 1990. *The Collected Economic Articles of Harold Hotelling.* New York: Springer Verlag.

Dasgupta, P., & David, P. 1994. "Towards a New Economics of Science." *Research Policy*, 23:487–521.

Daston, Lorraine. 1983. "The Theory of Will vs. the Science of the Mind." In W. Woodward & M. Ash, eds., *The Problematic Science: Psychology in the 19th Century*, pp. 88–118. New York: Praeger.

Dauben, Joseph. 1979. *Georg Cantor: His Mathematics and the Philosophy of the Infinite.* Princeton: Princeton University Press.

D'Autume, A., & Cartelier, J., eds. 1997. *Is Economics Becoming a Hard Science?* Aldershot: Elgar.

Davidson, Donald, & Marschak, Jacob. 1962. "Experimental Tests of a Stochastic Decision Theory." In C. Churchman & P. Ratoosh, eds., *Measurement*, pp. 233–70. New York: Wiley.

Davidson, Donald, & Suppes, Patrick. 1957. *Decision Making: An Experimental Approach.* Stanford: Stanford University Press.

Davies, Max, & Verhulst, Michel, eds. 1958. *Operational Research in Practice.* London: Pergamon.

Davis, Douglas, & Holt, Charles. 1993. *Experimental Economics.* Princeton: Princeton University Press.

Davis, Douglas, & Williams, Arlington. 1991. "The Hayek Hypothesis." *Economic Inquiry*, 29:261–74.

Davis, Harold. 1941. *The Analysis of Economic Time Series.* Bloomington, Ind.: Principia Press.

Davis, John, ed. 1998. *The New Economics and Its History.* Durham: Duke University Press.

Davis, M., Sigal, R., & Weyuker, E. 1994. *Computability, Complexity and Languages.* San Diego: Academic Press.

Davis, Paul. 1984. "RAND's Experience in Applying Artificial Intelligence Techniques to Strategic-Level Military-Political War Gaming." RAND P-6977. RAND Corporation, Santa Monica, Calif.

Davis, Paul, & Schwabe, William. 1985. "Search for a Red Agent to Be Used in War Games and Simulations." RAND P-7107. RAND Corporation, Santa Monica, Calif.

Davis, Paul, Stair, P., & Bennett, B. 1983. "Automated War Gaming as a Technique for Exploring Strategic Command and Control." RAND N-2044-NA. RAND Corporation, Santa Monica, Calif.

Davis, Philip, & Hersh, Reuben. 1981. *The Mathematical Experience*. Boston: Houghton Mifflin.

Dawkins, Richard. 1976. *The Selfish Gene.* Oxford: Oxford University Press.

Dawson, John. 1997. *Logical Dilemmas: The Life and Work of Kurt Gödel.* Wellesley, Mass.: AK Peters.

Day, Richard, & Cigno, A. 1978. *Modeling Economic Changes.* Amsterdam: North Holland.

Debreu, Gerard. 1959. *Theory of Value.* Cowles Monograph 17. New Haven: Yale University Press.

1974. "Excess Demand Functions." *Journal of Mathematical Economics*, 1:15–23.

1983a. "Mathematical Economics at Cowles." Unpublished manuscript. Rev. ed. in Cowles Foundation, 1991.

1983b. *Mathematical Economics*. Cambridge: Cambridge University Press.

1991. "The Mathematization of Economic Theory." *American Economic Review*, 81:1–7.

Degler, Carl. 1991. *In Search of Human Nature*. New York: Oxford University Press.

DeGroot, Morris. 1989. "A Conversation with David Blackwell." In Duren, 1989.

de Landa, Manuel. 1991. *War in the Age of Intelligent Machines*. New York: Zone Books.

1997. *A Thousand Years of Nonlinear History*. New York: Zone Books.

De Long, J. B., & Froomkin, A. M. 2000. "Speculative Microeconomics for Tomorrow's Economy." In Brian Kahin & Hal Varian, eds., *Internet Publishing and Beyond*, pp. 6–44. Cambridge: MIT Press.

de Marchi, Neil, ed. 1993. *Non-Natural Economics: Reflections on the Project of More Heat than Light*. Durham: Duke University Press.

de Marchi, Neil, & Gilbert, Christopher, eds. 1989. *History and Methodology of Econometrics*. Oxford: Clarendon Press.

de Marchi, Neil, & Hirsch, A. 1990. *Milton Friedman*. Ann Arbor: University of Michigan Press.

Demchak, Chris. 1991. *Military Organizations, Complex Machines*. Ithaca: Cornell University Press.

Denbigh, K., & Denbigh, J. 1985. *Entropy in Relation to Incomplete Knowledge*. Cambridge: Cambridge University Press.

Dennett, Daniel. 1978. *Brainstorms*. Cambridge: MIT Press.

1995. *Darwin's Dangerous Idea*. New York: Simon & Schuster.

1996. *Kinds of Minds*. New York: Basic.

Denning, D. 1999. *Information Warfare and Security*. Reading, Mass.: Addison-Wesley.

Dennis, Michael. 1994. "Our First Line of Defense: Two University Laboratories in the Postwar American State." *Isis*, 85:427–55.

Depew, David, & Weber, Bruce. 1995. *Darwinism Evolving*. Cambridge: MIT Press.

Dertouzos, Michael. 1997. *What Will Be*. New York: Harper Edge.

Desmond, Adrian, & Moore, James. 1991. *Darwin*. London: Michael Joseph.

Dewdney, A. 1990. *The Magic Machine*. New York: W. H. Freeman.

Diamond, Peter, & Rothschild, Michael, eds. 1989. *Uncertainty in Economics*. San Diego: Academic Press.

Digby, James. 1989. "Operations Research and Systems Analysis at RAND, 1948–67." RAND Note N-2936-RC. RAND Corporation, Santa Monica, Calif.

1990. "Strategic Thought at RAND, 1948–63." RAND N-3096-RC. RAND Corporation, Santa Monica, Calif.

Dimand, Mary Ann, & Dimand, Robert. 1996. *The History of Game Theory*. Vol. 1. London: Routledge.

Domowitz, Ian. 1990. "The Mechanics of Automated Trade Execution Systems." *Journal of Financial Intermediation*, 1.2 (June):167–94.

1992. "Automating the Price Discovery Process." *Journal of Financial Services Research*, 6.4 (January):305–26.

1993a. "Automating the Continuous Double Auction in Practice." In Friedman & Rust, 1993, pp. 27–60.

1993b. "A Taxonomy of Automated Trade Execution Systems." *Journal of International Money and Finance*, 12:607–31.

Domowitz, Ian, & Wang Jianxin. 1994. "Auctions as Algorithms." *Journal of Economic Dynamics and Control*, 18:29–60.

Dore, Mohammed. 1995. "The Impact of John von Neumann's Method." In Rima, 1995, pp. 436–52.

Dore, Mohammed, Chakravarty, Sukhamoy, & Goodwin, Richard, eds. 1989. *John von Neumann and Modern Economics.* Oxford: Clarendon.

Dorfman, Robert. 1960. "Operations Research." *American Economic Review*, 50:575–623.

1984. "The Discovery of Linear Programming." *Annals of the History of Computing*, 6:283–95.

1997. *Economic Theory and Public Decisions.* Cheltenham: Elgar.

Dorfman, Robert, Samuelson, Paul, & Solow, Robert. 1958. *Linear Programming and Economic Analysis.* New York: McGraw-Hill.

Doti, James. 1978. "The Response of Economic Literature to Wars." *Economic Inquiry*, 16:616–25.

Dowling, Deborah. 1999. "Experimenting on Theories." *Science in Context*, 12:261–74.

Downey, Gary, & Dumit, Joseph, eds. 1997. *Cyborgs and Citadels.* Santa Fe: School of American Research Press.

Dresden, Max. 1987. *H. A. Kramers.* Berlin: Springer.

Dresher, Melvin. 1950. "Methods of Solution in Game Theory." *Econometrica*, 18:179–81.

1951. "Theory and Applications of Games of Strategy." RAND R-216. RAND Corporation, Santa Monica, Calif.

1961. *Games of Strategy*. Englewood, Cliffs, N.J.: Prentice-Hall.

Dreyfus, Herbert. 1988. *Mind over Machine.* Rev. ed. New York: Free Press.

Dreyfus, Stuart. 1961. "Dynamic Programming." In Ackoff, 1961, pp. 211–42.

Dupuy, Jean-Pierre. 1994. *Aux origines des sciences cognitives.* Paris: Decouverte.

1996. "The Autonomy of Social Reality." In Khalil & Boulding, 1996.

ed. 1998. *Self-Deception and Paradoxes of Rationality.* Stanford: CSLI.

Duren, Peter, ed. 1988. *A Century of Mathematics in America, Part I.* Providence: American Mathematical Society.

ed. 1989. *A Century of Mathematics in America, Part III.* Providence: American Mathematical Society.

Dyson, George. 1997. *Darwin among the Machines.* Reading, Mass.: Addison-Wesley.

Dzuback, Mary Ann. 1991. *Robert M. Hutchins.* Chicago: University of Chicago Press.

Earman, John, & Norton, John. 1998. "The Wrath of Maxwell's Demon." *Studies in the History and Philosophy of Modern Physics*, 4:435–71.

1999. "Exorcist IV: The Wrath of Maxwell's Demon, Pt. 2." *Studies in the History and Philosophy of Science*, B, 30:1–40.

Economides, Nicholas, & Schwartz, Robert. 1995. "Electronic Call Market Trading." *Journal of Portfolio Management*, 21 (Spring):10–18.

Edgeworth, Francis. 1881. *Mathematical Psychics.* London: C. K. Paul.
Edwards, Paul. 1996. *The Closed World.* Cambridge: MIT Press.
Egidi, Massimo, Marris, R., & Viale, R. eds. 1992. *Economics, Bounded Rationality and the Cognitive Revolution.* Aldershot: Elgar.
Eigen, Manfred. 1971. "Self-Organization of Matter and the Evolution of Biological Macromolecules." *Naturwissenschaften*, 58:465–528.
1995. "What Will Endure in 20th Century Biology?" In Murphy & O'Neill, 1995.
Eldridge, Niles. 1997. "Evolution in the Marketplace." *Structural Change and Economic Dynamics*, 8:471–88.
Ellerman, David. 1980. "Property Theory and Orthodox Economics." In E. Nell, ed., *Growth, Property and Profits*, pp. 250–63. Cambridge: Cambridge University Press.
Elliott, Kevin. 2000. "Analytic Philosophy and Post-WWII Military Research." Unpublished manuscript, University of Notre Dame.
Ellsberg, Daniel. 1956. "The Theory of the Reluctant Duelist." *American Economic Review*, 46:909–23.
1961a. "Risk, Ambiguity and the Savage Axioms." *Quarterly Journal of Economics*, 75:643–69.
1961b. "The Crude Analysis of Strategic Choices." *American Economic Review*, 51.2:472–78.
Elster, Jon. 1998. "Emotions and Economic Theory." *Journal of Economic Literature*, 36:47–74.
Emmeche, Claus. 1994. *The Garden in the Machine.* Princeton: Princeton University Press.
Emmett, Ross. 1998. "What Is Truth in Capital Theory?" In Davis, 1998, pp. 231–50.
Engelbrecht, Hans-Jurgen. 1997. "A Comparison and Critical Assessment of Porat and Rubin and Wallace and North." *Information Economics and Policy*, 9:271–90.
Englander, E. 1988. "Technology and Oliver Williamson's Transactions Cost Economics." *Journal of Economic Behavior and Organization*, 10:339–53.
Enke, Stephen. 1951. "Equilibrium among Spatially Separated Markets: Solution by Electric Analogue." *Econometrica*, 19:40–47.
1954. "More on the Misuse of Mathematics in Economics." *Review of Economics and Statistics*, 36:131–33.
1958a. "An Economist Looks at Air Force Logistics." *Review of Economics and Statistics*, 40:230–39.
1958b. "On the Economic Management of Large Organizations: A Case Study in Military Logistics Involving Laboratory Simulation" RAND P-1368. RAND Corporation, Santa Monica, Calif.
1965. "Using Costs to Select Weapons." *American Economic Review, Papers and Proceedings*, 55.
ed. 1967. *Defense Management.* Englewood Cliffs, N.J.: Prentice-Hall.
Enthoven, Alain. 1963a. "Economic Analysis in the Department of Defense." *American Economic Review*, 53:413–23.
1963b. "Systems Analysis and Decision Making." *Military Review*, 43:7–8.
1969. "The Planning, Programming and Budgeting System in the Department of Defense: Some Lessons from Experience." In *The Analysis and Evaluation*

of Public Expenditures, 3:901–8. Joint Economic Committee. Washington, D.C.: GPO.

1995. "Tribute to Charles Hitch." *OR/MS Today.*

Enthoven, Alain, & Rowen, Henry. 1961. "Defense Planning and Organization." In *Public Finances: Sources, Needs and Utilization*, pp. 365–420. Princeton: Princeton University Press.

Epstein, Joshua, & Axtell, Robert. 1996. *Growing Artificial Societies.* Washington, D.C.: Brookings.

Epstein, Richard, & Carnielli, Walter. 1989. *Computability.* Pacific Grove, Calif.: Wadsworth & Brooks.

Epstein, Roy. 1987. *A History of Econometrics*. Amsterdam: North Holland.

Ewing, Katherine. 1990. "The Illusion of Wholeness." *Ethos*, 18:251–78.

Fagin, Ronald, Halpern, Joseph, Moses, Yoram, & Vardi, Moshe. 1996. *Reasoning about Knowledge.* Cambridge: MIT Press.

Fama, Eugene. 1980. "Agency Problems and the Theory of the Firm." *Journal of Political Economy*, 88:288–307.

Fama, Eugene, & Laffer, Arthur. 1971. "Information and Capital Markets." *Journal of Business*, 9:44.

Farmer, J. D. 1991. "A Rosetta Stone for Connectionism." In Forrest, 1991, pp. 153–87.

Feigenbaum, Edward. 1989. "What Hath Simon Wrought?" In D. Klahr & K. Kotovsky, eds., *Complex Information Processing*, pp. 165–82. Hillsdale, N.J.: Erlbaum.

Feigenbaum, Mitchell. 1995. "Computer-Generated Physics." In Brown et al., 1995, pp. 1823–54.

Feiwel, George, ed. 1982. *Samuelson and Neoclassical Economics.* Boston: Kluwer.

ed. 1987a. *Arrow and the Ascent of Modern Economics.* New York: New York University Press.

ed. 1987b. *Arrow and the Foundations of the Theory of Economic Policy.* New York: New York University Press.

Ferstman, Chaim, & Kalai, Ehud. 1993. "Complexity Considerations and Market Behavior." *RAND Journal of Economics*, 24:224–34.

Feyerabend, Paul. 1955. Review of *Mathematical Foundations of Quantum Mechanics*, by John von Neumann. *British Journal for the Philosophy of Science*, 8:343–47.

1999. *Knowledge, Science and Relativism.* Cambridge: Cambridge University Press.

Feynman, Richard. 1996. *The Feynman Lectures on Computation.* Ed. A. Hey & R. Allen. Reading, Mass.: Addison-Wesley.

Fine, Terrence. 1973. *Theories of Probability.* New York: Academic Press.

Fishburn, Peter. 1989. "Retrospective on the Utility Theory of von Neumann and Morgenstern." *Journal of Risk and Uncertainty*, 2:127–58.

1991. "Decision Theory: The Next 100 Years." *Economic Journal*, 101:27–32.

Fisher, Donald. 1993. *Fundamental Development of the Social Sciences: Rockefeller Philanthropy and the US Social Science Research Council.* Ann Arbor: University of Michigan Press.

Fisher, Franklin. 1989. "Games Economists Play: A Noncooperative View." *RAND Journal of Economics*, 20:113–24.

Fisher, Gene, & Walker, Warren. 1994. "Operations Research and the RAND Corporation." RAND P-7857. RAND Corporation, Santa Monica, Calif.

Fisher, Irving. 1965. *Mathematical Investigations into the Theory of Value and Prices.* New York: Kelley.

Fisher, R. A. 1922. "On the Mathematical Foundations of Theoretical Statistics." *Philosophical Transactions of the Royal Society of London*, A222:309–68.

Flagle, Charles, Huggins, William, & Roy, Robert, eds. 1960. *Operations Research and Systems Engineering.* Baltimore: Johns Hopkins Press.

Flamm, Kenneth. 1987. *Targeting the Computer.* Washington, D.C.: Brookings Institution.

1988. *Creating the Computer*. Washington, D.C.: Brookings Institution.

Fleck, James. 1987. "The Development and Establishment of AI." In B. Bloomfield, ed., *The Question of AI.* London: Croom Helm.

1994. "Knowing Engineers." *Social Studies of Science*, 24:105–13.

Flood, Merrill. 1951a. "A Preference Experiment." RAND P-256, P-258, P-263. RAND Corporation, Santa Monica, Calif.

1951b. Review of *Cybernetics*, by Norbert Wiener. *Econometrica*, 19:477–78.

1952a. "Some Experimental Games." RAND RM-798. RAND Corporation, Santa Monica, Calif.

1952b. "Testing Organization Theories." RAND P-312. RAND Corporation, Santa Monica, Calif.

1954. "Application of Transportation Theory to Scheduling a Military Tanker Fleet." *Journal of the Operations Research Society of America*, 2:150–62.

1956. "The Travelling Salesman Problem." *Journal of the Operations Research Society of America*, 4:61–75.

1958a. "Some Experimental Games." *Management Science*, 5:5–26.

1958b. "Operations Research and Automation Science." *Journal of Industrial Engineering*, 9:239–42.

1963. "What Future Is There for Intelligent Machines?" *General Systems*, 8:219–26.

1978. "Let's Redesign Democracy." *Behavioral Science*, 23:429–40.

Flood, Merrill, & Savage, Leonard. 1948. "A Game Theoretic Study of the Tactics of Area Defense." RAND D-287. RAND Corporation, Santa Monica, Calif.

Fogel, David, ed. 1998. *Evolutionary Computation: The Fossil Record.* New York: IEEE Press.

Foley, Duncan. 1970. "Economic Equilibrium with Costly Marketing." *Journal of Economic Theory*, 2:276–91.

1994. "A Statistical Theory of Markets." *Journal of Economic Theory*, 62:321–45.

1998. Introduction to Albin, 1998, pp. 3–72.

1999. "The Ins and Outs of Late 20th Century Economics." In Heertje, 1999, pp. 70–118.

Fontana, Walter, & Buss, Leo. 1996. "The Barrier of Objects: From Dynamical Systems to Bounded Organizations." In John Casti & Anders Karlqvist, eds., *Barriers and Boundaries*, pp. 55–115. Reading: Perseus Books.

Forrest, Stephanie, ed. 1991. *Emergent Computation.* Cambridge: MIT Press.
Forman, Paul. 1987. "Beyond Quantum Electronics." *Historical Studies in the Physical and Biological Sciences.* 18:149–229.
1997. "Recent Science: Late Modern and Postmodern." In Soderqvist, 1997, pp. 179–214.
Forsythe, Diana. 1993a. "Engineering Knowledge." *Social Studies of Science*, 23:445–77.
1993b. "The Construction of 'Work' in AI." *Science, Technology and Human Values*, 18:460–79.
Fortun, Michael, & Schweber, Silvan. 1993. "Scientists and the Legacy of WWII." *Social Studies of Science*, 23:595–642.
Foucault, Michel. 1977. *Discipline and Punish.* Trans. Alan Sheridan. New York: Pantheon.
Fouraker, Lawrence, & Siegel, Sidney. 1963. *Bargaining Behavior.* New York: McGraw-Hill.
Fox, Robert, & Guagnini, E. 1998–99. "Laboratories, Workshops and Sites." *Historical Studies in the Physical and Biological Sciences*, 29.1:55–140; 29.2:193–294.
Frank, Robert. 1988. *Passions within Reason.* New York: Norton.
Frank, Thomas, & Weiland, Matt, eds. 1997. *Commodify Your Dissent.* New York: Norton.
Friedman, Daniel. 1991. "Evolutionary Games in Economics." *Econometrica*, 59:637–66.
1998. "On the Economic Applications of Evolutionary Game Theory." *Journal of Evolutionary Economics*, 8:15–43.
Friedman, Daniel, & Rust, John, eds. 1993. *The Double Auction Market.* Reading, Mass.: Addison-Wesley.
Friedman, Daniel, & Sunder, Shyam. 1994. *Experimental Economics: A Primer.* Cambridge: Cambridge University Press.
Friedman, Matthew. 1997. *Fuzzy Logic.* Montreal: Vehicle Press.
Friedman, Milton. 1949. "The Marshallian Demand Curve." *Journal of Political Economy*, 57:463–95.
1953. "Choice, Chance and the Personal Distribution of Income." *Journal of Political Economy*, 61:277–90.
1966. *Price Theory: A Provisional Text.* Rev. ed. Chicago: Aldine.
Friedman, Milton, & Kuznets, Simon. 1945. *Income from Independent Professional Practice.* New York: NBER.
Friedman, Milton, & Wallis, W. A. 1942. "The Empirical Derivation of Indifference Functions." In Lange et al., 1942, pp. 175–89.
Fulkerson, D., & Johnson, S. 1957. "A Tactical War Game." *Operations Research*, 5:704–12.
Fuller, Steve. 1993. *Philosophy, Rhetoric and the End of Knowledge.* Madison: University of Wisconsin Press.
2000. *Thomas Kuhn: A Philosophical History for Our Time.* Chicago: University of Chicago Press.
Futia, Carl. 1977. "The Complexity of Economic Decision Rules." *Journal of Mathematical Economics*, 4:289–99.

Gabor, Denis. 1946. "Theory of Communication." *Journal of the Institute of Electrical Engineers*, 93:429–57.

Gale, David. 1953. "A Theory of N-Person Games with Perfect Information." *Proceedings of the National Academy of Sciences*, 39:496–501.

1955. "The Law of Supply and Demand." *Mathematica Scandinavica*, 5:155–69.

1963. "A Note on Global Instability of General Equilibrium." *Naval Research Logistics Quarterly*, 10:81–87.

Gale, David, & Stewart, F. 1953. "Infinite Games with Perfect Information." In Kuhn & Tucker, 1953, pp. 245–66.

Galison, Peter. 1994. "The Ontology of the Enemy." *Critical Inquiry*, 21:228–66.

1996. "Computer Simulations in the Trading Zone." In Peter Galison & David Stump, eds., *The Disunity of Science*, pp. 118–57. Stanford: Stanford University Press.

1997. *Image and Logic: A Material Culture of Microphysics.* Chicago: University of Chicago Press.

1998. "The Americanization of Unity." *Daedalus*, 127:45–71.

Galison, Peter, & Hevly, Bruce, eds. 1992. *Big Science.* Stanford: Stanford University Press.

Gamow, George. 1970. *My World Line.* New York: Viking.

Gandy, Robin. 1988. "The Confluence of Ideas in 1936." In Rolf Herken, ed., *The Universal Turing Machine: A Half Century Survey*, pp. 51–102. Oxford: Oxford University Press.

1996. "Human versus Machine Intelligence." In Millican & Clark, 1996.

Gani, J., ed. 1986. *The Craft of Probabilistic Modeling.* New York: Springer.

Gardner, Howard. 1987. *The Mind's New Science.* New York: Basic.

Gardner, Martin. 1968. *Logic Machines, Diagrams and Boolean Algebra.* New York: Dover.

Garey, M., & Johnson, D. 1979. *Computers and Intractability*. New York: Freeman.

Garner, W. 1962. *Uncertainty and Structure*. New York: Wiley.

Gass, S. 1989. "Comments on the History of Linear Programming." *Annals of the History of Computing*, 11:147–51.

Geanakoplos, John. 1992. "Common Knowledge." *Journal of Economic Perspectives*, 6:53–82.

Geiger, Roger. 1992. "Science, Universities and National Defense, 1945–70." In Thackray, 1992.

Geissler, M. 1959. "The Use of Monte Carlo Methods, Man-Machine Simulation and Analytical Methods for Studying Large Human Organizations." RAND P-1634. RAND Corporation, Santa Monica, Calif.

George, Peter. 1998. *Dr. Strangelove.* New York: Barnes & Noble.

Georgescu-Roegen, Nicholas. 1967. *Analytical Economics*. Cambridge: Harvard University Press.

1969. "A Critique of Statistical Principles in Relation to Social Phenomena." *Revue Internationale de Sociologie*, 5:347–70.

1971. *The Entropy Law and the Economic Process.* Cambridge: Harvard University Press.

1975. "The Measure of Information: a critique." In J. Rose & C. Bilciu, eds., *Modern Trends in Cybernetics and Systems*, 3:187–217. Berlin: Springer Verlag.

1976. *Energy and Economic Myths.* Oxford: Pergamon.

1987. "Entropy." In J. Eatwell et al., eds., *The New Palgrave*, pp. 153–56. London: Macmillan.

Getting, Ivan. 1989. *All in a Lifetime.* New York: Vantage Press.

Gibson, William. 1984. *Neuromancer*. New York: Ace.

Gieryn, Thomas. 1995. "Boundaries of Science." In Sheila Jasanoff et al., eds., *Handbook of Science and Technology Studies*, pp. 393–443. Thousand Oaks, Calif.: Sage.

Gigerenzer, Gerd. 1992. "Discovery in Cognitive Psychology." *Science in Context*, 5:329–50.

2000. *Adaptive Thinking.* Oxford: Oxford University Press.

Gigerenzer, Gerd, & Murray, David. 1987. *Cognition as Intuitive Statistics.* Hillsdale, N.J.: Erlbaum.

Gigerenzer, Gerd, et al. 1989. *The Empire of Chance*. Cambridge: Cambridge University Press.

Gilles, Donald. 1996. *Artificial Intelligence and the Scientific Method.* Oxford: Oxford University Press.

Gilboa, Itzhak. 1988. "The Complexity of Computing a Best Response Automaton in Repeated Games." *Journal of Economic Theory*, 45:342–52.

Gilman, S. 1956. "Operations Research in the Army." *Military Review*, 36:54–64.

Girshick, M., & Haavelmo, T. 1947. "Statistical Analysis of the Demand for Food." *Econometrica*, 15:79–110.

Glass, Robert, ed. 1998. *In the Beginning: Personal Recollections of Software Engineers.* Los Alamitos, Calif.: IEEE Press.

Glimm, James, Impagliazzo, John, & Singer, Isadore, eds. 1990. *The Legacy of John von Neumann.* Providence: American Mathematical Society.

Glosten, Lawrence. 1994. "Is the Electronic Open Limit Order Book Inevitable?" *Journal of Finance*, 49:1127–61.

Gode, D., & Sunder, S. 1993. "Allocative Efficiency of Markets with Zero Intelligence Traders: Markets as a Partial Substitute for Individual Rationality." *Journal of Political Economy*, 101:119–37.

1997. "What Makes Markets Allocatively Efficient?" *Quarterly Journal of Economics*, 112:603–30.

Goldstein, Andrew. 1991. "Interview with Charles and Katherine Fowler." <www.ieee.org/history_ce...tories/transcripts/fowler.html>.

Goldstein, J. R. 1961. "RAND: The History, Operations and Goals of a Nonprofit Corporation." RAND P-2236-1. RAND Corporation, Santa Monica, Calif.

Goldstein, William, & Hogarth, Robin, eds. 1997. *Research on Judgment and Decision Making.* Cambridge: Cambridge University Press.

Goldstine, Herman. 1972. *The Computer from Pascal to von Neumann.* Cambridge: MIT Press.

Goodwin, Craufurd, ed. 1991. *Economics and National Security.* Supplement to HOPE, vol. 23. Durham: Duke University Press.

1999. "The Patrons of Economics in a Time of Transition." In M. Rutherford, & M. Morgan, eds., *The Transformation of American Economics*, pp. 53–81. Durham: Duke University Press.

Goodwin, Craufurd, & Rogers, Alex. 1997. "Cyberpunk and Chicago." In J. Henderson, ed., *The State of the History of Economics*, pp. 39–55. London: Routledge.

Goodwin, Irwin. 1995. "Is It True, as 60 Minutes Confidently Argues, Physicists Are to Blame for Derivative Debacles?" *Physics Today*, 48.5 (May):55.

Goodwin, Richard. 1951. "Iteration, Automatic Computers, and Economic Dynamics." *Metroeconomica*, 3:1–17.

Gordon, Lincoln. 1956. "Economic Aspects of Coalition Diplomacy – the NATO Experience." *International Organization*, 10:529–43.

Gorn, Michael. 1988. *Harnessing the Genie.* Washington, D.C.: GPO.

Gottinger, Hans-Werner. 1983. *Coping with Complexity.* Boston: Reidel.

Gould, Stephen Jay. 1995. "What Is Life? As a Problem in History." In Murphy & O'Neill, 1995.

Granger, Clive, & Morgenstern, Oskar. 1970. *Predictability of Stock Market Prices.* Lexington, Mass.: Heath.

Grattan-Guinness, Ivor. 1990. "Work for the Hairdressers." *Annals of the History of Computing*, 12:177–85.

1991. "The Correspondence between George Boole and Stanley Jevons, 1863–64." *History and Philosophy of Logic*, 12:15–35.

ed. 1994a. *Companion Encyclopaedia of the History and Philosophy of the Mathematical Sciences.* 2 vols. London: Routledge.

1994b. "From Virtual Velocities to Economic Action." In Mirowski, 1994a, pp. 91–108.

2000. *The Search for Mathematical Roots, 1870–1940.* Princeton: Princeton University Press.

Gray, Chris Hables, ed. 1995. *The Cyborg Handbook*. New York: Routledge.

1997. *Postmodern War: The New Politics of Conflict.* New York: Guilford Press.

Green, Philip. 1966. *Deadly Logic.* New York: Schocken.

Greenlaw, Raymond, Hoover, H. J., & Ruzzo, Walter. 1995. *Limits to Parallel Computation.* New York: Oxford University Press.

Grossman, Sanford. 1989. *The Informational Role of Prices.* Cambridge: MIT Press.

Grossman, Sanford, & Stiglitz, Joseph. 1980. "On the Impossibility of Informationally Efficient Markets." *American Economic Review*, 70:393–402.

Gruenberger, F. 1979. "History of the Johnniac." *Annals of the History of Computing*, 1:49–64.

Grunberg, Emile, & Modigliani, Franco. 1954. "The Predictability of Social Events." *Journal of Political Economy*, 62:465–78.

Grunbichler, A., Longstaff, F., & Schwartz, E. 1994. "Electronic Screen Trading and the Transmission of Information." *Journal of Financial Intermediation*, 3:166–87.

Guedj, Denis. 1985. "Nicolas Bourbaki, Collective Mathematician." *Mathematical Intelligencer*, 7:18–22.

Guerlac, Henry. 1987. *Radar in World War II.* American Institute of Physics. Los Angeles: Tomash.

Guice, Jon. 1998. "Controversy and the State: Lord ARPA and Intelligent Computing." *Social Studies of Science*, 28:103–18.

Guillaume, Georges, & Guillaume, E. 1932. *Sur le fondements de l'economie rationelle avec une technique de la prevision.* Paris: Gauthier-Villars.

Gusterson, Hugh. 1996. *Nuclear Rites.* Berkeley: University of California Press.

Guttman, R., & Maes, P. 1998. "Cooperative vs. Competitive Multiagent Negotiations in Retail Electronic Commerce." In Klusch & Weiss, 1998.

Guttman, R., Moukas, A., & Maes, P. 1998. "Agent-Mediated Electronic Commerce: A Survey." *Knowledge Engineering Review*, 13:143–46.

Hacker, B. 1993. "Engineering a New Order: Military Institutions and Technical Education." *Technology and Culture*, 34:1–27.

Hacking, Ian. 1986. "Weapons Research and the Form of Scientific Knowledge." In D. Copp, ed., *Nuclear Weapons, Deterrence and Disarmament*, pp. 237–60. Boulder, Colo.: Westview.

1995. *Rewriting the Soul.* Princeton: Princeton University Press.

1998. "Canguillhem amid the Cyborgs." *Economy and Society*, 27:202–16.

1999. *The Social Construction of What?* Cambridge: Harvard University Press.

Hafner, Katie, & Lyon, Matthew. 1996. *Where Wizards Stay Up Late.* New York: Touchstone.

Hagen, Joel. 1992. *The Entangled Bank.* New Brunswick: Rutgers University Press.

Hagstrom, K. 1938. "Pure Economics as a Statistical Theory." *Econometrica*, 6:40–47.

Hailperin, Thomas. 1986. *Boole's Logic and Probability.* Amsterdam: North Holland.

Hakansson, Nils, Beja, Avraham, & Kale, Jivendra. 1985. "On the Feasibility of Automated Market Making by Programmed Specialists." *Journal of Finance*, 60:1–20.

Hakansson, Nils, Kunkel, J., & Ohlson, J. 1982. "Sufficient and Necessary Conditions for Information to Have Social Value in Pure Exchange." *Journal of Finance*, 37:1169–81.

Halford, G., Wilson, W., & Phillips, S. 1998. "Processing Capacity Defined by Relational Complexity." *Behavioral and Brain Sciences*, 21:803–64.

Hall, Brian. 1998. "Overkill Is Not Dead." *New York Times Sunday Magazine*, March 15, pp. 42–49.

Halmos, Paul. 1973. "The Legend of John von Neumann." *American Mathematical Monthly*, 80:382–94.

Halpern, J., & Moses, Y. 1984. "Knowledge and Common Knowledge in a Distributed Environment." IBM Research Report RJ4421. San Jose.

Hamilton, William. 1963. "The Evolution of Altruistic Behavior." *American Naturalist*, 97:354–56.

1964. "The Genetical Evolution of Social Behavior." *Journal of Theoretical Biology*, 7:1–16, 17–32.

Hamming, R. W. 1988. "The Use of Mathematics." In Duren, 1988.

Hammond, Dan. 1993. "An Interview with Milton Friedman." In Bruce Caldwell, ed., *Philosophy and Methodology of Economics I*, pp. 214–38. Aldershot: Elgar.
1996. *Theory and Measurement*. Cambridge: Cambridge University Press.
Hammond, Peter. 1990. "The Role of Information in Economics." European University Institute Paper, Eco no. 90/10, May.
Hands, D. Wade. 1985. "The Structuralist View of Economic Theories." *Economics and Philosophy*, 1:303–35.
1990. "Grunberg and Modigliani, Public Predictions and the New Classical Macroeconomics." *Research in the History of Economic Thought and Methodology*, 7:207–23.
1993. *Testing, Rationality and Progress.* Lanham, Md.: Rowman & Littlefield.
1994. "Restabilizing Dynamics." *Economics and Philosophy*, 10:243–83.
1995. "Blurred Boundaries: Recent Changes in the Relationship between Economics and the Philosophy of Natural Science." *Studies in the History and Philosophy of Science*, 26:1–22.
2000. *Reflection without Rules.* Cambridge: Cambridge University Press.
Hands, D. Wade, & Mirowski, P. 1998. "Harold Hotelling and the Neoclassical Dream." In R. Backhouse, D. Hausman, U. Maki, & A. Salanti, eds., *Economics and Methodology: Crossing Boundaries*, pp. 322–97. London: Macmillan.
1999. "A Paradox of Budgets." In Mary Morgan & Malcolm Rutherford, eds., *The Transformation of American Economics*, pp. 260–92. Durham: Duke University Press.
Hannon, Bruce. 1997. "The Uses of Analogy in Biology and Economics." *Structural Change and Economic Dynamics*, 8:471–88.
Haraway, Donna. 1981–82. "The High Cost of Information in Post WWII Evolutionary Biology." *Philosophical Forum*, 13:244–28.
1991. *Simians, Cyborgs and Women*. New York: Routledge.
1997. *Modest Witness@Second Millennium*. London: Routledge.
2000. *How Like a Leaf.* New York: Routledge.
Hardin, Garrett. 1961. *Nature and Man's Fate.* New York: Mentor
Hargreaves-Heap, Shaun, & Varoufakis, Yanis. 1995. *Game Theory: A Critical Introduction.* London: Routledge.
Harms, William. 1998. "The Use of Information Theory in Epistemology." *Philosophy of Science*, 65:472–501.
Harries-Smith, Peter. 1995. *A Recursive Vision.* Toronto: University of Toronto Press.
Harrison, Alexander. 2000. "Alice's Springs." *Times Literary Supplement*, June 9, pp. 14–15.
Harsanyi, John. 1967/8. "Games with Incomplete Information Played by Bayesian Players." *Management Science*, 14:159–82, 320–24, 486–502.
1977. *Rational Behavior and Bargaining Equilibrium in Games and Social Situations.* Cambridge: Cambridge University Press.
1982. "Noncooperative Bargaining Models." In *Games, Economic Dynamics and Time Series Analysis.* Wein: Physica Verlag.
Hart, David. 1998. *Forged Consensus.* Princeton: Princeton University Press.
Hart, Sergiu, & Neyman, Abraham, eds. 1995. *Games and Economic Theory: Selected Contributions in Honor of Robert Aumann.* Ann Arbor: University of Michigan Press.

Hartcup, Guy, 2000. *The Effect of Science on the Second World War*. London: Macmillan.

Hartley, R. 1928. "Transmission of Information." *Bell System Technical Journal*, 7:535–63.

Hartmanis, J. 1979. "Observations about the Development of Theoretical Computer Science." In *IEEE Foundations of Computer Science*. Los Angeles: IEEE Press.

Hausrath, Alfred. 1971. *Venture Simulation in War, Business and Politics*. New York: McGraw-Hill.

Hawkins, Mike. 1997. *Social Darwinism in European and American Thought, 1860–1945*. Cambridge: Cambridge University Press.

Hayek, F. ed. 1935. *Collectivist Economic Planning*. London: Routledge.

1944. *The Road to Serfdom*. Chicago: University of Chicago Press.

1945. "The Use of Knowledge in Society." *American Economic Review*, 35:519–30.

1948. *Individualism and Economic Order*. Chicago: University of Chicago Press.

1952. *The Sensory Order*. Chicago: University of Chicago Press.

1967. *Studies in Philosophy, Politics and Economics*. Chicago: University of Chicago Press.

1976. *The Pure Theory of Capital*. London: Routledge Kegan Paul.

1982. "The Sensory Order after 25 Years." In W. Weimer & D. Palermo, eds., *Cognition and the Symbolic Process*. Hillsdale, N.J.: Erlbaum.

1994. *Hayek on Hayek*. Ed. S. Kresge & L. Wenar. Chicago: University of Chicago Press.

Hayles, N. Katherine. 1990a. *Chaos Bound*. Ithaca: Cornell University Press.

1990b. "Designs on the Body: Wiener, Cybernetics and the Play of Metaphor." *History of the Human Sciences*, 3:211–28.

1994. "Boundary Disputes: Homeostasis, Reflexivity and the Boundaries of Cybernetics." *Configurations*, 2:441–48.

1995a. "Making the Cut: What Systems Theory Can't See." *Cultural Critique*, 31:71–100.

1995b. "Simulated Nature and Natural Simulations." In Cronon, 1995, pp. 409–25.

1999. *How We Became Posthuman*. Chicago: University of Chicago Press.

Haywood, O. 1951. *The Military Doctrine of Decision and the von Neumann Theory of Games*. RAND RM-258. RAND Corporation, Santa Monica, Calif.

1954. "Military Decision and Game Theory." *Journal of the Operations Research Society of America*, 2:365–85.

Healy, David. 1998. *The Antidepressant Era*. Cambridge: Harvard University Press.

Heertje, Arnold, Ed. 1999. *Makers of Modern Economics*. Vol. 4. Cheltenham: Elgar.

Heims, Steve. 1980. *John von Neumann and Norbert Wiener*. Cambridge: MIT Press.

1991. *The Cybernetics Group*. Cambridge: MIT Press.

Helmer, Olaf. 1949. "The Conference on Applications of Game Theory to Tactics." RAND D-444. RAND Corporation, Santa Monica, Calif.

1958. "The Prospects for a Unified Theory of Organizations." *Management Science*, 4:172–77.

1963. "The Game-Theoretical Approach to Organization Theory." *Synthese*, 15:245–53.

Helmer, Olaf, & Blackwell, David. 1949. "Continuous Three Move Games." RAND D-435. RAND Corporation, Santa Monica, Calif.

Helmer, Olaf, & Shapley, Lloyd. 1957. "Brief Description of the SWAP Game." RM-2058. RAND Corporation, Santa Monica, Calif.

Henderson, James. 1996. *Early British Mathematical Economics.* Savage, Md.: Rowman & Littlefield.

Herken, Gregg. 1987. *Counsels of War.* Rev. ed. Oxford: Oxford University Press.

Herken, Rolf, ed. 1988. *The Universal Turing Machine: A Half Century Survey.* Oxford: Oxford University Press.

Hersh, Seymour. 1991. *The Sampson Option.* New York: Random House.

Hershberg, James. 1993. *James Conant.* New York: Knopf.

Herschlag, Miriam, & Zwick, Rami. 2000. "Internet Auctions – Popular and Professional Literature Review." <http://members.theglobe.com/herschlag/default.html>.

Heusner, Beatrice. 1998. *Nuclear Mentalities?* Basingstoke: Macmillan.

Hilbert, D., & Ackerman, W. 1928. *Grundzuge der theoretischian Logik.* Berlin: Springer.

Hildenbrand, Werner. 1994. *Market Demand.* Princeton: Princeton University Press.

Hildreth, Clifford. 1953. "Alternative Conditions for Social Orderings." *Econometrica*, 21:81–94.

1986. *The Cowles Commission in Chicago.* Berlin: Springer Verlag.

Hillis, W. Daniel. 1992. "What Is Massively Parallel Computing?" *Daedalus*, 121 (Winter):1–15.

Hilton, Peter. 1986. "M. H. A. Newman." *Bulletin of the London Mathematical Society*, 18:67–72.

Hilton, Ronald. 1981. "The Determinants of Information Value." *Management Science*, 27:57–64.

Hirsch, A., & de Marchi, N. 1990. *Milton Friedman.* Ann Arbor: University of Michigan Press.

Hirshleifer, Jack. 1977. "Economics from a Biological Viewpoint." *Journal of Law and Economics*, 20:1–54.

1978. "Natural Economy vs. Political Economy." *Journal of Social and Biological Structures*, 1:319–37.

Hirshleifer, Jack, & Riley, John. 1992. *The Analytics of Uncertainty and Information.* Cambridge: Cambridge University Press.

Hitch, Charles. 1953. "Sub-Optimization in Operations Problems." *Journal of the Operations Research Society of America*, 1:87–99.

1958. "Economics and Military Operations Research." *Review of Economics and Statistics*, 40:199–209.

Hitch, Charles, & McKean, Roland. 1960. *Economics of Defense in the Nuclear Age.* Cambridge: Harvard University Press.

Hitchcock, Frank. 1941. "The Distribution of a Product from Several Sources and Numerous Localities." *Journal of Mathematical Physics*, 20:224–30.

Hodges, Andrew. 1983. *The Enigma.* New York: Simon & Schuster.

Hodgson, Geoffrey. 1993. *Economics and Evolution*. Ann Arbor: University of Michigan Press.

Hogarth, Robin, & Reder, Melvin, eds. 1987. *Rational Choice*. Chicago: University of Chicago Press.

Hoggatt, Austin. 1959. "An Experimental Business Game." *Behavioral Science*, 4:153–203.

1969. "Response of Paid Student Subjects to Differential Behavior of Robots in Bifurcated Duopoly Games." *Review of Economic Studies*, 36:417–32.

Hoggatt, A., Brandstatter, H., & Blatman, P. 1978. "Robots as Instrumental Functions in the Study of Bargaining Behavior." In H. Sauermann, ed., *Bargaining Behavior*. Tubingen: Mohr.

Holland, John. 1995. *Hidden Order*. Reading, Mass.: Addison-Wesley.

Holland, John, & Miller, John. 1991. "Artificial Adaptive Agents in Economic Theory." *American Economic Review Papers and Proceedings*, 81:365–70.

Holley, I. B. 1970. "The Evolution of Operations Research and Its Impact on the Military Establishment." In Wright & Paszek, 1970.

Hollis, Martin, & Sugden, Robert. 1993. "Rationality in Action." *Mind*, 102:1–34.

Holm, Hakan. 1992. *Complexity in Economic Theory: An Automata Theory Approach*. Lund Economic Studies no. 53. Lund.

Hood, W., & Koopmans, T., eds. 1953. *Studies in Econometric Method*. New Haven: Yale University Press.

Horgan, John. 1995. "From Complexity to Perplexity." *Scientific American*, 262 (June):104–9.

1996. *The End of Science*. Reading, Mass.: Addison-Wesley.

1999. *The Undiscovered Mind*. New York: Free Press.

Hotelling, Harold. 1932a. "Edgeworth's Taxation Paradox and the Nature of Demand and Supply Functions." *Journal of Political Economy*, 40:577–616.

1932b. Review of *New Methods of Measuring Marginal Utility*, by Ragnar Frisch. *Journal of the American Statistical Association*, 27:451–52.

1949. "Discussion of Paper by Roy." *Econometrica*, suppl., 17:188–89.

Hounshell, David. 1997a. "The Medium Is the Message, or How Context Matters." Unpublished manuscript, Carnegie Mellon University. Later version in Hughes & Hughes, 2000.

1997b. "The Cold War, RAND and the Generalization of Knowledge, 1946–62." *Historical Studies in the Physical and Biological Sciences*, 27:237–67.

Houthakker, Hendrik. 1956. "Economics and Biology." Technical Report no. 30, Office of Naval Research, Stanford University.

Huberman, Bernardo. 1988. *The Ecology of Computation*. Amsterdam: North Holland.

1998. "Computation as Economics." *Journal of Economic Dynamics and Control*, 22:1169–86.

Huberman, Bernabo, & Glance, N. 1995. "The Dynamics of Collective Action." *Computational Economics*, 8:49–66.

Huberman, Bernabo, & Hogg, T. 1995. "Distributed Computation as an Economic System." *Journal of Economic Perspectives*, 9:141–52.

Hughes, Agatha, & Hughes, Thomas, eds. 2000. *Systems, Experts and Computers*. Cambridge: MIT Press.

Hughes, Thomas. 1998. *Rescuing Prometheus.* New York: Pantheon.

Huhns, Michael, & Singh, Munindar, eds. 1998. *Readings in Agents.* San Francisco: Morgan Kaufmann.

Hume, David. 1966. *A Treatise on Human Nature.* La Salle, Ill.: Open Court.

Hunsacker, Jerome, & Mac Lane, Saunders. 1973. "Edwin Bidwell Wilson." *Biographical Memoirs of the National Academy of Sciences*, 43:285–320.

Hurwicz, Leonid. 1945. "The Theory of Economic Behavior." *American Economic Review*, 35:909–25.

1951. "Theory of Economic Organization." *Econometrica*, 19:54.

1953. "What Has Happened to the Theory of Games?" *American Economic Review*, 65:398–405.

1960. "Optimality and Informational Efficiency in Resource Allocation Processes." In K. Arrow et al., eds., *Mathematical Methods in the Social Sciences*, pp. 27–46. Stanford: Stanford University Press.

1969. "On the Concept and Possibility of Informational Decentralization." *American Economic Review*, 59:513–34.

1986a. "On Informationally Decentralized Systems." In McGuire & Radner, 1986, pp. 297–309.

1986b. "Incentive Aspects of Decentralization." In K. Arrow & M. Intrilligator, eds., *Handbook of Mathematical Economics*, vol. 3. Amsterdam: North Holland.

Husbands, Phil, & Harvey, Inman, eds. 1997. *Fourth European Conference on Artificial Life.* Cambridge: MIT Press.

Huxley, Thomas. 1874. "On the Hypothesis That Animals Are Automata." *Fortnightly Review*, 22:556–89.

Hyman, Anthony. 1982. *Charles Babbage: Pioneer of Computers.* Princeton: Princeton University Press.

Ingrao, Bruna, & Israel, Giorgio. 1990. *The Invisible Hand.* Cambridge: MIT Press.

Innocenti, Alessandro. 1995. "Oskar Morgenstern and the Heterodox Possibilities of Game Theory in Economics." *European Journal of the History of Economic Thought*, 17:205–27.

Issacs, Rufus. 1975. "The Past and Some Bits of the Future." In J. Grote, ed., *The Theory and Application of Differential Games*, pp. 1–11. Boston: Reidel.

James, William. 1879. "Are We Automata?" *Mind*, 4:1–22.

Janssen, Marten. 1993. *Microfoundations.* London: Routledge.

1998. "Individualism and Equilibrium Coordination in Games." In Backhouse et al., 1998, pp. 1–27.

Jardini, David. 1998. "Economics and the Science of War at RAND, 1946–52." Paper presented to the 1999 AEA meetings, New York.

Jaynes, Edwin. 1957. "Information Theory and Statistical Mechanics." *Physical Review*, 106:620–30; 108:171–90.

Jeffress, Lloyd, ed. 1951. *Cerebral Mechanisms in Behavior: The Hixon Symposium.* New York: Wiley.

Jeffreys-Jones, Rhoda, & Andrew, Christopher, eds. 1997. *Eternal Vigilance?* London: Cass.

Jennings, Nicholas, Sycara, Katia, & Wooldridge, Michael. 1998. "A Roadmap of Agent Research and Development." *Autonomous Agents and Multi-Agent Systems*, 1:7–38.

Jennings, Nicholas, & Wooldridge, Michael, eds. 1998. *Agent Technology.* Berlin: Springer.

Jevons, William Stanley. 1870. "On the Natural Laws of Muscular Exertion." *Nature*, 2 (June 30):158–60.

1890. *Pure Logic and Other Minor Works.* London: Macmillan.

1905. *The Principles of Science.* 2nd ed. London: Macmillan.

1970. *Theory of Political Economy*. 1871. Reprint, Baltimore: Penguin.

1972–81. *Papers and Correspondence of William Stanley Jevons.* Ed. R. D. C. Black. London: Macmillan.

Johansen, Leif. 1982. "On the Status of the Nash Type of Noncooperative Equilibrium in Economic Theory." *Scandinavian Journal of Economics*, 84:421–41.

Johnson, Stephen. 1997b. "Three Approaches to Big Technology." *Technology and Culture*, 38:891–919.

Jones, Charles, Enros, Philip, & Tropp, Henry. 1984. "Kenneth O. May, 1915–77: His Early Life." *Historia Mathematica*, 11:359–79.

Jones, J. P. 1982. "Some Undecidable Determined Games." *International Journal of Game Theory*, 11:63–70.

Jopling, David. 2000. *Self-Knowledge and the Self.* New York: Routledge.

Judd, Kenneth. 1997. "Computational Economics and Economic Theory: Substitutes or Complements?" *Journal of Economic Dynamics and Control*, 21:907–42.

1998. *Numerical Methods in Economics.* Cambridge: MIT Press.

Judson, Horace. 1979. *The Eighth Day of Creation.* New York: Simon & Schuster.

Junnarkar, Sandeep. 1997. "Making Boolean Logic Work on the Molecular Level." *New York Times Cybertimes*, May 21 <www.nytimes.com>.

Kafka, Franz. 1961. *Parables and Paradoxes.* New York: Schocken.

Kagel, John, Battalio, Raymond, & Green, Leonard. 1995. *Economic Choice Theory: An Experimental Analysis of Animal Behavior.* Cambridge: Cambridge University Press.

Kagel, John, & Roth, Alvin. eds. 1995. *Handbook of Experimental Economics.* Princeton: Princeton University Press.

Kahan, James, & Helwig, Richard. 1971. "Coalitions: A System of Programs for Computer-Controlled Bargaining Games." *General Systems*, 16:31–41.

Kahn, Herman. 1962. *Thinking about the Unthinkable.* New York: Avon.

Kahn, Herman, & Marshall, Andrew. 1953. "Methods of Reducing Sample Size in Monte Carlo." *Operations Research*, 1:263–78.

Kahneman, Daniel, & Tversky, Amos. 1982. "Prospect Theory." *Econometrica*, 24:178–91.

1996. "On the Reality of Cognitive Illusions." *Psychological Review*, 103:582–91.

Kalai, Ehud. 1990. "Bounded Rationality and Strategic Complexity in Repeated Games." In T. Ichiishi, A. Neyman, & Y. Tauman, eds., *Game Theory and Applications*, pp. 131–57. San Diego: Academic Press.

Kalai, Ehud, & Stanford, William. 1988. "Finite Rationality and Interpersonal Complexity in Repeated Games." *Econometrica*, 56:397–410.

Kalakota, Ravi, & Whinston, Andrew. 1996. *Frontiers of Electronic Commerce.* Reading, Mass.: Addison-Wesley.

Kalish, G., Milnor, J., Nash, J., & Nering, E. 1952. "Some Experimental N-Person Games." RAND RM-948. RAND Corporation, Santa Monica, Calif.

Kanwisher, N. 1989. "Cognitive Heuristics and American Security Policy." *Journal of Conflict Resolution*, 33:652–75.

Kaplan, Fred. 1983. *The Wizards of Armageddon.* New York: Simon & Schuster.

Kaplansky, I. 1945. "A Contribution to von Neumann's Theory of Games." *Annals of Mathematics*, 46:474–79.

Kari, Lila. 1998. "DNA Computing." <www.csd.uwo.ca/faculty/lila>.

Karmack, Elaine, & Nye, Joseph, Jr., eds. 1999. *Democracy.com?* Hollis, N.H.: Hollis Publishing.

Karnow, Curtis. 1996. "Liability for Distributed Artificial Intelligences." *Berkeley Technology Law Journal*, 11.1.

Kassel, Simon. 1971. *Soviet Cybernetics Research.* Santa Monica: RAND R-909. RAND Corporation, Santa Monica, Calif.

Kathren, Ronald, et al., eds. 1994. *The Plutonium Story: The Journals of Glenn T. Seaborg, 1939–46.* Columbus: Battelle Press.

Katz, Barry. 1989. *Foreign Intelligence: OSS, 1942–45.* Cambridge: Harvard University Press.

Kauffman, Stuart. 1993. *The Origins of Order.* New York: Oxford University Press.

1995. *At Home in the Universe.* Oxford: Oxford University Press.

Kay, Lily. 1993. *The Molecular Vision of Life.* Oxford: Oxford University Press.

1995. "Who Wrote the Book of Life?" *Science in Context*, 8.4:609–34.

1997a. "Cybernetics, Information, Life: The Emergence of Scriptural Representations of Heredity." *Configurations*, 5:23–91.

1997b. "Rethinking Institutions: Philanthropy as an Historiographic Problem." *Minerva*, 35:283–93.

1998. "Problematizing Basic Research in Molecular Biology." In Arnold Thackray, ed., *Private Science?*, pp. 20–38. Philadelphia: University of Pennsylvania Press.

2000. *Who Wrote the Book of Life?* Stanford: Stanford University Press.

Kaysen, Carl. 1947. "A Revolution in Economic Theory?" *Review of Economic Studies*, 14:1–15.

Keegan, Paul. 1999. "A Game Boy in the Crosshairs." *New York Times Magazine*, May 23, pp. 36–41.

Keller, Evelyn Fox. 1985. *Reflections on Gender and Science.* New Haven: Yale University Press.

1994. "The Body of a New Machine: Situating the Organism between Telegraphs and Computers." *Perspectives on Science*, 2:302–22.

1995. *Refiguring Life*. New York: Columbia University Press.

Kelley, J. 1956. "A New Interpretation of the Information Rate." *Bell System Technical Journal*, 35:917–26.

Kemeny, John. 1955. "Man Viewed as a Machine." *Scientific American*, 192 (April):58–67.
Kendrick, David. 1995. "Ten Wishes." *Computational Economics*, 8:67–80.
Ketabagian, Tamara. 1997. "The Human Prosthesis." *Critical Matrix*, 11:4–32.
Keunne, Robert. 1967. Review of *Defense Management*, by Stephen Enke. *American Economic Review*, 57:1399–1402.
Kevles, Daniel. 1995. *The Physicists*. New York: Knopf.
Khalil, Elias, & Boulding, Kenneth, eds. 1996. *Evolution, Order and Complexity.* London: Routledge.
Khalilzad, Zalmay, & White, John, eds. 1999. *The Changing Role of Information in Warfare.* Santa Monica, Calif.: RAND.
Kincaid, Harold. 1997. *Individualism and the Unity of Science.* Lanham, Md.: Rowman & Littlefield.
Kirman, Alan. 1992. "Whom or What Does the Representative Individual Represent?" *Journal of Economic Perspectives*, 6:117–36.
1993. "Ants, Rationality and Recruitment." *Quarterly Journal of Economics*, 108:137–56.
1997. "The Economy as an Evolving Network." *Journal of Evolutionary Economics*, 7:339–53.
1998. "Self-Organization and Evolution in Economics." *Jahrbuch für Komplexität*, 9:13–45.
1999. "The Future of Economic Theory." In Alan Kirman & L. Gerard-Varet, eds., *Economics beyond the Millennium*, pp. 8–22. Oxford: Oxford University Press.
Kitcher, Philip. 1985. *Vaulting Ambition.* Cambridge: MIT Press.
Kjeldsen, Tinne. 2000. "A Contextualized Historical Analysis of the Kuhn-Tucker Theorem in Nonlinear Programming." *Historia Mathematica*, 4:331–61.
Klausner, S., & Lidz, V., eds. 1986. *The Nationalization of the Social Sciences.* Philadelphia: University of Pennsylvania Press.
Klein, Judy. 1998a. "The Normative Economics of Statistical Quality Control, 1924–52." Working paper.
1998b. "Controlling Gunfire, Inventory and Expectations with the Exponentially Weighted Moving Average." Paper presented to the annual meeting of HES, Montreal.
1999a. "Economic Stabilization Policies and the Military Art of Control Engineering." Paper presented to the conference on Economists at War, Erasmus University, Rotterdam.
1999b. "An Economics of Experimentation: Investigations of the US Statistical Research Group in World War II." Paper presented to the annual meeting of ECHE, Paris.
Klein, Lawrence. 1991. "Econometric Contributions of the Cowles Commission, 1944–47." *Banca Nazionale del Lavoro Quarterly Review*, 177:107–17.
Kline, Morris. 1980. *Mathematics: The Loss of Certainty.* New York: Oxford University Press.
Klusch, M., & Weiss, G., eds. 1998. *Cooperative Information Agents II.* Berlin: Springer.
Knapik, Michael, & Johnson, Jay. 1998. *Developing Intelligent Agents for Distributed Systems.* New York: McGraw-Hill.

Knight, Frank. 1940. *Risk, Uncertainty and Profit*. Reprint no. 16. 1921. Reprint, London: LSE.

Knoblauch, Vicki. 1994. "Computable Strategies for Repeated Prisoner's Dilemma." *Games and Economic Behavior*, 7:381–89.

Koch, Sigmund, ed. 1985. *A Century of Psychology as a Science*. New York: McGraw-Hill.

Koestler, Arthur, & Smythies, J. R., eds. 1969. *Beyond Reductionism: New Perspectives in the Life Sciences*. New York: Macmillan.

Kohler, Robert. 1991. *Partners in Science*. Chicago: University of Chicago Press.

Kolata, Gina. 1998. *Clone*. New York: Morrow.

2001. "The Key Vanishes: Scientist Outlines Unbreakable Codes." *New York Times*, February 20, pp. D1, D4.

Koller, D., & Megiddo, N. 1992. "The Complexity of Two-Person Zero-Sum Games in Extensive Form." *Games and Economic Behavior*, 4:528–52.

Kolmogorov, A. 1965. "Three Approaches to the Quantitative Definition of Information." *Problems of Information Transmission*, 1:1–7.

Koopmans, Tjalling. 1939. *Tanker Freight Rates and Tankship Building*. Haarlem: Bohm.

1947. "Measurement without Theory." *Review of Economic Statistics*, 29:161–72.

1948a. "Systems of Linear Production Functions." RAND RM-46. RAND Corporation, Santa Monica, Calif.

1948b. "Remarks on Reduction and Aggregation." RAND RM-47. RAND Corporation, Santa Monica, Calif.

ed. 1950. *Statistical Inference in Dynamic Economic Models*. Cowles Commission Monograph 10. New York: Wiley.

ed. 1951. *Activity Analysis of Production and Allocation*. Cowles Commission Monograph 13. New York: Wiley.

1954. "On the Use of Mathematics in Economics." *Review of Economic Statistics*, 36:377–79.

1970. *Scientific Papers*. Berlin: Springer Verlag.

1985. *Scientific Papers of Tjalling C. Koopmans*. Vol. 2. Cambridge: MIT Press.

1991. *Three Essays on the State of Economic Science*. 1957. Reprint, New York: Kelley.

Koriat, A., & Goldsmith, M. 1996. "Memory Metaphors and the Real-Life/Laboratory Controversy." *Behavioral and Brain Sciences*, 19:167–228.

Korman, Gerd, & Klapper, Michael. 1978. "Game Theory's Wartime Connections and the Study of Industrial Conflict." *Industrial and Labor Relations Review*, 32:24–38.

Kragh, Helge. 1999. *Quantum Generations*. Princeton: Princeton University Press.

Kramer, Gerald. 1967. "An Impossibility Result in the Theory of Decision Making." In J. Bernd, ed., *Mathematical Applications in Political Science III*, pp. 39–51. Charlottesville: University Press of Virginia. Reissued as Cowles reprint no. 274.

Krentel, W., McKinsey, J., & Quine, W. V. 1951. "A Simplification of Games in Extensive Form." *Duke Mathematical Journal*, 18:885–900.

Kreps, David. 1990. *A Course in Microeconomic Theory*. Princeton: Princeton University Press.

Krige, John, & Pestre, Dominique, eds. 1997. *Science in the 20th Century*. Amsterdam: Harwood.

Kuenne, Robert. 1967. Review of *Defense Management*, by Stephen Enke. *American Economic Review*, 57:1399–1402.

Kuhn, H. 1953. "Extensive Games and the Problem of Information." In Kuhn & Tucker, 1953.

ed. 1995. *Festschrift for John Nash. Duke Mathematics Journal*, special suppl.

ed. 1997. *Classics in Game Theory.* Princeton: Princeton University Press.

Kuhn, H., & Tucker, A., eds. 1950. *Contributions to the Theory of Games.* Princeton: Princeton University Press.

eds. 1953. *Contributions to the Theory of Games*. Vol. 2. Princeton: Princeton University Press.

1958. "John von Neumann's Work in the Theory of Games and Mathematical Economics." *Bulletin of the American Mathematical Society*, 64:100–22.

Kurz, Heinz, & Salvadori, Neri. 1993. "Von Neumann's Growth Model and the Classical Tradition." *European Journal of the History of Economic Thought*, 1:129–60.

Lacey, Michael, & Furner, Mary, eds. 1993. *The State and Social Investigation in Britain and the US.* Washington, D.C.: Wilson Center Press.

Lagemann, Ellen. 1989. *The Politics of Knowledge.* Middletown: Wesleyan University Press.

Lamberton, Donald, ed. 1971. *Economics of Information and Knowledge*. Harmondsworth: Penguin.

1984. "The Emergence of Information Economics." In M. Jussawalla & H. Ebenfield, eds., *Communication and Information Economics*. Amsterdam: North Holland.

ed. 1996. *The Economics of Communication and Information.* Cheltenham: Elgar.

1998. "Information Economics Research: Points of Departure." *Information Economics and Policy*, 10:325–30.

Landauer, Rolf. 1991. "Information is Physical." *Physics Today*, 44 (May):23–29.

Lane, David. 1993. "Artificial Worlds and Economics, I and II." *Journal of Evolutionary Economics*, 3:89–107, 177–97.

Lane, David, et al. 1996. "Choice and Action." *Journal of Evolutionary Economics*, 6:43–76.

Lange, Oskar. 1965. *Wholes and Parts: A General Theory of System Behavior.* Oxford: Pergamon.

1967. "The Computer and the Market." In Charles Feinstein, ed., *Socialism and Economic Growth*, pp. 158–61. Cambridge: Cambridge University Press.

Lange, Oskar, McIntyre, Francis, & Yntema, Theodore, eds. 1942. *Studies in Mathematical Economics and Econometrics in Honor of Henry Schultz.* Chicago: University of Chicago Press.

Lange, Oskar, & Taylor, Frederick. 1938. *On the Economic Theory of Socialism.* Minnesota: University of Minnesota Press.

Langley, Pat, Simon, Herbert, Bradshaw, Gary, & Zytkow, Jan. 1987. *Scientific Discovery*. Cambridge: MIT Press.

Langton, Christopher, ed. 1994. *Artificial Life III.* Redwood City, Calif.: Addison-Wesley.

ed. 1995. *Artificial Life: An Overview.* Cambridge: MIT Press.

Langton, C., Taylor, C., Farmer, J., & Rasmussen, S., eds. 1992. *Artificial Life II.* Reading, Mass.: Addison-Wesley.

Lanouette, William. 1992. *Genius in the Shadows.* New York: Scribners.

Larson, Otto. 1992. *Milestones and Millstones: Social Science at the National Science Foundation.* New Brunswick, N.J.: Transaction Press.

Latour, Bruno. 1993. *We Have Never Been Modern.* Cambridge: Harvard University Press.

1999. *Pandora's Hope.* Cambridge: Harvard University Press.

Lavoie, Don. 1985. *Rivalry and Central Planning.* Cambridge: Cambridge University Press.

1990. "Computation, Incentives and Discovery." In J. Prybyla, ed., *Privatizing and Market Socialism*, pp. 72–79. London: Sage.

Lax, Peter. 1989. "The Flowering of Applied Mathematics in America." In Duran, 1989.

Ledyard, John. 1987. "Incentive Compatibility." In J. Eatwell et al., eds., *New Palgrave.* London: Macmillan.

Lee, Ron. 1985. "Bureaucracy as Artificial Intelligence." In L. Methlie & R. Sprague, eds., *Knowledge Representation for Decision Support Systems*, pp. 125–32. New York: Elsevier.

Leebaert, Derek. Ed. 1998. *The Future of the Electronic Marketplace.* Cambridge: MIT Press.

Leff, Harvey, & Rex, Andrew, eds. 1990. *Maxwell's Demon: Entropy, Information, Computing.* Princeton: Princeton University Press.

Legendi, Tomas, & Szentivanyi, Tibor, eds. 1983. *Leben und Werk von John von Neumann.* Mannheim: BI Wissenschaftsverlag.

Lehrer, E. 1988. "Repeated Games with Stationary Bounded Recall Strategies." *Journal of Economic Theory*, 46:130–44.

Lenoir, Timothy. 2000. "All but War Is Simulation: The Military Entertainment Complex." *Configurations*, 8:289–335.

Lenstra, J., Kan, A., & Schrijver, A., eds. 1991. *History of Mathematical Programming.* Amsterdam: North Holland.

Leonard, Robert. 1991. "War as a Simple Economic Problem." In Goodwin, 1991, pp. 261–83.

1992. "Creating a Context for Game Theory." In Weintraub, 1992, pp. 29–76.

1994a. "Reading Cournot, Reading Nash." *Economic Journal*, 104:492–511.

1994b. "Laboratory Strife." In Neil de Marchi & Mary Morgan, eds., *Higgling*, pp. 343–69. Suppl. to vol. 26 of *History of Political Economy.* Durham: Duke University Press.

1995. "From Parlor Games to Social Science: Von Neumann, Morgenstern, and the Creation of Game Theory." *Journal of Economic Literature*, 33:730–61.

1997. "Value, Sign and Social Structure: The Game Metaphor in Modern Social Science." *European Journal for the History of Economic Thought*, 4:299–326.

1998. "Ethics and the Excluded Middle." *Isis*, 89:1–26.

Leontief, Wassily. 1966. *Essays in Economics*. New York: Oxford University Press.

Leslie, Stuart. 1993. *The Cold War and American Science*. New York: Columbia University Press.

1999. "The Scientists and the Generals." *Studies in the History and Philosophy of Modern Physics*, 30:569–77.

Leunbach, G. 1948. "Theory of Games and Economic Behavior." *Nordisk Tidsskrift fur Teknisk Okonomi.*

Levins, Richard, & Lewontin, Richard. 1985. *The Dialectical Biologist.* Cambridge: Harvard University Press.

Levite, Ariel. 1989. *Offense and Defense in Israeli Military Doctrine.* Boulder, Colo.: Westview.

Levy, Pierre. 1997. *L'intelligence collective.* Paris: la Decouverte.

Levy, Stephen. 1992. *Artificial Life.* New York: Vintage.

Lewis, Alain. 1979. "On the Formal Character of Plausible Reasoning." RAND P-6462. RAND Corporation, Santa Monica, Calif.

1981. "The Use of Command/Task Structures in the Comparative Assessment of Organizational Components of the Strategic Nuclear Force System." RAND Internal Report. RAND Corporation, Santa Monica, Calif.

1985a. "On Effectively Computable Realizations of Choice Functions." *Mathematical Social Sciences*, 10:43–80.

1985b. "The Minimum Degree of Recursively Representable Choice Functions." *Mathematical Social Sciences*, 10:179–88.

1985c. "Complex Structures and Composite Models – An Essay on Methodology." *Mathematical Social Sciences*, 10:211–46.

1986. "Structure and Complexity: The Use of Recursion Theory in the Foundations of Neoclassical Mathematical Economics and the Theory of Games." Unpublished manuscript, Department of Mathematics, Cornell University.

1988. "An Infinite Version of Arrow's Theorem in an Effective Setting." *Mathematical Social Sciences*, 16:41–48.

1990. "On the Independence of Core-Equivalence Results from Zermelo-Fraenkel Set Theory." *Mathematical Social Sciences*, 19:55–95.

1991. "On the Effective Content of Asymptotic Verifications of Edgeworth's Conjecture." *Mathematical Social Sciences*, 22:275–324.

1992a. "On Turing Degrees of Walrasian Models and a General Impossibility Result in the Theory of Decision Making." *Mathematical Social Sciences*, 24:141–71.

1992b. "Some Aspects of Effectively Constructive Mathematics That Are Relevant to the Foundations of Neoclassical Mathematical Economics." *Mathematical Social Sciences*, 24:209–35.

Lewis, David. 1986. *Convention: A Philosophical Study.* 1969. Reprint, Oxford: Basil Blackwell.

Lewis, Harry, & Papadimitriou, Christos. 1981. *Elements of the Theory of Computation.* Englewood, Cliffs, N.J.: Prentice-Hall.

Libicki, Martin. 1997. *Defending Cyberspace and Other Metaphors.* Washington, D.C.: National Defense University Press.

Licklider, J. 1960. "Man-Computer Symbiosis." In *IRE Transactions on Human Factors in Electronics*, vol. HFE-1:4–11.

Lilienfeld, Robert. 1978. *The Rise of Systems Theory*. New York: Wiley.

Lindley, David. 1993. *The End of Physics.* New York: Basic.

Linial, N. 1994. "Game-Theoretic Aspects of Computing." In R. Aumann & S. Hart, eds., *Game Theory with Applications*, vol. 2. Amsterdam: North Holland.

Linster, Bruce. 1992. "Evolutionary Stability in the Infinitely-Repeated Prisoner's Dilemma Played by Two-State Moore Machines." *Southern Economic Journal*, 58:880–903.

1994. "Stochastic Evolutionary Dynamics in Repeated Prisoner's Dilemma." *Economic Inquiry*, 32:342–57.

Lipartito, Kenneth. 1994. "When Women Were Switches: Technology, Work and Gender in the Telephone Industry, 1890–1920." *American Historical Review*, 99:1075–1111.

Lipman, Barton. 1991. "How to Decide How to Decide How to . . ." *Econometrica*, 59:1105–25.

1995. "Information Processing and Bounded Rationality: A Survey." *Canadian Journal of Economics*, 28:42–67.

Lipton, R., & Baum, E., eds. 1996. *DNA Based Computers*. Providence: American Mathematics Society.

Loader, Brian, ed. 1998. *The Cyberspace Divide.* London: Routledge.

Lohr, Steve. 1996. "A New Battlefield: Rethinking Warfare in the Computer Age." *New York Times Cybertimes*, September 30 <www.nytimes.com>.

Lorenz, Edward. 1963. "Deterministic Nonperiodic Flow." *Journal of the Atmospheric Sciences*, 82:985–92.

Lotka, Alfred. 1956. *Elements of Mathematical Biology.* New York: Dover.

Lowen, Rebecca. 1997. *Creating the Cold War University.* Berkeley: University of California Press.

Lucas, Robert. 1972. "Expectations and the Neutrality of Money." *Journal of Economic Theory*, 4:103–24.

Lucas, William. 1967. "A Counterexample in Game Theory." *Management Science*, 13:766–67.

1969. "The Proof That a Game May Not Have a Solution." *Transactions of the American Mathematical Society*, 137.402 (March):219–29.

Luce, R. Duncan, & Raiffa, Howard. 1957. *Games and Decisions.* New York: Wiley.

Lucky, Robert. 1989. *Silicon Dreams.* New York: St. Martin's.

Lyon, David, & Zureik, Elia, eds. 1996. *Computers, Surveillance and Privacy.* Minneapolis: University of Minnesota Press.

Lyons, Gene. 1969. *The Uneasy Partnership.* New York: Russell Sage.

Maas, Harro. 1999. "Mechanical Rationality: Jevons and the Making of Economic Man." *Studies in the History and Philosophy of Science*, A, 30:587–619.

Maasoumi, F. 1987. "Information Theory." In J. Eatwell et al., eds., *The New Palgrave*, pp. 846–50. London: Macmillan.

Machlup, Fritz. 1962. *The Production and Distribution of Knowledge in the US.* Princeton: Princeton University Press.

Machlup, Fritz, & Mansfield, U., eds. 1983. *The Study of Information*. New York: Wiley.

MacKenzie, Donald. 1996. *Knowing Machines.* Cambridge: MIT Press.
1999. "Physics and Finance." Unpublished manuscript, University of Edinburgh.
2001. "Physics and Finance: S-Terms and Modern Finance as a Topic for Science Studies." *Science, Technology and Human Values*, 26:115–44.
MacKenzie, Donald, & Pottinger, Garrel. 1997. "Mathematics, Technology and Trust." *IEEE Annals of the History of Computing*, 19:41–59.
Mac Lane, Saunders. 1989. "The Applied Mathematics Group at Columbia in World War II." In Duren, 1989.
Macrae, N. 1992. *John von Neumann.* New York: Pantheon.
Madhavan, Ananth. 1992. "Trading Mechanisms in Securities Markets." *Journal of Finance*, 47:607–41.
Mahoney, Michael. 1988. "The History of Computing in the History of Technology." *Annals of the History of Computing*, 10:113–24.
1990. "Cybernetics and Information Technology." In *Companion Encyclopedia of the History of Science*, pp. 537–53. London: Routledge.
1992. "Computers and Mathematics: The Search for a Discipline of Computer Science." In J. Echeverria et al., eds., *The Space of Mathematics*, pp. 349–64. New York: de Gruyter.
1997. "Computer Science." In Krige & Pestre, 1997, pp. 617–34.
Mailath, George. 1998. "Do People Really Play Nash Equilibria?" *Journal of Economic Literature*, 36:1347–74.
Makowski, Louis, & Ostroy, Joseph. 1993. "General Equilibrium and Market Socialism." In P. Bardhan & J. Roemer, eds., *Market Socialism: The Current Debate.* New York: Oxford University Press.
Mandelbrot, Benoit. 1989. "Chaos, Bourbaki and Poincaré." *Mathematical Intelligencer*, 11:10–12.
Manktelow, K., & Over, D., eds. 1993. *Rationality: Psychological and Philosophical Perspectives.* London: Routledge.
Mantel, Rolf. 1976. "Homothetic Preferences and Community Excess Demand Functions." *Journal of Economic Theory*, 12:197–201.
Marglin, Stephen. 1964. "Decentralization with a Modicum of Increasing Returns." RAND RM-3421-PR. RAND Corporation, Santa Monica, Calif.
1969. "Information in Price and Command Systems." In J. Margolis & H. Guitton, eds., *Public Economics*, pp. 54–77. London: Macmillan.
Marimon, Ramon, McGrattan, Ellen, & Sargent, Thomas. 1990. "Money as a Medium of Exchange in an Economy with Artificially Intelligent Agents." *Journal of Economic Dynamics and Control*, 14:329–74.
Markowitz, Harry. 1959. *Portfolio Selection.* New Haven: Yale University Press.
Markowsky, George. 1997. "An Introduction to Algorithmic Information Theory." *Complexity*, 2.4:14–37.
Marschak, Jacob. 1923. "Wirtschaftsrechnung und Gemeinwirtschaft." *Archiv für Sozialwissenschaft*, 51.
1939. Review of *Theory and Measurement of Demand*, by Henry Schultz. *Economic Journal*, 49:486–89.
1940. "Peace Economics." *Social Research*, 7.

1941. "A Discussion of Methods in Economics." *Journal of Political Economy*, 49:441–48.

1946. "Neumann's and Morgenstern's New Approach to Static Economics." *Journal of Political Economy*, 54:97–115.

1971. "Optimal Symbol Processing: A Problem in Individual and Social Economics." *Behavioral Science*, 16:202–17.

1973. "Limited Role of Entropy in Information Economics." In R. Conti & A. Roberti, eds., *Fifth Conference on Optimization Techniques*, pt. 2, pp. 264–71. Berlin: Springer Verlag.

1974. *Economic Information, Decision and Prediction: Selected Essays*. 3 vols. Dordrecht: Reidel.

Marschak, Jacob, & Davidson, Donald. 1957. "Experimental Tests of Stochastic Decision Theory." Cowles Foundation Discussion Paper 22. New Haven.

Marschak, Jacob, & Miyasawa, K. 1968. "Economic Comparability of Information Systems." *International Economic Review*, 9:137–74.

Marschak, Jacob, & Radner, Roy. 1972. *Economic Theory of Teams*. New Haven: Yale University Press.

Marshall, Alfred. 1994. "Ye Machine." 1867. *Research in the History of Economic Thought and Method*. Archival Supplement 4. Greenwich, Conn.: JAI Press.

Marshall, Andrew. 1950. "A Test of Klein's Model III." Abstract. *Econometrica*, 18:241.

1958. "Experimental Simulation and Monte Carlo." RAND P-1174. RAND Corporation, Santa Monica, Calif.

Martin, L. W. 1962. "The Market for Strategic Ideas in Britain." *American Political Science Review*, 56:23–41.

Masani, Pesi. 1990. *Norbert Wiener*. Basel: Birkhauser.

Mayberry, John, Nash, John, & Shubik, Martin. 1953. "A Comparison of Treatments of the Duopoly Situation." *Econometrica*, 21:141–54.

Mayr, Otto. 1986. *Authority, Liberty, and Automatic Machinery in Early Modern Europe*. Baltimore: Johns Hopkins University Press.

Mays, Wolf. 1962. "Jevons' Conception of Scientific Method." *Manchester School*, 30:223–50.

Mays, Wolf, & Henry, D. 1973. "Jevons and Logic." *Mind*, 62:484–505.

McArthur, Charles. 1990. *Operations Analysis in the US Army Eighth Air Force in WWII*. Providence: American Mathematical Society.

McCabe, Kevin, Rassenti, Stephen, & Smith, Vernon. 1989. "Designing Smart Computer-Assisted Markets in an Experimental Auction." *European Journal of Political Economy*, 5:259–83.

McCarthy, John. 1956. "Measure of the Value of Information." *Proceedings of the National Academy of Sciences*, 42:654–55.

McCloskey, J., & Trefethen, F., eds. 1954. *Operations Research for Management*. Baltimore: Johns Hopkins University Press.

McCorduck, Pamela. 1979. *Machines Who Think*. San Francisco: Freeman.

McCue, Brian. 1990. *U Boats in the Bay of Biscay*. Washington, D.C.: National Defense University.

McCulloch, Warren, & Pitts, Walter. 1943. "A Logical Calculus of the Ideas Immanent in Nervous Activity." *Bulletin of Mathematical Biophysics*, 5:115–33.

McCumber, John. 2001. *Time in the Ditch: American Philosophy in the McCarthy Era.* Evanston: Northwestern University Press.
McDonald, John. 1950. *Strategy in Poker, Business and War.* New York: Norton.
McGee, V. 1991. "We Turing Machines Aren't Expected Utility Maximizers (Even Ideally)." *Philosophical Studies*, 64:115–23.
McGuckin, William. 1984. *Scientists, Society and the State.* Columbus: Ohio State University Press.
McGuire, C. 1986. "Comparisons of Information Structures." In McGuire & Radner, 1986, pp. 100–12.
McGuire, C., & Radner, R., eds. 1986. *Decision and Organization.* Minneapolis: University of Minnesota Press.
McKelvey, R., & McLennan, A. 1996. "Computation of Equilibria in Finite Games." In H. Amman et al., eds., *Handbook of Computational Economics*, pp. 87–142. Amsterdam: North Holland.
McKenney, James. 1995. *Waves of Change.* Cambridge: Harvard Business School Press.
McKinsey, John. 1952a. "Some Notions and Problems of Game Theory." *Bulletin of the American Mathematical Society*, 58:591–611.
1952b. *Introduction to the Theory of Games.* New York: McGraw-Hill.
McPherson, Michael. 1984. "On Schelling, Hirschman and Sen." *Partisan Review*, 51:236–47.
Megiddo, Nimrod. 1986. "Remarks on Bounded Rationality." IBM Research paper RJ-5270.
Megiddo, N., & Widgerson, A. 1986. "On Play by Means of Computing Machines." In Joseph Halpern, ed., *Aspects of Reasoning about Knowledge.* Los Altos: Morgan Kaufman.
Mehra, Jagdish, & Rechenberg, Helmut. 1982. *The Historical Development of Quantum Theory.* Vol. 3. New York: Springer.
Melman, Seymour. 1971. *Pentagon Capitalism.* New York: Pantheon.
1974. *The Permanent War Economy.* New York: Pantheon.
Mendelsohn, Everett. 1997. "Science, Scientists and the Military." In Krige & Pestre, 1997, pp. 75–202.
Mendelsohn, Everett, et al., eds. 1988. *Science Technology and the Military.* Boston: Kluwer.
Mendelson, M., Peake, J., & Williams, R. 1979. "Towards a Modern Exchange." In Ernest Bloch & Robert Schwartz, eds., *Impending Changes for Securities Markets.* Greenwich, Conn.: JAI Press.
Menger, Karl. 1934. *Moral, Wille und Weltgestaltung.* Vienna: Springer.
Mensch, A., ed. 1966. *Theory of Games.* New York: American Elsevier.
Merz, Martina. 1999. "Multiplex and Unfolding." *Science in Context*, 12:293–316.
Metakides, G., & Nerode, Anil. 1971. "Recursively Enumerable Vector Spaces." *Annals of Mathematical Logic*, 3:147–71.
Metropolis, Nicholas. 1992. "The Age of Computing: A Personal Memoir." *Daedalus*, 121:119–30.
Metropolis, N., Howlett, J., & Rota, G., eds. 1980. *A History of Computing in the 20th Century.* New York: Academic Press.
Mettrie, Julien de la. 1961. *Man a Machine.* La Salle, Ill.: Open Court.

Meyer, David. 1999. "Quantum Strategies." *Physical Review Letters*, 64.

Mihara, H. 1997. "Arrow's Theorem and Turing Computability." *Economic Theory*, 10:257–76.

Milgrom, Paul. 1984. "Axelrod's Evolution of Cooperation." *RAND Journal of Economics*, 15:305–9.

Milgrom, Paul, & Stokey, Nancy. 1982. "Information, Trade and Common Knowledge." *Journal of Economic Theory*, 26:17–27.

Miller, George. 1956. "The Magic Number Seven, Plus or Minus Two." *Psychological Review*, 63:81–97.

Miller, L., et al. 1989. "Operations Research and Policy Analysis at RAND, 1968–88." RAND N-2937-RC. RAND Corporation, Santa Monica, Calif.

Miller, Ross. 1986. "Markets as Logic Programs." In L. Pau, ed., *Artificial Intelligence in Economics and Management*, pp. 129–36. Amsterdam: North Holland.

1990. "The Design of Decentralized Auction Mechanisms That Coordinate Continuous Trade in Synthetic Securities." *Journal of Economic Dynamics and Control*, 14:237–53.

1996. "Smart Market Mechanisms." *Journal of Economic Dynamics and Control*, 20:967–78.

Millhauser, Steven. 1998. "The New Automaton Theatre." In *The Knife Thrower*, pp. 89–111. New York: Vintage.

Millikan, P., & Clark, A. eds. 1996. *Machines and Thought: The Legacy of Alan Turing.* Oxford: Clarendon Press.

Millman, S., ed. 1984. *A History of Engineering and Science in the Bell System: Communication Sciences, 1925–80.* Indianapolis: AT&T Bell.

Milnor, John. 1995. "A Nobel Prize for John Nash." *Mathematical Intelligencer*, 17:11–17.

Minas, J., et al. 1960. "Some Descriptive Aspects of Two Person Non-Zero Sum Games." *Journal of Conflict Resolution*, 4:193–97.

Minsky, Marvin. 1979. "Computer Science and the Representation of Knowledge." In M. Dertouzos & J. Moses, eds., *The Computer Age: A 20 Year View.* Cambridge: MIT Press.

1985. *Society of Mind.* New York: Simon & Schuster.

Mirowski, Philip. 1988. *Against Mechanism.* Totawa, N.J.: Rowman & Littlefield.

1989a. *More Heat than Light.* Cambridge: Cambridge University Press.

1989b. "The Measurement without Theory Controversy." *Economies et Sociétés*, 27:65–87.

1989c. "The Probabilistic Counter-Revolution." *Oxford Economic Papers*, 41: 217–35.

1990. "Problems in the Paternity of Econometrics: Henry Ludwell Moore." *History of Political Economy*, 22:587–609.

1991. "When Games Grow Deadly Serious." In Goodwin, 1991, pp. 227–60.

1992. "What Were von Neumann and Morgenstern Trying to Accomplish?" In Weintraub, 1992, pp. 113–47.

1993. "The Goalkeeper's Anxiety at the Penalty Kick." In de Marchi, 1993, pp. 305–49.

ed. 1994a. *Natural Images in Economic Thought*. Cambridge: Cambridge University Press.

1994b. *Edgeworth on Chance, Economic Hazard and Statistics*. Savage, Md.: Rowman & Littlefield.

1994c. "What Are the Questions?" In Roger Backhouse, ed., *New Directions in Economic Methodology*, pp. 50–74. London: Routledge.

1994d. "A Visible Hand in the Marketplace of Ideas." *Science in Context*, 7:563–89.

1996. "Do You Know the Way to Santa Fe?" In Steve Pressman, ed., *Interactions in Political Economy*, pp. 13–40. London: Routledge.

1997. "On Playing the Economics Trump Card in the Philosophy of Science: Why It Did Not Work for Michael Polanyi." *PSA 96*, suppl. to *Philosophy of Science*, 64:S127–38.

1998a. "Economics, Science and Knowledge: Polanyi vs. Hayek." *Tradition and Discovery*, 25:29–42.

1998b. "Machine Dreams: Economic Agent as Cyborg." In John Davis, ed., *The New Writing in the History of Economics. History of Political Economy* suppl., pp. 13–40. Durham: Duke University Press.

1998c. "Operationalism." In J. Davis et al., eds., *Handbook of Economic Methodology*, pp. 346–49. Cheltenham: Elgar.

1999. "Cyborg Agonistes." *Social Studies of Science*, 29:685–718.

Forthcoming. "On the Evolution of Some Species of American Evolutionary Economics." Paper presented to HES conference, Vancouver, June 2000.

Mirowski, Philip, & Sent, Esther-Mirjam, eds. 2001. *Science Bought and Sold*. Chicago: University of Chicago Press.

Mirowski, Philip, & Sklivas, Steven. 1991. "Why Econometricians Don't Replicate (Although They Do Reproduce)." *Review of Political Economy*, 3:146–63.

Mirowski, Philip, & Somefun, Koye. 1998. "Markets as Evolving Computational Entities." *Journal of Evolutionary Economics*, 8:329–56.

Mirrlees, James. 1997. "Information and Incentives." *Economic Journal*, 107: 1311–29.

Miser, Hugh. 1991. "A Structure for the Intellectual History of Operations Research." In *Operational Research '90*, pp. 49–62. Oxford: Pergamon Press.

Mises, Ludwig von. 1920. "Die Wirtschaftsrechnung im sozialistischen Gemeinwesen." *Archiv für Sozialwissenschaften und Sozialpolitik*, 47. Translated in Hayek, 1935.

Molander, R., Riddile, A., & Wilson, P. 1996. *Strategic Information Warfare*. RAND MR-661. RAND Corporation, Santa Monica, Calif.

Monod, Jacques. 1971. *Chance and Necessity*. New York: Knopf.

Mood, Alexander. 1990. "Miscellaneous Reminiscences." *Statistical Science*, 5:35–43.

Moore, Edward. 1956. "Gedanken Experiments on Sequential Machines." In Shannon & McCarthy, 1956.

Moore, Gregory. 1988. "The Emergence of First-Order Logic." In William Aspray & Philip Kitcher, eds., *History and Philosophy of Modern Mathematics*, pp. 95–135. Minneapolis: University of Minnesota Press.

Moore, Walter. 1989. *Schrödinger: Life and Thought*. Cambridge: Cambridge University Press.

Moravec, Hans. 1988. *Mind Children.* Cambridge: Harvard University Press.

Morawski, Jill, ed. 1988. *The Rise of Experimentation in American Psychology.* New Haven: Yale University Press.

Morgan, Mary. 1990. *The History of Econometric Ideas.* Cambridge: Cambridge University Press.

Morgan, Mary, & Rutherford, Malcolm, eds. 1998. *The Transformation of American Economics.* Durham: Duke University Press.

Morgenstern, Oskar. 1937. *The Limits of Economics.* Trans. V. Smith. London: W. Hodge.

1949. "Economics and the Theory of Games." *Kyklos*, 3:294–308.

1954. "Experiments and Large-Scale Computation in Economics." In *Economic Activity Analysis*, pp. 483–549. New York: Wiley.

1959. *The Question of National Defense.* New York: Random House.

1966. "The Compressibility of Economic Systems and the Problem of Economic Constants." *Zeitschrift fur Nationalökonomie*, 26:190–203.

1976a. "Collaborating with John von Neumann." *Journal of Economic Literature*, 14.

1976b. *Selected Economic Writings.* Ed. Andrew Schotter. New York: New York University Press.

Morin, Alexander. 1993. *Science Policy and Politics.* Englewood Cliffs, N.J.: Prentice-Hall.

Morse, Philip. 1948. "Mathematical Problems in Operations Research." *Bulletin of the American Mathematical Society*, 54:602–21.

1977. *In at the Beginnings: A Physicist's Life.* Cambridge: MIT Press.

Morse, Philip, & Kimball, George. 1951. *Methods of Operations Research.* New York: Wiley.

Morse, Philip, et al. 1981. "Where the War Stories Began." *Interfaces*, 11 (October):37–44.

Morton, Alan. 1994. "Packaging History: Emergence of the UPC in the US, 1970–75." *History and Technology*, 11:101–11.

Mosselmans, Bert. 1998. "W. S. Jevons and the Extent of Meaning in Logic and Economics." *History and Philosophy of Logic*, 19:83–99.

Mosteller, Frederick. 1978. "Samuel Wilks." *International Encyclopaedia of Statistics*, pp. 1250–53.

Mosteller, F., & Nogee, P. 1951. "An Experimental Measurement of Utility." *Journal of Political Economy*, 59:288–318.

Mowery, David, & Rosenberg, Nathan. 1998. *Paths of Innovation.* Cambridge: Cambridge University Press.

Murphy, M., & O'Neill, L., eds. 1995. *What Is Life? The Next 50 Years.* Cambridge: Cambridge University Press.

Muth, John. 1961. "Rational Expectations and the Theory of Price Movements." *Econometrica*, 29:315–35.

Myerson, Roger. 1996. "John Nash's Contribution to Economics." *Games and Economic Behavior*, 14:287–95.

1998. "Work on Game Theory." In Michael Szenberg, ed., *Passion and Craft*, pp. 227–33. Ann Arbor: University of Michigan Press.

1999. "Nash Equilibrium and the History of Economic Theory." *Journal of Economic Literature*, 107:1067–82.

Nachbar, John. 1992. "Evolution in Finitely Repeated Prisoner's Dilemma." *Journal of Economic Behavior and Organization*, 19:307–26.
1997. "Prediction, Optimization and Learning in Repeated Games." *Econometrica*, 65:275–309.
Nagy, Denes, Horvath, Peter, & Nagy, Ferenc. 1989. "The von Neumann-Ortvay Connection." *Annals of the History of Computing*, 11.183–88.
Nasar, Sylvia. 1994. "The Lost Years of a Nobel Laureate." *New York Times*, November 13, sec. 3, pp. 1, 8.
1998. *A Beautiful Mind.* New York: Simon and Schuster.
Nash, John. 1950a. "The Bargaining Problem." *Econometrica* 18:155–62.
1950b. "Equilibrium Points in N-Person Games." *Proceedings of the National Academy of Science*, 36:48–49.
1951. "Non-Cooperative Games." *Annals of Mathematics*, 54:286–95.
1952. "Some Games and Machines for Playing Them." RAND D-1164. RAND Corporation, Santa Monica, Calif.
1954. "Parallel Control." RAND RM-1361. RAND Corporation, Santa Monica, Calif.
1994. "Autobiography." <www.nobel.sdsc.edu/laureates/economy-1994-2-autobio.html>.
1996. *Essays on Game Theory*. Cheltenham: Elgar.
Nash, John, & Thrall, R. M. 1952. "Some War Games" RAND D-1379. RAND Corporation, Santa Monica, Calif.
National Research Council. 1997. *Modeling and Simulation: Linking Entertainment and Defense.* Washington, D.C.: National Academy Press.
1999. *Funding a Revolution: Government Support for Computing Research.* Washington, D.C.: National Academy Press.
National Science Board. 1996. *Science and Engineering Indicators*. Washington, D.C.: GPO.
1998. *Science and Engineering Indicators.* Arlington, Va.: National Science Foundation.
National Science Foundation. 1995. *Science, Technology and Democracy in the Cold War and After.* Washington, D.C.
Neisser, Hans. 1952. "The Strategy of Expecting the Worst." *Social Research*, 19:346–63.
Neisser, Ulric. 1963. "The Imitation of Man by Machine." *Science*, 139:193–97.
Nelson, Richard, & Winter, Sidney. 1982. *An Evolutionary Theory of Economic Change.* Cambridge: Harvard University Press.
Nersessian, Nancy. 1995. "Opening the Black Box: Cognitive Science and the History of Science." *Osiris*, 10:194–214.
Neufeld, Jacob. 1990. *The Development of the Ballistic Missile.* Washington, D.C.: Office of Air Force History.
ed. 1993. *Reflections on R&D in the US Air Force.* Washington, D.C.: Center for Air Force History.
Newell, Allen. 1991. "Metaphors of Mind, Theories of Mind." In J. Sheehan & M. Sosna, eds., *The Boundaries of Humanity*. Berkeley: University of California Press.

Newell, Allen, Shaw, J., & Simon, H. 1957. "Empirical Explorations of the Logic Theory Machine." *Proceedings of the Western Joint Computer Conference.* New York: Institute of Radio Engineers.

Newell, Allen, & Simon, H. 1956. "Current Developments in Complex Information Processing." RAND P-850. RAND Corporation, Santa Monica, Calif.

1972. *Human Problem Solving.* Englewood Cliffs, N.J.: Prentice-Hall.

Neyman, Abraham. 1985. "Bounded Complexity Justifies Cooperation in the Finitely Repeated Prisoner's Dilemma." *Economics Letters*, 19:227–30.

Neyman, Jerzy, & Pearson, Egon. 1933. "On the Problem of the Most Efficient Tests of a Statistical Hypothesis." *Philosophical Transactions of the Royal Society*, A231:289–337.

Niehans, Jurg. 1990. *A History of Economic Theory*. Baltimore: Johns Hopkins University Press.

Nietzsche, Friedrich. 1966. *Beyond Good and Evil.* Ed. W. Kaufmann. New York: Vintage.

1974. *The Gay Science.* New York: Vintage.

Noble, David. 1979. *America by Design.* New York: Oxford University Press.

Norberg, Arthur. 1996. "Changing Computing: The Computing Community and DARPA." *IEEE Annals of the History of Computing*, 18:40–53.

Norberg, Arthur, & O'Neill, Judy. 1996. *Transforming Computer Technology*. Baltimore: Johns Hopkins Press.

Norman, A. 1987. "A Theory of Monetary Exchange." *Review of Economic Studies*, 54:499–517.

1994. "Computability, Complexity and Economics." *Computational Economics*, 7:1–21.

Novick, David. 1947. "Research Opportunities in the War Production Board Records." *American Economic Review, Papers and Proceedings*, 37.2:690–99.

1954. "Mathematics: Logic, Quantity and Method." *Review of Economics and Statistics*, 36:357–58; and symposium 36:359–85; 37:297–300.

1965. *Program Budgeting.* Cambridge: Harvard University Press.

1988. "Beginning of Military Cost Analysis." RAND P-7425. RAND Corporation, Santa Monica, Calif.

Nyquist, H. 1924. "Certain Factors Affecting Telegraph Speed." *Bell System Technical Journal*, 3:324–46.

O'Brien, Ellen. 1994. "How the G Got into the GNP." In Karen Vaughn, ed., *Perspectives in the History of Economic Thought*, 10:241–55. Aldershot: Elgar.

O'Brien, Tim. 1990. *The Things They Carried.* Boston: Houghton Mifflin.

Office of Technology Assessment. U.S. Congress. 1995a. *Distributed Interactive Simulation of Combat.* Washington, D.C.

1995b. *Electronic Surveillance in a Digital Age.* Washington, D.C.

1995c. *A History of Department of Defense Federally Funded R&D Centers.* Washington, D.C.

O'Hara, Maureen. 1995. *Market Microstructure Theory.* Malden: Blackwell.

O'Leary, Daniel. 1991. "Principia Mathematica and the Development of Automated Theorem Proving." In Thomas Drucker, ed., *Perspectives on the History of Mathematical Logic*, pp. 47–53. Boston: Birkhauser.

O'Neill, Barry. 1989. "Game Theory and the Study of the Deterrence of War." In Paul Stern et al., eds., *Perspectives in Deterrence*, pp. 134–56. New York: Oxford University Press.

1992. "A Survey of Game Theory Models of Peace and War." In R. Aumann & S. Hart, eds., *Handbook of Game Theory*, vol. 1. Amsterdam: North Holland.

Oreskes, Naomi. 2000. "Laissez-tomber: Military Patronage and Women's Work." *Historical Studies in the Physical Sciences*, 30.

Osborne, Martin, & Rubinstein, Ariel. 1998. "Games with Procedurally Rational Players." *American Economic Review*, 88:834–47.

Ossowski, Sascha. 1999. *Coordination in Artificial Agent Societies.* Berlin: Springer.

Owens, Larry. 1989. "Mathematicians at War." In D. Rowe & J. McCleary, eds., *The History of Modern Mathematics*, vol. 2. Boston: Academic Press.

1994. "The Counterproductive Management of Science in WWII." *Business History Review*, 68:515–76.

1996. "Where Are We Going, Phil Morse?" *Annals of the History of Computing*, 18:34–41.

Page, T. 1952–53. "Tank Battle." *Journal of the Operations Research Society of America*, 1:85–86.

Palmer, R., Arthur, W. B., Holland, J., LeBaron, B., & Tayler, P. 1994. "Artificial Economic Life: A Simple Model of a Stockmarket." *Physica D*, 75:264–74.

Papadimitriou, C. 1992. "On Players with a Bounded Number of States." *Games and Economic Behavior*, 4:122–31.

Parsegian, V. 1997. "Harness the Hubris: Useful Things Physicists Could Do in Biology." *Physics Today*, 50.7 (July):23–27.

Patinkin, Donald. 1965. *Money, Interest and Prices.* 2nd ed. New York: Harper & Row.

1995. "The Training of an Economist." *Banca Nazionale del Lavoro Quarterly*, 48:359–95.

Paxson, Edwin. 1949. "Recent Developments in the Mathematical Theory of Games." *Econometrica*, 17:72–73.

1963. "War Gaming." RAND RM-3489-PR. RAND Corporation, Santa Monica, Calif.

Peat, F. D. 1997. *Infinite Potential.* Reading, Mass.: Addison-Wesley.

Peck, Merton, & Scherer, Frederick. 1962. *The Weapons Acquisition Process.* Cambridge: Harvard University Business School Press.

Pelaez, Eloina. 1999. "The Stored Program Computer: Two Conceptions." *Social Studies of Science*, 29:359–89.

Penrose, Roger. 1989. *The Emperor's New Mind.* Baltimore: Penguin.

Perelman, Michael. 1998. *Class Warfare in the Information Age.* New York: St. Martin's.

Pfeiffer, John. 1949. "The Office of Naval Research." *Scientific American*, 180 (February):11–15.

Pickering, Andrew, ed. 1992. *Science as Practice and Culture.* Chicago: University of Chicago Press.

1995a. "Cyborg History and the WWII Regime." *Perspectives in Science*, 3:1–45.

1995b. *The Mangle of Practice*. Chicago: University of Chicago Press.

1997. "The History of Economics and the History of Agency." In James Henderson, ed., *The State of the History of Economics*, pp. 6–18. London: Routledge.

1998. "A Gallery of Monsters." Unpublished manuscript, Dibner Institute.

1999. "Units of Analysis: Notes on WWII as a Discontinuity in the Social and Cyborg Sciences." Paper presented to the conference on Economists at War, Erasmus University, Rotterdam.

Pierides, Stella. 1998. "Machine Phenomena." In J. Berke et al., eds., *Even Paranoids Have Enemies*, pp. 177–88. London: Routledge.

Pikler, Andrew. 1951. "The Quanta-Kinetic Model of Monetary Theory." *Metroeconomica*, 3:70–95.

1954. "Utility Theories in Field Physics and Mathematical Economics." *British Journal for the Philosophy of Science*, 5:47–58, 303–18.

Pimbley, Joseph. 1997. "Physicists in Finance." *Physics Today*, 50.1 (January): 42–46.

Pinch, Trevor. 1977. "What Does a Proof Do If It Does Not Prove?" In E. Mendelsohn et al., eds., *The Social Production of Scientific Knowledge*. Dordrecht: Reidel.

Pitt, Joseph, ed. 1981. *Philosophy of Economics*. Dordrecht: Reidel.

Plotkin, Henry. 1994. *Darwin Machines and the Nature of Knowledge*. Cambridge: Harvard University Press.

Plott, Charles. 1986. "Laboratory Experiments in Economics: The Implications of Posted Price Institutions." *Science*, 232:732–38.

1991. "Will Economics Become an Experimental Science?" *Southern Economic Journal*, 57:901–19.

Plott, Charles, & Smith, Vernon. 1978. "An Experimental Examination of Two Exchange Institutions." *Review of Economic Studies*, 45:133–53.

Plott, Charles, & Sunder, Shyam. 1988. "Rational Expectations and the Aggregation of Diverse Information in the Laboratory Security Markets." *Econometrica*, 56:1085–1118.

Polanyi, Michael. 1958. *Personal Knowledge*. Chicago: University of Chicago Press.

Porat, Marc. 1977. *The Information Economy*. Washington, D.C.: U.S. Department of Commerce.

Porat, Marc, & Rubin, M. 1977. *The Information Economy*. Vol. 1. Washington, D.C.: GPO.

Porter, Theodore. 1986. *The Rise of Statistical Thinking*. Princeton: Princeton University Press.

Poundstone, William. 1992. *Prisoner's Dilemma*. New York: Anchor.

Pour-El, Marian, & Richards, Jonathan. 1989. *Computability in Analysis and Physics*. Berlin: Springer.

Prasad, Kislaya. 1991. "Computability and Randomness of Nash Equilibria in Infinite Games." *Journal of Mathematical Economics*, 20:429–51.

1997. "On the Computability of Nash Equilibria." *Journal of Economic Dynamics and Control*, 21:943–53.

Price, R. 1985. "A Conversation with Claude Shannon." *Cryptologia*, 9:167–75.

Prietula, Michael, Corley, Kathleen, & Gasser, Les, eds. 1998. *Simulating Organizations*. Cambridge: MIT Press.

Putnam, Hilary. 1960. "Minds and Machines." In Sidney Hook, ed., *Dimensions of Mind.* New York: New York University Press.

1967. *Mind, Language and Reality.* Cambridge: Cambridge University Press.

Pylyshyn, Zenon. 1984. *Computation and Cognition.* Cambridge: MIT Press.

Quade, Edwin, ed. 1964. *Analysis for Military Decisions.* Chicago: Rand McNally.

Quastler, Henry, ed. 1955. *Information Theory in Psychology.* Glencoe, Ill.: Free Press.

Quesnay, François. 1972. *Quesnay's tableau économique.* Ed. Marquerite Kuczynski and Ronald Meek. London. Macmillan.

Quine, Willard, McKinsey, J. J., & Krentel, W. 1951. "A Simplification of Games in Extensive Form." *Duke Mathematical Journal*, 18:885–900.

Rabin, Matthew. 1998. "Psychology and Economics." *Journal of Economic Literature*, 36:11–46.

Rabin, Michael. 1957. "Effective Computability of Winning Strategies." In *Contributions to the Theory of Games*, 3:147–57. Annals of Mathematical Studies no. 39. Princeton: Princeton University Press.

1967. "Mathematical Theory of Automata." In *Proceedings of the Symposium on Applied Mathematics*, vol. 19. Providence: American Mathematical Society.

1977. "Complexity of Computations." *Communications of the ACM*, 20:625–33.

Rabin, Michael, & Scott, Dana. 1959. "Finite Automata." *IBM Journal of Research*, 3:114–25.

Rabinbach, Anson, ed. 1985. *The Austrian Socialist Experiment, 1918–1934.* Boulder, Colo.: Westview.

1990. *The Human Motor.* New York: Basic.

Radner, Roy. 1959. "The Application of Linear Programming to Team Decision Problems." *Management Science*, 5:143–50.

1961. "The Evaluation of Information in Organizations." In *Proceedings of the Fourth Berkeley Symposium on Mathematical Statistics*, vol. 1. Berkeley: University of California.

1984. "Radner on Marschak." In Henry Spiegel and Warren Samuels, eds., *Contemporary Economists in Perspective*, pp. 443–80. Greenwich, Conn.: JAI Press.

1986. "Can Bounded Rationality Solve the Prisoner's Dilemma?" In W. Hildenbrand & A. Mas-Collel, eds., *Contributions to Mathematical Economics in Honor of Gerar Debeu.* Amsterdan: North Holland.

1987. "Jacob Marschak." In J. Eatwell et al., eds., *New Palgrave*, pp. 348–50. London: Macmillan.

1989. "Uncertainty and General Equilibrium." In J. Eatwell et al., eds., *General Equilibrium*, pp. 305–23. New York: Norton.

1993. "The Organization of Decentralized Information Processing." *Econometrica*, 61:1109–46.

Raffaelli, Tiziano. 1994. "Marshall's Analysis of the Human Mind." *Research in the History of Economic Thought and Analysis*, suppl., 4:57–93.

RAND Corporation. 1955. *A Million Random Digits with 100,000 Normal Deviates.* Santa Monica, Calif.

1963. *The Rand Corporation: The First 15 Years.* Santa Monica, Calif.

Various dates. *RAND Research Review.* <www.rand.org/publications/RRR>.

Rapoport, Anatol. 1959. "Critiques of Game Theory." *Behavioral Science*, 4: 49–66.
1960. *Fights, Games and Debates.* Ann Arbor: University of Michigan Press.
1964. *Strategy and Conscience.* New York: Schocken.
1968. "The Promise and Pitfalls of Information Theory." In Walter Buckley, ed., *Modern Systems Research for the Behavioral Sciences*, pp. 137–42. Chicago: Aldine.
1997. "Memories of Kenneth Boulding." *Review of Social Economy*, 60:416–31.
2000. *Certainties and Doubts.* Detroit: Black Rose Press.
Rapoport, Anatol, & Chammah, Albert. 1965. *Prisoner's Dilemma.* Ann Arbor: University of Michigan Press.
Rapoport, Anatol, & Orwant, Cynthia. 1962. "Experimental Games: A Review." *Behavioral Science*, 7:1–37.
Rashid, Salim. 1994. "John von Neumann, Scientific Method, and Empirical Economics." *Journal of Economic Methodology*, 1:279–93.
Raymond, Arthur. 1974. *Who? Me?* Privately printed.
Redei, Miklos. 1996. "Why John von Neumann Did Not Like the Hilbert Space Formalism of Quantum Mechanics (and What He Liked Instead)." *Studies in the History and Philosophy of Modern Physics*, 27:493–510.
Reder, Melvin. 1982. "Chicago Economics: Permanence and Change." *Journal of Economic Literature*, 20:1–38.
Redman, Deborah. 1998. *The Rise of Political Economy as a Science.* Cambridge: MIT Press.
Rees, Mina. 1987. "Warren Weaver (1894–1978)." *Biographical Memoirs of the National Academy of Sciences*, 57:493–530.
1988. "Mathematical Sciences and WWII." In Peter Duren, ed., *A Century of Mathematics in America*. Providence: AMS.
Reich, Leonard. 1985. *The Making of American Industrial Research.* Cambridge: Cambridge University Press.
Reingold, Nathan. 1991. *Science American Style.* New Brunswick: Rutgers University Press.
1995. "Choosing the Future." *Historical Studies in the Physical and Biological Sciences*, 25:2:301–27.
Reiter, Stanley. 1977. "Information and Performance in the $(\text{New})^2$ Welfare Economics." *American Economic Review Papers and Proceedings*, 67:226–34.
Rellstab, Urs. 1992. "New Insights into the Collaboration between John von Neumann and Oskar Morgenstern." In Weintraub, 1992, pp. 77–93.
Rescher, Nicholas. 1997. "H2O: Hempel-Helmer-Oppenheim." *Philosophy of Science*, 64:334–60.
Rhodes, Richard. 1986. *The Making of the Atomic Bomb.* New York: Simon & Schuster.
Richards, Graham. 1992. *Mental Machinery.* Baltimore: Johns Hopkins University Press.
Richardson, G. P. 1991. *Feedback Thought in the Social Sciences and Systems Theory.* Philadelphia: University of Pennsylvania Press.
Richter, Marcel, & Wong, Kam Chan. 1998. "What Can Economists Compute?" Economics discussion paper, University of Minnesota.

1999a. "Non-Computability of Competitive Equilibrium." *Economic Theory*, 14:1–27.

1999b. "Computable Preferences and Utility." *Journal of Mathematical Economics*, 31.

1999c. "Bounded Rationalities and Computable Economics." Economics discussion paper, University of Minnesota.

Rider, Robin. 1992. "Operations Research and Game Theory." In Weintraub, 1992, pp. 225–39.

1994. "Operational Research." In I. Grattan-Guinness, ed., *Companion Encyclopaedia of the History and Philosophy of the Mathematical Sciences*, pp. 837–42. London: Routledge.

Rima, Ingrid, ed. 1995. *Measurement, Quantification and Economic Analysis.* London: Routledge.

Riordan, Michael, & Hoddeson, Lillian. 1997. *Crystal Fire.* New York: Norton.

Riskin, Jessica. 1998. "The Defecating Duck." Paper presented to the American Society for 18th Century Studies, Notre Dame.

Rizvi, S. Abu. 1991. "Specialization and the Existence Problem in General Equilibrium Theory." *Contributions to Political Economy*, 10:1–10.

1994. "Game Theory to the Rescue?" *Contributions to Political Economy*, 13:1–28.

1997. "The Evolution of Game Theory." Paper presented to the seminar in philosophy and economics, Erasmus University, Rotterdam.

1998. "Responses to Arbitrariness in Contemporary Economics." In John Davis, ed., *The New Economics and Its History*, pp. 273–88. Durham: Duke University Press.

Robertson, George, et al., eds. 1996. *Future Natural.* London: Routledge.

Rochlin, Gene. 1997. *Trapped in the Net.* Princeton: Princeton University Press.

Romano, Richard. 1982. "The Economic Ideas of Charles Babbage." *History of Political Economy*, 14:385–405.

Rorty, Richard. 1989. *Contingency, Irony and Solidarity.* Cambridge: Cambridge University Press.

Rose, Nikolas. 1990. *Governing the Soul.* London: Routledge.

1996. *Inventing Our Selves.* Cambridge: Cambridge University Press.

1999. *Powers of Freedom.* Cambridge: Cambridge University Press.

Rosenberg, Nathan. 1994. *Exploring the Black Box.* Cambridge: Cambridge University Press.

Rosenblatt, Frank. 1958. *The Perceptron.* Washington, D.C.: U.S. Department of Commerce.

1962. *Principles of Neurodynamics.* Washington, D.C.: Spartan Books.

Rosenbleuth, A., Wiener, N., & Bigelow, J. 1943. "Behavior, Purpose and Teleology." *Philosophy of Science*, 10:18–24.

Rosenschein, Jeffrey, & Zlotkin, Gilead. 1994. *Rules of Encounter.* Cambridge: MIT Press.

Roszak, Theodore. 1994. *The Cult of Information.* 2nd ed. Berkeley: University of California Press.

Rota, Gian-Carlo. 1993. "The Lost Cafe." *Contention*, 2:41–61.

1997. *Indiscrete Thoughts.* Boston: Birkhauser.

Roth, Alvin. 1993. "On the Early History of Experimental Economics." *Journal of the History of Economic Thought*, 15:184–209. See also <http://www.pitt.edu/~alroth/history.html>.

1994. "Lets Keep the Con Out of Experimental Economics." *Empirical Economics*, 19:279–89.

Roth, Alvin, & Murnighan, J. 1982. "The Role of Information in Bargaining." *Econometrica*, 50:1123–42.

Roussas, S., & Hart, A. 1951. "Empirical Verification of a Composite Indifference Map." *Journal of Political Economy*, 59:288–318.

Rubinstein, Ariel. 1982. "Perfect Equilibrium in a Bargaining Model." *Econometrica*, 50:97–108.

1986. "Finite Automata Play the Repeated Prisoner's Dilemma." *Journal of Economic Theory*, 39:83–96.

1991. "Comments on the Interpretation of Game Theory." *Econometrica*, 59:909–24.

1993a. "On Price Recognition and Computational Complexity in a Monopolistic Model." *Journal of Political Economy*, 101:473–85.

1993b. "A Subjective Perspective on the Interpretation of Economic Theory." In A. Heertje, ed., *The Makers of Modern Economics*, vol. 1. Hemel Hempstead: Harvester Wheatsheaf.

1998a. *Models of Bounded Rationality*. Cambridge: MIT Press.

1998b. "Economics and Language." Nancy Schwartz lecture, Northwestern University.

Ruhla, Charles. 1992. *The Physics of Chance.* Oxford: Oxford University Press.

Russell, Bertrand, & Whitehead, Alfred. 1910. *Principia Mathematica.* Cambridge: Cambridge University Press.

Rust, John. 1997. "Dealing with the Complexity of Economic Calculations." Working paper, Yale University.

Rustem, Bert, & Velupillai, K. 1990. "Rationality, Computability and Complexity." *Journal of Economic Dynamics and Control*, 14:419–32.

Ryerson, James. 2000. "Games People Play: Rational Choice with a Human Face." *Lingua Franca*, 9.9 (January):61–71.

Saari, D. 1995. "Mathematical Complexity of Simple Economies." *Notices of the American Mathematical Society*, 42:222–30.

Samuelson, Larry. 1997. *Evolutionary Games and Equilibrium Selection.* Cambridge: MIT Press.

Samuelson, Paul. 1938. "A Note on the Pure Theory of Consumer's Behavior." *Economica*, 5:61–71.

1944a. "A Warning to the Washington, D.C. Expert." *New Republic*, 111 (September 11):297–99.

1944b. "Analysis of Tracking Data." Cambridge: MIT Radiation Laboratory.

1965. *Foundations of Economic Analysis.* New York: Atheneum.

1966. *Collected Economic Papers.* Vols. 1, 2. Cambridge: MIT Press.

1977. *Collected Economic Papers.* Vol. 3. Cambridge: MIT Press.

1986. *Collected Economic Papers.* Vol. 5. Cambridge: MIT Press.

1998. "How *Foundations* Came to Be." *Journal of Economic Literature*, 36: 1375–86.

Sandler, Todd, & Hartley, Keith. 1995. *The Economics of Defense.* Cambridge: Cambridge University Press.
Sapolsky, Harvey. 1990. *Science and the Navy.* Princeton: Princeton University Press.
Sargent, Thomas. 1993. *Bounded Rationality in Macroeconomics.* Oxford: Clarendon Press.
Saunders, Frances. 1999. *Who Paid the Piper?* London: Granta.
Savage, Leonard. 1954. *The Foundations of Statistics.* New York: Wiley.
Scarf, Herbert. 1959. "Bayes' Solutions of the Statistical Inventory Problem." *Annals of Mathematical Statistics*, 30:490–508.
1960. "Some Examples of Global Instability in General Equilibrium." *International Economic Review*, 1:157–72.
1969. "An Example of an Algorithm for Calculating General Equilibrium Prices." *American Economic Review*, 59:669–77.
1973. *The Computation of Economic Equilibria.* New Haven: Yale University Press.
1997. "Tjalling Koopmans." *Biographical Memoirs of the National Academy of Sciences*, 67:263–91.
Scarf, Herbert, & Shapley, Lloyd. 1954. "Games with Information Lags." RAND RM-1320. RAND Corporation, Santa Monica, Calif.
Schabas, Margaret. 1990. *A World Ruled by Number.* Princeton: Princeton University Press.
Schacter, Daniel. 1996. *Searching for Memory.* New York: Basic.
Schaffer, Simon. 1994. "Babbage's Intelligence: Calculating Engines and the Factory System." *Critical Inquiry*, 21:203–27.
1996. "Babbage's Dancer and the Impresarios of Mechanism." In Spufford & Uglow, 1996, pp. 53–80.
1999. "Enlightened Automata." In William Clark et al., eds., *The Sciences in Enlightened Europe*, pp. 126–68. Chicago: University of Chicago Press.
Schelling, Thomas. 1958. "Comment." *Review of Economics and Statistics*, 40:221–24.
1960. *The Strategy of Conflict.* Cambridge: Harvard University Press.
1964. Review of *Strategy and Conscience*, by Anatol Rapoport. *American Economic Review*, 54:1082–88.
1969. "Models of Segregation." *American Economic Review, Papers and Proceedings*, 59:2:488–93.
1971. "Dynamic Models of Segregation." *Journal of Mathematical Sociology*, 1:143–86.
1980. "The Intimate Contest for Self-Command." *Public Interest*, 60:94–118.
1984. "Self-Command in Practice and in Policy." *American Economic Review*, 74:1–11.
Schement, Jorge, & Ruben, Brent, eds. 1993. *Between Communication and Information.* New Brunswick, N.J.: Transactions.
Schiller, Dan. 1999. *Digital Capitalism.* Cambridge: MIT Press.
Schlesinger, Kurt. 1933. "Uber die Produktionsgleichungen der Okonomischen Wertlehre." In Karl Menger, ed., *Ergebnisse eines Mathematischen Kolloquiums*, 6:10–11.

Schmemann, Serge. 1998. "Israel Clings to Its Nuclear Ambiguity." *New York Times*, June 21.

Schmidt, Christian. 1995a. "Nash versus von Neumann et Morgenstern." *Revue economique*.

ed. 1995b. *Theorie des jeux et analyse economique 50 ans après*. Paris: Sirey.

Schrecker, Ellen. 1986. *No Ivory Tower: McCarthyism and the Universities*. Oxford: Oxford University Press.

Schrödinger, Erwin. 1945. *What Is Life?* Cambridge: Cambridge University Press.

Schultz, Andrew. 1970. "The Quiet Revolution: From Scientific Management to Operations Research." *Cornell Engineering Quarterly*, 4:2–10.

Schultz, Henry. 1938. *Theory and Measurement of Demand*. Chicago: University of Chicago Press.

Schurr, Sam, & Marschak, Jacob. 1950. *Economic Aspects of Atomic Power*. Princeton: Princeton University Press.

Schwartz, Benjamin. 1989. "The Invention of Linear Programming." *Annals for the History of Computing*, 11:145–51.

Schwartz, Peter. 1995. "War in the Age of Intelligent Machines" *Wired*. <www.hotwired.com/wired/3.04/features/pentagon.html>.

Schweber, Sylvan. 1988. "The Mutual Embrace of Science and the Military." In Mendelsohn et al., 1988, pp. 3–46.

1994. *QED and the Men Who Made It*. Princeton: Princeton University Press.

Scodel, A., et al. 1960. "Some Descriptive Aspects of Two Person Non-Zero Sum Games, II." *Journal of Conflict Resolution*, 4:114–19.

Searle, John. 1980. "Minds, Brains and Programs." *Behavioral and Brain Sciences*, 3:417–57.

1992. *The Rediscovery of Mind*. Cambridge: MIT Press.

1995. "The Myth of Consciousness." *New York Review of Books*, November 2.

Selcraig, James. 1982. *The Red Scare in the Midwest, 1945–55*. Ann Arbor: UMI Research Press.

Selten, Reinhard. 1990. "Bounded Rationality." *Journal of Institutional and Theoretical Economics*, 146:649–58.

1993. "In Search of a Better Understanding of Economic Behavior." In A. Heertje, ed., *Makers of Modern Economics*, 1:115–39. Hemel Hempstead: Harvester Wheatsheaf.

Sen, Amartya. 1987. *On Ethics and Economics*. Oxford: Basil Blackwell.

1999. "The Possibility of Social Choice." *American Economic Review*, 89:349–78.

Sent, Esther-Mirjam. 1997a. "Sargent versus Simon: Bounded Rationality Unbound." *Cambridge Journal of Economics*, 21:323–38.

1997b. "Bounded Rationality on the Rebound." Working paper, University of Notre Dame.

1998. *The Evolving Rationality of Rational Expectations*. Cambridge: Cambridge University Press.

1999. "John Muth's Rational Expectations." Economics working paper, University of Notre Dame. Forthcoming in *History of Political Economy*.

Forthcoming. "Sent Simulating Simon Simulating Scientists." *Studies in the History and Philosophy of Science*.

Shanker, Stuart, ed. 1989. *Gödel's Theorem in Focus*. London: Routledge.

1995. "Turing and the Origins of AI." *Philosophia Mathematica*, 3:52–85.

Shannon, Claude. 1948. "The Mathematical Theory of Communication." *Bell System Technical Journal*, 27:379–423, 623–56.

1949. "Communication Theory of Secrecy Systems." *Bell System Technical Journal*, 28:656–715.

1950. "Programming a Computer for Playing Chess." *Philosophical Magazine*, 41:256–74.

1987. "Interview." *Omni*, 10.8 (August):61–66, 110.

1993. *Claude Shannon: Collected Papers.* Piscataway, N.J.: IEEE Press.

Shannon, Claude, & McCarthy, J. eds. 1956. *Automata Studies.* Princeton: Princeton University Press.

Shannon, Claude, & Weaver, Warren. 1949. *Mathematical Theory of Communication.* Urbana: University of Illinois Press.

Shapiro, Carl, & Varian, Hal. 1999. *Information Rules.* Boston: Harvard Business School Press.

Shapley, Deborah. 1993. *Promise and Power.* Boston: Little Brown.

Shasha, Dennis, & Lazere, Cathy. 1995. *Out of Their Minds*. New York: Copernicus.

Sheehan, J., & Sosna, M., eds. 1991. *The Boundaries of Humanity*. Berkeley: University of California Press.

Shenker, Orly. 1999. "Maxwell's Demon and Baron Munchausen." *Studies in the History and Philosophy of Modern Physics*, 30:347–72.

Sherry, Michael. 1995. *In the Shadow of War.* New Haven: Yale University Press.

Shils, Edward, ed. 1991. *Remembering the University of Chicago.* Chicago: University of Chicago Press.

Shlapak, David. 1985. "The Rand Strategy Assessment Center and the Future of Simulation and Gaming." RAND P-7162. RAND Corporation, Santa Monica, Calif.

Shoham, Yoav, ed. 1996. *Theoretical Aspects of Rationality and Knowledge.* San Francisco: Morgan Kaufman.

Shrecker, Ellen. 1986. *No Ivory Tower: McCarthyism and the Universities.* New York: Oxford University Press.

Shubik, Martin. 1952. "Information, Theories of Competition and the Theory of Games." *Journal of Political Economy*, 60:145–50.

ed. 1964. *Game Theory and Related Approaches to Social Behavior.* New York: Wiley.

1969. "A Note on a Simulated Stock Market." *Decision Sciences*, 1:129–41.

1970. "A Curmudgeon's Guide to Microeconomics." *Journal of Economic Literature*, 8:405–34.

1975. *The Uses and Methods of Gaming.* Amsterdam: Elsevier.

ed. 1976. *Essays in Mathematical Economics in Honor of Oskar Morgenstern.* Princeton: Princeton University Press.

1981. "Perfect or Robust Noncooperative Equilibrium: Search for the Philosopher's Stone?" In *Essays in Game Theory*. Mannheim: BI Wissenschaftsverlag.

1982. *Game Theory in the Social Sciences*. Cambridge: MIT Press

1992. "Game Theory at Princeton, 1949–55." In Weintraub, 1992, pp. 151–63.

1996. "Why Equilibrium? A Note on the Noncooperative Equilibrium of Some Matrix Games." *Journal of Economic Behavior and Organization*, 29:537–39.

1997. "On the Trail of a White Whale." In A. Heertje, ed., *Makers of Modern Economics III*. Cheltenham: Elgar.

1998. "Game Theory, Complexity and Simplicity, Pt. II." *Complexity*, 3.3:36–45.

Shubik, Martin, Brewer, G., & Savage, E. 1972. *The Literature of Gaming, Simulation and Model Building.* RAND R-620-ARPA. RAND Corporation, Santa Monica, Calif.

Shubik, Martin, & Wolf, G. 1971. "An Artificial Player for a Business Market Game." *Simulation and Games*, 2:27–43.

Sichel, Dan. 1997. *The Computer Revolution.* Washington, D.C.: Brookings.

Siegelmann, Hava. 1999. *Neural Networks and Analog Computation.* Boston: Birkhauser.

Simon, Herbert. 1945. Review of *Theory of Games and Economic Behavior*. *American Journal of Sociology*, 27:558–60.

1947. *Administrative Behavior.* New York: Free Press.

1954. "Bandwagon and Underdog Effects and the Possibility of Election Predictions." *Public Opinion Quarterly*, 18:245–53.

1955. "A Behavioral Model of Rational Choice." *Quarterly Journal of Economics*, 69:99–118.

1956a. "A Comparison of Game Theory and Learning Theory." *Psychometrika*, 21:267–72.

1956b. "Rational Choice and the Structure of the Environment." *Psychological Review*, 63:129–38.

1956c. "Dynamic Programming under Uncertainty with a Quadratic Criterion Function." *Econometrica*, 24:74–81.

1957. *Models of Man.* New York: Wiley.

1960. *The New Science of Management Decison.* New York: Harper and Row.

1973. "The Organization of Complex Systems." In Howard Patee, ed., *Hierarchy Theory*, pp. 3–27. New York: Braziller.

1979. "Rational Decision Making in Business Organizations." *American Economic Review*, 69:493–513.

1980. "Cognitive Science: The Newest Science of the Artificial." *Cognitive Science*, 4:33–46.

1981. *The Sciences of the Artificial.* Rev. ed. Cambridge: MIT Press.

1982. *Models of Bounded Rationality*. 2 vols. Cambridge: MIT Press.

1983. *Reason in Human Affairs.* Stanford: Stanford University Press.

1987. "Two Heads Are Better than One: The Collaboration between AI and OR." *Interfaces*, 17 (July):8–15.

1988. "Remembering John von Neumann." *OR/MS Today*, 15:10.

1989a. *Models of Thought*. New Haven: Yale University Press.

1989b. "The Scientist as Problem Solver." In D. Klahr & K. Kotovsky, eds., *Complex Information Processing*, pp. 375–98. Hillsdale, N.J.: Erlbaum.

1991a. *Models of My Life.* New York: Basic.

1991b. "Comments on the Symposium on Computer Discovery." *Social Studies of Science*, 21:143–48.

1992a. *Economics, Bounded Rationality and the Cognitive Revolution*. Ed. Massimo Egidi & Robin Marris. Aldershot: Elgar.

1992b. Review of *John von Neumann and the Origins of Modern Computing*, by William Aspray. *Minerva*, 30:570–77.

1994. "Interview." *Omni*, 16.

1995. "Comment on Kagel." In G. Wolters & J. Lennox, eds., *Concepts, Theories and Rationality in the Biological Sciences*, pp. 359–66. Pittsburgh: University of Pittsburgh Press.

1996. "Machine as Mind." In Milikan & Clark, 1996, pp. 81–102.

Simon, Herbert, & Newell, A. 1958. "Heuristic Problem Solving." *Operations Research*, 6:1–10.

Singh, M., Rao, A., & Wooldridge, M., eds. 1998. *Intelligent Agents IV.* Berlin: Springer.

Sippel, R. 1997. "An Experiment on the Pure Theory of Consumer's Behavior." *Economic Journal*, 107:1431–44.

Sismondo, Sergio. 1997. "Modeling Strategies: Creating Autonomy for Biology's Theory of Games." *History and Philosophy of the Life Sciences*, 19:147–61.

Sklar, Lawrence. 1993. *Physics and Chance.* Cambridge: Cambridge University Press.

Skryms, Brian. 1996. *Evolution of the Social Contract.* Cambridge: Cambridge University Press.

Slaughter, Sheila, & Rhoads, Gary. 1996. "The Emergence of a Competitiveness R&D Policy Coalition." *Science, Technology and Human Values*, 21:303–39.

Slezak, Peter. 1989. "Scientific Discovery by Computer as Empirical Refutation of the Strong Programme." *Social Studies of Science*, 19:563–600.

Smith, Adam. 1976. *An Inquiry into the Nature and Causes of the Wealth of Nations*. 1776. Reprint, Chicago: University of Chicago Press.

Smith, Barry. 1997. "The Connectionist Mind: A Study of Hayekian Psychology." In S. Frowen, ed., *Hayek the Economist and Social Philosopher*, pp. 9–30. London: Macmillan.

Smith, Bruce. 1966. *The Rand Corporation.* Cambridge: Harvard University Press.

Smith, Carl. 1994. *A Recursive Introduction to the Theory of Computation*. Berlin: Springer Verlag.

Smith, Crosbie, & Wise, M. Norton. 1989. *Energy and Empire*. Cambridge: Cambridge University Press.

Smith, James Allen. 1991. *The Idea Brokers.* New York: Free Press.

Smith, John Maynard. 1982. *Evolution and the Theory of Games.* Cambridge: Cambridge University Press.

1988. *Did Darwin Get it Right?* London: Penguin.

1990. "What Can't the Computer Do?" *New York Review of Books*, March 15, pp. 21–25.

Smith, John Maynard, & Szathmary, Eors. 1995. *The Major Transitions in Evolution.* Oxford: W. H. Freeman.

Smith, Merrit, ed. 1985. *Military Enterprise and Technological Change.* Cambridge: MIT Press.

Smith, Roger. 1997. *Norton History of the Human Sciences.* New York: Norton.

Smith, Vernon. 1962. "An Experimental Study of Competitive Market Behavior." *Journal of Political Economy*, 70:111–37.

1982. "Microeconomic Systems as an Experimental Science." *American Economic Review*, 72:923–55.

1991a. *Papers in Experimental Economics.* Cambridge: Cambridge University Press.

1991b. "Rational Choice: The Contrast between Economics and Psychology." *Journal of Political Economy*, 99:877–96.

1992. "Game Theory and Experimental Economics." In Weintraub, 1992, pp. 241–82.

Snyder, Frank. 1993. *Command and Control.* Washington, D.C.: National Defense University.

Sober, Elliott, ed. 1984. *Conceptual Issues in Evolutionary Biology.* Cambridge: MIT Press.

Soderqvist, Thomas, ed. 1997. *The Historiography of Contemporary Science and Technology.* Amsterdam: Harwood.

Solberg, W., & Tomilson, R. 1997. "Academic McCarthyism and Keynesian Economics." *History of Political Economy*, 29:55–81.

Sommerhalder, R., & van Westrehen, S. 1988. *The Theory of Computability.* Wokingham: Addison-Wesley.

Sonnenschein, Hugo. 1972. "Market Excess Demand Functions." *Econometrica*, 40:549–63.

1973. "Do Walras' Identity and Continuity Characterize the Class of Community Excess Demand Functions?" *Journal of Economic Theory*, 6:345–54.

Sorokin, Pitrim. 1956. *Contemporary Sociological Theories*. New York: Harper Torchbooks.

Soule, M., & Lease, Gary, eds. 1995. *Reinventing Nature?* Washington, D.C.: Island Press.

Spear, S. 1989. "Learning Expectations under Computability Constraints." *Econometrica*, 57:889–910.

Spence, A. M. 1974. "An Economist's View of Information." In C. Canadia, A. Luke, & J. Harris, eds., *Annual Review of Information Science and Technology*, 9:57–78.

Spiro, Melford. 1993. "Is the Western conception of the Self Peculiar within the Context of World Cultures?" *Ethos*, 21:107–53.

Spitzer, N., & Sejnowski, T. 1997. "Biological Information Processing." *Science*, 277 (September 26):1060–61.

Spitzer, R., ed. 1979. *Diagnostic and Statistical Manual of Mental Disorders.* Washington, D.C.: American Psychiatric Association.

Spufford, Francis, & Uglow, Jenny, eds. 1996. *Cultural Babbage*. London: Faber and Faber.

Spulber, Daniel. 1999. *Market Microstructure.* Cambridge: Cambridge University Press.

Statistical Research Group. 1945. *Sequential Analysis of Statistical Data: Applications.* SRG Report 255, AMP Report 30.2R.

Steele, David. 1992. *From Marx to Mises: Post-Capitalist Society and the Challenge of Economic Calculation.* La Salle, Ill.: Open Court.

Stein, George. 1996. "Information Attack: Information Warfare in 2025." <www.fas.org/spp/military/docops/usaf/2025/v3c3-1.htm>.

Steinberg, Jessica. 1998. "The Military-Technological Complex Is Thriving in Israel." *New York Times*, December 6, sec. 3, p. 5.

Steinhaus, Hugo. 1960. "Definitions for a Theory of Games of Pursuit." 1925. *Naval Research Logistics Quarterly*, 7.2 (June):105–8.

Sterelny, Kim. 1995. "Understanding Life: Recent Work in the Philosophy of Biology." *British Journal for the Philosophy of Science*, 46:155–83.

Sterling, Bruce. 1993. "War Is Virtual Hell." *Wired.* <www.wired.com/wired/1.1/features/virthell>.

Stewart, Iain. 1993. "Logical and Schematic Characterization of Complexity Classes." *Acta Informatica*, 30:61–87.

Stigler, George. 1955. "Mathematics in Economics: Further comment." *Review of Economics and Statistics*, 37:299–300.

1961. "The Economics of Information." *Journal of Political Economy*, 69:213–25.

1968. *The Economics of Industry.* Homewood, Ill.: Irwin.

1982. *The Economist as Preacher.* Chicago: University of Chicago Press.

1988. *Memoirs of an Unregulated Economist.* New York: Basic.

Stigler, George, & Becker, Gary. 1977. "De Gustibus non est Dispudandum." *American Economic Review*, 67.2:76–90.

Stigler, Stephen. 1986. *The History of Statistics.* Cambridge: Harvard University Press.

Stiglitz, Joseph. 1985. "Information and Economic Analysis." *Economic Journal*, 95:21–41.

1993. "Reflections on Economics and on Being and Becoming an Economist." In A. Heertje, ed., *Makers of Modern Economics*, 1:140–84. Hemel Hempstead: Harvester Wheatsheaf.

2000. "The Contributions of the Economics of Information to 20th Century Economics." *Quarterly Journal of Economics*, 140:1441–78.

Stockfisch, Jack. 1987. "The Intellectual Foundations of Systems Analysis." RAND P-7401. RAND Corporation, Santa Monica, Calif.

Stoll, Hans. 1992. "Principles of Trading Market Structure." *Journal of Financial Services Research*, 6:75–107.

Stone, D., Denham, A., & Garnett, M., eds. 1998. *Think Tanks across Nations.* Manchester: University of Manchester Press.

Stone, L. 1989. "What's Happened in Search Theory?" *Operations Research*, 37:501–6.

Stone, Richard. 1948. "The Theory of Games." *Economic Journal*, 58:185–201.

Strathern, Marilyn. 1995. "Nostalgia and the New Genetics." In D. Battaglia, ed., *Rhetorics of Self-Making*, pp. 97–120. Berkeley: University of California Press.

Sunder, Shyam. 1992. "The Market for Information: Experimental Evidence." *Econometrica*, 60:667–95.

Swade, Doron. 1991. *Charles Babbage and His Calculating Engines.* London: Science Museum.

Swedberg, Richard, ed. 1990. *Economics and Sociology.* Princeton: Princeton University Press.

Szenberg, Michael, ed. 1992. *Eminent Economists.* Cambridge: Cambridge University Press.

Szilard, Leo. 1929. "Uber die Entropieverminderung in einem Thermodynamischen System bei Eingriffen Intelligenter Wesen." *Zeitschrift für Physik*, 53:840–56.

Takacs, David. 1996. *The Idea of Biodiversity.* Baltimore: Johns Hopkins University Press.

Taubes, Gary. 1997. "Computer Design Meets Darwin." *Science*, 277 (September 26):1931–32.

1998. "Evolving a Conscious Machine." *Discover*, 19.6 (June):73–79.

Taylor, Charles. 1989. *Sources of the Self.* Cambridge: Harvard University Press.

Taylor, Peter. 1988. "Technocratic Optimism, H. T. Odum and the Partial Transformation of the Ecological Metaphor." *Journal of the History of Biology*, 21:213–44.

Taylor, R. G. 1998. *Models of Computation and Formal Languages.* New York: Oxford University Press.

Thackray, Arnold, ed. 1992. "Science after '40." *Osiris*, 7.

Thagard, Paul. 1993. "Computational Tractability and Conceptual Coherence: Why Do Computer Scientists Believe That P ≠ NP?" *Canadian Journal of Philsophy*, 23:349–64.

Thaler, Richard. 1987. "The Psychology of Choice and the Assumptions of Economics." In Alvin Roth, ed., *Laboratory Experiments in Economics.* Cambridge: Cambridge University Press.

Thaler, Richard, & Shefrin, Hersh. 1981. "An Economic Theory of Self-Control." *Journal of Political Economy*, 89:392–410.

Theil, Henri. 1967. *Economics and Information Theory.* Amsterdam: North Holland.

Theusneyer, Lincoln. 1947. *Combat Scientists.* Boston: Little Brown.

Thomas, Clayton, & Deemer, Walter. 1957. "The Role of Operational Gaming in Operations Research." *Journal of the Operations Research Society of America*, 5:1–27.

Thomas, J. 1993. "Non-Computable Rational Expectations Equilibria." *Mathematical Social Sciences*, 25:133–42.

Thompson, E. P. 1967. "Time, Work Discipline and Industrial Capitalism." *Past and Present*, 38:56–97.

Thompson, G. F. 1998. "Encountering Economics and Accounting." *Accounting, Organization and Society*, 23:283–323.

Thompson, F. 1952. "Equivalence of Games in Extensive Form." RAND RM-759. RAND Corporation, Santa Monica, Calif.

Thomsen, Esteban. 1992. *Prices and Knowledge.* London: Routledge.

Thrall, R., Coombs, C., & Davis, R., eds. 1954. *Decision Processes.* New York: Wiley.

Thresmeyer, Lincoln, & Burchard, John. 1947. *Combat Scientists.* Boston: Little Brown.

Tidman, Keith. 1984. *The Operations Evaluation Group.* Annapolis: Naval Institute Press.

Torretti, Roberto. 1999. *The Philosophy of Physics.* Cambridge: Cambridge University Press.

Trachtenberg, Marc. 1999. *A Constructed Peace.* Princeton: Princeton University Press.

Trefethen, Florence. 1954. "A History of Operations Research." In J. McCloskey & F. Trefethen, eds., *Operations Research for Management*, pp. 3–35. Baltimore: Johns Hopkins University Press.

Tucker, A., & Luce, R., eds. 1959. *Contributions to the Theory of Games.* Vol. 4. Princeton: Princeton University Press.

Tucker, Samuel. ed. 1966. *A Modern Design for Defense Decisions: A McNamara-Hitch-Enthoven Anthology.* Washington, D.C.: Armed Forces College.

Turing, Alan. 1936–37. "On Computable Numbers." *Proceedings of the London Mathematical Society*, ser. 2, 42:230–65.

1950. "Computing Machinery and Intelligence." *Mind*, 59:433–60.

1953. "The Chemical Basis of Morphogenesis." *Philosophical Transactions of the Royal Society*, B237:37–72.

1992. *Mechanical Intelligence.* Ed. D. C. Ince. Amsterdam: North Holland.

1996. "Intelligent Machinery: A Heretical Theory." *Philosophia Mathematica*, 4:256–60.

Turkle, Sherry. 1984. *The Second Self.* New York: Simon & Schuster.

1995. *A Life on the Screen.* New York: Simon & Schuster.

Turner, Stephen. 1999. "Does Funding Produce Its Effects?" In Teresa Richardson and Donald Fisher, eds., *The Development of the Social Sciences in the US and Canada: The Role of Philanthropy*, pp. 215–27. Stamford, Conn.: Ablex.

Turner, Stephen, & Turner, Jonathan. 1990. *The Impossible Science.* Newbury Park, Calif.: Sage.

Tustin, Arnold. 1957. *The Mechanism of Economic Systems.* London: Heinemann.

Tversky, Amos. 1969. "Intransitivities of Preferences." *Psychological Review*, 76:31–48.

Tversky, Amos, & Kahneman, Daniel. 1981. "The Framing of Decisions and the Psychology of Choice." *Science*, 211:453–58.

Twomey, Christopher. 1999. "The McNamara Line and the Turning Point for Civilian-Scientist Advisors in America." *Minerva*, 37:235–58.

Ulam, Stanislaw. 1952. "Random Processes and Transformations." In *Proceedings of the International Congress of Mathematicians, 1950.* Providence: American Mathematical Society.

1956. "Applications of Monte Carlo Methods to Tactical Games." In Herbert Meyer, ed., *Symposium on Monte Carlo Methods*, p. 63. New York: Wiley.

1958. "John von Neumann, 1903–57." *Bulletin of the American Mathematical Society*, 64:1–49.

1976. *Adventures of a Mathematician.* New York: Scribners.

1980. "Von Neumann: The Interaction of Mathematics and Computing." In Metropolis et al., 1980, pp. 93–100.

1986. *Science, Computers and Mathematics.* Ed. Mark Reynolds & Gian-Carlo Rota. Boston: Birkhauser.

Ullman, Ellen. 1997. *Close to the Machine.* San Francisco: City Lights.
Uzawa, Hirofumi. 1988. *Preference, Production and Capital.* Cambridge: Cambridge University Press.
Van Boening, Mark, & Wilcox, Nathaniel. 1996. "Avoidable Cost: Ride a Double Auction Roller Coaster." *American Economic Review*, 86:461–77.
van Creveld, Martin. 1985. *Command in War.* Cambridge: Harvard University Press.
van der Lubbe, Jan. 1997. *Information Theory.* Cambridge: Cambridge University Press.
Varian, Hal. 1978. *Microeconomic Theory.* New York: Norton.
Varnholt, Burkhard. 1997. "Electronic Securities Trading: The Options for Designing Electronic Markets." *Swiss Journal of Economics and Statistics*, 133:165–74.
Vatin, François. 1998. *Economie politique et economie naturelle chez Antoine-Augustin Cournot.* Paris: Presses Universitaires de France.
Velupillai, K. 1995. "Synopsis of the Ryde Lectures" Working paper, University of California, Los Angeles.
1996. "The Computable Alternative to the Formalization of Economics." *Kyklos*, 49:251–73.
1997. "Expository Notes on Computability and Complexity in Arithmetical Games." *Journal of Economic Dynamics and Control*, 21:955–79.
2000. *Computable Economics: The Arne Ryde Lectures.* Oxford: Oxford University Press.
von Foerster, Heinz, ed. 1952. *Cybernetics: Transactions of the 8th Macy Conference.* New York: Macy Foundation.
von Furstenberg, George, ed. 1990. *Acting under Uncertainty.* Boston: Kluwer.
von Neumann, John. 1928. "Zur Theorie der Gesellschaftsspiele." *Mathematische Annalen*, 100:295–320. Translated in von Neumann, 1959, 4:13–42.
1937. "Uber ein Okonomischen Gleichungssystem." In Karl Menger, ed., *Ergebnisse eines Mathematischen Kolloquiums*, 8.
1945. "A Model of General Economic Equilibrium." *Review of Economic Studies*, 13:1–9.
1949. Review of *Cybernetics*, by N. Wiener. *Physics Today*, 2:33–34.
1955. *Mathematical Foundations of Quantum Mechanics.* Trans. Robert Beyer. 1932. Reprint, Princeton: Princeton University Press.
1956. "The Impact of Recent Developments in Science on the Economy and on Economics." *Looking Ahead*, 4:11.
1958. *The Computer and the Brain.* New Haven: Yale University Press.
1959. "On the Theory of Games of Strategy." Trans. Sonya Bargmann. In A. Tucker & R. Luce, eds., *Contributions to the Theory of Games*, 4:13–42. Princeton: Princeton University Press. Originally published as von Neumann, 1928.
1961–63. *The Collected Works of John von Neumann.* Ed. A. Taub. 6 vols. New York: Pergamon Press.
1966. *The Theory of Self-Reproducing Automata.* Ed. Arthur Burks. Urbana: University of Illinois Press.

1987. *The Papers of John von Neumann on Computing and Computer Theory.* Ed. A. Burks & W. Aspray. Cambridge: MIT Press.

von Neumann, John, & Morgenstern, Oskar. 1964. *Theory of Games and Economic Behavior.* 1944. Reprint, New York: Wiley.

Vroman, Jack. 1995. *Economic Evolution.* London: Routledge.

Vulkan, Nir. 1999. "Economic Implications of Agent Technology and E-Commerce." *Economic Journal*, 109:F67–90.

Waddington, Conrad. 1973. *OR in World War II.* London: Elek.

Wald, Abraham. 1936. "Uber einige Gleichungssysteme der mathematischen Okonomie." *Zeitschrift für Nationalökonomie*, 7:637–70.

1951. "On Some Systems of Equations of Mathematical Economics." 1934. *Econometrica*, 19:368–403.

Wald, Abraham, Dvoretsky, A., & Wolfowitz, J. 1951. "Elimination of Randomization in Certain Statistical Decision Procedures and Zero-Sum Two Person Games." In Anderson et al., 1957, pp. 602–22.

Waldrop, Mitchell. 1992. *Complexity*. New York: Simon & Schuster.

Wallis, W. A. 1980. "The Statistical Research Group." *Journal of the American Statistical Association*, 75:320–30.

1991. "Leonard Jimmie Savage." In Shils, 1991, pp. 436–51.

Wallis, W. A., & Friedman, Milton. 1942. "The Empirical Derivation of Indifference Functions." In Lange et al., pp. 175–89.

Wang, Jessica. 1999a. *American Science in an Age of Anxiety.* Chapel Hill: University of North Carolina Press.

1999b. "Merton's Shadow: Perspectives on Science and Democracy." *Historical Studies in the Physical Sciences*, 30:279–306.

Ware, Willis. 1976. "Project RAND and Air Force Decision Making." RAND P-5737. RAND Corporation, Santa Monica, Calif.

Waring, Steven. 1991. *Taylorism Transformed.* Chapel Hill: University of North Carolina Press.

Weagle, Matt. 1995 "Computational Complexity vs. Rational Choice." Unpublished manuscript, University of Notre Dame.

1996. "Budget Constraints and Experimental Double Auctions." Working paper, University of Notre Dame.

Weatherford, Roy. 1982. *Philosophical Foundations of Probability Theory.* London: Routledge Kegan Paul.

Weaver, Warren. 1946. "Summary." In *Summary Technical Report of the Applied Mathematics Panel*, vol. 1. Washington: NDRC.

1947. "Science and Complexity." *American Scientist*, 36:536–44.

1970. *Scenes of Change.* New York: Scribners.

Webster, Frank. 1995. *Theories of the Information Society.* London: Routledge.

Weibull, Jorgen. 1997. *Evolutionary Game Theory.* Cambridge: MIT Press.

Weimer, Walter, & Palermo, David, eds. 1982. *Cognition and Symbolic Processes.* Vol. 2. Hillsdale, N.J.: Erlbaum.

Weinberg, Steven. 1998. "The Revolution That Didn't Happen." *New York Review of Books*, October 8, pp. 48–52.

Weintraub, E. R. 1985. *General Equilibrium Analysis: Studies in Appraisal.* Cambridge: Cambridge University Press.

1991. *Stabilizing Dynamics*. Cambridge: Cambridge University Press.

ed. 1992. *Toward a History of Game Theory*. Suppl. to vol. 24 of *History of Political Economy*. Durham: Duke University Press.

1998. "Axiomatisches Misunderstandnis." *Economic Journal*, 108:1837–47.

Weintraub, E. R., & Gayer, Ted. 2000. "Negotiating at the Boundary." *History of Political Economy*, 32:441–71.

2001. "Equilibrium Proofmaking." *Journal of the History of Economic Thought*.

Weintraub, E. R., & Mirowski, P. 1994. "The Pure and the Applied: Bourbakism Comes to Mathematical Economics." *Science in Context*, 7:245–72.

Weirs, L., & Jungenfelt, K. 1984. "Tjalling Koopmans." In W. Samuels & H. Spiegel, eds., *Contemporary Economists in Perspective*, pp. 351–71. Greenwich, Conn.: JAI Press.

Weiss, Rick. 1998. "Patent Sought on the Making of Part-Human Creatures." *Washington Post*, April 2, p. A12.

Weissert, Thomas. 1997. *The Genesis of Simulation in Dynamics*. New York: Springer Verlag.

Weizenbaum, Joseph. 1976. *Computer Power and Human Reason*. San Francisco: W. H. Freeman.

Wellman, Michael. 1994. Review of *The Ecology of Computation*, by Bernardo Huberman. In W. Clancey et al., eds., *Contemplating Minds*. Cambridge: MIT Press.

1995. "Rationality in Decision Machines." <http://ai.eecs.umich.edu/people/wellman>.

Werskey, Gary. 1988. *The Visible College*. London: Free Association.

Westhoff, F., & Yarbrough, B. 1996. "Complexity, Organization and Stuart Kauffman's *Origins of Order*." *Journal of Economic Behavior and Organization*, 29:1–25.

Westwick, Peter. 2000. "Secret Science: A Classified Community in the National Laboratories." *Minerva*, 38:363–91.

Wheeler, David. 1996. "Evolutionary Economics." *Chronicle of Higher Education*, July 15, pp. A8–A12.

White, Bebo. 1998. "The World Wide Web and High Energy Physics." *Physics Today*, 51.11 (November):30–35.

White, Michael. Forthcoming. "Where Did Jevens' Energy Come From?" *History of Political Economy*.

Whittle, Peter. 1983. *Prediction and Regulation by Linear Least-Square Methods*. 2nd ed. Minneapolis: University of Minnesota Press.

Wiener, Norbert. 1949. *Extrapolation, Interpolation and Smoothing of Stationary Time Series*. Cambridge: MIT Press.

1953. *Ex-Prodigy*. New York: Simon & Schuster.

1954. *The Human Use of Human Beings*. 2nd ed. New York: Doubleday.

1956. *I am a Mathematician*. New York: Doubleday.

1961. *Cybernetics*. 1948. 2nd ed. Cambridge: MIT Press.

1964. *God and Golem*. Cambridge: MIT Press.

Williams, Arlington. 1980. "Computerized Double Auction Markets." *Journal of Business*, 53:235–58.

Williams, Nigel. 1997. "Cold War Thaw Leads to Turmoil in NATO Science." *Science*, 278:795.
Winsberg, Eric. 1999. "Sanctioning Models: The Epistemology of Simulation." *Science in Context*, 12:275–92.
Winston, Wayne. 1991. *Operations Research: Applications and Algorithms.* Boston: PWS-Kent.
Winter, Sidney. 1987. "Natural Selection and Evolution." In J. Eatwell et al., eds., *The New Palgrave*, pp. 108–16. London: Macmillan.
Wise, M. Norton. 1993. "Enlightenment Balancing Acts." In Paul Horwich, ed., *World Changes.* Cambridge: MIT Press.
Wiseman, Alan. 2000. *The Internet Economy*. Washington, D.C.: Brookings.
Wittgenstein, Ludwig. 1978. *Remarks on the Foundations of Mathematics.* Cambridge: MIT Press.
Wohlstetter, Albert. 1961. "Nuclear Sharing: NATO and the N + 1 Country." *Foreign Affairs*, 39 (April):355–87.
Wolff, Robert Paul. 1970. "Maximization of Expected Utility as a Criterion of Rationality in Military Strategy and Foreign Policy." *Social Theory and Practice*, 1:99–111.
Wong, Stanley. 1978. *The Foundations of Paul Samuelson's Revealed Preference Theory*. Boston: Routledge Kegan Paul.
Wooldridge, Michael, & Jennings, Nicholas, eds. 1995. *Intelligent Agents*. Berlin: Springer.
Worster, Donald. 1994. *Nature's Economy*. 2nd ed. Cambridge: Cambridge University Press.
1995. "Nature and the Disorder of History." In Soule & Lease, 1995, pp. 65–86.
Wright, Monte, & Paszek, Lawrence, eds. 1970. *Science, Technology and Warfare.* Washington, D.C.: GPO.
Yockey, Hubert, ed. 1958. *Proceedings of a Symposium on Information Theory in Biology.* New York: Pergamon.
1992. *Information Theory and Molecular Biology.* Cambridge: Cambridge University Press.
Yonay, Yuval. 1998. *The Struggle over the Soul of Economics.* Princeton: Princeton University Press.
Young, Jeffrey. 1998. "Using Computer Models to Study the Complexities of Human Society." *Chronicle of Higher Education*, July 24, pp. A17–A20.
Young, Oran. 1975. *Bargaining: Formal Theories of Negotiation.* Urbana: University of Illinois Press.
Zachary, G. 1997. *Endless Frontier: Vannevar Bush.* New York: Free Press.
Zappia, Carlo. 1996. "The Notion of Private Information in a Modern Perspective: A Reappraisal of Hayek's Contribution." *European Journal of the History of Economic Thought*, 3:107–31.
Zemel, E. 1989. "Small Talk and Cooperation: A Note on Bounded Rationality." *Journal of Economic Theory*, 49:1–9.
Zermelo, Ernst. 1913. "Uber eine Answendung der Mengenlehre auf die Theorie des Schachspiels." *Proceedings of the Fifth International Congress of Mathematicians*, 2:501–4. Cambridge: Cambridge University Press. Translated in *Games and Economic Behavior* 24 (2001):123–37.

Ziman, John. 1994. *Prometheus Bound.* Cambridge: Cambridge University Press.
Zimmerman, David. 1996. *Top Secret Exchange.* Montreal: McGill-Queens University Press.
Zizek, Slavoj. 1999. *The Ticklish Subject.* London: Verso.
Zupan, M. 1990. "Why Nice Guys Finish Last: A Comment on Axelrod's Evolution of Cooperation." *Public Choice*, 65:291–93.
Zurek, Wojciech, ed. 1990. *Complexity, Entropy and the Physics of Information.* Santa Fe Institute Studies, vol. 8. Redwood City, Calif.: Addison-Wesley.

Index

Printed in the United States
61182LVS00003B/1-30

9 780521 775267